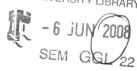

MICROWAVE FILTERS FOR COMMUNICATION SYSTEMS

THE WILEY BICENTENNIAL–KNOWLEDGE FOR GENERATIONS

*E*ach generation has its unique needs and aspirations. When Charles Wiley first opened his small printing shop in lower Manhattan in 1807, it was a generation of boundless potential searching for an identity. And we were there, helping to define a new American literary tradition. Over half a century later, in the midst of the Second Industrial Revolution, it was a generation focused on building the future. Once again, we were there, supplying the critical scientific, technical, and engineering knowledge that helped frame the world. Throughout the 20th Century, and into the new millennium, nations began to reach out beyond their own borders and a new international community was born. Wiley was there, expanding its operations around the world to enable a global exchange of ideas, opinions, and know-how.

For 200 years, Wiley has been an integral part of each generation's journey, enabling the flow of information and understanding necessary to meet their needs and fulfill their aspirations. Today, bold new technologies are changing the way we live and learn. Wiley will be there, providing you the must-have knowledge you need to imagine new worlds, new possibilities, and new opportunities.

Generations come and go, but you can always count on Wiley to provide you the knowledge you need, when and where you need it!

WILLIAM J. PESCE
PRESIDENT AND CHIEF EXECUTIVE OFFICER

PETER BOOTH WILEY
CHAIRMAN OF THE BOARD

MICROWAVE FILTERS FOR COMMUNICATION SYSTEMS: FUNDAMENTALS, DESIGN, AND APPLICATIONS

RICHARD J. CAMERON
CHANDRA M. KUDSIA
RAAFAT R. MANSOUR

BICENTENNIAL
BICENTENNIAL
1807
⚛WILEY
2007
BICENTENNIAL
BICENTENNIAL

WILEY-INTERSCIENCE
A JOHN WILEY & SONS, INC., PUBLICATION

Library of Congress Cataloging-in-Publication Data:

Cameron, Richard J.
  Microwave filters for communication systems: fundamentals, design and applications / Richard J.
Cameron, Chandra M. Kudsia, Raafat R. Mansour.
       p. cm.   1005219509
    Includes index.
    ISBN 978-0-471-45022-1 (cloth)
    1. Microwave filters. 2. Telecommunication systems. I. Kudsia, Chandra M.
    II. Mansour, Raafat R. III. Title.
    TK7872.F5C35 2007
    621.381'33--dc22

                                                          2006036928

Printed in the United States of America

10 9 8 7 6 5 4 3 2 1

To our wives, Patricia, Wendy, and Miranda

CONTENTS

2 FUNDAMENTALS OF CIRCUIT THEORY APPROXIMATION   83

FOREWORD

It gives me great pleasure to write a foreword for this monumental work by three of the most prominent contemporary experts in the field of microwave filters. For more than three decades, significant research and development efforts have been focused on improving performance, introducing new concepts, and developing new technologies for microwave filters. As a result, microwave filters available today are significantly better than what was available 30 years ago: much smaller size, lower loss, flat pass bands, very high rejections, extremely steep transition response from pass band to stop band, stop band transmission zeros, symmetrical or asymmetrical responses, and possibility of self equalized group delay. In addition, the design techniques employed today are dramatically different. In the past, microwave filter designers used approximate circuit models and idealized components to produce a design. Such designs required many iterations and empirical adjustments to achieve the desired performance. Today, synthesis techniques coupled with very accurate electromagnetic simulations and sophisticated optimization software algorithms, operating on very fast computers, allow the filter designers to go from the drawing boards almost immediately to the final product, without the need for empirical adjustments or manual tuning. This cuts dramatically the time required for product development cycle. It also allows implementation of much more efficient and enhanced capacity communication systems.

This book captures the fundamentals and practical aspects of modern microwave filter design. It brings in one place a wealth of information that has not been unified before. It is a reference book for the experienced filter designer, and could be an excellent text book for senior level undergraduate or graduate courses on microwave filters. It can also be a guide for researchers in their efforts to advance the state of the art of microwave filters design. And finally, it is quite useful for communications systems designers and planners to learn the current capabilities and limitations of modern microwave filters and multiplexing networks. This book is well balanced between theory and practical implementation.

The vast experience of the authors in industries ranging from satellite communications to cellular base stations is clearly reflected in the book. This experience will no doubt immensely benefit the coming generations of engineering students, practitioners, and the microwave industry as a whole.

ALI ATIA

PREFACE

The book begins with a simple model of a communication system. It addresses the
issues: (1) whether there is a limitation on the available bandwidth for a wireless
communication system, (2) what are the limitations for transmitting information
in the available bandwidth, and (3) what are the cost sensitive parameters of a
communication system? Each issue is then addressed to gain understanding of
various system parameters with emphasis on the role and requirements of filter networks in different parts of the communication system. This sets the stage to address
the fundamentals of filter design based on circuit theory approximation. It continues
with a description of classical filters. This is followed by the development of
computer aided techniques to generate general class of prototype filter functions,
exhibiting symmetrical or asymmetrical frequency response. This general formulation is accomplished by incorporating hypothetical *frequency invariant reactive*
(FIR) elements in the lowpass prototype filter design. The FIR elements show up
as frequency offsets of resonant circuits in real bandpass or bandstop filters.
Absence of FIR elements represents the classical filter function that gives rise to
symmetrical frequency response. From this general formulation of the filter function, synthesis techniques are described to realize the equivalent lumped parameter
circuit model of filter networks. Next step in the synthesis procedure is to translate
the circuit model of the filter into its equivalent microwave structure. As a first
approximation, this can be achieved by making use of the extensive existing data
that relates circuit models to the physical dimensions and properties of structures
used for microwave filters. For more accurate determination of physical dimensions,
modern EM based techniques and tools are described to determine filter dimensions
with near arbitrary accuracy. This knowledge is carried through in the design of
multiplexing networks having arbitrary bandwidths and channel separations.

Separate chapters are devoted to computer-aided tuning and high power considerations in filter design. Our goal has been to give the reader a broad view of filter
requirements and design, and sufficient depth to follow continuing advances in
this field. Throughout the book, emphasis has been on fundamentals and practical
considerations in filter design. Distinct features of the book include (i) system
considerations in the design of filters, (ii) the general formulation and synthesis of
filter functions including the FIR elements, (iii) synthesis techniques for lowpass
prototype filters exhibiting symmetrical or asymmetrical frequency response in a
variety of topologies, (iv) application of EM techniques to optimize physical dimensions of microwave filter structures, (v) design and tradeoffs of various multiplexer
configurations, (vi) computer-aided filter tuning techniques and, (vii) high power
considerations for terrestrial and space applications. The material in the book is
organized in twenty chapters as follows:

- Chapter 1 is devoted to an overview of communication systems; more specifically to the relationship between the communication channel and other
  elements of the system. The intent here is to provide the reader with sufficient
  background to be able to appreciate the critical role and requirements of RF
  filters in communication systems.
- The principles that unify communication theory and circuit theory approximations are explained in Chapter 2. It highlights the essential assumptions
  and the success of the frequency analysis approach that we take for granted
  in analyzing electrical networks.
- Chapter 3 describes synthesis of the characteristic polynomials to realize the
  classical maximally flat, Chebyshev, and elliptic function lowpass prototype
  filters. It includes a discussion of FIR elements and their inclusion to generate
  filter functions with asymmetrical frequency response. This leads to transfer
  function polynomials (with certain restrictions) with complex coefficients, a
  distinct departure from the more familiar characteristic polynomials with
  rational and real coefficients. This provides a basis to analyze the most
  general class of filter functions in the lowpass prototype domain, including
  minimum and non-minimum phase filters, exhibiting symmetrical or
  asymmetrical frequency response.
- Chapter 4 presents synthesis of characteristic polynomials of lowpass prototype
  filters with arbitrary amplitude response using computer-aided optimization
  technique. The key lies in making sure that the optimization procedure is
  highly efficient. This is accomplished by determining the gradients of the objective function analytically, and linking it directly to the desired amplitude
  response shape. It includes minimum phase and non-minimum phase filters
  exhibiting symmetrical or asymmetrical frequency response. To demonstrate
  the flexibility of this method, examples of some unconventional filters are
  included.
- Chapter 5 provides a review of the basic concepts used in the analysis of multiport microwave networks. These concepts are important for filter designers

since any filter or multiplexer can be divided into smaller two-, three- or N-port networks connected together. Five matrix representations of microwave networks are described, namely, [Z], [Y], [ABCD], [S] and [T] matrices. These matrices are interchangeable, where the elements of any matrix can be written in terms of those of the other four matrices. Familiarity with the concepts of these matrices is essential in understanding the material presented in this book.

• Chapter 6 begins with a review of some important scattering parameter relations that are relevant for the synthesis of filter networks. This is followed by a discussion of the general kind of Chebyshev function and its application in generating the transfer and reflection polynomials for the equi-ripple class of filter characteristics with an arbitrary distribution of the transmission zeros. In the final part of this chapter, the special cases of pre-distorted and dual-band filtering functions are discussed.

• In Chapter 7, filter synthesis based on the [ABCD] matrix is presented. The synthesis procedure is broken into two stages. The first stage involves lumped element lossless inductors, capacitors and FIR elements. The second stage includes the immitance inverters. Use of such inverters allows for the prototype electrical circuit in a form suitable for realization with inter-coupled microwave resonators. The technique is applicable for synthesising lowpass prototype filters with symmetrical or asymmetrical response, in ladder form, as well as cross-coupled topologies. A further generalization is introduced to allow the synthesis of singly-terminated filters. The synthesis process described in this chapter represents the most general technique for synthesizing lumped element lowpass prototype filter networks.

• In Chapter 8, the concept of $N \times N$ coupling matrix for the synthesis of band-pass prototype filters is introduced. The procedure is modified by including FIR elements to allow synthesis of asymmetric filter response as well. The procedure is then extended to $N + 2$ coupling matrix by separating out the purely resistive and purely reactive portions of the $N \times N$ matrix. The $N + 2$ coupling matrix allows multiple couplings with respect to the input and output ports, in addition to the main input/output couplings to the first and last resonator as envisaged in the $N \times N$ coupling matrix. This allows synthesis of fully canonical filters and simplifies the process of similarity transformations to realize other filter topologies. This synthesis process yields the general coupling matrix with finite entries for all the couplings. The next step in the process is to derive topologies with a minimum number of couplings, referred to as canonical forms. This is achieved by applying similarity transformations to the coupling matrix. Such transformations preserve the eigen values and eigen vectors of the matrix, thus ensuring that the desired filter response remains unaltered. There are two principal advantages of this synthesis technique. Once the general coupling matrix with all the permissible couplings has been synthesized, it allows matrix operations on the coupling matrix to realize a variety of filter topologies. The second advantage is that the coupling matrix represents the practical bandpass filter topology. Therefore, it is possible to identify each element of the practical filter uniquely, including its Q value, dispersion characteristics and sensitivity. This permits a more accurate determination of the practical filter characteristics and an insight into ways to optimize filter performance.

- Chapter 9 develops methods of similarity transformations to realize a wide range of topologies appropriate for dual-mode filter networks. Dual-mode filters make use of two orthogonally polarized degenerate modes, supported in a single physical resonator, be it a cavity, a dielectric disc, or a planar structure, thereby allowing a significant reduction in the size of filters. Besides the longitudinal and folded configurations, structures referred to as cascade quartets and cul-de sac filters are also included. The chapter concludes with examples and a discussion of the sensitivity of the various dual-mode filter topologies.

- In Chapter 10, we introduce two unusual circuit sections: the extracted pole section and the trisection. These sections are capable of realizing one transmission zero each. They can be cascaded with other circuit elements in the filter network. Application of these sections extends the range of topologies for realizing microwave filters. This is demonstrated by synthesizing filters that include cascaded quartets, quintets and sextets filter topologies. Lastly, the synthesis of the box section, and its derivative, the extended box configuration is explained. Examples are included to illustrate the intricacies of this synthesis procedure.

- Theoretical and experimental techniques for evaluating the resonant frequency and unloaded Q-factor of microwave resonators are described in Chapter 11. Resonators are the basic building blocks of any bandpass filter. At microwave frequencies, resonators can take many shapes and forms. The chapter includes two approaches for calculating the resonant frequency of arbitrarily-shaped resonators: the eigen-mode analysis and the S-parameter analysis. Examples are given illustrating the implementation of these two techniques using EM-based commercial software tools such as HFSS. It also includes a step by step procedure for measuring the loaded and unloaded Q values using either the polar display of a vector network analyzer or the linear display of a scalar network analyzer.

- Chapter 12 addresses the synthesis techniques for the realization of lowpass filters at microwave frequencies. Typical bandwidth requirements for lowpass filters in communication systems are in the GHz range. As a consequence, prototype models based on lumped elements are not suitable for realization at microwave frequencies. It requires use of distributed elements for the prototype filters. The chapter begins with a description of the commensurate line elements and their suitability for realizing distributed lowpass prototype filters. It then goes into a discussion of characteristic polynomials that are best suited for modeling practical lowpass filters, and methods to generate such polynomials. This is followed by a detailed description of the synthesis techniques for the stepped impedance and the lumped/distributed lowpass filters.

- Chapter 13 deals with the practical design aspects of dual-mode bandpass filters. It includes use of dual-mode resonators that operate in the dominant mode, as well as in the higher order propagation modes. A variety of examples are included to illustrate the design procedure. These examples include longitudinal and canonical configurations, the extended box design, the extracted pole filter, and filters with all inductive couplings. The examples also include symmetrical and asymmetrical response filters. The steps involved in the simultaneous optimization of amplitude and group delay response of a dual-mode linear phase filter are described. Examples in this chapter span the analysis and synthesis techniques described in Chapters 3 to 11.

- Chapter 14 presents the use of EM simulator tools for designing microwave filters. It is shown how one can couple the filter circuit models with EM simulation tools to synthesize the physical dimensions of microwave filters with near arbitrary accuracy. The starting point for such computations is usually the physical dimensions derived from the best circuit model of the filter. Methods are described to compute, with much greater accuracy, the input/output and inter-resonator couplings by using the commercially available EM simulator software. The techniques can be adapted for a direct approach to determine the physical dimensions of filters from the elements of the coupling matrix [M], using K-impedance inverter, or J-admittance inverter models. Numerical examples are given in this chapter to illustrate, step by step, the application of this approach to the design of dielectric resonator, waveguide and microstrip filters. For simple geometries with negligible coupling between non-adjacent resonators, this approach yields excellent results. Use of EM tools represents a major advance in the physical realization of microwave filters.

- Chapter 15 presents several techniques for EM-based design of microwave filters. The most direct approach is to combine an accurate EM simulation tool with an optimization software package, and then optimize the physical dimensions of the filter to achieve the desired performance. This is effectively a tuning process where the tuning is done by the optimization package rather than a technologist. The starting point for this technique is the filter dimensions obtained using methods described in Chapter 14. Direct optimization approach, without any simplifying assumptions can be still very computation intensive. A number of optimization strategies including adaptive frequency sampling, neural networks, and multi-dimensional Cauchy technique are described to reduce optimization time. Two advanced EM based techniques, the space mapping technique (SM), and the coarse model technique (CCM) are described in detail offering a significant reduction in computation time. The chapter concludes with examples of filter dimensions obtained by using aggressive space mapping (ASM) and CCM techniques.

- Chapter 16 develops the design of dielectric resonator filters in a variety of configurations. Commercial software packages such as HFSS and CST Microwave Studio can be readily utilized to calculate the resonant frequency, field distribution, and resonator Q of dielectric resonators having any arbitrary shape. Using such tools, mode charts, along with plots, illustrating the field distribution of the first four modes in dielectric resonators are included. It also addresses the computation of the resonant frequency, and the unloaded Q (Q_0) of cylindrical resonators, including the support structure. Design considerations in terms of Q_0, spurious response, temperature drift and power handling capability are described. The chapter concludes with a detailed description of the design and tradeoffs for cryogenic dielectric resonator filters. Dielectric resonator filters are widely employed in wireless and satellite applications. Continuing advances in the quality of dielectric materials is a good indication of the growing application of this technology.

- Chapter 17 deals with the analysis and synthesis of allpass networks, often referred to as equalizers. Such external allpass equalizers can be cascaded with filters to improve the phase and group delay response of filter networks. The chapter concludes with a discussion of the practical tradeoffs between the linear phase filters and externally equalized filter networks.

- Chapter 18 presents the design and tradeoffs for multiplexing networks for a variety of applications. It begins with a discussion of tradeoffs among the various types of multiplexing networks, including circulator-coupled, hybrid-coupled and manifold-coupled multiplexers, employing single mode or dual mode filters. It also includes multiplexers based on using directional filters. This is followed by the detailed design considerations for each type of multiplexer. The design methodology and optimization strategy are dealt with in depth for the manifold-coupled multiplexer, by far the most complex microwave network. Numerous examples and photographs are included to illustrate the designs. The chapter is concluded with a brief discussion of the high power capability of diplexers for cellular applications.

- Chapter 19 is devoted to the computer-aided techniques for tuning microwave filters. From a theoretical standpoint, the physical dimensions of a microwave filter can be perfected using EM based techniques with near arbitrary accuracy. In practice, the use of EM based tools can be very time consuming and prohibitively so for higher order filters and multiplexing networks. Moreover, owing to manufacturing tolerances and variations in material characteristics, practical microwave filters cannot duplicate the theoretical design. These problems are further exacerbated by the very stringent performance requirements for applications in the wireless and satellite communication systems. As a result, filter tuning is deemed an essential post-production process. Techniques discussed in this chapter include, (i) sequential tuning of coupled resonator filters, (ii) computer-aided tuning based on circuit model parameter extraction, (iii) computer-aided tuning using poles/zeros of the input reflection coefficient, (iv) time domain tuning, and (v) fuzzy logic tuning. The relative advantages of each technique are described.

- Chapter 20 deals with the high power considerations in the design of microwave filters and multiplexing networks. It includes an overview of the phenomenon of microwave gaseous breakdown for terrestrial applications. It highlights the importance of various factors that can severely degrade the performance of high power equipment. The phenomenon of multipaction, applicable for space applications, is described in some depth. The topic of design margins, especially when the equipment must handle a number of high power carriers, is discussed in detail. Methods to prevent multipaction are highlighted. Another phenomenon that becomes significant in designing high power equipment is that of passive inter-modulation (PIM). PIM is difficult to analyze and depends upon the choice of materials and workmanship standards. Guidelines to minimize PIM in the design of high power equipment are included.

The book is aimed at senior undergraduate and graduate students as well as practitioners of microwave technology. In writing this book, we have borrowed heavily from our industrial experience, giving seminars and teaching courses at universities and interactions with the engineering community at large at various conferences. It reflects a lifetime of experiences in advancing the state-of-the-art in microwave filters and multiplexing networks.

ACKNOWLEDGMENTS

Authors had the benefit of working with a large number of very dedicated and talented colleagues at COM DEV over many years. We appreciate their contribution to our knowledge. Authors would like to pay a special tribute to the late Dr. Val O'Donovan, co-founder of COM DEV, whose leadership and encouragement created an environment that allowed us and others to make many contributions towards advancing the state-of-the-art in microwave technology.

We would like to thank many people for their help in the preparation of this book: Dietmar Schmitt of ESA and Ming Yu of COM DEV for providing an early review of the book and many useful suggestions, Alastair Malarky (formerly COM DEV) for providing a critique of Chapter 1, Santiago Cogollos (University of Valencia), Soren Peik (University of Applied Sciences Bremen), Michael Earle (Consultant) for providing feedback on selected chapters. We would also like to thank many graduate students in the E & CE department at the University of Waterloo for providing feedback and help in the preparation of the book. In particular, we wish to thank Michel Elnagger, Vahid Miraftab, Hamid Salehi, Joe Salfi, George Shaker, and Winter Yan. Our thanks also go to the administrative staff that took care of numerous tasks associated with the book; specifically, Louise Green and Bill Jolley at the University of Waterloo, Emma Shanks at COM DEV, and Rosaline Wong at AMI.

Authors have been fortunate in interfacing with so many engineers on a large number of national and international satellite programs. Our appreciation goes to numerous colleagues and others in the communication satellite and cellular communication industry worldwide with whom we had the pleasure of working and, in the process, gaining wide ranging experience and expertise. It was a pleasure to bring this knowledge into this book.

CHAPTER 1

RADIO FREQUENCY (RF) FILTER NETWORKS FOR WIRELESS COMMUNICATIONS—THE SYSTEM PERSPECTIVE

This chapter is devoted to an overview of communication systems, especially the relationship between the communication channel and other elements of the system. The intent here is to provide the reader with sufficient background to be able to appreciate the critical role and requirements of radio frequency (RF) filters in communication systems. A number of standard texts [1–8] have been referred to in developing a significant portion of this chapter.

The chapter is divided into three parts: Part I presents a simple model of a communication system, the radio spectrum and its utilization, the concept of information, and system link budgets. Part II describes the noise and interference environment in a communication channel, the nonideal amplitude and phase characteristics of the channel, the choice of modulation–demodulation schemes, and how these parameters impact the efficient use of the allocated bandwidths. Part III discusses the impact of system design on the requirements, and the specifications of microwave filter networks in satellite and cellular communication systems.

Microwave Filters for Communication Systems: Fundamentals, Design, and Applications,
by Richard J. Cameron, Chandra M. Kudsia, and Raafat R. Mansour
Copyright © 2007 John Wiley & Sons, Inc.

PART I INTRODUCTION TO A COMMUNICATION SYSTEM, RADIO SPECTRUM, AND INFORMATION

1.1 MODEL OF A COMMUNICATION SYSTEM

Communication refers to the process of conveying information-bearing signals from one point to another that is physically separate. In ancient times, people communicated over long distances by a variety of means, such as smoke signals, drum beating, homing pigeons, and horseback riders. All such means were slow in the transmission of information over any appreciable distance. It was the invention of electricity that changed all that. Communication became almost instantaneous by the transmission of electrons through wires, or electromagnetic waves through empty space or fibers, limited only by the speed of light—a fundamental constraint of our universe.

At the highest (simplest) level, communication involves an information source, a transmitter, a communication medium (or channel), a receiver, and an information destination (sink), as depicted in Figure 1.1. Until the 1980s, most information was communicated in an analog format, called *analog communication*. Today, most information is communicated in a digital format, called *digital communication*. Even analog information is routinely converted to digital format for transmission and then converted back to analog information at the destination.

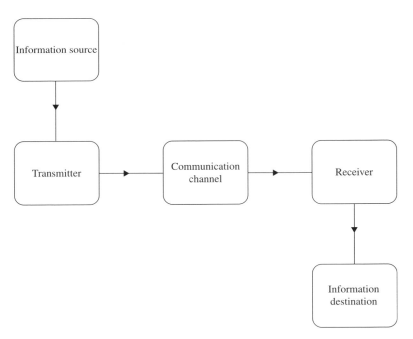

Figure 1.1 Simple model of a communication system.

All communication systems are required to be linear. For such systems, the law of superposition holds. It allows the use of common media for the transmission and reception for an arbitrary number of independent signals, subject only to the constraints of the available bandwidths and adequate power levels. However, the components of a communication system do not need to be all linear, as long as the overall system is linear over the specified range of bandwidth within an acceptable degree of nonlinearity. In fact, all active devices are inherently nonlinear, which is essential for the purposes of frequency generation, modulation, demodulation, and the amplification of signals. However, such intentional nonlinearities can be controlled for a specific application. The undesirable nonlinearities, though, are present to some degree in any communication system, and often are critical in optimizing the overall performance. For broadband wireless communication systems such as line-of-sight (LOS) or satellite systems used for long distance networks, the frequency spectrum is divided into a number of RF channels, often referred to as *transponders*. Within each RF channel, there can be more than one RF carrier, depending on the system requirements. The channelization of the frequency spectrum provides the flexibility for the communication traffic flow in a multiuser environment. High-power amplifiers can operate with relatively high efficiency, since they are required to amplify a single carrier, or a limited number of signals, on a channel-by-channel basis, incurring a minimum of distortion.

Irrespective of the communication format, it should be recognized that the passage of the transmitted signal through the communication channel is a strictly analog operation. The communication channel is a nonideal, lossy medium, and the reception of signals at the receiver involves recovery of the transmitted signal in the presence of impairments, in particular, thermal noise at the receiver, signal distortions (within the channel, and from nonideal transmitter and receiver), and interference from other signals or echoes (multipath) seen by the receiver.

1.1.1 Building Blocks of a Communication System

In this section, the building blocks shown in Figure 1.2 are described for both analog and digital communication systems.

Information Source   The information source consists of a large number of individual signals that are combined in a suitable format for transmission over the communication medium. Such a signal is referred to as the *baseband signal*. The transducers in Figure 1.2 are required to convert the energy of the individual information sources, either acoustic (voice) or electrical, into an appropriate electrical signal suitable for transmission. For an analog system, all the individual signals, as well as the combined baseband signal, are in an analog format as illustrated in Figure 1.2*a*.

For digital systems, the baseband signal is a digital datastream, whereas the individual signals constituting the baseband can be digital or analog. Consequently, the individual analog signals need to be converted to their equivalent digital format via

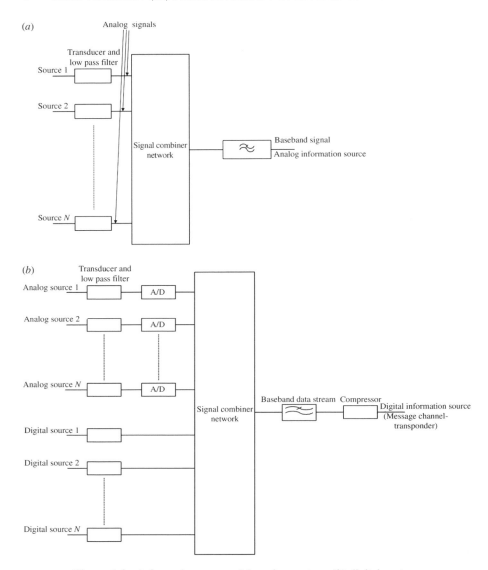

Figure 1.2 Information source: (*a*) analog system; (*b*) digital system.

analog-to-digital (A/D) converters. Another feature of the information source in a digital system is the use of data compression to conserve bandwidth. A compressor takes the digital data and exploits their redundancy and other features to reduce the amount of data that need to be transmitted but still permits the information to be recovered. The information source for a digital communication system is described in Figure 1.2*b*.

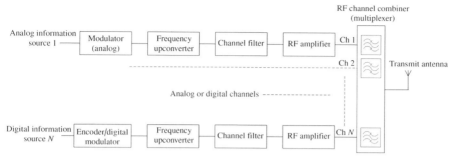

Figure 1.3 Transmitter block diagram.

Transmitter The block diagram of a transmitter is presented in Figure 1.3, and the functionality of each element is described:

Encoder—in digital systems, an encoder introduces error correction data into the baseband information stream that permits recovery of the digital information even after significant impairments are created in the communication channel.

Modulator—it transcribes the baseband signal into a carrier frequency that is more suitable for transmission over the communication medium. The modulator can shift the signal frequencies for ease of transmission, change the bandwidth occupancy, or materially alter the form of the signal to optimize the noise or distortion performance.

Upconverter—often the modulator and demodulator cannot work at the RF communication frequency. Instead they operate at a lower, intermediate frequency (IF). Thus, the transmitter and receiver contain circuits that convert between the communication RF frequency and the modulator–demodulator IF frequencies.

RF amplifier—since the communication channel is a lossy medium, it requires amplification of the carrier frequency. RF power amplifiers provide this function, which has a direct bearing on the range of the communication link.

RF multiplexer—used to combine the power of a number of RF channels into a composite broadband signal for transmission via a common antenna.

Transmit antenna—launches the RF power into space and focuses it towards the receiving station. The "gain" of the transmit and receive antennas has a direct bearing on the link budget and thus on the range of the communication system.

Communication Channel For wireless systems, the communication channel is free space. As a consequence, the properties of space, including the atmosphere, play a critical role in system design.

Receiver The block diagram of a receiver is shown in Figure 1.4, and the functionality of each element is described:

Receive antenna—it intercepts the RF power and focuses it to a transmission line connected to the low-noise amplifier (LNA).

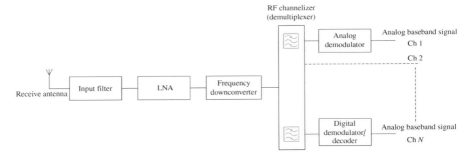

Figure 1.4 Receiver block diagram.

LNA—amplifies the very weak received signal with a minimum addition of noise to it.

Downconverter—serves the function of frequency conversion, which is similar to that required in the transmitter chain. The downconverter converts the uplink frequency band to the downlink frequency band.

Demodulator—extracts the baseband signal from the RF carrier by a process opposite that of a modulator.

Decoder—exploits the error correction data previously inserted into the information stream, and uses it to correct the errors made during digital demodulator's recovery of the data.

Information Destination The functionality of the information destination block is opposite that of the information source. In the case of digital systems, an expander is used to reverse the operation of the compressor as denoted in Figure 1.5.

Since a communication system is costly to install, its commercial viability is critically dependent on the number of users who must share the transmission medium. As a result, the information source usually consists of a large number of signals, occupying a finite range of frequencies. The width of the range of frequencies is called the *bandwidth* of the signal.

Some key questions then arise. Is there a limitation on the available bandwidth for a communication system? What are the limitations of transmitting information over the chosen transmission media in the available bandwidth? What are the cost-sensitive parameters in a communications system?

Figure 1.5 Information sink for a digital system.

1.2 RADIO SPECTRUM AND ITS UTILIZATION

To understand the limitations on the available bandwidths for a communication system, it is necessary to understand the radio spectrum and its utilization [4].

Electromagnetic waves cover an extremely broad spectrum of frequencies, from a few cycles per second to gamma rays with frequencies of up to 10^{23} cycles per second. The radio spectrum is that portion of the electromagnetic spectrum that can be electronically and effectively radiated from one point in space and received at another. This includes frequencies anywhere from 9 kHz to 400 GHz. Although most of the commercial use takes place between 100 kHz and 44 GHz, some experimental systems reach as high as 100 GHz. The signals employing these same frequencies can also be transmitted over long distances by wire, coaxial cables, and glass fibers. However, since such signals are not intended to be radiated, they are not considered part of the radio spectrum. A communication system is allowed to access only a portion the radio spectrum, due not only to technology limitations but also for regulatory reasons. The radio spectrum is subdivided into smaller frequency "bands" by national and international agencies, where each band is restricted to a limited set of types of operation. In addition, each band is a controlled commodity that often has a license fee associated with its use. Obviously, this represents a huge incentive to minimize the bandwidth required to communicate a signal.

1.2.1 Radio Propagation at Microwave Frequencies

There are many sources of energy loss in the free-space communication medium. The most serious ones include the rainfall and oxygen in the atmosphere. The atmospheric losses as a function of frequency are portrayed in Figure 1.5. The radio energy is absorbed and scattered by the raindrops, and this effect becomes more intense as the wavelength approaches the size of the raindrops. Consequently, rainfall and water vapor produce intense attenuation effects at the higher microwave frequencies. The first absorption band, due to water vapor, peaks at approximately 22 GHz, and the first absorption band, due to the oxygen in the atmosphere, peaks at about 60 GHz.

For fixed-LOS terrestrial microwave radio links, multipath fading is another major cause of signal loss. Fading results from the variations in the refractive index of air for the first few tenths of a kilometer above the earth's surface. Such gradients in refraction bend the rays, which, on reflection from the ground or other layers, combine with the direct rays, causing coherent interference.

Mobile communication adds a new dimension to the propagation problem. Besides the requirements of omnidirectional coverage and the mobility of end users, the communication system must deal with non LOS and multipath problems caused by the signal reflections from tall buildings, trees, valleys, and other large objects in an urban environment. In addition, the coverage area for mobile services includes the area inside buildings. Also, mobility incurs the *Doppler shift*, a change in the frequency of the received signal, further complicating the issue.

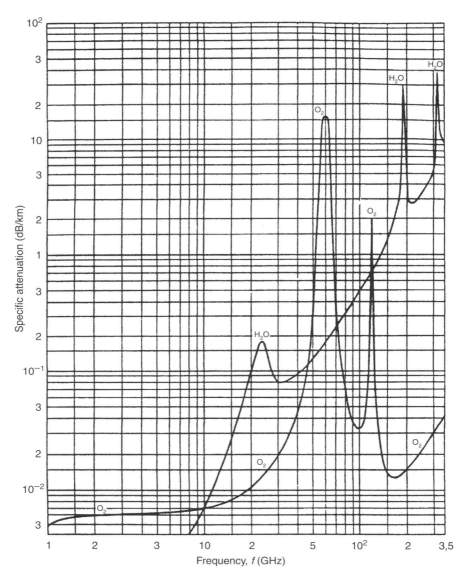

Figure 1.6 Atmospheric losses as a function of frequency. (From CCIR Rep. 719-3, Vol. V Annex, 1990).

From the foregoing analysis, it is evident that the attenuation due to rain and other atmospheric effects, coupled with multipath fading in an urban environment, severely constrains the available frequency spectrum suitable for commercial communications.

1.2.2 Radio Spectrum as a Natural Resource

The radio spectrum is a natural resource unlike any other. It is a completely renewable resource that can never be permanently depleted. It is also universally available. The limitation comes with its usage. The radio spectrum has a finite capacity that, if exceeded, results in interference, incapacitating the system. For this reason, national governments grant users the privilege of using radio spectra in exchange for agreements to abide by usage rules. Because RF signals often cross national borders where they can interfere with radio frequencies allocated in another country, nations must cooperate to find ways to coordinate their individual allocations. Since communication is vital to all nations, there are national and international agencies that regulate the allocation and usage of the RF spectrum.

The International Telecommunications Union (ITU), part of the United Nations, is the international body that determines the worldwide radio spectrum allocations. The ITU accomplishes this through World Administrative Radio Conferences (WARC), where ITU member nations attempt to reach a consensus on proposals by different countries. These meetings require a consensual approach to decisionmaking, rendering the process very tedious, and often resulting in delays. Once a consensus is reached, the ITU publishes tables of the frequency allocations, deemed as the *radio regulations* of ITU. Each country then makes its own detailed frequency allocation plan consistent with the radio regulations tables. In addition, there are other consultative committees such as the International Radio Consultative Committee (CCIR), a part of the ITU assigned to study and recommend the standards for interoperability and guidelines and the control of interference from various services. On the domestic front, most countries have government agencies such as the Federal Communications Commission (FCC) of the United States that regulate all the nonfederal government use of the frequency spectrum. There are other agencies that control frequency allocations for government use, including the military.

In most countries, the priority followed by the regulators for allocating radio spectrum is as follows:

- Military
- Public safety functions such as aeronautical and marine emergency communications, police, fire, and other emergency services
- National telecommunication companies for telephone
- Broadcast radio and television
- Private users such as mobile systems and other services

Because of the complex web of procedures, priorities, government policies, and so on, once a service has been established and uses a particular portion of the spectrum, it is seldom changed. Incumbency tends to be a big advantage, and often an obstacle in the efficient use of the radio spectrum. The emergence of widespread wireless communications and services has added enormous pressure in both domestic and international regulatory bodies, as well as in original equipment manufacturers (OEMs) to ensure that the frequency allocations and spectrum usage represent the most efficient use of this natural resource.

1.3   CONCEPT OF INFORMATION

What are the limitations for transmitting information over the chosen transmission media? At the fundamental level, the answer lies in the seminal work of Shannon [9], who developed the concepts of information theory.

In 1948, Claude Shannon of the Bell Labs pointed out that "for a signal to carry information, the signal must be changing; and to convey information, the signal must resolve uncertainty."

Thus, the measure of the amount of information is a probabilistic one. Shannon defined the uncertainty of the outcome between two equally likely message probabilities as a unit of information; thereby, he used, as his measure, the binary digit or bit. Also, he showed that the information capacity of a system is fundamentally limited by a very few parameters. Specifically, he demonstrated that the maximum information capacity C of a channel is limited by the channel bandwidth B and the signal-to-noise ratio (S/N) in the channel as described by the relationship

$$C = B \log_M \left(1 + \frac{S}{N} \right) \quad \text{information messages per second} \qquad (1.1)$$

where M is the number of possible source information message states. Here, S/N defines the conditions at the demodulator input, where S is the signal power in watts, and N is white Gaussian noise with a mean power of N watts distributed uniformly over the channel bandwidth at the same location. In analog communication, M is difficult to define; however, in digital communication, the binary data $(M = 2)$ is considered and the capacity limit is given by

$$C = B \log_2 \left(1 + \frac{S}{N} \right) \quad \text{information bits per second} \qquad (1.2)$$

For digital communication, this capacity is defined in the context of information bits, measured at the output of the compressor function before the encoding function, and not at the data source as might be expected; B and S/N are defined within the channel at the input of the receiver section.

Note that the information is not defined at the data source, since there is a significant distinction between "data" and "information":

Data is the raw output from the source, for example, a digitized voice, digitized video, or a sequence of text characters, defining a text document.

Information is the salient content in the raw data. This content is often much less than the data used to represent the raw source output.

For example, consider a voice digitizer. It must continually sample the voice analog source and produce output, even when a person pauses between the words or sentences. Similarly, in a video signal, much of the picture does not change from frame to frame. In a text document, some characters or words are repeated more often than others, but the same number of bits is used to represent each character or word at the data source.

Digital compression exploits the knowledge about the characteristics of the data source type to map the data from one format of representation into another format that reduces the number of digital bits required to provide the salient information. The details of this are not addressed in this book. However, it is important to realize that there can be substantial reductions of raw data prior to their transmission with minimal or no loss of the useful information. For example, the current generation digital TV uses MPEG2 compression that provides at least one order of magnitude reduction in the average data rate required to represent the source data compared with the digitization of the raw video. Digital compression has certainly attracted the interest of the research community over the last few decades, resulting in continual improvement.

Shannon's information theorem proves that as long as the information transmission rate R is less than C, it is possible to limit the error in transmission to an arbitrarily small value. The technique for approaching this limit in digital communication is called *coding*, another favored area of research over the last few decades. The current technology comes within a few tenths of a decibel of the Shannon limit. The theorem states that for $R < C$, transmission can be accomplished without error in the presence of noise. This is a surprising result in the presence of Gaussian noise, since its probability density extends to infinity.

This theorem, although restricted to the Gaussian channel, is fundamentally important for two reasons: (1) the channels encountered in physical systems are generally Gaussian, and (2) the results obtained for the Gaussian channel often provide a *lower bound* on the performance of a system, indicating that it has the highest probability of error. Thus, if a particular encoder–decoder is used with a Gaussian channel and an error probability P_e results, then, with a non-Gaussian channel, another encoder–decoder can be designed so that P_e is smaller. Similar channel capacity equations have been derived for a number of non-Gaussian channels.

Shannon's theorem indicates that a noiseless Gaussian channel ($S/N = \infty$) has an infinite capacity, irrespective of the bandwidth, but the channel capacity does not become infinite as the bandwidth becomes infinite. This is due to the increase in the noise power as the bandwidth is increased. As a result, for a fixed signal power and in the presence of white Gaussian noise, the channel capacity approaches the upper limit with the increasing bandwidth. By using $N = \eta B$ in equation (1.2), where η is the noise density (watts/Hz), we obtain

$$C = B \log_2\left(1 + \frac{S}{\eta B}\right) = \frac{S}{\eta} \log_2\left[1 + \frac{S}{\eta B}\right]^{\eta B / S} \tag{1.3}$$

and

$$\lim_{B \to \infty} C \approx \frac{S}{\eta} \log_2 e = 1.44 \frac{S}{\eta} \tag{1.4}$$

It is evident that there is also an absolute lower bound on the received signal power in a system for a given capacity, irrespective of the bandwidth utilized. For a given capacity

$$S \geq \frac{C\eta}{1.44} \tag{1.5}$$

In accordance with equation (1.2), once minimum received signal power is exceeded, the bandwidth is traded off with the S/N ratio and vice versa. For example, if $S/N = 7$ and $B = 4$ kHz, $C = 12 \times 10^3$ bps. If the SNR is increased to $S/N = 15$ and B is decreased to 3 kHz, the channel capacity remains the same. With a 3 kHz bandwidth the noise power is three-fourth as large as with a 4 kHz bandwidth. As a result, the signal power must be increased by the factor $\frac{3}{4} \times \frac{15}{7} = 1.6$. Therefore, this 25% reduction in bandwidth requires a 60% increase in signal power.

Once the basic parameters of radio spectrum and channel bandwidths have been established, the next question is: *What are the cost-sensitive parameters in the communication system?* This leads to an evaluation of the communication channel and system link parameters.

1.4 COMMUNICATION CHANNEL AND LINK BUDGETS

Communication capacity of a channel is bounded by signal power S, bandwidth B, and noise N in the channel. The parameters and the resultant tradeoffs are investigated in the subsequent sections.

1.4.1 Signal Power in a Communication Link

Wireless communication systems require transmission of signals as radio frequencies in space. RF signals travel through earth's atmosphere and are collected by the receiver. The direction of the propagation can vary from a horizontal direction for terrestrial fixed and mobile systems to a vertical direction for satellite systems. The received signal power is primarily a function of four elements of the communication system:

- The transmitter output electrical power at the radio frequency
- The fraction of the transmitter electrical power directed at the receiver as an RF free-space wave, defined by the transmit antenna gain
- The loss of energy in the communication medium, including the loss due to spherical spreading of the energy
- The fraction of the received free-space RF power at the receiver converted into electrical energy, defined by the receive antenna gain

RF Power For a given allocation of bandwidth and a certain level of noise, the communication capacity is governed by the available power for radio transmission. At first glance, it appears that the capacity can be arbitrarily increased by increasing the level of the RF power. There are two problems with this approach: (1) the generation of RF power is expensive and represents a cost-sensitive parameter, and (2) a substantial amount of the RF power is wasted in the process of radio transmission, rendering the power the most costly portion of a wireless system. It behooves us to examine the basic limitations of power transmission in a wireless system.

1.4.2 Transmit and Receive Antennas

Antennas are used to launch EM energy from a transmission line into space and viceversa. Antennas are linear, reciprocal elements, and, as a consequence, the properties of an antenna are the same in transmission and in reception. The reciprocity theorem holds for all the antenna characteristics. A physical antenna (reflector type) consists of a reflecting surface and a radiating or absorbing feed network. The reflector is used to focus the energy in a given direction. The feed element converts electric currents to EM waves in a transmitting system or converts EM waves to electric currents in a receiving system. A typical parabolic microwave antenna is depicted in Figure 1.7. A transmit antenna focuses energy in the direction of a receive station or over a specific geographic area. For a receive antenna, the aperture collects the power and focuses it on the input feed element of the receive system. The general properties of the antennas described here can be applied to a broad range of frequency spectra.

Antenna Gain   The gain of an antenna is defined with respect to an isotropic antenna. An isotropic transmitting antenna is equivalent to a point source, radiating uniform spherical waves equally in all directions as denoted in Figure 1.8a. The power p_0 in watts, radiated from such a source has a uniform power flux density of $(p_0/4\pi r^2)$ watts per square meter at distance r from the source. Assume that the power source is located at the input terminal of an antenna. The power radiated

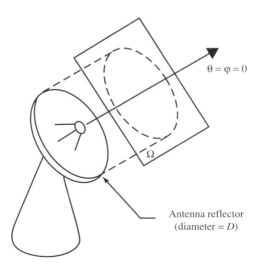

$\theta = \varphi = 0$

Ω

Antenna reflector
(diameter = D)

Figure 1.7   Parabolic antenna.

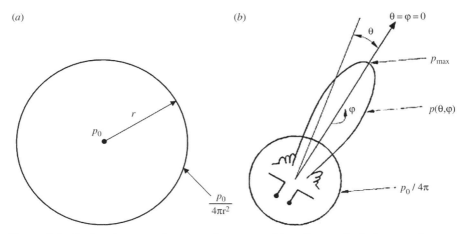

Figure 1.8   Radiated power of a transmit antenna: (*a*) uniform spherical waves from an isotropic source; (*b*) a directional antenna beam.

from this source is proportional to $p_0/4\pi$ in any direction (θ,ϕ) of the surrounding space. A directional antenna radiates power $p(\theta,\phi)$ in direction θ,ϕ as shown in Figure 1.8*b*. The gain of the directional antenna, with respect to an isotropic source, is then given by

$$g(\theta, \phi) = \frac{p(\theta, \phi)}{(p_0/4\pi)} \tag{1.6a}$$

The gain varies, depending on the specific values of θ and ϕ. The maximum value of the gain, as determined by the maximum value of power within the (θ,ϕ) envelope, is simply defined as the antenna gain g. It is usually expressed in decibels:

$$G = 10\log g \qquad \text{dBi or simply dB} \tag{1.6b}$$

Note that this definition is independent of the physical attributes of the antenna and depends solely on the geometry of the radiation patterns of the antennas. For most applications employing uniform parabolic antennas, this maximum gain occurs along the boresight of the antenna, where $\theta = \phi = 0$.

Effective Aperture and Antenna Gain   Physical antennas are designed to radiate and capture energy in the desired directions with a minimum of loss and spillage of energy outside the region of interest. Effective aperture A_e of an antenna is defined as the equivalent physical area of the antenna that captures or radiates energy within the desired direction of θ and ϕ. It is defined by

$$A_e = \eta A \tag{1.7}$$

where $\eta(<1)$ is the efficiency of the antenna and A is the physical aperture used to radiate or capture energy. The effective aperture represents the projected area in the direction of the beam to achieve the maximum gain, and includes the degradation due to the losses and nonuniformities of the structure, and the nonuniformity in the illumination of the aperture. If an antenna is perfect and the energy propagation is uniform, A_e equals the actual projected area A, and η is unity. For practical applications, η varies between 0.5 and 0.8.

The next question is how to determine the amount of power that an effective aperture is able to focus in a given frequency band. This is a complex problem, but the result is simple and elegant. From the theory of antennas [10], the gain of an antenna is related to the effective aperture by

$$g = \frac{4\pi A_e}{\lambda^2} \qquad (1.8)$$

where λ is the wavelength of the radio frequency. This demonstrates that the gain of an antenna depends on its effective aperture and operating frequency. The higher the frequency, the higher the gain of the antenna for a given aperture size. This relationship indicates that there are inherent limitations in terms of the achievable gain for a given size of the aperture. Moreover, the surface of the aperture needs to be accurate to within small fractions of λ, making higher-frequency structures more expensive. Another useful relationship can be derived by considering the solid angle Ω within which an antenna focuses the power. This can be readily derived from equation (1.8) as

$$\Omega = \frac{\lambda^2}{A_e} \text{ rad}^2 \qquad (1.9)$$

and

$$g = 4\pi/\Omega$$

For a satellite-based system, once the antenna footprint is specified (i.e., the coverage area is delineated), the solid angle to cover that area from the satellite gets fixed. This implies that the antenna gain (and therefore the aperture size) is determined by the coverage area. In other words, the limit on the achievable gain from the satellite is set by the coverage area and not by the design or physical structure of the antenna. This relationship also shows that in order to achieve a high directivity, that is, a smaller solid angle, the area of the aperture must be much larger than the operating wavelength. In the microwave region, the physical apertures required to achieve relatively high gains are reasonable and readily implemented. For such antennas, the radiated fields do not interact strongly with nearby objects that are not in the direction of the transmission. Thus, the ground effects, which play a major role in the design of megahertz region (AM, FM, and TV broadcasts) antennas, are generally not important above 1 GHz.

Power Flux Density The radiation of EM energy follows the fundamental inverse square law. Consequently, energy, radiated by the isotropic antenna to a distance r, is uniformly distributed over a sphere of area $4\pi r^2$, as shown in Figure 1.8a. Note that the energy is independent of the frequency of the radiation.

The power received per unit area is simply $p_0/4\pi r^2$, where p_0 is the amount of power at the input of the isotropic source. This value is defined as the power flux density (pfd);

$$\text{pfd} = \frac{p_0}{4\pi r^2} \qquad \text{W/m}^2 \qquad (1.10)$$

Power Radiated and Received by Antennas in a Communication System

Consider the power transmitted and received via radio transmission in an ideal communication channel. Let

p_T	be the power of the transmitter in watts
g_T	be the transmit antenna gain
g_R	be the receive antenna gain
A_{eT}, A_{eR}	be the effective apertures of the transmit and receive antennas in meters
r	be the distance between the transmitter and receiver in meters

Then power transmitted $= p_T \times g_T$ watts. The power flux density in the desired direction at distance r is given by

$$\text{pfd} = \frac{p_T g_T}{4\pi r^2} \qquad \text{W/m}^2 \qquad (1.11a)$$

The product $p_T g_T$ is called the *equivalent isotropic radiated power* (EIRP), an important factor in link calculations. The pfd represents the received power that is intercepted by an ideal antenna with an aperture of 1 m^2, expressed in decibels

$$PFD = P_T + G_T - 10\log 4\pi r^2$$

$$= EIRP - 10\log 4\pi r^2 \qquad \text{dBW/m}^2 \qquad (1.11b)$$

where $P_T = 10\log p_T$, $G_T = 10\log g_T$ and $PFD = 10\log$ (pfd). This equation relates that no matter what, in a radio transmission, a very large portion of the transmitted energy, represented by the term $10\log 4\pi r^2$, is not seen by the receiver. Also, this loss is independent of the frequency, a point often lost if the equation is manipulated in terms of the gains of the transmit and receive antennas. To clarify this point, equation (1.11) can be expressed in terms of power p_r received at distance r as follows:

$$p_r = (\text{pfd})_t \times A_{eR} = \frac{p_T g_T}{4\pi r^2} g_R \frac{\lambda^2}{4\pi}$$

$$= p_T g_T g_R \left(\frac{\lambda}{4\pi r}\right)^2 \qquad \text{watts} \qquad (1.12a)$$

In decibels, this is

$$P_R = P_T + G_T + G_R - 20 \log \frac{4\pi r}{\lambda} \quad \text{dBW} \qquad (1.12\text{b})$$

This equation indicates that the power received has frequency-and distance-dependent terms. The combined dependence arises because of the receive antenna, required to focus the received energy to a point source. In other words, the term $20 \log (4\pi r/\lambda)$, often referred to as the *range loss*, represents the power loss between two isotropic antennas within a particular range and at a particular frequency. As far as spreading of the radio energy from a point source is concerned, the energy follows the classic inverse square law, independent of the frequency, described by equation (1.11). Regardless, either equation, (1.11) or (1.12), can be used for link calculations.

Example 1.1 In a microwave radio relay link, repeater stations are located 50 km apart. They are equipped with 10-W high-power amplifiers (HPAs) and antennas with a gain of 30 dB. Assuming the transmission-line and filter losses to be 2 dB, compute (1) the EIRP of the transmitter and the power flux density at the receive antenna and (2) given an antenna efficiency of 0.8, calculate the diameter of a circular parabolic antenna to realize the 30 dB gain in the 4 GHz frequency band.

Solution

1. EIRP $= 10 \text{ dBW} + (-2 \text{ dB}) + 30 \text{ dB}$
   $= 38 \text{ dBW}$

   PFD $=$ EIRP $- 10 \log 4\pi r^2$
   $= 38 - 105$
   $= -67 \text{ dBW/m}^2 \text{ (or } 200 \text{ nW/m}^2)$

   Thus, the aperture of a square meter area, located 50 km from the transmitter, intercepts 200 nW of RF power, radiated by the transmit antenna.

2. The aperture to realize a specific antenna gain is derived from

$$A = \frac{G \lambda^2}{\eta \, 4\pi} \quad \text{m}^2$$

The power received by the aperture depends on the radio frequency even though the power flux density remains unchanged. For a circular parabolic antenna with a diameter D, the aperture area A equals $(\pi/4) \times D^2$. By expressing the antenna gain in decibels, and the diameter in meters, we obtain

$$G = 20 \log D + 20 \log f_{\text{MHz}} + 10 \log \eta - 39.6 (\text{dB})$$

For $f = 4000$ MHz, $G = 30$ dB, and $\eta = 0.55$, D is calculated to be one meter.

PART II NOISE IN A COMMUNICATION CHANNEL

1.5 NOISE IN COMMUNICATION SYSTEMS

Noise in its broadest definition consists of any undesired signal in a communication circuit. Noise represents the fundamental constraint in the transmission capacity of a communication system. It is also of interest to the ITU and domestic radio regulation agencies. An essential element of radio regulations is the specification of allowable levels of radiation to other existing or proposed systems. Such a restriction to control the interference between communication systems is essential in a multisystem environment. Without the restriction, it is not possible to design a reliable communication system. Typically, the regulating agencies specify the allowable level of radiation outside the intended geographic area and allocated frequency spectrum. Also, the regulations include the interference guidelines of competing services and systems. The noise sources external to a communication system are:

1. Cosmic radiation, including that from the sun
2. Anthropogenic (synthetic; caused by humans) noise such as powerlines, electrical machinery, consumer electronics, and other terrestrial sources
3. Interference from other communications systems

There is little to be done about the cosmic noise except to ensure that it is considered in the system design. Generally, anthropogenic noise occurs at low frequencies and is rarely a problem for communication systems operating above 1 GHz. Interference from other communication systems is strictly controlled by ITU and domestic agencies. Typically, this requires control of transmitter powers, antenna patterns, generation, and suppression of frequencies outside the allocated spectrum. The regulations ensure that these sources of unwanted radiation are kept to a minimum, and significantly below the levels of noise inherent in the design of a communication system. The dominant sources of noise in a communication system include the following:

- Interference from adjacent copolarized channels
- Interference from adjacent cross-polarized channels
- Multipath interference
- Thermal noise
- Intermodulation (IM) noise
- Noise due to channel imperfections

These noise sources are now reviewed.

1.5.1 Adjacent Copolarized Channel Interference

One thing that is common to all communication systems is the channelization of the frequency spectrum. Channelization allows flexibility in meeting multidestination

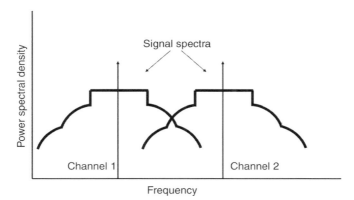

Figure 1.9   Adjacent channel interference.

traffic requirements, as well as in maximizing the communication capacity of the system. In a subsequent section, modulation schemes employed for long-haul transmission are described. The most commonly used modulation scheme is frequency modulation (FM). One property of FM is that it produces sidebands of decreasing magnitudes extending to infinity. Consequently, there is no escape from some energy spilling over in the adjacent bands causing distortion. To a degree, this spillage can be controlled by the choice of channelization filters, highlighting the criticality of filter networks in a communication system. Figure 1.9 is a pictorial representation of adjacent channel interference.

1.5.2   Adjacent Cross-Polarized Channel Interference

The nature of EM radiation allows the polarization of energy in a given direction. This property is exploited in communication systems by employing orthogonal polarization of antenna beams. It allows frequency reuse that doubles the available bandwidth. Polarization can be linear or circular. The primary limitation is the level of cross-polarization isolation that is achievable in practical antenna networks. Therefore, this interference is totally dependent on and controlled by the antenna design. Typical cross-polarization isolation in practical systems is of the order of 27–30 dB. It should be noted that polarization can be altered, when EM radiation propagates through the atmosphere. This also needs to be accounted for in the design.

1.5.3   Multipath Interference

This distortion mechanism is caused by obstacles in the spread energy field of the transmitter where the energy is reflected in the direction of the receiver. Such obstacles can be tall buildings or trees or foliage in an urban environment. Also, distortion can stem from refractive-index gradients and other atmospheric effects. Interference occurs when the reflected rays from obstacles or ground are received by the receiver. These interfering signals are time-displaced echos of the original transmission. Figure 1.10 is a pictorial description of multipath interference. In a

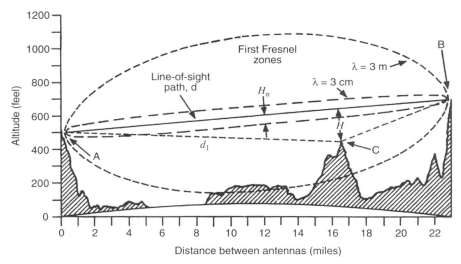

Figure 1.10   Multipath interference. (From Ref. 1.)

fixed-LOS environment, the radio paths are optimized by taking such obstructions into account. This is not the case when it comes to mobile communication systems. The mobile terminal can be fixed or in motion. Even a handheld terminal might have micromotion. When a terminal is in motion, the path characteristics are constantly changing and multipath propagation tends to become the limiting factor. Mobile units experience multipath scattering, reflection, and diffraction by various obstructions and buildings in the vicinity. Although constructive and destructive fading can become quite complex, there are ways to deal with this problem, including frequency and space diversity and forward error correction. Regardless of these compensation techniques, multipath interference continues to be the principal cause of missed calls, fading, and disruptions in mobile communications. The limitation of the available bandwidth for this service further exacerbates the problem. This topic and other issues related to radio propagation are described by Freeman [7].

1.5.4   Thermal Noise

Thermal noise permeates all communication systems and is the ultimate limit to their performance. For this reason, thermal noise is addressed in more depth. A fundamental source of noise is due to the constant agitation of molecules and its constituents in a conductor. This agitation at the atomic level is a universal characteristic of all matter. Molecules consist of a nucleus and a cloud of electrons around it. The nucleus consists of neutrons and protons, and the number of protons is equal to the number of electrons, thereby neutralizing the molecules. Electrons and positive ions (nuclei with the electrons removed) are, on average, distributed uniformly in a conductor, rendering the structure electrically neutral. The electrons in a conductor are in continual random motion, and in thermal equilibrium with the molecules. However, the random agitation of the electrons creates statistical fluctuations from the neutrality and that translates into electrical noise. The mean square velocity of electrons is directly proportional to the absolute temperature. The equipartition law due to Boltzmann and Maxwell (and the works of Johnson and Nyquist) states that for a thermal noise source, the available power in a 1 Hz bandwidth is given by

$$p_n(f) = kT \qquad W/Hz \qquad (1.13)$$

where k is the Boltzmann constant = 1.3805×10^{-23} J/K and T is the absolute temperature of the thermal noise source in degrees Kelvin. At a room temperature of 17°C or 290 K, the available power is $p_n(f) = 4.0 (10^{-21})$ W/Hz or -174.0 dBm/Hz. The result given by the equipartition theory is one of a constant power density spectrum versus frequency. Because of this property, a thermal noise source is referred to as a *white noise source*, an analogy to white light that contains all the visible wavelengths of light. In all the reported measurements, the available power of a thermal noise source has been found to be proportional to the bandwidth over any range, from direct current to the highest microwave frequencies. If the bandwidths are unlimited, the results of the equipartition theory states that the available power of a thermal noise source should also be unlimited. Obviously, this is not possible. The reason is that the equipartition theory is based on classical mechanics, which tends to break down as the higher frequencies are approached. If the principles of quantum mechanics are applied to the problem, kT must be replaced by $hf/[\exp(hf/kT) - 1]$, where h is Planck's constant = 6.626×10^{-34} J-s. By applying this result to the expression for available power of a thermal noise source [1], we obtain

$$p_n(f) = \frac{hf}{\exp(hf/kT) - 1} \qquad W/Hz \qquad (1.14)$$

This relationship demonstrates that at arbitrarily high frequencies, the thermal noise spectrum eventually drops to zero, but this does not mean that noiseless devices can be built at these frequencies. A quantum noise term equal to hf needs to be added to equation (1.14) in this case. Figure 1.11 offers a plot of the noise density as a function of frequency. The transitional region occurs at about 40 GHz for $T = 3$ K, at 400 GHz for $T = 30$ K, and at 4000 GHz at room temperature. For most practical purposes, the available noise power of a thermal noise source can be assumed to be directly proportional to the product of the bandwidth of the system or detector, and the absolute temperature of the source.

Thus

$$p_a = kTB \quad W \qquad (1.15a)$$

where B is the noise bandwidth of the system or detector in hertz and p_a is the available noise power in watts. By assuming an ambient temperature of 290 K and expressing the available noise power in dBm gives

$$p_a = -174 + 10 \log B \qquad dBm \qquad (1.15b)$$

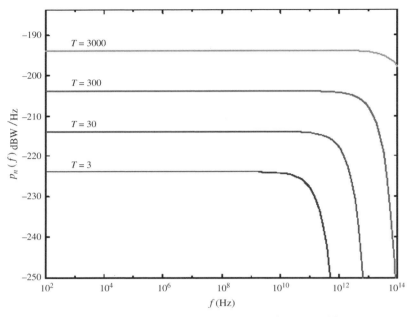

Figure 1.11 Thermal noise power densities as a function of frequency.

This represents the minimum amount of noise power that must ultimately limit the signal-to-noise ratio (SNR) of a signal. This relationship represents an average value, it does not tell us anything about the statistical distribution. As stated earlier, thermal noise is attributed to the random motions of electrons in the conductors. Therefore, thermal noise might be regarded as the superposition of an exceedingly large number of independent electronic contributions. It is well known in the field of statistics that the limiting form for the distribution function of the sum of a large number of independent quantities that can have various distributions is a Gaussian function. This result is known in statistics as the *central limit theorem*. Consequently, thermal noise satisfies the theoretical conditions for a Gaussian function and must follow this distribution. The Gaussian probability density function for the zero mean is reflected in Figure 1.12*a*, and its equation is

$$p(V) = \frac{1}{\sigma_n \sqrt{2\pi}} \exp\left(\frac{-V^2}{2\sigma_n^2}\right) \tag{1.16}$$

where V represents the instantaneous voltage and σ_n the standard deviation. The Gaussian distribution function is shown in Figure 1.12*b* and is given by the integral of equation (1.16)

$$P(V) = \frac{1}{\sigma_n \sqrt{2\pi}} \int_{-\infty}^{V} \exp\left(\frac{-x^2}{2\sigma_n^2}\right) dx \tag{1.17}$$

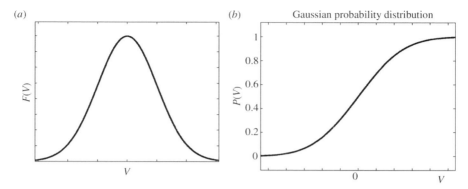

Figure 1.12 (*a*) Gaussian density and (*b*) distribution functions.

It can be readily shown that the mean square voltage (the expected value of V^2) is equal to the variance, σ_n^2. Thus, the rms (root mean square) voltage of the Gaussian distributed noise source is given by σ_n, the standard deviation.

Gaussian noise has a probability greater than zero of exceeding any finite magnitude no matter how large. As a result, the peak factor, given by the ratio of peak to rms voltage, does not exist for a thermal noise signal. For this particular case, it is convenient to modify the definition of the peak factor to be the ratio of the value, exceeded by the noise a certain percentage of the time to the rms noise value. This percentage of time is commonly chosen to be 0.01%. A table of the normal distribution indicates that signal magnitudes greater than $3.89\sigma_n$ (i.e., $|V| > 3.89\sigma_n$) occur less than 0.01% of the time.

Since σ_n is the rms value of the noise signal, the peak factor for a thermal noise signal is 3.89, or 11.80 dB. Inclusion of 0.001% peaks increases the peak factor by 1.1 dB, raising it to a value of 12.9 dB. The fact that thermal noise is white, as well as Gaussian, has led many engineers into carelessly viewing white and Gaussian noise as synonymous, which is not always the case. For example, passing Gaussian noise through a linear network such as a filter will leave it Gaussian but can drastically change the frequency spectrum. A single impulse does not exhibit a Gaussian amplitude distribution, but has a flat or white frequency spectrum.

Effective Input Noise Temperature Since the available noise power of a thermal noise source is directly proportional to the absolute temperature of the source, an equivalent noise temperature can be attributed to it [1]. For resistive elements, the noise temperature is equal to the physical temperature of the resistor; that is, if a given noise source produces an available power of p_a watts in a small frequency interval of df hertz, the noise temperature of the noise source is given by $T = p_a/k\,df$. It should be emphasized that the concept of noise temperature does not have to be restricted to the noise sources alone, and that the noise temperature does not have to equal the physical temperature of the source. For example, consider an antenna. The output noise is simply the noise collected by the aperture due to the radiating elements in the antenna's field of view. The physical temperature of the antenna has no bearing on this, and the noise temperature can still be used to define the noise power from the antenna.

Consider a two-port network with an available gain of $g_a(f)$. When it is connected to a noise source having a noise temperature of T, the available noise power in a small band df at the output of the network is $p_{no} = g_a(f)kT\,df + p_{ne}$. This power consists of two components: (1) power due to the external noise source, $g_a(f)kT\,df$, and (2) the power due to the internal noise sources of the network p_{ne}, which is the available noise power of the network when the input of the network is connected to a noise-free source. The effective noise temperature T_e of this equivalent representation of the internal noise source of the noisy network is then given by

$$T_e = \frac{p_{ne}}{g_a(f)k\,df} \tag{1.18}$$

The available noise power at the output of the network in terms of the effective input noise temperature now becomes

$$p_{no} = g_a(f)\,k\,(T + T_e)df \tag{1.19}$$

The effective input noise temperature T_e can vary as a function of the frequency, depending on how g_a and p_{ne} vary. The concept of the effective input noise temperature is useful when the source noise temperature differs from that of the standard temperature. It has a distinct advantage when the noise performance of a complete communications system is being evaluated. Another concept that is useful in analyzing communication systems is the noise figure.

Noise Figure   The IRE (Institute of Radio Engineers, precursor to IEEE) definition of the noise factor for a two-port network is as follows: "The noise figure (noise factor) at a specified input frequency is the ratio of the total noise power per unit bandwidth at a corresponding output frequency, available at the output when the noise temperature of the input source is standard (290 K); to that portion of this output power engendered at the input frequency by the input source." In terms of this definition

$$\text{Noise figure} = n_F = \frac{p_{no}}{g_a(f)kT_0\,df} \tag{1.20}$$

where $T_0 = 290$ K and is referred to as the standard temperature. A noise figure such as that described for a narrowband df is called a *spot noise figure*, which can vary as a function of frequency. The noise figure can also be related to the effective noise temperature. In equation (1.19), if T is replaced by T_0, as required in the definition of the noise figure, the output noise power p_{no} is then given by $g_a(f)k(T_0 + T_e)df$. The output noise in terms of the noise figure is given by equation (1.20). Equating these two expressions yields the relationship between the noise figure and the effective noise temperature as

$$n_F = 1 + \frac{T_e}{T_0} \tag{1.21}$$

and

$$T_e = T_0(n_F - 1) \tag{1.22}$$

The concept of the noise figure is most useful when the input source has a noise temperature approximately equal to the standard temperature. Consider the expression for the noise power at the output of the network as given by equation (1.20). By rewriting this equation in terms of dBm (reference 1 mW), we have

$$P_{no} = N_F + G_a + 10\log df - 174 \qquad \text{dBm} \tag{1.23}$$

The symbols are defined as follows:

$$P_{no} = \frac{10 \log p_{no}}{10^{-3}}$$

$$N_F = 10\log n_F$$

$$G_a = 10\log g_a \tag{1.24}$$

Thus, the available noise power of the two-port network in dBm can be written as the sum of the noise power of the thermal noise source in dBm, the available gain of the network in dB, and the noise figure of the network in dB. As a result, the effects of internal noise sources of a two-port network are accounted for by adding the noise figure in dB to the available noise power of the source in dBm.

Noise of a Lossy Element   Any signal is attenuated by the lossy elements in its path, regardless of the physical temperature of the element. This is equivalent to saying that any lossy element in the signal path contributes noise. By adopting arguments similar to the preceding sections, the effective input noise temperature of a lossy element is derived as [1]

$$T_e = T(l_a - 1) \tag{1.25}$$

where l_a is the loss ratio of the element. The loss L_a of the element in dB is given by $L_a = 10\log l_a$. The noise figure of the lossy element is then given by

$$n_F = 1 + \frac{T}{T_0}(l_a - 1) \tag{1.26}$$

If the lossy element is at the standard temperature T_0, then

$$n_F = l_a \quad \text{and} \quad N_F = L_a \text{ dB} \tag{1.27}$$

As an example, for a transmission line with a loss of 1 dB at room temperature, the effective input noise temperature is given by $T_e = 75$ K, and the noise figure is simply 1 dB.

Example 1.2   What is the amount of thermal noise power radiated by a source at a room temperature of 290 K in a bandwidth of 50 MHz? What are the peak values of this noise that do not exceed 1% and 0.1% of the time?

Solution   Using equation (1.14), the thermal noise is given by

$$N_T = kTB$$
$$= -228.6 + 10 \log T + 10 \log B \quad \text{watts}$$

For $T = 290$ K and $B = 50 \times 10^6$ Hz, we have

$$N_T = -127 \, \text{dBW}$$
$$= 0.2 \, \text{pW}$$

As a result, an antenna that is pointed at a 290 K source, where the source fills the antenna beamwidth, receives 0.2 pW of noise power in a bandwidth of 50 MHz. As described in Section 1.5.4, the peak factors for thermal noise not exceeding 1% and 0.1% of the time are 11.8 and 12.9 dB, respectively. As a consequence, the peak value of the noise can be as high as -115.2 dBW (3.0 pW) for 1% of the time and -114.1 dBW (3.9 pW) for 0.1% of the time.

Example 1.3   What is the noise power at the output of a low noise amplifier (LNA) with a noise figure of 2 dB and a gain of 30 dB over a bandwidth of 500 MHz? What is the equivalent noise temperature of the LNA?

Solution   Equations (1.20)–(1.22) lead to

$$N_{\text{LNA}} = N_F + G_a + 10 \log df - 174 \, \text{dBm}$$
$$= (2 \, \text{dB}) + (30 \, \text{dB}) + 10 \log 500 \times 10^6 - 174$$
$$= -55 \, \text{dBm}$$

The equivalent noise temperature is given by

$$T_{\text{LNA}} = T_0(n_F - 1)$$
$$= 290(1.585 - 1)$$
$$= 169.6 \, \text{K}$$

1.5.5   Noise in Cascaded Networks

The two networks connected in tandem in Figure 1.13 have effective input temperatures T_{e1} and T_{e2} and available gains g_1 and g_2.

Suppose that these two tandem networks are connected to a noise source with a noise temperature of T. In a small frequency band df the noise power at the output due to the noise source alone is $g_1 g_2 kT \, df$. The noise power due to noise sources in

Figure 1.13   Noise in cascaded networks.

the first network is $g_1g_2kT_{e1}\,df$ and in the second network is $g_2kT_{e2}\,df$. The total noise appearing at the output of the second network is $kg_2(g_1T + g_1T_{e1} + T_{e2})df$. Since the portion of this noise due to the noise sources internal to the two networks is $kg_2(g_1T_{e1} + T_{e2})df$, the effective input temperature T_e of the two networks in tandem is therefore given by

$$T_e = \frac{kg_2(g_1T_{e1} + T_{e2})df}{g_1\,g_2\,k\,df}$$

$$= T_{e1} + \frac{T_{e1}}{g_1} \tag{1.28}$$

This result can be easily generalized to n networks in tandem. The resulting effective input noise temperature is

$$(T_e)_1 = T_{e1} + \frac{T_{e2}}{g_1} + \frac{T_{e3}}{g_1g_2} + \cdots + \frac{T_{en}}{g_1g_2\cdots g_{n-1}} \tag{1.29a}$$

It should be noted in equation (1.29a) that the reference point for the effective noise temperature is the input terminal of the first element in the chain. This relationship can be readily modified to compute the effective noise temperature with reference to any terminal within the chain. For example, if the chosen reference is the input port of element $3(T_{e3})$, then the effective noise temperature is given by

$$(T_e)_3 = T_{e1}\,g_1\,g_2 + T_{e2}\,g_2 + T_{e3} + \frac{T_{e4}}{g_3} + \frac{T_{e5}}{g_3g_4} + \cdots \tag{1.29b}$$

Equation (1.29) is known as Friis formula, named in honor of H. T. Friis.

   This relationship is advantageous in system calculations where the selected reference is typically the input terminal of the LNA. Such a reference provides the evaluation of the figure of merit defined by the ratio of the effective antenna gain to the noise temperature (G/T).

   From the relationship between the noise figure and effective input noise temperature, it can be easily demonstrated that the resulting noise figure of n stages in tandem is

$$(n_F)_1 = n_{F1} + \frac{n_{F2} - 1}{g_1} + \cdots + \frac{n_{Fn} - 1}{g_1g_2\cdots g_{n-1}} \tag{1.30}$$

The significance of these two relationships becomes apparent when a multistage amplifier in which each stage has an available gain of at least 20 dB is considered.

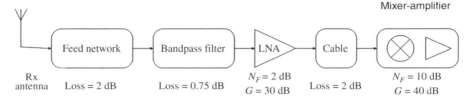

Figure 1.14   Example of noise calculations of a receive network.

If each stage has the same effective input noise temperature, then the noise contribution of only the first stage is significant. Only the noise sources occurring before or in the first stage of the amplifier are important so far as the noise calculations are concerned. However, if the gain of the first stage is small or if the noise contribution of the second stage is large, then it is the first two stages that essentially determine the overall noise temperature of the network.

Example 1.4   Calculate the noise temperature contributions in a 6 GHz, 500 MHz bandwidth receive section as denoted in Figure 1.14. Assume that the receive antenna has a noise temperature of 70 K.

Solution   The noise temperatures and associated gain/loss ratios for the various elements are computed as follows:

Feed Network:

$$l_1 = 1.5849$$

$$g_1 = \frac{l}{l_1} = 0.631$$

$$T_{e1} = 290\,(l_1 - 1) = 169.6\,\text{K}$$

Bandpass Filter:

$$l_2 = 1.1885$$

$$g_2 = \frac{l}{l_2} = 0.8414$$

$$T_{e2} = 290\,(l_2 - 1) = 54.7\,\text{K}$$

LNA:

$$n_{F3} = 1.5849$$
$$g_3 = 1000$$
$$T_{e3} = 290(n_{F3} - 1) = 169.6\,\text{K}$$

Cable:

$$l_4 = 1.5849$$
$$g_4 = \tfrac{1}{14} = 0.631$$
$$T_{e4} = 290\,(l_4 - 1) = 169.6\,\mathrm{K}$$

Mixer-Amplifier:

$$n_{F5} = 10$$
$$g_5 = 10,000$$
$$T_{e5} = 290\,(n_{F5} - 1) = 2610\,\mathrm{K}$$

The total system noise temperature referred to the input of the LNA is

$$(T_e)_{\text{sys}} = (T_{\text{ant}} + T_{e1})\,g_1 g_2 + T_{e2}\,g_2 + T_{e3} + \frac{T_{e4}}{g_3} + \frac{T_{e5}}{g_3 g_4}$$
$$= 127.2 + 46.0 + 169.6 + 0.17 + 4.14$$
$$= 347.1\,\mathrm{K}$$

It is interesting to note the relatively large contribution to the overall system noise temperature by the losses prior to the LNA and the relatively insignificant noise contributions by components after the LNA. These are important considerations in the design of communication systems, emphasizing the need to mitigate the feeder and input filter losses.

1.5.6   Intermodulation (IM) Noise

IM noise is generated by the nonlinearities present in communication systems. Like thermal noise, nonlinearities are present to some degree in all electrical networks. Nonlinearities can limit the useful signal levels in a system and thus become an important design consideration. The device that is the primary source of IM is the nonlinear high-power amplifier (HPA). It is an essential element of a communication system. The HPA's efficiency and the degree of nonlinearity are inversely related; thus the HPA characteristics and operating power levels are important design parameters.

Consider the voltage transfer characteristics of a general two-port, which can be a device, network, or system, as depicted in Figure 1.15. For a two-port memoryless nonlinear network, the transfer function is described by the Taylor series expansion:

$$e_0 = a_1 e_i + a_2 e_i^2 + a_3 e_i^3 + \cdots \tag{1.31}$$

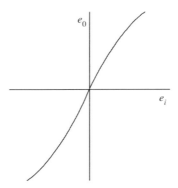

Figure 1.15 Transfer characteristics of a nonlinear two-port network.

For a single-frequency sinusoid input, $e_1 = A \cos(ax)$, we obtain

$$e_0 = a_1 A \cos ax + a_2 A^2 \cos^2 ax + a_3 A^3 \cos^3 ax + \cdots$$
$$= K_0 + K_1 \cos(ax) + K_2 \cos(2ax) + K_3 \cos(3ax) + \cdots \qquad (1.32)$$

where the K values are constants related to $a_1, a_2, a_3, \ldots$. Therefore, a single sinusoid input leads to an output containing the fundamental frequency and its harmonics. Similarly, if $e_i = A \cos \omega_{1i} + B \cos \omega_{2i} + \cos \omega_{3i} + \cdots$, then by using trigonometric identities, we obtain

$$e_0 = K_0 + K_{1i} f(\omega_i) + K_{2i} f(2\omega_i) + K_{3i} f(3\omega_i) + \cdots \qquad (1.33)$$

where

$f(\omega_i) =$ collection of first-order terms containing ω

$f(2\omega_i) =$ second-order terms such as $2\omega_1, \omega_1 + \omega_2, \omega_1 - \omega_2$

$f(3\omega_i) =$ third-order terms such as $3\omega_1, 2\omega_1 \pm \omega_2, 2\omega_2 \pm \omega_3,$
$\qquad \omega_1 + \omega_2 + \omega_3, \omega_1 + \omega_2 - \omega_3, \ldots$

$K =$ a distinct constant associated with each term

Consequently, the output contains harmonics and all possible combinations of the sum and difference frequencies of the input signals. The dc (direct-current) term is of no interest and is filtered out. The desired output corresponds to the linear case, given by the first-order products K_{1i} due to term a_1. All the other outputs are spurious and contribute to the objectionable noise and interference. As the number of input signals increases, the number of IM products grows rapidly. These IM products fall either inside or outside of the RF channel, depending on the order of the product and the separations between the carrier frequencies. In the classic case of a single channel per carrier (SCPC), frequency-division multiplex (FDM-FM) system, IM products are so widespread that they resemble noise. As a result, for RF channels employing a large number of carrier frequencies, there is a tradeoff between thermal noise and IM noise. As the number of carriers increases,

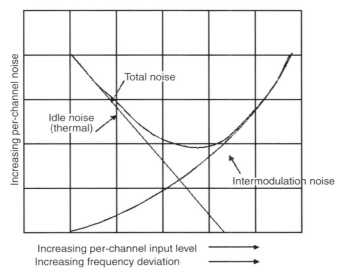

Figure 1.16   Effect of loading and thermal noise in a communication channel.

thermal noise per carrier is smaller, whereas the IM noise level is larger. There is some point of optimum loading of the wideband FM transmitter where the thermal noise and the IM noise in the channel when combined represent a minimum of the total system noise, as described in Figure 1.16. This provides guidance for the best operating power level for the amplifier. Typically, the third-order IM product is dominant and often forms part of the specification. Amplifiers are designed to have a carrier to third-order product values exceeding 20 dB. This topic is discussed further in Sections 1.8.3 and 1.8.4.

1.5.7   Distortion Due to Channel Imperfections

An ideal transmission channel transmits all signals without distortion over a certain bandwidth, and completely attenuates all signals outside this bandwidth. Such a performance is characterized by a channel that provides a constant loss (ideally zero) and linear phase (i.e., constant delay) over the passband and infinite attenuation outside it, as shown in Figure 1.17a,b. Such a filter performance is not feasible. The unit impulse response of such a filter exists for negative values of time, violating the condition of causality [10].

Although ideal filter characteristics are not feasible, it is possible to approach these characteristics as closely as desired. The characteristics of a practical filter are plotted in Figure 1.17c,d. For practical filters, there exists a tradeoff between the design complexity and its departure from ideal filter characteristics. Other components in the communication system, including amplifiers, frequency converters, modulators, cables, and waveguides, are wideband devices, exhibiting a minimal deviation from the flat amplitude and group delays over the narrow channel

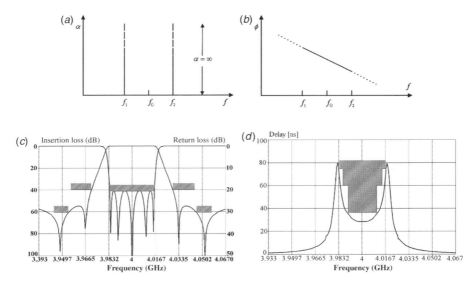

Figure 1.17 Transmission channel characteristics: (*a*) an ideal amplitude response; (*b*) an ideal phase response; (*c*) a practical amplitude respons; (*d*) a practical group delay response.

bandwidths. In other words, filter networks represent the controlling features of amplitude and phase characteristics of a transmission channel. This has been a key driver for filter R&D (research and development) for many decades.

All practical filters exhibit transmission deviations. Filters are passive and linear devices. Their response is time-invariant and has amplitude and phase shapes that are prescribed functions of frequency. Unlike nonlinear devices such as HPAs, no new frequencies are produced when an FM signal is passed through the filter. However, filters do change the relative amplitude and phase of the carrier and sidebands, which, in turn, is interpreted by the demodulator as additional modulation, causing distortion in the received signal. For insight into this process, consider an FM signal with many sidebands applied to a network with ideal transmission characteristics except at the frequency of one of the sideband components. The amplitude of this component gets altered, and is equivalent to adding an extraneous signal to the applied FM signal. As a result, the demodulated output signal consists of the desired signal, which is proportional to the input modulating signal, and the undesired signal. When many carriers are present, it can be demonstrated that the transmission deviations in an FM system can introduce baseband frequency components at the output that do not exist at the input. In this sense, the transmission deviations in an FM system have an effect similar to that of an amplifier nonlinearity. For that reason, the distortion introduced by transmission deviations is often called *IM noise*. There is no exact solution to the problem of determining IM noise produced by transmission deviations. An approximate approach for analog systems that has proved successful is described in Ref. 1 and is based on previous works [12–14].

The analysis assumes that the transmission characteristics of the channel are sufficiently smooth so that the gain and phase as a function of frequency can be represented by

$$Y_n(\omega) = \left[1 + g_1(\omega - \omega_c) + g_2(\omega - \omega_c)^2 + g_3(\omega - \omega_c)^3 + g_4(\omega - \omega_c)^4\right]$$
$$\times\ e^{i\left[b_2(\omega - \omega_c)^2 + b_3(\omega - \omega_c)^3 + b_4(\omega - \omega_c)^4\right]}$$

(1.34)

where

$$\omega_c = \text{carrier frequency in radians per second (rad/s)}$$
$$g_1, g_2, g_3, g_4 = \text{linear, parabolic, cubic, and quartic gain coefficients,}$$
$$\text{respectively}$$
$$b_2, b_3, b_4 = \text{parabolic, cubic, and quartic phase coefficients, respectively}$$

This assumption is consistent with the frequency response that can be readily achieved by microwave filter networks. The second assumption is that the FM signal has a sufficiently low modulation index, and that the bandwidth of the channel is much smaller than the carrier frequency. This is indeed the case for most communication systems. From these assumptions, it is possible to compute the distortion due to transmission deviations. Another related source of distortion is introduced by the HPAs that follow filter networks. The nonlinearity of amplifiers converts the amplitude variations introduced by filters to phase variations, causing distortion of the FM signal. If a limiter can successfully remove the amplitude variations before the signal reaches the device, it can be dismissed. A summary of these distortion terms and noise power (included in Appendix A1) is described Ref. 15. These values are applicable for analog FM transmission.

For digital systems, the impact of transmission deviations is relatively small, unless the data rate is very high. For advanced digital modulation schemes, sophisticated simulation tools are needed to compute distortion as a function of the amplitude and phase deviations in the RF channel. It leads to tradeoffs between in-band response (transmission deviations) and out-of-band attenuation, consistent with microwave filter technology and other system design parameters [16]. Such tradeoffs characterize the RF channels. In effect, RF channel filters control the amplitude and phase characteristics of a communication channel; that is, the channel filters define the effective usage of the available channel bandwidths.

It should be noted that for analog FM transmission, variations in the amplitude slope of a filter, followed by a nonlinear amplifier causes AM-to-PM conversion, resulting in intelligible (coherent) crosstalk between FM carriers. With digital modulation, crosstalk is unintelligible (noncoherent) but there is still a modulation transfer, causing an increase in the required E_b/N_0 (ratio of energy per bit to noise density) in the link. Digital transmission is relatively insensitive to variations in group delay unless, of course, it is very large and the data rate is high (short symbol duration). In most cases, the effect of the group delay slope is small on the link E_b/N_0.

1.5.8 RF Link Design

A communication link is characterized by a number of RF links in tandem. In this section, we describe the carrier-to-noise ratio for a single link, and the impact of cascading a number of such links.

Carrier-to-Noise Ratio (C/N) The carrier-to-noise ratio (C/N) of a signal is simply given by C/N, where C is the carrier power and N is the total noise present in a given bandwidth. If N_0 is the noise density as defined by the noise power in a bandwidth of 1 Hz, then, by virtue of equation (1.15), we have

$$N_0 = kT_s \quad \text{and} \quad N = N_0 B = kT_s B \tag{1.35}$$

where k is the Boltzmann constant, T_s is the total effective noise temperature, and B is the bandwidth of the carrier frequency. In a communication system, T_s is chosen as the total effective temperature referenced to the input of the LNA of the receive system. It includes the receive antenna noise temperature, lossy transmission lines, bandpass filter prior to the LNA, and the noise temperature of the LNA itself. If G_R is the receive antenna gain at the same reference (i.e., including the losses up to LNA), the carrier power at the reference point is [see equation (1.12b)]

$$C = \text{EIRP} + G_R - P_L \tag{1.36}$$

where $P_L = 20\log(4\pi r/\lambda)$ is referred to as the path loss between the transmit and receive antennas. Therefore

$$\frac{C}{N} = \text{EIRP} + G_R - P_L + 228.6 - 10\log T_s - 10\log B$$

$$= \text{EIRP} + \frac{G}{T_s} - P_L + 228.6 - 10\log B \tag{1.37}$$

This represents the link equation for the thermal noise of the system. The factor $G_R - 10\log T_s$ or G/T_s represents the figure of merit of the receiving system. The link equation can also be expressed in the alternative form

$$\frac{C}{N_0} = \text{EIRP} + \frac{G}{T_s} - P_L + 228.6 \tag{1.38}$$

$$\frac{C}{T_s} = \text{EIRP} + \frac{G}{T_s} - P_L \tag{1.39}$$

Multiple RF Links in Tandem A communication channel, consisting of a number of RF links, is described in Figure 1.18.

Since the n links in tandem are physically different, the noise power generated in each of the links is not coherent with those generated in the other links. Thus, the noise power of all the links can be summed up as independent power sources to attain the total end-to-end noise power contribution and the overall C/N. The noise power of each link can be referenced to the carrier level, which can be

(a)

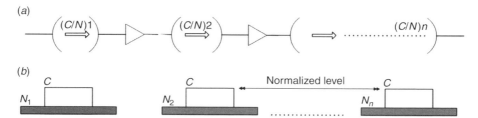

(b)

Figure 1.18 A communication channel depicting (a) RF Links in tandem and (b) normalized thermal noise of each link.

normalized to unity. Therefore, by adding all the noise floors N_i, the overall carrier-to-noise ratio $(C/N)^*$ is given by

$$\left[\left(\frac{C}{N}\right)^*_T\right]^{-1} = \left[\left(\frac{C}{N}\right)^*_1\right]^{-1} + \left[\left(\frac{C}{N}\right)^*_2\right]^{-1} + \cdots + \left[\left(\frac{C}{N}\right)^*_n\right]^{-1} \qquad (1.40)$$

The terms $(C/N)^*_1$, $(C/N)^*_2$ signify that the C/N is expressed in terms of ratios, as opposed to decibels.

Example 1.5 A transmit earth station has a 1 kW output power and an antenna with a gain of 55 dB. The transmission line and filter losses between the amplifier and antenna amount to 2 dB. Using Example 1.4 as reference

1. Compute the value of EIRP, the power flux density, and the C/N_0 for the 6 GHz uplink for a satellite located at 40,000 km from the surface of the earth. Assume that the receive antenna at the satellite has a gain of 25 dB and includes a margin of 3 dB due to losses through the atmosphere and antenna pointing accuracy,
2. Compute the value of the carrier to thermal noise at the satellite for a 36-MHz RF channel.
3. Compute the value of the combined uplink–downlink thermal noise, assuming that the power radiated by the satellite to the earth station in the 4-GHz downlink beam results in a carrier to noise ratio of 20 dB. What is the impact on the total noise, if the uplink power is reduced by 10 dB?

Solution
1. Uplink $\text{EIRP} = 30\,\text{dBW} + (-2\,\text{dB}) + 55$
$$= 83\,\text{dBW}$$

Power flux density, $\text{PFD} = \text{EIRP} - 10\log 4\pi r^2$
$$= 83 - 163$$
$$= -80\,\text{dBW/m}^2$$

This represents a power density of 10 nW per square meter at the satellite located in the geostationary orbit. From equation (1.38), we have

$$\frac{C}{N_0} = \text{EIRP} - 3 - 20 \log \frac{4\pi r}{\lambda} + G_R - L + 228.6 - 10 \log T_s$$

$$= 83 - 3 - 20 \log \frac{4\pi \times 40 \times 10^6}{(3 \times 10^8)/6 \times 10^9} + 25 - 2.75 + 228.6 - 10 \log 346.9$$

$$= 83 - 3 - 200 + 25 - 2.75 + 228.6 - 25.4$$
$$= 108 \text{ dBHz}$$

2. $\dfrac{C}{N} = \dfrac{C}{N_0} - 10 \log 36 \times 10^6 = 32.44 \text{ dB}$

3. The combined value of the carrier to thermal noise, in terms of ratios, is given by

$$\left[\left(\frac{C}{N}\right)^*_T\right]^{-1} = \left[\left(\frac{C}{N}\right)^*_{\text{UL}}\right]^{-1} + \left[\left(\frac{C}{N}\right)^*_{\text{DL}}\right]^{-1}$$

$$\left(\frac{C}{N}\right)^*_{\text{UL}} = 10^{3.244} \quad \text{and} \quad \left(\frac{C}{N}\right)^*_{\text{DL}} = 10^2$$

The total thermal noise ratio is, therefore, given by

$$\left[\left(\frac{C}{N}\right)^*_T\right]^{-1} = 10^{-3.244} + 10^{-2}$$

$$= 0.00057 + 0.01$$

$$\approx 0.01$$

For the ratio in decibels (dB), we obtain $(C/N)_T = 20$ dB.

If the uplink power is reduced by 10 dB, the $(C/N)_{\text{UL}}$ is reduced to 22.44 dB, and the carrier to total thermal noise power is

$$\left[\left(\frac{C}{N}\right)^*_T\right]^{-1} = 10^{-2.244} + 10^{-2}$$

$$= 0.0057 + 0.01$$

$$= 0.0157$$

Since this represents a carrier-to-noise ratio of 18.04 dB, the impact on the total noise is no longer negligible.

1.6 MODULATION–DEMODULATION SCHEMES IN A COMMUNICATION SYSTEM

In a communication system, the baseband signal, consisting of a large number of individual message signals, needs to be transmitted over the communication medium, separating the transmitter from the receiver. The efficiency of the transmission requires that this information be processed in some manner before it is transmitted. Modulation is the process whereby the baseband signal is suitably impressed on a carrier signal to increase its efficiency for transmission over the medium. Modulation can shift the signal frequencies to facilitate transmission or change the bandwidth occupancy, or it can alter the form of the signal to optimize the noise or distortion performance. At the receiver, a demodulation scheme reverses this process.

Modulation techniques are categorized as linear or nonlinear, depending on whether the modulated signal varies linearly (i.e., the superposition holds) or varies nonlinearly with the message.

There are two forms of modulation: amplitude modulation and angle (phase or frequency) modulation. The process of modulation is represented by

$$M(t) = a(t) \cos[(\omega_c t + \phi(t)] \tag{1.41}$$

Here $a(t)$ represents the amplitude of the sinusoidal carrier and $\omega_c t + \phi(t)$ is the phase angle. Although both the amplitude and angle modulation can be present simultaneously, an amplitude-modulated system is one in which $\phi(t)$ is a constant and $a(t)$ is made proportional to the modulating signal. Similarly, in an angle-modulated system, $a(t)$ is held constant and $\phi(t)$ is rendered proportional to the modulating signal.

1.6.1 Amplitude Modulation

For the amplitude modulated wave, we have

$$M(t) = a(t) \cos \omega_c t \tag{1.42}$$

where the carrier is at frequency f_c and $a(t)$ is the modulating time function. If $a(t)$ is a single-frequency sinusoid of unit amplitude at frequency f_m, then $a(t) = \cos \omega_m t$, and the modulated wave is

$$M(t) = \cos \omega_m t \cos \omega_c t$$

This can be expanded to

$$M(t) = \frac{1}{2} \cos (\omega_c - \omega_m)t + \frac{1}{2} \cos (\omega_c + \omega_m)t \tag{1.43}$$

The modulated wave contains no component at the original carrier and only a sideband on either side of the carrier, spaced f_m hertz from the carrier, as reflected in Figure 1.19.

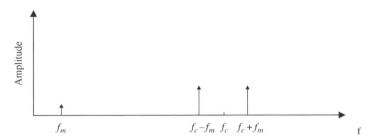

Figure 1.19   Amplitude-modulated carrier with a single sinusoid modulating frequency.

The effect of the product modulation is to translate $a(t)$ in the frequency domain so that it is reflected symmetrically about f_c. It can be shown that this is true for a complex waveform as well. If either sideband is rejected by a filter, the result is a single-sideband (SSB) signal, representing a pure frequency translation. A more general representation of the amplitude modulation is

$$M(t) = [1 + ma(t)] \cos \omega_c t \tag{1.44}$$

This expression is equivalent to adding a dc term of magnitude unity. Also, it is imperative that

$$|ma(t)| < 1 \tag{1.45}$$

so that the envelope of modulated wave remains undistorted, as denoted in Figure 1.20.

Here, m is defined as the modulation index with a maximum value of unity, representing 100% modulation. It represents the relative magnitude of the modulating wave for a carrier frequency of unity magnitude. The modulated wave is derived as from

$$M(t) = \cos \omega_c t + \frac{m}{2} \cos (\omega_c - \omega_m)t + \frac{m}{2} \cos (\omega_c + \omega_m)t \tag{1.46}$$

The average power in each sideband is $m^2/4$ or a total sideband power of $m^2/2$ watts. Thus, for 100% modulation, only one third of the total power is in the information bearing sideband. For complex signals, the power in the sidebands is considerably less, amounting to a few percent of the total power. A second drawback of the AM wave is its sensitivity to amplitude deviations in the signal path. Any such deviation can result in distortion of the signal. Furthermore, for AM signals, linear HPAs are necessary for signal amplification to avoid excessive distortion of the signal. There are limits to obtaining practical linear amplifiers with an adequate gain and power, rendering AM impractical for long-haul transmissions. However, one substantial advantage of AM, is the preservation of the bandwidth. As a consequence, it finds application for frequency translation and in multiplexing individual message channels by translating them to higher frequencies. A good example of amplitude modulation is in the formation of a baseband signal with a large number of voice channels.

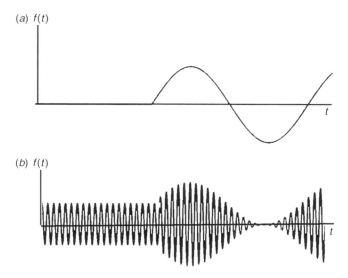

Figure 1.20 Amplitude modulation of a carrier: (*a*) modulating signal; (*b*) amplitude-modulated carrier.

1.6.2 Formation of a Baseband Signal

A message channel is composed of a large number of independent signals multiplexed to form a composite signal that occupies a continuous spectral range, referred to as the *bandwidth* of the signal. A hierarchy has been established for North American telephony communication systems to allow standardization and a high degree of commonality. The basic message channel, although originally intended for voice transmission, has also been adopted for data transmission. The basic group consists of 12 channels, each 4 kHz wide extending over the band 60–108 kHz. The equipment can be considered as a series of modulators with distinct carrier frequencies, followed by the appropriate bandpass filters and then multiplexed to form the composite signal as shown in Figure 1.21.

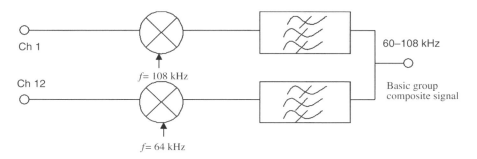

Figure 1.21 Formation of a message signal.

Here, modulation is amplitude modulation, and the single sideband is extracted via the bandpass filters. Amplitude modulation is a linear process, although the mixer to modulate the carrier is inherently a nonlinear device. Here, the mixer is used in a quasi-linear mode, and simply translates the modulation frequency symmetrically about the carrier frequency. This group of 12 channels, occupying a 48 kHz bandwidth, is a basic building block in the Bell system. The next step in the frequency-division multiplex (FDM) hierarchy is the combination of five groups in a 60-channel supergroup, for a bandwidth of 240 kHz, occupying a 312–552 kHz frequency band. Modern broadband transmission systems are capable of larger and larger systems, extending to groups of up to 3600 message channels. Although the grouping described here is generally used for long-haul Bell systems, it is not a universal standard. Along with telephony channels, television channels occupy bandwidths of 4–6 MHz or data channels ranging from kilobits to several megabits per second range, or Internet traffic. Therefore, a typical composite signal, referred to as a *baseband signal*, can vary from a simple basic group to a combination of groups of voice channels, television signals, and data channels. The signal composition is determined by the traffic requirements of the system. The hierarchy of data rates exists all the way up to optical fiber rates of 40 Gbps (gigabits per second).

1.6.3   Angle-Modulated Signals

The modulated wave for angle modulated signals is represented by

$$M(t) = A \cos[\omega_c t + \phi(t)] \tag{1.47}$$

Phase modulation (PM) is defined as angle modulation in which the instantaneous phase deviation $\phi(t)$ is proportional to the modulating signal voltage. Frequency modulation is angle modulation (FM) in which the carrier varies with the integral of the modulating signal. The average power for a PM or an FM wave is proportional to the square of the voltage and results in

$$P(t) = A_c^2 \cos^2[\omega_c t + \phi(t)]$$
$$= A_c^2 \left[\frac{1}{2} + \frac{1}{2} \cos(2\omega_c t + 2\phi(t)) \right] \tag{1.48}$$

The second term is assumed to consist of a large number of sinusoids about the carrier frequency $2f_c$, with the average value of zero and

$$P_{av} = \frac{A_c^2}{2} \tag{1.49}$$

Thus, the average power of a FM wave is the same as the average power in the absence of modulation. This represents a major advantage, compared to the AM, where the average power of information bearing sidebands is a third or less than that of the average power of the carrier frequency. However, there are no free lunches, and this power advantage comes at the expense of increased bandwidths. FM frequency analysis is quite complicated and beyond the scope of this book. A brief account of the key results and underlying assumptions are now described.

Spectra of Analog Angle-Modulated Signals

Narrowband FM Phase and frequency modulation are special cases of angle modulation, and one form is easily derived from the other. The discussion is confined to the frequency modulation, owing to its wider range of applications for both analog and digital communications systems. A frequency-modulated carrier is shown in Figure 1.22.

The modulating signal is assumed to be a repetitive sawtooth of period T, where $(2\pi/T) \ll \omega_c$. As the sawtooth modulating signal increases in magnitude, the FM signal oscillates more rapidly, resulting in the widening of the frequency spectrum. Note that its amplitude remains unchanged.

Analysis of the FM process is inherently much more complicated than that for AM, due to the nonlinearity of the FM process. Assume a sinusoidal modulating signal $v(t)$ at frequency f_m:

$$v(t) = a \cos \omega_m t \tag{1.50}$$

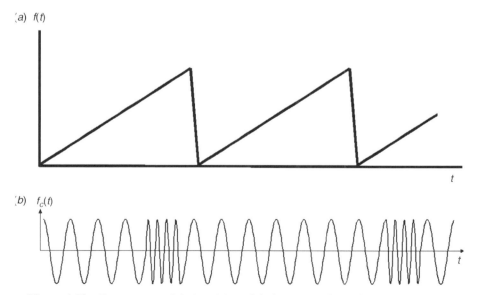

(a) $f(t)$

(b) $f_c(t)$

Figure 1.22 Frequency modulation: (*a*) modulating wave; (*b*) modulated FM carrier.

The instantaneous radian frequency ω_i is then

$$\omega_i = \omega_c + \Delta\omega \cos \omega_m t, \quad \Delta\omega \ll \omega_c \tag{1.51}$$

where $\Delta\omega$ is a constant, depending on the amplitude a. As a result, the instantaneous radian frequency varies about the unmodulated carrier frequency ω_c at the rate ω_m of the modulating signal and with a maximum deviation of $\Delta\omega$ radians. The phase variation $\theta(t)$ for this special case is computed by

$$\theta(t) = \int \omega_i dt = \omega_c t + \frac{\Delta\omega}{\omega_m} \sin \omega_m t + \theta_0 \tag{1.52}$$

where θ_0 is a constant, providing a reference for the phase. By choosing it to be zero, we see that the frequency-modulated carrier is

$$M(t) = \cos(\omega_c t + m\sin \omega_m t) \tag{1.53}$$

with

$$m = \frac{\Delta\omega}{\omega_m} = \frac{\Delta f}{f_m}$$

where m is the modulation index and is given by the ratio of the frequency deviation to the baseband bandwidth. The modulated carrier frequency is written in its expanded form as $M(t) = \cos \omega_c t \cos(m\sin \omega_m t) - \sin \omega_c t \sin(m\sin \omega_m t)$. For $m \ll \pi/2$, we have

$$M(t) \approx \cos \omega_c t - m\sin \omega_m t \, \sin \omega_c t$$

$$\approx \cos \omega_c t - \frac{m}{2}[\cos(\omega_c - \omega_m)t - \cos(\omega_c + \omega_m)t] \tag{1.54}$$

Under this condition, the system is referred to as *narrowband FM* and has a form similar to that of AM carriers. It contains the original unmodulated carrier and two sideband frequencies, displaced $\pm \omega_m$ radians from ω_c. The bandwidth of the narrowband FM signal is thus $2f_m$, like that of an AM signal. Despite this similarity, there is a clear distinction between AM and the narrowband FM; in the FM case, the carrier amplitude is constant whereas for the AM, the amplitude varies in accordance with the modulating signal. The carrier and sideband terms are in phase for the AM, whereas the sidebands are in phase quadrature with the carrier for narrowband FM case.

Wideband FM   The case where $m > (\pi/2)$ is referred to as *wideband FM*. Its analysis requires the expansion of $M(t)$ given by equation (1.53). In the equation, the terms $\cos(m\sin \omega_m t)$ and $\sin(m\sin \omega_m t)$ are periodic functions of ω_m and are expanded in a Fourier series of period $2\pi/\omega_m$. This expansion can be expressed in terms of the Bessel functions [3]:

$$
\begin{aligned}
M(t) = {} & J_0(m) \cos \omega_c t - J_1(m)[\cos{(\omega_c - \omega_m)t} - \cos{(\omega_c + \omega_m)t}] \\
& + J_2(m)[\cos{(\omega_c - 2\omega_m)t} + \cos{(\omega_c + 2\omega_m)t}] \\
& - J_3(m)[\cos{(\omega_c - 3\omega_m)t} - \cos{(\omega_c + 3\omega_m)t}] \\
& + \cdots
\end{aligned}
\tag{1.55}
$$

Equation (1.55) represents a time function, consisting of a carrier and an infinite number of sidebands, spaced at frequencies, $\omega_c \pm \omega_m$, $\omega_c \pm 2\omega_m$, and so on.

These successive sets of sidebands are called *first-order sideband*, *second-order sideband*, ..., the magnitudes of which are determined by the coefficients $J_1(m)$, $J_2(m)$, ... respectively.

When two or more sinusoids are present in the modulating signal, the spectrum contains not only multiples of individual modulating frequencies but also all the possible sums and differences of multiples of the modulation frequencies. As more and more modulating signals are added, the complexity of the solution increases rapidly. Eventually, the baseband signal is represented by random noise, extending uniformly across the baseband from 0 to f_m hertz. The frequency spectrum of the corresponding FM signal appears as a continuum of sidebands. The magnitudes of the carrier and sideband terms depend on m, the modulation index, expressed by the appropriate Bessel function. From a theoretical viewpoint, the bandwidth required for 100% of the signal energy is infinite. For practical systems, only significant sidebands with a magnitude of at least 1% of the magnitude of the unmodulated carrier are considered. The number of significant sidebands varies with m and can be determined from the tabulated values of the Bessel function. For the large value of m (>10), the minimum bandwidth is given by $2\Delta f$, where Δf is the peak deviation. A general rule, postulated by J. R. Carson in 1939 for the minimum bandwidths of an FM signal, is

$$
B_T \approx 2[f_m + \Delta f]
\tag{1.56}
$$

This is an approximate rule and is appropriate for most applications. The actual bandwidth required is, to some extent, a function of the waveform of the modulating signal and the quality of transmission desired. It can be seen from this expression that for $m < 1$, the minimum bandwidth is given by $2f_m$, and for $m > 10$, the minimum bandwidth is $2\Delta f$. A key point is that wideband FM systems require greater bandwidth compared with that of AM systems, the extent of which is determined by the modulation index.

1.6.4 Comparison of FM and AM Systems

So far, only ideal AM and FM systems have been investigated. It was shown that an AM system is a linear process, and conserves the bandwidth but the modulation process transfers only one third of the input power to the information bearing sidebands, the remainder power remains at the carrier frequency. An FM system, however, is a nonlinear process. Modulation produces new frequencies, and the modulated signal requires much larger bandwidths. From a theoretical standpoint, the energy of an FM signal is dispersed over an infinite bandwidth. However, the bulk of the energy is contained in the first few sidebands.

For most FM signals, 99% of the energy is contained in a bandwidth of $2(f_m + \Delta f)$ where Δf is the peak deviation and f_m is the baseband frequency. The advantage of the FM system is that all of the input power is transferred to the information-bearing sidebands. The average power at the carrier frequency, after modulation, is zero. Another advantage is that the amplitude of the FM envelope is nearly constant and, thus, the signals are amplified with a minimum of distortion by the practical non-linear amplifiers. The critical parameter for any modulation scheme is the S/N ratio under different traffic conditions. For an AM system, this ratio is expressed by [3]

$$\frac{S}{N} = \frac{A_c^4}{8N^2 + 8NA_c^2} \tag{1.57}$$

where A_c is the voltage amplitude of the unmodulated carrier and N is the mean noise power. The carrier-to-noise ratio (CNR) is

$$\frac{C}{N} = \frac{A_c^2}{2N} \tag{1.58}$$

Therefore

$$\frac{S}{N} = \frac{1}{2} \frac{(C/N)^2}{1 + 2(C/N)} \tag{1.59}$$

For $C/N \ll 1$, the output SNR drops as the square of the CNR. This is the suppression that is characteristic of envelope detection. For CNR $\gg 1$

$$\frac{S}{N} = \frac{1}{4} \frac{C}{N} \tag{1.60}$$

The output SNR is then linearly dependent on the C/N, another characteristic of envelope detectors. In addition, the relation shows that no SNR improvement is possible for AM systems. An increase in the transmission bandwidth $2f_m$, needed to pass the AM signals, serves only to increase the noise N, decreasing the output SNR.

For a FM system, the SNR is given by [3],

$$\frac{S}{N} = 3\left(\frac{\Delta f}{B}\right)^2 \frac{C}{2 N_0 B} \tag{1.61}$$

Here, C is the average power of the FM carrier and $\Delta f/B$ is the modulation index m. Denoting $2N_0 B = N$, the average noise power in the sidebands, we obtain

$$\frac{S}{N} = 3 m^2 \frac{C}{N} \tag{1.62}$$

Compare the FM and AM systems by assuming the same unmodulated carrier power and noise spectral density n_0 for both. For a 100% modulated FM signal, the C/N corresponds to the CNR of an AM system as described by equation (1.49); that is, $(S/N)_{AM} = C/N$. Equation (1.62) can be modified to read

$$\left(\frac{S}{N}\right)_{FM} = 3\,m^2 \left(\frac{S}{N}\right)_{AM} \tag{1.63}$$

For a large modulation index (corresponding to a wide transmission bandwidth, with $m \gg 1$), the SNR can be increased significantly over that of the AM case. For example, if $m = 5$, the FM output SNR is 75 times that of an equivalent AM system. Alternatively, for the same SNR at the output in both receivers the power of the FM carrier can be reduced 75 times, but this requires an increase in the transmission bandwidth from $2B$ (AM case) to $16B$ (FM case). Frequency modulation provides a substantial improvement in the SNR, but at the expense of increased bandwidth. This is, of course, is a characteristic of all noise improvement systems.

Is it possible to continue increasing the output SNR indefinitely by increasing the frequency deviation, and the corresponding bandwidth? If the transmitter power is fixed, increasing the frequency deviation increases the required bandwidth with it, incurring more noise. Eventually, the noise power at the limiter becomes comparable to the signal power and the noise is found to "take over" the system. The output SNR falls off much more sharply than the input CNR. This effect is called

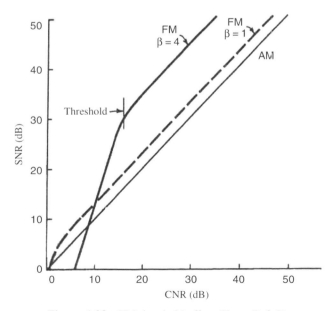

Figure 1.23   FM threshold effect (From Ref. 3).

the threshold effect, as described in Figure 1.23. For proper operation of FM systems, the CNR must be kept above the threshold value, typically >13 dB.

1.7 DIGITAL TRANSMISSION

The widespread use of digital communication systems is the result of many factors, including the relative simplicity of digital circuit design, the ease with which integrated circuit techniques can be applied to digital circuitry, and the rapid advances in digital signal processing techniques. Above all, it is the ruggedness of the digital signals, achieved via coding to minimize the effects of noise, that makes digital transmission the preferred means of communication. This ruggedness comes at a price. Initially, the price paid was that of the increased bandwidth requirement. However, the development of compression and coding techniques has had a major impact on digital communication systems. Compression has matured to the point where digital systems now require less bandwidth than do analog signals, for example, in television channels. A 36-MHz transponder that once carried one analog TV channel can now carry 10 digital channels Not only that; the 10 digital channels can be combined into one signal to allow the transmitters to operate near saturation. However, the advances in compression and coding techniques require increasingly sophisticated signal processing. In a way, that is the price paid in achieving rugged and bandwidth efficient digital communication systems.

1.7.1 Sampling

The well-known Nyquist criterion states that "If a message that is magnitude–time function is sampled instantaneously at regular intervals, and at a rate at least twice the highest significant message frequency, then the samples contain all of the information of the original message."

This rather amazing result is fundamental to the ability to digitize analog signals without the loss of information. As an example, a message bandlimited to f_m hertz is completely specified by its amplitudes at any set of points in time spaced T seconds apart, where $T = \frac{1}{2}f_m$, as shown in Figure 1.24. The result implies that the bandwidth required to convert an analog signal into a digital representation is at least twice the bandwidth of the highest frequency in the signal. The amplitude-modulated pulse signal can then be transmitted to a receiver in any form that is suitable from a transmission standpoint. At the receiving end, a reverse process is followed to recreate the original pulse amplitude modulated signal. To recover the original message, it is necessary to pass the impulses through an ideal lowpass filter with a cutoff of f_m. The output of this filter is a replica of the original message, delayed in time. Since information can be digitized without loss of accuracy, the challenge is then to exploit the potential for digital signal processing (DSP) for communications systems. The first step in this process is the quantization of the sampled signals.

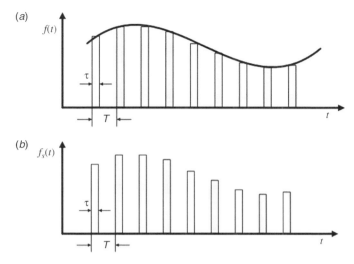

Figure 1.24   The sampling process (τ = sampling time; $T = 1/f_c$: sampling interval): (*a*) input function $f(t)$ and (*b*) sampled output $f_s(t)$.

1.7.2 Quantization

The process of quantization consists of breaking the amplitudes of the signals into a prescribed number of discrete amplitude levels. Then, when the message is sampled in a pulse-amplitude-modulated (PAM) system, the discrete amplitude nearest the true amplitude, is sent. Consequently, the process of quantization introduces an error in the representation of the amplitudes of the sampled signals. Moreover, this error is irretrievable. This seemingly purposeful distortion of the signal is kept below the noise introduced during transmission and at the receiver. In essence, the uncertainty introduced by the fundamental thermal noise, and noise due to imperfections in circuits and devices, limits the ability to distinguish between all the possible amplitude levels; thus making quantization possible. The advantage of quantization is that once the number of discrete amplitude levels is established, each level can then be coded in some arbitrary form before transmission. Thus, quantization makes it possible to deploy the full potential of DSP techniques to optimize the information flow in a communication system.

1.7.3 PCM Systems

Systems that embody the transmission of digitized and coded signals are commonly called *pulse-code-modulated* (PCM) systems. Binary digital systems constitute the most frequently encountered form of PCM systems. A quantized sample can be sent as a simple pulse with certain possible discrete amplitudes. However, if many discrete samples are required, design circuits become quite complicated and uneconomical. On the contrary, if several pulses are used as a code group to describe the amplitude of a sample, then each pulse can have only two states. In a binary system, a code group of m on/off pulses can be used to represent 2^m amplitudes. For example, eight pulses yield 2^8 or 256 amplitude levels. The m pulses must be transmitted in the general sampling interval allotted for the quantized sample. This constraint necessitates an increase in the bandwidth by a factor of m.

For the digital transmission of a 4-kHz voice channel, using 256 levels of quantization and binary code requires a bandwidth of $4 \times 2 \times 8$ or 64 kHz, a factor of 16 times greater than that of an analog system. The required bandwidth can be reduced by choosing a smaller number of amplitude levels for sampling or by using a code where the pulses can be represented by more than two amplitude levels. A quantized signal sample can be coded into a group of m pulses, each with n possible amplitude levels. Thus, if the signal is quantized into M possible amplitude levels, then, $M = n^m$. Each combination of n^m must correspond to one of the M levels. Consider an example with four amplitude levels ($n = 4$) to represent a pulse. The 256 amplitude levels of the sampled signal in the previous example can now be represented by four pulses per sample ($m = 4$). This requires a bandwidth of $2 \times n$ or 8 times the bandwidth of the analog system. Similarly, if a smaller value is selected for M, the bandwidth is again reduced. For all such schemes, there is a tradeoff involved in terms of system noise and signal power. This ability to trade the bandwidth and SNR, back and forth via coding and signal processing, is a characteristic of all PCM systems.

1.7.4 Quantization Noise in PCM Systems

The process of quantization introduces noise at the transmitting terminal. This noise depends on the number of amplitude intervals chosen to represent a signal. For a signal quantized with a uniform interval, the peak signal to rms noise ratio is given by [2,3]

$$\left(\frac{S}{N}\right)^* = 3 M^2$$

$$= 3\, n^{2m}$$

For the binary case, $n = 2$ and the SNR in dB is

$$\left(\frac{S}{N}\right) = 4.8 + 6\, m \text{ dB} \tag{1.64}$$

This relation provides the tradeoffs between the SNR and the relative bandwidth, summarized in Table 1.1.

The frequency spectrum of the quantizing noise with a uniform interval is essentially flat over the range of interest. Quantization does not need to be uniform. In fact, few message signals possess uniform amplitude distribution, and most have a large dynamic range. To overcome such constraints, quantization, typically,

TABLE 1.1 SNR Versus Relative Bandwidth for Binary Transmission

Number of Quantizing Levels	Binary Digits for Coding	Relative Bandwidth	Peak SNR (dB)
8	3	6	22.8
16	4	8	28.8
32	5	10	34.8
64	6	12	40.8
128	7	14	46.8
256	8	16	52.8
512	9	18	58.8
1024	10	20	64.8

has nonuniform spacing and is optimized to achieve a relatively uniform signal-to-distortion ratio over a wide dynamic range. Such nonuniform quantization is referred to as *companding*. Quantization noise can be reduced to any desired degree by choosing increasing finer quantization levels. However, the larger the number of quantizing steps, the greater the required bandwidth is. It is, therefore, desirable to choose as few steps as necessary to meet the objectives of the transmission. Needless to say, many subjective tests have been carried out to determine the acceptable number of levels for voice, video, and data signals. High-quality speech transmission is readily achieved with 128 levels or a 7-bit PCM. Good quality television requires 9- or 10-bit PCM.

1.7.5 Error Rates in Binary Transmission

The deliberate quantization error or noise imparted to the PCM signal is a major source of signal impairment and originates only at the transmitting (or coding) end of the system. This noise can be made arbitrarily small at the expense of increased bandwidth. The other type of noise that is always present is the thermal noise generated by dissipative elements in the path of the signals, and the noise generated by active devices. These noises are random and follow a Gaussian distribution function. They are added to the incoming group of pulses at the receiver. The noise density and distribution function of such noise are the same as those given in equations (1.19) and (1.20).

To detect the presence or absence of a pulse in a binary system, a minimum SNR is required on the digital line. If the pulse power is too low compared to the noise power, the detector will make occasional errors, indicating a pulse where there is none or vice versa. However, if the signal power is increased, the error can be rendered arbitrarily small. To determine the probability of error quantitatively, assume the amplitude of the pulse to be V_p when it is present and zero when it is absent, designated by 1 and 0, respectively. The composite sequence of the binary symbols and the received noise is sampled once every binary interval and decision must be made as to whether a 1 or 0 is present. A simple and perhaps obvious way to decide is to say that if the voltage plus noise sample exceeds $V_p/2$, it is a 1 and if it is less than $V_p/2$, it is a zero. Error then occurs if, with a pulse present, the composite voltage sample is less than $V_p/2$ or, with a pulse absent, if the noise alone exceeds $V_p/2$. To calculate the probability of error, let us assume that a 0 is sent. The probability of error is then just the probability that noise will exceed $V_p/2$ volts and be mistaken for a 1. The probability of error is therefore the voltage that appears somewhere between $V_p/2$ and ∞. Assuming noise to be Gaussian with an rms value of σ, the probability of error is given by [3]

$$P_{e0} = \frac{1}{\sqrt{2\pi\sigma^2}} \int_{V_{p/2}}^{\infty} e^{-v^2/2\sigma^2} \, dv \qquad (1.65)$$

In a similar manner, the probability of error when a pulse is sent and interpreted as zero:

$$P_{e1} = \frac{1}{\sqrt{2\pi\sigma^2}} \int_{-\infty}^{V_{p/2}} e^{-(v-V_p)^2/2\sigma^2} \, dv \qquad (1.66)$$

The two types of error are mutually exclusive and are equivalent. If it is further assumed that the two binary signals are equally likely, then the system probability P_e is the same as P_{e0} or P_{e1}: $P_e = P_{e0} = P_{e1}$.

The probability function P_e is well known and available in various mathematical tables. It is plotted in decibels as a function of V_p/σ in Figure 1.25. It should be noted that P_e depends solely on V_p/σ, the ratio of peak signal to rms noise ratio. It is interesting that P_e has a maximum value of $\frac{1}{2}$. Thus, even if the signal is entirely lost in the noise, the receiver cannot be wrong by more than half the time on the average. The probability curve indicates a sharp drop around the 16 dB level. Below this level, the error rate increases sharply, and is called the *threshold effect*. For this reason, for the transmission of binary digits, the threshold level is chosen to be somewhere between 16 and 18 dB.

The implicit assumptions in this probability curve are (1) the statistics of the receive signal plus noise are Gaussian, and (2) the transmission system is transparent, with no effect on the signal statistics or noise prior to detection. These assumptions allow the chosen decision level to be midway within a pulse amplitude.

1.7.6 Digital Modulation and Demodulation Schemes

Advances in data compression, digital modulation, and coding techniques coupled with the spectacular reduction in the costs of digital circuits are diverting more and more traffic to the digital domain. It is a matter of time before most, if not all, traffic is carried by using digital communications system. A brief overview of digital modulation schemes, aspiring to achieve ever greater power and bandwidth efficiencies, approaching the Shannon limit, is outlined. The frequency spectrum of various modulation schemes has an impact on the desired filter characteristics, required to extract and process information in the communication channel.

Digital baseband signals can be modulated on a sinusoidal carrier by modulating one or more of its three basic parameters: amplitude, frequency, and phase.

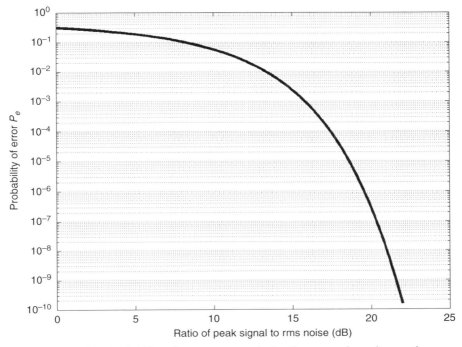

Figure 1.25 Probability of error versus peak signal to rms noise voltage ratio.

Accordingly, there are three basic modulation schemes: amplitude shift keying (ASK), frequency shift keying (FSK), and phase shift keying (PSK).

Amplitude Shift Keying (ASK) This type of modulation is characterized by the amplitude of the carrier wave switching between zero (OFF state) and some predetermined amplitude level (ON state). The amplitude modulated ASK signal is therefore given by

$$M(t) = Af(t)\cos \omega_c t \qquad (1.67)$$

where $f(t) = 1$ or 0, over intervals T seconds long. This is analogous to the amplitude modulation in analog systems. The Fourier transform of the ASK signal is then given by

$$F(\omega) = \frac{A}{2}[F(\omega - \omega_c) + F(\omega + \omega_c)] \qquad (1.68)$$

The binary signal simply shifts the frequency spectrum to the carrier frequency f_c with the signal energy distributed between the upper and lower sidebands. The required transmission bandwidth is twice the baseband bandwidth. For a pulse of amplitude

A and width T (the binary interval), the spectrum is given by

$$A\frac{T}{2}\left[\frac{\sin(\omega - \omega_c)T/2}{(\omega - \omega_c)T/2} + \frac{\sin(\omega + \omega_c)T/2}{(\omega + \omega_c)T/2}\right] \tag{1.69}$$

This is the well-known $\sin(x/x)$ response of a pulse with a finite width, as illustrated in Figure 1.26.

Frequency Shift Keying (FSK)   The frequency-modulated signal by a binary pulse is represented by

$$M(t) = A\cos\omega_1 t \quad \text{or} \quad M(t) = A\cos\omega_2 t \tag{1.70}$$

where $-T/2 \leq t \leq T/2$.

In this scheme, one of the frequencies, say, f_1, can represent a one and the other frequency f_2, a zero. An alternative representation of FSK is, by letting $f_1 = f_c - \Delta f$ and $f_2 = f_c + \Delta f$:

$$M(t) = A\cos(\omega_c \pm \Delta\omega)t \tag{1.71}$$

The frequency deviates $\pm\Delta f$ about f_c, and Δf represents the frequency deviation. The frequency spectrum of the FSK is complex and has a form similar to that for analog FM.

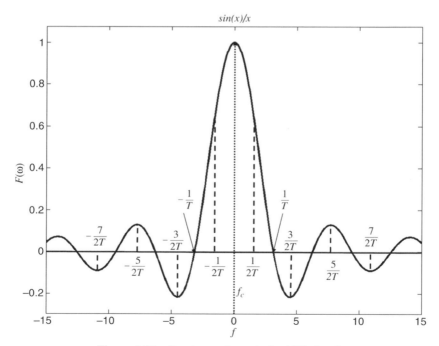

Figure 1.26   Spectrum of a periodic ASK signal.

Phase Shift Keying (PSK) This form of modulation is characterized by a change of phase of the carrier frequency. Owing to the binary nature of the baseband signal, this simply implies a change of polarity. A PSK signal is represented in a form similar to ASK

$$M(t) = f(t) \cos \omega_c t, \quad \frac{-T}{2} < t < \frac{T}{2} \tag{1.72}$$

where $f(t) = \pm 1$. Here a 1 in the baseband binary stream corresponds to positive polarity and a 0 to negative polarity. The PSK signal has the same double-sideband characteristic as the ASK transmission. This result is similar to low index phase modulated analog systems. A comparison of the basic ASK, FSK, and PSK modulations is provided in Figure 1.27.

1.7.7 Advanced Modulation Schemes

As stated in Section 1.4, the two primary resources in a communication system are the signal power and the available transmission bandwidth. Advanced modulation schemes are geared to achieving higher efficiencies in either or both of these resources. Before proceeding further, let us examine the bandwidth and power efficiencies of a digital system.

Bandwidth Efficiency As described in Section 1.3, Shannon's Theorem States

$$C = B \log_2 \left(1 + \frac{S}{N} \right)$$

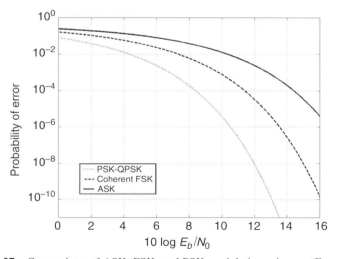

Figure 1.27   Comparison of ASK, FSK, and PSK modulation schemes (From Ref. 3.)

If the information rate R is equal to C, then

$$\frac{R}{B} = \log_2\left(1 + \frac{S}{N}\right) \tag{1.73}$$

defines the ultimate limit of the bandwidth efficiency, or the so-called Shannon limit. As described in Section 1.7.3, the quantized state of a signal can be represented by pulses of varying amplitudes or phases. Each state of a signal sample may be represented by [17]

$$M = 2^m$$
$$m = \log_2 M \tag{1.74}$$

Each state of M is referred to as a *symbol*, and consists of m bits. Each symbol is transmitted as an electric voltage or current waveform. If the duration of the transmission of a symbol is T_s, then the data rate R is given by

$$R = \frac{m}{T_s} = \frac{\log_2 M}{T_s} \tag{1.75}$$

If T_b represents the duration of a bit ($=T_s/m$) and B is the allocated bandwidth, then the transmission bandwidth efficiency is expressed as

$$\frac{R}{B} = \frac{\log_2 M}{BT_s} = \frac{1}{BT_b} \tag{1.76}$$

The smaller the BT_b product, the higher is the bandwidth efficiency of the communication system. By assuming ideal Nyquist filtering, bandwidth B is simply given by $1/T_s$ and

$$\frac{R}{B} = \log_2 M \quad \text{bps/Hz} \tag{1.77}$$

(where bps = bits per second). This represents the Shannon limit of bandwidth efficiency. As M increases, so does R/B. However, this comes at the cost of increased E_b/N_0. Figure 1.28 depicts the tradeoff between the bandwidth and required E_b/N_0 for multiphase or M-ary (MPSK) modulation schemes. As expected, each modulation scheme leads to its own unique frequency spectrum. For example, Figure 1.29 traces the frequency spectrum for a binary (BPSK) and quadrature (QPSK), and offset QPSK modulation schemes. QPSK and offset QPSK are the most widely used modulation schemes in satellite communication systems.

Power Efficiency   Power efficient modulation schemes are best suited for FSK modulation systems. In the binary FSK, the required bandwidth is twice the symbol rate and has a bandwidth efficiency of 0.5 bps/Hz. This is similar to the

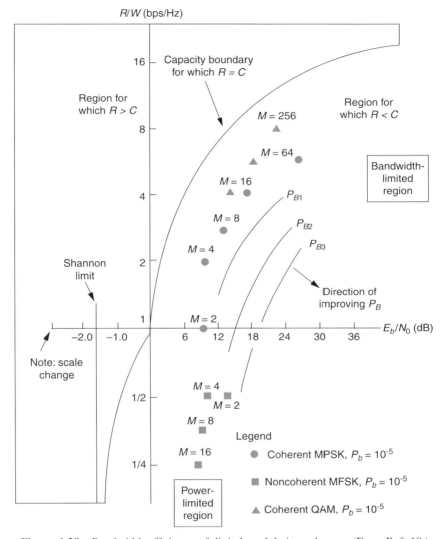

Figure 1.28   Bandwidth efficiency of digital modulation schemes. (From Ref. 18.)

case of a narrowband analog FM that requires twice the bandwidth of the baseband signal. Again, analogous to the wideband analog FM, the power efficiency is improved by trading off the bandwidth. For MFSK scheme, the minimum bandwidth, required in accordance with the Nyquist criterion, is provided in [17]

$$B = \frac{M}{T_s} = MR_s \tag{1.78}$$

where R_s $(=1/T_s)$ is the symbol rate.

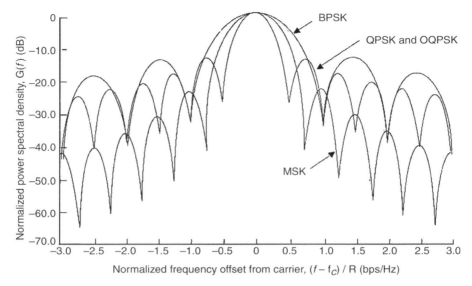

Figure 1.29 Normalized power spectral densities for PSK modulation schemes. (From Ref. 18.)

By using M different orthogonal waveforms, each requiring a bandwidth of $1/T_s$, the bandwidth efficiency of incoherent orthogonal MFSK signals with Nyquist filtering is given by

$$\frac{R}{B} = \frac{\log_2 M}{M} \quad \text{bps/Hz} \tag{1.79}$$

Bandwidth–power tradeoffs for MFSK modulation schemes are depicted in Figure 1.28, which shows how it is possible to achieve lower E_b/N_0 at the expense of increased bandwidth.

Bandwidth- and Power-Efficient Modulation Schemes Table 1.2 describes a wide variety of advanced digital modulation schemes [8]. They can be classified into two large categories: constant-envelope and non-constant-envelope. Generally, the constant envelope class is considered as the most suitable, where the effects of nonlinear amplification in high power amplifiers are an important system consideration.

The PSK schemes (see Fig. 1.29 for PSK scheme power spectral densities) have a constant envelope but discontinuous phase transitions from symbol to symbol. Classic PSK techniques include the BPSK and QPSK. More generally, modulation schemes with M-ary PSK (MPSK) and M-ary FSK (MFSK) signals can be used with a variety of tradeoffs between power and bandwidth efficiencies.

TABLE 1.2 Advanced Digital Modulation Schemes

Abbreviation	Alternate Abbreviation	Definition
ASK	—	Amplitude shift keying
FSK	—	Frequency shift keying (generic name)
BFSK	FSK	Binary frequency shift keying
MFSK	—	M-ary frequency shift keying
PSK	—	Phase shift keying (generic name)
BPSK	2PSK	Binary phase shift keying
DPSK	—	Differential BPSK
QPSK	4PSK	Quadrature phase shift keying
DPQSK	—	Differential QPSK (with differential demodulation)
DEQPSK	—	Differential QPSK (with coherent demodulation)
OQPSK	SQPSK	Offset QPSK, staggered QPSK
$\lambda/4$-QPSK	—	$\lambda/4$-quadrature phase shift keying
$\lambda/4$-DQPSK	—	$\lambda/4$-differential QPSK
CTPSK	—	$\lambda/4$-controlled transition PSK
MPSK	—	M-ary phase shift keying
CPM	—	Continuous phase modulation
SHPM	—	Single-h (modulation index) phase modulation
MHPM	—	Multi-h phase modulation
LREC	—	Rectangular pulse of length L
CPFSK	—	Continuous phase frequency shift keying
MSK	FFSK	Minimum shift keying, fast frequency shift keying
DMSK	—	Differential MSK
MSK	—	Gaussian MSK
SMSK	—	Serial MSK
TFM	—	Timed frequency shift keying
CORPSK	—	Correlative PSK
QAM	—	Quadrature amplitude modulation
SQAM	—	Superposed QAM
Q^2PSK	—	Quadrature phase shift keying
QPSK	—	Differential Q^2PSK
IJF OQPSK	—	Intersymbol -interference jitter-free OQPSK
SQORC	—	Staggered quadrature-overlapped raised-cosine modulation

Source: Ref. 8.

The continuous phase modulation (CPM) schemes have not only a constant envelope but also continuous phase transitions from symbol to symbol. They have less sidelobe energy in their spectra in comparison with the PSK schemes. A great variety of CPM schemes can be obtained by varying the modulation index and pulse frequency [8].

1.7.8 Quality of Service and S/N Ratio

The quality of a wireless communication link depends not only on its design but also on random effects of the propagation environment, such as rain attenuation, tropospheric and ionospheric scintillation, Faraday rotation, Doppler effect, and antenna pointing errors. As a consequence, transmission performance is defined probabilistically, in terms of the specific quality of the signal over a percentage of time. This has given rise to a variety of standards, some agreed on and others still open for discussion [8]. Conventionally, the quality of the signal is expressed in terms of SNR in the case of analog transmission and bit error rate (BER) in the case of digital transmission. For most applications, the specification calls for SNR to remain above a certain value or BER to remain below a certain value over 99% and 99.9% of the time, averaged over a year. Over these time periods, typical SNR for analog TV is specified to be 53 and 45 dB, respectively. At the present time, no consensus has emerged on the standards for digital transmission, although more and more traffic is moving toward it. It is expected that nearly all traffic will be digital by 2010. With the widespread applications and advances in coding technology, it is becoming possible to have virtually error-free transmission for digital signals. For digital TV, satellite systems in the 14/11 GHz band are targeting a performance objective of one uncorrected error per transmission hour, specifically, BER $\leq 10^{-10}$ or 10^{-11}, depending on user bit rate [8].

It should be noted that SNR depends on two parameters, CNR and signal modulation. CNR is a measure of the efficiency of radio transmission at RF, whereas modulation provides the conversion of CNR into SNR. The variety of modulation and coding techniques provide the tradeoffs between CNR and SNR. Systems are always designed to ensure that CNR levels are significantly larger than the threshold value of FM of digital demodulators.

PART III   IMPACT OF SYSTEM DESIGN ON THE REQUIREMENTS OF FILTER NETWORKS

1.8   COMMUNICATION CHANNELS IN A SATELLITE SYSTEM

Communication satellites (Fig. 1.30) are radio relay stations in space. They serve much the same purpose as the microwave towers one sees spread over populated landmass areas. Satellite communication has evolved since the 1970s and represents a mature niche in the field of telecommunications [5–8].

The satellites receive radio signals transmitted from the ground, amplify them, translate them in frequency, and retransmit them back to the ground. Since the satellites are at high altitude, they can see all the microwave transmitters and receivers (earth stations) on almost one-third of the earth. Thus, they can connect any pair of stations or provide point-to-multipoint services, such as television. The coverage can be extended to any part of the earth via intersatellite links or interconnection with

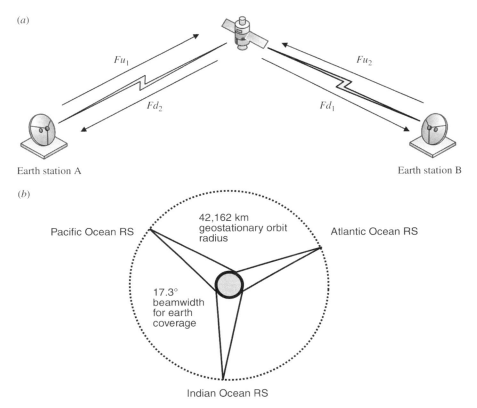

Figure 1.30 Satellite communications: (*a*) typical satellite link; (*b*) worldwide coverage using a three-geostationary-satellite system.

long-distance fiber networks, giving satellite systems an inherent advantage of being insensitive to distance. Satellites are unique in their capability to provide global, seamless, and ubiquitous coverage, including mobile services via handheld units. The frequency plans for commercial satellite systems are listed in Table 1.3.

It should be noted that frequency allocations are periodically addressed and revised by domestic and international regulatory agencies in order to accommodate new services. To determine the characteristics of communication channels, let us examine the typical block diagram of a communication subsystem of a satellite repeater, as depicted in Figure 1.31. Most commercial satellite systems employ dual orthogonal polarization (linear or circular) providing twofold increase in the available bandwidth. Advance satellite systems employ multiple beams, which allow further reuse of the available bandwidth. However, such advance architectures come at the cost of higher complexity for the spacecraft.

Regardless of the satellite architectures, the block diagram of transponders on a given beam and polarization is essentially the same as the one in Figure 1.31. A receive antenna is connected to a wideband filter, followed by a low-noise receiver.

TABLE 1.3a   **Frequency Allocations for Satellite Systems**

Approximate Frequency Range (GHz)	Letter	Typical Usage
1.5–1.6	L	Mobile satellite service (MSS)
2.0–2.7	S	Broadcasting-satellite service (BSS)
3.7–7.25	C	Fixed-satellite service (FSS)
7.25–8.4	X	Government satellites
10.7–18	Ku	Fixed-satellite service (FSS)
18–31	Ka	Fixed-satellite service (FSS)
44	Q	Government satellites

TABLE 1.3b   **Intersatellite Frequency Allocations**

Allocation Frequency (GHz)	Total Bandwidth (MHz)	Satellite Services
22.55–23.55	1000	Fixed, mobile, broadcasting
59–64	5000	Fixed, mobile, radio location
126–134	8000	Fixed, mobile, radio location

After that, the signal is channelized into its various transponders via an input multiplexing network. The allocated frequency band in a satellite system is divided into a number of RF channels, often referred to as *transponders*. Typical channel separation and bandwidths in Ku- and C-band satellite systems are listed in Table 1.4.

Each RF channel is amplified separately and recombined by an output multiplexer (OMUX) network into a composite wideband signal that feeds into the transmit antennas. Input and output switch matrices (ISM and OSM) are there to provide onboard reconfiguration of traffic flow from one transponder to another, from one beam to another, or various combinations thereof. Often, the ISM and OSM consist of mechanical switches. For ISM, loss is not a constraint, and, as a consequence, solid-state switches can be utilized to conserve mass and volume. For the OSM, loss is critical, and there is no substitute for mechanical switches by virtue of their very low loss. From the standpoint of the filtering requirements, the block

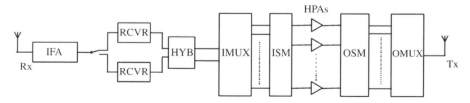

Figure 1.31   Communication sub-system block diagram.

TABLE 1.4   Typical Channel Separations and Bandwidths in C and Ku Bands for Fixed-Satellite Services

Channel Separation (MHz)	Desired Usable Bandwidth (MHz)
27	24
40	36
61	54
80	72

diagram can be separated into three distinct groups: the front end, the channelizer section, and the high-power output circuits. Such a breakdown is typical of most communication repeaters. We now deal with each of the three subgroups separately.

1.8.1 Receive Section

The receive section consists of a wideband input receive filter, the LNA, frequency downconverter, and driver amplifier, as shown in Figure 1.32. For high reliability, satellites invariably employ a redundant receiver that requires a switch prior to the receiver. At the receive antenna, the signal strength is at its lowest level, and therefore, it is imperative to minimize the energy loss prior to the LNA. The wideband receive filter is required to ensure that only the signals in the 500 MHz bandwidth are fed to the LNA and all other signals outside this range are attenuated. For the filters and transmission lines prior to the LNA, the designs are dictated by a need for low insertion loss in the passband. Typical requirement calls for an insertion loss of no more than a few tenths of a decibel.

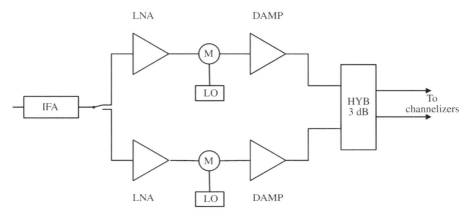

Figure 1.32 Satellite receiver block diagram.

The frequency downconverter consists of a mixer–local oscillator (LO) assembly. The separation between the transmit and receive frequencies is required to minimize the interference between them. A driver amplifier (DAMP) is used to attain the required power levels prior to channelization and final amplification of the signals. It also allows LNA to operate at low enough power level consistent with the lowest possible noise figure. The hybrid is a type of 3 dB power divider that provides two separate paths for the channelization of the transponders.

1.8.2   The Channelizer Section

A detailed block diagram of the channelizer or input multiplexer (IMUX) section is shown in Figure 1.33.

Once the signal has been amplified by the LNA, the loss in the subsequent equipment is no longer critical. This is due to the fact that the receive system noise temperature is reduced by the gain of the LNA, and the post-LNA losses have less impact (as seen in Section 1.5.5). Instead, the design driver is the efficient channelization of the composite signal into its various RF channels or transponders with a minimum loss of bandwidth and, at the same time, providing enough isolation to control interference from other channels. In other words, channelizing filters largely determine the usable bandwidth of each RF channel, with a minimum of guard band. The design of such channel filters must be close to that of an ideal filter. Moreover, the bandwidth requirements for these filters range from 0.3% to 2%. Such narrowband filters incur relatively large transmission deviations, and often require a degree of phase and group delay equalization. As an example of a 6/4-GHz satellite system, the typical specifications for a channelizing filter are described in Table 1.5.

The defining feature of the channelizing filter networks is the very stringent out-of-band rejection requirement. It necessitates the use of filters with transmission zeros just outside the passband to achieve a steep isolation response. Although the insertion loss at the band center is not critical, the amplitude variation across the passband is, and can be minimized only by employing filters with high unloaded Q. An interesting feature of the channelization scheme is the use of channel dropping ferrite circulators and 3-dB power dividers with coaxial cables to ease the layout of the equipment. Circulators are three-port nonreciprocal devices, constructed from ferrite materials, biased with external magnets, to achieve the desired property of the unidirectional flow of energy. Such devices incur a few tenths of a decibel loss in coaxial or planar structures, and less than a tenth of a decibel loss in the waveguide realization. The reverse isolation is 30–40 dB. Circulators are inherently wideband devices, whose bandwidths range from 10% to 20%. It is possible to optimize the performance over narrower bandwidths. Circulators are widely used in radar and communication systems. As shown in Figure 1.33, use of 3-dB hybrid and channel dropping circulators enable design simplicity and flexibility for the layout of the multiplexing networks. However, this design simplicity comes at the expense of increased loss, which, as described earlier, is of little consequence for this portion of the network. Consequently, the narrowband channelizing filters need to be high-Q structures, whereas the wideband ancillary equipment,

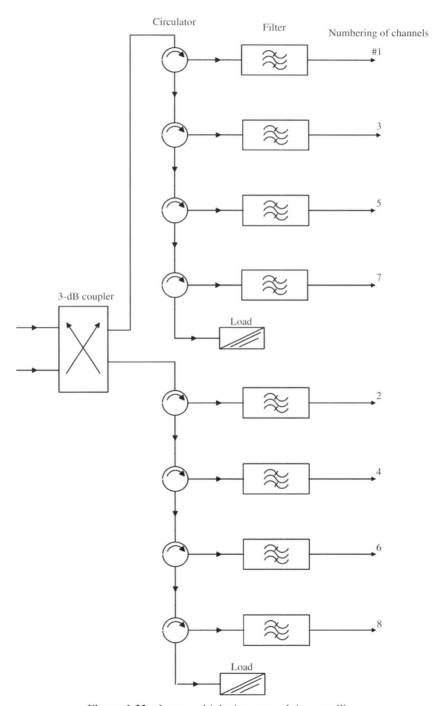

Figure 1.33   Input multiplexing network in a satellite.

TABLE 1.5   Typical Specifications for Channelizing Filters in a 6/4-GHz Satellite System

Frequency band	3.7–4.2 GHz
Number of RF channels	12
Channel separation	40 MHz
Passband bandwidth	36 MHz
Rejection over narrowband	10–15 dB at band edge of adjacent copolarized channels, rising to 30–40 dB within 10–15% of channel bandwidth (2–3 MHz) beyond band edge over frequency band 3.7–4.2 GHz
Rejection over wideband	>40–45 dB over receive 5.925–6.425 GHz band
Insertion loss in passband	Not critical
Passband loss variation	<1 dB
Passband relative group delay	<1–2 ns over middle 70%, rising to 20–30 ns toward band edges
Operating temperature range	0–50°C

namely, circulators, isolators, and hybrid, use the conventional lower-Q but more compact coax designs.

Narrowband filters incur relatively high group delay variation across its passband. Group delay varies inversely with the absolute bandwidth and is further constrained by the steepness of the required amplitude response of the filter. Group delay can be equalized using allpass external equalizer network or by using higher-order self-equalized linear phase filters. Either way, group delay equalization increases the design complexity and cost. Most satellite systems employ some degree of equalization. An added benefit of group delay equalization is that it also reduces the amplitude variation across the passband at the expense of a small increase in midband loss. Tradeoffs for the channelizing filter are critical in establishing the usable passband bandwidth, and hence the efficiency in the utilization of the frequency spectrum.

1.8.3   High-Power Amplifiers (HPAs)

High-power amplifiers are required to raise the power levels of RF signals prior to their transmission back to earth. System level tradeoffs determine the gain and maximum RF power of HPAs to ensure the desired communication capacity for a transponder. The traveling-wave-tube amplifier (TWTA) is the dominant power amplifier for communications satellites, although many satellites use solid-state power amplifiers (SSPA) as well. Over the years, the available power, reliability, and efficiency of TWTAs has significantly improved, allowing its continued dominance of the satellite market. However, solid-state power amplifiers (SSPAs) tend to exhibit better nonlinear characteristics for moderate power levels for C- and Ku-band satellite systems. As always, system-level tradeoffs are required to select the appropriate HPA for a given system. HPAs are power-hungry, requiring its operation at a high efficiency. HPAs are inherently nonlinear devices and their characteristics involve tradeoff between efficiency, output power level, and the amount of nonlinearity. Typical characteristics of a TWTA are depicted in Figure 1.34.

To obtain maximum RF power, the amplifier must be operated under the condition of saturation. However, at this power level, the amplifier is highly nonlinear, and is suitable only for amplification of a single carrier. For multicarrier operation, it is essential to operate the TWTA in a backoff mode to keep the carrier-to-intermodulation (C/I) ratio to an acceptable level. As described in Section 1.5.6, as the number of carriers in an RF channel increases, the number of IM products grows rapidly. These products fall inside as well as outside the RF channel. It is usual to specify the IM performance by the intercept point defined as the theoretical output power $\overline{IP}$, where the extrapolated linear single carrier output power is equal to the extrapolated two-carrier IM power, as represented in Figure 1.35. The carrier to third order $(2f_1 \pm f_2$ or $2f_2 \pm f_1)$ intermodulation (IM_3) is then given by [3,19]

$$\frac{C}{IM_3} = 2(\overline{IP} - P_0), \text{ dB}$$

where P_0 is the single carrier output power level. The value of the intercept point is specified by the tube manufacturer.

This relationship shows that carrier to third-order IM decreases 2 dB for every dB backoff of the output power of the HPA. Also, the greater the intercept point, the larger the value of C/IM. The intercept point is a characteristic of the amplifier and provides a measure of its linearity. There are good reasons for specifying HPA performance with respect to two carriers. It represents the minimum number

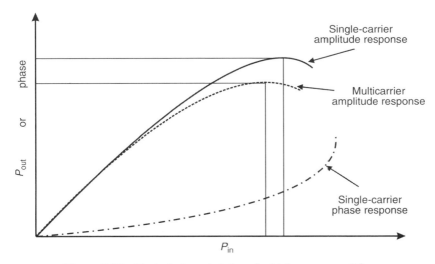

Figure 1.34 Typical characteristics of a high-power amplifier.

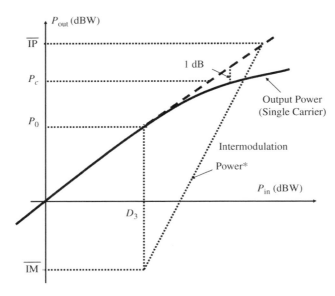

Figure 1.35   Third-order intermodulation (IM) intercept point in amplifiers. (From Ref. 8.)

of carriers to generate an IM product. In addition, third-order products have the highest power level and are critical in determining C/IM performance. The level of odd-order IM products tends to decrease between 10 and 20 dB with increasing order. IM products that fall inside the RF channel can be controlled only by the operating power level and linearity of the HPAs. There exists a tradeoff between the efficiency and linearity of HPAs. Highly linear HPAs tend to have lower efficiency and vice versa.

The typical way to control IM is to operate it in the linear region of the power curve, which is usually 2–3 dB lower than its saturated power level. For two-carrier operation, HPAs are backed off by 2–3 dB below saturation power level. For multi-carrier operation, the backoff can be as much as 10–12 dB to achieve acceptable performance. This represents a significant penalty in the output power to ensure acceptable IM performance. Obviously, linearity of amplifiers is a major issue. Often a "linearizer" is employed to improve the linearity of amplifiers, incurring extra hardware and cost. Typically, amplifiers are designed to have carrier to third-order product values exceeding 20 dB. It should be noted that the prime reason for operating HPAs in a backoff mode for multicarrier operation is to control those IM products that fall inside the RF channel; those outside the RF channel are of no consequence, unless they lie within the transmit or receive frequency bands. High-power filters or multiplexing networks that follow HPAs are designed to provide adequate attenuation to such interference. The isolation between the transmit and receive antennas also provide protection to the IM products falling in the receive band. The other frequency band of significance corresponds to the second and third harmonics of carriers generated by HPAs. These harmonics, if not suppressed, can interfere with ground-based systems as well as with military- and science-based space systems. The harmonic suppression must be achieved prior to the transmit antenna. Harmonic suppression is part of the output multiplexing subsystem.

1.8.4 Transmitter Section Architecture

The transmit section of the satellite combines the outputs of various high-power channel amplifiers in an output multiplexing network for transmission via a common antenna. Once the signals have been amplified by the final amplifier, conservation of power becomes critical. Consequently, the challenge is how to achieve the lowest loss for each RF channel, preserve the useable bandwidth, and, at the same time, keep the design of multiplexer and antenna subsystems simple. This tradeoff has given rise to two alternative multiplexing schemes for satellite systems.

Early satellite systems employed the noncontiguous multiplexing scheme. In such an architecture, alternate channels are combined in the so-called noncontiguous multiplexer, as depicted in Figure 1.36a. The output power of each noncontiguous multiplexer is then combined in a 3-dB hybrid as shown in Figure 1.36b. Power from each input port of the hybrid is divided equally between the two output ports with a phase difference of 90°. Thus, each output port of the hybrid contains all the RF channels at half the power. Such a scheme requires an antenna feed network with two input ports, and is called *a dual-mode feed network*. This architecture was employed extensively for satellites in the 1970s and 1980s. Its advantage is the design simplicity of the multiplexer network. The disadvantage, though, is that the architecture requires a more complex beamforming network. Also, it is difficult to optimize the antenna gain, resulting in some loss of EIRP. The alternative is to combine the powers of all the transponders on a given polarization in a single device, the contiguous band multiplexer as shown in Figure 1.36c. Its main advantage is that it allows a simpler beamforming network and it is easier to optimize the antenna gain. Furthermore, owing to its inherent sharper amplitude characteristics, it reduces the effect of multipath in the satellite, resulting in improved RF channel characteristics [16]. Its disadvantage is that the design of such a multiplexer is more complex and it incurs slightly higher loss and group delay variation across the passband of RF channels. Over the years, technology advances have overcome the disadvantage of design complexity. Higher antenna gain achievable with this scheme more than compensates for the disadvantage of slightly higher loss in the multiplexer. As a result, most modern satellite systems employ contiguous multiplexing scheme since it yields better performance in terms of the overall transponder channel characteristics and EIRP. For some applications, especially for systems with narrower-band transponders ($\sim$20 MHz), a noncontiguous scheme offers a better design. Comparison of satellite architectures employing alternate multiplexing schemes is described in more depth in Ref. 16.

Output Multiplexer   The output multiplexer (OMUX) performs the function that is reverse to that of channelizing section. It combines the power of the individual RF channels to form a single composite signal for transmission back to earth via a

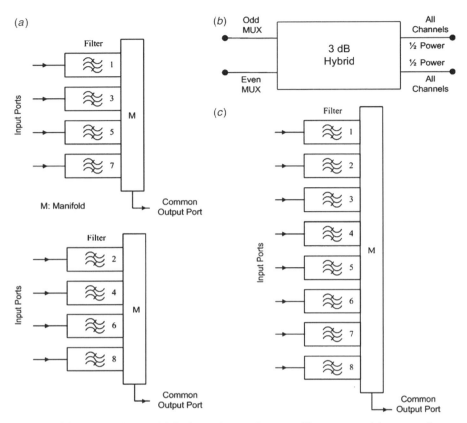

Figure 1.36 Alternative multiplexing schemes in a satellite system: (*a*) noncontiguous multiplexing scheme; (*b*) hybrid combiner; (*c*) contiguous multiplexing scheme.

common antenna. OMUX consists of a number of bandpass filters whose outputs are connected to a common manifold. Each filter corresponds to a particular transponder channel and is optimized to accept the amplified signal within the passband of that channel and reject frequencies corresponding to other transponders. In addition, the performance of channel combining filters and overall multiplexer design is further optimized to achieve low loss and provide rejection over the receive band. TWTAs produce not only the desired amplified signals but also IM products and harmonics that must be suppressed. The OMUX provides part of that suppression. The typical requirements imposed for a channel in the OMUX of a 6/4-GHz satellite system are described in Table 1.6.

Harmonic Rejection Filters The purpose of such filters is to provide a high rejection at the second and third harmonic incurring a minimum of loss in the passband of the transponder. This is accomplished by a low-loss lowpass filter design. The power handling capability of such filters is a critical requirement. Two

TABLE 1.6   Typical Specification of a Channel in the High-Power Combining Network in a 6/4-GHz Satellite System

Frequency band	3.7–4.2 GHz
Number of RF channels	12
Channel separation	40 MHz
Passband bandwidth	36 MHz
Passband insertion loss	Minimum (typically <0.2–0.3 dB)
Rejection over narrowband[a]	Shaped isolation response >5–10 dB at band edges of adjacent copolarized channels, rising to 20–30 dB within 15–20% of channel bandwidth (3–4 MHz) beyond the band edge and over the frequency band 3.7–4.2 GHz
Rejection over wideband	>30–35 dB over the receive 5.925–6.425 GHz band
Passband relative group delay	<1–2 ns over middle 70%, rising to 10–20 ns toward band edges
Power handling	10–100 W per channel
Operating temperature range	0–50°C

[a]The isolation response shape is dictated by the filter and multiplexer technology and the requirements of low loss and high power handling capacity.

alternative designs are feasible to meet the power constraints as shown in Figure 1.37. One way is to use a single-harmonic filter connected after the multiplexer as shown in Figure 1.37*a*. The advantage is that it uses a single filter, incurring a lower mass and volume. However, it must be capable of handling the combined power of all the RF channels on the multiplexer, representing a major disadvantage. The alternative is to use a harmonic filter on a channel by channel basis, as depicted in Figure 1.37*b*. In this scheme, the harmonic filter is required to handle

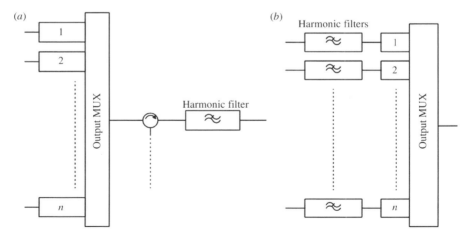

Figure 1.37 High-power output multiplexing network in a satellite, using (*a*) a single-harmonic filter and (*b*) individual channel harmonic filters.

the power of a single channel, and not the combined power of all the channels. Its disadvantage is the increase in the number of harmonic filters required.

Early satellite systems employed a single harmonic filter in OMUX. As the power of satellites has increased, there seems to be a shift toward employing harmonic filters for each individual channel. The particular choice is dictated by system level tradeoffs.

PIM Requirements for High-Power Output Circuits All active devices are inherently nonlinear and therefore generate IM products. What is not well understood is that all devices, including passive components such as filters, multiplexers, and antennas, have a degree of nonlinearity, and are thus capable of generating IM products [20]. Such IM is known as *passive intermodulation* (PIM). It arises owing to imperfections in the structure of materials. For most applications, the level of PIM is sufficiently low, and hence not an issue in system design. It can be an issue if a common antenna is used for the transmission and the reception of signals in a communication system. PIM is also problematic if the high-power transmitter and low-level receivers are in close proximity, allowing coupling between the two. For example, the difference in the power levels between the transmit and receive section in a satellite repeater can be as high as 130–140 dB. This implies that the level of PIM should be 170 dB or more below the power level of the transmitter to ensure that there is no interference with the low-level received signals. It has implications for the system design, as well as for the quality of materials and work standards for high-power equipment. One important impact on the system design pertains to the allocation of the transmit and receive frequency bands. As described in Section 1.8.3, the third-order IM is the dominant product. Consequently, the separation between the transmit and receive frequency bands should be such to avoid at least the third-order IM from falling into the receive band, preferably the fifth-order IM as well. If F_1 and F_2 represent the lower and upper band edge of the transmit frequency band, then

$$\text{Frequency location of IM products} = |m F_2 \pm n F_1|$$

$$\text{IM order} = m + n, \ m \ \& \ n \ \text{are integers}$$

The frequency spectrum of IM products generated by two carriers is illustrated in Figure 1.38.

It is evident that even-order IM products are of no consequence since they lie well outside the bands of interest. Also, the odd order IM products that lie below F_1 are of no consequence since the receive frequency band is invariably chosen to be higher than the transmit band. The reason for this choice is the higher efficiency of equipment at the lower transmit frequency band. The IM products that are of interest are the ones that are above F_2. The frequency locations of the upper odd order IM products are given by $2F_2 - F_1$ for the third order, $3F_2 - 2F_1$ for the fifth order, and so on. As an example, transmit and receive bands for a 6/4-GHz satellite system are 3.7–4.2 GHz and 5.925–6.425 GHz, respectively. In this case, the minimum odd-order IM product

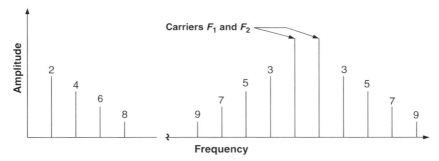

Figure 1.38   Frequency spectrum of IM products generated by two carriers.

that falls in the receive band is the ninth-order product. This represents a relatively safe separation between the transmit and receive frequency bands. Some of the earlier military satellite systems had much closer separation that allowed third-order PIM to fall in the receive band, and that proved to be very problematic. It was only then that the system designers were made aware of the PIM phenomenon.

High-power output circuits are often subjected to a specification on PIM. It depends on the antenna design for the communication system. If a common antenna is used for transmission and reception, then any PIM products generated in the receive band can couple via the antenna feed networks into the receive section and interfere with the received signals. This can also occur for systems with independent transmit and receive antennas located in close proximity. A most conservative specification of PIM is to assume Tx–Rx isolation to be zero and the PIM level to be equivalent to that of third-order IM, regardless of the order of the PIM that is permissible to fall into the receive band. Such a specification can put excessive constraints on hardware designers, and seldom can be met or demonstrated with a high degree of confidence. In practical systems, the Tx–Rx isolation is of the order of 20–30 dB for systems with a common antenna. It is much higher (60–100 dB) if separate transmit and receive antennas are used. If the permissible PIM falling into the receive band is of fifth order, its value is typically 10–20 dB lower than that of third-order PIM. Higher-order PIMs go down in level by 10–20 dB with each increasing order. This has been demonstrated by experiments for a number of satellite systems [21]. A typical PIM specification for the 6/4-GHz satellite system with a common antenna is −120 dBm for the third-order IM product.

1.9   RF FILTERS IN CELLULAR SYSTEMS

Cellular radio evolved from mobile communication systems. The earliest application of mobile radio communications was to ocean-going vessels. This was followed by radio services to aircraft, and subsequently, on land for public safety services such as police, fire departments, and hospital ambulances. However, it was not until the early 1990s that mobile communications took hold as a means for personal communications, extending the reach of fixed-telephony networks to people just about

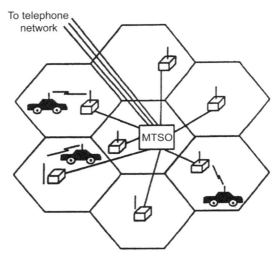

Figure 1.39 Conceptual layout of a cellular system. (From Ref. 7.)

anywhere: in their backyards, in automobiles, urban centers or remote regions. It is indeed possible to remain connected anywhere on the planet with access to open space. The conceptual layout of a cellular radio system is shown in Figure 1.39.

The geographic area served by a cellular system is divided into small geographic cells, ideally in a hexagonal arrangement. Typically, the cell centers are spaced 6–12 km, depending on the terrain and local climate. The system consists of cell sites, a switching center, and mobile units. Each cell represents a microwave radio repeater. The radio facility is housed in a building or shelter, and can connect and control any mobile unit within its own allocated area. The switching facility, referred to as the *mobile telephone switching office* (MTSO), provides switching and control functions for a group of cell sites. In addition, the MTSO provides connectivity with the public switched telecommunications network (PSTN), adding the distant reach of public telephony to mobile users.

The proliferation of cellular communications has posed a major problem for bandwidth allocations for wireless systems. The present cellular radio bandwidth assignments cannot support the burgeoning demand for cellular services, especially in urban areas. As more and more people are able to afford cellular service, demand is creeping up for additional data and video services, severely constraining the bandwidth assignments. The efficient use of the frequency spectrum has therefore become paramount in the design of cellular systems. Table 1.7 describes the frequency allocations for wireless PCS systems.

Frequency allocations are regulated by domestic and international agencies and are reviewed on a regular basis.

The architecture of cellular systems requires many tradeoffs, especially with respect to the size (coverage area) of the cells, the number of channels allocated

TABLE 1.7 Frequency Allocations for Cellular Systems

	Uplink (MHz)	Downlink (MHz)
America	824–849	869–894
	1850–1910	1930–1990
Japan	810–826	940–956
	1429–1453	1477–1501
Europe, other regions	453–458	463–468
	871–904	916–949
	890–915	935–960
	880–915	925–960
	1710–1785	1805–1880

to each cell, the layout of the cell sites, and the traffic flow [7]. The total available (one-way) bandwidth is split up into N sets of channel groups. The channels are then allocated to cells, one channel set per cell in a regular pattern that repeats to fill the required number of cells. As N increases, the distance between channel sets (D) increases, reducing the level of interference. As the number of channels sets (N) increases, the number of channels per cell decreases, reducing the system capacity. Selecting the optimum number of channel sets is a compromise between capacity and quality. Note that only certain values of N lead to regular repeat patterns without gaps. These are $N = 3,4,7,9,12$, and then multiples thereof.

The use of directional antennas can improve performance by allowing smaller cells and higher operating capacity. Regardless of the architecture of the system, each cell site requires highly efficient RF filters to ensure maximum use of the available frequency spectrum. The range and key specifications of RF filters required in a cellular system are summarized in Table 1.8.

Cellular base stations combine the functions of a microwave repeater and a switching network. As a consequence, filtering requirements and constraints are very similar to those encountered in a satellite repeater. Receive and transmit

TABLE 1.8 Filter Requirements for Cellular Systems

Frequency Band (MHz)	Bandwidth (%)	Key Requirements
	Base Stations	
453–468	0.25–1	Low loss (typically <2 dB)
824–960	0.25–3	High Tx–Rx rejection (60–90 dB)
1429–1501	0.25–1.5	High power handling (tens to hundreds of watts) for Tx filters
1710–1990	0.25–4	Compact size and low cost
	Handsets	
70	0.05–0.3	Compact size and low cost

filters must exhibit low loss; elsewhere, loss is not a constraint. Typically, a base station requires a diplexer to separate, transmit, and receive frequencies. Low loss is a key requirement for such diplexers. Since the majority of cell sites are spread throughout urban areas, real estate is expensive and, as a consequence, size of equipment is also a constraint. For filters required for handsets, minimum mass and size are the governing criteria. Owing to large volumes and a highly competitive environment, cost is an overarching constraint for all filters in cellular base stations.

1.10   IMPACT OF SYSTEM REQUIREMENTS ON RF FILTER SPECIFICATIONS

The emergence and widespread applications of wireless communication systems has stretched the required performance of all components and subsystems that constitute such systems. The scarcity of the available frequency spectrum for wireless communications has put new demands in signal processing and filtering in order to maximize the efficiency of the usage of the frequency spectrum. The factors affecting the specification of RF filters can be categorized as follows:

Frequency plan
Interference environment
Modulation–demodulation schemes
Operating environment
Location of filters in the communication link
HPA characteristics
Limitations of microwave filter technology

Each of these factors can be related to the appropriate area of system design.

Frequency Plan   As described in Section 1.2, the available frequency spectrum for commercially viable wireless communication is a limited natural resource. It is imperative that the communication system optimize the use of the available bandwidth. RF filters perform the function of protecting the overall frequency band from outside interference, efficient channelization of the available bandwidth into various transponders to meet the traffic demands, and efficient power combining to form a common feed, minimizing the cost of the antennas. Consequently, the achievable performance of practical filters largely determines the effective usable bandwidth of the allocated frequency spectrum. The frequency of operation and channel bandwidths has a significant impact on the achievable performance. At higher frequencies, filter design is more complex and sensitive to manufacturing tolerances. Choice of narrower channel bandwidths implies higher loss, higher transmission deviations, and sensitivity. Typical allocation of the guard band between RF channels is 10%. It is based on the assumption that the channel filters can provide an isolation of greater than 30 to 40 dB to the interference from adjacent channels. It implies an efficiency of 90% in the usage of the available frequency band. However, the allocation of 10% bandwidth to the guard band is arbitrary, and is rarely achieved with an acceptable passband performance. Achievable guard band is governed by the constraints imposed by microwave filter technology. For practical systems, an efficiency of around 80% in the effective usage of the available bandwidth is typically realized [16]. As a result, there is continued pressure to come up with filters to reduce the guard band without degrading the system performance or undue increase in the system cost.

Interference Environment   The interference environment of an RF channel in a communication system is described in Section 1.3. The prime purpose of RF filters is to channelize (or combine) the allocated bandwidth with a minimum of guard band while maintaining a good passband performance. The near-out-of-band isolation requirements are dictated almost entirely by the interference from adjacent (copolarized) channels. The level of interference depends on the energy spectrum of signals in the adjacent channels. It is possible, therefore, to specify near isolation in accordance with the shape of the interfering spectrum. This specification is the most severe constraint imposed on channel filters (see Sections 1.8.2 and 1.8.4). It typically drives the tradeoff between the out-of-band amplitude response and the in-band amplitude and group delay variations. The level of such deviations depends on the filter technology, design complexity, and cost. Interference from other sources, such as cross-polarized channels, rain fades, multipath effects, and ionosperic scintillation, fall inside the RF channel and are therefore controlled by the system design. Intermodulation interference caused by nonlinear HPAs (Sections 1.5.6 and 1.8.3) is controlled by the operating power level of HPAs.

Modulation Schemes   Modulation schemes play a critical role in maximizing communication capacity of RF channels for different traffic requirements. They also provide a tradeoff between power and bandwidth required for transmission of signals. Most systems require a high degree of flexibility in being able to handle different types of modulation schemes as well as a varying number of carriers in an RF channel. Such channel characteristics are best achieved by microwave filters exhibiting equiripple pass- and stopbands; such a response represents the optimal approach to a "brick wall" amplitude response. If necessary, a degree of phase and group delay equalization can be incorporated to achieve improved passband characteristics. For dedicated RF channels, it is possible to optimize the channel characteristics in accordance with the energy spectrum of the modulation scheme.

HPA Characteristics   High-power amplifiers are inherently wideband and nonlinear devices. In addition to amplifying the RF power, HPAs also generate harmonics of the fundamental frequency, and wideband thermal noise. Another consequence of nonlinear characteristics is the generation of IM noise. As a result, HPAs control specification of harmonic suppression and wide out-of-band rejection response, as described in Sections 1.8.2 and 1.8.4.

Location of RF Filter in the Communication Link   A communication link consists of a cascade of transmitters and receivers. RF filters are required after HPAs, prior to LNAs and in between LNA and HPAs as presented in Section 1.5. The impact on the specification of filters, based on their location in the communication link, is as follows:

Filter prior to LNA	Low loss
Filters after HPAs	Low loss and high power
Filters between LNA	handling capability
and HPAs	High selectivity

In addition to electrical requirements, there are always the constraints of size, mass, and cost. Low loss and high power handling capability imply larger size. However, if loss is not a constraint, then it is possible to tradeoff smaller size at the expense of higher loss.

Operating Environment   For wireless communication, there are two operating environments: outer space or terrestrial space.

Space Environment   Key drivers for space applications are mass, size, reliability, and performance. In addition, the equipment must withstand the launch environment, radiation in outer space, and higher temperature range. These requirements have been drivers for the many innovations in filter technology over the last three decades. Typically, the microwave filters used in space are composed of air or dielectric-loaded metal structures, and are therefore immune to the radiation environment. However, if high-temperature superconductors (HTS) are adopted to realize such filters, the radiation environment becomes a design constraint, which can be readily overcome by enclosing filter networks in a metal enclosure of appropriate thickness. Another phenomenon that is peculiar to this environment is multipactor breakdown, which can occur in the hard vacuum of outer space. It is dealt with in Chapter 20.

Terrestrial Environment   For terrestrial systems, cost is the main driver, along with the requirement of relatively large quantities of filters. This has spawned innovations in the area of volume production.

Limitations of Microwave Filter Technology   An ideal filter is characterized by a brick-wall amplitude response, and linear phase (constant group delay) in the passband. Such a response can only be approximated by a very high-order filter composed of lossless materials, an impractical proposition. Practical filters exhibit transmission deviations, relative to an ideal response, which depends on the design complexity and the quality of the materials used in the realization of the filter networks. Microwave filters are distributed structures. As a result, the electrical, mechanical, and thermal properties of materials play an important role in the design and performance of microwave filters. The dissipation of energy in the materials impairs the filter performance. In addition, the performance of filters, especially those with narrow bandwidths (<5%), is very sensitive to the operating temperature range. The thermal stability of materials becomes a critical parameter as well. Often, low loss and thermal stability in materials are competing requirements. For any microwave structure, especially narrowband devices such as filters, there is always a tradeoff between size (and mass) and the dissipation of energy. The bigger the structure, the lower the loss and vice versa. Another factor that is crucial in determining the size of filters is the dielectric constant of materials. The higher the dielectric constant, the smaller the size, but is often at the cost of higher loss and poorer thermal stability. Suffice it to say that advances in materials are the key in realizing improved performance in ever more compact structures. The design aspect of microwave filters is primarily driven by two factors:

1. The stringent electrical requirements imposed to maximize the usage of the available bandwidths
2. The sensitivity of microwave structures to design tolerances, discontinuities, operating environment, and the inherent dispersion characteristics

At lower microwave frequencies (<1 GHz) and when specifications are not overly stringent, the use of circuit theory approximation to design a wide range of filters is possible. At higher frequencies or when specifications are close to what is theoretically possible, EM theory must be applied to circuit models to achieve a higher level of accuracy. The miniaturization of equipment is another driver of innovative design driven by advances in manufacturing technology and EM techniques.

1.11 IMPACT OF SATELLITE AND CELLULAR COMMUNICATIONS ON FILTER TECHNOLOGY

In the 1970s and 1980s, the advent of satellite communications was the key R&D driver for microwave filter networks [22]. For space applications, the size and mass of onboard equipment have a critical impact on cost. In addition, power generation in space is costly, and thus, low loss for high power equipment is equally critical. Lastly, the equipment must be ultrareliable to operate in space, without failure, for the lifetime of the spacecraft. In that timeframe, there were many advances and innovations in the field of filters and multiplexing networks. These advances included the development of dual- and triple-mode waveguide filters with arbitrary amplitude and phase response, dielectric resonator filters, contiguous and non-contiguous multiplexing networks, surface acoustic wave (SAW) filters, high-temperature superconductor (HTS) filters, and a variety of coaxial, finline, and microstrip filters. The evolution of widespread wireless cellular communication systems in the 1990s capitalized on this R&D and pushed the envelope further in developing materials and processes for large-scale and cost-effective production techniques for microwave filters for ground-based systems.

Computer-aided design and tuning by EM techniques have played a major role in this endeavor. In an interesting twist of circumstances, advances in production techniques are now being adopted for space microwave equipment to achieve lower costs. The exploitation of the nascent technology of microelectromechanical systems (MEMs) and high-temperature superconductors (HTSs) will continue to fuel R&D for the miniaturization of microwave filters.

SUMMARY

This chapter is devoted to an overview of communication systems; more specifically, the relationship between the communication channel and other elements of the system. The intent here is to provide the reader with sufficient background to be able to appreciate the critical role and requirements of RF filters in communication systems.

Starting with a description of a model of a communication system, the key issues addressed in this chapter are (1) whether there is a limitation to the available bandwidth for a communication system, (2) the limitations for transmitting information over a chosen transmission medium in the available bandwidth, and (3) the cost-sensitive parameters in a communication system.

To understand the limitations on the available bandwidths for a communication system, the radio spectrum and its utilization must be understood. This has been examined extensively, from the perspective of the limitations imposed by the earth's atmosphere for wireless communications, technology limitations in making efficient use of the frequency spectrum, and regulatory constraints.

The concept of information and its transmission over a medium is explored by reviewing the seminal work of Shannon, highlighting the fundamental tradeoffs between signal-to-noise ratio and bandwidth of a communication channel. This leads to a discussion of power sources and antennas required to establish a radio link. The topic of noise is considered with emphasis on the thermal noise, and how it sets a fundamental noise floor for the communication channel. The topic of modulation is described for both analog and digital systems. It includes the tradeoffs for signal to noise ratio for analog systems, and bit error rates for digital systems. The chapter concludes with an overview of the filtering requirements in typical cellular and satellite communication systems, and how the various system requirements impact the specifications of filter networks.

REFERENCES

1. Members of Technical Staff, *Transmission Systems for Communications*, 4th ed. (revised), Bell Telephone Laboratories, Inc., 1971.
2. H. Taub and D. L. Schilling, *Principles of Communications Systems*, 3rd ed., McGraw-Hill, New York, 1980.

3. M. Schwartz, *Information, Transmission, Modulation, and Noise*, 3rd ed., McGraw-Hill, New York, 1980.

4. R. A. Meyers, Ed., *Encyclopedia of Telecommunications*, Academic Press, San Diego, 1989.

5. G. D. Gordon and W. L. Morgan, *Principles of Communications Satellites*, Wiley, New York, 1993.

6. S. Haykin, *Communication Systems*, 4th edition, Wiley, New York, 2001.

7. R. L. Freeman, *Radio System Design for Telecommunications*, 2nd ed., Wiley, New York, 1997.

8. ITU-*Handbook on Satellite Communications*, 3rd ed., Wiley, New York, 2002.

9. C. E. Shannon, A mathematical theory of communications, *BSTJ* (*Br. Syst. Telecommun. J.*) **27**, 1948; C. E. Shannon, Communication in the presence of noise, *Proc. IRE* **37**, 10–21 (1949).

10. J. N. D. Krauss, *Antennas*, McGraw-Hill, New York, 1950.

11. B. P. Lathi, *Signals, Systems and Communication*, Wiley, New York, 1965.

12. T. G. Cross, Intermodulation noise in FM systems due to transmission deviations and AM/FM conversion, *BSTJ* **45**, 1749–1773 (Dec. 1966).

13. W. R. Bennett, H. E. Curtis, and S. O. Rice, Interchannel interference in FM and PM systems under noise loading conditions, *BSTJ* **34** 601–636 (May 1955).

14. G. J. Garrison, Intermodulation distortion in frequency-division multiplex FM systems—a tutorial summary, *IEEE Trans. Commun. Technol.* **Com-16**(2) (April 1968).

15. C. M. Kudsia and M. V. O'Donovan, *Microwave Filters for Communications Systems*, Artech House, Norwood, MA, 1974.

16. R. Tong and C. Kudsia, Enhanced performance and increased EIRP in communications satellites using contiguous multiplexers, *Proc. 10th AIAA Communication Satellite Systems Conf.*, Orlando, FL, March 19–22, 1984.

17. B. Sklar, Defining, designing, and evaluating digital communication systems, *IEEE Commun. Mag.*, (Nov. 1993).

18. F. Xiong, Modem technologies in satellite communications, *IEEE Commun. Mag.*, (Aug. 1994).

19. G. Maral and M. Bousquet, *Satellite Communications Systems*, Fourth edition, Wiley, New York, 2002.

20. R. C. Chapman et al., Hidden threat: Multicarrier passive component IM generation, Paper 76–296, *AIAA/CASI 6th Communications Satellite Systems Conf.*, Montreal, April 5–8, 1976.

21. C. Kudsia, COM DEV, and J. Fiedzuisko, LORAL, Organizers, High power passive equipment for satellite applications, *IEEE MTT-S Workshop Proc.*, Long Beach, CA, June 13–15, 1989.

22. C. Kudsia, R. Cameron, and W. C. Tang, Innovations in microwave filters and multiplexing networks for communications satellite systems, *IEEE Trans. Microwave Theory Tech* **40** (June 1992).

APPENDIX 1A INTERMODULATION DISTORTION SUMMARY

TABLE 1A.1   Direct Transmission Deviations

Transmission Deviation	Order of Distortion	NPR at Top Modulating Frequency (Without Preemphasis)
Parabolic gain, A_2 (dB/MHz^2)	Third	$\dfrac{1.72 \times 10^4}{A_2^4\, \sigma^4 f_m^4}$
Cubic gain, A_3 (dB/MHz^3)	Second	$\dfrac{33.6}{A_3^2\, \sigma^2 f_m^4}$
Quartic gain, A_4 (dB/MHz^4)	Third	$\dfrac{6.32}{A_3^2\, \sigma^2 f_m^4}$
Linear delay, B_1 (ns/MHz)	Second	$\dfrac{10^6}{\pi^2 B_1^2\, \sigma^2 f_m^2}$
	Third	$\dfrac{7.5 \times 10^5}{\pi^4 B_1^4\, \sigma^4 f_m^4}$
Parbolic delay, B_2 (ns/MHz^2)	Third	$\dfrac{7.5 \times 10^5}{\pi^2 B_2^2\, \sigma^4 f_m^2}$
Cubic delay, B_3 (ns/MHz^3)	Second	$\dfrac{1.19 \times 10^6}{\pi^2 B_3^2\, \sigma^2 f_m^6}$

Key: σ—multichannel RMS frequency deviation in MHz; f_m—top modulating frequency in MHz; A_n—is the nth-order amplitude coefficient in dB/(MHz)^n; B_n—is the nth-order group delay coefficient in ns/(MHz)^n; NPR—noise power ratio; a measure of intermodulation (IM) noise, given by ratio of white noise power spectral density to IM noise power spectral density in a communication channel.

TABLE 1A.2 Coupled Transmission Deviations—Significant Distortion Terms Only

Transmission Deviation	Order of Distortion	NPR at Top Modulating Frequency (Without Preemphasis)
Linear gain + linear AM/PM, A_1 (dB/MHz) + $K_{P1}(\bullet/\text{dB/MHz})$	Second	$\dfrac{3.28 \times 10^3}{K_{P1}^2 A_1^2 \sigma^2 f_m^2}$
Parabolic gain + constant AM/PM, A_2 (dB/MHz^2) + K_{P0} ($\bullet$/dB)	Second	$\dfrac{3.28 \times 10^3}{K_{P0}^2 A_2^2 \sigma_m^2 f_m^2}$
Quartic gain + constant AM/PM, A_4 (dB/MHz^4) + K_{P0} ($\bullet$/dB)	Second	$\dfrac{9.75 \times 10^2}{K_{P0}^2 A_4^2 \sigma^2 f_m^6}$
Linear delay + linear AM/PM, B_1 (ns/MHz) + K_{P1} ($\bullet$/dB/MHz)	Second	$\dfrac{1.73 \times 10^3}{\pi^2 K_{P1}^2 B_1^2 \sigma^2 f_m^4}$
Parabolic delay + constant AM/PM, B_2 (ns/MHz^2) + K_{P0} ($\bullet$/dB)	Second	$\dfrac{4.33 \times 10^7}{\pi^2 K_{P0}^2 B_2^2 \sigma^2 f_m^4}$

CHAPTER 2

FUNDAMENTALS OF CIRCUIT THEORY APPROXIMATION

Engineering problems represent exercises in approximation. The same can be said when we analyze any physical phenomena. At the most fundamental level, quantum mechanics can describe the properties of matter only in terms of probabilities. The point being made here is that with any engineering problem, there is always a tradeoff lurking somewhere, whether explicit or implicit. With that understanding, we can always optimize the parameters that are available within the constraints of a given system. The principles that unify communication theory and circuit theory approximations are explained in this chapter, which highlights the essential assumptions and the success of the frequency analysis approach that we take for granted in analyzing electrical networks.

2.1 LINEAR SYSTEMS

In nature, most systems, if not all, are nonlinear. Such systems are highly complex and not amenable to any general solution. Each problem and boundary condition requires an individual solution. However, most engineering problems can be approximated by linear systems over a limited range; herein lies our first assumption. It implies that we must ensure that all equipment used in the communication system satisfy the assumption of linearity over the operating range. Analysis of linear systems is well developed and covered in many textbooks [1].

Microwave Filters for Communication Systems: Fundamentals, Design, and Applications,
by Richard J. Cameron, Chandra M. Kudsia, and Raafat R. Mansour
Copyright © 2007 John Wiley & Sons, Inc.

Figure 2.1 Block diagram of an electrical system.

2.1.1 Concept of Linearity

Any system can be described as a blackbox that receives an input signal (or driving function), processes this signal, and then produces an output signal or response function, as depicted in Figure 2.1.

A system whose output is proportional to its input constitutes a linear system. This does not imply that the response is linearly proportional to the driving signal, although this constitutes a special case of linearity. Thus, if $r(t)$ is the response to $f(t)$, then $kr(t)$ is the response to $kf(t)$, where k is an arbitrary constant. Symbolically, this implies

$$f(t) \rightarrow r(t)$$
$$kf(t) \rightarrow kr(t) \tag{2.1}$$

However, linearity implies more than this. The defining property of a linear system is the law of superposition. This property states that if several causes are acting on a linear system, then the total effect on the system can be determined by considering each cause separately, and assuming all other causes to be zero. The total effect is then the sum of all the individual effects. Mathematically, this is expressed as follows:

$$f_1(t) \rightarrow r_1(t)$$
$$f_2(t) \rightarrow r_2(t)$$
$$f_1(t) + f_2(t) \rightarrow r_1(t) + r_2(t) \tag{2.2}$$

In more general terms, this can be stated as

$$a_1 f_1(t) + a_2 f_2(t) \rightarrow a_1 r_1(t) + a_2 r_2(t) \tag{2.3}$$

whereas this condition holds irrespective of the choice of a_1, f_1, a_2, f_2, and so on. Equation (2.3) represents the defining equation of a linear system.

We have restricted the discussion to a single-input/single-output system. This can be extended to multiple-input/multiple output systems.

2.2 CLASSIFICATION OF SYSTEMS

For linear and nonlinear systems alike, there can be a further classification:

- Time-invariant and time-variant systems
- Lumped-parameter and distributed-parameter systems
- Instantaneous (memoryless) and dynamic (with memory) systems
- Analog and digital systems

2.2.1 Time-Invariant and Time-Variant Systems

Systems whose parameters do not change with time are called *constant-parameter* or *time-invariant systems*. Linear time-invariant systems are characterized by linear equations with constant coefficients. Systems whose parameters change with time are called *variable-parameter* or *time-variant systems*. Linear time-variant systems are characterized by linear equations with time-dependent coefficients. Strictly speaking, all physical objects or systems change over time, when measured in eons. However, when systems are considered over finite durations of time as measured in tens or even hundreds of years, most fall into the category of time-invariant systems. All our discussions are confined to such systems.

2.2.2 Lumped and Distributed Systems

Any system is a collection of individual elements that are interconnected in a specific way. A system is deemed as lumped if all the elements in the system are assumed to behave as lumped elements. In a lumped model, the energy in the system is considered to be stored in distinct isolated elements such as inductors and capacitors in an electrical system, and springs in a mechanical system. Also, it is assumed that the disturbance initiated at any point is instantaneously propagated to any other point in the system. In electrical systems, this implies that the dimensions of the elements are very small compared to the wavelength of the signals to be transmitted. For a lumped electrical component, the voltage across its terminals or current flowing through it is related by a lumped parameter.

In contrast to lumped element systems, distributed systems are characterized by elements where the energy is distributed over the element's physical dimensions. For these systems, voltages and currents are distributed over the physical dimensions of the various elements. This implies that we need to deal with the variables of time as well as space. Moreover, in a distributed system, the propagation of energy takes a finite time to travel from a point in the system to any other point. The descriptive equations for distributed systems are partial differential equations, in contrast to the ordinary differential equations that describe lumped systems.

2.2.3 Instantaneous and Dynamic Systems

A system's output at any given instant depends on the past input. However, for many systems, the past history is irrelevant in determining the response. Such systems are said to be instantaneous or memoryless systems. More precisely, a system is said to be instantaneous if its output at any instant of time depends on its inputs at the same instant, and not on any past or future values of the inputs. A system whose output does depend on past inputs is said to be a system with memory or dynamic. In this book, we focus on instantaneous systems.

2.2.4 Analog and Digital Systems

A signal whose amplitude can take any value in a continuous range is an analog signal. Systems that deal with analog signals are referred to as *analog systems*. Signals, whether continuous or discrete, can also be represented by a finite number of values, referred to as *digital signals*. Systems that deal with digital signals are referred to as *digital systems*.

2.3 EVOLUTION OF ELECTRICAL CIRCUITS—A HISTORICAL PERSPECTIVE

In 1800, Volta published a paper that described, for the first time, the continuous flow of current by connecting a wire across a pair of dissimilar metals immersed in brine or a weak-acid electrolyte [2]. This important discovery represented the beginning of electrical circuits and circuit theory. Davy, Ampere, Coulomb, Ohm, and many others extended the concepts of electrical circuits to applications of direct current. Such circuits are characterized by currents, voltages, and electromotive force, which are all constants and not functions of time.

Two major discoveries on which the theory of time-varying currents and electromagnetic fields is based were made by Michael Faraday and James Clerk Maxwell. Faraday's discovery was experimental. He demonstrated that a changing magnetic field induces an electric field. Maxwell's discovery was intuitive and a leap in imagination. He postulated, by analogy, that a changing electric field would induce a magnetic field. In addition, he introduced the concept of displacement current so that time-varying electric fields play the same role as conduction currents, and combined the two into a total current that he showed to be continuous. This generalization cannot be overstated. It allows propagation of electric disturbances through dielectric media, including air and free space. Without this generalization, electrical disturbance can only propagate through a conductive medium. Maxwell went one step further and calculated the velocity of such electromagnetic disturbance and found it to be very close to the velocity of light, as measured by M. Fizeau. He concluded that the two phenomena are the same; specifically, light waves also represent an electromagnetic disturbance. The profound nature of this intuitive leap of imagination is evident in the following quotation [3]:

> It should be remembered that at this time no one had ever wittingly generated or detected electromagnetic waves. The concept was completely new, as was the notion of a displacement current. To link light to these hypothetical phenomena was a flash of brilliance seldom equalled in the history of science. It was not to be until eight years after Maxwell's death that these hypotheses would receive substantiation through the experiments of Hertz.

Maxwell combined all the fundamental relations of electrodynamics into four compact, beautiful equations, the Maxwell equations. This set of equations is the foundation of the whole field of electrical engineering.

2.3.1 Circuit Elements

Passive lumped electric circuits are characterized by three basic elements: capacitor (C), inductor (L), and resistor (R). The electromagnetic energy shuttles back and forth between the magnetic field of an inductor and the electric field of the capacitor,

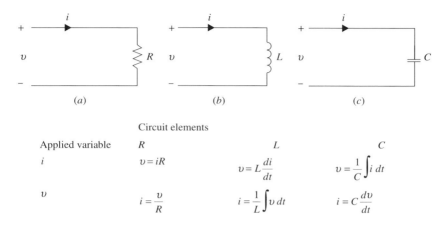

Figure 2.2   Relationships between the idealized elements in an electrical network: (*a*) a resistor; (*b*) an inductor; (*c*) a capacitor.

while it is being gradually dissipated as thermal energy in the resistor. The electrical variables that occur in networks are voltages or potential v, and currents i, where v and i are the instantaneous values of the voltage and current, respectively. The elements R, L, and C are defined in terms of two variables v and i as described in Figure 2.2. The elements are assumed to be linear, lumped, finite, passive, bilateral (LLFPB), and time-invariant. Systems composed of such elements conform to linear equations (algebraic, differential, or difference equations) with constant coefficients.

When the values of variables v and i are related in terms of the element values, time becomes an important factor. Currents and voltages must therefore be expressed as functions of time $i(t)$ and $v(t)$, respectively. Therefore, the applied signals are also functions of time, usually voltages. The response of the network to these signals is a voltage or a current in some other part of the network. It should be noted that we are still dealing with time-invariant systems, that is, the values of R, L, and C remain independent of time.

2.4   NETWORK EQUATION OF LINEAR SYSTEMS IN THE TIME DOMAIN

To analyze electrical networks, we can write loop equations to obtain the relationships for the complete network. Figure 2.3 reflects a simple network with the applied signal $v(t)$ and the required response is the current flowing in the circuit.

The system equation is

$$iR + L\frac{di}{dt} + \frac{1}{C}\int i\,dt = v(t) \tag{2.4}$$

By differentiating this expression in a second-order differential equation, we obtain

$$L\frac{d^2i}{dt^2} + R\frac{di}{dt} + \frac{1}{C}i(t) = \frac{dv(t)}{dt} \tag{2.5}$$

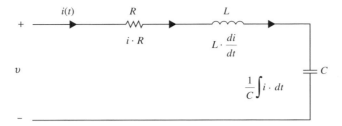

Figure 2.3   Time-domain analysis of a series RLC circuit.

Equation (2.5) indicates the nature of a differential equation characterizing lumped, linear, and time-invariant systems. For systems comprising of many more elements, branches, and loops, the general form of the differential equation is

$$a_m \frac{d^m i}{dt^m} + a_{m-1} \frac{d^{m-1} i}{dt^{m-1}} + \cdots + a_i \frac{di}{dt} + a_0 i(t)$$

$$= b_n \frac{d^n f}{dt^n} + b_{n-1} \frac{d^{n-1} f}{dt^{n-1}} + \cdots + b_1 \frac{df}{dt} + b_0 f(t) \qquad (2.6)$$

For convenience, we can replace differential operator d/dt by algebraic operator p. Consequently, the operation of integration with respect to t is replaced by $1/p$;

$$p \equiv \frac{d}{dt}, \quad pf(t) = \frac{df(t)}{dt}, \quad \ldots p^m f(t) = \frac{d^m f}{dt^m}$$

$$\frac{1}{p} f(t) = \int f(t) dt, \ldots$$

With this notation, the system equation is

$$(a_m p^m + a_{m-1} p^{m-1} + \cdots + a_1 p + a_0) i(t)$$

$$= (b_n p^n + b_{n-1} p^{n-1} + \cdots + b_1 p + b_0) f(t) \qquad (2.7)$$

This is a linear differential equation with constant coefficients and is solved by classical methods. The solution consists of two components: a source-free component and the component due to source. For a stable system, the source-free component always decays with time. For this reason, the source-free component is designated as the *transient component*, and the component due to source is called the *steady-state component*. The source-free component is obtained by setting the driving function to zero. It is independent of the driving function, and depends only on the nature of the system. This response is a characteristic of the system, and is also known as the *natural response* of the system. We are concerned with circuits where the transient component has vanished, leaving the system response entirely due to the steady-state source. For most applications, the transient response vanishes quickly and is not a factor in analyzing circuits powered by steady-state sources.

Several methods are used to obtain the solution of a linear differential equation due to a source. Assuming that the driving function $f(t)$ yields only a finite number of derivatives, we can represent the component due to source to be a linear combination of $f(t)$ and all its higher derivatives [1]

$$i(t) = h_1 f(t) + h_2 \frac{df}{dt} + \cdots + h_{r+1} \frac{d^r f}{dt^r} \tag{2.8}$$

where only the first r derivatives of $f(t)$ are independent. Coefficients $h_1, h_2, \ldots, h_r$ are obtained by substituting equation (2.8) in equation (2.7), and equating the coefficients with similar terms on both sides. The solutions of such a set of simultaneous differential equations are very time-consuming and sometimes quite formidable. The problem of solving differential equations can be overcome by limiting the problem to a reduced class of functions of time for which simple differential relationships exist. This leads to the simpler world of frequency domain.

2.5 NETWORK EQUATION OF LINEAR SYSTEMS IN THE FREQUENCY-DOMAIN EXPONENTIAL DRIVING FUNCTION

If the driving function $f(t)$ is an exponential function of time, it simplifies analysis of electrical circuits. For an exponential function, all the derivatives are also exponential functions of the same form. None of the derivatives is independent. Thus, the component due to source must be the same form as the driving function itself. Thus, if driving function $f(t)$ is exponential, the response function $i(t)$ is given by

$$i(t) = h f(t)$$
$$f(t) = e^{st}$$
$$i(t) = h e^{st} \tag{2.9}$$

where s is the complex frequency variable $\sigma + j\omega$. The value of c is obtained by substituting equation (2.9) in equation (2.7) as follows:

$$(a_m p^m + a_{m-1} p^{m-1} + \cdots + a_1 p + a_0) h e^{st}$$
$$= (b_n p^n + b_{n-1} p^{n-1} + \cdots + b_1 p + b_0) e^{st} \tag{2.10}$$

Since

$$p e^{st} = \frac{d}{dt} e^{st} = s e^{st} \quad \text{and} \quad p^r e^{st} = s^r e^{st}$$

equation (2.10) can be simplified to

$$(a_m s^m + a_{m-1} s^{m-1} + \cdots + a_1 s + a_0) h$$
$$= (b_n s^n + b_{n-1} s^{n-1} + \cdots + b_1 s + b_0)$$

Therefore

$$h = \frac{b_n s^n + b_{n-1} s^{n-1} + \cdots + b_1 s + b_0}{a_m s^m + a_{m-1} s^{m-1} + \cdots + a_1 s + a_0} \qquad (2.11)$$

The quantity h represents the system transfer function, usually denoted as $H(s)$, which characterizes the system in the frequency domain. This result forms the basis of frequency-domain analysis.

2.5.1   Complex Frequency Variable

At this point, we will describe the significance of the complex frequency variable in the analysis of electrical circuits. The complex frequency variable is given by $s = \sigma + j\omega$. A signal $f(t)$ may then be represented as

$$f(t) = e^{st}$$
$$= e^{\sigma t}(\cos \omega t + j \sin \omega t)$$

As a consequence, when s is complex, function e^{st} has real and imaginary parts. Furthermore, $e^{\sigma t} \cos \omega t$ and $e^{\sigma t} \sin \omega t$ represent functions oscillating at angular frequency ω, with the amplitude increasing or decreasing exponentially. This depends on whether σ is positive or negative. In the graphical representation of the complex frequency variable in Figure 2.4, the horizontal axis is the real axis represented by σ, and the vertical axis is the imaginary axis represented by ω.

Thus, the imaginary axis is associated with real frequencies. It should be mentioned that the left half-plane represents exponentially decaying functions ($\sigma < 0$), and the right half-plane represents exponentially growing functions ($\sigma > 0$). Each point on the complex frequency plane corresponds to a certain mode of the exponential function. This leads to the interesting question as to the significance of the frequencies that lie along the negative portion of the $j\omega$ axis.

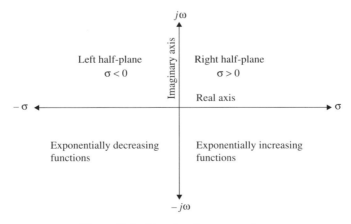

Figure 2.4   Complex frequency plane.

What is a negative frequency? By definition, frequency is an inherently positive quantity. The confusion arises because we are defining the frequency, not as a number of cycles per second of a particular waveform, but as an index of the exponential function. As a result, the negative frequencies are associated with negative exponents. Signals of negative and positive frequencies can be combined to obtain real functions, as follows: $e^{j\omega t} + e^{-j\omega t} = 2\cos \omega t$.

Similarly, the signals of two complex conjugate frequencies can form real signals. Any real function of time encountered in practice can be expressed as a continuous sum of the exponential functions that occur in pairs with complex conjugate frequencies. Thus, the use of a complex variable for the exponential functions represents a complete and powerful tool for analysing electrical systems.

2.5.2 Transfer Function

A system with an exponential input is displayed in Figure 2.5.

In the frequency-domain analysis, the response of a system is characterized by its transfer function defined as

$$H(s) = \frac{\text{response function}}{\text{driving function}} \tag{2.12}$$

For an exponential driving function, the response of a linear time-invariant system is given by $H(s)e^{st}$. By definition, the transfer function of a system is defined with respect to an exponential driving function. Equally, it should be remembered that the concept of transfer function is meaningful for linear systems alone. Furthermore, a system or electrical network can consist of a single-port, two-port, or multiport terminals. Thus there are a number of positions where a driving function can be applied and where a response can be observed. The transfer function is not a unique quantity unless the input and output terminals are specified. Filter networks are typically two-port networks, and $H(s)$ is usually specified in terms of the output to input voltage ratio. Using the more familiar notation

$$H(s) = \frac{b_n s^n + b_{n-1} s^{n-1} + \cdots + b_1 s + b_0}{a_m s^m + a_{m-1} s^{m-1} + \cdots + a_1 s + a_0}$$

$$= \frac{n(s)}{d(s)} \tag{2.13}$$

where $n(s)$ and $d(s)$ are respectively the numerator and denominator polynomials of the transfer function. In addition, this establishes the link between the system's time-domain and frequency-domain representations. For the system in differential equation (2.10), $H(s)$ is a rational function in s. It is a quotient of polynomials

Figure 2.5   Electrical network with an exponential input function.

with real coefficients. This is a very significant result and represents a huge simplification in the analysis of electrical circuits. In other words, for an exponential driving function, the response component due to source is also an exponential function of the same form as the driving function. Moreover, in a system, the transfer function is not unique unless the location of the excitation and response are specified.

2.5.3 Signal Representation by Continuous Exponentials

In order to exploit the simplicity of equation (2.13) to analyze real-life systems, two conditions must be satisfied:

1. Any signal can be represented in terms of exponential waveforms.
2. The responses due to the individual exponential functions can be added up to obtain the overall response.

If these two conditions are satisfied, then the simple expression of the transfer function $H(s)$ is all that is needed to obtain the response of the system for any signal in the frequency domain.

The second condition represents the principle of superposition of linear systems. As stated earlier, a fundamental premise of network analysis is that it is confined to linear systems. We need to preserve the linearity of the systems and deal with the inevitable nonlinearities as distortions that must be controlled within specified bounds. As a consequence, for linear systems, the success of the frequency-domain analysis approach depends on our ability to represent a given function of time as a sum of various exponential functions to satisfy the first condition. Nearly all practical signals or waveforms are classified as either periodic or nonperiodic. A periodic function is expressed as a sum of discrete exponential functions by the well-known Fourier series. Similarly, a nonperiodic function is represented as a continuous sum of exponential functions by making use of the well-known Fourier and Laplace transforms. As a consequence, any function, periodic or nonperiodic, is amenable to representation as a sum of exponential functions. More detailed and rigorous analyses of signals and systems can be found in the text by Lathi [1]. Suffice it to say here that the basis of frequency-domain analysis rests in the assumptions of the system's linearity and the power of exponential functions to represent signals and waveforms encountered in practice.

2.5.4 Transfer Functions of Electrical Networks

Lumped electrical networks are composed of individual elements R, L, and C. An exponential voltage applied across each of these elements produces an exponential response (i.e., current) as a consequence of the properties of linearity and the time invariance of these elements. Thus, if current i through the element is given by $i(t) = e^{st}$, then the voltage v across the element is $v(t) = Z(s)e^{st}$, where $Z(s) = v(t)/i(t)$ is defined as the impedance of the element. If we consider the exponential voltage as the driving function and current as the resulting response

TABLE 2.1   Properties of Positive Real Functions

1. An impedance or admittance function has the form $Z(s)$ or $Y(s) = n(s)/d(s)$, where the coefficients in the numerator and denominator polynomials are rational, real, and positive; thus

   $Z(s)$ is real when s is real
   Complex poles and zeros of $Z(s)$ occur in conjugate pairs

2. The poles and zeros of $Z(s)$ have either negative or zero real parts
3. The poles of $Z(s)$ on the imaginary axis must be simple, and their residues must be real and positive
4. The degrees of the numerator and the denominator polynomials in $Z(s)$ differ, at most, by 1; thus the number of finite poles and finite zeros of $Z(s)$ differ, at most, by 1; $Z(s)$ has neither multiple poles nor zeros at the origin

through the element, then $v(t) = e^{st}$ and $i(t) = Y(s)e^{st}$, where $Y(s) = i(t)/v(t)$ is defined as the admittance of the element. It is evident that $Y(s) = 1/Z(s)$. Together, these functions are referred to as *immitance functions* and are easily evaluated as follows:

$$Z(s) = Y^{-1}(s) = R \quad \text{for a resistor}$$
$$= sL \quad \text{for an inductor}$$
$$= \frac{1}{sC} \quad \text{for a capacitor} \qquad (2.14)$$

The transfer function of a system is the ratio of the response to an exponential excitation across a pair of specified terminals. It is composed of various combinations of the immitance functions of individual elements, R, L, and C, giving rise to a quotient of polynomial in s, the complex frequency variable. The coefficients of the polynomial are real, since R, L, and C are physical elements and have real values. As a consequence, the transfer function of an electrical network has the same form as that in by equation (2.11). Do all quotients of polynomials represent a physical network? Intuitively, the answer is "No." Physical networks must satisfy the law of the conservation of energy and must yield a real response to real stimuli. Such functions describe physical networks and are called *positive real functions*. The input impedance or driving point impedance function of any network, seen in Figure 2.3, is obviously a positive real function. The analysis of such functions has been described in many books [4]. Table 2.1 lists the properties of positive real functions.

2.6   STEADY-STATE RESPONSE OF LINEAR SYSTEMS TO SINUSOIDAL EXCITATIONS

The steady-state response of a system to the exponential driving function e^{st} was found to be $H(s)e^{st}$. It follows that the steady-state response of the system to a function $e^{j\omega t}$ will be $H(j\omega)e^{j\omega t}$. Since $\cos \omega t + j\sin \omega t = e^{j\omega t}$ and the principle of superposition holds for linear systems, we can conclude that the steady-state response of the system to driving functions $\cos \omega t$ and $\sin \omega t$ is given by $\text{Re}[H(j\omega)e^{j\omega t}]$ and $\text{Im}[H(j\omega)e^{j\omega t}]$, respectively. $H(j\omega)$ is complex in general and is expressed in polar form as

$$H(j\omega) = |H(j\omega)|e^{j\theta(\omega)}$$

and

$$H(j\omega)e^{j\omega t} = |H(j\omega)|e^{j(\omega t + \theta)} \tag{2.15}$$

Therefore

$$\text{Re}\left[H(j\omega)e^{j\omega t}\right] = |H(j\omega)| \cos(\omega t + \theta)$$

and

$$\text{Im}\left[H(j\omega)e^{j\omega t}\right] = |H(j\omega)| \sin(\omega t + \theta) \tag{2.16}$$

The term $|H(j\omega)|$ signifies the magnitude of the response to a unit sinusoidal function. The response function is shifted in phase by angle θ with respect to the driving function. The angle θ is the phase angle of $H(j\omega)$, specifically, $\angle H(j\omega) = \theta$.

2.7 CIRCUIT THEORY APPROXIMATION

There are three basic assumptions implicit in circuit theory [5]:

1. The system is physically small enough that the propagation effects can be ignored; that is, the electrical effects occur instantaneously throughout the entire system. Ignoring the spatial dimensions results in what we refer to as lumped-parameter elements and systems.
2. The net charge on every component in the system is always zero. Thus, no component can collect a net excess of charge, although some components can hold equal but opposite charges.
3. There is no coupling between the components in a system.

Assumption 1 implies that the wavelengths of the signals are much larger than the physical dimensions of the electric components. At low frequencies, say, 1 MHz, the free-space wavelength is 300 m. Components measuring in the range of a few centimeter are well within the range of the assumption. At microwave frequencies, say, 1 GHz, the wavelength is 0.3 m. The dimensions of circuits in this range are comparable to the wavelength, and, therefore, outside the range of the assumption. How can we then justify the use of circuit theory on the basis of the lumped-element representation at microwave frequencies? The answer lies in the modeling of microwave circuits such that their behavior can be predicted accurately using lumped-element representation. Such is indeed the case with distributed microwave structures over a limited frequency range. Consequently, circuit theory approximation is widely used in the analysis and synthesis for a large class of microwave filters.

Assumption 2 reflects the law of conservation of the charge or current in an electrical system. For time-varying signals, the concept of displacement current allows preservation of this principle when circuits include capacitors. Gustav Kirchhoff formalized these concepts, now referred to as *Kirchhoff's laws:*

1. The algebraic sum of all the currents at any node in a circuit equals zero.
2. The algebraic sum of all the voltages around any closed path in a circuit equals zero.

Assumption 3 implies that each element in circuit theory is independent of the other elements in the network. It allows for mutual coupling between two inductors that form a transformer. Such a mutual coupling is directly related to the inductances of the two separate elements forming the transformer.

In summary, the circuit theory is based on five ideal elements:

- An ideal voltage source is a circuit element that maintains a prescribed voltage across its terminals regardless of the device.
- An ideal current source is a circuit element that maintains a prescribed current within its terminals, regardless of the voltage across it.
- An ideal resistor is linear, time-invariant, bilateral, lumped element that follows Ohm's law.
- An ideal inductor is a linear, time-invariant, lumped circuit parameter that relates the voltage induced by a time-varying magnetic field to the current, producing the time-varying magnetic field.
- An ideal capacitance is a linear, time-invariant, lumped circuit parameter that relates the current induced by a time-varying electric field to the voltage, producing the time-varying electric field.

The application of the lumped-element circuit theory at microwave frequencies is an approximation, albeit a very good one over a limited frequency range. It is important to understand that lumped circuit techniques represent an approximation of the more encompassing EM theory. Consequently, It is possible to apply both the lumped circuit approximation and EM theory to a given problem. Circuit theory allows a good first approximation and, if necessary, EM techniques can be applied to further refine the approximation, especially over wider frequency bands. Such is indeed the case for a large number of practical microwave engineering applications. The benefits of circuit theory are

1. It provides simple solutions (of significant accuracy) to problems that would otherwise become hopelessly complicated if we were to use electromagnetic field theory. We can analyze and build practical circuits with circuit theory.
2. The analysis and design of many useful electrical systems are less complicated if we divide them into subsystems, called *components*. We can then use the terminal behavior of each component to predict the behavior of the

interconnection. The ability to derive the circuit models of physical devices makes circuit theory an attractive approach.

3. Circuit analysis introduces a methodology for solving large networks of linked linear differential equations, prevalent throughout engineering and technology. This allows access to a common pool of knowledge that exists in other engineering disciplines.

4. Circuit theory is an interesting area of study in its own right. Much of the remarkable development of electrical power systems can be attributed to the development of circuit theory as a separate discipline of study.

SUMMARY

The principles that unify communication theory and circuit theory approximation are explained in this chapter. It highlights the necessity for communication systems to be linear over the specified range of bandwidth. When a signal, composed of a number of time varying functions is applied to a linear system, the overall response is given by the sum of the responses due to the individual functions. This follows from the principle of superposition, an inherent property of a linear system. As is well known, any function, periodic or nonperiodic is amenable to representation by the sum of exponential functions via Fourier series or Laplace transforms. Thus, a communication channel can be characterized by an exponential driving function, and that represents a major simplification in the analysis of communication systems. This approach is called the *frequency analysis approach* or the *frequency-domain analysis.*

The next step in the simplification process for analyzing a communication channel is the circuit theory approximation. It assumes that the basic passive electrical components, namely, resistors, inductors, and capacitors, are discrete, lumped, linear elements. Such modeling, when combined with exponential driving functions, leads to the transfer functions of circuits in terms of a rational quotient of polynomials with real coefficients. This is a huge simplification in the analysis and synthesis of filter networks. This chapter highlights the fundamental assumptions and the success of the frequency analysis approach that we take for granted in analyzing electrical networks.

REFERENCES

1. B. P. Lathi, *Signals, Systems and Communications*, Wiley, New York, 1965.
2. A. Volta, On the electricity excited by the mere contact of conducting substances of different kinds, *Phil. Trans. Roy. Soc. (Lond.)* **90** (1800).
3. R. S. Elliot, *Electromagnetics—History, Theory, and Applications*, IEEE Press, 1993.
4. M. E. Van Valkenburg, *Modern Network Synthesis*, Wiley, New York, 1960.
5. J. W. Nilsson, *Electric Circuits*, 4th ed., Addison-Wesley, Reading, MA, 1993.

CHAPTER 3

CHARACTERIZATION OF LOSSLESS LOWPASS PROTOTYPE FILTER FUNCTIONS

This chapter describes the synthesis process for the characteristic polynomials to realize the ideal, classical, prototype filters: the maximally flat, Chebyshev, and elliptic function filters. The chapter includes a discussion of filters that are not symmetric with respect to their center frequency. This leads to transfer function polynomials (with certain restrictions) with complex coefficients, a distinct departure from the more familiar characteristic polynomials with rational and real coefficients. This provides a basis for analysis of the most general class of filter functions in the lowpass prototype domain, minimum and nonminimum phase filters, exhibiting a symmetric or an asymmetric frequency response.

3.1 THE IDEAL FILTER

In communication systems, filter networks are required to transmit and attenuate signals in specified frequency bands. Ideally, this must be accomplished with the minimum of distortion and loss of energy of the transmitted signal.

3.1.1 Distortionless Transmission

A signal is characterized by the amplitude and phase of its frequency components, referred to as the *waveform* of the signal. For distortionless transmission, the waveform must be preserved; that is, the output signal must be an exact replica of the

Microwave Filters for Communication Systems: Fundamentals, Design, and Applications,
by Richard J. Cameron, Chandra M. Kudsia, and Raafat R. Mansour
Copyright © 2007 John Wiley & Sons, Inc.

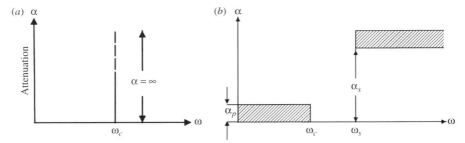

Figure 3.1 Lowpass filter amplitude response: (*a*) ideal filter; (*b*) practical filter. (Note that ω_c represents the passband cutoff frequency, and ω_s represents the lower edge of stopband; α_p and α_s represent the maximum and minimum attenuation in the passband and stopband, respectively.)

input signal. This can be achieved only if the filter network has a constant amplitude and time delay for all the frequencies in the passband. A constant time delay implies that the phase shift is directly proportional to the frequency and is given by $-\omega t$. Outside the passband, the filter network is required to attenuate all frequencies, implying a brick-wall amplitude response, exhibiting infinite attenuation and a zero transition band. The unit impulse response of a brick-wall filter exists for the negative values of time, thus violating the causality condition [1]. A causal function starts at some finite time, say, $t = 0$, and has zero value for $t < 0$. The causality condition means that a physical system cannot anticipate the driving function. Causality condition gives rise to two fundamental constraints for the filter function: (1) the amplitude function can be zero at some discrete frequencies but cannot be zero over a finite band of frequencies, and (2) the amplitude function cannot fall off to zero faster than a function of an exponential order. The implication is that the zero transition band is not allowed. However, it is possible to approach the ideal characteristic as closely as desired. Needless to say, it is the practical constraints that set the limit of the achievable performance. The ideal and the practical response of a lowpass prototype filter is depicted in Figure 3.1.

3.1.2 Maximum Power Transfer in Two-Port Networks

Communication systems consist of a tandem connection of several two-port networks. Energy is transferred from the source to load via linear transducers such as amplifiers, modulators, filters, and cables. It is to be understood that although components such as modulators and amplifiers are inherently nonlinear, their operation is controlled to be linear with a minimal and acceptable amount of nonlinearity. For practical systems, it is imperative that all the available power is transferred at each stage so as to realize the highest levels of signal to noise ratios in the system. To achieve this condition, let us examine an ideal lossless network connected between a source and a load, as reflected in Figure 3.2.

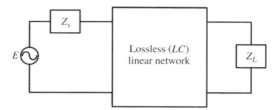

Figure 3.2 Power transfer in a two-port network.

The available power of a source is defined as the maximum power that can be drawn from the source. The maximum available power from a source of internal impendence Z_s is obtained when the load connected to its terminals is the complex conjugate of Z_s. Therefore, if $Z_S = R_S + jX_S$, then $Z_L = R_S - jX_S$ for a maximum power transfer.

Therefore, the available power from the source is $E^2/4R_s$. The power delivered to the load is maximized if the output impedance of the two-port network is the conjugate of Z_L. It should be noted that the available power depends only on the resistive component of the source. Similarly, the maximum power that can be delivered to the load depends solely on its resistive component. This leads us to the lossless two-port networks, terminated in resistors of equal value at each end, as the desired filter networks for communications systems. Such filter networks are referred to as the doubly terminated filters.

3.2 CHARACTERIZATION OF POLYNOMIAL FUNCTIONS FOR DOUBLY TERMINATED LOSSLESS LOWPASS PROTOTYPE FILTER NETWORKS

The synthesis of filter networks is carried out by developing lumped lossless lowpass filters that are normalized in terms of frequency and impedance and terminated in resistors of equal value. By scaling in frequency and amplitude, it is then possible to derive filter networks over any desired frequency range and impedance levels. This simplifies the design of practical filters regardless of their frequency range and how they are realized physically. The lowpass prototype filter is normalized to a cutoff frequency of 1 radian per second and terminated in resistors of 1 Ω. The cutoff frequency of unity implies that the passband extends from ω = 0 to ω = 1.

As described in the previous section, even for an ideal filter, we cannot have a zero loss over the entire passband without violating the causality condition. However, it is possible to have a zero loss at a finite number of frequencies. Such frequencies are referred to as *reflection zeros*, for which no power is reflected, and thus, the signal incurs zero loss. An obvious consequence of this is that all the reflection zeros of the filter function are confined to the passband. Furthermore,

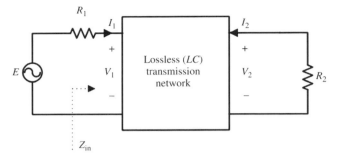

Figure 3.3 Doubly terminated lossless transmission network.

the magnitude of maximum reflected power over the passband serves as a design parameter of the filter prototype networks. The use of lossless elements makes the synthesis process much simpler. The effect of dissipation can be readily incorporated after the synthesis process with little loss of accuracy, provided the dissipation is small. Filter design based on lossless prototype networks is quite adequate for a variety of applications. This chapter deals with the circuit theory approximation for the design of lossless lowpass prototype filter functions.

Figure 3.3 illustrates a lossless two-port network terminated in resistors. This is a general representation of a doubly terminated filter network capable of maximum power transfer.

The maximum available power P_{max} from the ideal voltage source E is $E^2/4R_1$. Power P_2, delivered to the load R_2, is $|V_2|^2/R_2$. Therefore

$$\frac{P_{max}}{P_2} = \left| \frac{1}{2} \sqrt{\frac{R_2}{R_1}} \frac{E}{V_2} \right|^2 \tag{3.1}$$

$$= \frac{1}{4} \left| \frac{E}{V_2} \right|^2 \quad \text{for} \quad R_1 = R_2$$

For passive lossless two-port networks, P_2 can only be equal to or less than P_{max}. For such networks, it is most convenient to define a characteristic function $K(s)$ by [2,3]

$$\frac{P_{max}}{P_2} = 1 + \left| K(s) \right|^2_{s=j\omega} \tag{3.2}$$

For the lumped, linear, and time-invariant circuits considered here, $K(s)$ is a rational function in s with real coefficients. The ratio P_{max}/P is defined by the transmission function $H(s)$, where

$$\left| H(s) \right|^2_{s=j\omega} = 1 + \left| K(s) \right|^2_{s=j\omega} \tag{3.3}$$

Also, $H(s)$ is referred to as the *transducer function* and can be written in the alternative form

$$H(s)H(-s) = 1 + K(s)K(-s) \tag{3.4}$$

since all network functions are rational in s with real coefficients. To characterize the polynomial $K(s)$, let us consider the reflection and transmission coefficients of the prototype network.

3.2.1 Reflection and Transmission Coefficients

In transmission-line theory, the reflection coefficient ρ is defined as

$$\rho(s) = \pm \frac{\text{reflected wave}}{\text{incident wave}} \tag{3.5}$$

The sign of ρ is positive when the ratio is defined in terms of voltages and negative when defined in terms of currents. The negative sign arises because the direction of current for the reflected wave is opposite that of incident wave. The use of symbol ρ for the reflection coefficient is quite common for analyzing two-port networks and their power relationships. The symbol Γ is used to denote the complex reflection coefficient, and generally used when analyzing circuits in terms of scattering parameters. For the lossless network in Figure 3.3 and with P_{max} as the reference or maximum power available, the reflection coefficient in terms of power is given by

$$\left|\rho(j\omega)\right|^2 = \frac{\text{reflected power}}{\text{available power}} = \frac{P_r}{P_{max}} \tag{3.6}$$

where P_r represents the reflected power. Since, the reflected power plus the transmitted power delivered to the load resistor R_2 must equal the available power

$$\frac{\text{Reflected power}}{\text{Available power}} + \frac{\text{transmitted power}}{\text{available power}} = 1$$

or

$$\left|\rho(j\omega)\right|^2 + \left|t(j\omega)\right|^2 = 1 \tag{3.7}$$

where $t(s)$ is defined as the transmission coefficient and represents the ratio of the transmitted wave to the incident wave. Therefore

$$\left|t(j\omega)\right|^2 = \frac{\text{transmitted power}}{\text{available power}} = \frac{P_2}{P_{max}} = 1 - \left|\rho(j\omega)\right|^2 \tag{3.8}$$

In terms of decibels, the transmission and reflection loss (often termed the *return loss*) are defined as

$$A = 10 \log \frac{1}{|t(j\omega)|^2} = -10 \log |t(j\omega)|^2 \quad \text{dB}$$

$$R = 10 \log \frac{1}{|\rho(j\omega)|^2} = -10 \log |\rho(j\omega)|^2 \quad \text{dB} \tag{3.9}$$

Making use of equation (3.7), we can relate the transmission and return loss by

$$A = -10 \log[1 - 10^{-R/10}] \quad \text{dB}$$

and

$$R = -10 \log[1 - 10^{-A/10}] \quad \text{dB} \tag{3.10}$$

From the transmission-line theory [2], the reflection coefficient ρ for a two-port network is given by

$$\rho(s) = \frac{Z_{\text{in}}(s) - R_1}{Z_{\text{in}}(s) + R_1} = \frac{z_{\text{in}}(s) - 1}{z_{\text{in}}(s) + 1}, \quad \text{where} \quad z_{\text{in}}(s) = \frac{Z_{\text{in}}(s)}{R_1} \tag{3.11}$$

The input impedance of a terminated network is a positive real function (Section 2.5), and its normalized impedance z can be expressed as

$$z(s) = \frac{n(s)}{d(s)} \tag{3.12}$$

where $n(s)$ and $d(s)$ represent the numerator and denominator polynomials and z(s) is a positive real function [2]. Thus

$$\rho(s) = \frac{z(s) - 1}{z(s) + 1} = \frac{n(s) - d(s)}{n(s) + d(s)} = \frac{F(s)}{E(s)} \tag{3.13}$$

Since $z(s)$ is positive real, we can conclude that

1. The denominator polynomial $n(s) + d(s) = E(s)$ must be the Hurwitz polynomial, with all its roots lying inside the left half of the s-plane.
2. The numerator polynomial $n(s) - d(s) = F(s)$ may or may not be a Hurwitz polynomial. However, the polynomial coefficient must be real and, as a result, its roots must be real or at the origin or occur in conjugate complex pairs.

The magnitude of ρ, along the imaginary axis, is therefore given by

$$\left|\rho(j\omega)\right|^2 = \frac{F(s)\,F^*(s)}{E(s)\,E^*(s)} \qquad (3.14)$$

where the asterisks indicate the complex conjugate of the respective functions. Furthermore, for $s = j\omega$, $F^*(s) = F(-s)$ and $E^*(s) = E(-s)$, and therefore

$$\left|\rho(j\omega)\right|^2 = \frac{F(s)\,F(-s)}{E(s)\,E(-s)}$$

$$\left|t(j\omega)\right|^2 = \frac{E(s)\,E(-s) - F(s)\,F(-s)}{E(s)\,E(-s)} = \frac{P(s)\,P(-s)}{E(s)\,E(-s)} \qquad (3.15)$$

where $P(s)P(-s) = E(s)E(-s) - F(s)F(-s)$. Expressing the right-hand side of equation (3.15) by $P(s)P(-s)$ is justified only if it is a perfect square. This is indeed the case since the root pattern has a quadrantal symmetry and the roots on the imaginary axis have even multiplicity. It is implicit here that polynomials $F(s)$ and $P(s)$ are normalized to ensure that the transmitted power is equal to or less than the available power.

The three polynomials are referred to as the *characteristic polynomials*. For lowpass prototype filter networks, their properties are summarized as follows [3]:

1. $F(s)$ is a polynomial with real coefficients, and its roots lie along the imaginary axis as conjugate pairs. $F(s)$ can have multiple roots only at the origin. The roots represent frequencies at which no power is reflected, often termed *reflection zeros*. At these frequencies, the filter loss is zero, and $F(s)$ is a pure odd or even polynomial.

2. $P(s)$ is a pure even polynomial with real coefficients. Its roots lie on the imaginary axis in conjugate pairs. Such roots represent frequencies at which no power is transmitted, and the filter loss is infinite. These frequencies are often referred to as *transmission zeros* or *attenuation poles*. Its roots can also occur as conjugate pairs on the real axis or as a complex quad in the s plane. Such roots lead to linear phase (nonminimum phase) filters.

3. $E(s)$ is a strict Hurwitz polynomial, as all its roots lie in the left half of the s plane.

In terms of the characteristic polynomials, we obtain

$$\rho(s) = \frac{F(s)}{E(s)} \quad \text{and} \quad t(s) = \frac{P(s)}{E(s)} \qquad (3.16)$$

Another interesting relationship that relates the transmitted to the reflected powers is

$$\left|K(s)\right|^2_{s=j\omega} = \frac{\left|\rho(j\omega)\right|^2}{\left|t(j\omega)\right|^2} = \frac{P_r}{P_t} \qquad (3.17)$$

where P_r is the reflected power and P_t is the transmitted power.

The transmission function and the characteristic function are then expressed as the ratio of the polynomials in the form

$$H(s) = \frac{E(s)}{P(s)} \tag{3.18}$$

and

$$K(s) = \frac{F(s)}{P(s)} \tag{3.19}$$

It should be noted that in terms of scattering parameters described in subsequent chapters, the reflection and transmission coefficients are synonymous with S_{11} and S_{21}, respectively, represented as

$$\rho(s) \equiv \Gamma(s) \equiv S_{11}(s)$$
$$t(s) \equiv S_{21}(s)$$

3.2.2 Normalization of the Characteristic Polynomials

The transmission function of a lossless prototype filter, defined in terms of the characteristic function $K(s)$ in equation (3.3), is

$$\left|H(s)\right|^2_{s=j\omega} = 1 + \left|K(s)\right|^2_{s=j\omega}$$

Without loss of generality, we can introduce an arbitrary constant factor ε and restate the relationship as

$$\left|H(s)\right|^2_{s=j\omega} = 1 + \varepsilon^2 \left|K(s)\right|^2_{s=j\omega} \tag{3.20}$$

where ε is the ripple factor. It is employed to normalize the maximum amplitude of the filter in the passband. Since the polynomial $K(s)$ represents the ratio of polynomials $F(s)$ and $P(s)$, it is more appropriate to represent $K(s)$ in the form

$$K(s) = \varepsilon \frac{F(s)}{P(s)} \tag{3.21}$$

In the synthesis procedure, polynomials $F(s)$ and $P(s)$ are normalized so that their highest coefficients are unity. This is accomplished by extracting the highest coefficients of these polynomials and representing their ratio as a constant factor that can be readily absorbed within the ripple factor. Thus, ε is used to normalize $F(s)$ and $P(s)$, as well as the prototype amplitude response given by equation (3.20).

3.3 CHARACTERISTIC POLYNOMIALS FOR IDEALIZED LOWPASS PROTOTYPE NETWORKS

Filter requirements call for a low loss in the passband and a high loss in other frequency bands. Such a requirement can best be achieved by assigning all the zeros of $F(s)$ to the $j\omega$ axis in the passband region, and all zeros of $P(s)$ to the $j\omega$ axis in the high-loss frequency bands. For some applications, the zeros of $P(s)$ have a non-$j\omega$-axis location. This results in an improved phase and group delay response in the passband at the expense of attenuation in the stopband. Such a tradeoff is sometimes beneficial for the overall system requirements. In order to ensure that the characteristic polynomials conform to be positive real functions (Section 2.5), the pole–zero locations of lossless lowpass prototype networks normalized to a unity cutoff frequency are restricted as follows:

1. All the zeros of $K(s)$ must lie symmetrically on the $j\omega$ axis in the passband region.
2. All the poles of $K(s)$ must lie symmetrically on the $j\omega$ axis or real axis, or be distributed symmetrically along the real and imaginary axes, forming a conjugate complex quad.

The distribution of zero locations for polynomials $F(s)$ and $P(s)$ are illustrated in Figure 3.4.

The polynomial $F(s)$ containing the reflection zeros has the form

$$F(s) = s(s^2 + a_1^2)(s^2 + a_2^2)\ldots \quad \text{for odd order filters}$$
$$= (s^2 + a_1^2)(s^2 + a_2^2)\ldots \quad \text{for even order filters} \quad (3.22)$$

The polynomial $P(s)$ can be formed by factors $(s^2 \pm b_i^2)$ or $(s^4 \pm c_i s^2 + d_i)$, or any combination of such factors. Therefore, the characteristic function $K(s)$ assume the following forms

$$K(s) = \varepsilon \frac{F(s)}{P(s)} = \varepsilon \frac{s\left(s^2 + a_1^2\right)\left(s^2 + a_2^2\right)\cdots}{\left(s^2 \pm b_1^2\right)\left(s^2 \pm b_2^2\right)\cdots(s^4 \pm c_1 s^2 + d_1)\cdots} \quad (3.23)$$

for odd-order networks and

$$K(s) = \varepsilon \frac{F(s)}{P(s)} = \varepsilon \frac{\left(s^2 + a_1^2\right)\left(s^2 + a_2^2\right)\left(s^2 + a_3^2\right)\cdots}{\left(s^2 \pm b_1^2\right)\left(s^2 \pm b_2^2\right)\cdots(s^4 \pm c_1 s^2 - d_1)\cdots} \quad (3.24)$$

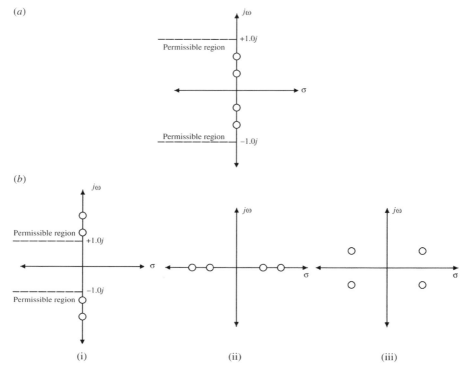

Figure 3.4   Permissible zero locations of lowpass prototype characteristic polynomials; (a) zero locations of $F(s)$; (b) zero locations of $P(s)$ on the $j\omega$ axis, real axis, and as a complex quad.

for even-order networks. The constraints on the a and b terms are

$$1 < a_1, a_2, \cdots \geq 0$$
$$b_1, b_2, \cdots > 1 \quad \text{for } j\omega\text{-axis zeros and no restriction for real-axis zeros}$$
$$c_1, c_2, \ldots, d_1, d_2, \ldots, > 0$$

These relationships clearly indicate that all the polynomials are rational with real coefficients and are either pure even or pure odd, reflecting the nature of positive real functions to realize lowpass prototype networks. From equations (3.23) and (3.24), it is evident that the frequencies $\pm ja_1, \pm ja_2, \ldots$ are zeros of reflection, that is, the frequencies at which all the power is transmitted and none is reflected. As a result, these frequencies are also called *attenuation zeros*. Similarly, frequencies $\pm jb_1, \pm jb_2, \ldots$ are zeros of transmission and no power is transmitted, and there is a total reflection of the power. For this reason, these frequencies are more commonly called *attenuation poles* or *transmission zeros*. Together, a's and b's are termed *critical frequencies*. The zeros that lie on the real axis or form a symmetric complex quad are used to improve the phase and group delay response at the expense of the amplitude response. Such filters are referred to as *linear phase filters*. The polynomial $E(s)$ is strictly Hurwitz, that is, all its zeros must lie in the left half of the s plane.

3.4 LOWPASS PROTOTYPE CHARACTERISTICS

Filters are specified in terms of their amplitude and phase or group delay response. Computation of these quantities is described in this section.

3.4.1 Amplitude Response

The insertion loss in dB is defined as

$$A\ (dB) = 10 \log \frac{P_{max}}{P_2} = 20 \log |H(j\omega)|$$

$$= 20 \log \left| \varepsilon \, \frac{E(s)}{P(s)} \right| \quad dB \tag{3.25}$$

If we denote $s_1, s_2, \ldots, s_n$ as the roots of the Hurwitz polynomial $E(s)$, and $p_1, p_2, \ldots, p_m$ as the roots of polynomial $P(s)$

$$|H(j\omega)| = \varepsilon \, \frac{|(s - s_1)(s - s_2) \cdots (s - s_n)|}{|(s - p_1)(s - p_2) \cdots (s - p_m)|} \tag{3.26}$$

where ε is an arbitrary constant for normalizing the amplitude response; ε is computed by evaluating $|H|$ at $\omega = 0$. At zero frequency, the value of H is either unity or exceeds unity by a small amount determined by the maximum permissible ripple in the passband.

The insertion loss is then computed as

$$A(dB) = 20 \log e[\ln|(s - s_1)|$$
$$+ \ln|(s - s_2| + \cdots - \ln|(s - p_1)| - \ln|(s - p_2)| - \cdots]$$
$$+ 20 \log e \ln \varepsilon \tag{3.27}$$

At real frequencies, $s = j\omega$, a typical term in the insertion loss is given by

$$|s - s_k|_{s=j\omega} = |s - \sigma_k \mp j\omega_k|_{s=j\omega}$$

$$= \sqrt{\sigma_k^2 + (\omega \mp \omega_k)^2} \tag{3.28}$$

where $s_k = \sigma_k + j\omega_k$ represents the kth root of the respective polynomial.

3.4.2 Phase Response

The phase response for the filter network is computed by

$$\beta(\omega) = -\text{Arg}\, H\,(j\omega)$$

$$= -\tan^{-1} \frac{\text{Im}\, H(j\omega)}{\text{Re}\, H(j\omega)}$$

$$= \sum_{poles} \tan^{-1} \left(\frac{\omega - \omega_k}{\sigma_k} \right) - \sum_{zeros} \tan^{-1} \left(\frac{\omega - \omega_k}{\sigma_k} \right) \tag{3.29}$$

The group delay τ of the filter is calculated by

$$\tau = -\frac{d\beta}{d\omega} = \sum_{\text{zeros}} \left(\frac{\sigma_k}{\sigma_k^2 + (\omega - \omega_k)^2} \right) \qquad (3.30)$$

The zeros of $P(s)$ are either located on the imaginary axis or distributed symmetrically about the imaginary axis. Consequently, they do not contribute to the phase variation (except for the integer multiples of π radians). Since the group delay is the derivative of the phase, it is entirely determined by the roots of $E(s)$. The slope of the group delay is given by

$$\frac{d^2\beta}{d\omega^2} = \sum_{\text{zeros}} 2\tau_k^2 \tan \beta_k - \sum_{\text{zeros}} 2\tau_k^2 \tan \beta_k \qquad (3.31)$$

The term τ_k is the group delay factor associated with the kth pole or zero such that

$$\tau_k = -\frac{d\beta_k}{d\omega} = \frac{\sigma_k}{\sigma_k^2 + (\omega - \omega_k)^2} \qquad (3.32)$$

3.4.3 Phase Linearity

The phase linearity over a specified frequency band is defined as the deviation of the phase from the ideal linear phase response. Usually, the ideal phase curve is defined by the slope of the phase at zero frequency for the lowpass prototype filter. However, we can choose any reference frequency over the band of interest and define the phase linearity with respect to the slope at that reference frequency.

Let ω_{ref} be the reference frequency. The linear phase ϕ_L at a frequency ω is then expressed by

$$\phi_L(\omega) = \left(\frac{d\beta}{d\omega} \right)_{\omega=\omega_{\text{ref}}} (\omega - \omega_{\text{ref}})$$

$$= \tau_{\text{ref}} (\omega - \omega_{\text{ref}}) \qquad (3.33)$$

where τ_{ref} is the absolute group delay at the reference frequency. The phase linearity $\Delta\phi$ at a given frequency ω is given by

$$\Delta\phi = \beta(\omega) - \phi_L \qquad (3.34)$$

where $\beta(\omega)$ is the actual phase at ω. If the reference frequency is chosen to be zero, then

$$\Delta\phi = \beta(\omega) - \tau_0\omega, \quad \text{where} \quad \tau_0 = \frac{d\beta}{d\omega}\bigg|_{\omega=0} \qquad (3.35)$$

From a practical standpoint, it is desirable to compute the minimum phase deviation from a straight-line approximation of the phase over the band of interest. This implies curve fitting of a straight line over the phase response so as to obtain an equiripple phase about the straight line.

3.5 CHARACTERISTIC POLYNOMIALS VERSUS RESPONSE SHAPES

In this section, we describe the possible forms of the characteristic function and the resulting response shapes for doubly terminated lowpass prototype filter networks.

3.5.1 All-Pole Prototype Filter Functions

Such filter functions are characterized by

$$t(s) = \frac{1}{E(s)} \tag{3.36}$$

with $P(s) \equiv 1$. There are no finite transmission zeros and the attenuation rises monotonically beyond the passband. All the attenuation poles are located at infinity. The response shape is determined by the form of the polynomial $F(s)$. The two basic forms of $F(s)$ are

$$F(s) \;\rightarrow\; s^n$$

or

$$F(s) \;\rightarrow\; s^m \left(s^2 + a_1^2\right) \left(s^2 + a_2^2\right) \cdots \tag{3.37}$$

The first form has all the zeros at the origin, and the transmission response is characterized by a maximally flat passband response, commonly known as the *Butterworth response*.

The more general second form of $F(s)$ is characterized by some of its zeros at the origin and others at the finite frequencies, $a_1, a_2, \ldots$. Figure 3.5 offers such an arbitrary response shape. By an appropriate choice of $a_1, a_2, \ldots$, we can obtain an equiripple response shape. A special case of the equiripple response exists when there are maximum permissible peaks in the passband. For such a case, $m = 1$ for odd-order filters and zero for even-order filters. The critical frequencies a_i are chosen via the Chebyshev polynomial to ensure the maximum number of equiripple peaks in the passband. For this reason, such filters are called *Chebyshev filters*.

3.5.2 Prototype Filter Functions with Finite Transmission Zeros

Such a filter function is characterized by

$$P(s) = \left(s^2 + b_1^2\right) \left(s^2 + b_2^2\right) \cdots$$
$$F(s) = s^m \left(s^2 + a_1^2\right) \left(s^2 + a_2^2\right) \cdots \tag{3.38}$$

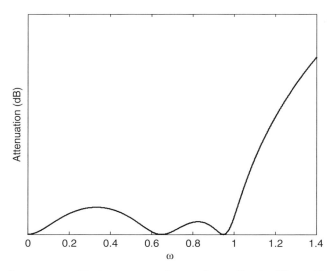

Figure 3.5 Lowpass amplitude response shape of an all-pole filter with an arbitrary distribution of reflection zeros.

The difference between this and the all-pole filter function is the introduction of transmission zeros, which tends to increase the selectivity of the amplitude response. Figure 3.6 illustrates a comparison of the amplitude response of a fourth-order equiripple filter both with and without a finite transmission zero in the stopband. When the transmission zeros are introduced, the increased selectivity is obtained at the expense of a poorer wideband amplitude response in the stopband.

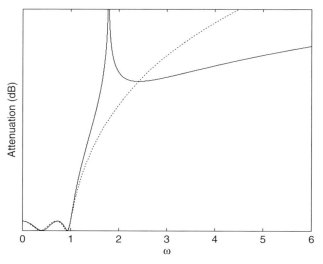

Figure 3.6 Lowpass amplitude response of a fourth-order filter with and without a finite transmission zero.

3.6 CLASSICAL PROTOTYPE FILTERS

This section describes the derivation of characteristic polynomials of the classical filters.

3.6.1 Maximally Flat Filters

Let us consider an nth-order polynomial $K_n(\omega)$ that satisfies the following conditions:

1. $K_n(\omega)$ is an nth-order polynomial.
2. $K_n(0) = 0$.
3. $K_n(\omega)$ is maximally flat at the origin.
4. $K_n(1) = \varepsilon$.

Condition 1 implies that

$$\varepsilon K_n(\omega) = c_0 + c_1\omega + c_2\omega^2 + \cdots + c_n\omega^n \tag{3.39}$$

Condition 2 requires that $c_0 = 0$. "Maximally flat at the origin" implies that as many derivatives as possible are zero at the origin. Thus

$$\varepsilon\frac{dK_n}{d\omega} = c_1 + 2\,c_2\omega + \cdots + nc_n\omega^{n-1} \tag{3.40}$$

For this polynomial to be zero at $\omega = 0$, it is required that $c_1 = 0$. Similarly, higher-order derivatives can be made to vanish by assigning zero to the higher-order coefficients. Thus, condition 3 requires

$$K_n(\omega) = c_n\omega^n \tag{3.41}$$

Finally, condition 4 yields $c_n = \varepsilon$. Summarizing, we can write

$$K_n(\omega) = \varepsilon\omega^n \tag{3.42}$$

where ε is the ripple factor, which defines the maximum amplitude response over the passband. The choice of ε as unity implies normalization with respect to the half-power point at the unity cutoff frequency. In terms of the characteristic polynomials, since all the poles of the filter are located at infinity, $P(s) = 1$. As a result, by virtue of equation (3.41), the polynomial $F(s)$ is given by

$$F(s) = s^n \tag{3.43}$$

The Hurwitz polynomial $E(s)$ is determined as follows. For the normalized (half-power) Butterworth lowpass prototype filter of degree n, we obtain

$$\left|K(j\omega)\right|^2 = \omega^{2n}$$

$$\left|H(j\omega)\right|^2 = 1 + \omega^{2n} = \left|E(j\omega)\right|^2 \tag{3.44}$$

By analytic continuation, we can replace ω^2 with $-s^2$ to obtain

$$E(s)E(-s) = 1 + (-s^2)^n \tag{3.45}$$

The $2n$ zeros of this function are located on the unit circle. Their positions are given by [8]

$$s_k = \begin{cases} \exp\left[\dfrac{j\pi}{2n}(2k-1)\right] & \text{for } n \text{ even} \\[2ex] \exp\left(\dfrac{j\pi k}{n}\right) & \text{for } n \text{ odd} \end{cases} \tag{3.46}$$

where $k = 1, 2, \ldots, 2n$. Since $E(s)$ represents the Hurwitz polynomial, the zeros obtained from equation (3.46) that lie in the left half-plane are assigned to $E(s)$. The amplitude response of a three-pole maximally flat filter is represented in Figure 3.7. The poles of the transfer function for this filter by using equation (3.46) are computed as $s_1 = -0.5 + 8660j$, $s_2 = -1.0$, and $s_3 = -0.5 - 0.8660j$.

3.6.2 Chebyshev Approximation

Let us define the characteristic function by

$$\left|K(j\omega)\right|^2 = \varepsilon^2 T_n^2\left(\frac{\omega}{\omega_c}\right) \tag{3.47}$$

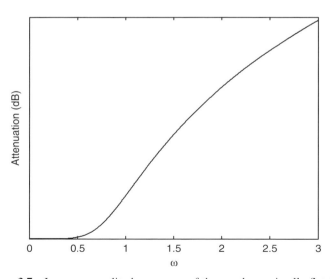

Figure 3.7 Lowpass amplitude response of three-pole maximally flat filter.

The polynomial $T_n(x)$ is the nth-degree polynomial that has the following properties:

1. T_n is even (odd) if n is even (odd).
2. T_n has all its zeros in the interval $-1 < x < 1$.
3. T_n oscillates between the values of ± 1 in the interval $-1 \le x \le 1$.
4. $T_n(1) = +1$.

The amplitude response, characterized by such a function, is shown in Figure 3.8. T_n can be derived as [4]

$$T_n(x) = \cos(n\cos^{-1} x) \qquad (3.48)$$

This is a transcendental function and represents the well-known Chebyshev polynomial. Filters derived from this polynomial are referred to as *Chebyshev filters*. Equation (3.48) is represented by the recursion relationship

$$T_{n+1}(x) = 2xT_n(x) - T_{n-1}(x) \qquad (3.49)$$

Since $T_0(x) = 1$ and $T_1(x) = x$, we can generate the Chebyshev polynomial for any order n. For example, a third-order Chebyshev polynomial is computed by

$$T_3(x) = 4x^3 - 3x \qquad (3.50)$$

The roots of this polynomial are given by 0 and $\pm j0.8660$. In terms of the characteristic polynomials, $P(s) = 1$, and by virtue of equation (3.47), the

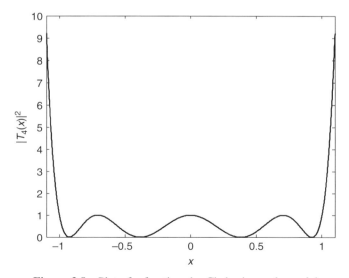

Figure 3.8   Plot of a fourth-order Chebyshev polynomial.

polynomial $F(s)$ is given by

$$F(s) = T_n\left(\frac{s}{j}\right) \tag{3.51}$$

The Hurwitz polynomial is determined as follows:

$$H(s)H(-s) = 1 + \left[\varepsilon T_n\left(\frac{s}{j}\right)\right]^2 = E(s)E(-s) \tag{3.52}$$

The roots s_k are the solutions to

$$T_n\left(\frac{s_k}{j}\right) = \pm\frac{j}{\varepsilon}\cos\left(n\cos^{-1}\frac{s_k}{j}\right) \tag{3.53}$$

This equation can be readily solved to yield [2]

$$\sigma_k = \pm\sinh\left(\frac{1}{n}\sinh^{-1}\frac{1}{\varepsilon}\right)\sin\frac{\pi}{2}\frac{2k-1}{n}$$

$$\omega_k = \cosh\left(\frac{1}{n}\sinh^{-1}\frac{1}{\varepsilon}\right)\cos\frac{\pi}{2}\frac{2k-1}{n} \tag{3.54}$$

where $k = 1, 2, \ldots, 2n$. The Hurwitz polynomial $E(s)$ is formed by the left half-plane roots obtained from equation (3.54). For example, for a 20 dB return loss, $\rho = 0.1$ and $\varepsilon = 0.1005$. This corresponds to an amplitude ripple of 0.0436 dB in the passband. By using equation (3.54), the roots of the Hurwitz polynomial are computed as -1.1714, $-0.5857 \pm j1.3368$. Figure 3.9 exhibits the amplitude response of a three-pole Chebyshev filter.

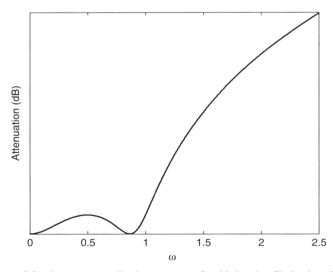

Figure 3.9 Lowpass amplitude response of a third-order Chebyshev filter.

3.6.3 Elliptic Function Filters

Let us examine a filter for which $|K(j\omega)| \leq \varepsilon$ in the passband and $|K(j\omega)| \geq K_{\min}$ in the stopband. The maximum passband attenuation and minimum stopband attenuation are then related as

$$A_{\max} = 10 \log \left(1 + \varepsilon^2\right) \tag{3.55}$$

$$A_{\min} = 10 \log \left(1 + K_{\min}^2\right) \tag{3.56}$$

There is an infinity of solutions to this approximation problem. However, the solution of most interest is when the response is equiripple in both passband and stopband. Such a response provides the sharpest rise in attenuation and is known as an *elliptic function response*, since its solution is dependent on the Jacobian elliptic functions. To examine this approximation problem, we express the characteristic function as [4]

$$\left|K(j\omega)\right|^2 = \left[\varepsilon R_n\left(\frac{\omega}{\omega_p}, L\right)\right]^2 \tag{3.57}$$

where $R_n\left(\frac{\omega}{\omega_p}, L\right)$ is a rational function. For filters having an equiripple passband and stopband, this function must have the following properties:

1. R_n is even (odd) if n is even (odd).
2. R_n has all its n zeros in the interval $-1 < (\omega/\omega_p) < 1$ and all its n poles outside that interval.
3. R_n oscillates between values of ± 1 in the interval $-1 < (\omega/\omega_p) < 1$.
4. $R_n (1, L) = +1$.
5. $1/R_n$ oscillates between values of $\pm 1/L$ in the interval $|\omega| > \omega_s$.

The problem now reduces to the determination of a rational function $R_n(x, L)$ that possesses the previous properties; in other words, all points in $|x| < 1$ and $|x| > x_L$, where $|R_n(x, L)| = 1$ and L, respectively, must represent maximum or minimum values of $R_n(x, L)$ at these points. Therefore

$$\frac{dR_n(x, L)}{dx} \bigg|_{|R_n(x,L)|=1} = 0 \text{ except } |x| = 1$$

$$\frac{dR_n(x, L)}{dx} \bigg|_{|R_n(x,L)|=L} = 0 \text{ except } |x| = x_L \tag{3.58}$$

where $x = (\omega/\omega_p)$ and $x_L(\omega_s/\omega_p)$. As a result, $R_n(x, L)$ is defined through the differential equation

$$\left(\frac{dR_n}{dx}\right)^2 = M^2 \frac{\left(R_n^2 - 1\right)\left(R_n^2 - L^2\right)}{\left(1 - x^2\right)\left(x^2 - x_L^2\right)}$$

or

$$\frac{C\,dR_n}{\sqrt{\left(1 - R_n^2\right)\left(L^2 - R_n^2\right)}} = \frac{M\,dx}{\sqrt{\left(1 - x^2\right)\left(x_L^2 - x^2\right)}} \tag{3.59}$$

where C and M are constants. This is the differential equation for the Chebyshev rational function, and its solution involves elliptic integrals. This rational function is derived and expressed as [3]

$$R_n(x, L) = C_1 x \prod_{v=1}^{(n-1)/2} \frac{x^2 - sn^2(2vK/n)}{x^2 - [x_L/sn(2vK/n)]^2} \quad \text{for } n \text{ odd} \tag{3.60}$$

$$R_n(x, L) = C_2 x \prod_{v=1}^{(n/2)} \frac{x^2 - sn^2[(2v-1)K/n]}{x^2 - \left\{[x_L/sn(2vk/n)K/n]\right\}^2} \quad \text{for } n \text{ even} \tag{3.61}$$

Here $sn[(vK/n)]$ is the Jacobian elliptic function and K is the complete elliptic integral of the first kind [not to be confused with the characteristic function $K(j\omega)$] defined by [5]

$$K = \int_0^{\pi/2} \frac{d\xi}{\sqrt{1 - k^2 \sin^2 \xi}} \tag{3.62}$$

where k is the modulus of the elliptic integral. The modulus is less than unity for K to be real, permiting an alternative definition of k where

$$k = \sin \theta = \frac{\omega_p}{\omega_s} \tag{3.63}$$

and θ is the modular angle. By substituting for x, x_L, and $\sin \theta = \omega_p/\omega_s$, the characteristic function for n odd is expressed as [6]

$$|K(j\omega)| = \varepsilon C_1 \frac{\omega}{\omega_p} \prod_{v=1}^{(n-1)/2} sn^2\left(\frac{2vK}{n}\right) \frac{\omega^2 - sn^2\left(\frac{2vK}{n}\right)\omega_s^2 \sin^2 \theta}{\omega^2 sn^2\left(\frac{2vK}{n}\right) - \omega_s^2} \tag{3.64}$$

If we choose the geometric mean of ω_p and ω_s as the normalizing frequency of unity, then

$$\sqrt{\omega_p \cdot \omega_s} = 1 \quad \text{or} \quad \omega_p = \frac{1}{\omega_s} = \sqrt{\sin \theta} \tag{3.65}$$

The characteristic function $|K(j\omega)|$ is rendered symmetric, such that the zeros of the numerator and denominator are reciprocal of each other

$$|K(j\omega)| = \varepsilon\omega \prod_{v=1}^{(n-1)/2} \left[\frac{\omega^2 - \left\{ \sqrt{\sin\theta}\ sn\left(\frac{2vK}{n}\right) \right\}^2}{\omega^2 \left\{ \sqrt{\sin\theta}\ sn\left(\frac{2vK}{n}\right) \right\}^2 - 1} \right]$$

where

$$\varepsilon \rightarrow \frac{\varepsilon C_1}{\sqrt{\sin\theta}} \prod_{v=1}^{(n-1)/2} \sin\theta\ sn^2\left(\frac{2vK}{n}\right) \tag{3.66}$$

In a similar manner, the symmetric form is obtained when n is even. We now express the characteristic function in the familiar symmetric form [6]

$$K(s) = \varepsilon s \prod_{v=1}^{(n-1)/2} \frac{(s^2 + a_{2v}^2)}{(s^2 a_{2v}^2 + 1)} \quad \text{for } n \text{ odd} \tag{3.67}$$

$$K(s) = \varepsilon \prod_{v=1}^{n/2} \frac{(s^2 + a_{2v-1}^2)}{(s^2 a_{2v-1}^2 + 1)} \quad \text{for } n \text{ even} \tag{3.68}$$

where $a_v = \sqrt{\sin\theta}\ [sn(vK)/n]$, $v = 1, 2, \ldots, n$. It is evident that this representation implies that the cutoff frequency ω_p is normalized to $\sqrt{\sin\theta}$. For prototype filters, ω_p is usually selected to be unity. Such a normalization is achieved by dividing all the critical frequencies by $\sqrt{\sin\theta}$. For a given n and θ, the critical frequencies are determined, and thus the characteristic function $K_n(\omega)$. The zeros of the Hurwitz polynomial $E(s)$ are obtained by the analytic continuation

$$1 + \varepsilon^2 K_n^2(s) = 0$$

or

$$P(s)P(-s) + \varepsilon^2 F(s)F(-s) = 0 \tag{3.69}$$

For the case where n is odd, $F(s)$ is an odd polynomial, and

$$[P(s) + \varepsilon F(s)][P(s) - \varepsilon F(s)] = 0 \tag{3.70}$$

Where n is even, both $F(s)$ and $P(s)$ are even polynomials, and

$$P^2(s) + \varepsilon^2 F^2(s) = 0 \tag{3.71}$$

The roots of this equation determine all the poles in the s plane. Only the left half-plane poles are used to form the polynomial $E(s)$. This completes the derivation of the transfer function for elliptic function filters.

3.6.4   Odd-Order Elliptic Function Filters

The characteristic function of odd-order elliptic filters is represented by equation (3.67). The amplitude response of a third-order elliptic function filter is plotted in Figure 3.10. It has a zero attenuation at $\omega = 0$ and an infinite attenuation at $\omega = \infty$. Here we consider an example of a third-order filter and compute its critical frequencies as follows.

Assuming $\theta = 40°$ as the modular angle, the frequencies corresponding to the equiripple passband and stopband by using equation (3.65) are calculated by

$$\omega_p = \sqrt{\sin \theta} = 0.801470$$

$$\omega_s = \frac{1}{\omega_p} = 1.247287$$

The complete elliptic integral K in equation (3.62) is evaluated as [5] $K = 1.786769$. The Jacobian elliptic function $sn(vK/3)$ is evaluated next. This is followed by the computation of critical frequencies via equation (3.67):

$$a_v = \sqrt{\sin \theta}\; sn\left(\frac{vK}{3}\right)$$

The critical frequencies are evaluated as $a_1 = 0.4407$, $a_2 = 0.7159$, and $a_3 = 0.8017$.

As described by equation (3.67), a_2 represents the attenuation zero and $1/a_2$, the attenuation pole. The attenuation maxima in the passband is given by a_1, whereas a_3

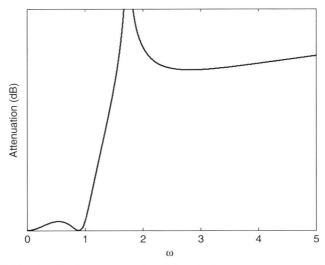

Figure 3.10   Amplitude response of a third-order lowpass prototype elliptic filter.

represents the normalizing frequency ω_p, given by $\sqrt{\sin\theta}$. The reciprocal of a_3 represents ω_s, the frequency which corresponds to the attenuation minima in the stopband. It is more convenient to have prototype filters whose passband is normalized to the unity cutoff. This is obtained by dividing the critical frequencies by $\sqrt{\sin\theta}$. Thus, the attenuation zeros and poles normalized to a cutoff frequency of unity are as follows:

Normalized attenuation zero $= 0.8929$

Normalized attenuation pole $= 1.7423$

Normalized $\omega_s = 1.5557$

3.6.5 Even-Order Elliptic Function Filters

The characteristic function for the even-order class A elliptic function filter is represented by the symmetric equation (3.68). The attenuation response of such a four-pole filter is shown in Figure 3.11a. This filter has a finite attenuation at $\omega = 0$ and $\omega = \infty$ and has the sharpest rise in attenuation in the stopband. The critical frequencies are given by $a_v = \sqrt{\sin\theta}\,sn[(v/4)K]$ and are normalized with respect to $\omega_p = \sqrt{\sin\theta}$. Critical frequencies can be normalized to a unity cutoff by dividing them by $\sqrt{\sin\theta}$. By following the same procedure, described in the example for the three-pole filter, for a modular angle $\theta = 40°$, K, the complete elliptic integral is evaluated to be 1.786769. The computed critical frequencies with unity cutoff are summarized as follows:

Normalized attenuation zeros $= 0.4267, 0.9405$

Normalized attenuation poles $= 1.6542, 3.6461$

Normalized $\omega_s = 1.5557$

Class B Elliptic Filters   Even-order class B elliptic filters are derived in order to push the highest attenuation pole toward infinity, enabling a ladder realization without mutual inductances. The critical frequencies for this class of filters are

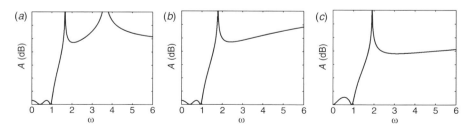

Figure 3.11   Lowpass amplitude response of four-pole elliptic filters: (a) class A; (b) class B; (c) class C four elliptic filters.

derived by the frequency transformation as given in [6], where

$$b_\gamma = \sqrt{\frac{w}{a_\gamma^{-2} - a_1^2}}$$

$$b_v' = \sqrt{\frac{w}{a_\gamma^2 - a_1^2}}$$

$$w = \sqrt{\left(1 - a_1^2 a_n^2\right)\left(1 - \frac{a_1^2}{a_n^2}\right)} \tag{3.72}$$

where a_γ are derived by equation (3.68).

The corresponding characteristic function for this class of filter is

$$\left|K(j\omega)\right| = \frac{\left(b_1^2 - \omega^2\right)\left(b_3^2 - \omega^2\right) \cdots \left(b_{n-1}^2 - \omega^2\right)}{\left(b_3'^2 - \omega^2\right)\left(b_5'^2 - \omega^2\right) \cdots \left(b_{n-1}'^2 - \omega^2\right)} \tag{3.73}$$

and the new cutoff frequency is given by

$$\omega_p = b_n = \sqrt{a_n a_{n-1}} < a_n \tag{3.74}$$

The response shape for this class of filter is plotted in Figure 3.11b. For such elliptic filters, the attenuation pole is shifted toward infinity. This type of filter provides the sharpest rise in attenuation, next only to the class A elliptic filters. This response has the advantage of not requiring any coupling between the input and output terminals. Consequently, the class B elliptic filter is the most widely used of even-order elliptic function filters.

By making use of the example of class A four-pole elliptic filter, the critical frequencies for the class B filter are computed by equations (3.72) as follows:

$$w = 0.8697$$
$$b_1 = 0.3212$$
$$b_3 = 0.7281$$
$$b_3' = 1.3879$$

The cutoff frequency is derived from $\omega_p = \sqrt{a_4 a_3} = 0.7775$, and the normalization with respect to a unity cutoff is obtained by dividing all critical frequencies by $\sqrt{a_4 a_2}$. Therefore

Normalized attenuation zeros = 0.4131, 0.9361
Normalized attenuation pole = 1.7851
Normalized $\omega_s = 1.6542$

Class C Elliptic Filters Even-order class C elliptic filters are characterized by a zero attenuation at $\omega = 0$ and an infinite attenuation at $\omega = \infty$. This class of filter function can be obtained through a frequency transformation [6]:

$$c_\gamma = \sqrt{\frac{a_\gamma^2 - a_1^2}{1 - a_\gamma^2 a_1^2}} = \sqrt{a_{\gamma-1} \cdot a_{\gamma+1}} \qquad (3.75)$$

The corresponding characteristic function is

$$\left| K(j\omega) \right| = \frac{\omega^2 \left(c_3^2 - \omega^2\right)\left(c_5^2 - \omega^2\right) \cdots \left(c_{n-1}^2 - \omega^2\right)}{\left(1 - c_3^2 \omega^2\right)\left(1 - c_5^2 \omega^2\right) \cdots \left(1 - c_{n-1}^2 \omega^2\right)} \qquad (3.76)$$

and the new cutoff frequency is given by $\omega_p = a_{n-1} < b_n < a_n$.

Using equations (3.68), (3.74), and (3.76), we can see that the cutoff frequency is progressively reduced from class A to class B and class C filters, implying a reduced sharpness in the rise of the attenuation in stopband. The response shape for a fourth-order filter using such a transformation is shown in Figure 3.11c. By following the same procedure as for classes A and B, the critical frequencies are computed using equations (3.75) and are summarized as follows:

Normalized attenuation zero = 0.9223
Normalized attenuation pole = 1.9069

3.6.6 Filters with Transmission Zeros and a Maximally Flat Passband

These filters are characterized by all the attenuation zeros at the origin and the arbitrary distribution of the attenuation poles.

The characteristic function $K(s)$ has the form

$$K(s) = \varepsilon \frac{F(s)}{P(s)} = \varepsilon \frac{s^n}{\left(s^2 + b_1^2\right)\left(s^2 + b_2^2\right) \cdots \left(s^2 + b_m^2\right)} \qquad (3.77)$$

where n is the order of the filter and m ($m \le n$) represents the number of transmission zeros. The conventional equiripple form of this class of filters is commonly referred to as *inverse Chebyshev filters*.

3.6.7 Linear Phase Filters

These filters are characterized by the presence of at least one pair of real-axis zeros or a complex quad of zeros or any combination thereof. Any type of minimum phase filter can be augmented by such zeros to improve the passband response. However, this improvement in group delay comes at the expense of a poorer amplitude response. The form of the characteristic function of a four-pole Chebyshev linear phase filter with a single pair of real-axis zeros is represented by

$$K(s) = \varepsilon \, \frac{\left(s^2 + a_1^2\right)\left(s^2 + a_2^2\right)}{\left(s^2 - b_1^2\right)} \tag{3.78}$$

where b_1 is the real-axis zero. Linear phase filters are described in more depth in Chapters 4 and 13.

3.6.8   Comparison of Maximally Flat, Chebyshev, and Elliptic Function Filters

The amplitude response of a four-pole lowpass prototype maximally flat, Chebyshev, and elliptic function (class B) filter is described in Figure 3.12. The response is normalized to a unity cutoff frequency. For the elliptic filter, the transmission zero is located at the normalized frequency of 1.5. A maximally flat filter exhibits monotonically rising attenuation, and attenuation slope. In the near-out-of-band region, the filter provides less attenuation than an equivalent Chebyshev and elliptic filter. However, in the far-out-of-band region, the maximally flat filter provides higher attenuation and has the highest asymptotic slope.

The Chebyshev filter has an equiripple passband and a monotonically rising attenuation band. It provides a higher attenuation than does the maximally flat filter in the near-out-of band region. The elliptic filter is characterized by an equiripple passband and stopband. This filter has the highest attenuation slope in the near-out-of-band region. However, in the far-out-of-band region, the filter provides less attenuation as compared to the maximally flat or Chebyshev filter.

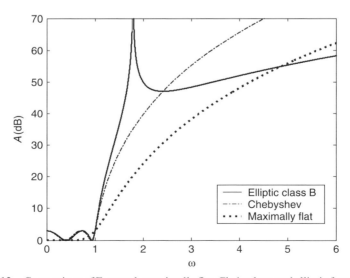

Figure 3.12   Comparison of Four-pole maximally flat, Chebyshev, and elliptic function filters.

This comparison highlights the tradeoffs that exist between the classical filters. For most applications, Chebyshev and elliptic function filters offer the best compromise between the inband response and out of band rejection. As often is the case, we must perform a tradeoff analysis to determine the best design for a given specification.

3.7 UNIFIED DESIGN CHART (UDC) RELATIONSHIPS

The characteristic function $K(s)$ is given by

$$\left| K(s) \right|^2_{s=j\omega} = \frac{P_r}{P_t} = \frac{\text{reflected power}}{\text{transmitted power}} \tag{3.79}$$

If we choose ω_1 as the bandwidth, corresponding to the maximum ripple in the passband, and ω_3 as the bandwidth, corresponding to the minimum attenuation in the stopband, then

$$\left(\frac{P_r}{P_t}\right)_{s=j\omega_1} \left(\frac{P_t}{P_r}\right)_{s=j\omega_3} = \frac{\left| K(s) \right|^2_{s=j\omega_1}}{\left| K(s) \right|^2_{s=j\omega_3}} = FF(\omega_1, \omega_3) \tag{3.80}$$

where $FF(\omega_1, \omega_3)$ represents the ratio of the reflected and transmitted powers at a pair of frequencies ω_1 and ω_3. In terms of the insertion and reflection loss in decibels, we have

$$(R_1 - A_1) + (A_3 - R_3) = 10 \log FF(\omega_1, \omega_3)$$

or

$$(R_1 + A_3) - (A_1 + R_3) = F \tag{3.81}$$

where F simply represents the value $10 \log FF(\omega_1, \omega_3)$. Note that

R_1, A_1 Return loss and transmission loss in dB, corresponding to maximum ripple in the passband

R_3, A_3 Return loss and transmission loss in dB, corresponding to minimum attenuation in stopband

F A dimensionless quantity with dB as the unit, referred to as the "characteristic factor"

Frequency F versus (ω_3/ω_1) leads to the unified design charts [7]. For most practical applications, the return loss in the passband is chosen to be >20 dB, implying $R_1 = 20$ dB and $A_1 = 0.0436$ dB. In the stopband, the attenuation required is typically >20 dB, implying $A_3 = 20$ dB and $R_3 = 0.0436$ dB. Thus, for practical applications, $F > 40$ dB, and A_1 and R_3 are negligible compared with F. Consequently

$$(R_1 + A_3) \approx F \tag{3.82}$$

Thus, a tradeoff exists, dB for dB, between the return loss in the passband and transmission loss in the stopband. It should be noted that the relationship $F = (R_1 + A_3) - (A_1 + R_3)$ is an exact relationship. Also, there are only two independent variables, since the return loss and transmission loss in dBs are related by

$$A = -10 \log \left[1 - 10^{-R/10} \right] \tag{3.83}$$

The UDCs provide, at first glance, the order of filter required for a given set of amplitude constraints. In addition, the charts provide the tradeoffs between the return loss in the passband and attenuation in the stopband or, alternatively, between the passband and stopband bandwidths for a given order of filter and the amplitude constraints. Appendix 3A includes the UDCs for the maximally flat, Chebyshev, elliptic, and quasielliptic function filters. These charts provide a quick and accurate first-cut design of a filter to meet a given amplitude specification.

3.7.1 Ripple Factor

UDCs are generated as a function of the ratio of the characteristic polynomial $K(s)$ at a pair of frequencies, from equation (3.81). The ripple factor ε cancels out in this relationship, and thus, the UDCs are independent of it. On the other hand, the choice of ε governs the tradeoff between the return loss in the passband and the attenuation in the stopband for a given F. More explicitly, from the relationship (3.81), the choice of ε alters the values of $R_1(A_1)$ and $A_3(R_3)$ within the constraint of constant F. In other words, the dB-for-dB tradeoff between the return loss in the passband and attenuation in the stopband is preserved and provided by the selection of the ripple factor. The ripple factor ε is used to normalize the maximum permissible amplitude in the passband. It is determined as follows:

$$\left| t(j\omega) \right|^2 = \frac{1}{1 + \varepsilon^2 \left| K(s) \right|^2_{s=j\omega}} \tag{3.84}$$

$$= 1 - \left| \rho(j\omega) \right|^2$$

In this formulation, the ripple factor ε is separated from the characteristic function $K(s)$. This alternative formulation is similar to that in equation (3.20). The formulation allows the evaluation of ε in terms of the practical parameters of transmission or the reflection loss in dB. As mentioned in Section 3.2.2, there are many ways to normalize the transfer function with ε within the constraint of satisfying the law of the conservation of energy. Regardless of how ε is used in the normalization process, ε must yield the same transfer function. If ω_1 is the bandwidth corresponding to the maximum ripple of A_1 dB in the passband, then

$$\varepsilon = \sqrt{\frac{\left| t(j\omega_1) \right|^{-2} - 1}{\left| K(j\omega_1) \right|^2}} \tag{3.85}$$

The values of the transmission and reflection coefficients in terms of decibels are given by equation (3.9);

$$A = 10\log = -10\log|t(j\omega)|^2 \quad \text{dB},$$

$$R = 10\log = -10\log|\rho(j\omega)|^2 \quad \text{dB}$$

For lowpass prototype networks, ω_1 is referred to as the *cutoff frequency* and is typically chosen to be unity. For Chebyshev and maximally flat filters, we obtain

$$\left|K(s)\right|_{s=j1} = 1$$

$$\varepsilon = \sqrt{\left|t(j\omega_1)\right|^{-2}-1}$$

$$= \frac{1}{\sqrt{\left|\rho(j\omega_1)\right|^{-2}-1}}$$

(3.86)

For example, for a Chebyshev filter with a return loss of 20 dB and $\rho = 0.1$, the value of ε is 0.1. For elliptic filters, ε must be determined through the relationship of equation (3.86). For instance, for the three-pole elliptic function filter discussed in Section 3.6.5, the value of the characteristic function $K(s)$ at $s = j$ is computed to be 0.099575. For a 20 dB return loss in the passband, $t^2 = 0.99$, and ε is evaluated to be 1.00933. By using this value of ε, the poles of the transfer function [the roots of the Hurwitz polynomial $E(s)$] are computed to be -1.6317 and $-0.3205 \pm j1.3193$.

3.8 LOWPASS PROTOTYPE CIRCUIT CONFIGURATIONS

The characteristic polynomials of the lossless lowpass prototype filters are based on lumped circuit elements. The circuit configuration of such a lowpass prototype filter as a ladder network is depicted in Figure 3.13a and its dual, in Figure 3.13b. The circuit model of the prototype filter provides the link to the physical realization of the filter networks.

The next step in the design process is to determine the element values g_k from the characteristic polynomial of a desired lowpass prototype filter. In Chapters 6–8, the synthesis procedures are presented to determine the element values from characteristic polynomials for any class of filter function. In the case of the classic maximally flat and Chebyshev filters, explicit formulae are derived in Ref. 8. The prototype ladder network of Figure 3.13 have all its transmission zeros located at infinity. As a result, only all-pole filters such as the maximally flat or the Chebyshev-type filter can be synthesized by such a configuration. For filters with finite transmission zeros, we need to introduce resonant circuits in the prototype network, as shown in Figure 3.14.

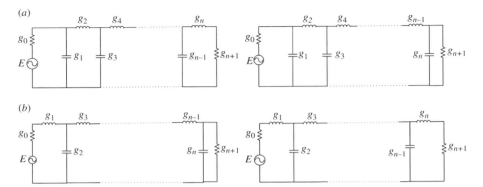

Figure 3.13 General forms of lowpass prototype lossless all-pole ladder filters: (*a*) shunt–series configuration; (*b*) series–shunt configuration.

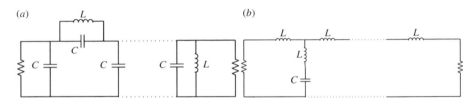

Figure 3.14 General forms of lowpass prototype lossless ladder filters with transmission zeros: (*a*) using parallel resonators; (*b*) using series resonators.

For such networks, some of the zeros of the polynomial $P(s)$ are located along the $j\omega$ axis and represent frequencies at which no power is transmitted. For the classic elliptic function filters, there is an analytic procedure to synthesize the lumped circuit elements [6]. The synthesis techniques for the general class of filters with arbitrary distribution of $P(s)$ are explained in Chapters 7 and 8.

3.8.1 Scaling of Prototype Networks

The prime purpose to derive the lowpass prototype circuits is to realize a basic model from which we can derive practical filters at any frequency with any bandwidth and impedance levels. The element values of prototype filters can be changed to other impedance levels and frequency bands by the appropriate transformations. Let us consider two networks, one an unscaled (prototype network) and the other a scaled network, as shown in Figure 3.15.

Suppose we desire to scale the driving-point impedance $Z'(s)$ of the prototype network by

$$Z(s) = bZ'\left(\frac{s}{a}\right) \tag{3.87}$$

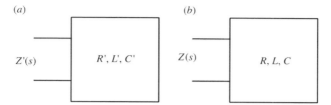

Figure 3.15 Scaling of prototype networks: (*a*) prototype network; (*b*) scaled network.

where $Z(s)$ is the equivalent impedance of the scaled network and a and b are dimensionless, real, and positive constants for scaling the frequency and impedance level, respectively. Both the impedance and frequency transformations are applied independently. If the two networks are identical in structure (topology), there is a simple relationship of the element values in the two networks [2]:

$$R = bR'$$

$$L = \frac{b}{a}L'$$

$$C = \frac{1}{ab}C \tag{3.88}$$

These relationships are restricted to transforming the lowpass prototype networks to practical lowpass filters. However, we must consider a more general frequency transformation to realize the highpass, bandpass, and bandstop filters.

Let us consider a general transformation $\phi(\omega)$, defined as a quotient of polynomials. Equation (3.88) represents the special case of $\phi(\omega)$, when the frequency is scaled by a constant factor. However, the most important case is one in which $\phi(\omega)$ is a reactance function [9]. In such a case, the inductors and capacitors are replaced by the networks of inductors and capacitors but not resistors. This topic has been dealt with in many books [10,11]. The reactance frequency transformations and equivalent element values are summarized Tables 3.1 and 3.2.

3.8.2 Frequency Response of Scaled Networks

Amplitude Response Under frequency transformation, the amplitude and phase response of the prototype network has a direct, one-to-one correspondence with the practical network derived from it. In other words, the amplitude and phase responses of the two networks at the same relative frequencies are identical. As an example, the amplitude response of the prototype filter at $\omega' = 0, 1$ is the same as the one for a lowpass filter at $\omega = 0$ and ω_c. For a corresponding bandpass filter, the amplitude response at the band center and band edge is the same as the prototype filter response at $\omega = 0$ and 1, respectively.

TABLE 3.1   Frequency Transformations

ω'	Normalized Frequency Variable of Lowpass Prototype Filter ($\omega'_c = 1$)
ω	Unnormalized frequency variable
ω_c	Unnormalized cutoff frequency
ω_0	Bandcenter frequency obtained as geometric mean of passband edges ω_2 and ω_1
$\omega_2 - \omega_1 = \Delta\omega$	Passband/stopband bandwidths
Transformation of lowpass prototype filter to	Equivalent frequency transformation variable
Lowpass filter	$\omega' = \dfrac{\omega}{\omega_c}$
Highpass filter	$\omega' = \dfrac{\omega_c}{\omega}$
Bandpass filter	$\omega' = \dfrac{\omega_0}{\Delta\omega}\left(\dfrac{\omega}{\omega_0} - \dfrac{\omega_0}{\omega}\right)$
Bandstop filter	$\omega' = \dfrac{1}{\dfrac{\omega_0}{\Delta\omega}\left(\dfrac{\omega}{\omega_0} - \dfrac{\omega_0}{\omega}\right)}$

Group Delay Response   The group delay response is represented by the derivative of the phase response. Consequently, the group delay response of the transformed network depends on the absolute bandwidths of the practical networks. Let us consider the group delay response for the lowpass and bandpass filters.

Lowpass Filter Group Delay   The lowpass prototype to lowpass filter transformation is given by $\omega' = \omega/\omega_c$, and the group delay τ of the lowpass filter is provided by

$$\tau = -\frac{d\beta}{d\omega} = -\frac{d\beta}{d\omega'}\frac{d\omega'}{d\omega} = \frac{1}{\omega_c}\tau' \tag{3.89}$$

where τ' is the group delay of the prototype network. Thus, for a lowpass filter with a cutoff frequency of 1 GHz, the group delay of the lowpass filter is simply given by $0.16\tau'$ nanoseconds. The group delay is, therefore, scaled by ω_c, the cutoff frequency of the lowpass filter.

Bandpass Filter Group Delay   By following the same procedure as for the lowpass filter, the group delay τ_{BPF} of the bandpass filter is readily derived as

$$\omega' = \frac{\omega_0}{\Delta\omega}\left(\frac{\omega}{\omega_0} - \frac{\omega_0}{\omega}\right)$$

TABLE 3.2   Lumped Circuit Transformations

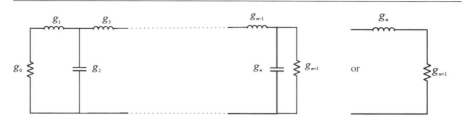

g_k: *Element Values of the Lowpass Prototype Circuit*

Circuit Transformations Lowpass Prototype to	Transformed Circuit Configuration and Equivalent Element Values

Lowpass filter

$$L_k = \frac{g_k}{\omega_c}$$

$$C_k = \frac{g_k}{\omega_c}$$

Highpass filter

$$L_k = \frac{1}{g_k\omega_c}$$

$$C_k = \frac{1}{g_k\omega_c}$$

Bandpass filter

$$L_k = \frac{g_k}{\Delta\omega}$$

$$C_k = \frac{\Delta\omega}{g_k\omega_0^2}$$

$$k = 1, 3, 5, \cdots$$

$$C_k = \frac{g_k}{\Delta\omega}$$

$$L_k = \frac{\Delta\omega}{g_k\omega_0^2}$$

$$k = 2, 4, \cdots$$

(Continued)

TABLE 3.2 *Continued*

Circuit Transformations Lowpass Prototype to	Transformed Circuit Configuration and Equivalent Element Values
Bandstop filter	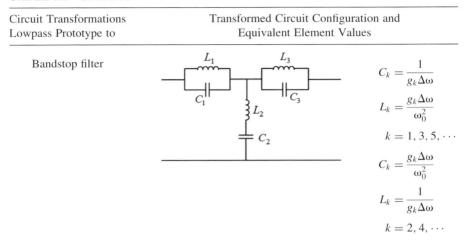

$$C_k = \frac{1}{g_k \Delta\omega}$$

$$L_k = \frac{g_k \Delta\omega}{\omega_0^2}$$

$$k = 1, 3, 5, \cdots$$

$$C_k = \frac{g_k \Delta\omega}{\omega_0^2}$$

$$L_k = \frac{1}{g_k \Delta\omega}$$

$$k = 2, 4, \cdots$$

Note: The dual of these circuits has the same equivalent element values.

and

$$\tau_{\text{BPF}} = -\frac{d\beta}{d\omega} = \frac{d\beta}{d\omega'} \frac{d\omega'}{d\omega}$$

$$= \frac{\omega_0}{\Delta\omega} \left(\frac{1}{\omega_0} + \frac{\omega_0}{\omega^2} \right) \tau'$$

$$\approx \frac{2}{\Delta\omega} \tau' \qquad (3.90)$$

At the band center, $\omega = \omega_0$, and

$$\tau_0 = \frac{2}{\Delta\omega} \tau_0' \qquad (3.91)$$

τ_0' is the delay of the prototype filter at $\omega' = 0$. These relationships demonstrate that the shape and the relative values of the group delay are determined by the choice of the prototype filter and the absolute bandwidth of the practical filter. It is interesting to note that the group delay is independent of the choice of the center frequency of the bandpass filter, a point that often goes unnoticed.

3.9 EFFECT OF DISSIPATION

In the physical realization of filter networks, the finite conductivities of the materials involved introduce dissipation of energy. As a consequence, the expected response of the transfer functions based on the lossless elements requires adjustments.

The approximate effect of incidental dissipation in an inductance is to associate with it a series resistance; in a capacitance element, the dissipation effectively produces a parallel conductance. To consider this effect on the transfer function, let us analyze the effect of changing the complex frequency variable s to $s + \delta$, where δ is a positive quantity, that is

$$s \rightarrow s + \delta$$
$$sL_k \rightarrow sL_k + \delta L_k \tag{3.92}$$
$$sC_k \rightarrow sC_k + \delta C_k$$

where the k subscript denotes the kth element. This conversion then implies that each inductance is associated with a proportionate amount of series resistance δL_k, and that each capacitance has a parallel conductance of the value δC_k. If r and r' are the associated series resistance and shunt conductance, then

$$\delta L_k = r \quad \text{or} \quad \delta = \frac{r}{L_k}$$

Similarly

$$\delta = \frac{r'}{C_k}$$

This common ratio δ is referred to as the *dissipation ratio* or factor. Since $s + \delta = s_k$ for the kth pole or zero, s assumes the value

$$s = s_k - \delta \tag{3.93}$$

Thus, all the zeros and poles of the transfer function are displaced horizontally to the left by an amount δ. It is assumed that the dissipation ratios of the inductances and capacitances are identical. Thus, by subtracting δ from the real part of the poles and zeros of the transfer function, we obtain the network response with finite unloaded Q values. Incorporating the displacement δ into the pole–zero configurations, equation (3.18) is modified to

$$A = 20 \log e \left[\frac{\ln |s + \delta - s_1| + \ln|s + \delta - s_2| + \cdots -}{\ln|s + \delta - p_1| + \ln|s + \delta - p_2| - \cdots} \right] + 20 \log e \ln \varepsilon \tag{3.94}$$

At $\omega = 0$, let $A = A_0$, and therefore

$$A_0 = 20 \log e \left[\frac{\ln |\delta - s_1| + \ln|\delta - s_2| + \cdots -}{\ln|\delta - p_1| + \ln|\delta - p_2| - \cdots} \right] + 20 \log e \ln \varepsilon \tag{3.95}$$

A typical term in the bracket is $\ln| \delta - s_k|$, and is computed by

$$\ln|\delta - s_k| = \ln\lfloor(\delta - \sigma_k)^2 + \omega_k^2\rfloor^{1/2}$$
$$\cong \ln |s_k| - \delta \frac{\sigma_k}{|s_k|^2} \quad \text{neglecting } \delta^2 \text{ terms} \tag{3.96}$$

Substituting, we write

$$A_0 = 20 \log e \, \delta \left[\sum \frac{\sigma_p}{|s_p|^2} - \sum \frac{\sigma_k}{|s_k|^2} \right] \tag{3.97}$$

Note that the term containing $\sum \ln|(s_k)| - \sum \ln|(s_p)|$ cancels out with $\ln \varepsilon$ in accordance with equation (3.26). Next, we determine the phase and group delay response. If β and τ are the phase and group delay, respectively, then

$$\beta = \sum_{zeros} \tan^{-1} \left(\frac{\omega - \omega_k}{\delta - \sigma_k} \right) - \sum_{poles} \tan^{-1} \left(\frac{\omega - \omega_p}{\delta - \sigma_p} \right)$$

Therefore

$$\tau = -\frac{d\beta}{d\omega} = \frac{d}{d\omega} \left[\sum_{poles} \tan^{-1} \left(\frac{\omega - \omega_p}{\delta - \sigma_p} \right) - \sum_{zeros} \tan^{-} \left(\frac{\omega - \omega_k}{\delta - \sigma_k} \right) \right] \tag{3.98}$$

At $\omega = 0$, let $\tau = \tau_0$, and therefore

$$\tau_0 = \sum \frac{\delta - \sigma_p}{(\delta - \sigma_p)^2 + \omega_p^2} - \sum \frac{\delta - \sigma_k}{(\delta - \sigma_k)^2 + \omega_k^2} \tag{3.99}$$

If we again assume a small amount of dissipation, $\delta \ll 1$ and $\sigma \gg \delta$, the approximate expression for group delay is

$$\tau_0 \approx \left[\sum \frac{\sigma_p}{|s_p|^2} - \sum \frac{\sigma_k}{|s_k|^2} \right] \tag{3.100}$$

Substituting in equation (3.97), we obtain

$$A_0 \approx 20 \log e \, \delta \, \tau_0 \tag{3.101}$$

This demonstrates that the insertion loss at $\omega = 0$ (or the band center for bandpass filters) is directly proportional to the absolute time delay at that frequency for a small dissipation.

3.9.1   Relationship of Dissipation Factor δ and Quality Factor Q_0

The dissipation factor δ represents the displacement of the poles and zeros along the real (σ) axis in the complex frequency plane. δ has the dimensions of radian frequency.

Unloaded $Q(Q_0)$ represents the radian frequency times the energy stored, divided by the energy dissipated per cycle, or

$$Q_0 = \omega \left(\frac{\text{energy stored}}{\text{average power loss}} \right)$$

$$= 2\pi \left(\frac{\text{energy stored}}{\text{energy dissipated per cycle}} \right) \tag{3.102}$$

where Q_0 is a dimensionless quantity, and provides an indication of the losses in the circuit. For an *LCR* circuit, we obtain

$$Q_0 = \omega_0 \, \frac{\mu}{\omega_R} \tag{3.103}$$

where μ is the energy stored in the circuit, ω_0 is the resonant frequency, and ω_R is the average power loss:

$$\omega_R = -\frac{d\mu}{dt} \tag{3.104}$$

The energy in the circuit can be calculated when the energy is all in the inductance

$$\mu = \frac{1}{2} L \, (I_{\text{max}})^2 = L \frac{A^2}{2} \tag{3.105}$$

where $I = A \cos(\omega_0 t + \phi)$. The average power loss in resistance r is

$$\omega_R = \frac{1}{2} r (I_{\text{max}})^2 = r \frac{A^2}{2}$$

$$\therefore Q_0 = \omega_0 \frac{LA^2/2}{rA^2/2} \tag{3.106}$$

$$= \omega_0 \frac{L}{r}$$

As introduced in the previous section, $\delta = r/L$ for an inductance. If we apply the change of variable $s \rightarrow s + \delta$, implying a uniform shift, then the pertinent value of δ equals $r/2L$. It is implied that the losses are divided equally between L and C. Therefore

$$Q_0 = \frac{\omega_0}{2} \frac{1}{\delta} \tag{3.107}$$

In terms of the lowpass prototype g_k parameters, we obtain

$$\delta = \frac{r}{g} \tag{3.108}$$

and

$$Q_0 = \omega' \frac{g}{r} \qquad (3.109)$$

where g is the element value and ω' is the normalized frequency variable. Typically, the normalization is chosen with respect to 1 radian per second, and therefore

$$Q_0 = \frac{1}{\delta} \qquad (3.110)$$

3.9.2   Equivalent δ for Lowpass and Highpass Filters

The frequency transformation for lowpass filters is

$$\omega' = \frac{\omega}{\omega_c}$$

If the normalizing, or cutoff frequency of the prototype structure, $\omega' = 1$ corresponds to $\omega = \omega_c$

$$\omega' g \to \omega \frac{g}{\omega_c}$$

or $g \to g/\omega_c$ and $r \to r$, and

$$\therefore \qquad \begin{aligned} \delta_{Lp} &= \frac{r}{g/\omega_c} \\ &= \omega_c \delta \end{aligned} \qquad (3.111)$$

where δ_{Lp} and δ are the dissipation factors for the lowpass filter and the prototype filter, respectively. The unloaded Q for the lowpass filter is calculated by

$$Q_0 = \frac{\omega_c}{\delta_{Lp}} = \frac{1}{\delta}$$

or

$$\delta = \frac{1}{Q_0} \qquad (3.112)$$

Thus, the equivalent displacement of the pole–zero pattern of the lowpass prototype structure is simply the reciprocal of the unloaded Q of the lowpass filter, determined at the cutoff frequency. Similarly, for the highpass filters, $\delta = 1/Q_0$.

3.9.3   Equivalent δ for Bandpass and Bandstop Filters

The frequency transformation for a bandpass filter is represented by

$$\omega' = \frac{\omega_0}{\Delta\omega} \left(\frac{\omega}{\omega_0} - \frac{\omega_0}{\omega} \right) \qquad (3.113)$$

Therefore

$$\omega' g \rightarrow \frac{g}{\Delta\omega} \quad \omega - g \frac{\omega_0^2}{\Delta\omega} \frac{1}{\omega} \tag{3.114}$$

that is, an inductive element value of the lowpass prototype structure gives rise to an LC tuned circuit where the equivalent inductance and capacitance are calculated by

$$L \rightarrow \frac{g}{\Delta\omega}, \quad C \rightarrow g \frac{\omega_0^2}{\Delta\omega}, \quad r \rightarrow r \tag{3.115}$$

Assuming that r is associated with L only (with no loss of generality, since the same result is obtained by assuming equal losses for L and C), we calculate

$$\delta_{BP} = \frac{r}{L} = \frac{r}{g/\Delta\omega} \tag{3.116}$$

The unloaded Q at the resonance frequency is expressed by

$$Q_0 = \frac{\omega_0}{\delta_{BP}} = \frac{\omega_0}{\Delta\omega} \frac{g}{r}$$

$$= \frac{\omega_0}{\Delta\omega} \frac{1}{\delta}$$

or

$$\delta = \frac{\omega_0}{\Delta\omega} \frac{1}{Q_0} = \frac{f_0}{\Delta f} \frac{1}{Q_0} \tag{3.117}$$

where δ is the equivalent displacement of the pole–zero pattern of the lowpass prototype structure. Similarly, it can be shown that for bandstop filters

$$\delta = \frac{f_0}{\Delta f} \frac{1}{Q_0} \tag{3.118}$$

Let us examine the example of an eight-pole 26-dB return loss Chebyshev filter. The group delay of the prototype network τ' is computed to be 5.73 s at $\omega = 0$. Let us now assume that the practical filters are derived and realized in a coaxial structure with an unloaded Q of 1000. The dissipation factor and loss for a lowpass and a bandpass filter are computed as follows.

Lowpass Filter   Let us assume an unloaded Q of 1000. The corresponding dissipation factor $\delta = (1/1000) = .001$. Loss α at $\omega = \omega_c$ is therefore given by, $\alpha = 20 \log e \, \delta \, \tau' = 8.686 \times 10^{-3} \times 5.73 = 0.05$ dB.

It should be noted that the loss is independent of the choice of the cutoff frequency of the lowpass filter. The loss depends solely on the choice of the prototype filter and the unloaded Q (defined at the cutoff frequency) of the practical structure.

Bandpass Filter Let us assume a 1% bandwidth filter, for which the dissipation factor results from

$$\delta = \frac{f_0}{\Delta f}\frac{1}{Q_0} = \frac{100}{Q_0} = 0.1$$

By substituting a value of 0.1 for δ, the loss for the bandpass filter at $\omega = \omega_0$ is computed as 5.0 dB. This loss is 100 times the loss of the lowpass filter. The loss of the bandpass or bandstop filter depends on the percentage bandwidth, in addition to the choice of the prototype network and the quality factor of the practical structure. The loss does not depend on the choice of the center frequency per se, although Q_0 for a given structure varies with the choice of the center frequency. The dissipative loss for bandpass filters can be very high for narrow bandwidths, and therefore, require high-Q structures for most practical applications.

3.10 ASYMMETRIC RESPONSE FILTERS

For some applications, it is desirable to have an asymmetric response in the amplitude or phase or both. This poses a challenge for the design of filter networks. The prototype filters are composed of lossless inductors and capacitors, terminated in a matching resistor. The result is a lowpass network that is inherently symmetric with respect to the zero frequency. Also, the frequency transformation necessarily leads to networks that are symmetric with respect to the center of the passband. Consequently, as long as we confine a prototype network to real elements (L, C, and R), it is not feasible to realize an asymmetric response. However, in a bandpass circuit, it is possible to arrange the resonant circuits in such a way to yield an asymmetric frequency response. The challenge therefore lies in devising an equivalent lowpass prototype network, which, when transformed into bandpass filters via frequency transformation, yields the appropriate asymmetric response. Then, the asymmetric prototype network is well within the scope of numerous and well-established general network synthesis techniques employed for symmetric lowpass prototype filters.

Baum [12] was the first to introduce a hypothetical frequency-invariant reactive (FIR) element in the design of filter networks. The element is used as a mathematical tool in the formulation of the lowpass prototype filter. This hypothetical element becomes physically realizable only after a transformation from lowpass to bandpass or to bandstop filter. These hypothetical elements, introduced in the lowpass domain, render the filter response to be asymmetric with respect to zero frequency. Such a prototype circuit allows the realization of an asymmetric bandpass response via lowpass to bandpass frequency transformation. In the process, the fictitious FIR elements disappear as the frequency offsets to the resonant circuits, required in the realization of bandpass filters.

The challenge was how to incorporate this imaginary element in the synthesis procedure without violating the rules of circuit theory. Baum applied the concept of FIR elements to a few elementary synthesis procedures. By following Baum's introduction of the fictitious FIR elements, progress on the synthesis procedures incorporating such elements was slow, partly due to limited practical applications, and equally due to the complexity of the problem. The textbooks [13,14] that have addressed this problem are difficult to follow. Nevertheless, over time, this concept has been picked up by others and is now fully incorporated in the most general network synthesis techniques [15–18].

3.10.1 Positive Functions

The foundation of the classical theory is based on the concept of a positive real function. A network made up of L, C, and R elements can be described by the ratio of two polynomials in the complex frequency variable s with positive real constants. All the positive real functions exhibit this characteristic. The roots of such polynomials yield complex conjugate zeros instigating an amplitude and phase response that exhibits an even or odd symmetry about zero frequency. The introduction of constant reactances (FIR elements) results in complex coefficients for the polynomials of the transfer function. This allows the unsymmetric distribution of the poles and zeros, giving rise to frequency response that is no longer symmetric with respect to the zero frequency. Such functions are called *positive functions* [14].

Ernst [19] has described the generalized two-port networks, including the hypothetical constant reactance element. The nomenclature for describing FIR elements is

Frequency-invariant inductive element: $\quad X \geq 0, \; V(s) = jXI(s)$

Frequency-invariant capacitive element: $\quad B \geq 0, \; I(s) = jBV(s)$

X and B represent the frequency invariant inductance and capacitance, respectively. By applying the Laplace transform to the voltages and currents by assuming zero initial conditions, the complex impedance $Z(s)$ in the complex variable $s = \sigma + j\omega$ is defined as

$$Z(s) = \frac{V(s)}{I(s)}$$

Thus the impedances of FIR elements are

FIR inductive element: $\quad X \geq 0, Z(s) = jX$

FIR capacitive element: $\quad B \geq 0, Z(s) = \dfrac{1}{jB}$ (3.119)

Pictorially, FIR elements have been portrayed by a circle, a rectangle, or a set of parallel lines connected at each end. The alternative representations of this element are shown in Figure 3.16.

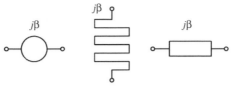

Figure 3.16   Different representations of the frequency-invariant reactance (FIR) element.

The impedance function of a loop or branch in a general network, consisting of R, L, C, and FIR elements, is

$$Z(s) = R + sL + \frac{1}{sC} + jX + \frac{1}{jB} \tag{3.120}$$

The driving point impedance of such a network is given by

$$Z(s) = \frac{V(s)}{I(s)}$$

$$= \frac{a_0 + a_1 s + a_2 s^2 + \cdots + a_n s^n}{b_0 + b_1 s^2 + b_2 s^2 + \cdots + b_m s^m} \tag{3.121}$$

where the coefficients a_i and b_i are now complex. Such functions are referred to as *positive functions*. If the FIR elements are not part of the network ($X = B = 0$), then all a_i and b_i are real, and $Z(s)$ becomes a positive real function. The question is, then, what constraints must be imposed on the positive functions to ensure physical realizability, when the lowpass prototype filter function is transformed to yield a real asymmetric bandpass filter. The imposed constraint is the same as in the case of the positive real functions, where the law of conservation of energy must hold. Invoking this constraint and working through the maze of complex power relationships [14], we find the following properties for the positive functions:

1. If $f(s)$ is a complex rational function in the complex variable $s = \sigma + j\omega$, then $f(s)$ is said to be a positive function if $\text{Re}\{f(s)\} \geq 0$ for $\text{Re}\{s\} \geq 0$.

If, in addition, $f(s)$ is real for real s (i.e., when $j\omega = 0$), then $f(s)$ is said to be a positive real function.

2. There are no poles or zeros of $f(s)$ in the right half-plane.
3. The poles and zeros of $f(s)$, along the $j\omega$ axis, are simple and have positive real residues.
4. The degrees of the numerator and denominator polynomials differ, at most, by 1. Thus, the number of finite poles and finite zeros differ, at most, by 1.
5. If $f(s)$ is a positive function, then $1/f(s)$ is also a positive function.
6. The linear combinations of the positive functions are a positive function.

These properties are very similar to those of positive real functions. The principal differences are

1. The coefficients of the numerator and denominator polynomials of $Z(s)$ or $Y(s)$ are real and positive for the positive real functions and complex for the positive functions.
2. The poles and zeros of $Z(s)$ or $Y(s)$ occur as complex conjugate pairs in the case of positive real functions, whereas there is no such restriction for positive functions.

This leads to the important question of how to represent the positive functions in terms of the lowpass prototype characteristic polynomials. Invoking the law of conservation of energy, it can be shown that the zeros of $P(s)$ must lie on the imaginary axis or appear as pairs, symmetrically located with respect to the imaginary axis. This derivation is detailed in Chapter 6. There is no requirement that such zeros have a corresponding complex conjugate pair to represent positive real functions. Consequently, the polynomial $P(s)$ has complex coefficients. As for polynomial $F(s)$, all its zeros must lie on the imaginary axis, but they need not be distributed symmetrically around zero frequency, leaving $F(s)$ to possess complex coefficients as well. Of course, this results in complex coefficients for $E(s)$. Thus, for the general case, that includes the realization of symmetric or asymmetric lowpass prototype filters, the characteristic polynomials have the following properties:

1. The roots of $P(s)$ lie on the imaginary axis or appear as pairs of zeros located symmetrically with respect to the imaginary axis. The roots' degree is $\leq n$, where n is the order of the filter.
2. The roots of $F(s)$ lie on the imaginary axis where the degree is n.
3. $E(s)$ is a Hurwitz polynomial of degree n. All its roots lie in the left half-plane of s.

As a result, $P(s)$ can be formed by any of the following factors or combinations of factors:

$$s \pm jb_i, \quad |b_i| > 1, \quad s^2 - \sigma_i^2, \quad (s - \sigma_i + j\omega_i)(s + \sigma_i + j\omega_i) \qquad (3.122)$$

The polynomial $P(s)$ has the form

$$P(s) = s^m + jb_{m-1}s^{m-1} + b_{m-2}s^{m-2} + jb_{m-3}s^{m-3} + \cdots + b_0 \qquad (3.123)$$

where m is the number of finite zeros of $P(s)$.

The polynomial $F(s)$ is formed by factors $(s + ja_i)$, where $|a_i| < 1$. It has the form

$$F(s) = s^n + ja_{n-1}s^{n-1} + a_{n-2}s^{n-2} + ja_{n-3}s^{n-3} + \cdots + a_0 \qquad (3.124)$$

where n is the order of the filter. The possible locations of roots of $P(s)$ for lowpass prototype filters, normalized to unity cutoff frequency, are illustrated in Figure 3.17. A four-pole asymmetric filter with a Chebyshev passband and an equiripple stopband in the upper half of the frequency band is exhibited in Figure 3.18. Asymmetric classes of filters are dealt with extensively in Chapters 4, 6, and 7.

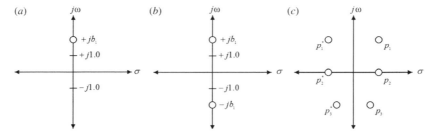

Figure 3.17   Permissible zero locations of $P(s)$ to realize asymmetric response filters: (*a*) asymmetric transmission zero location in the upper band; (*b*) asymmetric transmission zero locations in the upper and lower bands; (*c*) asymmetric zero locations to realize linear phase response.

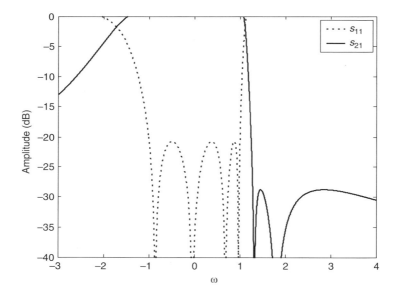

Figure 3.18   Lowpass amplitude response of a four-pole asymmetric filter.

SUMMARY

This chapter describes the synthesis process for the characteristic polynomials to realize the ideal classical prototype filters: maximally flat, Chebyshev, and elliptic function filters. The effect of uniform dissipation in filter elements is included by shifting the pole–zero pattern of the transfer function by the dissipation factor for the practical filter. The dissipation factor is derived in terms of the quality factor (unloaded Q) of the proposed structure and the bandwidth requirements of the practical filter. The computed response, including the effect of dissipation, gives remarkably accurate results for narrow-bandwidth ($<5\%$) microwave filters.

It so happens that this range includes many practical applications. A brief review of the scaling of frequency and impedance of the prototype networks to realize practical filter networks is also included.

The chapter concludes with a discussion of filters that are not symmetric with respect to their center frequency. Such bandpass or bandstop filters are referred to as *asymmetric filters*. The synthesis of such filters in the lowpass prototype domain is accomplished by introducing two fictitious network elements, namely, the constant positive or negative reactance, referred to as a (*frequency-invariant reactance*) (FIR) element. This leads to transfer function polynomials with complex coefficients, implying that the characteristic polynomials no longer require their roots to be complex conjugate pairs. However, the law of the conservation of energy requires that the roots exhibit symmetry with respect to the imaginary axis to ensure physical realizability. The existing synthesis procedures can then be applied to synthesize asymmetric lowpass prototype filters. In the process of frequency transformation, the FIR elements disappear as the frequency offsets to the resonant circuits of bandpass and bandstop filters. Consequently, transfer functions with complex coefficients, within the constraint of polynomial roots exhibiting symmetry with respect to the imaginary axis, represent the most general lowpass prototype networks.

From the relationships described in this chapter, we can perform tradeoffs to meet the filtering requirements of a communication channel for a wide range of applications, independent of the physical realization of the filter networks. In other words, it provides a reliable tool for system engineers and filter designers alike, to assess and specify realistic requirements for filters.

REFERENCES

1. B. P. Lathi, *Signals, Systems and Communications*, Wiley, New York, 1965.

2. M. E. Van Valkenburg, *Modern Network Synthesis*, Wiley, New York, 1960.

3. G. C. Temes and S. Mitra, *Modern Filter Theory and Design*, Wiley, New York, 1973.

4. R. W. Daniels, *Approximation Methods for Electronic Filter Design*, McGraw-Hill, New York, 1974.

5. M. Abramowitz and I. A. Stegun, *Handbook of Mathematical Functions*, Dover Publications, New York, 1972.

6. R. Saal and E. Ulbrich, On the design of filters by synthesis, *IRE Trans. Circuit Theory* (Dec. 1958).

7. C. M. Kudsia and V. O'Donovan, *Microwave Filters for Communications Systems*, Artech House, Norwood, MA, 1974.

8. L. Weinberg, Explicit formulas for Tschebyshev and Butterworth ladder networks, *J. Appl. Phys.* **28**, 1155–1160 (1957).

9. A. Papoulis, Frequency transformations in filter design, *IRE Trans.* **CT-3**, 140–144 (1956).

10. G. Matthaei, L. Young, and E. M. T. Jones, *Microwave Filters, Impedance Matching Networks and Coupling Structures*, Artech House, Norwood, MA, 192.

11. I. Hunter, *Theory and Design of Microwave Filters*, IEE, 2001.

12. R. F. Baum, Design of unsymmetrical bandpass filters, *IRE Trans. Circuit Theory* 33–40 (June 1957).

13. J. D. Rhodes, *Theory of Electrical Filters*, Wiley, New York, 1976.

14. V. Belevitch, *Classical Network Theory*, Holden-Day, San Francisco, 1968.

15. R. J. Cameron, Fast generation of Chebyshev filter prototypes with asymmetrically-prescribed zeros, *ESA J.* **6**, 83–95 (1982).

16. R. J. Cameron, General prototype network synthesis methods for microwave filters, *ESA J.* **6**, 193–207 (1982).

17. R. J. Cameron and J. D. Rhodes, Asymmetrical realizations for dual-mode bandpass filters, *IEEE Trans. Microwave Theory Tech.* **MTT-29**, 51–58 (Jan. 1981).

18. H. C. Bell, Canonical asymmetric coupled-resonator filters, *IEEE Trans. Microwave Theory Tech.* **MTT-30**, 1335–1340 (Sept. 1982).

19. C. Ernst, *Energy Storage in Microwave Cavity Filter Networks*, Ph.D. thesis, Univ. Leeds, Aug. 2000.

APPENDIX 3A   UNIFIED DESIGN CHARTS

This appendix contains unified design charts (UDCs) for the Butterworth, Chebyshev, and quasielliptic filters with a single pair of transmission zeros. Also, the UDC for asymmetric filters with a single zero in the upper half of the frequency band is included. It also includes critical frequencies for the quasielliptic and asymmetric filters in Tables 3A.3 and 3A.4. An example of a channelizing filter in a satellite transponder is provided to illustrate the versatility of these charts in determining the filter tradeoffs for a given specification.

Example 3A.1   Channelizing filters in a 6/4-GHz communication satellite system are required to separate 36-MHz channels with a 40 MHz spacing between them. By using UDCs, determine the tradeoffs with respect to the order of the filter, and available passband bandwidths using an isolation constraint of 30 dB, or greater at $f_0 \pm 25$ MHz.

Solution   For this example, we shall assume a minimum return loss of, say, 20 dB in the passband, which is typical for such applications. Allowing for the operating

TABLE 3A.1   Filter Tradeoffs Using Unified Design Charts (UDCs)

n	ω_3/ω_1	Chebyshev Response Bandwidth (MHz)	F (dB)	Quasielliptic Response (Single Pair of Transmission Zeros) ω_3/ω_1	Bandwidth (MHz)	F (dB)
6		Not applicable		1.39	36	58
				1.26	39.7	50
7		Not applicable		1.39	36	67
				1.18	42.4	50
8	1.39	36	54	1.39	36	77
	1.34	37.3	50	1.125	44.45	50
9	1.39	36	62	Not necessary		
	1.265	39.5	50			

Key: n—order of filter; F—characteristic factor in dB; ω_3/ω_1—ratio of isolation bandwidth to passband bandwidth; $F \geq 50$ dB; $\omega_3/\omega_1 \leq 1.39$; ≥ 1.265.

TABLE 3A.2   Filter Tradeoffs Summary

Filter Design	Passband Bandwidth (MHz)	Return Loss in Passband	Isolation at $f_0 \pm 25$ MHz (dB)
9th-order Chebyshev	38	22	33
6th-order quasielliptic	37.7	22	33
7th-order quasielliptic	39	23	37

TABLE 3A.3 Critical Frequencies for Quasielliptic Filters with a Single Pair of Transmission Zeros

N	F (dB)	Critical Frequencies				
		Attenuation Zeros				Attenuation Poles
4	40	0.4285	0.9419			1.5169
	50	0.4085	0.9343			1.9139
	60	0.3973	0.9268			2.4692
5	40	0.6567	0.9672			1.2763
	50	0.6321	0.9615			1.4921
	60	0.6161	0.9577			1.7870
6	40	0.3054	0.7785	0.9794		1.1695
	50	0.2917	0.7568	0.9752		1.3042
	60	0.2818	0.7415	0.9723		1.4869
7	40	0.5037	0.8476	0.9860		1.1137
	50	0.4863	0.8301	0.9807		1.2052
	60	0.4728	0.8170	0.9807		1.3293
8	40	0.2299	0.6327	0.8896	0.9899	1.0812
	50	0.2227	0.6158	0.8757	0.9876	1.1472
	60	0.2168	0.6021	0.8649	0.9859	1.2367

environment and manufacturing tolerances, it is reasonable to assume a margin of ± 0.25 MHz in the alignment of center frequency, implying that the passband bandwidth is no more than 39.5 MHz. Using the nomenclature of UDC, the criteria for filters tradeoffs is $F \geq 20 + 30$ or ≥ 50 dB and

$$\frac{\omega_3}{\omega_1} \leq \frac{50}{36} \quad \text{or} \quad \leq 1.39$$

$$\text{and} \quad \geq \frac{50}{39.5} \quad \text{or} \quad \geq 1.265$$

With this, the filter tradeoffs are derived for the Chebyshev and quasielliptic (single pair of transmission zeros) filters using the UDCs. Tradeoffs are summarized in Table 3A.1.

For the Chebyshev response filters, the minimum order of filter to satisfy the requirements is eight. However, such a design provides very little margin to allow for the operating environment, primarily the operating temperature range. Therefore, the realistic design is a Chebyshev filter of the ninth order. For such a design, the available bandwidth can be chosen between 36 and 39.5 MHz with the corresponding values of F varying between 62 and 50 dB. Therein lies the tradeoff between bandwidth and isolation, or the return loss or a combination thereof. Greater bandwidth results in a better passband response, whereas higher isolation minimizes

TABLE 3A.4 Critical Frequencies for Asymmetric Filters with a Single Zero in the Upper Half of the Frequency Band

		Critical Frequencies				
N	F (dB)	Attenuation Zeros				Attenuation Poles
4	40	−0.8943	−0.1994			1.3378
		0.5829	0.9600			
	50	−0.9008	−0.2415			1.6321
		0.5336	0.9509			
	60	−0.9060	−0.2748			2.0587
		0.4961	0.9440			
5	40	−0.9340	−0.4631			1.1925
		0.2145	0.7488	0.9766		
	50	−0.9367	−0.4843			1.3578
		0.1743	0.7151	0.9709		
	60	−0.9392	−0.5028			1.5918
		0.1406	0.6881	0.9665		
6	40	−0.9554	−0.6246	−0.0871		1.124
		0.4593	0.8339	0.9847		
	50	−0.9568	−0.6354	−0.1115		1.2297
		0.4267	0.8101	0.9809		
	60	−0.9580	−0.6454	−0.1337		1.3776
		0.3985	0.7907	0.9779		
7	40	−0.9681	−0.7259	−0.3059		1.0864
		0.1809	0.6090	0.8825	0.9893	
	50	−0.9688	−0.7318	−0.3203		1.198
		0.1582	0.5834	0.8651	0.9866	
	60	−0.9695	−0.7374	−0.3339		1.2618
		0.1372	0.5608	0.8507	0.9844	
8	40	−0.9761	−0.7923	−0.4610	−0.474	1.0636
		0.3677	0.7056	0.9127	0.9921	
	50	−0.9765	−0.7958	−0.4967	−0.0622	1.1176
		0.3480	0.6853	0.8995	0.9900	
	60	−0.9769	−0.7991	−0.4782	−0.0764	1.1922
		0.3296	0.6672	0.8885	0.9884	

interchannel interference. A choice of $F = 55$ dB provides a margin of 5 dB, which can be shared between the return loss and isolation, say, a nominal return loss of 22 dB and a nominal isolation of 33 dB. The corresponding ω_3/ω_1 is 1.32, which yields a passband bandwidth of nearly 38 MHz. This represents a reasonable compromise and first-cut design.

Similar reasoning for the quasielliptic case yields a minimum order of six to satisfy the requirements. Here again, we choose a value of $F = 55$ dB, and as in the previous case, $\omega_3/\omega_1 = 1.325$. This yields a passband bandwidth of 37.7 MHz. Thus, a sixth-order quasielliptic filter with a single pair of

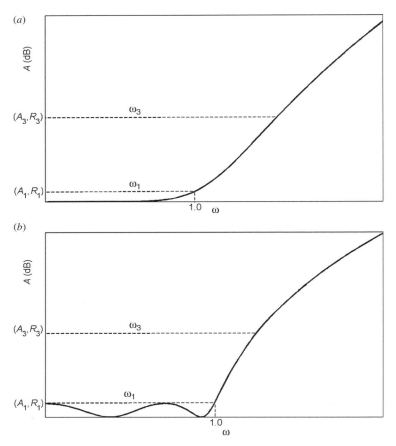

Figure 3A.1 UDC nomenclature for (*a*) Butterworth and (*b*) Chebyshev filters.

transmission zeros satisfies the requirement. By going a step further, a seventh-order quasielliptic filter offers wider margins at the cost of an increase in the order of filter by one. For example, a choice of 39 MHz passband bandwidth implies ω_3/ω_1 of 1.38. The corresponding value of F is 60 dB. This represents a 10 dB margin, shared between return loss and isolation. A good choice is to use a return loss of 23 dB and isolation of 37 dB. The three choices are summarized in Table 3A.2.

A further tradeoff can be made, dB for dB, between the return loss and isolation. This example exhibits the simplicity of UDCs in developing filter tradeoffs. Further optimization can be carried out by simulating these first-cut designs for both amplitude and phase/group delay response and then making minor adjustments. For many applications, use of UDC alone can be quite adequate for the design.

(a)

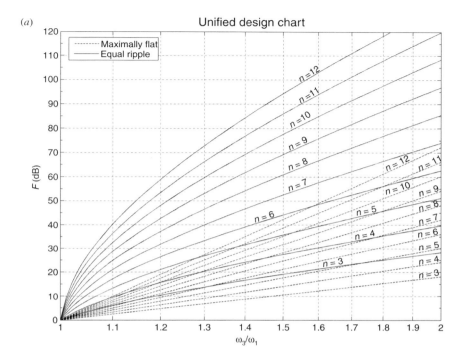

(b)

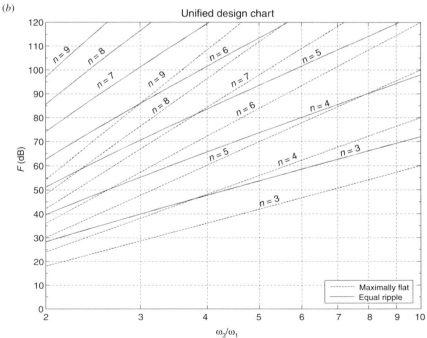

Figure 3A.2   Unified design chart for Butterworth and Chebyshev filters.

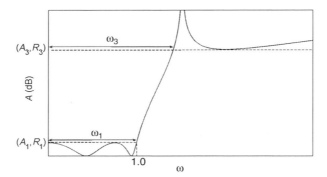

Figure 3A.3 UDC nomenclature for Quasielliptic filter with a single pair of transmission zeros.

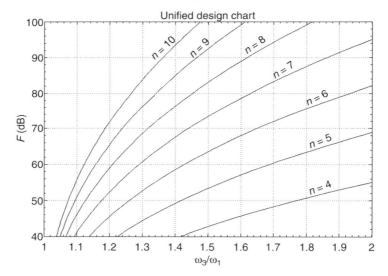

Figure 3A.4 Unified design chart for quasielliptic filters with a single pair of transmission zeros.

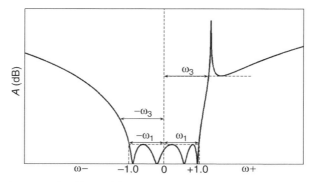

Figure 3A.5 Asymmetric filter with a single transmission zero in the upper half of frequency band.

(a)

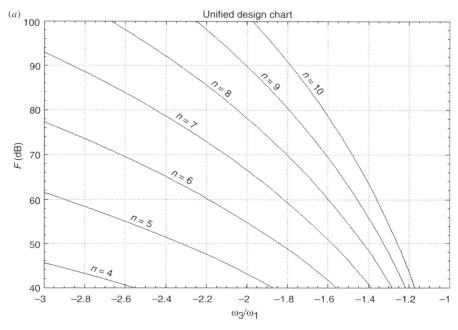

(b)

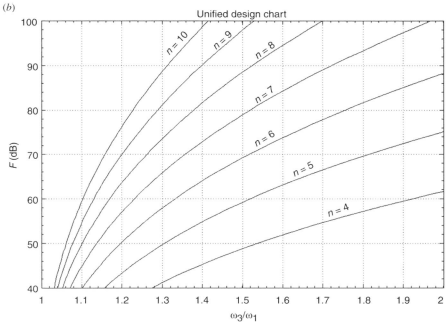

Figure 3A.6   Unified design chart for asymmetric filters with a single zero in the upper half of the frequency band: (a) lower half of frequency band; (b) upper half of frequency band.

CHAPTER 4

COMPUTER-AIDED SYNTHESIS OF CHARACTERISTIC POLYNOMIALS

A lowpass prototype filter can be completely characterized in terms of its critical frequencies, the poles and zeros of the transfer function. This easily leads to the possibility of generating such critical frequencies for a desired response shape by adopting computer-aided optimization techniques. For most types of practical filters, analytic techniques already exist to compute the critical frequencies. However, such techniques are restricted to the well-defined filter functions such as the equiripple Chebyshev and elliptic functions, or the monotonically increasing maximally flat function. Any departure from these techniques necessitates a computer-aided optimization procedure.

This chapter is devoted to the synthesis of the characteristic polynomials of lowpass lossless prototype filters using an efficient computer-aided optimization technique. It includes minimum phase as well as linear phase filters exhibiting symmetric or asymmetric response. The technique is completely general and can be readily adapted to synthesize characteristic polynomials with an arbitrary response. Classical filters such as Chebyshev or elliptic function filters can be derived as special cases of the more general characteristic polynomials.

Microwave Filters for Communication Systems: Fundamentals, Design, and Applications,
by Richard J. Cameron, Chandra M. Kudsia, and Raafat R. Mansour
Copyright © 2007 John Wiley & Sons, Inc.

4.1   OBJECTIVE FUNCTION AND CONSTRAINTS FOR SYMMETRIC LOWPASS PROTOTYPE FILTER NETWORKS

The transmission coefficient of a symmetric minimum phase lowpass prototype filter network is given by (Section 3.3)

$$|t(s)|^2_{s=j\omega} = \frac{1}{1 + \varepsilon^2 |K(s)|^2_{s=j\omega}}$$

where

$$K(s) = \varepsilon \frac{s(s^2 + a_1^2)(s^2 + a_2^2)\cdots(s^2 + a_i^2)\cdots}{(s^2 + b_1^2)(s^2 + b_2^2)\cdots(s^2 + b_i^2)\cdots}$$

for odd-order filter functions and

$$K(s) = \varepsilon \frac{(s^2 + a_1^2)(s^2 + a_2^2)\cdots(s^2 + a_i^2)\cdots}{(s^2 + b_1^2)(s^2 + b_2^2)\cdots(s^2 + b_i^2)\cdots} \tag{4.1}$$

for even-order filters.

The a and b terms are the attenuation zeros and poles, respectively (referred to as the *critical frequencies*), and ε is the ripple factor. The number of independent variables is represented by the finite critical frequencies. Attenuation zeros at origin or poles at infinity represent fixed frequencies and are therefore, not independent. ε provides a tradeoff between the passband and the stopband responses for a given characteristic factor F (Section 3.7) and again, does not represent an independent variable. If we want a specific value of F, we can take it as an equality constraint. Alternatively, critical frequencies can be generated with F as a parameter. This is more efficient since it dispenses with the equality constraint in the optimization process, and leads to the generation of unified design charts (Section 3.7).

Next comes the question as to which filter performance parameters are essential to determine a practical design. All practical applications require a maximum permissible ripple in the passband, and a minimum of isolation in the stopband. Phase or group delay response, in the passband and elsewhere, is uniquely related to the amplitude response for minimum phase filters by Hilbert's transforms. For such filters, it is sufficient to optimize the amplitude response and accept the resulting phase response. We are concerned primarily with minimum phase filters, although the subsequent analysis is general enough to analyze and optimize non-minimum phase types of filters as well.

To constrain the amplitude ripples in the passband or stopband, we must determine the frequencies at which these amplitude maxima and minima occur. This is accomplished by differentiating the equation for the characteristic function $K(s)$

and equating it to zero:

$$\frac{\partial}{\partial s}|K(s)|^2 = 0, \quad \frac{\partial}{\partial s}\left|\frac{F(s)}{P(s)}\right| = 0, \quad \text{or} \quad \left|\dot{F}(s)P(s)\right| - \left|F(s)\dot{P}(s)\right| = 0 \qquad (4.2)$$

Here, $\dot{F}(s)$ and $\dot{P}(s)$ represent the differentials with respect to the complex variable s. This determines the frequencies corresponding to the amplitude maxima in the passband and minima in the stopband, for an arbitrary choice of ϵ. The transmission and reflection coefficients, or the transmission and return loss in decibels, are computed as described by the equations in Section 3.4. The cutoff frequency ω_c, corresponding to the maximum ripple in the passband, is determined by solving for the highest frequency at which $t(s)$ is equal to the maximum ripple in the passband. Alternatively, $\omega_c = 1$ can be included as a constraint in the optimization process. All the other frequencies are then normalized with respect to $\omega_c = 1$.

For practical applications, the most convenient parameters for specifying the amplitude response are the return loss in passband and the transmission loss in the stopband, both parameters having units of decibels (dB).

By denoting the return loss and transmission loss at the ith attenuation maxima or minima as $R(i)$ and $T(i)$, respectively, the objective function U has terms of the following three types

$$U = U_1 + U_2 + U_3$$

where

$$U_1 = ABS\left[\left|R(i) \sim R(j)\right| + A_{ij}\right]$$
$$U_2 = ABS\left[\left|T(i) \sim T(j)\right| + B_{ij}\right]$$
$$U_3 = ABS\left[\left|R(i) \sim T(j)\right| + C_{ij}\right] \qquad (4.3)$$

where A_{ij}, B_{ij}, and C_{ij} represent arbitrary constants that constrain the differences in the amplitudes at the attenuation maxima and minima to some arbitrary values. For the equiripple passband, $A_{ij} = 0$, and for the equiripple stopband, $B_{ij} = 0$.

The term U_3 represents the characteristic factor F, and is best utilized as an equality constraint. Thus, in a generalized form, the objective function U is represented by

$$U = \sum_{i \neq j} ABS\left[\left|R(i) - R(j)\right| - A_{ij}\right] + \sum_{k \neq \ell} ABS\left[|T(k) - T(\ell)| - B_{k\ell}\right] + U_3 \qquad (4.4)$$

where the summations extend over all the attenuation maxima in the passband for the first term, and over all the attenuation minima in the stopband for the second term. The return loss and transmission loss are calculated by

$$T = -10 \log |t(s)|^2_{s=j\omega} \mathrm{dB}$$

$$R = -10 \log \left[1 - |t(s)|^2_{s=j\omega} \right] \mathrm{dB} \qquad (4.5)$$

For the prototype filter normalized to the unity cutoff, the constraints on the independent variables (the critical frequencies) are

$$0 \le a_1, a_2, \ldots, < 1$$
$$b_1, b_2, \ldots, > 1 \qquad (4.6)$$

These constraints simply imply that the attenuation zeros lie in the passband and the transmission zeros lie in the stopband.

4.2   ANALYTIC GRADIENTS OF THE OBJECTIVE FUNCTION

For the optimization process, we need to define an unconstrained artificial function that includes the objective function, as well as the inequality and equality constraints. Such an artificial function U_{art} is given by [1,2]

$$U_{art} = U + \frac{r}{\sum_i |\phi_i|} + \frac{\sum_k |\psi_k|^2}{\sqrt{r}} \qquad (4.7)$$

where $r =$ a variable of optimization to control the imposed constraints. (it is a positive number with a typical range of r between 10^{-4} and 1)
    $U =$ is the unconstrained objective function
    $\phi_i = i$th inequality constraint
    $\psi_k = k$th equality constraint

At the optimum value, $U_{art} = 0$. This implies that each individual term comprising U_{art} must be zero. The unconstrained function U tends to zero at the optimum value. To ensure the convergence of the constraint terms to zero, r is made smaller such that

$$\frac{r}{\sum_i |\phi_i|} \to 0 \quad \text{and} \quad \frac{\sum_k |\psi_k|^2}{\sqrt{r}} \to 0 \quad \text{as} \quad r \to 0$$

This is done in steps to ensure the smooth convergence of U_{art} to the desired level of accuracy. The gradient of U_{art}, with respect to the attenuation zeros a_i, is computed by

$$g_i = \frac{\partial U_{\text{art}}}{\partial a_i} = \frac{\partial U}{\partial a_i} + r \frac{\partial}{\partial a_i} \frac{1}{\sum_i |\phi_i|} + \frac{1}{\sqrt{r}} \frac{\partial}{\partial a_i} \sum_k |\psi_k|^2$$

$$= \frac{\partial U}{\partial a_i} + \Phi + \Psi \tag{4.8}$$

The same holds true for the attenuation poles. Each term on the right-hand side is evaluated independently in the following sections.

4.2.1 Gradient of the Unconstrained Objective Function

A typical form of the unconstrained function U is

$$U = |R(1) - R(2)| + |R(2) - R(3)| + \cdots + |T(4) - T(5)|$$
$$+ |T(5) - T(6)| + \cdots \tag{4.9}$$

where $R(k)$ and $T(k)$ are the values of the return loss and the transmission loss in dB at the kth attenuation maxima in the passband or minima in the stopband. Considering a typical term, we write

$$U_{ik} = |R(i) - R(k)| = R(i) - R(k) \quad \text{if} \quad R(i) \geq R(k)$$
$$= -(R(i) - R(k)) \quad \text{if} \quad R(i) \leq R(k) \tag{4.10}$$

$$\therefore \frac{\partial U_{ik}}{\partial a_i} = \pm \left[\frac{\partial R(i)}{\partial a_i} - \frac{\partial R(k)}{\partial a_i} \right] \tag{4.11}$$

The positive sign holds when $R(i) \geq R(k)$ and vice versa. The partial derivatives $\partial R / \partial a_i$ are determined by considering the transmission function

$$|t|^2 = \frac{1}{1 + |K(s)|^2} = 1 - |\rho|^2 \tag{4.12}$$

where ρ is the reflection coefficient. The critical frequencies are independent of the choice of the ripple factor. As a consequence, it is convenient to assume ε to be unity. By making use of equation (4.1), we obtain

$$\therefore \frac{\partial |t|^2}{\partial a_i} = -\frac{1}{[1 + |K(s)|^2]^2} 2K(s) \frac{\partial |K(s)}{\partial a_i}$$

$$= -4 \frac{a_i}{(s^2 + a_1^2)} \frac{K^2(s)}{[1 + K^2(s)]^2} \tag{4.13}$$

Making use of equation (4.5), the gradients in terms of transmission and reflection loss in dB are given by [3]

$$\therefore \frac{\partial T}{\partial a_i} = 40 \log e \frac{K^2(s)}{1 + K^2(s)} \frac{a_i}{(s^2 + a_i^2)}$$

$$= 40 \log e |\rho|^2 \frac{a_i}{(s^2 + a_i^2)} \tag{4.14}$$

and

$$\frac{\partial R}{\partial a_i} = -40 \log e \frac{1}{1 + K^2(s)} \frac{a_i}{(s^2 + a_i^2)}$$

$$= -40 \log e |t|^2 \frac{a_i}{(s^2 + a_i^2)} \tag{4.15}$$

where e is the natural base for logarithms. In a similar manner, it can be demonstrated that the gradients, with respect to the attenuation poles b_i, are computed by

$$\frac{\partial T}{\partial b_i} = -40 \log e |\rho|^2 \frac{b_i}{(s^2 + b_i^2)} \tag{4.16}$$

and

$$\frac{\partial R}{\partial b_i} = 40 \log e |t|^2 \frac{b_i}{(s^2 + b_i^2)} \tag{4.17}$$

Thus, the partial derivatives of the transmission or return loss in dB, at any frequency with respect to the independent variables, are determined analytically, and thus the gradient of the unconstrained function U.

4.2.2   Gradient of the Inequality Constraint

The inequality constraint ϕ has the form

$$\Phi = \frac{r}{|\phi_1|} + \frac{r}{|\phi_2|} + \cdots \tag{4.18}$$

where r is a positive number, and the ϕ terms have the following forms

$$\phi_k = \begin{cases} a_i \\ 1 - a_i \\ b_i - 1 \\ N - b_i, \ N \text{ is a positive number greater than 1} \end{cases} \tag{4.19}$$

$$\therefore \frac{\partial}{\partial a_i} \frac{1}{|\phi_k|} = -\frac{1}{\phi_k^2} \frac{\partial \phi_k}{\partial a_i} \qquad \phi_k > 0$$

$$= +\frac{1}{\phi_k^2} \frac{\partial \phi_k}{\partial a_i} \qquad \phi_k < 0 \tag{4.20}$$

TABLE 4.1   Gradients of the Inequality Constraint

Form of ϕ	Gradient of ϕ	Constraints
a_i	$\mp \dfrac{1}{a_i^2}$	$a_i \gtrless 0$
$1 - a_i$	$\pm \dfrac{1}{(1.0 - a_i)^2}$	$(1.0 - a_i) \gtrless 0$
$b_i - 1$	$\mp \dfrac{1}{(b_i - 1)^2}$	$(b_i - 1) \gtrless 0$
$N - b_i$	$\pm \dfrac{1}{(N - b_i)^2}$	$(N - b_i) \gtrless 0$

where a_i represents the ith independent variable. The gradients for the various forms of ϕ are described in Table 4.1 [3].

As a result, the gradient of Φ has the following form:

$$\frac{\partial \Phi}{\partial a_i} + \frac{\partial \Phi}{\partial b_i} = r \cdot \left[\sum \left(\mp \frac{1}{a_i^2} \right) + \sum \pm \left[\frac{1}{(1 - a_i^2)} \right] \right.$$

$$\left. + \sum \left(\mp \frac{1}{(b_i - 1)^2} \right) + \sum \left(\pm \frac{1}{(N - b_i)} \right) \right] \qquad (4.21)$$

This includes the inequalities associated with the attenuation zeros, as well as the poles.

4.2.3   Gradient of the Equality Constraint

The gradient of the equality constraint is computed by

$$\Psi = \frac{1}{\sqrt{r}} \frac{\partial}{\partial a_i} \sum |\psi_k|^2$$

where ψ_k has the form

$$\psi_k = A + R(1) + T(2) - T(1) - R(2) + \cdots \qquad (4.22)$$

where A is a constant. $R(\ )$ and $T(\ )$ represent the return loss and transmission loss in dB at a maxima in the passband and minima in the stopband respectively. The gradient is evaluated as follows:

$$\frac{\partial}{\partial a_i} |\psi_k|^2 = 2\psi_k \cdot \frac{\partial \psi_k}{\partial a_i} \qquad (4.23)$$

$$\Psi = \frac{2}{\sqrt{r}} \sum_k \psi_k \left[\frac{\partial R(1)}{\partial a_i} + \frac{\partial T(2)}{\partial a_i} + \frac{\partial T(1)}{\partial a_i} - \frac{\partial R(2)}{\partial a_i} \right.$$

$$\left. + \cdots + \frac{\partial R(1)}{\partial b_i} + \frac{\partial T(1)}{\partial b_i} + \cdots \right] \qquad (4.24)$$

The values of $R(\ )$ or $T(\ )$ and their partial derivatives are derived by equations (4.14)–(4.17). Thus, the gradients of the unconstrained objective function U_{art} are determined analytically.

4.3   OPTIMIZATION CRITERIA FOR CLASSICAL FILTERS

In this section, we develop the appropriate objective functions for optimization to generate the known classes of filter functions.

4.3.1   Chebyshev Function Filters

This function is characterized by an equiripple response in the passband and a monotonically rising stopband. The objective function and constraints can be represented as

$$U = |R(1) - R(2)| + |R(2) - R(3)| + \cdots$$
$$1 < a_1, a_2, \ldots, > 0 \tag{4.25}$$

where $R(k)$ is the return loss at the kth attenuation maxima and a_i are the critical frequencies. For example, a sixth-order filter has the objective function

$$U = |R(1) - R(2)| + |R(2) - R(3)| + |R(3) - R(4)| \tag{4.26}$$

where $R(4)$ is characterized by $s = j$, as depicted in Figure 4.1. This term is used to normalize the critical frequencies to a unity cutoff frequency in the passband. By adopting the analytic gradients described by equations (4.21) and (4.24) in the

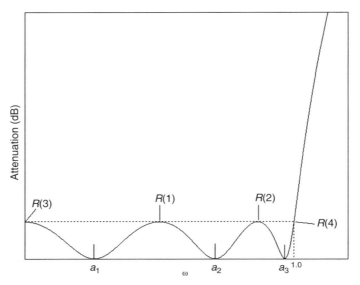

Figure 4.1   Attenuation zeros and maxima of a sixth-order Chebyshev filter.

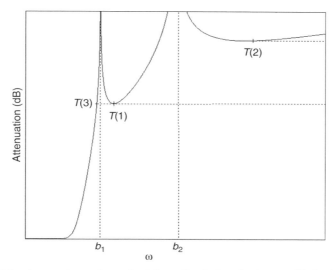

Figure 4.2 Attenuation poles and minima of a sixth-order inverse Chebyshev filter.

optimization process, the critical frequencies are determined to be 0.2588, 0.7071, and 0.9659.

4.3.2 Inverse Chebyshev Filters

These filters are characterized by a monotonically increasing passband (maximally flat) and an equiripple stopband:

$$K(s) = \frac{s^6}{(s^2 + b_1^2)(s^2 + a_2^2)}$$

The response of a six-pole filter with arbitrary locations of attenuation poles is illustrated in Figure 4.2. The objective function for an equiripple stopband, normalized to unity cutoff frequency, and inequality constraints are as follows:

$$U = |T(1) - T(2)| + |T(2) - T(3)|, \quad b_1, b_2, > 1 \tag{4.27}$$

The critical frequencies are readily evaluated as 1.0379 and 1.4679.

4.3.3 Elliptic Function Filters

These filters are characterized by an equiripple passband and equiripple stopband. For a six-pole class C elliptic function filter, displayed in Figure 4.3, the characteristic function has the form

$$K(s) = \frac{s^2(s^2 + a_1^2)(s^2 + a_2^2)}{(s^2 + b_1^2)(s^2 + b_2^2)}$$

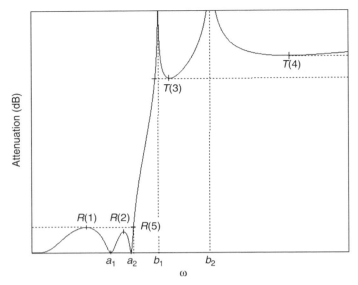

Figure 4.3   Attenuation maxima and minima of a sixth-order lowpass prototype class C elliptic function filter.

and the objective function has the form

$$U = |R(1) - R(2)| + |R(2) - R(5)| + |T(3) - T(4)| \qquad (4.28)$$

$R(5)$ is characterized by $s = j$ to incorporate the unity cutoff frequency. For filters with transmission zeros, it is necessary to impose some constraint on the characteristic factor F (Section 3.7), which, for the example under consideration, is expressed as

$$F = |R(2) + T(3)| - |T(2) + R(3)|$$
$$= A(\text{say}) \qquad (4.29)$$

As an equality constraint, this is written as follows:

$$\Psi = A - |R(2) + T(3)| + |T(2) + R(3)| \qquad (4.30)$$

In general, equality constraints increase the computation time significantly. One way to bypass this equality constraint and at the same time decrease the number of independent variables by one, is to fix one of the attenuation pole frequencies. This automatically constrains the characteristic factor F. From a practical standpoint, this is most useful since tradeoffs are generated by selecting the pole frequencies in the neighborhood where a high isolation is specified. With this approach, computations can be performed very efficiently. For example, we chose b_1 to be fixed at 1.2995. The remaining critical frequencies are then determined as $a_1 = 0.7583$, $a_2 = 0.9768$, and $b_2 = 1.6742$.

4.4 GENERATION OF NOVEL CLASSES OF FILTER FUNCTIONS

Now that the generation of classical filter functions has been demonstrated, we will examine an unconventional eighth-order filter with the characteristic polynomial:

$$K(s) = \frac{s^4(s^2 + a_1^2)(s^2 + a_2^2)}{(s^2 + b_1^2)(s^2 + b_2^2)} \tag{4.31}$$

This function is characterized by a pair of attenuation poles, a double zero at the origin, and two other nonorigin reflection zeros, described in Figure 4.4. Now, we will introduce the following three cases for optimization.

4.4.1 Equiripple Passbands and Stopbands

Assuming a fixed attenuation pole $b_1 = 1.25$, we express the objective function and constraints as

$$U = |R(1) - R(2)| + |R(2) - R(5)| + |T(3) - T(4)|$$
$$1 < a_1, a_2, > 0, \quad b_2 > 1 \tag{4.32}$$

$R(5)$ is used to constrain the cutoff frequency to unity.

The computed critical frequencies are summarized in Table 4.2.

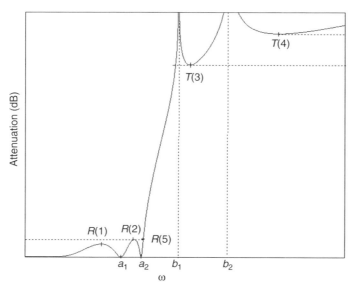

Figure 4.4 Lowpass prototype response of an eighth-order filter with two attenuation poles and a double zero at the origin.

TABLE 4.2   Critical Frequencies for an Optimized Eight-Pole Filter with Two Attenuation Poles and a Double Attenuation Zero at the Origin, and Fixed Attenuation Pole at $b_1 = 1.25$

B_{34} (dB)	Computed a_1	Critical a_2	Frequencies b_2	Characteristic Factor F (dB)
0	0.8388	0.9845	1.4671	67.4
10	0.8340	0.9839	1.6211	63.1
−10	0.8636	0.9878	1.1541	63.25

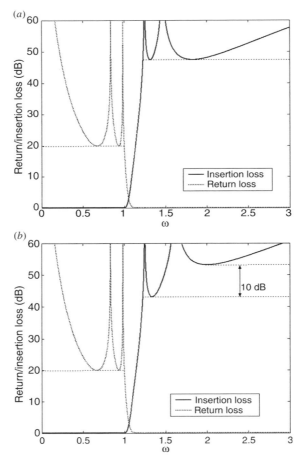

Figure 4.5   Frequency response of an optimized lowpass prototype eight-pole filter with two attenuation poles and a double zero at the origin: (*a*) an equiripple stopband; (*b*) a 10 dB differential in stopband minima.

4.4.2 Nonequiripple Stopband with an Equiripple Passband

Here, we consider an equiripple passband and a nonequiripple stopband. As before, we fix the first attenuation pole at $b_1 = 1.25$, and the objective function and constraints are then calculated by

$$U = |R(1) - R(2)| + |R(2) - R(5)| + |T(3) - T(4) - B_{34}| \tag{4.33}$$
$$1 < a_1, a_2, > 0, \quad b_2 > 1$$

The term B_{34} is an arbitrary constant. It represents the difference between the attenuation minima at $T(3)$ and $T(4)$ in dB. The computed critical frequencies for $B_{34} = \pm 10\,\text{dB}$ are listed in Table 4.2. $B_{34} = 10\,\text{dB}$ implies that the second minima in the stopband is greater than the first one by 10 dB, and $B_{34} = -10\,\text{dB}$ implies the reverse.

For the case $B_{34} = 0$ and 10 dB, the frequency response is exhibited in Figure 4.5.

4.5 ASYMMETRIC CLASS OF FILTERS

The characteristic polynomials for the asymmetric class of filters are discussed in Section 3.10. For the sake of simplicity, we first deal with the minimum phase filters for which the zeros of $P(s)$ are confined to the imaginary axis. For such filters, the polynomial $K(s)$ is expressed as

$$K(s) = \frac{(s + ja_1)(s + ja_2) \cdots (s + ja_n)}{(s + jb_1)(s + jb_2) \cdots (s + jb_m)} \tag{4.34}$$

where $|a_i| < 1$, $|b_i| > 1$, $m \le n$ and n is the order of the filter network. Since the ripple factor ε does not alter the critical frequencies, it is assumed to be unity in the formulation of $K(s)$. The gradient of the transmission coefficient with respect to the reflection or transmission zeros is computed by equation (4.13). As an illustration, we compute the gradient with respect to b_i by

$$|K(s)| = \frac{|s + ja_1|\,|s + ja_2| \cdots}{|s + jb_1|\,|s + jb_2| \cdots} \tag{4.35}$$

$$\ln |K(s)| = -\ln |s + jb_i| + \text{other terms}$$

$$\therefore \frac{1}{|K(s)|} \frac{\partial}{\partial b_i} |K(s)| = -\frac{1}{|s + jb_i|} \cdot \frac{\partial}{\partial b_i} |s + jb_i| + \cdots \tag{4.36}$$

since $|s + jb_i| = (s^2 + b_i^2)^{1/2}$, the gradient is computed to be

$$\frac{\partial}{\partial b_i} |K(s)| = -|K(s)| \frac{b_i}{s^2 + b_i^2} \tag{4.37}$$

In a similar way

$$\frac{\partial}{\partial a_i}|K(s)| = |K(s)|\frac{a_i}{s^2 + a_i^2} \tag{4.38}$$

By substituting the values of

$$\frac{\partial}{\partial a_i}|K(s)| \quad \text{or} \quad \frac{\partial}{\partial b_i}|K(s)|$$

in equation (4.13), we obtain

$$\frac{\partial |t|^2}{\partial a_i} = -2\frac{a_i}{(s^2 + a_i^2)}\frac{|K(s)|^2}{\left[1 + |K(s)|^2\right]^2} \tag{4.39}$$

$$\frac{\partial |t|^2}{\partial b_i} = 2\frac{b_i}{(s^2 + b_i^2)} \cdot \frac{|K(s)|^2}{\left\{1 + |K(s)|^2\right\}^2} \tag{4.40}$$

These relationships can be expressed in a more compact form in terms of transmission and reflection loss in dB, described by equations (4.14) and (4.15):

$$\frac{\partial T}{\partial a_i} = 20\log e|\rho|^2\frac{a_i}{s^2 + a_i^2} \tag{4.41}$$

$$\frac{\partial R}{\partial a_i} = -20\log e|t|^2\frac{a_i}{s^2 + a_i^2} \tag{4.42}$$

In a similar manner

$$\frac{\partial T}{\partial b_i} = -20\log e \cdot |\rho|^2\frac{b_i}{s^2 + b_i^2} \tag{4.43}$$

$$\frac{\partial R}{\partial b_i} = 20\log e|t|^2\frac{b_i}{s^2 + b_i^2} \tag{4.44}$$

The gradients differ by a factor of 2 for the asymmetric case as compared to the symmetric case. This is expected since we are dealing with a single zero as opposed to a pair of zeros for the symmetric case.

4.5.1 Asymmetric Filters with Chebyshev Passband

Let us consider a Four-pole filter with four finite attenuation zeros in the passband and two attenuation poles in the upper half of the stopband, as shown in Figure 4.6.

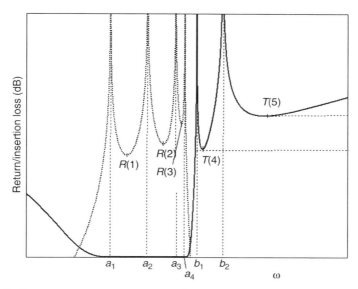

Figure 4.6 Lowpass prototype response of a Four-pole asymmetric filter with two attenuation poles in the upper half of the stopband.

For a Chebyshev passband and arbitrary stopband, the objective function is derived by

$$U = |R(1) - R(2)| + |R(2) - R(3)| + |T(4) - T(5) - B_{45}| \qquad (4.45)$$

Here, B_{45} is an arbitrary constant and represents the difference between the attenuation minima at $T(4)$ and $T(5)$. For an equiripple stopband, $B_{45} = 0$. For filters requiring an arbitrary value of B_{45}, it is desirable to generate the unified design charts (Section 3.7) as a function of the close in attenuation minima, often a critical specification requirement. This can be accomplished by choosing the attenuation pole frequency closest to the passband as a parameter to generate the unified design charts for an arbitrary value of B_{34}. This has the merit of reducing the number of variables for optimization by one. Sometimes, it is beneficial to fix both attenuation pole frequencies, in which case the objective function is simply given by

$$U = |R(1) - R(2)| + |R(2) - R(3)| \qquad (4.46)$$

Such a function is much simpler to optimize. It provides a good indication for the starting values to optimize the objective function for specific values of B_{45}. For this example, b_1 and b_2 are chosen to be 1.25 and 1.5, respectively. Following the optimization procedure, the remaining critical frequencies for an equiripple Chebyshev passband are determined as -0.8466, 0.0255, 0.7270, and 0.9757, with a cutoff frequency of ± 1.0. Figure 4.7 describes the filter response.

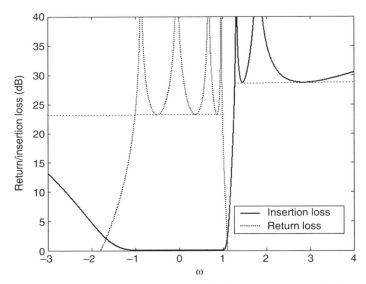

Figure 4.7 Lowpass prototype response of a fourth-order asymmetric filter with two attenuation poles in the upper half of the frequency band and optimized for an equiripple response.

4.5.2   Asymmetrical Filters with Arbitrary Response

To demonstrate the flexibility of optimization approach, we consider a four-pole asymmetric filter with a double zero at the origin and two attenuation poles in the upper half of the stopband, as described in Figure 4.8. The characteristic polynomial

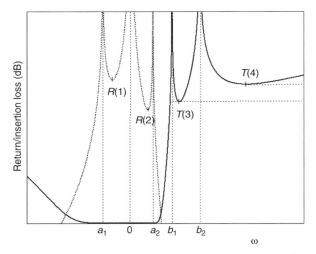

Figure 4.8 Lowpass prototype response of a fourth–order asymmetric filter with a double zero at the origin and two attenuation poles in the upper half of the frequency band.

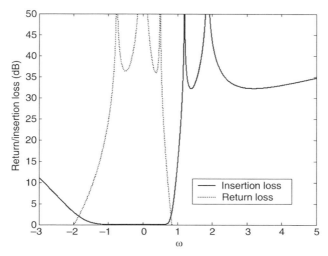

Figure 4.9 Lowpass prototype response of a four-pole asymmetric filter with a double zero at the origin, two attenuation poles in the upper half of the frequency band, and optimized for the equiripple response.

for such a filter is

$$K(s) = \frac{s^2(s + ja_1)(s + ja_2)}{(s + jb_1)(s + jb_2)} \quad (4.47)$$

The objective function for this class of filter can be described as

$$U = |R(1) - R(2) + A_{12}| + |T(3) - T(4) + B_{34}| \quad (4.48)$$

where A_{12} and B_{34} are constants that define the departure of the passband and stopband from an equiripple response; $A_{12} = 0$ represents an equiripple passband, and $B_{34} = 0$ represents an equiripple stopband. For this example, we optimize the objective function by choosing A_{12} and B_{34} to be zero. The optimized frequency response is plotted in Figure 4.9; the optimized characteristic polynomials and attenuation minima and maxima are listed in Table 4.3. An interesting aspect of such an

TABLE 4.3 Critical Frequencies and Polynomial Characteristics for Optimized Four-Pole Asymmetric Prototype Filter with Three Equiripple Peaks in Passband and Two Transmission Zeros in Upper Stopband

a_i	b_i	Roots of $E(s)$	Frequency Maxima in Passband	Attenuation Minimum in Stopband
−0.745	1.2	$-0.8881 - 1.6837j$	—	—
0	1.865	$-1.4003 - 0.1067j$	−0.4963	1.3944
0	∞	$-0.6811 + 0.7139j$	0.3764	3.2005
0.5	∞	$-0.1689 + 0.8315j$	—	—

asymmetric filter is that its cutoff frequency is no longer symmetric with respect to the zero frequency of the prototype filter. The band center must be calculated by computing the geometric mean of the equiripple passband. As a result, the filter response can be renormalized to the new band center frequency and passband bandwidth.

4.6   LINEAR PHASE FILTERS

The technique described for the computer-aided optimization of characteristic polynomials for minimum phase filters is equally applicable for linear phase filters, both symmetric and asymmetric. Linear phase filters, when the zeros of $P(s)$ are located symmetrically on the real axis, are readily derived from equation (4.17). The equations for the gradients are modified as

$$\frac{\partial T}{\partial b_i} = -40 \log e \cdot |\rho|^2 \cdot \frac{b_i}{(s^2 \pm b_i^2)} \tag{4.49}$$

and

$$\frac{\partial R}{\partial b_i} = 40 \log e \cdot |t|^2 \cdot \frac{b_i}{(s^2 \pm b_i^2)} \tag{4.50}$$

where the negative sign is applicable to the case of a linear phase filter with zero location at $\pm b_1$ along the real axis. The complex zeros of the transfer function are located symmetrically about the imaginary axis, as noted in Section 3.10. Let us define a complex zero by

$$Q(s) = (s + \sigma_1 \pm j\omega_1)(s - \sigma_1 \pm j\omega_1)$$

$$= s^2 \pm j2s\omega_1 - (\sigma_1^2 + \omega_1^2) \tag{4.51}$$

A single pair of complex zeros gives rise to an asymmetric response, whereas a complex quad yields a symmetrical response. Following equation (4.51), we obtain

$$|Q(s)|^2 = \{(s^2 + \sigma_1^2 + \omega_1^2)^2 - 4s^2\sigma_1^2\} \tag{4.52}$$

$$\frac{\partial |Q(s)|}{\partial \omega_1} = -\frac{2\omega_1(3s^2 - \sigma_1^2 - \omega_1^2)}{|Q(s)|} \tag{4.53}$$

$$\frac{\partial |Q(s)|}{\partial \sigma_1} = -\frac{2\sigma_1(s^2 - \sigma_1^2 - \omega_1^2)}{|Q(s)|} \tag{4.54}$$

In terms of characteristics polynomial $K(s)$, we have4

$$\frac{\partial}{\partial \sigma_1}|K(s)| = -\frac{|K(s)|}{|Q(s)|} \cdot \frac{\partial |Q(s)|}{\partial \sigma_1}$$

$$\frac{\partial}{\partial \omega_1}|K(s)| = -\frac{|K(s)|}{|Q(s)|} \cdot \frac{\partial |Q(s)|}{\partial \omega_1}$$

Finally, the gradient in terms of transmission and reflection loss in dB are

$$\frac{\partial T}{\partial \sigma_1} = -40 \log e \cdot |\rho|^2 \sigma_1 \frac{(s^2 - \sigma_1^2 - \omega_1^2)}{|Q(s)|^2} \qquad (4.55)$$

$$\frac{\partial T}{\partial \omega_1} = -40 \log e |t|^2 \omega_1 \frac{(3s^2 - \sigma_1^2 - \omega_1^2)}{|Q(s)|^2} \qquad (4.56)$$

Since linear phase filters require the simultaneous optimization of the amplitude and phase (or group delay), the problem becomes quite complex. For practical reasons, it is far more expeditious to assume specific values of a complex zero as well as $j\omega$-axis zeros [i.e., zeros of $P(s)$] and then optimize for an equiripple passband. By selecting zeros of $P(s)$ judiciously, and then tinkering with these values, we can optimize the characteristic polynomials for both amplitude and phase to achieve the desired response.

4.7 CRITICAL FREQUENCIES FOR SELECTED FILTER FUNCTIONS

With the optimization techniques presented in the preceding section, the critical frequencies are computed for an eight-pole lowpass prototype filter function, not amenable to analytical techniques. Characteristic factor F is used as a parameter in these computations, and it includes the following filter functions:

1. An eight-pole filter with two transmission zeros having a 10 dB difference in the attenuation minima and an equiripple passband.
2. An eight-pole filter with a double zero at the origin, two transmission zeros, and an equiripple passband and stopband.

The computed critical frequencies are included in Appendix 4A. These data add to the range of unified design charts and also provide guidelines for developing software to synthesize characteristic polynomials for filters with arbitrary amplitude and phase response.

SUMMARY

The transfer function of a lossless lumped-element lowpass prototype filter is completely characterized in terms of its poles and zeros. Such a characterization is well suited for computer-aided design techniques. The key lies in making sure that the optimization procedure is highly efficient. This is accomplished by determining the gradients of the objective function analytically, and linking it directly to the desired amplitude response shape. In this chapter, we address this issue by deriving analytical gradients with respect to the critical frequencies for the most general lowpass prototype filter. This includes minimum phase and nonminimum phase filters exhibiting symmetric or asymmetric frequency response. The efficiency of this procedure is demonstrated by including examples of the classical Chebyshev and elliptic function filters as special cases of the design procedure. The technique is completely general and can be readily adapted to synthesize the characteristic polynomials for filters with arbitrary responses. To demonstrate the flexibility of this method, examples of some unconventional filters are included.

REFERENCES

1. J. Kowalik and M. R. Osbourne, *Methods for Unconstrained Optimization Problems*, Elsevier, New York, 1968.
2. J. W. Bandler, Optimization methods for computer-aided design, *IEEE Trans. MTT* 533–552 (Aug. 1969).
3. C. M. Kudsia and M. N. S. Swamy, Computer-aided optimization of microwave filter networks for space applications, *IEEE MTT-S Int. Microwave Symp. Digest*, May 28–30, 1980, Washington, DC.

APPENDIX 4A CRITICAL FREQUENCIES FOR AN UNCONVENTIONAL 8-POLE FILTER

TABLE 4A.1   Critical Frequencies and Unified Design Chart Data for an Eight-Pole Filter Having Two Transmission Zeros with 10 dB Difference in Attenuation Minima and an Equiripple Passband

$$K(s) = \frac{(s^2 + a_1^2)(s^2 + a_2^2)(s^2 + a_3^2)(s^2 + a_4^2)}{(s^2 + b_1^2)(s^2 + b_2^2)}$$

a_1	a_2	a_3	a_4	b_1	b_2	b_s	F^a
0.2702	0.7049	0.9255	0.9938	1.0460	1.2070	1.0384	40
0.2618	0.6885	0.9159	0.9925	1.0664	1.2514	1.0568	45
0.2539	0.6732	0.9067	0.9913	1.0914	1.3037	1.0802	50
0.2403	0.6464	0.8903	0.9891	1.1601	1.4321	1.1437	60
0.2295	0.6250	0.8768	0.9873	1.2520	1.5911	1.2306	70
0.2210	0.6081	0.8660	0.9857	1.3719	1.7876	1.3449	80
0.2147	0.5954	0.8577	0.9846	1.5192	2.0186	1.4860	90
0.2099	0.5857	0.8514	0.9837	1.6972	2.2901	1.6572	100

[a]F: characteristic factor in dB—$F = (A_1 + R_3) - (A_3 + R_1) \approx (R_1 + A_3)$.

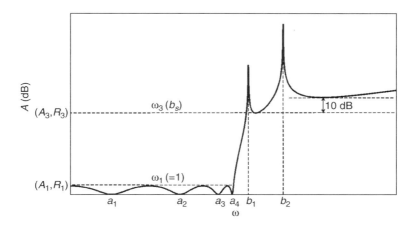

TABLE 4A.2   Critical Frequencies and Unified Design Chart Data for an Eight-Pole Filter with a Double Zero at the Origin, Two Transmission Zeros, and Equiripple Passband and Stopband

$$K(s) = \frac{s^4(s^2 + a_1^2)(s^2 + a_2^2)}{(s^2 + b_1^2)(s^2 + b_2^2)}$$

a_1	a_2	b_1	b_2	F
0.9012	0.9926	1.0501	1.1566	40
0.8871	0.9909	1.0734	1.1990	45
0.8742	0.9893	1.1023	1.2473	50
0.8628	0.9878	1.1364	1.3011	55
0.8522	0.9863	1.1774	1.3626	60
0.8429	0.9851	1.2249	1.4315	65
0.8347	0.9839	1.2788	1.5074	70
0.8213	0.9820	1.4057	1.6807	80
0.8109	0.9804	1.5640	1.8911	90
0.8029	0.9793	1.7571	2.1430	100

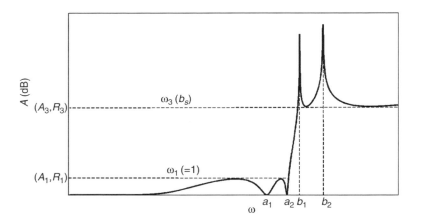

CHAPTER 5

ANALYSIS OF MULTIPORT MICROWAVE NETWORKS

Any filter or multiplexer circuit can be viewed as a combination of multiport subnetworks. Figure 5.1 illustrates a circuit that consists of a five-port network connected to several two-port networks. The overall circuit in this case is a four-channel multiplexer. The five-port network is the combining manifold junction. Each filter could be also viewed as a cascade of two-port or three-port subnetworks. In this chapter, we discuss several matrix representations of multiport microwave networks. Also, we present various techniques to analyze linear passive microwave circuits that are formed by connecting any number of multiport networks. As an example we apply these techniques, at the end of this chapter, to demonstrate step by step the evaluation of the overall scattering matrix of a three-channel multiplexer.

The most commonly used matrices to describe a network are the $[Z]$, $[Y]$, $[ABCD]$, and $[S]$ matrices [1–3]. The $[Z]$, $[Y]$, and $[ABCD]$ matrices relate the voltage and current at the various ports by describing the network at the lumped-element level. The scattering matrix $[S]$ and the transmission matrix $[T]$ relate the incident and reflected normalized voltages at the various ports. While at microwave frequencies, the voltage and currents cannot be measured, the $[Z]$, $[Y]$, and $[ABCD]$ matrices are often used to provide a physical insight into the equivalent circuit of the microwave network. In RF design, the most commonly quoted parameters are the S parameters. They represent parameters that can be quantified and measured. In addition, the S matrix can be easily extended to represent the behavior of the network for the dominant mode, as well as the higher-order modes. It should be however noted that the $[Z]$, $[Y]$, $[ABCD]$, $[S]$, and $[T]$ matrices are

Microwave Filters for Communication Systems: Fundamentals, Design, and Applications,
by Richard J. Cameron, Chandra M. Kudsia, and Raafat R. Mansour
Copyright © 2007 John Wiley & Sons, Inc.

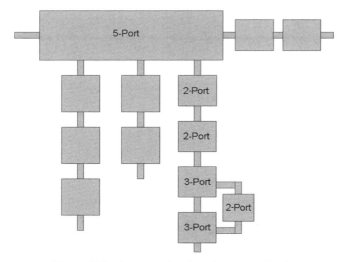

Figure 5.1   An example of a microwave circuit.

interchangeable and can fully describe the characteristics of any microwave network. Familiarity with the concepts of these matrices is crucial for understanding the materials presented in this book. These matrices are also essential to comprehend the various techniques presented in this chapter for analysis of microwave filter networks.

5.1   MATRIX REPRESENTATION OF TWO-PORT NETWORKS

5.1.1   Impedance [Z] and Admittance [Y] Matrices

Let us consider the two-port network shown in Figure 5.2. The current and the voltages at the two ports are related by [Z] and [Y] matrices as follows:

$$\begin{bmatrix} V_1 \\ V_2 \end{bmatrix} = \begin{bmatrix} Z_{11} & Z_{12} \\ Z_{21} & Z_{22} \end{bmatrix} \begin{bmatrix} I_1 \\ I_2 \end{bmatrix} \tag{5.1}$$

$$\begin{bmatrix} I_1 \\ I_2 \end{bmatrix} = \begin{bmatrix} Y_{11} & Y_{12} \\ Y_{21} & Y_{22} \end{bmatrix} \begin{bmatrix} V_1 \\ V_2 \end{bmatrix} \tag{5.2}$$

$$[V] = [Z][I] \tag{5.3}$$

$$[I] = [Y][V] \tag{5.4}$$

It is clear that the [Z] and [Y] matrices are related as

$$Z = [Y]^{-1} \tag{5.5}$$

Figure 5.2   A two-port microwave network.

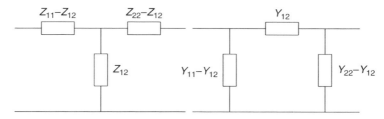

Figure 5.3   Equivalent T and π networks of a two-port microwave network.

It can be easily shown that Z_{11} is the input impedance at port 1 when port 2 is an open circuit: $I_2 = 0$. Y_{11} is the input admittance at port 1 when port 2 is a short circuit: $V_2 = 0$. Similarly, one can derive the physical meaning of the rest of the parameters from equations (5.1) and (5.2). For microwave networks that are lossless, the elements of the $[Z]$ and $[Y]$ matrices are pure imaginary. For networks that do not have ferrites, plasmas or active devices, that is, networks that are reciprocal, one can show that $Z_{12} = Z_{21}$ and $Y_{12} = Y_{21}$.

In filter design, the $[Z]$ and $[Y]$ matrices are often used to describe the $[T]$ or π lumped-element equivalent circuit of the microwave network as shown in Figure 5.3. For example, in dealing with a coupling iris in filters, the $[T]$ or π equivalent circuit of the iris can easily identify whether the coupling is inductive or capacitive. As demonstrated in Chapter 14, the equivalent circuit can also identify the loading effects of the iris on the adjacent resonators.

5.1.2   The [ABCD] Matrix

The $[ABCD]$ matrix representation of the two-port depicted in Figure 5.4 is given in equation (5.6). Note that the current in port 2 is defined as $-I_2$; that is, the current flows out of port 2. This definition is adopted to ease the evaluation of the overall $[ABCD]$ matrix in dealing with cascaded networks. For symmetrical networks $A = D$, and for reciprocal networks one can show that the $ABCD$ parameters

Figure 5.4   The $ABCD$ representation of a two-port network.

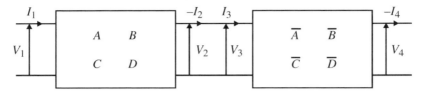

Figure 5.5 Two networks cascaded in series.

satisfy equation (5.7). Figure 5.5 shows two networks connected in cascade. In this case, $I_3 = -I_2$, and the overall $[ABCD]$ matrix is calculated by simple matrix multiplication of the two $[ABCD]$ matrices

$$\begin{bmatrix} V_1 \\ I_1 \end{bmatrix} = \begin{bmatrix} A & B \\ C & D \end{bmatrix} \begin{bmatrix} V_2 \\ -I_2 \end{bmatrix} \tag{5.6}$$

$$AD - CB = 1 \tag{5.7}$$

$$\begin{bmatrix} V_1 \\ I_1 \end{bmatrix} = \begin{bmatrix} A & B \\ C & D \end{bmatrix} \begin{bmatrix} V_2 \\ -I_2 \end{bmatrix}, \qquad \begin{bmatrix} V_3 \\ I_3 \end{bmatrix} = \begin{bmatrix} \overline{A} & \overline{B} \\ \overline{C} & \overline{D} \end{bmatrix} \begin{bmatrix} V_4 \\ -I_4 \end{bmatrix} \tag{5.8}$$

But V_2 and I_2 are related to V_3 and I_3 as:

$$\begin{bmatrix} V_2 \\ -I_2 \end{bmatrix} = \begin{bmatrix} V_3 \\ I_3 \end{bmatrix} \tag{5.9}$$

Substituting equations (5.8) and (5.9) in equation (5.6) we get

$$\begin{bmatrix} V_1 \\ I_1 \end{bmatrix} = \begin{bmatrix} A & B \\ C & D \end{bmatrix} \begin{bmatrix} \overline{A} & \overline{B} \\ \overline{C} & \overline{D} \end{bmatrix} \begin{bmatrix} V_4 \\ -I_4 \end{bmatrix} \tag{5.10}$$

The relation between the $[Z]$ matrix and $[ABCD]$ matrix can be obtained by setting $I_2 = 0$ and $V_2 = 0$ in both matrices as shown below:

$$\begin{bmatrix} V_1 \\ V_2 \end{bmatrix} = \begin{bmatrix} Z_{11} & Z_{12} \\ Z_{21} & Z_{22} \end{bmatrix} \begin{bmatrix} I_1 \\ I_2 \end{bmatrix} \tag{5.11}$$

$$\begin{bmatrix} V_1 \\ I_1 \end{bmatrix} = \begin{bmatrix} A & B \\ C & D \end{bmatrix} \begin{bmatrix} V_2 \\ -I_2 \end{bmatrix} \tag{5.12}$$

Setting $I_2 = 0$ in equations (5.11) and (5.12) gives

$$V_1 = Z_{11}I_1, \qquad V_2 = Z_{21}I_1 \tag{5.13}$$

$$V_1 = AV_2, \qquad I_1 = CV_2 \tag{5.14}$$

C and D can then be related to the elements of the $[Z]$ matrix as follows:

$$C = \frac{I_1}{V_2} = \frac{1}{Z_{21}}, \qquad A = \frac{V_1}{V_2} = \frac{Z_{11}}{Z_{21}} \tag{5.15}$$

TABLE 5.1 [*ABCD*] Parameters of the Commonly Used Circuit Configurations

Circuit	*ABCD* Parameters	
Z_0, β	$A = \cos \beta l$ $C = jY_0 \sin \beta l$	$B = jZ_0 \sin \beta l$ $D = \cos \beta l$
Z	$A = 1$ $C = 0$	$B = Z$ $D = 1$
Y	$A = 1$ $C = Y$	$B = 0$ $D = 1$
Y_1, Y_2	$A = 1 + \dfrac{Y_2}{Y_1}$ $C = Y_2$	$B = \dfrac{1}{Y_1}$ $D = 1$
Y_2, Y_1	$A = 1$ $C = Y_1$	$B = \dfrac{1}{Y_2}$ $D = 1 + \dfrac{Y_1}{Y_2}$
Z_1, Z_2, Z_3	$A = 1 + \dfrac{Z_1}{Z_3}$ $C = \dfrac{1}{Z_3}$	$B = Z_1 + Z_2 + \dfrac{Z_1 Z_2}{Z_3}$ $D = 1 + \dfrac{Z_2}{Z_3}$
Y_3, Y_1, Y_2	$A = 1 + \dfrac{Y_2}{Y_3}$ $C = Y_1 + Y_2 + \dfrac{Y_1 Y_2}{Y_3}$	$B = \dfrac{1}{Y_3}$ $D = 1 + \dfrac{Y_1}{Y_3}$
$N{:}1$	$A = N$ $C = 0$	$B = 0$ $D = \dfrac{1}{N}$

Similarly, by setting $V_2 = 0$ in both equations, one can easily show that

$$B = \frac{Z_{11}Z_{22} - Z_{12}Z_{21}}{Z_{21}}, \qquad D = \frac{Z_{22}}{Z_{21}} \qquad (5.16)$$

Table 5.1 depicts the $[ABCD]$ matrix of several circuit components that are often used in the design of microwave circuits [3].

5.1.3 The Scattering $[S]$ Matrix

The S parameters are important in microwave design because they are easier to measure and work with at high frequencies [4–6]. They are conceptually simple, analytically convenient, and capable of providing a great insight into the transmission and reflection of RF energy in microwave circuits.

The S parameters of the two-port network shown in Figure 5.6 relate the incident and the reflected normalized voltage waves at the two ports as described by equation (5.17). The relationships between the incident and reflected voltages V^+ and V^- and port currents and voltages I and V are explained in details in other publications [1, 4–6].

$$\begin{bmatrix} V_1^- \\ V_2^- \end{bmatrix} = \begin{bmatrix} S_{11} & S_{12} \\ S_{21} & S_{22} \end{bmatrix} \begin{bmatrix} V_1^+ \\ V_2^+ \end{bmatrix} \qquad (5.17)$$

$$V_1^- = S_{11}V_1^+ + S_{12}V_2^+ \qquad (5.18)$$

$$V_2^- = S_{21}V_1^+ + S_{22}V_2^+ \qquad (5.19)$$

Terminating port 2 with a matched load, that is, $V_2^+ = 0$, results in

$$V_1^- = S_{11}V_1^+, \qquad V_2^- = S_{21}V_1^+ \qquad (5.20)$$

It is clear from equation (5.20) that the parameter S_{11} is the reflection coefficient at port 1, when port 2 is terminated with a matched load, and the parameter S_{21} is the transmission coefficient from port 1 to port 2. Similarly, one can show that S_{22} is the reflection coefficient at port 2 when port 1 is terminated by a matched load. The S parameters are characteristics of the network itself and are defined assuming

Figure 5.6   The S matrix of a two-port network.

that all ports are matched. Having a different termination will change the reflection seen at the ports.

In reciprocal networks, $S_{12} = S_{21}$. For lossless networks, the S matrix satisfies a condition known as the *unitary condition*. This condition can be easily derived by equating the average power dissipated inside the network to zero. In lossless networks the energy is conserved. Thus the energy incident to the network is equal to the energy propagating away from it, which can be written as

$$[V^+]_t[V^+]^* = [V^-]_t[V^-]^* \tag{5.21}$$

where $[V^-]$ and $[V^+]$ are column vectors. $[\quad]^*$ denotes complex conjugate, and $[\quad]_t$ denotes transpose:

$$[V^+]_t[V^+]^* = \{[S][V^+]\}_t \cdot \{[S][V^+]\}^* \tag{5.22}$$

$$[V^+]_t[V^+]^* = [V^+]_t[S]_t[S]^*[V^+]^* \tag{5.23}$$

$$0 = [V^+]_t \cdot \{[U] - [S]_t[S]^*\}[V^+]^* \tag{5.24}$$

Equation (5.24) can hold only if the following equation is satisfied:

$$[S]_t[S]^* = [U] \tag{5.25}$$

This equation, where $[U]$ is the unitary matrix, is known as the *unitary condition*. It states that multiplication of S transpose by S complex conjugate equals the unity matrix U. For a two-port network it can be written in the following form:

$$\begin{bmatrix} S_{11} & S_{21} \\ S_{12} & S_{22} \end{bmatrix} \begin{bmatrix} S_{11}^* & S_{12}^* \\ S_{21}^* & S_{22}^* \end{bmatrix} = \begin{bmatrix} 1 & 0 \\ 0 & 1 \end{bmatrix} \tag{5.26}$$

$$|S_{11}|^2 + |S_{21}|^2 = 1 \tag{5.27}$$

$$|S_{12}|^2 + |S_{22}|^2 = 1 \tag{5.28}$$

$$S_{11}S_{12}^* + S_{21}S_{22}^* = 0 \tag{5.29}$$

In defining the S parameters by using equation (5.17) [1], it is assumed that the characteristic impedances of all ports are equal. This is case in the great majority of circuits. However, for cases where the ports have different characteristic impedances. one must use the normalized incident and reflected voltages in defining the scattering matrix. The incident and reflected voltages are normalized as follows:

$$a_1 = \frac{V_1^+}{\sqrt{Z_{01}}}, \qquad a_2 = \frac{V_2^+}{\sqrt{Z_{02}}} \tag{5.30}$$

$$b_1 = \frac{V_1^-}{\sqrt{Z_{01}}}, \qquad b_2 = \frac{V_2^-}{\sqrt{Z_{02}}} \tag{5.31}$$

Thus, in the general case the scattering matrix is defined as [1,4–6]

$$\begin{bmatrix} b_1 \\ b_2 \end{bmatrix} = \begin{bmatrix} S_{11} & S_{12} \\ S_{21} & S_{22} \end{bmatrix} \begin{bmatrix} a_1 \\ a_2 \end{bmatrix} \tag{5.32}$$

To illustrate the importance of the normalization when dealing with a network with different characteristic impedances, consider the problem of calculating the scattering matrix of a shunt admittance jB between two ports in two cases. In case 1, the ports have the same characteristic impedance, whereas in case 2 the two ports have different characteristic impedances.

Case 1.   Refer to Figure 5.7.

When port 2 is terminated in the characteristic admittance Y_0, the load Y_L seen at the input is then $Y_L = jB + Y_0$. The reflection coefficient Γ seen at the input port equals S_{11} and is then given by

$$S_{11} = \Gamma = \frac{Y_0 - Y_L}{Y_0 + Y_L} \tag{5.33}$$

$$S_{11} = \frac{Y_0 - (Y_0 + jB)}{2Y_0 + jB} = \frac{-jB}{2Y_0 + jB} \tag{5.34}$$

Because of the shunt configuration, the voltage V_1 at port 1 equals the voltage V_2 at port 2. The port voltages V_1 and V_2 are related to the incident and reflected voltages as follows:

$$V_1 = V_1^+ + V_1^-, \qquad V_2 = V_2^- \tag{5.35}$$

Note that V_2^+ is zero since port 2 is terminated in a matched load.

$$V_1^+ + V_1^- = V_2^- \tag{5.36}$$
$$[1 + S_{11}]V_1^+ = V_2^- \tag{5.37}$$

$$S_{21} = \frac{V_2^-}{V_1^+} = [1 + S_{11}] = \frac{2Y_0}{2Y_0 + jB} \tag{5.38}$$

Similarly, by assuming that port 1 is terminated in a matched load, one can show that

$$S_{22} = \frac{-jB}{2Y_0 + jB}, \qquad S_{12} = \frac{2Y_0}{2Y_0 + jB} \tag{5.39}$$

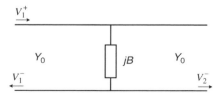

Figure 5.7   A shunt admittance connected between two ports having the same characteristic impedance.

Note that $S_{12} = S_{21}$ since the network is reciprocal and $S_{11} = S_{22}$ since the network is symmetric.

Case 2. Refer to Figure 5.8.

First we assume that port 2 is terminated in a matched load, that is, it is terminated in Y_{02}. S_{11} is then given by

$$S_{11} = \frac{V_1^-}{V_1^+} = \frac{Y_{01} - (jB + Y_{02})}{Y_{01} + (jB + Y_{02})} \tag{5.40}$$

$$S_{11} = \frac{Y_{01} - jB - Y_{02}}{Y_{01} + jB + Y_{02}} \tag{5.41}$$

Following the same approach as in case 5.1, the S parameters are derived by relating the voltages at the two ports, resulting in

$$V_1^+ + V_1^- = V_2^- \Rightarrow [1 + S_{11}] = S_{21} \tag{5.42}$$

$$S_{21} = \frac{V_2^-}{V_1^+} = 1 + \frac{Y_{01} - jB - Y_{02}}{Y_{01} + jB + Y_{02}} = \frac{2Y_{01}}{Y_{01} + jB + Y_{02}} \tag{5.43}$$

Similarly, if we terminate port 1 in a matched load we can follow the same procedure to derive expressions for S_{22} and S_{12} as follows:

$$S_{22} = \frac{V_2^-}{V_2^+} = \frac{Y_{02} - jB - Y_{01}}{Y_{01} + jB + Y_{02}} \tag{5.44a}$$

$$S_{12} = \frac{V_1^-}{V_2^+} = \frac{2Y_{02}}{Y_{01} + jB + Y_{02}} \tag{5.44b}$$

Note that even though the circuit is reciprocal, the use of nonnormalized incident and reflected voltages leads to a solution that violates the reciprocity condition. However, if normalized voltages are used as shown below, a proper solution is obtained. When port 2 is terminated in a matched load, equation (5.43) is modified to reflect the normalization:

$$S_{21} = \frac{V_2^- / \sqrt{Z_{02}}}{V_1^+ / \sqrt{Z_{01}}} = \frac{V_2^-}{V_1^+} \sqrt{\frac{Z_{01}}{Z_{02}}} \tag{5.45}$$

$$S_{21} = \frac{2\sqrt{Y_{01}Y_{02}}}{Y_{01} + jB + Y_{02}} \tag{5.46}$$

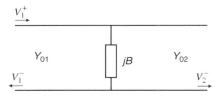

Figure 5.8 A shunt admittance connected between two ports having different characteristic impedances.

Similarly, when port 2 is terminated in a matched load, equation (5.45) is modified to read

$$S_{12} = \frac{V_1^- / \sqrt{Z_{01}}}{V_2^+ / \sqrt{Z_{02}}} = \frac{V_1^-}{V_2^+} \sqrt{\frac{Z_{02}}{Z_{01}}} \tag{5.47}$$

$$S_{12} = \frac{2\sqrt{Y_{01} Y_{02}}}{Y_{01} + jB + Y_{02}} \tag{5.48}$$

It is noted that in this case the use of normalized incident and reflected voltage leads to a solution that satisfies the reciprocity condition. The preceding example highlights the importance of adopting normalized voltages when we deal with networks having ports with different characteristic impedances.

The relationship between the scattering matrix $[S]$ and the impedance matrix $[Z]$ can be derived by relating the voltage and current to incident and reflected voltages:

$$[V^-] = [S][V^+] \tag{5.49}$$

$$[V] = [Z][I] \tag{5.50}$$

By using transmission-line theory [1,3] and assuming unity characteristic impedance for all ports, we get

$$[V] = [V^+] + [V^-] \tag{5.51}$$

$$[I] = ([V^+] - [V^-]) \tag{5.52}$$

$$[V^+] = \frac{1}{2}([V] + [I]) \Rightarrow \quad [V^+] = \frac{1}{2}([Z] + [U])[I] \tag{5.53}$$

$$[V^-] = \frac{1}{2}([V] - [I]) \Rightarrow \quad [V^-] = \frac{1}{2}([Z] - [U])[I] \tag{5.54}$$

$$[V^-] = ([Z] - [U])([Z] + [U])^{-1}[V^+] \tag{5.55}$$

$$[S] = ([Z] - [U])([Z] + [U])^{-1} \tag{5.56}$$

Similarly, the $[S]$ matrix is related to the $[Y]$ matrix as follows:

$$[S] = ([U] - [Y])([U] + [Y])^{-1} \tag{5.57}$$

Conversely, the $[Z]$ and $[Y]$ matrices are written in terms of the $[S]$ matrix as

$$[Z] = ([U] + [S])([U] - [S])^{-1} \tag{5.58}$$

$$[Y] = ([U] - [S])([U] + [S])^{-1} \tag{5.59}$$

These formulas are valid where all ports have the same characteristic impedance values. In the general case, where the ports have different impedances, the following relations are applied [2]:

$$[S] = [\sqrt{Y_0}]([Z] - [Z_0])([Z] + [Z_0])^{-1}[\sqrt{Z_0}] \tag{5.60}$$

$$[S] = [\sqrt{Z_0}]([Y_0] - [Y])([Y_0] + [Y])^{-1}[\sqrt{Y_0}] \tag{5.61}$$

$$[Z] = [\sqrt{Z_0}]([U] + [S])([U] - [S])^{-1}[\sqrt{Z_0}] \tag{5.62}$$

$$[Y] = [\sqrt{Y_0}]([U] - [S])([U] + [S])^{-1}[\sqrt{Y_0}] \tag{5.63}$$

where

$$[Z_0] = \begin{bmatrix} Z_{01} & 0 \\ 0 & Z_{02} \end{bmatrix} \tag{5.64}$$

$$[Y_0] = \begin{bmatrix} Y_{01} & 0 \\ 0 & Y_{02} \end{bmatrix} \tag{5.65}$$

$$[\sqrt{Z_0}] = \begin{bmatrix} \sqrt{Z_{01}} & 0 \\ 0 & \sqrt{Z_{02}} \end{bmatrix} \tag{5.66}$$

$$[\sqrt{Y_0}] = \begin{bmatrix} \sqrt{Y_{01}} & 0 \\ 0 & \sqrt{Y_{02}} \end{bmatrix} \tag{5.67}$$

5.1.4 The Transmission Matrix [T]

Another matrix that relates the incident and reflected voltages at the ports is known as the *transmission matrix* [T] (see Figure 5.9). It is defined by equation (5.68). The [T] matrix is similar to the [ABCD] matrix in the sense that it facilitates calculation of the overall transmission matrix of two cascaded networks by a

Figure 5.9 *T*-matrix representation of a two-port network.

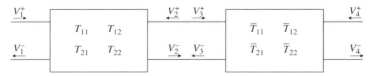

Figure 5.10 Two networks connected in cascade.

simple matrix multiplication. For reciprocal networks the determinant of the [T] matrix equals unity as given in equation (5.69)

$$\begin{bmatrix} V_1^+ \\ V_1^- \end{bmatrix} = \begin{bmatrix} T_{11} & T_{12} \\ T_{21} & T_{22} \end{bmatrix} \begin{bmatrix} V_2^- \\ V_2^+ \end{bmatrix} \tag{5.68}$$

$$T_{11}T_{22} - T_{12}T_{21} = 1 \tag{5.69}$$

Consider the two cascaded networks shown in Figure 5.10, the overall [T] matrix of the cascaded network is given as

$$\begin{bmatrix} V_1^+ \\ V_1^- \end{bmatrix} = \begin{bmatrix} T_{11} & T_{12} \\ T_{21} & T_{22} \end{bmatrix} \begin{bmatrix} V_2^- \\ V_2^+ \end{bmatrix}, \quad \begin{bmatrix} V_3^+ \\ V_3^- \end{bmatrix} = \begin{bmatrix} \overline{T}_{11} & \overline{T}_{12} \\ \overline{T}_{21} & \overline{T}_{22} \end{bmatrix} \begin{bmatrix} V_4^- \\ V_4^+ \end{bmatrix} \tag{5.70}$$

$$\begin{bmatrix} V_1^+ \\ V_1^- \end{bmatrix} = \begin{bmatrix} T_{11} & T_{12} \\ T_{21} & T_{22} \end{bmatrix} \begin{bmatrix} \overline{T}_{11} & \overline{T}_{12} \\ \overline{T}_{21} & \overline{T}_{22} \end{bmatrix} \begin{bmatrix} V_4^- \\ V_4^+ \end{bmatrix} \tag{5.71}$$

The relation between [S] and [T] matrices is obtained by setting V_2^+ and V_2^- in both matrices such that

$$\begin{bmatrix} V_1^- \\ V_2^- \end{bmatrix} = \begin{bmatrix} S_{11} & S_{12} \\ S_{21} & S_{22} \end{bmatrix} \begin{bmatrix} V_1^+ \\ V_2^+ \end{bmatrix} \tag{5.72a}$$

$$\begin{bmatrix} V_1^+ \\ V_1^- \end{bmatrix} = \begin{bmatrix} T_{11} & T_{12} \\ T_{21} & T_{22} \end{bmatrix} \begin{bmatrix} V_2^- \\ V_2^+ \end{bmatrix} \tag{5.72b}$$

Letting $V_2^+ = 0$, from the [S] matrix, we have

$$V_1^- = S_{11}V_1^+, \qquad V_2^- = S_{21}V_1^+ \tag{5.73}$$

From the [T] matrix we get

$$V_1^+ = T_{11}V_2^-, \qquad V_1^- = T_{21}V_2^- \tag{5.74}$$

Using equations (5.73)–(5.74), we get

$$T_{11} = \frac{1}{S_{21}} \tag{5.75a}$$

$$T_{21} = \frac{S_{11}}{S_{21}} \tag{5.75b}$$

Letting $V_2^- = 0$, in equations (5.72a) and (5.72b) we then have

$$V_1^- = S_{11}V_1^+ + S_{12}V_2^+ \tag{5.76}$$
$$0 = S_{21}V_1^+ + S_{22}V_2^+ \tag{5.77}$$
$$V_1^+ = T_{12}V_2^+ \tag{5.78}$$
$$V_1^- = T_{22}V_2^+ \tag{5.79}$$

From equations (5.77) and (5.78), we get

$$T_{12} = -\frac{S_{22}}{S_{21}} \tag{5.80}$$

Substituting (5.80) in (5.78) and then using equation, (5.82) gives

$$V_1^- = -\frac{S_{11}S_{22}}{S_{21}} V_2^+ + S_{12}V_2^+ \tag{5.81}$$

$$T_{22} = \frac{V_1^-}{V_2^+} = S_{12} - \frac{S_{11}S_{22}}{S_{21}} \tag{5.82}$$

Similarly, one can show that

$$S_{11} = \frac{T_{21}}{T_{11}} \tag{5.83}$$

$$S_{12} = T_{22} - \frac{T_{21}T_{12}}{T_{11}} \tag{5.84}$$

$$S_{21} = \frac{1}{T_{11}} \tag{5.85}$$

$$S_{22} = -\frac{T_{12}}{T_{11}} \tag{5.86}$$

Table 5.2 summarizes the relationships between the $[Z]$, $[Y]$, $[ABCD]$, and $[S]$ matrices [3].

5.1.5 Analysis of Two-Port Networks

Consider the two-port network shown in Figure 5.11. If port 2 is terminated by a load, the network is reduced to a one-port network. The input impedance or the reflection coefficient at this port can be calculated from knowledge of the scattering matrix of the two-port network and the reflection coefficient of the terminating load:

$$V_1^- = S_{11}V_1^+ + S_{12}V_2^+ \tag{5.87}$$

$$V_2^- = S_{21}V_1^+ + S_{22}V_2^+ \tag{5.88}$$

The reflection coefficient seen at the output port is given by

$$\Gamma = \frac{Z_L - Z_0}{Z_L + Z_0} = \frac{V_2^+}{V_2^-} \tag{5.89}$$

TABLE 5.2 Relationship Between [S], [Z], [Y], and [ABCD] Matrices

	S	Z	Y	$ABCD$
S_{11}	S_{11}	$\dfrac{(Z_{11}-Z_0)(Z_{22}+Z_0)-Z_{12}Z_{21}}{(Z_{11}+Z_0)(Z_{22}+Z_0)-Z_{12}Z_{21}}$	$-\dfrac{(Y_{11}-Y_0)(Y_{22}+Y_0)-Y_{12}Y_{21}}{(Y_{11}+Y_0)(Y_{22}+Y_0)-Y_{12}Y_{21}}$	$\dfrac{A+B/Z_0-CZ_0-D}{A+B/Z_0+CZ_0+D}$
S_{12}	S_{12}	$\dfrac{2Z_{12}Z_0}{(Z_{11}+Z_0)(Z_{22}+Z_0)-Z_{12}Z_{21}}$	$-\dfrac{2Y_{12}Y_0}{(Y_{11}+Y_0)(Y_{22}+Y_0)-Y_{12}Y_{21}}$	$\dfrac{2(AD-BC)}{A+B/Z_0+CZ_0+D}$
S_{21}	S_{21}	$\dfrac{2Z_{21}Z_0}{(Z_{11}+Z_0)(Z_{22}+Z_0)-Z_{12}Z_{21}}$	$-\dfrac{2Y_{21}Y_0}{(Y_{11}+Y_0)(Y_{22}+Y_0)-Y_{12}Y_{21}}$	$\dfrac{2}{A+B/Z_0+CZ_0+D}$
S_{12}	S_{12}	$\dfrac{(Z_{11}+Z_0)(Z_{22}-Z_0)-Z_{12}Z_{21}}{(Z_{11}+Z_0)(Z_{22}+Z_0)-Z_{12}Z_{21}}$	$-\dfrac{(Y_{11}+Y_0)(Y_{22}-Y_0)-Y_{12}Y_{21}}{(Y_{11}+Y_0)(Y_{22}+Y_0)-Y_{12}Y_{21}}$	$\dfrac{-A+B/Z_0-CZ_0+D}{A+B/Z_0+CZ_0+D}$
Z_{11}	$Z_0\dfrac{(1+S_{11})(1-S_{22})+S_{12}S_{21}}{(1-S_{11})(1-S_{22})-S_{12}S_{21}}$	Z_{11}	$\dfrac{Y_{22}}{Y_{11}Y_{22}-Y_{12}Y_{21}}$	$\dfrac{A}{C}$
Z_{12}	$Z_0\dfrac{2S_{12}}{(1-S_{11})(1-S_{22})-S_{12}S_{21}}$	Z_{12}	$\dfrac{-Y_{12}}{Y_{11}Y_{22}-Y_{12}Y_{21}}$	$\dfrac{AD-BC}{C}$
Z_{21}	$Z_0\dfrac{2S_{21}}{(1-S_{11})(1-S_{22})-S_{12}S_{21}}$	Z_{21}	$\dfrac{-Y_{21}}{Y_{11}Y_{22}-Y_{12}Y_{21}}$	$\dfrac{1}{C}$
Z_{22}	$Z_0\dfrac{(1-S_{11})(1+S_{22})+S_{12}S_{21}}{(1-S_{11})(1-S_{22})-S_{12}S_{21}}$	Z_{22}	$\dfrac{Y_{11}}{Y_{11}Y_{22}-Y_{12}Y_{21}}$	$\dfrac{D}{C}$

	S-parameters	Z-parameters	Y-parameters	ABCD-parameters
Y_{11}	$Y_0 \dfrac{(1-S_{11})(1+S_{22})+S_{12}S_{21}}{(1+S_{11})(1+S_{22})-S_{12}S_{21}}$	$\dfrac{Z_{22}}{Z_{11}Z_{22}-Z_{12}Z_{21}}$	Y_{11}	$\dfrac{D}{B}$
Y_{12}	$Y_0 \dfrac{-2S_{12}}{(1+S_{11})(1+S_{22})-S_{12}S_{21}}$	$\dfrac{-Z_{12}}{Z_{11}Z_{22}-Z_{12}Z_{21}}$	Y_{12}	$\dfrac{BC-AD}{B}$
Y_{21}	$Y_0 \dfrac{-2S_{21}}{(1+S_{11})(1+S_{22})-S_{12}S_{21}}$	$\dfrac{-Z_{21}}{Z_{11}Z_{22}-Z_{12}Z_{21}}$	Y_{21}	$\dfrac{-1}{B}$
Y_{22}	$Y_0 \dfrac{(1+S_{11})(1-S_{22})+S_{12}S_{21}}{(1+S_{11})(1+S_{22})-S_{12}S_{21}}$	$\dfrac{Z_{11}}{Z_{11}Z_{22}-Z_{12}Z_{21}}$	Y_{22}	$\dfrac{A}{B}$
A	$\dfrac{(1+S_{11})(1-S_{22})+S_{12}S_{21}}{2S_{21}}$	$\dfrac{Z_{11}}{Z_{21}}$	$-\dfrac{Y_{22}}{Y_{21}}$	A
B	$Z_0 \dfrac{(1+S_{11})(1+S_{22})-S_{12}S_{21}}{2S_{21}}$	$\dfrac{Z_{11}Z_{22}-Z_{12}Z_{21}}{Z_{21}}$	$-\dfrac{1}{Y_{21}}$	B
C	$\dfrac{1}{Z_0}\dfrac{(1-S_{11})(1-S_{22})-S_{12}S_{21}}{2S_{21}}$	$\dfrac{1}{Z_{21}}$	$-\dfrac{Y_{11}Y_{22}-Y_{12}Y_{21}}{Y_{21}}$	C
D	$\dfrac{(1-S_{11})(1+S_{22})+S_{12}S_{21}}{2S_{21}}$	$\dfrac{Z_{22}}{Z_{21}}$	$-\dfrac{Y_{11}}{Y_{21}}$	D

$T_{11}=1/S_{21}, \; T_{12}=-S_{22}/S_{21}, \; T_{21}=S_{11}/S_{21}, \; T_{22}=(S_{12}S_{21}-S_{11}S_{22})/S_{21}.$

$S_{11}=T_{21}/T_{11}, \; S_{12}=(T_{11}T_{22}-T_{12}T_{21})/T_{11}, \; S_{21}=1/T_{11}, \; S_{22}=-T_{12}/T_{11}.$

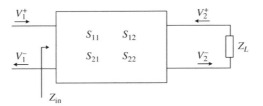

Figure 5.11 A two-port network terminated in load Z_L.

Substituting equation (5.89) in equation (5.88) gives V_2^+ in terms of V_1^+:

$$V_2^+ = \Gamma V_2^- = \Gamma S_{21} V_1^+ + \Gamma S_{22} V_2^+ \tag{5.90}$$

$$(1 - \Gamma S_{22}) V_2^+ = \Gamma S_{21} V_1^+ \tag{5.91}$$

$$V_2^+ = \frac{\Gamma S_{21}}{1 - \Gamma S_{22}} V_1^+ \tag{5.92}$$

$$V_1^- = S_{11} V_1^+ + \frac{S_{12} \Gamma S_{21}}{1 - \Gamma S_{22}} V_1^+ \tag{5.93}$$

The reflection coefficient and input impedance at port 1 are then expressed as

$$\Gamma_{in} = \frac{V_1^-}{V_1^+} \tag{5.94}$$

$$\Gamma_{in} = S_{11} + \frac{S_{12} \Gamma S_{21}}{1 - \Gamma S_{22}} \tag{5.95}$$

$$Z_{in} = \frac{1 + \Gamma_{in}}{1 - \Gamma_{in}} \tag{5.96}$$

The scattering matrix of the two-port network with the added lengths of transmission lines as shown in Figure 5.12 (i.e., extending the reference planes) can be calculated as follows:

$$\begin{bmatrix} V_1^- \\ V_2^- \end{bmatrix} = [S] \begin{bmatrix} V_1^+ \\ V_2^+ \end{bmatrix} \tag{5.97}$$

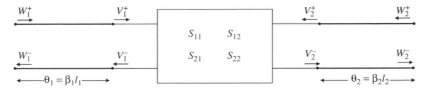

Figure 5.12 A two-port network with extended reference planes.

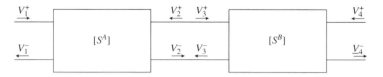

Figure 5.13 Two networks connected in cascade.

We also have

$$\begin{bmatrix} V_1^- \\ V_2^- \end{bmatrix} = \begin{bmatrix} e^{j\theta_1} & 0 \\ 0 & e^{j\theta_2} \end{bmatrix} \begin{bmatrix} W_1^- \\ W_2^- \end{bmatrix} \qquad (5.98)$$

$$\begin{bmatrix} V_1^+ \\ V_2^+ \end{bmatrix} = \begin{bmatrix} e^{-j\theta_1} & 0 \\ 0 & e^{-j\theta_2} \end{bmatrix} \begin{bmatrix} W_1^+ \\ W_2^+ \end{bmatrix} \qquad (5.99)$$

Substituting equations (5.98) and (5.99) in equation (5.97) gives

$$\begin{bmatrix} W_1^- \\ W_2^- \end{bmatrix} = \begin{bmatrix} e^{-j\theta_1} & 0 \\ 0 & e^{-j\theta_2} \end{bmatrix} [S] \begin{bmatrix} e^{-j\theta_1} & 0 \\ 0 & e^{-j\theta_2} \end{bmatrix} \begin{bmatrix} W_1^+ \\ W_2^+ \end{bmatrix} \qquad (5.100)$$

The resultant $[S]$ matrix of the network shown in Figure 5.13 is then given by

$$\text{Resultant Matrix} = \begin{bmatrix} e^{-j\theta_1} & 0 \\ 0 & e^{-j\theta_2} \end{bmatrix} [S] \begin{bmatrix} e^{-j\theta_1} & 0 \\ 0 & e^{-j\theta_2} \end{bmatrix} \qquad (5.101)$$

5.2 CASCADE OF TWO NETWORKS

Different methods are used to deal with cascaded networks. Let us consider first the case when using the S parameters directly. The overall scattering matrix of the cascaded networks $[S^{\text{cascade}}]$ can be obtained from the scattering matrices of the two networks, $[S^A]$ and $[S^B]$ as follows:

$$V_1^- = S_{11}^A V_1^+ + S_{12}^A V_2^+ \qquad (5.102)$$

$$V_2^- = S_{21}^A V_1^+ + S_{22}^A V_2^+ \qquad (5.103)$$

$$V_3^- = S_{11}^B V_3^+ + S_{12}^B V_4^+ \qquad (5.104)$$

$$V_4^- = S_{21}^B V_3^+ + S_{22}^B V_4^+ \qquad (5.105)$$

In view of Figure 5.13, we also have

$$V_2^+ = V_3^- \qquad (5.106)$$

$$V_2^- = V_3^+ \qquad (5.107)$$

Solving the six equations (5.105)–(5.110) provides V_1^- and V_4^- in terms of V_1^+ and V_4^+ as

$$\begin{bmatrix} V_1^- \\ V_4^- \end{bmatrix} = \left[S^{\text{cascade}} \right] \begin{bmatrix} V_1^+ \\ V_4^+ \end{bmatrix} \tag{5.108}$$

where

$$S_{11}^{\text{cascade}} = S_{11}^A + \frac{S_{12}^A S_{11}^B S_{21}^A}{1 - S_{22}^A S_{11}^B} \tag{5.109}$$

$$S_{12}^{\text{cascade}} = \frac{S_{12}^A S_{12}^B}{1 - S_{22}^A S_{11}^B} \tag{5.110}$$

$$S_{21}^{\text{cascade}} = \frac{S_{21}^A S_{21}^B}{1 - S_{22}^A S_{11}^B} \tag{5.111}$$

$$S_{22}^{\text{cascade}} = S_{22}^B + \frac{S_{21}^B S_{22}^A S_{12}^B}{1 - S_{22}^A S_{11}^B} \tag{5.112}$$

In analyzing symmetric networks, the problem can be simplified by making use of symmetry. Consider the symmetric network shown in Figure 5.14. The network can be divided into two identical halves. The problem reduces to the analysis of one half of the circuit with an electric wall termination (i.e., a short circuit) and magnetic wall termination (i.e., an open circuit).

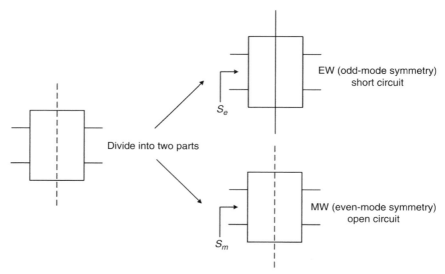

Figure 5.14 A two-port symmetric network divided into two identical halves (EW, MW = electric, magnetic walls).

By evaluating the reflection coefficients S_e and S_m of the reduced circuits, the scattering matrix of the overall circuit can be written as

$$S_{11} = \frac{1}{2}[S_m + S_e] = S_{22} \tag{5.113}$$

$$S_{12} = \frac{1}{2}[S_m - S_e] = S_{21} \tag{5.114}$$

Equations (5.113) and (5.114) are derived by considering Figure 5.15. The network is decomposed into the superposition of two networks, one with odd-mode excitation, $V_1^+ = V/2$, $V_2^+ = -V/2$, and another with even-mode excitation, $V_1^+ = V/2$, $V_2^+ = V/2$. These two excitations will provide the electric wall and magnetic wall, respectively, at the plane of symmetry. Note also that the superposition of the two networks produces the original excitation, $V_1^+ = V$, $V_2^+ = 0$. Addition of the reflected signals and the transmitted signals from both excitations provides the reflected and transmitted signals from which S_{11} and S_{21} can be calculated.

$$V_R = \frac{V}{2}S_e + \frac{V}{2}S_m = \frac{V}{2}(S_e + S_m) \tag{5.115}$$

$$V_T = \frac{V}{2}(S_m - S_e) \tag{5.116}$$

$$S_{11} = \frac{V_R}{V} = \frac{1}{2}[S_e + S_m] \tag{5.117}$$

$$S_{21} = \frac{V_T}{V} = \frac{1}{2}[S_m - S_e] \tag{5.118}$$

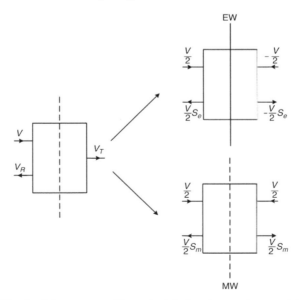

Figure 5.15 Applying superposition principle to a two-port symmetric network.

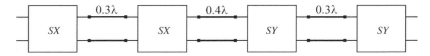

Figure 5.16 Four two-port networks connected in cascade.

Example 5.1 Consider the cascaded four networks shown in Figure 5.16, calculate the overall scattering matrix of the network assuming that the scattering matrices of networks *SX* and *SY* are given by

$$SX = \begin{pmatrix} \frac{1}{3} + j\frac{2}{3} & j\frac{2}{3} \\ j\frac{2}{3} & \frac{1}{3} - j\frac{2}{3} \end{pmatrix}, \qquad SY = \begin{pmatrix} \frac{1}{3} - j\frac{2}{3} & j\frac{2}{3} \\ j\frac{2}{3} & \frac{1}{3} + j\frac{2}{3} \end{pmatrix}$$

Solution The *S* matrix of a lossless interconnecting transmission line (*TL*) is written as follows:

$$S = \begin{pmatrix} 0 & e^{-j\theta} \\ e^{-j\theta} & 0 \end{pmatrix} = \begin{pmatrix} 0 & e^{-j\beta\ell} \\ e^{-j\beta\ell} & 0 \end{pmatrix}$$

For the 0.3λ line, $\theta = (2\pi/\lambda) \cdot (0.3\lambda) = 0.6\pi$, while for the 0.4λ line

$$\theta = \frac{2\pi}{\lambda} \cdot (0.4\lambda) = 0.8\pi$$

$$S_{0.3\lambda} = \begin{pmatrix} 0 & e^{-j(0.6\pi)} \\ e^{-j(0.6\pi)} & 0 \end{pmatrix}, \quad S_{0.4\lambda} = \begin{pmatrix} 0 & e^{-j(0.8\pi)} \\ e^{-j(0.8\pi)} & 0 \end{pmatrix}$$

The overall network can be analyzed by a variety of techniques; we will solve this example using four different methods.

Method 1. Using the [*ABCD*] matrix approach, the solution can be summarized as follows:

1. Convert each *S* matrix into its *ABCD* equivalent matrix.
2. Multiply all *ABCD* matrices to obtain the overall *ABCD* matrix.
3. Convert the *ABCD* matrix into an *S* matrix.

Step 1. Convert each *S* matrix in to its *ABCD* equivalent matrix.
In view of Table 5.2, the conversion can be as follows:

$$\begin{pmatrix} A & B \\ C & D \end{pmatrix} = \begin{pmatrix} \dfrac{(1 + S_{11})(1 - S_{22}) + S_{12}S_{21}}{2S_{21}} & \dfrac{(1 + S_{11})(1 + S_{22}) - S_{12}S_{21}}{2S_{21}} \\ \dfrac{(1 - S_{11})(1 - S_{22}) - S_{12}S_{21}}{2S_{21}} & \dfrac{(1 - S_{11})(1 + S_{22}) + S_{12}S_{21}}{2S_{21}} \end{pmatrix}$$

Thus

$$\begin{pmatrix} A & B \\ C & D \end{pmatrix}_X = \begin{pmatrix} 1 & -j2 \\ -j & -1 \end{pmatrix} \qquad \begin{pmatrix} A & B \\ C & D \end{pmatrix}_Y = \begin{pmatrix} -1 & -j2 \\ -j & 1 \end{pmatrix}$$

and

$$\begin{pmatrix} A & B \\ C & D \end{pmatrix}_{0.3\lambda} = \begin{pmatrix} -0.309 & j0.951 \\ j0.951 & -0.309 \end{pmatrix}$$

$$\begin{pmatrix} A & B \\ C & D \end{pmatrix}_{0.4\lambda} = \begin{pmatrix} -0.809 & j0.588 \\ j0.588 & -0.809 \end{pmatrix}$$

Before we proceed, we should stop here and check for the *ABCD* matrices, and apply the simple checks for reciprocity ($AD - BC = 1$). This can be a good guide for the accuracy of the calculations.

Step 2. Multiply all *ABCD* matrices to get the overall *ABCD* matrix:

$$\begin{pmatrix} A & B \\ C & D \end{pmatrix}_{\text{overall}} = \begin{pmatrix} A & B \\ C & D \end{pmatrix}_X \times \begin{pmatrix} A & B \\ C & D \end{pmatrix}_{0.3\lambda} \times \begin{pmatrix} A & B \\ C & D \end{pmatrix}_X \times \begin{pmatrix} A & B \\ C & D \end{pmatrix}_{0.4\lambda}$$

$$\times \begin{pmatrix} A & B \\ C & D \end{pmatrix}_Y \times \begin{pmatrix} A & B \\ C & D \end{pmatrix}_{0.3\lambda} \times \begin{pmatrix} A & B \\ C & D \end{pmatrix}_Y$$

$$= \begin{pmatrix} 10.248 & j16.911 \\ -j6.151 & 10.248 \end{pmatrix}$$

Again, before proceeding further, check for reciprocity.

Step 3. Using Table 5.2, the *ABCD* matrix is converted into an *S* matrix as follows:

$$\begin{pmatrix} S_{11} & S_{12} \\ S_{21} & S_{22} \end{pmatrix} = \begin{pmatrix} \dfrac{A+B-C-D}{A+B+C+D} & \dfrac{2(AD-BC)}{A+B+C+D} \\ \dfrac{2}{A+B+C+D} & \dfrac{-A+B-C+D}{A+B+C+D} \end{pmatrix}$$

Again note that the relation above holds when the characteristic impedance of both the input and the output lines are equal. Hence

$$(S)_{\text{overall}} = \begin{pmatrix} 0.996e^{j1.087} & 0.086e^{-j0.483} \\ 0.086e^{-j0.483} & 0.996e^{j1.087} \end{pmatrix}$$

The resultant *S* matrix exhibits the following:

a. $S_{12} = S_{21}$, or the matrix is in the form of $[S] = [S]^T$, which means that the circuit is reciprocal. This should be expected since each of the blocks forming the network was reciprocal.

b. $[S]_t[S]^* = [U]$ (the unitary condition). This is expected since each of the network blocks is lossless.

Method 2. Converting each S matrix to its T equivalent matrix. This can be of much interest if the modeling technique leads directly to the transmission matrix [7]. The solution can be summarized as follows:

1. Convert each S matrix into its T equivalent matrix.
2. Multiply all T matrices to get the overall T matrix.
3. Convert the T matrix into S matrix.

Step 1. Convert each S matrix into its T equivalent matrix.

Generally for normalized characteristic impedance at both the input and the output ports, the conversion can proceed as follows:

$$\begin{pmatrix} T_{11} & T_{12} \\ T_{21} & T_{22} \end{pmatrix} = \begin{pmatrix} \dfrac{1}{S_{21}} & \dfrac{-S_{22}}{S_{21}} \\ \dfrac{S_{11}}{S_{21}} & \dfrac{S_{12}S_{21} - S_{11}S_{22}}{S_{21}} \end{pmatrix}$$

Thus

$$\begin{pmatrix} T_{11} & T_{12} \\ T_{21} & T_{22} \end{pmatrix}_X = \begin{pmatrix} -j1.5 & 1+j0.5 \\ 1-j0.5 & j1.5 \end{pmatrix}$$

$$\begin{pmatrix} T_{11} & T_{12} \\ T_{21} & T_{22} \end{pmatrix}_Y = \begin{pmatrix} -j1.5 & -1+j0.5 \\ -1-j0.5 & j1.5 \end{pmatrix}$$

$$\begin{pmatrix} T_{11} & T_{12} \\ T_{21} & T_{22} \end{pmatrix}_{0.3\lambda} = \begin{pmatrix} e^{j0.6\pi} & 0 \\ 0 & e^{-j0.6\pi} \end{pmatrix}$$

$$\begin{pmatrix} T_{11} & T_{12} \\ T_{21} & T_{22} \end{pmatrix}_{0.4\lambda} = \begin{pmatrix} e^{j0.8\pi} & 0 \\ 0 & e^{-j0.8\pi} \end{pmatrix}$$

Before we proceed, it is worthwhile to check the resultant T matrices for reciprocity (determinant of the matrix $T = 1$). This can be a good guide for the accuracy of our calculations.

Step 2. Multiply all T matrices to get the overall T matrix

$$(T)_{\text{overall}} = (T)_X \times (T)_{0.3\lambda} \times (T)_X \times (T)_{0.4\lambda} \times (T)_Y \times (T)_{0.3\lambda} \times (T)_Y$$

$$= \begin{pmatrix} 11.577e^{j0.483499} & -j11.534 \\ j11.534 & 11.577e^{-j0.483499} \end{pmatrix}$$

Note that the determinant of this matrix is one.

Step 3. Convert the T matrix into an S matrix. This can be done using

$$
\begin{pmatrix} S_{11} & S_{12} \\ S_{21} & S_{22} \end{pmatrix} = \begin{pmatrix} \dfrac{T_{21}}{T_{11}} & \dfrac{T_{11}T_{22} - T_{12}T_{21}}{T_{11}} \\ \dfrac{1}{T_{11}} & \dfrac{-T_{12}}{T_{11}} \end{pmatrix}
$$

$$
(S)_{\text{overall}} = \begin{pmatrix} 0.996e^{j1.087} & 0.086e^{-j0.483} \\ 0.086e^{-j0.483} & 0.996e^{j1.087} \end{pmatrix}
$$

It is clear that solutions of methods 1 and 2 are identical.

Method 3. Cascading directly using the scattering matrices as given in Section 5.3. From equations (5.112) and (5.115) we have

$$
S_{11}^{\text{cascade}} = S_{11}^{A} + \frac{S_{12}^{A}S_{11}^{B}S_{21}^{A}}{1 - S_{22}^{A}S_{11}^{B}}
$$

$$
S_{21}^{\text{cascade}} = \frac{S_{12}^{A}S_{12}^{B}}{1 - S_{22}^{A}S_{11}^{B}}
$$

$$
S_{21}^{\text{cascade}} = \frac{S_{21}^{A}S_{21}^{B}}{1 - S_{22}^{A}S_{11}^{B}}
$$

$$
S_{22}^{\text{cascade}} = S_{22}^{B} + \frac{S_{21}^{B}S_{22}^{A}S_{12}^{B}}{1 - S_{22}^{A}S_{11}^{B}}
$$

Applying this concept successively to the network given in Figure 5.16, starting first with the scattering matrix SY to represent output network B, and the scattering matrix of the 0.3λ section representing the input network A as shown in Figure 5.17, we get

$$
S^{\text{cas1}} = \begin{pmatrix} 0.1222 + j0.7353 & 0.6340 - j0.2060 \\ 0.6340 - j0.2060 & 0.3333 + j0.6667 \end{pmatrix}
$$

Now S^{cas1} represents output network B while network SY represents input network A:

$$
S^{\text{cas2}} = \begin{pmatrix} 0.3460 - j0.8893 & 0.0277 + j0.2979 \\ 0.0277 + j0.2979 & 0.5038 + j0.8104 \end{pmatrix}
$$

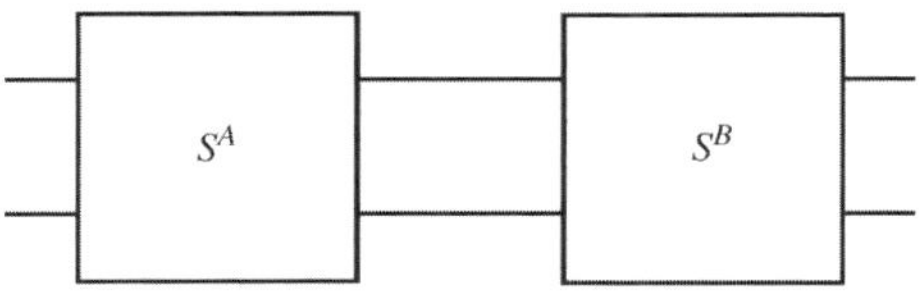

Figure 5.17 Two-port networks A and B connected in series.

S^{cas2} represents the output network B while the scattering matrix of the 0.4λ line represents network A:

$$S^{cas3} = \begin{pmatrix} 0.9527 + j0.0543 & 0.1527 - j0.2572 \\ 0.1527 - j0.2572 & 0.5038 + j0.8104 \end{pmatrix}$$

With S^{cas3} representing output network B and scattering matrix SX representing network A we get.

$$S^{cas4} = \begin{pmatrix} -0.0281 + j0.9744 & 0.2175 - j0.0501 \\ 0.2175 - j0.0501 & 0.4517 + j0.8638 \end{pmatrix}$$

Similarly, we get S^{cas5} as

$$S^{cas5} = \begin{pmatrix} -0.5500 - j0.8048 & -0.1149 - j0.1914 \\ -0.1149 - j0.1914 & 0.4517 + j0.8638 \end{pmatrix}$$

As a final step, with S^{cas5} representing output network B and scattering matrix SX representing network A, we get the overall S matrix as follows:

$$(S)_{\text{Overall}} = \begin{pmatrix} 0.4631 + j0.8821 & 0.0765 - j0.0402 \\ 0.0765 - j0.0402 & 0.4631 + j0.8821 \end{pmatrix}$$
$$= \begin{pmatrix} 0.996e^{j1.087} & 0.086e^{-j0.483} \\ 0.086e^{-j0.483} & 0.996e^{j1.087} \end{pmatrix}$$

The solution is identical to that given by methods 1 and 2. Method 3 can be easily programmed to deal with the cascade of a large number of two-port networks.

Method 4. In this method, we make use of network symmetry. In view of the scattering matrices SX and SY, it is clear that the network is symmetric, that is, it has a mirror-image symmetry at a vertical axis through the middle of the 0.4λ section (most Chebyshev filters are symmetric structures). The problem can be simplified by analyzing only half of the network terminated in magnetic and electric walls. It is worthy to note that the use of symmetry can considerably reduce the computation time. One needs first to calculate the scattering matrix of half of the original network, which is a cascade of the network SX, a 0.3λ transmission line section, network SX, and a 0.2λ transmission line section as shown in Figures 5.18 and 5.19. The S matrix of the reduced network S^{common} can be calculated using any of the above three methods and is given by

$$S^{\text{common}} = \begin{pmatrix} 0.5038 + j0.8104 & 0.2918 + j0.0657 \\ 0.2918 + j0.0657 & -0.8026 + j0.5160 \end{pmatrix}$$

The system is then reduced to the calculation of the reflection coefficient of the reduced two networks shown in Figures 5.18 and 5.19.

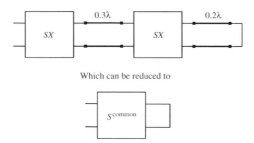

Figure 5.18 Half-network terminated by electric wall.

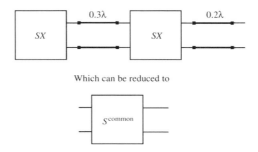

Figure 5.19 Half-network terminated by electric wall.

Electric Wall Case With the output port of S^{common} terminated in the reflection coefficient Γ_L, in view of equation (5.98) the input reflection coefficient Γ_{in} can be written as follows:

$$\Gamma_{\text{in}} = S_{11}^{\text{common}} + \frac{S_{12}^{\text{common}} S_{21}^{\text{common}} \Gamma_L}{1 - S_{22}^{\text{common}} \Gamma_L}$$

For short circuit $\Gamma_L = -1$, then $S_e = \Gamma_{\text{in}} = 0.3867 + j0.9222$.

Magnetic Wall Case For open circuit $\Gamma_L = 1$, then $S_m = \Gamma_{\text{in}} = 0.5396 + j0.8419$.

Hence, for the overall S matrix, we obtain

$$S_{11} = S_{22} = \frac{1}{2}(S_m + S_e) = 0.4631 + j0.8821$$

$$S_{12} = S_{21} = \frac{1}{2}(S_m - S_e) = 0.0765 - j0.0402$$

$$(S)_{\text{overall}} = \begin{pmatrix} 0.4631 + j0.8821 & 0.0765 - j0.0402 \\ 0.0765 - j0.0402 & 0.4631 + j0.8821 \end{pmatrix}$$

$$= \begin{pmatrix} 0.996e^{j1.087} & 0.086e^{-j0.483} \\ 0.086e^{-j0.483} & 0.996e^{j1.087} \end{pmatrix}$$

5.3 MULTIPORT NETWORKS

Consider the multiport microwave network shown in Figure 5.20. It consists of N ports having incident voltages $V_1^+, V_2^+, V_3^+, \ldots, V_N^+$, and reflected voltages $V_1^-, V_2^-, V_3^-, \ldots, V_N^-$. The scattering matrix is computed by

$$
\begin{bmatrix} V_1^- \\ V_2^- \\ \cdot \\ \cdot \\ V_N^- \end{bmatrix} = \begin{bmatrix} S_{11} & S_{12} & \cdot & \cdot & S_{1N} \\ S_{21} & S_{22} & \cdot & \cdot & S_{2N} \\ \cdot & \cdot & \cdot & \cdot & \cdot \\ \cdot & \cdot & \cdot & \cdot & \cdot \\ \cdot & \cdot & \cdot & \cdot & \cdot \\ S_{N1} & \cdot & \cdot & \cdot & S_{NN} \end{bmatrix} \cdot \begin{bmatrix} V_1^+ \\ V_2^+ \\ \cdot \\ \cdot \\ V_N^+ \end{bmatrix}
\tag{5.119}
$$

The reciprocity and lossless unitary conditions can be extended to the N-port network. The reciprocity condition is $S_{ij} = S_{ji}$. The unitary condition is $[S]^t[S]^* = [U]$ and can be written as follows:

$$
\begin{bmatrix} S_{11} & S_{21} & \cdot & \cdot & S_{N1} \\ S_{12} & S_{22} & \cdot & \cdot & S_{N2} \\ \cdot & \cdot & \cdot & \cdot & \cdot \\ \cdot & \cdot & \cdot & \cdot & \cdot \\ \cdot & \cdot & \cdot & \cdot & \cdot \\ S_{1N} & \cdot & \cdot & \cdot & S_{NN} \end{bmatrix} \cdot \begin{bmatrix} S_{11}^* & S_{12}^* & \cdot & \cdot & S_{1N}^* \\ S_{21}^* & S_{22}^* & \cdot & \cdot & S_{2N}^* \\ \cdot & \cdot & \cdot & \cdot & \cdot \\ \cdot & \cdot & \cdot & \cdot & \cdot \\ \cdot & \cdot & \cdot & \cdot & \cdot \\ S_{N1}^* & \cdot & \cdot & \cdot & S_{NN}^* \end{bmatrix}
$$
$$
= \begin{bmatrix} 1 & 0 & \cdot & \cdot & \cdot & 0 \\ 0 & 1 & \cdot & \cdot & \cdot & 0 \\ \cdot & \cdot & \cdot & \cdot & \cdot & \cdot \\ \cdot & \cdot & \cdot & \cdot & \cdot & \cdot \\ \cdot & \cdot & \cdot & \cdot & \cdot & \cdot \\ 0 & \cdot & \cdot & \cdot & \cdot & 1 \end{bmatrix}
\tag{5.120}
$$

If the terminal planes are shifted by adding lengths outward by $l_1, l_2, \ldots, l_N$, then the resultant scattering matrix $[S^T]$ is given by

$$
\begin{bmatrix} e^{-j\beta_1 l_1} & 0 & \cdot & \cdot & \cdot & 0 \\ 0 & e^{-j\beta_2 l_2} & \cdot & \cdot & \cdot & 0 \\ \cdot & \cdot & \cdot & \cdot & \cdot & \cdot \\ \cdot & \cdot & \cdot & \cdot & \cdot & \cdot \\ \cdot & \cdot & \cdot & \cdot & \cdot & \cdot \\ 0 & 0 & \cdot & \cdot & \cdot & e^{-j\beta_N l_N} \end{bmatrix} [S] \begin{bmatrix} e^{-j\beta_1 l_1} & 0 & \cdot & \cdot & \cdot & 0 \\ 0 & e^{-j\beta_2 l_2} & \cdot & \cdot & \cdot & 0 \\ \cdot & \cdot & \cdot & \cdot & \cdot & \cdot \\ \cdot & \cdot & \cdot & \cdot & \cdot & \cdot \\ \cdot & \cdot & \cdot & \cdot & \cdot & \cdot \\ 0 & 0 & \cdot & \cdot & \cdot & e^{-j\beta_N l_N} \end{bmatrix}
\tag{5.121}
$$

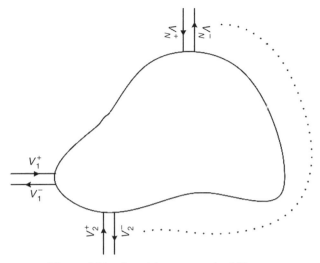

Figure 5.20   A multiport network of N ports.

In general, an N-port network, with M-port terminated by loads, reduces to a network of $(N - M)$ ports, and the resultant network matrix has a size of $(N - M) \times (N - M)$.

The same symmetry concept, applied to the two-port network, can be also extended to multiport networks. Consider, for example, the four-port network shown in Figure 5.21a. An analysis of this network reduces to analysis of two networks, each having two ports with electric wall (EW) and magnetic wall

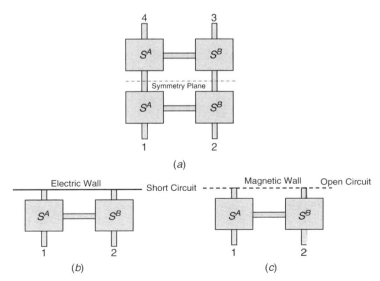

Figure 5.21   A four-port network (a) decomposed into two 2-port networks (b) with electric wall and (c) magnetic wall termination.

(MW) terminations as shown in Figure 5.21*b*.

$$\begin{bmatrix} V_1^- \\ V_2^- \\ V_3^- \\ V_4^- \end{bmatrix} = \begin{bmatrix} [S_{11}^h] & [S_{12}^h] \\ [S_{21}^h] & [S_{22}^h] \end{bmatrix} \begin{bmatrix} V_1^+ \\ V_2^+ \\ V_3^+ \\ V_4^+ \end{bmatrix} \tag{5.122}$$

$$[S_{11}^h] = [S_{22}^h] = \frac{1}{2}[[S_m] + [S_e]] \tag{5.123}$$

$$[S_{12}^h] = [S_{21}^h] = \frac{1}{2}[[S_m] - [S_e]] \tag{5.124}$$

where $[S_m]$ and $[S_e]$ are (2×2) scattering matrices of the reduced two-port network with magnetic wall and electric wall terminations, respectively.

5.4 ANALYSIS OF MULTIPORT NETWORKS

Consider the generalized network shown in Figure 5.22. It consists of several multiport networks connected together. The overall circuit has P external ports and C internal ports. Some of the internal ports are terminated by loads. We will follow the analysis given in [2] to determine the scattering matrix of the resultant network, specifically to determine the scattering matrix that describes the relationships between external ports 1, 2, 3, . . . , P. The main steps are presented below:

Step 1. Identify the external ports P and the internal ports C. The internal ports include those ports that are terminated in loads and ports that are connected to each other. The external ports are the ports that define the scattering matrix of the resultant network.

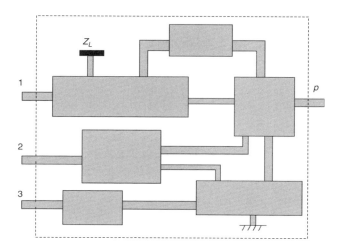

Figure 5.22 An arbitrary microwave network consisting of several multiport networks connected together.

Step 2. Rewrite the scattering matrices of the networks in the form of a large matrix of size $(P + C) \times (P + C)$. The incident and reflected voltages are arranged into two groups: one for the external P ports (V_P^+, V_P^-) and the other group for the internal C ports (V_C^+, V_C^-). The overall matrix then be viewed as consisting of four sub-matrices: S_{PP} of size $(P \times P)$, S_{PC} of size $(P \times C)$, S_{CP} of size $(C \times P)$, and S_{CC} of size $(C \times C)$.

Step 3. Find the connection matrix Γ. This matrix relates the internal ports to each other. That is, it relates V_C^- to V_C^+ by the equation $V_C^- = \Gamma V_C^+$.

To calculate the scattering matrix of the overall network $[S^P]$, we need to relate V_P^- to V_P^+. By dividing the ports into P ports and C ports one can write the overall network scattering matrix as follows:

$$\begin{bmatrix} V_P^- \\ V_C^- \end{bmatrix} = \begin{bmatrix} S_{PP} & S_{PC} \\ S_{CP} & S_{CC} \end{bmatrix} \begin{bmatrix} V_P^+ \\ V_C^+ \end{bmatrix} \tag{5.125}$$

The vectors V_C^-, V_C^+ are related to each other by the connectivity matrix Γ as

$$V_C^- = \Gamma V_C^+ \tag{5.126}$$

Substituting equation (5.129) into equation (5.128) gives

$$\Gamma V_C^+ = S_{CP} V_P^+ + S_{CC} V_C^+ \tag{5.127}$$

$$[\Gamma - S_{CC}] V_C^+ = S_{CP} V_P^+ \tag{5.128}$$

$$V_C^+ = [\Gamma - S_{CC}]^{-1} S_{CP} V_P^+ \tag{5.129}$$

$$V_P^- = [S_{PP} V_P^+ + S_{PC} [\Gamma - S_{CC}]^{-1} S_{CP}] V_P^+ \tag{5.130}$$

The resultant S matrix $[S^P]$ that relates the external P ports can then be written as

$$[S^P] = [S_{PP} + S_{PC} [\Gamma - S_{CC}]^{-1} S_{CP}] \tag{5.131}$$

To illustrate this method, we consider the following two examples.

Example 5.2   Find the scattering matrix of the network shown in Figure 5.23. Assume that the scattering matrix of the four-port network is given by

$$\begin{bmatrix} V_1^- \\ V_2^- \\ V_3^- \\ V_4^- \end{bmatrix} = \begin{bmatrix} S_{11} & S_{12} & S_{13} & S_{14} \\ S_{21} & S_{22} & S_{23} & S_{24} \\ S_{31} & S_{32} & S_{33} & S_{34} \\ S_{41} & S_{42} & S_{43} & S_{44} \end{bmatrix} \begin{bmatrix} V_1^+ \\ V_2^+ \\ V_3^+ \\ V_4^+ \end{bmatrix}$$

In this problem, ports 1 and 2 are the external ports and are denoted by P, whereas ports 3 and 4 are internal ports and are denoted by C. The scattering matrix of the four-port network is then rewritten as

$$\begin{bmatrix} V_P^- \\ V_C^- \end{bmatrix} = \begin{bmatrix} S_{PP} & S_{PC} \\ S_{CP} & S_{CC} \end{bmatrix} \begin{bmatrix} V_P^+ \\ V_C^+ \end{bmatrix}$$

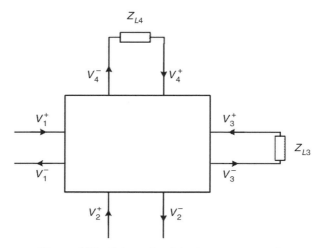

Figure 5.23   Schematic of the four-port network.

where

$$S_{PP} = \begin{bmatrix} S_{11} & S_{12} \\ S_{21} & S_{22} \end{bmatrix}, \quad S_{PC} = \begin{bmatrix} S_{13} & S_{14} \\ S_{23} & S_{24} \end{bmatrix},$$

$$S_{CP} = \begin{bmatrix} S_{31} & S_{32} \\ S_{42} & S_{42} \end{bmatrix}, \quad S_{CC} = \begin{bmatrix} S_{33} & S_{34} \\ S_{43} & S_{44} \end{bmatrix}$$

$$V_P^- = S_{PP} V_P^+ + S_{PC} V_C^+ \tag{5.132}$$

$$V_C^- = S_{CP} V_P^+ + S_{CC} V_C^+ \tag{5.133}$$

In view of Figure 5.23 we also have

$$V_3^+ = \Gamma_3 V_3^-, \qquad \Gamma_3 = \frac{Z_{L3} - Z_{03}}{Z_{L3} + Z_{03}}$$

$$V_4^+ = \Gamma_4 V_4^-, \qquad \Gamma_4 = \frac{Z_{L4} - Z_{04}}{Z_{L4} + Z_{04}}$$

$$[V_C^-] = \begin{bmatrix} \dfrac{1}{\Gamma_3} & 0 \\ 0 & \dfrac{1}{\Gamma_4} \end{bmatrix} [V_C^+]$$

$$V_C^- = \Gamma V_C^+$$

Substituting equation (5.137) into equation (5.133) yields:

$$\Gamma_i V_C^+ = S_{CP} V_P^+ + S_{CC} V_C^+ \Rightarrow V_C^+ = [\Gamma - S_{CC}]^{-1} S_{CP} V_P^+ \tag{5.138}$$

By substituting equation (5.138) into equation (5.132), the resultant scattering matrix between the external ports 1 and 2 is given as

$$V_P^- = S_{PP} V_P^+ + S_{PC} [\Gamma - S_{CC}]^{-1} S_{CP} V_P^+ \tag{5.139}$$

$$\begin{bmatrix} V_1^- \\ V_2^- \end{bmatrix} = [S_{PP} + S_{PC}[\Gamma - S_{CC}]^{-1}S_{CP}]\begin{bmatrix} V_1^+ \\ V_2^+ \end{bmatrix} \tag{5.140}$$

Example 5.3 Evaluate the S parameters of the three-channel multiplexer shown in Figure 5.24.

The multiplexer consists of a combining junction (manifold) whose S matrix is $[S^M]$ and three channel filters whose S matrices are $[S^A]$, $[S^B]$, and $[S^C]$. The objective is to evaluate the reflection coefficient at the manifold input, reflection coefficients at the channel filter outputs, and the transmission coefficients between the input of the manifold and the channel filters; thus, we need to drive the scattering matrix among ports 1, 8, 9, and 10. These ports are the external ports and will be denoted by P. The internal ports, in this problem are 2, 3, 4, 5, 6, and 7 and are denoted as ports C. Assume that $[S^M]$, $[S^A]$, $[S^B]$, and $[S^C]$ are given by

$$\begin{pmatrix} V_1^- \\ V_2^- \\ V_3^- \\ V_4^- \end{pmatrix} = \begin{pmatrix} S_{11}^M & S_{12}^M & S_{13}^M & S_{14}^M \\ S_{21}^M & S_{22}^M & S_{23}^M & S_{24}^M \\ S_{31}^M & S_{32}^M & S_{33}^M & S_{34}^M \\ S_{41}^M & S_{42}^M & S_{43}^M & S_{44}^M \end{pmatrix} \begin{pmatrix} V_1^+ \\ V_2^+ \\ V_3^+ \\ V_4^+ \end{pmatrix} \tag{5.141}$$

$$\begin{pmatrix} V_5^- \\ V_8^- \end{pmatrix} = \begin{pmatrix} S_{11}^A & S_{12}^A \\ S_{21}^A & S_{22}^A \end{pmatrix} \begin{pmatrix} V_5^+ \\ V_8^+ \end{pmatrix}, \quad \begin{pmatrix} V_6^- \\ V_9^- \end{pmatrix} = \begin{pmatrix} S_{11}^B & S_{12}^B \\ S_{21}^B & S_{22}^B \end{pmatrix} \begin{pmatrix} V_6^+ \\ V_9^+ \end{pmatrix} \tag{5.142}$$

$$\begin{pmatrix} V_7^- \\ V_{10}^- \end{pmatrix} = \begin{pmatrix} S_{11}^C & S_{12}^C \\ S_{21}^C & S_{22}^C \end{pmatrix} \begin{pmatrix} V_7^+ \\ V_{10}^+ \end{pmatrix} \tag{5.143}$$

We will rewrite equations (5.141)–(5.143) into one equation as

$$\begin{bmatrix} V_P^- \\ V_C^- \end{bmatrix} = \begin{bmatrix} S_{PP} & S_{PC} \\ S_{CP} & S_{CC} \end{bmatrix} \begin{bmatrix} V_P^+ \\ V_C^+ \end{bmatrix} \tag{5.144}$$

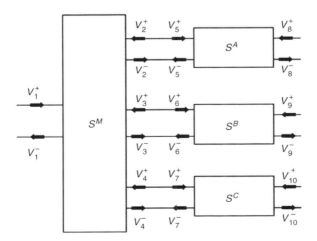

Figure 5.24 A three-channel multiplexer.

where

$$V_P^- = \begin{pmatrix} V_1^- \\ V_8^- \\ V_9^- \\ V_{10}^- \end{pmatrix}, \quad V_P^+ = \begin{pmatrix} V_1^+ \\ V_8^+ \\ V_9^+ \\ V_{10}^+ \end{pmatrix},$$

$$V_C^- = \begin{pmatrix} V_2^- \\ V_3^- \\ V_4^- \\ V_5^- \\ V_6^- \\ V_7^- \end{pmatrix}, \quad V_C^+ = \begin{pmatrix} V_2^+ \\ V_3^+ \\ V_4^+ \\ V_5^+ \\ V_6^+ \\ V_7^+ \end{pmatrix}$$

(5.145)

The matrices S_{PP}, S_{PC}, S_{CP}, and S_{CC} can then be written as follows:

$$S_{PP} = \begin{pmatrix} S_{11}^M & 0 & 0 & 0 \\ 0 & S_{22}^A & 0 & 0 \\ 0 & 0 & S_{22}^B & 0 \\ 0 & 0 & 0 & S_{22}^C \end{pmatrix}$$

(5.146)

$$S_{PC} = \begin{pmatrix} S_{12}^M & S_{13}^M & S_{14}^M & 0 & 0 & 0 \\ 0 & 0 & 0 & S_{21}^A & 0 & 0 \\ 0 & 0 & 0 & 0 & S_{21}^B & 0 \\ 0 & 0 & 0 & 0 & 0 & S_{21}^C \end{pmatrix}$$

(5.147)

$$S_{CP} = \begin{pmatrix} S_{21}^M & 0 & 0 & 0 \\ S_{31}^M & 0 & 0 & 0 \\ S_{41}^M & 0 & 0 & 0 \\ 0 & S_{12}^A & 0 & 0 \\ 0 & 0 & S_{12}^B & 0 \\ 0 & 0 & 0 & S_{12}^C \end{pmatrix}$$

(5.148)

$$S_{CC} = \begin{pmatrix} S_{22}^M & S_{23}^M & S_{24}^M & 0 & 0 & 0 \\ S_{32}^M & S_{33}^M & S_{34}^M & 0 & 0 & 0 \\ S_{42}^M & S_{43}^M & S_{44}^M & 0 & 0 & 0 \\ 0 & 0 & 0 & S_{11}^A & 0 & 0 \\ 0 & 0 & 0 & 0 & S_{11}^B & 0 \\ 0 & 0 & 0 & 0 & 0 & S_{11}^C \end{pmatrix}$$

(5.149)

The internal reports are related as

$$V_2^- = V_5^+, \quad V_2^+ = V_5^- \tag{5.150}$$

$$V_3^- = V_6^+, \quad V_3^+ = V_6^- \tag{5.151}$$

$$V_4^- = V_7^+, \quad V_4^+ = V_7^- \tag{5.152}$$

The connection matrix Γ relates the internal ports V_C^- to V_C^+ as $V_C^- = \Gamma V_C^+$, where Γ is given by

$$\Gamma = \begin{pmatrix} 0 & 0 & 0 & 1 & 0 & 0 \\ 0 & 0 & 0 & 0 & 1 & 0 \\ 0 & 0 & 0 & 0 & 0 & 1 \\ 1 & 0 & 0 & 0 & 0 & 0 \\ 0 & 1 & 0 & 0 & 0 & 0 \\ 0 & 0 & 1 & 0 & 0 & 0 \end{pmatrix} \tag{5.153}$$

The scattering matrix of the three-channel multiplexer is expressed as

$$S^P = S_{PP} + S_{PC}[\Gamma - S_{CC}]^{-1} S_{CP} \tag{5.154}$$

It is noted that the connection matrix in this case consists of 1s and 0s. If there are transmission lines connecting the filters to the manifold, these transmission lines can be included in $[S^A]$, $[S^B]$, and $[S^C]$ by shifting the reference planes. Alternatively, one can take the transmission lines into consideration in writing the elements of the connection matrix Γ. In this case, the 1s in the connection matrix will be replaced by exponential terms of the form of $e^{j\theta}$.

It should be noted that while we considered the case of only a three-channel multiplexer in Example 5.3, one can easily extend the procedure to formulate a computer program to calculate the scattering matrix of a multiplexer of N channels.

SUMMARY

This chapter provides a review of the basic concepts used in the analysis of multiport microwave networks. These concepts are important for filter designers, since any filter or multiplexer circuit can be divided into smaller two-, three- or N-port networks connected together. Five matrix representations of microwave networks are described; namely, $[Z]$, $[Y]$, $[ABCD]$, $[S]$, and $[T]$ matrices. These matrices are interchangeable, where the elements of any matrix can be written in terms of those of the other four matrices. A table is given summarizing the relationship between these matrices.

The $[Z]$, $[Y]$, and $[ABCD]$ matrices are useful in understanding and dealing with the lumped-element filter circuit models. The $[Z]$ and $[Y]$ matrices are useful in providing a physical insight of the circuit itself. For example, the measured S-parameters of an iris connected between two waveguides cannot directly identify the nature of the coupling provided by the iris. However, by converting the S-parameters to Z- or Y-parameters, the coupling can be readily identified, whether it is inductive (magnetic) or capacitive (electric) coupling. For filter or multiplexer circuits that have ports with unequal impedances, it is important to take into consideration the proper impedance normalization when switching between [S] to [Z] or [Y] matrices. The concept of using symmetry to simplify the analysis of large symmetrical circuits is also described. This concept is particularly useful when using computation-intensive commercial software tools in designing filter circuits. An example is included to demonstrate the flexibility in evaluating the overall S-parameters of a cascaded network four different ways, by utilizing the equivalent [ABCD], [T], and [S] matrices, as well as by invoking the symmetry of the network.

An analysis is presented for evaluating the scattering parameters of microwave circuits formed by connecting several multiport networks together. The analysis is quite general and can be employed to analyze any filter or multiplexer circuit. The chapter concludes with an example showing the evaluation of the scattering parameters of a 3-channel multiplexer. The reader can easily generalize the concept to evaluate the scattering matrix of multiplexers with any number of channels.

REFERENCES

1. R. E. Collin, *Foundation for Microwave Engineering*, McGraw-Hill, New York, 1966.
2. K. C. Gupta, *Computer-Aided Design of Microwave Circuits*, Artech House, Dedham, MA, 1981.
3. D. Pozar, *Microwave Engineering*, 2nd ed., Wiley, New York, 1998.
4. H. J. Carlin, The scattering matrix in network theory, *IRE Trans. Circuit Theory* **CT-3**, 88–96 (June 1956).
5. D. C. Youla, On scattering matrices normalized to complex port numbers, *Proc. IRE* **49**, 1221 (July 1961).
6. K. Kurokawa, Power waves and the scattering matrix, *IEEE Trans. Microwave Theory Tech.* **MTT-13**, 194–202 (March 1965).
7. R. R. Mansour and R. H. MacPhie, An improved transmission matrix formulation of cascaded discontinuities and its application to E-plane circuits, *IEEE Trans. Microwave Theory Tech.* **MTT-34**, 1986–1498 (Dec. 1986).

CHAPTER 6

SYNTHESIS OF A GENERAL CLASS OF THE CHEBYSHEV FILTER FUNCTION

In this chapter, we review some important scattering parameter relations that are relevant for the synthesis of filter networks. This is followed by a discussion of the general kind of Chebyshev function and its application in generating the transfer and reflection polynomials for equiripple filter characteristics with an arbitrary distribution of the transmission zeros. In the final part of this chapter, the special cases of predistorted and dual-band filtering functions are discussed.

6.1 POLYNOMIAL FORMS OF THE TRANSFER AND REFLECTION PARAMETERS $S_{21}(s)$ AND $S_{11}(s)$ FOR A TWO-PORT NETWORK

For the great majority of the filter circuits, we shall initially consider two-port networks; a "source port" and a "load port" (Fig. 6.1).

For a two-port network, the scattering matrix is represented by a 2×2 matrix

$$\begin{bmatrix} b_1 \\ b_2 \end{bmatrix} = \begin{bmatrix} S_{11} & S_{12} \\ S_{21} & S_{22} \end{bmatrix} \cdot \begin{bmatrix} a_1 \\ a_2 \end{bmatrix} \tag{6.1}$$

where b_1 and b_2 are the power waves propagating away from ports 1 and 2, respectively, and a_1 and a_2 are the power waves incident at ports 1 and 2, respectively.

Microwave Filters for Communication Systems: Fundamentals, Design, and Applications, by Richard J. Cameron, Chandra M. Kudsia, and Raafat R. Mansour
Copyright © 2007 John Wiley & Sons, Inc.

Figure 6.1   Two-port network.

If the network is passive, lossless, and reciprocal, its 2×2 S-parameter matrix yields two conservation of energy equations

$$S_{11}(s)S_{11}(s)^* + S_{21}(s)S_{21}(s)^* = 1 \tag{6.2}$$

$$S_{22}(s)S_{22}(s)^* + S_{12}(s)S_{12}(s)^* = 1 \tag{6.3}$$

and one unique orthogonality equation[1]

$$S_{11}(s)S_{12}(s)^* + S_{21}(s)S_{22}(s)^* = 0 \tag{6.4}$$

where the S parameters are now assumed to be functions of s $(= j\omega)$, the frequency variable.

As described in Section 3.10, the reflection parameter $S_{11}(s)$ at port 1 of the network is expressed as the ratio of two finite-degree polynomials $E(s)$ and $F(s)$, and real constant ε_R

$$S_{11}(s) = \frac{F(s)/\varepsilon_R}{E(s)} \tag{6.5}$$

---

[1]For an Nth degree polynomial $Q(s)$ with purely imaginary variable $s = j\omega$ and complex coefficients q_i, $i = 0, 1, 2, \ldots, N$, $Q(s)^*$ is the same as $Q^*(s^*)$ or $Q^*(-s)$, which might be a little more familiar to the reader.

$$\text{If}\quad Q(s) = q_0 + q_1 s + q_2 s^2 + \cdots + q_N s^N$$
$$\text{then}\quad Q(s)^* = Q^*(-s) = q_0^* - q_1^* s + q_2^* s^2 - \quad\cdots + q_N^* s^N \quad (N \text{ even})$$
$$\cdots - q_N^* s^N \quad (N \text{ odd})$$

The effect of this conjugating operation (*paraconjugation*) is to reflect the complex singularities of $Q(s)$ symmetrically about the imaginary axis [as opposed to reflecting about the real axis for the operation $Q(s) \rightarrow Q^*(s)$ (*conjugation*)]. If the N complex-plane singularities of $Q(s)$ are s_{0k}, $k = 1, 2, \ldots, N$, then the singularities of $Q(s)^*$ or $Q^*(-s)$ will be $-s_{0k}^*$. When generating the polynomial $Q(s)^*$ or $Q^*(-s)$ from the singularities of $Q(s)$ using the transformed variable $s_{0k} \rightarrow -s_{0k}^*$, the resultant polynomial must be multiplied through by $(-1)^N$ to give the correct sign to the new leading coefficient:

$$Q(s)^* = Q^*(-s) = (-1)^N \prod_{k=1}^{N} (s + s_{0k}^*)$$

where $E(s)$ is an Nth-degree polynomial with complex coefficients $e_0, e_1, e_2, \ldots, e_N$, where N is the degree of the filter network under consideration. Also, $F(s)$ is an Nth degree polynomial with complex coefficients $f_0, f_1, f_2, \ldots, f_N$. ε_R allows the normalization of the highest degree coefficients of $E(s)$ and $F(s)$ to unity (i.e., e_N and $f_N = 1$). Because this is a lossless passive network, $E(s)$ is strictly Hurwitz; that is, all the roots of $E(s)$ [poles of $S_{11}(s)$] are in the left half of the complex plane. These poles need not be symmetric about the real axis. $F(s)$, the numerator polynomial of $S_{11}(s)$, for lowpass and bandpass filters is also of degree N. For band-stop filters the degree of $F(s)$ can be $<N$. The roots of $F(s)$ [zeros of $S_{11}(s)$] are the points of zero reflected power ($b_i = 0$), or points of perfect transmission.

By reorganizing equation (6.2) and substituting for $S_{11}(s)$, we obtain

$$S_{21}(s)S_{21}(s)^* = 1 - \frac{F(s)F(s)^*/\varepsilon_R^2}{E(s)E(s)^*}$$

$$= \frac{P(s)P(s)^*/\varepsilon^2}{E(s)E(s)^*}$$

Thus, the transfer parameter $S_{21}(s)$ can be expressed as the ratio of two polynomials

$$S_{21}(s) = \frac{P(s)/\varepsilon}{E(s)} \tag{6.6}$$

where $P(s)P(s)^*/\varepsilon^2 = E(s)E(s)^* - F(s)F(s)^*/\varepsilon_R^2$.

From equations (6.5) and (6.6) it is seen that $S_{11}(s)$ and $S_{21}(s)$ share a common denominator polynomial $E(s)$. The numerator of $S_{21}(s)$ is a polynomial $P(s)/\varepsilon$ whose zeros are the transmission zeros (Tx zeros or TZs) of the filtering function. The degree n_{fz} of the polynomial $P(s)$ corresponds to the number of finite-position Tx zeros that the transfer function incorporates. This also implies that $n_{fz} \leq N$; otherwise, $S_{21}(s)$ exceeds unity as $s \to j\infty$, which is, of course, impossible for a passive network.

A transmission zero can be realized in two ways. The first case occurs when the degree n_{fz} of $P(s)$ is less than the degree N of the denominator polynomial $E(s)$, and $s \to j\infty$. At $s = j\infty$, $S_{21}(s) = 0$ and this is known as a *transmission zero at infinity*. When there are no finite-position zeros ($n_{fz} = 0$), the filtering function is known as an *all-pole response*. When $0 < n_{fz} < N$, the number of Tx zeros at infinity is $N - n_{fz}$.

The second case occurs when the frequency variable s coincides with an imaginary-axis root of its numerator polynomial $P(s)$, that is, $s = s_{0i}$, where s_{0i}, is a purely imaginary root of $P(s)$. The zeros are not necessarily on the imaginary axis, and if there is one root s_{0i}, that is complex, there must be a second root $-s_{0i}^*$ to make up a pair having symmetry about the imaginary axis. This ensures that polynomial $P(s)$ has coefficients that alternate between purely real and purely imaginary as the power of s increases. This is a condition that must hold if the filter is to be realized with purely reactive components (see Chapter 7).

Consideration of the orthogonality unitary condition provides an important relation between the phases of the $S_{11}(s)$, $S_{22}(s)$, and $S_{21}(s)$ polynomials, and between the zeros of $S_{11}(s)$ and $S_{22}(s)$ in the complex s plane. If we apply the

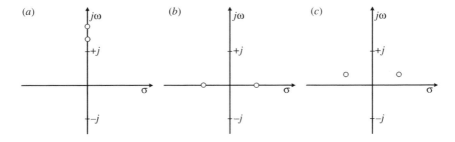

Figure 6.2   Possible positions for the finite-position roots of $P(s)$, the denominator of $S_{21}(s)$: (a) two asymmetric imaginary axis zeros; (b) real axis pair; and (c) complex pair.

reciprocity condition $S_{12}(s) = S_{21}(s)$ to the unitary conditions described in equations (6.2), (6.3), and (6.4), we obtain

$$S_{11}(s)S_{11}(s)^* + S_{21}(s)S_{21}(s)^* = 1 \tag{6.7a}$$

$$S_{22}(s)S_{22}(s)^* + S_{21}(s)S_{21}(s)^* = 1 \tag{6.7b}$$

$$S_{11}(s)S_{21}(s)^* + S_{21}(s)S_{22}(s)^* = 0 \tag{6.7c}$$

If we write these vector quantities in polar coordinates and drop the s for the moment for presentational simplicity, $S_{11} = |S_{11}| \cdot e^{j\theta_{11}}$, $S_{22} = |S_{22}| \cdot e^{j\theta_{22}}$, and $S_{21} = |S_{21}| \cdot e^{j\theta_{21}}$ [1]. From equations (6.7a) and (6.7b), we note that $|S_{11}| = |S_{22}|$, and so, from equation (6.7a), we have

$$|S_{21}|^2 = 1 - |S_{11}|^2$$

By rewriting equation (6.7c) in the polar form, we obtain

$$|S_{11}|e^{j\theta_{11}} \cdot |S_{21}|e^{-j\theta_{21}} + |S_{21}|e^{j\theta_{21}} \cdot |S_{11}|e^{-j\theta_{22}} = 0$$

$$|S_{11}| \, |S_{21}|\left(e^{j(\theta_{11}-\theta_{21})} + e^{j(\theta_{21}-\theta_{22})}\right) = 0 \tag{6.8}$$

This equation can be satisfied only if

$$e^{j(\theta_{11}-\theta_{21})} = -e^{j(\theta_{21}-\theta_{22})} \tag{6.9}$$

By replacing the minus sign in equation (6.9) with $e^{j(2k \pm 1)\pi}$, where k is an integer, we have

$$e^{j(\theta_{11}-\theta_{21})} = e^{j((2k \pm 1)\pi + \theta_{21}-\theta_{22})}$$

or

$$\theta_{21} - \frac{(\theta_{11} + \theta_{22})}{2} = \frac{\pi}{2}(2k \pm 1) \tag{6.10}$$

Since the vectors $S_{11}(s)$, $S_{22}(s)$, and $S_{21}(s)$ are represented by rational polynomials in the variable s and share a common denominator polynomial $E(s)$, their phases are

$$\theta_{21}(s) = \theta_{n21}(s) - \theta_d(s)$$

$$\theta_{11}(s) = \theta_{n11}(s) - \theta_d(s) \tag{6.11}$$

$$\theta_{22}(s) = \theta_{n22}(s) - \theta_d(s)$$

where $\theta_d(s)$ is the angle of the common denominator polynomial $E(s)$, and $\theta_{n21}(s)$, ... are the angles of the numerator polynomials of $S_{21}(s)$ etc.

By substituting these separated phases into equation (6.10), the $\theta_d(s)$ cancel and the following important relationship is obtained:

$$-\theta_{n21}(s) + \frac{(\theta_{n11}(s) + \theta_{n22}(s))}{2} = \frac{\pi}{2}(2k \pm 1) \tag{6.12}$$

This equation states that at any value of the frequency variable s, the difference between the angle of the $S_{21}(s)$ numerator vector, and the average of phases of the $S_{11}(s)$ and $S_{22}(s)$ numerator vectors must be an odd multiple of $\pi/2$ radians, that is, orthogonal. The frequency independence of the right-hand side of equation (6.12) allows us to observe two important properties of $\theta_{n21}(s)$, $\theta_{n11}(s)$, and $\theta_{n22}(s)$:

- The numerator polynomial of $S_{21}(s)$ $(P(s)/\varepsilon)$ must have zeros that are either on the imaginary axis of the complex s plane or in mirror-image pairs symmetrically arranged about the imaginary axis. In this way, the value of $\theta_{n21}(s)$ [the angle of $P(s)$] remains at an integral number of $\pi/2$ radians at any position of s, along the imaginary axis from $-j\infty$ to $+j\infty$.
- Similarly, the average of the phases of the $S_{11}(s)$ and $S_{22}(s)$ numerator polynomials, $(\theta_{n11}(s) + \theta_{n22}(s))/2$, must be evaluated to an integral number of $\pi/2$ radians at any value of s between $-j\infty$ and $+j\infty$. This means that the zeros of the numerator polynomial of $S_{11}(s)$ [i.e., the roots of $F(s)$]² and the zeros of the numerator polynomial of $S_{22}(s)$ [the roots of $F_{22}(s)$], must either be coincident on the imaginary axis or arranged in mirror-image pairs about the imaginary axis.

The definition of these angles is illustrated in Figure 6.3.

The second statement allows us to relate the numerator polynomials of $S_{11}(s)$ and $S_{22}(s)$, namely, $F(s)$ and $F_{22}(s)$, respectively. If their zeros (roots) are coincident the

---

²Formally, if the numerator polynomial of $S_{22}(s)$ is $F_{22}(s)$, then the numerator polynomial of $S_{11}(s)$ should be denoted $F_{11}(s)$. However, common usage in this book and elsewhere has meant that $F(s) \equiv F_{11}(s)$.

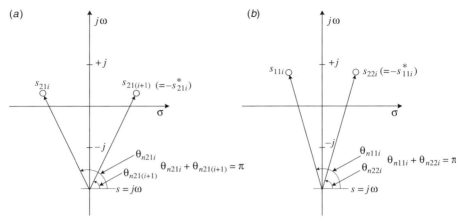

Figure 6.3   Angles of $S_{21}(s)$, $S_{11}(s)$, and $S_{22}(s)$ numerator polynomials for an arbitrary position of the frequency variable s on the imaginary axis: (*a*) for a transmission zero pair; (*b*) for an $S_{11}(s)$ zero and the complementary $S_{22}(s)$ zero.

imaginary axis or are symmetrically arranged about the imaginary axis, then the *i*th zero of $F_{22}(s)$, namely s_{22i}, is related to the corresponding zero of $F(s)$, (s_{11i}) as follows

$$s_{22i} = -s_{11i}^* \tag{6.13}$$

where $i = 1, 2, \ldots, N$.

By forming the $F_{22}(s)$ polynomial from the s_{22i} zeros, we obtain

$$F_{22}(s) = \prod_{i=1}^{N} (s - s_{22i})$$

$$= \prod_{i=1}^{N} (s + s_{11i}^*) \tag{6.14a}$$

$$= (-1)^N \prod_{i=1}^{N} (s - s_{11i})^* \tag{6.14b}$$

Since we already know that the polynomial $F(s) = \prod_{i=1}^{N} (s - s_{11i})$, it is obvious that

$$F_{22}(s) = (-1)^N F(s)^* \tag{6.15}$$

This leads to the first of two rules associated with the orthogonality condition:

- The numerator polynomial of $S_{22}(s)$, $F_{22}(s)$, can be formed from the product of the negative conjugate zeros of $F(s)$ [equation (6.14a)], or from the conjugate of the $F(s)$ polynomial multiplied by -1 when N is an odd integer [equations (6.14b) and (6.15)].

Multiplication by -1, when N is odd, gives an expression for the angle of $F_{22}(s)$, equation (6.15);

$$\theta_{n22}(s) = -\theta_{n11}(s) + N\pi \qquad (6.16)$$

By substituting equation (6.16) into equation (6.12), we have

$$-\theta_{n21}(s) + \frac{N\pi}{2} = \frac{\pi}{2}(2k \pm 1) \qquad (6.17)$$

Finally, let us turn to $\theta_{n21}(s)$. This is the angle of the numerator polynomial of $S_{21}(s)$, $P(s)$, which is of degree n_{fz}, the number of finite-position transmission zeros in the transfer function. Because of the symmetry of these transmission zeros about the imaginary axis, $\theta_{n21}(s)$ is an integral number of $\pi/2$ radians regardless of the values of s and n_{fz} (Fig. 6.3a), represented as

$$\theta_{n21}(s) = \frac{n_{fz}\pi}{2} + k_1\pi \qquad (6.18)$$

where k_1 is an integer. Now by substituting into equation (6.17), we obtain

$$-\frac{n_{fz}\pi}{2} - k_1\pi + \frac{N\pi}{2} = \frac{\pi}{2}(2k \pm 1)$$

$$(N - n_{fz})\frac{\pi}{2} - k_1\pi = \frac{\pi}{2}(2k \pm 1) \qquad (6.19)$$

Equation (6.19) implies that the integer quantity $(N - n_{fz})$ must be odd so that its right-hand side (RHS) is satisfied. For the cases where $(N - n_{fz})$ is even, then an extra $\pi/2$ radians must be added to the left-hand side (LHS) of equation (6.19) to maintain the orthogonality condition, which is the same as adding $\pi/2$ to $\theta_{n21}(s)$ on the LHS of (6.18). Adding $\pi/2$ to $\theta_{n21}(s)$ is the equivalent of multiplying the polynomial $P(s)$ by j. Table 6.1 summarizes these conclusions.

This is the second of the two rules associated with the orthogonality condition:

- When the highest-degree coefficients of the Nth-degree polynomials $E(s)$ and $F(s)$, and the n_{fz} ($\leq N$)-degree polynomial $P(s)$ are normalized to unity (i.e., monic polynomials), then the $P(s)$ polynomial must be multiplied by j when the integer quantity $(N - n_{fz})$ is even.

In practice the polynomials $E(s)$, $F(s)$, and $P(s)$ are usually formed as products of their singularities, such as $F(s) = \prod_{i=1}^{N}(s - s_{11i})$; so, their leading coefficients are automatically equal to unity in the majority of cases.

TABLE 6.1   Multiplication of $P(s)$ by j to Satisfy the Orthogonality Condition

N	n_{fz}	$N - n_{fz}$	Multiply $P(s)$ by j
Odd	Odd	Even	Yes
Odd	Even	Odd	No
Even	Odd	Odd	No
Even	Even	Even	Yes

By applying the orthogonality unitary condition to the scattering matrix, we obtain

For $(N - n_{fz})$ odd,
$$F(s)P(s)^* + P(s)F_{22}(s)^* = 0 \qquad (6.20a)$$

For $(N - n_{fz})$ even,
$$F(s)[jP(s)]^* + [jP(s)]F_{22}(s)^* = 0$$

or
$$F(s)P(s)^* - P(s)F_{22}(s)^* = 0 \qquad (6.20b)$$

For $(N - n_{fz})$ even or odd,
$$F(s)P(s)^* - (-1)^{(N-n_{fz})}P(s)F_{22}(s)^* = 0 \qquad (6.20c)$$

which is represented in the following S-matrix form:

$$
\begin{bmatrix} S_{11} & S_{12} \\ S_{21} & S_{22} \end{bmatrix} = \frac{1}{E(s)} \begin{bmatrix} \dfrac{F(s)}{\varepsilon_R} & \dfrac{P(s)}{\varepsilon} \\ \dfrac{P(s)}{\varepsilon} & -(-1)^{(N-n_{fz})}\dfrac{F_{22}(s)}{\varepsilon_R} \end{bmatrix} \qquad (6.21)
$$

Since $F_{22}(s) = (-1)^N F(s)^*$ [equation (6.15)], the polynomial S matrix is rewritten as

$$
\begin{bmatrix} S_{11} & S_{12} \\ S_{21} & S_{22} \end{bmatrix} = \frac{1}{E(s)} \begin{bmatrix} \dfrac{F(s)}{\varepsilon_R} & \dfrac{P(s)}{\varepsilon} \\ \dfrac{P(s)}{\varepsilon} & (-1)^{(n_{fz}+1)}\dfrac{F(s)^*}{\varepsilon_R} \end{bmatrix} \qquad (6.22)
$$

which is perhaps a more familiar form [2]. In a similar way, we can obtain alternative forms to represent the S matrix by manipulating equation (6.20c). Whatever the form, the unitary conditions will always be satisfied.

From unitary conditions, two additional relationships are derived:

- The relationship between the real constants ε and ε_R used to normalize the polynomials $P(s)$ and $F(s)$, respectively.
- The alternating pole-principle, which allows the accurate determination of either $E(s)$, $F(s)$, or $P(s)$ while knowing the other two.

6.1.1 Relationship Between ε and ε_R

The transfer and reflection functions $S_{21}(s)$ and $S_{11}(s)$ were defined earlier in terms of their rational polynomials

$$S_{21}(s) = \frac{P(s)/\varepsilon}{E(s)}, \qquad S_{11}(s) = \frac{F(s)/\varepsilon_R}{E(s)} \tag{6.23}$$

respectively, where $E(s)$ and $F(s)$ are Nth degree, $P(s)$ is a polynomial of degree n_{fz}, the number of finite-position transmission zeros, and ε and ε_R are real constants normalizing $P(s)$ and $F(s)$ such that $|S_{21}(s)|$ and $|S_{11}(s)|$ are ≤ 1 at any value of s, the frequency variable. It is assumed that the three polynomials have been normalized such that their highest-power coefficients are unity. Supplemental to this, for the cases where $N - n_{fz}$ is an even integer, the polynomial $P(s)$ is multiplied by j to satisfy the unitary condition (6.7c) as described above.

The real constant ε is determined by evaluating $P(s)/E(s)$ at a convenient value of s, where $|S_{21}(s)|$ or $|S_{11}(s)|$ are known, for instance, at $s = \pm j$, where the equiripple return loss level for Chebyshev filters or the 3 dB (half-power point) for Butterworth filters is known. This sets the maximum level for $|S_{21}(s)|$ at 1, and if $n_{fz} < N$, $|S_{21}(s)| = 0$ at infinite frequency $s = \pm j\infty$. When $|S_{21}(s)| = 0$, the conservation of energy condition (6.7(a)) dictates that

$$S_{11}(j\infty) = \frac{1}{\varepsilon_R} \left| \frac{F(j\infty)}{E(j\infty)} \right| = 1 \tag{6.24}$$

Because the highest-degree coefficients (e_N and f_N) of $E(s)$ and $F(s)$, respectively, are unity, it is easily seen that $\varepsilon_R = 1$. When $n_{fz} = N$, that is, where all N available transmission zeros are at finite positions in the complex plane and therefore $P(s)$ is an Nth-degree polynomial (fully canonical function), then the attenuation at $s = \pm j\infty$ is finite and ε_R is derived from the conservation of energy condition equation (6.7a) by

$$S_{11}(j\infty)S_{11}(j\infty)^* + S_{21}(j\infty)S_{21}(j\infty)^* = 1$$

$$\frac{F(j\infty)F(j\infty)^*}{\varepsilon_R^2 E(j\infty)E(j\infty)^*} + \frac{P(j\infty)P(j\infty)^*}{\varepsilon^2 E(j\infty)E(j\infty)^*} = 1 \tag{6.25}$$

For the fully canonical case, $E(s)$, $F(s)$, and $P(s)$ are all Nth-degree polynomials with the highest-power coefficients $= 1$. Therefore, at $s = \pm j\infty$, we obtain

$$\frac{1}{\varepsilon_R^2} + \frac{1}{\varepsilon^2} = 1 \quad \text{or} \quad \varepsilon_R = \frac{\varepsilon}{\sqrt{\varepsilon^2 - 1}} \tag{6.26}$$

which is slightly greater than unity since ε has to be >1.

Also, it is evident that for the fully canonical case, the insertion loss at $s = \pm j\infty$ is

$$S_{21}(\pm j\infty) = \frac{1}{\varepsilon} = 20\log_{10}\varepsilon \text{ dB} \tag{6.27a}$$

and the return loss is

$$S_{11}(\pm j\infty) = \frac{1}{\varepsilon_R} = 20\log_{10}\varepsilon_R \text{ dB} \tag{6.27b}$$

Although the fully canonical characteristics are rarely used for the synthesis of bandpass filters, they are sometimes used for bandstop filters, which will be considered in a later chapter.

6.2   ALTERNATING POLE METHOD FOR DETERMINATION OF THE DENOMINATOR POLYNOMIAL $E(s)$

For the polynomial synthesis method to be described later, transmission zeros are prescribed in the complex plane, which defines the $S_{21}(s)$ numerator polynomial $P(s)$. Then the coefficients of the $S_{11}(s)$ numerator polynomial $F(s)$ are found using an analytic or recursive method. It then remains to find the $S_{11}(s)$ and $S_{21}(s)$ denominator polynomial $E(s)$ to complete the design of the filtering function.

If two of the three polynomials are known, the third may be derived from the conservation of energy equation (6.7a) as follows:

$$S_{11}(s)S_{11}(s)^* + S_{21}(s)S_{21}(s)^* = 1 \text{ or } \frac{F(s)F(s)^*}{\varepsilon_R^2} + \frac{P(s)P(s)^*}{\varepsilon^2} = E(s)E(s)^* \tag{6.28}$$

The LHS of equation (6.28) is constructed using polynomial multiplications to find polynomial $E(s)E(s)^*$, which must be a scalar quantity. This means that the $2N$ roots of $E(s)E(s)^*$, must form a symmetric pattern about the imaginary axis in the complex plane, so that at any frequency s product $E(s)E(s)^*$ is scalar (see Fig. 6.4).

Since we know that the roots of $E(s)$ are strictly Hurwitz, those roots of $E(s)E(s)^*$ that are in the left-half plane must belong to $E(s)$ and those in the right-half plane, to $E(s)^*$. By choosing the N roots in the left-half plane, the polynomial $E(s)$ is formed.

Although this method is completely general, it does mean working with double-degree polynomials, and sometimes, with higher degree filter functions, the roots of $E(s)E(s)^*$ tend to cluster around $s = \pm j$, leading to inaccurate root-finding. An alternative method, due to Rhodes and Alseyab [3] is presented here, with which the roots of $E(s)$ can be found without having to determine the roots of polynomials of degree $2N$.

By expanding equation (6.28) in two different ways, we obtain

$$\varepsilon^2\varepsilon_R^2 E(s)E(s)^* = [\varepsilon_R P(s) + \varepsilon F(s)][\varepsilon_R P(s)^* + \varepsilon F(s)^*]$$
$$- \varepsilon\varepsilon_R[P(s)^*F(s) + P(s)F(s)^*] \tag{6.29a}$$

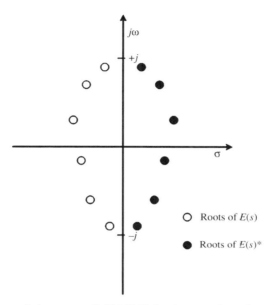

Figure 6.4 Pattern of the roots of $E(s)E(s)^*$ in the complex plane (symmetric about imaginary axis).

and

$$\varepsilon^2 \varepsilon_R^2 E(s)E(s)^* = [\varepsilon_R(jP(s)) + \varepsilon F(s)][\varepsilon_R(jP(s))^* + \varepsilon F(s)^*]$$
$$- \varepsilon\varepsilon_R[(jP(s))^*F(s) + (jP(s))F(s)^*] \qquad (6.29b)$$

Equation (6.29a) is in the correct form for the cases where the integer quantity $(N - n_{fz})$ is odd. For the term on the extreme right to equal zero

$$P(s)^*F(s) = -P(s)F(s)^* \qquad (6.30)$$

If the orthogonality unitary condition for $(N - n_{fz})$ odd [equation (6.20a)] is recalled, namely

$$F(s)P(s)^* + P(s)F_{22}(s)^* = 0$$

it is obvious that equation (6.30) is vaild only if $F(s) = F_{22}(s)$. This can happen only when all the zeros of $F(s)$ lie on the imaginary axis and are coincident with those of $F_{22}(s)$.

Equation (6.29b) is in the correct form for the cases where $(N - n_{fz})$ is even, where the polynomial $P(s)$ must be multiplied by j. For the term on the extreme right to equal zero, the following condition must hold:

$$P(s)^*F(s) = P(s)F(s)^* \qquad (6.31)$$

By recalling the orthogonality unitary condition, described in equation (6.20b) for $(N - n_{fz})$ even, we have

$$F(s)P(s)^* - P(s)F_{22}(s)^* = 0$$

and once more it can be seen that equation (6.31) can be satisfied only if $F(s) = F_{22}(s)$. Again, this can happen only when all the zeros of $F(s)$ lie on the imaginary axis and are coincident with those of $F_{22}(s)$.

If the zeros of $F(s)$ and $F_{22}(s)$ fulfill this condition, equations (6.29a) and (6.29b) for $(N - n_{fz})$ odd are reduced to

$$\varepsilon^2 \varepsilon_R^2 E(s)E(s)^* = [\varepsilon_R P(s) + \varepsilon F(s)][\varepsilon_R P(s)^* + \varepsilon F(s)^*]$$
$$= [\varepsilon_R P(s) + \varepsilon F(s)][\varepsilon_R P(s) + \varepsilon F(s)]^* \qquad (6.32a)$$

and for $(N - n_{fz})$ even

$$\varepsilon^2 \varepsilon_R^2 E(s)E(s)^* = [\varepsilon_R(jP(s)) + \varepsilon F(s)][\varepsilon_R(jP(s))^* + \varepsilon F(s)^*]$$
$$= [\varepsilon_R(jP(s)) + \varepsilon F(s)][\varepsilon_R(jP(s)) + \varepsilon F(s)]^* \qquad (6.32b)$$

In the ω plane, $P(\omega)$ and $F(\omega)$ will have purely real coefficients. Use (6.32(b)) modified as follows to find the ω-plane singularities for $(N - n_{fz})$ even or odd:

$$\varepsilon^2 \varepsilon_R^2 E(\omega)E(\omega)^* = [\varepsilon_R P(\omega) - j\varepsilon F(\omega)][\varepsilon_R P(\omega) - j\varepsilon F(\omega)]^* \qquad (6.32c)$$

Rooting one of the two terms on the RHSs of equation (6.32) results in a pattern of singularities alternating between the left-half and right-half planes, as depicted in Figure 6.5.

Rooting the other term will give the complementary set of singularities, completing the symmetry of the pattern about the imaginary axis and ensuring that the RHS of (6.32) is properly scalar as the LHS demands.

As a result, it is necessary to form only one of the two terms in equation (6.32) from the known $P(s)$ and $F(s)$ polynomials, and then root the resultant Nth-degree polynomial with complex coefficients to obtain the singularities. Knowing that the polynomial $E(s)$ must be Hurwitz, any singularity in the right half-plane (RHP) plane can be reflected about the imaginary axis to lie in the image position in the left half-plane (LHP) plane. Now, knowing the positions of the N singularities in the LHP, we can form the polynomial $E(s)$. In finding $E(s)$, we only have to deal with Nth-degree polynomials; also because the alternating singularities tend to be less clustered about $s = \pm j$, greater accuracy is guaranteed.

In practice, for most classes of filtering function, such as Butterworth or Chebyshev, the reflection zeros lie on the imaginary axis and the alternating singularity method can be applied to find $E(s)$. For certain specialized cases, such

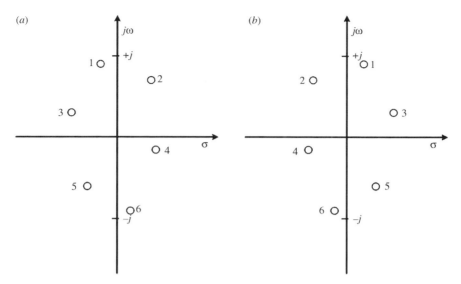

Figure 6.5 Singularities of polynomials for the case $(N - n_{fz})$ even: (*a*) $[\varepsilon_R(jP(s)) + \varepsilon F(s)]$ and (*b*) $[\varepsilon_R(jP(s)) + \varepsilon F(s)]^*$.

as predistorted filters, where some or all of the reflection zeros lie in the complex plane and not on the imaginary axis, the conservation of energy method for finding $E(s)$ [equation (6.28)] will have to be used.

6.3 GENERAL POLYNOMIAL SYNTHESIS METHODS FOR CHEBYSHEV FILTER FUNCTIONS

The methods presented in this section include the design of the transfer functions and the synthesis of the prototype filter networks with characteristics belonging to the general class of Chebyshev filter function:

- Even or odd degree
- Prescribed transmission and/or group delay equalization zeros
- Asymmetric or symmetric characteristics
- Singly or doubly terminated networks

In this section, an efficient recursive technique for generating the Chebyshev transfer and reflection polynomials, given the arbitrary location of the transmission zeros, is described [4,5]. In the next chapter, the methods for generating the corresponding circuits, and the $N \times N$ and $N + 2$ coupling matrices from these polynomials, is described.

6.3.1 Polynomial Synthesis

To simplify the mathematics in this section, we shall work in the ω plane, where ω is the real frequency variable related to the more familiar complex frequency variable s by $s = j\omega$.

For any two-port lossless filter network composed of a series of N intercoupled resonators, the transfer and reflection functions can be expressed as a ratio of two Nth-degree polynomials

$$S_{11}(\omega) = \frac{F(\omega)/\varepsilon_R}{E(\omega)}, \qquad S_{21}(\omega) = \frac{P(\omega)/\varepsilon}{E(\omega)} \tag{6.33}$$

$$\varepsilon = \frac{1}{\sqrt{10^{RL/10} - 1}} \left| \frac{P(\omega)}{F(\omega)/\varepsilon_R} \right|_{\omega = \pm 1} \tag{6.34}$$

where RL is the prescribed return loss level in dB at $\omega = \pm 1$. It is assumed that the polynomials $P(\omega)$, $F(\omega)$, and $E(\omega)$ are normalized such that their highest-degree coefficients are unity. $S_{11}(\omega)$ and $S_{21}(\omega)$ share a common denominator $E(\omega)$, and the polynomial $P(\omega) = \prod_{n=1}^{n_{fz}} (\omega - \omega_n)$ carries n_{fz} transfer function finite-position transmission zeros.[3] For a Chebyshev filter function ε is a constant normalizing $S_{21}(\omega)$, to the equiripple level at $\omega = \pm 1$.

Since, by definition, a Chebyshev function has all its reflection zeros on the real axis of the ω plane, the alternating pole formula for a lossless network, as described by equation (6.32c), is as follows:

$$S_{21}(\omega)S_{21}(\omega)^* = \frac{P(\omega)P(\omega)^*}{\varepsilon^2 E(\omega)E(\omega)^*} = \frac{1}{\left[1 - j\dfrac{\varepsilon}{\varepsilon_R} k\, C_N(\omega) \right]\left[1 + j\dfrac{\varepsilon}{\varepsilon_R} k\, C_N(\omega)^* \right]}$$

where $k\, C_N(\omega) = \dfrac{F(\omega)}{P(\omega)}$ and k is a constant[4]   (6.35)

$C_N(\omega)$ is known as the filter function of degree N, and its poles and zeros are the roots of $P(\omega)$ and $F(\omega)$, respectively. For the general Chebyshev characteristic, it has the form [4]

$$C_N(\omega) = \cosh\left[\sum_{n=1}^{N} \cosh^{-1}(x_n(\omega)) \right] \tag{6.36a}$$

---

[3]$P(\omega) = 1$ for $n_{fz} = 0$.
[4]k is an unimportant normalizing constant, inserted here to account for the fact that in general the polynomials of $C_N(\omega)$ have nonunity highest-degree coefficients, whereas the polynomials $P(\omega)$ and $F(\omega)$ in this text are assumed to be monic.

or, by using the identity $\cosh\theta = \cos j\theta$, the alternative expression for $C_N(\omega)$ is given by

$$C_N(\omega) = \cos\left[\sum_{n=1}^{N} \cos^{-1}(x_n(\omega))\right] \tag{6.36b}$$

where $x_n(\omega)$ is a function of the frequency variable ω. For analyzing $C_N(\omega)$, equation (6.36a) can be adopted for $|\omega| \geq 1$ and equation (6.36b), for $|\omega| \leq 1$.

To properly represent a Chebyshev function, $x_n(\omega)$ requires the following properties:

- At $\omega = \omega_n$, where ω_n is a finite-position prescribed transmission zero, or where ω_n is at infinite frequency ($\omega_n = \pm\infty$), $x_n(\omega) = \pm\infty$
- At $\omega = \pm 1$, $x_n(\omega) = \pm 1$
- Between $\omega = -1$ and $\omega = 1$ (in-band), $1 \geq x_n(\omega) \geq -1$

The first condition is satisfied if $x_n(\omega)$ is a rational function with its denominator $= \omega - \omega_n$:

$$x_n(\omega) = \frac{f(\omega)}{\omega - \omega_n} \tag{6.37}$$

The second condition states that at $\omega = \pm 1$

$$x_n(\omega)|_{\omega=\pm 1} = \frac{f(\omega)}{\omega - \omega_n}\bigg|_{\omega=\pm 1} = \pm 1 \tag{6.38}$$

This condition is satisfied if $f(1) = 1 - \omega_n$ and $f(-1) = 1 + \omega_n$, giving $f(\omega) = 1 - \omega\,\omega_n$. Therefore

$$x_n(\omega) = \frac{1 - \omega\,\omega_n}{\omega - \omega_n} \tag{6.39}$$

Differentiating $x_n(\omega)$ with respect to ω shows that there are no turning points or inflection points between $\omega = -1$ and $\omega = 1$. If $x_n(\omega) = -1$ at $\omega = -1$ and $x_n(\omega) = +1$ at $\omega = +1$, then $|x_n(\omega)| \leq 1$ while $|\omega| \leq 1$, thus satisfying the third condition. Figure 6.6 illustrates the behavior of $x_n(\omega)$ for a case where the finite transmission zero $\omega_n = +1.3$.

By dividing through by ω_n to deal with any transmission zeros at $\omega_n = \pm\infty$, we obtain the final form for the function $x_n(\omega)$ as

$$x_n(\omega) = \frac{\omega - 1/\omega_n}{1 - \omega/\omega_n} \tag{6.40}$$

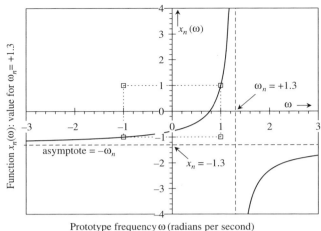

Figure 6.6 Function $x_n(\omega)$ in the ω plane with a prescribed transmission zero at $\omega_n = +1.3$.

In equation (6.40), $\omega_n = s_n/j$ is the position of the nth transmission zero in the complex frequency plane, and it may be easily verified from equations (6.38) and (6.39) that $C_N(\omega)|_{\omega=\pm 1} = 1$, when $|\omega| < 1$, $C_N(\omega) < 1$, and when $|\omega| > 1$, $C_N(\omega) > 1$, all of which are necessary conditions for a Chebyshev response. Also, as all N of the prescribed transmission zeros ω_n approach infinity, $C_N(\omega)$ degenerates to the familiar pure Chebyshev function:

$$C_N(\omega)\bigg|_{\omega_n \to \infty} = \cosh\left[N\cosh^{-1}(\omega)\right] \qquad (6.41)$$

The rule in prescribing the positions of the transmission zeros are that symmetry must be preserved about the imaginary ($j\omega$) axis of the complex s plane, to ensure that the unitary conditions are preserved. Also, for the polynomial synthesis methods about to be described, the number of transmission zeros with finite positions in the s plane n_{fz} must be $\leq N$. If $n_{fz} < N$, those zeros without finite positions must be placed at infinity. Although the $N + 2$ coupling matrix (Chapter 8) can accommodate fully canonical filtering functions (i.e., the number of finite-position transmission zeros n_{fz} equals the degree N of the filter), the $N \times N$ matrix can accommodate a maximum of only $N - 2$ finite-position zeros (minimum path rule.[5]) In synthesizing the polynomials for the $N \times N$ coupling matrix, at least two of the transmission zeros must be placed at infinity.

The next step is to solve equation (6.36a) to determine the coefficients of the numerator polynomial of $C_N(\omega)$. Normalizing this polynomial to set its

---

[5]The minimum path rule is a simple formula for calculating the number of finite-position transmission zeros (i.e., those that are not at $\omega = \pm\infty$) that an Nth-degree direct-coupled resonator network can realize. If the number of resonators in the shortest path (along nonzero interresonator couplings) between the input (source) and output (load) terminals of a network equals $n_{\min}$, then the maximum number of finite-position TZs (n_{fz}) that the network may realize is $n_{fz} = N - n_{\min}$.

highest-degree coefficient to unity will yield the polynomial $F(\omega)$. By knowing ε from equation (6.34) and the prescribed polynomial $P(\omega) = \prod_{n=1}^{n_{fz}} (\omega - \omega_n)$, the Hurwitz polynomial $E(\omega)$ common to the denominators of $S_{21}(\omega)$ and $S_{11}(\omega)$ can be evaluated from equation (6.35). It then becomes possible to proceed onward to prototype network synthesis, from which a real electrical network with the transfer and reflection characteristics $S_{21}(\omega)$ and $S_{11}(\omega)$ may be synthesized.

The first step in the polynomial synthesis procedure is to replace the $\cosh^{-1}$ term in equation (6.36a) with its identity

$$C_N(\omega) = \cosh \left[\sum_{n=1}^{N} \ln (a_n + b_n) \right]$$

where

$$a_n = x_n(\omega) \quad \text{and} \quad b_n = \left(x_n^2(\omega) - 1 \right)^{1/2} \tag{6.42}$$

Then

$$C_N(\omega) = \frac{1}{2} \left[e^{\sum \ln (a_n + b_n)} + e^{-\sum \ln (a_n + b_n)} \right]$$

$$= \frac{1}{2} \left[\prod_{n=1}^{N} (a_n + b_n) + \frac{1}{\prod_{n=1}^{N} (a_n + b_n)} \right] \tag{6.43}$$

By multiplying the second term in equation (6.43) top and bottom by $\prod_{n=1}^{N} (a_n - b_n)$ yields

$$C_N(\omega) = \frac{1}{2} \left[\prod_{n=1}^{N} (a_n + b_n) + \prod_{n=1}^{N} (a_n - b_n) \right] \tag{6.44}$$

From equation (6.42), it is evident that $\prod_{n=1}^{N}(a_n + b_n) \cdot \prod_{n=1}^{N}(a_n - b_n) = \prod_{n=1}^{N}(a_n^2 - b_n^2)$ will always be unity. Equation (6.44) is expressed in its final form by substituting for a_n, b_n, and $x_n(\omega)$ using equations (6.40) and (6.42) as

$$a_n = \frac{\omega - 1/\omega_n}{1 - \omega/\omega_n}$$

$$b_n = \frac{\sqrt{(\omega - 1/\omega_n)^2 - (1 - \omega/\omega_n)^2}}{1 - \omega/\omega_n}$$

$$= \frac{\sqrt{(\omega^2 - 1)(1 - 1/\omega_n^2)}}{1 - \omega/\omega_n} \tag{6.45}$$

$$= \frac{\omega' \sqrt{(1 - 1/\omega_n^2)}}{1 - \omega/\omega_n}$$

where $\omega' = \sqrt{(\omega^2 - 1)}$, a transformed frequency variable.

Equation (6.44) then becomes

$$C_N(\omega) = \frac{1}{2}\left[\frac{\prod_{n=1}^{N}[(\omega - 1/\omega_n) + \sqrt{(1 - 1/\omega_n^2)}\omega'] + \prod_{n=1}^{N}[(\omega - 1/\omega_n) - \sqrt{(1 - 1/\omega_n^2)}\omega']}{\prod_{n=1}^{N}(1 - \omega/\omega_n)}\right]$$

$$= \frac{1}{2}\left[\frac{\prod_{n=1}^{N}(c_n + d_n) + \prod_{n=1}^{N}(c_n - d_n)}{\prod_{n=1}^{N}(1 - \omega/\omega_n)}\right]$$

where

$$c_n = \left(\omega - \frac{1}{\omega_n}\right) \quad \text{and} \quad d_n = \omega'\sqrt{1 - \frac{1}{\omega_n^2}} \tag{6.46}$$

In comparison with equation (6.35), it is obvious that the denominator of $C_N(\omega)$ has the same zeros as $P(\omega)$ the numerator polynomial of $S_{21}(\omega)$ generated from the prescribed transmission zeros ω_n. From equation (6.35) it is also observed that the numerator of $C_N(\omega)$ has the same zeros as the numerator $F(\omega)$ of $S_{11}(\omega)$, and appears at first to be a mixture of two finite-degree polynomials, one in the variable ω purely, whilst the other has each coefficient multiplied by the transformed variable ω'.

However, the coefficients multiplied by ω' cancel each other when equation (6.46) is multiplied out. This is best proven by demonstration, by multiplying out the left- and right-hand product terms in the numerator of $C_N(\omega)$ in (6.46) for a few low values of N:

For $N = 1$, $\text{Num}[C_1(\omega)] = \frac{1}{2}\left[\prod_{n=1}^{1}(c_n + d_n) + \prod_{n=1}^{1}(c_n - d_n)\right] = c_1$.

For $N = 2$, $\text{Num}[C_2(\omega)] = c_1 c_2 + d_1 d_2$.

For $N = 3$, $\text{Num}[C_3(\omega)] = (c_1 c_2 + d_1 d_2)c_3 + (c_2 d_1 + c_1 d_2)d_3$

(and so on).

At each stage, the expansions result in a sum of factors, where each factor is composed of multiples of c_n and d_n elements. Because of the positive sign in the LH product term in equation (6.46), the c_n, d_n factors, resulting from multiplying out the LH term, are always positive in sign. Multiplying out the RH product term in equation (6.46) produces the same c_n, d_n factors; however, the negative sign indicates that those factors containing an odd number of d_n elements are negative in sign and are canceled with the corresponding factors from the LH product term.

Now, the remaining factors contain only the even numbers of d_n elements. Consequently, $\omega' = \sqrt{\omega^2 - 1}$, which is a common multiplier for all the d_n elements [equation (6.46)], is raised by even powers only, producing subpolynomials in the variable ω. As a result, the numerator of $C_N(\omega)$ will be a polynomial in the variable ω only.

These relationships can be used to develop a simple algorithm to determine the coefficients of the numerator polynomial of $C_N(\omega)$. Normalizing $\text{Num}[C_N(\omega)]$ to set its highest-degree coefficient to unity yields $F(\omega)$, the numerator of $S_{11}(\omega)$.

6.3.2 Recursive Technique

The numerator of equation (6.46) may be written as

$$\text{Num}\,[C_N(\omega)] = \frac{1}{2}\left[G_N(\omega) + G'_N(\omega)\right] \tag{6.47}$$

where

$$G_N(\omega) = \prod_{n=1}^{N}[c_n + d_n] = \prod_{n=1}^{N}\left[\left(\omega - \frac{1}{\omega_n}\right) + \omega'\sqrt{\left(1 - \frac{1}{\omega_n^2}\right)}\right] \tag{6.48a}$$

and

$$G'_N(\omega) = \prod_{n=1}^{N}[c_n - d_n] = \prod_{n=1}^{N}\left[\left(\omega - \frac{1}{\omega_n}\right) - \omega'\sqrt{\left(1 - \frac{1}{\omega_n^2}\right)}\right] \tag{6.48b}$$

The method for computing the coefficients of $\text{Num}[C_N(\omega)]$ is a recursive technique, where the solution for the nth degree is built up from the results of the $(n-1)$th degree. First, let us consider the polynomial $G_N(\omega)$ [equation (6.48a)]. This polynomial can be rearranged into two polynomials $U_N(\omega)$ and $V_N(\omega)$, where the $U_N(\omega)$ polynomial contains the coefficients of the terms in the variable ω only, whereas each coefficient of the auxiliary polynomial $V_N(\omega)$ is multiplied by the transformed variable ω'

$$G_N(\omega) = U_N(\omega) + V_N(\omega)$$

where

$$U_N(\omega) = u_0 + u_1\omega + u_2\omega^2 + \cdots + u_N\omega^N$$

and

$$V_N(\omega) = \omega'(v_0 + v_1\omega + v_2\omega^2 + \cdots + v_N\omega^N) \tag{6.49}$$

The recursion cycle is initiated with the terms corresponding to the first prescribed transmission zero ω_1, that is, by setting $N = 1$ in equations (6.48a) and (6.49):

$$G_1(\omega) = [c_1 + d_1]$$

$$= \left(\omega - \frac{1}{\omega_1}\right) + \omega'\sqrt{\left(1 - \frac{1}{\omega_1^2}\right)}$$

$$= U_1(\omega) + V_1(\omega) \tag{6.50}$$

For the first cycle of the process, $G_1(\omega)$ has to be multiplied by the terms corresponding to the second prescribed zero ω_2 [equation (6.48a)], given by

$$G_2(\omega) = G_1(\omega) \cdot [c_2 + d_2]$$

$$= [U_1(\omega) + V_1(\omega)] \cdot \left[\left(\omega - \frac{1}{\omega_2} \right) + \omega' \sqrt{\left(1 - \frac{1}{\omega_2^2} \right)} \right] \qquad (6.51)$$

$$= U_2(\omega) + V_2(\omega)$$

Let us multiply out this expression for $G_2(\omega)$, and again allocating terms purely in ω to $U_2(\omega)$, terms multiplied by ω' to $V_2(\omega)$, and recognizing that $\omega' V_n(\omega)$ results in $(\omega^2 - 1) \cdot (v_0 + v_1\omega + v_2\omega^2 + \cdots + v_n\omega^n)$ [equation (6.49)], a polynomial purely in ω and therefore to be allocated to $U_n(\omega)$:

$$U_2(\omega) = \omega\, U_1(\omega) - \frac{U_1(\omega)}{\omega_2} + \omega' \sqrt{\left(1 - \frac{1}{\omega_2^2} \right)} V_1(\omega) \qquad (6.52a)$$

$$V_2(\omega) = \omega\, V_1(\omega) - \frac{V_1(\omega)}{\omega_2} + \omega' \sqrt{\left(1 - \frac{1}{\omega_2^2} \right)} U_1(\omega) \qquad (6.52b)$$

Having obtained these new polynomials $U_2(\omega)$ and $V_2(\omega)$, the cycle is repeated with the third prescribed zero, and so on until all N of the prescribed zeros (including those at $\omega_n = \infty$) are used, specifically, $(N - 1)$ cycles.

This procedure may be very easily programmed—a compact Fortran subroutine is given below. The N ω-plane transmission zeros (including those at infinity) are contained in the complex array XP.

```
        X=1.0/XP(1)                   initialize with first prescribed zero ω₁
        Y=CDSQRT(1.0-X**2)
        U(1)=-X
        U(2)=1.0
        V(1)=Y
        V(2)=0.0
C
        DO 10 K=3, N+1                 multiply in second and subsequent
                                          prescribed zeros
        X=1.0/XP(K-1)
        Y=CDSQRT(1.0-X**2)
        DO 11 J=1, K-1                multiply by constant terms
        U2(J)=-U(J)*X-Y*V(J)
```

```
11    V2(J)=-V(J)*X+Y*U(J)
         DO 12 J=2, K                    multiply by terms in ω
         U2(J)=U2(J)+U(J-1)
12    V2(J)=V2(J)+V(J-1)
         DO 13 J=3, K                    multiply by terms in ω²
13    U2(J)=U2(J)+Y*V(J-2)
         DO 14 J=1, K                    update Uₙ and Vₙ
         U(J)=U2(J)
14    V(J)=V2(J)
10    CONTINUE                          reiterate for 3rd, 4th, ..., Nth
                                        prescribed zero.
```

If the same process is repeated for $G'_N(\omega) = U'_N(\omega) + V'_N(\omega)$ [equation (6.48b)], then $U'_N(\omega) = U_N(\omega)$ and $V'_N(\omega) = -V_N(\omega)$. Therefore, from equations (6.47) and (6.49), we obtain

$$\text{Num}[C_N(\omega)] = \frac{1}{2}\left[G_N(\omega) + G'_N(\omega)\right] = \frac{1}{2}((U_N(\omega) + U'_N(\omega))$$

$$+ (V_N(\omega) + V'_N(\omega))) = U_N(\omega) \tag{6.53}$$

Equation (6.53) demonstrates that the numerator of $C_N(\omega)$ [which has the same zeros as $F(\omega)$] is equal to $U_N(\omega)$ after $(N - 1)$ cycles of this recursion method. Now the zeros of $F(\omega)$ are revealed by finding roots of $U_N(\omega)$, and along with the prescribed zero polynomial $P(\omega)/\varepsilon$, the denominator polynomial $E(\omega)$ is constructed by using the alternating singularity principle:

- Construct the complex polynomial $P(\omega)/\varepsilon - jF(\omega)/\varepsilon_R$ [equation (6.32c)], and find its zeros (which will alternate between the lower and upper halves of the ω plane).
- Take the conjugate of any zero that is in the lower half of the ω plane (this is equivalent to reflecting zeros in the right half of the s plane about the imaginary axis to satisfy the Hurwitz condition).
- Reconstruct the polynomial to obtain $E(\omega)$.

To illustrate the procedure, the recursions are applied to a 4th-degree example with an equiripple return loss level of 22 dB and prescribed zeros at $+j1.3217$ and $+j1.8082$, chosen to give two attenuation lobe levels of 30 dB each on the upper side of the passband.

Initializing equation (6.50) with $\omega_1 = 1.3217$ yields

$$U_1(\omega) = -0.7566 + \omega$$
$$V_1(\omega) = \omega'(0.6539)$$

After the first cycle, with $\omega_2 = 1.8082$, we obtain

$$U_2(\omega) = -0.1264 - 1.3096\omega + 1.5448\omega^2$$
$$V_2(\omega) = \omega'(-0.9920 + 1.4871\omega)$$

and after the second cycle, with $\omega_3 = \infty$

$$U_3(\omega) = 0.9920 - 1.6134\omega - 2.3016\omega^2 + 3.0319\omega^3$$

$$V_3(\omega) = \omega'(-0.1264 - 2.3016\omega + 3.0319\omega^2)$$

After the third cycle, with $\omega_4 = \infty$, we obtain

$$U_4(\omega) = 0.1264 + 3.2936\omega - 4.7717\omega^2 - 4.6032\omega^3 + 6.0637\omega^4$$

$$V_4(\omega) = \omega'(0.9920 - 1.7398\omega - 4.6032\omega^2 + 6.0637\omega^3$$

At this stage, the polynomial $U_4(\omega)$ is the unnormalized numerator polynomial of the reflection function $S_{11}(\omega)$ (the polynomial $F_4(\omega)$), and finding its roots yields the N in-band reflection zeros. Roots of $V_4(\omega)$ yield the $N-1$ in-band reflection maxima. The s-plane coordinates of the zeros together with the corresponding transmission poles are listed in Table 6.2a, and plots of the transfer and reflection characteristics are shown in Figure 6.7.

TABLE 6.2   (4–2) Asymmetric Chebyshev Filter Function with Two Prescribed Transmission Zeros

	a. Singularities of Transfer and Reflection Function			
	Reflection Zeros [Roots of $U_4(s) = F_4(s)$]	Transmission Zeros (Prescribed)	Transmission/ Reflection Poles [Roots of $E_4(s)$]	In-Band Reflection Maxima [Roots of $V_4(s)$]
1	$-j0.8593$	$+j1.3217$	$-0.7437 - j1.4178$	$-j0.4936$
2	$-j0.0365$	$+j1.8082$	$-1.1031 + j0.1267$	$+j0.3796$
3	$+j0.6845$	$j\infty$	$-0.4571 + j0.9526$	$+j0.8732$
4	$+j0.9705$	$j\infty$	$-0.0977 + j1.0976$	—
	$\varepsilon_R = 1.0$	$\varepsilon = 1.1548$		

	b. Corresponding Polynomials		
$s^i, i =$	$E(s)$	$F(s)$	$P(s)$
0	$-0.1268 - j2.0658$	$+0.0208$	$-j2.3899$
1	$+2.4874 - j3.6255$	$-j0.5432$	$+3.1299$
2	$+3.6706 - j2.1950$	$+0.7869$	$j1.0$
3	$+2.4015 - j0.7591$	$-j0.7591$	
4	$+1.0$	$+1.0$	

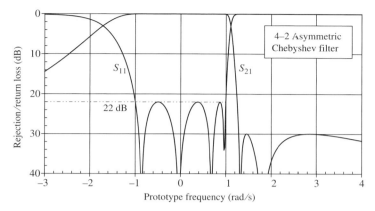

Figure 6.7 Lowpass prototype transfer and reflection characteristics of the (4–2) asymmetric Chebyshev filter with two prescribed transmission zeros at $s_1 = +j1.3217$ and $s_2 = +j1.8082$.

6.3.3 Polynomial Forms for Symmetric and Asymmetric Filtering Functions

Table 6.2b lists the coefficients of the $E(s)$, $F(s)$, and $P(s)$ polynomials formed from the singularities in Table 6.2a, normalized in each case to set the highest-degree coefficient to unity. In addition, the coefficients of $P(s)$ are multiplied by j to preserve the unitary conditions, because in this case, $N - n_{fz} = 2$ is an even integer (see Section 6.1). With these polynomials, we are now ready to proceed with the synthesis of a prototype electrical network, which, under analysis, results in exactly the same performance characteristics as the original polynomials.

At this stage it is worth studying the form of the polynomials of the preceding 4th degree asymmetric filter to help understand the distinctive characteristics of the electrical network that will be synthesized from them:

- Because the zeros of $E(s)$ [the poles of the filter function] are asymmetrically located about both the real and the imaginary axes, all except the leading coefficient of $E(s)$ are complex.
- Zeros of $F(s)$ [the zeros of the reflection function $S_{11}(s)$] are distributed asymmetrically along the imaginary axis. This means that the coefficients of $F(s)$ alternate between purely real and purely imaginary as the power of s increases.
- Similarly, because the zeros of $P(s)$ [the zeros of the transfer function $S_{21}(s)$] all lie on the imaginary axis (or can be distributed symmetrically about the imaginary axis), the coefficients of $P(s)$ also alternate between purely real and purely imaginary as the power of s increases.

By analyzing these polynomials with frequency variable s, ranging from $-j\infty$ to $+j\infty$, the asymmetric transfer and reflection characteristics are as shown in Figure 6.7.

In Chapter 7, we examine the network synthesis techniques. After the polynomials of the $[ABCD]$ transfer matrix are formed from the coefficients of $E(s)$ and $F(s)$, an electrical network is synthesized from these matrix polynomials. As the network progressively builds up component by component, the polynomials are correspondingly reduced in degree one by one. This procedure is referred to as "extracting" the components from the polynomials. At the end of a successful synthesis cycle, when the electrical network has been fully synthesized, the coefficients of the polynomials are zero apart from one or two constants.

It can be demonstrated that if the coefficients of the polynomials are complex, it is necessary to extract components known as *frequency-invariant reactances* (FIRs) from the polynomials, in addition to the capacitors and inductors in the course of building up the network and reducing the polynomial coefficients to zero. Eventually, the FIRs translate into frequency offsets from the bandpass filter's nominal center frequency ("asynchronously tuned"), giving the required asymmetric performance.

If the transmission zeros of the filter function are prescribed symmetrically about the real axis, then the pattern of the singularities of the polynomials $E(s)$ and $F(s)$, which are generated from these transmission zeros, are also symmetric about the real axis. Symmetry about the real and imaginary axes means that the polynomials are composed of purely real coefficients:

- $E(s)$ is an Nth-degree polynomial with real coefficients.
- $P(s)$ is an even polynomial of degree n_{fz} with real coefficients, where n_{fz} is the number of prescribed finite-position transmission zeros ranging from zero ("all pole" function) to N ("fully canonical").
- $F(s)$ is an Nth-degree even polynomial if N is even and odd if N is odd, again with purely real coefficients.

The purely real coefficients indicate that any FIR that is extracted during the synthesis process is zero in value, and that all the resonators of the filter are synchronously tuned at centre frequency.

6.4 PREDISTORTED FILTER CHARACTERISTICS

Synthesis techniques for the filter functions described so far assume that the filter is composed of lossless (nondissipative) components. In practice, all components dissipate some energy. As described in Section 3.9, the response of the filter with dissipative components is less selective and has a more rounded appearance as compared to the lossless prototype filter. Transmission and reflection zeros are less distinct, and in bandpass filters the shoulders of the response near the band edges are more rounded. Figure 6.8 shows the amplitude response of the $(4-2)$ asymmetric 40-MHz wide bandpass filter at 12 GHz as a function of the unloaded Q. Filter response based on lossless elements ($Q_u = $ infinity) is also included. The effects of progressively worsening Q_u are clear.

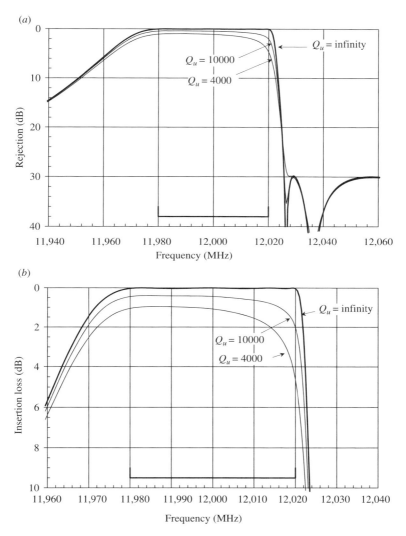

Figure 6.8 Transfer characteristics of a (4–2) bandpass filter with Q_u = infinity, 10,000, and 4000: (*a*) rejection; (*b*) in-band insertion loss.

To analyze the transfer and reflection responses of a filter with a finite Q_u, a positive real factor σ is added to the purely imaginary frequency variable $s = j\omega$, such that $s = \sigma + j\omega$ (see Fig. 6.9). As described in Section 3.9, σ is calculated from

$$\sigma = \frac{f_0}{\text{BW}} \cdot \frac{1}{Q_u} \tag{6.54}$$

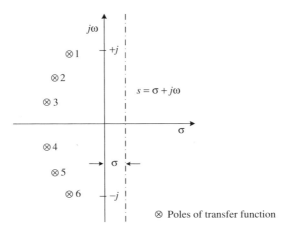

Figure 6.9   Track of the frequency variable s in the complex plane with a finite-value Q_u.

where Q_u is the unloaded Q factor of the filter's resonators, and f_0 and BW are respectively the center frequency and design bandwidth of the bandpass filter. The effect of adding σ to the frequency variable causes s to track to the right of the $j\omega$ axis as it moves from $-j\infty$ to $+j\infty$ in frequency.

The same analysis results as for the lossless case will be obtained if all the transfer/reflection singularities, based on the lossless prototype filter case, are also shifted to the right by the amount σ. By doing this, we find that the magnitudes and phases of the vectors drawn between the frequency variable position and the positions of the poles and zeros of the filter function remain unchanged. Now, we are effectively analyzing a lossless circuit, but the rightward shift of the singularities has resulted in an analyzed response equal to the lossy case.

Predistortion synthesis involves synthesizing an ideal transfer function (e.g., Chebyshev) and then moving all the poles of the filter function [the zeros of the polynomial $E(s)$] to the right by the amount σ, derived from the anticipated unloaded Q of the filter structure [6]. With the poles in these new positions, a *lossless* analysis gives the predistorted response. In the presence of the finite Q_u that has been anticipated for the resonators of the real filter, the frequency variable $s = \sigma + j\omega$ tracks to the right of the imaginary axis by the amount σ. However, since the poles have also been pre-shifted to the right by this same amount, they are now in the correct relative positions in relation to s for the ideal lossless response to be recovered. Figure 6.10 illustrates the procedure.

For filter functions with finite transmission zeros, it is not possible to recover the ideal response with dissipative elements. This is because in order to be able to synthesize a realizable network, there must be symmetry of the TZs about the imaginary axis; it is not possible to give them a one-way shift by the amount σ. However, the poles dominate the in-band response, and the deviation of the transfer characteristic $S_{21}(s)$ from ideal will not be significant if the TZs are left in their original positions.

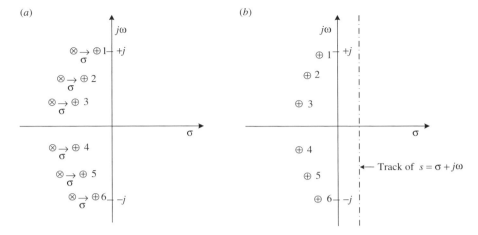

Figure 6.10 Predistortion synthesis: (a) poles shifted to right by amount σ before synthesis; (b) poles are now in correct nominal positions relative to the track of the frequency variable with loss ($s = \sigma + j\omega$) to give equivalent lossless response under analysis.

So, the synthesis of a predistorted network might proceed as follows:

- Derive the $S_{21}(s)$ polynomials $E(s)$ and $P(s)$ for the ideal response.
- Compute the loss factor σ by using equation (6.54), knowing the center frequency and design bandwidth of the realized bandpass filter.
- Shift the positions of the poles by the real amount $+\sigma$, $p_k \rightarrow p_k + \sigma$, $k = 1, 2, \ldots, N$, that is, the poles are shifted to the right by an amount σ. The real part of p_k should always be negative to satisfy the Hurwitz condition.
- With the original $P(s)$ polynomial, recalculate ε. This is relatively arbitrary provided $S_{21}(s) = (P(s)/\varepsilon)/E(s) \leq 1$ at any value of s, usually chosen so that $S_{21}(s) = 1$ at its maximum value.
- Recalculate the numerator polynomial $F(s)$ of reflection function $S_{11}(s)$. This must be done using the conservation of energy equation (6.2), because in general the zeros of $S_{11}(s)$ and $S_{22}(s)$ will be off-axis and as mentioned in Section 6.3, the alternating singularity method cannot be used.

However, to fully compensate for a finite Q_u, effectively restoring it to infinity usually incurs a rather large insertion loss penalty. In practice, it is usual to only partly compensate for the finite Q_u, such as by compensating Q_u of 4000 to an effective Q (Q_{eff}) of 10,000. For partial compensation the equation for σ [see equation (6.54)] becomes

$$\sigma = \frac{f_0}{\text{BW}} \left(\frac{1}{Q_{\text{act}}} - \frac{1}{Q_{\text{eff}}} \right) \tag{6.55}$$

Partial compensation tends to reduce insertion loss and improve return loss without seriously degrading the in-band linearity performance from that of the ideal filter.

Yu et al. [7] introduced a method whereby the amounts that the poles are shifted to the right are weighted, for example, in a sinusoidal fashion such that those poles closest to the imaginary axis (usually in the proximity of the band edges) are moved less than those near the centre of the band. This also helps reduce insertion loss and improve return loss (easier tuning).

Figure 6.11 reflects the effects on the Four-pole filter intended for realization in rectangular waveguide resonators with a Q_u of 4000. Figure 6.11a depicts the response of the predistorted network analyzed with Q_u = infinity. This reveals

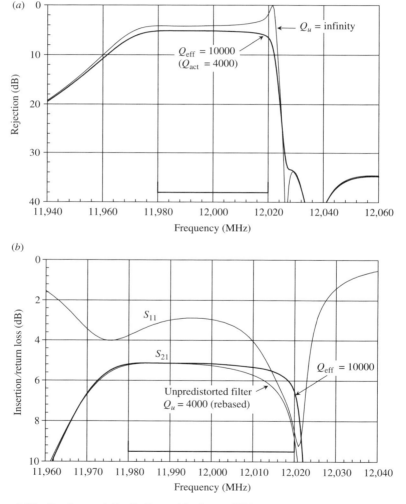

Figure 6.11 Predistorted (4–2) filter with $Q_{act} = 4000$ and $Q_{eff} = 10000$: (a) predistorted rejection characteristics with Q_u = infinity and $Q_{act} = 4000$ (effective Q of 10,000); (b) in-band return loss, and insertion loss compared with that of filter without predistortion.

the "cat's ear" predistortion of the characteristic. Figure 6.11*b* shows the filter response with the $Q_u = 4000$ that is expected from the real filter cavities. The in-band amplitude linearity is effectively improved to that of a filter with resonators of $Q_u = 10,000$, but with a standing insertion loss of about 5 dB. This loss is due partially to internal dissipative losses caused by finite Q_u values, but mostly by reflected RF energy (return loss) which is now tailored to compensate for the in-band roundings caused by the finite Q_u (Fig. 6.11*b*). The large in-band insertion loss implies that the predistortion technique can be applied only to filters in low-power (input) subsystems where RF gain is readily available to compensate, and isolators can be installed at the input and output to capture the reflected energy.

Effectively improving the Q_u of a filter means that the same in-band performance can be achieved with a filter with smaller and lighter cavities than a high-Q_u equivalent. Figure 6.12 offers an example of two C-band (10−4−4) filters, one made from dielectric resonator cavities and the smaller filter with an equivalent in-band performance from smaller and less complex coaxial resonator

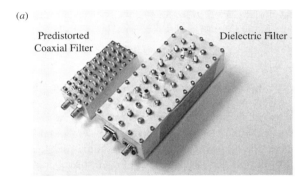

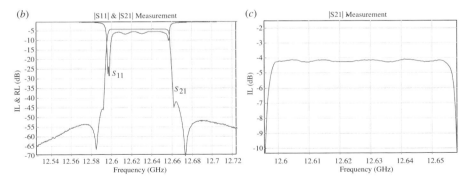

Figure 6.12 Predistorted C-band (10−4−4) coaxial filter: (*a*) comparison with unpredistorted dielectric filter; (*b*) rejection performance; (*c*) in-band insertion loss performance. (Courtesy COM DEV.)

cavities. The in-band amplitude and rejection performances of this predistorted filter are also given.

6.4.1   Synthesis of the Predistorted Filter Network

We now know that the zeros of reflection functions $S_{11}(s)$ and $S_{22}(s)$ [the roots of $F(s)$ and $F_{22}(s)$] of the predistorted filter are not located on the imaginary axis. However, for a synthesisable network, the zeros from $F(s)$ and $F_{22}(s)$ must form mirror-image pairs about the imaginary axis, to make up a symmetric overall pattern about the imaginary axis. Unlike the poles of the transfer and reflection function, there is no Hurwitz condition on the reflection zeros, and the individual zeros that make up the polynomial $F(s)$ can be arbitrarily chosen from the LHS or RHS of each pair. $F_{22}(s)$ is composed of the remaining zeros from each pair to form the complementary function to $F(s)$.

Thus, there are 2^N combinations of zeros that can be chosen to form $F(s)$ and $F_{22}(s)$, half of which are not simply exchanges of $F(s)$ and $F_{22}(s)$, that is, network reversals. Each combination leads to different values for the coefficients of $F(s)$ and, after the synthesis process, different values for the network elements.

A parameter that may be used to broadly classify the solutions is μ, where

$$\mu = \left| \sum_{k=1}^{N} \mathrm{Re}(s_k) \right| \tag{6.56}$$

and s_k are the N zeros that have been chosen from the LHS or RHS of each pair to form $F(s)$. It can be demonstrated that if the zeros are chosen so as to minimize μ, then the values of the elements of the folded-array network are maximally symmetric about the physical center of the network; however, the network is maximally asynchronously tuned. The opposite holds true if μ is maximized, that is all the zeros for $F(s)$ are chosen from only the LHS or RHS of each mirror-image pair. Here, the network is as synchronously tuned as possible, but the network element values are maximally asymmetric about the center of the network.

Although this principle holds true for any filtering characteristic, symmetric or otherwise, the effect is most convincingly demonstrated with an even-degree symmetric characteristic. The example chosen here is a 6th-degree 23-dB Chebyshev filter. It has a pair of transmission zeros at $\pm j1.5522$ to produce two rejection lobes on either side of the passband of 40 dB. The actual Q_u of the cavities is 4000, but predistortion has been applied to effectively raise the effective Q_u to 20,000.

Three possible arrangements for the $S_{11}(s)$ zeros are considered, one where the value of μ is maximized and two for the case where $\mu = 0$. There is only one arrangement for $\mu = \max$ where the zeros of $F(s)$, s_k are all in the left half (or right half) of the complex plane (Fig. 6.13a). Two possible solutions for $\mu = 0$ are portrayed in Figure 6.13b,c, where zeros s_k alternate between the LHP- and RHP as $j\omega$ increases, and another possibility where the zeros with positive imaginary parts are all in the LHP, and those with negative imaginary parts are in the RHP. The characteristic has been normalized such that two of the zeros lie on the imaginary axis. Note

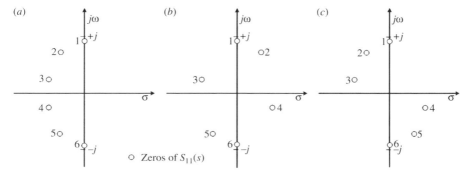

Figure 6.13 Possible arrangements for the zeros s_k of $F(s)$ for the symmetric (6–2) quasielliptic characteristic: (a) $\mu = $ max; (b) $\mu = 0$; (c) $\mu = 0$.

that the zeros of $F_{22}(s)$ assume mirror-image positions about the imaginary axis, to make up an overall symmetric pattern about the imaginary axis.

Let us apply the predistortion procedure to the filter function to increase its effective Q_u to 20,000. By using methods described in Chapter 8, the $N + 2$ coupling matrices for the three cases are synthesized, yielding the values as shown in Table 6.3. It is evident that in the first case ($\mu = $ max) that the filter is synchronously tuned; that is, the diagonal entries of the coupling matrix, which represent the resonant frequency offsets at each resonator node, are all zero. However, for this case the coupling elements in the matrix are asymmetric in value about the center of the filter (i.e., asymmetric about the cross-diagonal). The two cases where

TABLE 6.3 Folded Coupling Matrix Values for the Case Where $\mu = $ max and Two Cases for $\mu = 0$

Couplings	Case A ($\mu = $ max) (Fig. 6.13a)	Case B ($\mu = 0$) (Fig. 6.13b)	Case C ($\mu = 0$) (Fig. 6.13c)
M_{S1}	0.3537	0.9678	0.9677
M_{12}	0.8417	0.7890	0.6457
M_{23}	0.5351	0.4567	0.7284
M_{34}	0.6671	0.7223	0.5299
M_{45}	0.6642	$= M_{23}$	$= M_{23}$
M_{56}	1.1532	$= M_{12}$	$= M_{12}$
M_{6L}	1.3221	$= M_{S1}$	$= M_{S1}$
M_{25}	-0.1207	-0.0940	-0.1403
	Tuning Offsets		
M_{11}	0.0	-0.0859	0.4616
M_{22}	0.0	-0.2869	0.1824
M_{33}	0.0	0.5334	-0.3541
M_{44}	0.0	$= -M_{33}$	$= -M_{33}$
M_{55}	0.0	$= -M_{22}$	$= -M_{22}$
M_{66}	0.0	$= -M_{11}$	$= -M_{11}$

$\mu = 0$ exhibit symmetry of coupling values about the cross-diagonal, but are asynchronously tuned about the center of the filter.

As far as the electrical performance is concerned, the choice of which case to use for a practical application ($\mu = $ max or $\mu = $ min) is arbitrary, because $S_{21}(s)$ is the same for all cases. An interesting point is that for the $\mu = $ min cases, $S_{11}(s)$ and $S_{22}(s)$ characteristics display conjugate symmetry about the band center of the filter in the presence of finite Q_u. The reason for this becomes clear when we remember that the loss is represented by an offset for $j\omega$ to the right of the imaginary axis as s tracks from $-j\infty$ to $+j\infty$. On the practical side, it is easier to manufacture the physically symmetric $\mu = $ min cases, such as case B or case C in Table 6.3. Of these, case C is probably preferred because of the rather larger value for M_{25}. However, in all cases, tuning is considerably more difficult than that for the regular Chebyschev filters because of the low value and amorphous shape of the return loss characteristic. A "lossy synthesis" approach for lumped-element filters is presented by Hunter [8]. It involves the inclusion of resistive elements among the reactive components. For these predistorted networks, the amplitude characteristic is shaped by absorption rather than reflection, and the network becomes easier to tune.

6.5 TRANSFORMATION FOR DUAL-BAND BANDPASS FILTERS

Occasionally, applications arise for filters that exhibit two passbands within the confines of the passband of the original regular single-passband filter. The dual-band characteristic can be generated from a regular single-passband lowpass prototype (LPP) filter characteristic by the application of a symmetric mapping formula. The mapping transforms the original equiripple behavior (if the original LP filtering characteristic is Chebyshev) of the single passband over two subbands. The lower subband ranges from the lower edge of the original passband at $s = -j$ to $s = -jx_1$, and the upper passband extends from $s = +jx_1$ to $s = +j$, where the fractional parameters $\pm jx_1$ are the prescribed frequency points within the main passband, defining the inner edges of the two subbands (Fig. 6.14).

The effect of applying the transform is to double the degree of the filtering characteristic and the number of any transmission zeros ($N \to 2N$, $n_{fz} \to 2n_{fz}$). Despite this, the original equiripple level and the levels of any rejection lobes (apart from the central lobe) are preserved. The dual-band characteristic is symmetric about zero frequency, even if the original lowpass prototype filtering function is asymmetric.

The dual-band characteristic is achieved by applying a symmetric frequency transform

$$s = as'^2 + b \tag{6.57}$$

where s' is the frequency variable mapped from the prototype s plane. Constants a and b are derived by considering the following boundary conditions

$$\begin{aligned}
\text{At} \quad s' &= \pm j, & s &= -a + b & = +j \\
\text{At} \quad s' &= \pm jx_1, & s &= -ax_1^2 + b & = -j
\end{aligned}$$

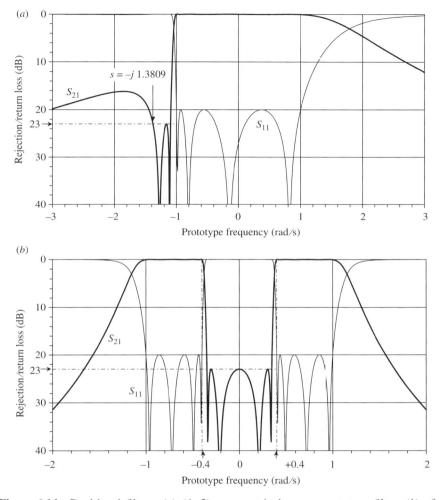

Figure 6.14 Dual-band filters: (a) (4-2) asymmetric lowpass prototype filter; (b) after transform to dual-band lowpass prototype with $x_1 = 0.4$.

giving

$$a = \frac{-2j}{1 - x_1^2} \quad \text{and} \quad b = \frac{-j(1 + x_1^2)}{1 - x_1^2} \tag{6.58}$$

Substituting equation (6.58) back into equation (6.57) yields the transform formula

$$s_i' = j\sqrt{\frac{1}{2}\left[(1 + x_1^2) - js_i(1 - x_1^2)\right]}$$

and

$$s'_{N+i} = s'^{*}_{i} \qquad (6.59)$$

for $i = 1, 2, \ldots, N$. With this transform, the Ns plane singularities of the $E(s)$, $F(s)$ polynomials and the n_{fz} singularities of the $P(s)$ polynomial of the lowpass prototype filter can be mapped to the $2N$ and $2n_{fz}$ singularities, respectively, of the dual-band characteristic. Network synthesis can then proceed as usual on the doublesized polynomials to realize the dual-band filter.

Any transmission zeros below the main passband of the lowpass prototype can be mapped to frequencies between the two passbands of the dual-band characteristic. An inspection of equation (6.59) reveals that a transmission zero at $s = -j(1 + x_1^2)/(1 - x_1^2)$ maps to $s' = 0$. It imposes a limit to the position of the TZ on the lower side of the LPP that can map into a pair of TZs in the dual-band characteristic. For example, where $x_1 = 0.4$, $s_{\max} = -j1.3809$ and any TZ with a position more negative than this in the LPP, will map into a pair of real-axis zeros in the dual-band response.

Figure 6.14b exhibits an 8th-degree dual-band filter with four transmission zeros and $x_1 = \pm 0.4$. They are formed by the transformation of a $(4-2)$ asymmetric filter with 20 dB return loss level and optimized such that the rejection lobe level nearest to the passband and the point at $s = -j1.3809$ (which transforms to $s' = 0$) are both 23 dB (Fig. 6.14a). After transformation, an equiripple rejection level of 23 dB in the interband region is obtained.

Table 6.4 displays the s-plane locations of the original $(4-2)$ asymmetric filtering function, as well as with their transformed positions for the $(8-4)$ symmetric dual-band filter. Figure 6.15 presents the coupling matrix (shown here in a folded array configuration), synthesized from the transformed singularities by employing the

TABLE 6.4 Singularities of Transfer and Reflection Polynomials

	Reflection Zeros	Transmission Zeros	Transmission/Reflection Poles
		a. (4−2) Asymmetric Lowpass Prototype—20 dB R_L	
1	$-j0.9849$	$-j1.2743$	$-0.0405 - j1.0419$
2	$-j0.8056$	$-j1.1090$	$-0.2504 - j1.0228$
3	$-j0.1453$	$j\infty$	$-1.0858 - j0.4581$
4	$+j0.8218$	$j\infty$	$-0.8759 + j1.4087$
	$\varepsilon_R = 1.0$	$\varepsilon = 0.6590$	
		b. After Transform to (8−4) Dual-Band Characteristic with $x_1 = 0.4$	
	$\pm j0.9619$	$\pm j0.3380$	$-0.1679 \pm j1.0954$
	$\pm j0.7204$	$\pm j0.2117$	$-0.3247 \pm j0.7022$
	$\pm j0.4916$	$\pm j\infty$	$-0.1287 \pm j0.4086$
	$\pm j0.4078$	$\pm j\infty$	$-0.0225 \pm j0.3780$
	$\varepsilon_R = 1.0$	$\varepsilon = 3.7360$	—

	S	1	2	3	4	5	6	7	8	L
S	0	0.8024	0	0	0	0	0	0	0	0
1	0.8024	0	0.8467	0	0	0	0	0	0	0
2	0	0.8467	0	0.4142	0	0	0	−0.2900	0	0
3	0	0	0.4142	0	0.2754	0	−0.4170		0	0
4	0	0	0	0.2754	0	0.2832		0	0	0
5	0	0	0	0	0.2832	0	0.2754	0	0	0
6	0	0	0	−0.4170	0	0.2754	0	0.4142	0	0
7	0	0	−0.2900	0	0	0	0.4142	0	0.8467	0
8	0	0	0	0	0	0	0	0.8467	0	0.8024
L	0	0	0	0	0	0	0	0	0.8024	0

Figure 6.15 Coupling matrix for (8–4) dual-band filtering characteristic (folded-array configuration).

methods described in Chapters 7 and 8. Note that the dual-band coupling matrix is synchronously tuned, even though the original prototype is asymmetric. Further similarity transforms (rotations) can be applied to the matrix to transform it to another configuration, such as a cascade quartet.

SUMMARY

In Chapter 3, we described the classical filters, namely, the Butterworth, Chebyshev, and elliptic function filters. In Chapter 4, we described an efficient procedure to generate completely arbitrary filter functions using computer-aided design techniques. Although this technique is quite efficient, it does require use of optimization to generate the desired filter function. In this chapter, we describe an important class of equiripple filter functions that can be derived by making use of a quasianalytic technique.

This chapter begins with a review of some of the important scattering parameter relationships (particularly the unitary conditions) that are relevant to the synthesis procedures introduced in later chapters. This is followed by a discussion of the general kind of Chebyshev polynomial and its application in filter design. The classical Chebyshev filter, described in Chapter 3, is derived by making use of the Chebyshev polynomial of the first kind. Some books and many articles use the expression "Chebyshev polynomial" to refer exclusively to the Chebyshev polynomial $T_n(x)$ of the first kind. Filters, obtained by such polynomials, have their transmission zeros located at infinity. The general Chebyshev polynomial described in this chapter can be utilized in generating filter functions with the maximum number of equiripple amplitude peaks in the passband for an arbitrary distribution of the zeros of the transfer function. To ensure physical realizability, the zeros of the transfer function must be located symmetrically about the imaginary axis. This constraint is related to satisfying the unitary condition. The derivation of this condition, along with a recursion technique to generate the transfer and reflection polynomials of this class of filter is described. The procedure is completely general and is applicable to realize symmetric as well as asymmetric response filters. Examples are included to illustrate the design procedure.

The final part of the chapter outlines the benefits of incorporating predistortion in the filter design for certain applications. The chapter concludes with a design example of a predistorted filter and a dual-band filter.

REFERENCES

1. R. E. Collin, *Foundations for Microwave Engineering*, McGraw-Hill, New York, 1966.
2. J. Helszajn, *Synthesis of Lumped Element, Distributed and Planar Filters*, McGraw-Hill, Basingstoke, UK, 1990.
3. J. D. Rhodes and S. A. Alseyab, The generalized Chebyshev low-pass prototype filter, *IEEE Trans. Circuit Theory* **8**, 113–125 (1980).
4. R. J. Cameron, Fast generation of Chebyshev filter prototypes with asymmetrically prescribed transmission zeros, *European Space Agency J.* **6**, 83–95 (1982).
5. R. J. Cameron, General coupling matrix synthesis methods for Chebyshev filtering functions, *IEEE Trans. Microwave Theory Tech.* **MTT-47**, 433–442 (April 1999).
6. A. E. Williams, W. G. Bush, and R. R. Bonetti, Predistortion technique for multicoupled resonator filters methods for Chebyshev filtering functions, *IEEE Trans. Microwave Theory Tech.* **MTT-33**, 402–407 (May 1985).
7. M. Yu, W.-C. Tang, A. Malarky, V. Dokas, R. J. Cameron, and Y. Wang, Predistortion technique for cross-coupled filters and its application to satellite communication systems, *IEEE Trans. Microwave Theory Tech.* **MTT-51**, 2505–2515, (Dec. 2003).
8. I. C. Hunter, *Theory and Design of Microwave Filters*, Electromagnetic Waves Series 48, IEE, London, 2001.

CHAPTER 7

SYNTHESIS OF NETWORK–CIRCUIT APPROACH

In the previous chapters, methods were established to derive the transfer and reflection polynomials for a broad class of lowpass prototype filter functions. The next step in the design process is to translate these polynomials into a prototype electrical circuit from which a real microwave filter may be developed. Two methods for doing this are available: the classical circuit synthesis method and the direct coupling matrix synthesis approach. In this chapter, the circuit synthesis approach, based on the [ABCD] transfer matrix, or the "chain" matrix, as it is sometimes known, is described.

There are many excellent and comprehensive contributions to the subject of circuit synthesis techniques in the literature [1–6]. It is not the intention to repeat the works here, but rather to utilize the theories to develop a general microwave filter synthesis technique. The techniques described here include both, the symmetric as well as asymmetric lowpass prototype filters. As described in Sections 3.10 and 6.1, such prototype networks require the hypothetical frequency-invariant reactive elements, referred to by the acronym FIR [7]. Inclusion of such elements leads to transfer and reflection polynomials to have complex coefficients. In bandpass or bandstop filters, FIR elements appear as frequency offsets to the resonant circuits. The various components used in the microwave synthesis process include the following:

- Frequency-dependent reactive elements: lumped capacitors and inductors. Their number in the lowpass prototype network determines the "degree" or "order" of

Microwave Filters for Communication Systems: Fundamentals, Design, and Applications,
by Richard J. Cameron, Chandra M. Kudsia, and Raafat R. Mansour
Copyright © 2007 John Wiley & Sons, Inc.

the filter network. Capacitors and inductors in a ladder network are interchangeable through the dual-network theorem.

- Frequency-invariant reactance or FIR[1] elements. The classical network synthesis theory is founded on the concept of the *positive real* condition for the driving point immittance function $Z(s)$ or $Y(s)$, that is, for a realizable network, $\text{Re}(Z(s))$ or $\text{Re}(Y(s)) > 0$, when $\text{Re}(s) > 0$, and $Z(s)$ or $Y(s)$ is real for s real [17]. With the introduction of FIRs into the network, the second condition (which implies symmetry of the frequency response) is no longer necessary and the driving-point function becomes a *positive* function [11]. The FIR represents an offset in the resonant frequency of a resonator from nominal (*nominal* here means zero for a lowpass resonator, and f_0 for a bandpass resonator), as described in Figure 7.1. For the asymmetric class of filters, the driving-point immittance polynomials possess coefficients that cannot be extracted as capacitors or inductors. However, these coefficients can be extracted as FIRs that are realized eventually by a microwave resonant element (a waveguide cavity or a dielectric resonator) that is tuned to be offset from the *nominal* center frequency of the filter; thus, the filter is *asynchronously* tuned.

- Frequency-independent transmission-line phase lengths, particularly the quarter-wave (90°) impedance or admittance inverters (collectively known as *immittance inverters*). In a microwave circuit, these elements act as transformers with a frequency-invariant 90° phase shift, and are approximately realized by a variety of microwave structures such as inductive irises and coupling probes. The use of inverters significantly simplify the physical realization of filters at microwave frequencies.

- Coupling inverters act as coupling elements between the resonant nodes of a network, representing a microwave filter. Those that couple directly between sequentially numbered resonators are known as *mainline couplings*; those between nonsequential resonators are known as cross-couplings, and those that couple between the source port and the load port and internal resonators of the filter are known as *input/output couplings*. There can be more than one coupling from the source or load ports to the internal resonators nodes in the filter, and there is a direct source–load coupling for canonical filter functions.

Figure 7.1 (*a*) $Y_{in} = 0$ when $s = s_0 = 0$—therefore, this lowpass resonator resonates at zero frequency; (*b*) $Y_{in} = sC_1 + jB_1 = 0$ when $s = s_0$. Thus the resonant frequency $s_0 = j\omega_0 = -jB_1/C_1$ that is, is offset by $\omega_0 = -B_1/C_1$ from the nominal center frequency.

[1]In common usage, the acronym "FIR" is used for both frequency-independent reactances *and* susceptances.

7.1 CIRCUIT SYNTHESIS APPROACH

For a two-port network operating between unity source and load terminations, the [ABCD] matrix representing the network has the following form [9]

$$[ABCD] = \frac{1}{jP(s)/\varepsilon} \cdot \begin{bmatrix} A(s) & B(s) \\ C(s) & D(s) \end{bmatrix} \qquad (7.1a)$$

where

$$S_{12}(s) = S_{21}(s) = \frac{P(s)/\varepsilon}{E(s)} = \frac{2P(s)/\varepsilon}{A(s) + B(s) + C(s) + D(s)} \qquad (7.1b)$$

$$S_{11}(s) = \frac{F(s)/\varepsilon_R}{E(s)} = \frac{A(s) + B(s) - C(s) - D(s)}{A(s) + B(s) + C(s) + D(s)} \qquad (7.1c)$$

$$S_{22}(s) = \frac{(-1)^N F(s)^*/\varepsilon_R}{E(s)} = \frac{D(s) + B(s) - C(s) - A(s)}{A(s) + B(s) + C(s) + D(s)} \qquad (7.1d)$$

The j that multiplies $P(s)$ in the denominator of equation (7.1) allows the cross-couplings to be extracted as inverters, and is there, in addition to the j that multiplies $P(s)$ when $N - n_{fz}$ is even, to preserve the orthogonality unitary condition (see Section 6.2). For nonunity source and load terminations R_S and R_L, the [ABCD] matrix is scaled as follows:

$$[ABCD] = \frac{1}{jP(s)/\varepsilon} \cdot \begin{bmatrix} \sqrt{\dfrac{R_L}{R_S}}A(s) & \dfrac{B(s)}{\sqrt{R_S R_L}} \\ \sqrt{R_S R_L}C(s) & \sqrt{\dfrac{R_S}{R_L}}D(s) \end{bmatrix} \qquad (7.2)$$

From equation (7.1), it is immediately clear that the $A(s)$, $B(s)$, $C(s)$, and $D(s)$ polynomials share a common denominator polynomial $P(s)/\varepsilon$. It is demonstrated that these polynomials can be built up from the elements of the circuit that models the real microwave filter, and then related to the coefficients of the $E(s)$ and $F(s)/\varepsilon_R$ polynomials, which, together with $P(s)/\varepsilon$, represent the desired transfer and reflection performance of the filter, respectively.

Our first task, therefore, is to relate the $A(s)$, $B(s)$, $C(s)$, and $D(s)$ polynomials to the S parameters that represent the transfer and reflection characteristics of the filter function. This is accomplished by building the [ABCD] matrix for a simple three-element ladder network, relating it to a real filter circuit and examining the form of the resulting $A(s)$, $B(s)$, $C(s)$, $D(s)$, and $P(s)$ polynomials. These polynomials are then compared directly with the coefficients of the polynomials that constitute $S_{21}(s)$ and $S_{11}(s)$. We also study the form of the [ABCD] polynomials for the more advanced circuits, namely, the asymmetric, cross-coupled, and single-ended types, in order to generalize the procedure.

Figure 7.2 depicts a simple lossless three-element prototype ladder circuit that is capable of realizing a lowpass filter. The three transmission zeros for this circuit all

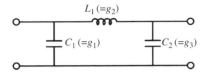

Figure 7.2   Basic lowpass prototype circuit.

lie at $s = j\infty$, which means that it is an all-pole filter. The network synthesis procedure is demonstrated by first building the $A(s)$, $B(s)$, $C(s)$, and $D(s)$ polynomials, corresponding to a third-order lowpass prototype ladder network with the elements C_1, L_1, and C_2 (sometimes known as the g values $g_1, g_2, g_3, \ldots$ of the prototype ladder network [8]). Then, the synthesis procedure is applied to demonstrate the recovery of the values of the original network elements C_1, L_1, and C_2 from the $A(s)$, $B(s)$, $C(s)$, and $D(s)$ polynomials.

7.1.1 Buildup of [*ABCD*] Matrix for the Third-Degree Network

Figure 7.2 represents a third-degree lowpass network.

Step A. Cascade Shunt C_1 with Series L_1

$$\begin{bmatrix} A & B \\ C & D \end{bmatrix} = \begin{bmatrix} 1 & 0 \\ sC_1 & 1 \end{bmatrix} \cdot \begin{bmatrix} 1 & sL_1 \\ 0 & 1 \end{bmatrix}$$

2nd-degree network ($N = 2$)

degree of polynomials :

$$= \begin{bmatrix} 1 & sL_1 \\ sC_1 & 1 + s^2L_1C_1 \end{bmatrix}$$

$A(s)$: $N - 2$

$B(s)$: $N - 1$

$C(s)$: $N - 1$

$D(s)$: N

$$\begin{bmatrix} A & B \\ C & D \end{bmatrix} = \begin{bmatrix} 1 & sL_1 \\ sC_1 & 1 + s^2C_1L_1 \end{bmatrix} \cdot \begin{bmatrix} 1 & 0 \\ sC_2 & 1 \end{bmatrix}$$

3rd-degree network ($N = 3$)

degree of polynomials:

$$= \begin{bmatrix} 1 + s^2L_1C_2 & sL_1 \\ s(C_1 + C_2) + s^3C_1C_2L_1 & 1 + s^2C_1L_1 \end{bmatrix}$$

$A(s)$: $N - 1$

$B(s)$: $N - 2$

$C(s)$: N

$D(s)$: $N - 1$

Note the form of the polynomials $A(s)$, $B(s)$, $C(s)$, and $D(s)$. For a lowpass prototype network of degree N, where N is even, $B(s)$ and $C(s)$ are of degree $N - 1$, $D(s)$ is of degree N, and $A(s)$ is of degree $N - 2$. For N odd, $A(s)$ and $D(s)$ are of degree $N - 1$, $C(s)$ is of degree N, and $B(s)$ is of degree $N - 2$. Thus, $A(s)$ and $C(s)$ are always even, and $B(s)$ and $D(s)$ are always odd.

7.1.2 Network Synthesis

Now we reverse the process. The coefficients of the polynomials $A(s)$, $B(s)$, $C(s)$, and $D(s)$ are known, and we need to find the values of C_1, L_1, and C_2 and extract them from the polynomials in sequence.

The values of the components nearest to the ports of the network are found through the network's short-circuit admittance (y) parameters or the open-circuit (z) parameters, and evaluating these parameters at $s = j\infty$. Knowing the value of the component, we can now extracted it by premultiplying the $[ABCD]$ matrix with its inverse, leaving a unit matrix in cascade with a remainder matrix.

Step A.1. Evaluate C_1   We can find C_1 from the short-circuit admittance parameter y_{11} or the open-circuit impedance parameter z_{11}. Starting at the left-hand side (LHS) of the network, we first find y_{11} or z_{11}, and evaluate C_1 at $s = j\infty$. First, we find y_{11} and z_{11}. For the general $[ABCD]$ matrix, we obtain

$$v_1 = A(s)v_2 + B(s)i_2$$
$$i_1 = C(s)v_2 + D(s)i_2$$

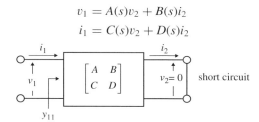

When

$$v_2 = 0, \quad y_{11} = \frac{i_1}{v_1} = \frac{D(s)}{B(s)}$$

$$i_2 = 0, \quad z_{11} = \frac{v_1}{i_1} = \frac{A(s)}{C(s)}$$

The overall $[ABCD]$ matrix is

$$\begin{bmatrix} A & B \\ C & D \end{bmatrix} = \begin{bmatrix} 1 + s^2 L_1 C_2 & sL_1 \\ s(C_1 + C_2) + s^3 C_1 C_2 L_1 & 1 + s^2 C_1 L_1 \end{bmatrix}$$

$$z_{11} = \frac{A(s)}{C(s)} = \frac{1 + s^2 L_1 C_2}{s(C_1 + C_2) + s^3 C_1 C_2 L_1}$$

$$sz_{11}|_{s \to \infty} = \frac{sA(s)}{C(s)}\bigg|_{s \to \infty} = \frac{1}{C_1}$$

Alternatively

$$y_{11} = \frac{D(s)}{B(s)} = \frac{1 + s^2 L_1 C_1}{s L_1}$$

$$\left. \frac{y_{11}}{s} \right|_{s \to \infty} = \left| \frac{D(s)}{s B(s)} \right|_{s \to \infty} = C_1$$

Step A.2. Extract C_1   The element extraction procedure is as follows. The [*ABCD*] matrix

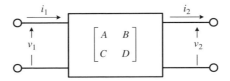

consists of a matrix for shunt C_1 in cascade with remainder [*ABCD*] matrix.

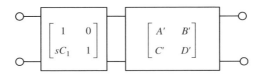

To extract C_1, premultiply by the inverse of the C_1 matrix

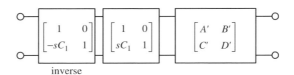

This leaves the remainder matrix in cascade with a unitary matrix, which can be ignored:

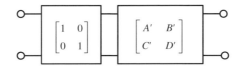

The overall [*ABCD*] matrix is given by

$$\begin{bmatrix} A & B \\ C & D \end{bmatrix} = \begin{bmatrix} 1 + s^2 L_1 C_2 & s L_1 \\ s(C_1 + C_2) + s^3 C_1 C_2 L_1 & 1 + s^2 C_1 L_1 \end{bmatrix} \tag{7.3}$$

Premultiply the matrix by the inverse matrix of shunt C_1 as follows:

$$\begin{bmatrix} A & B \\ C & D \end{bmatrix} = \begin{bmatrix} 1 & 0 \\ -sC_1 & 1 \end{bmatrix} \cdot \begin{bmatrix} 1 + s^2 L_1 C_2 & sL_1 \\ s(C_1 + C_2) + s^3 C_1 C_2 L_1 & 1 + s^2 C_1 L_1 \end{bmatrix} = \begin{bmatrix} A' & B' \\ C' & D' \end{bmatrix}$$

$$\therefore A'(s) = A(s)$$
$$B'(s) = B(s)$$
$$C'(s) = C(s) - sC_1 A(s)$$
$$D'(s) = D(s) - sC_1 B(s)$$

The remainder matrix then becomes

$$\begin{bmatrix} A' & B' \\ C' & D' \end{bmatrix} = \begin{bmatrix} 1 + s^2 L_1 C_2 & sL_1 \\ sC_2 & 1 \end{bmatrix} \tag{7.4}$$

Step B. Evaluate the Series Element L_1 Using the Remainder Matrix from Step A.2.

$$z_{11} = \frac{A(s)}{C(s)} = \frac{1 + s^2 L_1 C_2}{sC_2}$$

$$\therefore L_1 = \frac{z_{11}}{s}\bigg|_{s \to \infty} = \frac{A(s)}{sC(s)}\bigg|_{s \to \infty} \tag{7.5}$$

Alternatively, we obtain

$$y_{11} = \frac{D(s)}{B(s)} = \frac{1}{sL_1}$$

$$\therefore \frac{1}{L_1} = \frac{sD(s)}{B(s)}\bigg|_{s \to \infty} \tag{7.6}$$

Step B.1. Extract L_1 Premultiply the remainder matrix from the previous extraction (step A.2), by the inverse of matrix of series L_1

$$\begin{bmatrix} 1 & -sL_1 \\ 0 & 1 \end{bmatrix} \cdot \begin{bmatrix} 1 + s^2 L_1 C_2 & sL_1 \\ sC_2 & 1 \end{bmatrix} = \begin{bmatrix} A' & B' \\ C' & D' \end{bmatrix} \tag{7.7}$$

$$\therefore A'(s) = A(s) - sL_1 C(s)$$
$$B'(s) = B(s) - sL_1 D(s)$$
$$C'(s) = C(s)$$
$$D'(s) = D(s)$$

The remainder matrix then becomes

$$\begin{bmatrix} A' & B' \\ C' & D' \end{bmatrix} = \begin{bmatrix} 1 & 0 \\ sC_2 & 1 \end{bmatrix} \tag{7.8}$$

which is the matrix of C_2 in shunt. Premultiplication by the inverse matrix of C_2 leaves the unit matrix $\begin{bmatrix} 1 & 0 \\ 0 & 1 \end{bmatrix}$. Now we know the values of C_1, L_1, and C_2, and so the synthesis of the network is complete.

At each stage of the synthesis procedure, two values for each of the elements C_1, L_1, and C_2 are derived, one from the $C(s)$ and $A(s)$ polynomials and the other from the $B(s)$ and $D(s)$ polynomials. Nominally, they should be the same value, but with higher-degree networks, cumulative errors can cause the values to diverge quite significantly. Various techniques exist to mitigate the errors, such as taking advantage of any symmetry in the network, or by using transformed variable (Z-plane) synthesis techniques [10–12]. With the numerical accuracies available from 32-bit computers, it is possible to attain six decimal places of accuracy for the values of components for moderate degrees of network, say up to 12th degree, without using special techniques. However, it is important to use *only* the $A(s)$ and $C(s)$ polynomials or *only* the $B(s)$ and $D(s)$ polynomials to evaluate the series of components in the ladder network. Sometimes it is informative to evaluate the series first with the $A(s)$ and $C(s)$ polynomials, and then with the $B(s)$ and $D(s)$ polynomials. When the values of the components, determined by the two methods, begin to diverge in the sixth decimal place, then it will be known that the cumulating errors are starting to become significant.

7.2 LOWPASS PROTOTYPE CIRCUITS FOR COUPLED-RESONATOR MICROWAVE BANDPASS FILTERS

The doubly terminated lowpass prototype circuits, connected to their terminating impedances or admittances [8], are depicted in Figure 7.3. The element values in the network are termed the "g parameters" [8], which may be the value of a shunt capacitor or a series inductor. The source termination g_0 is resistive if g_1 is a capacitor, and conductive if g_1 is an inductor, and similarly for the load termination g_{N+1}. This gives an alternating impedance–admittance–impedance (or vice versa) reading, left to right along the ladder network. The circuits are formulated in this way so that identical responses are obtained from whichever form is used.

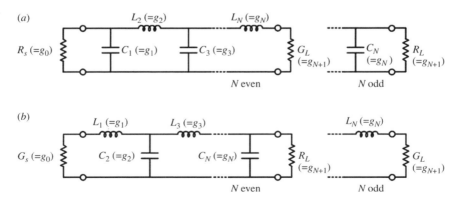

Figure 7.3 Lowpass prototype ladder networks: (*a*) where the leading component is a shunt capacitor; (*b*) the dual of the network.

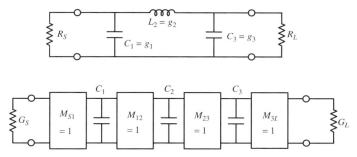

Figure 7.4 Third-degree lowpass prototype ladder networks: the form after addition of the unit inverters.

At this stage, the dual-network theorem is applied to add unit inverters to the network to transform the series components (inductors) to shunt capacitors. Figure 7.4 shows the series inductor L_2 of a third-degree network that is transformed to a shunt capacitor (with the same value as that of the inductor, g_2). Also, two unit inverters M_{S1} and M_{3L} are added to transform the terminating impedances into conductances. Now the network is all admittance or conductances, and is known as a *parallel resonator* lowpass prototype circuit. These inverters have a one-to-one relationship with the physical coupling elements in the finally realized filter structure.

A coupled-cavity microwave bandpass filter is realized directly from lowpass prototype circuits of the type shown in Figure 7.4. After the addition of the series and parallel resonating components [to form the *bandpass prototype* (BPP)], and the appropriate scaling to the center frequency and bandwidth of the RF bandpass filter, the design of the microwave structures or devices used to realize the elements of the bandpass circuit becomes apparent. These resonators are realized in the microwave structure by elements such as coaxial resonators. Figure 7.5 illustrates the procedure for a fourth-degree all-pole filter.

In this chapter, we concentrate on the synthesis of parallel-connected components in an open-wire transmission environment, because many waveguide and coaxial elements are modeled directly as shunt-connected elements. These include the inductive or capacitive apertures in rectangular waveguides [13]. If the series element is required, it can be obtained through the use of the dual-network theorem, where a shunt-connected capacitor becomes a series-connected inductor of the same value, or an admittance (J) inverter becomes an impedance (K) inverter [14–16].

7.2.1 Synthesis of the [*ABCD*] Polynomials for Circuits with Inverters

We use the third-degree network in Figure 7.4 to derive the general form of the polynomials for an inverter-coupled parallel resonator network:

Cascade the first inverter M_{S1} and the first shunt capacitor C_1:

$$\begin{bmatrix} 0 & j \\ j & 0 \end{bmatrix} \cdot \begin{bmatrix} 1 & 0 \\ sC_1 & 1 \end{bmatrix} = j\begin{bmatrix} 0 & 1 \\ 1 & 0 \end{bmatrix} \cdot \begin{bmatrix} 1 & 0 \\ sC_1 & 1 \end{bmatrix} = j\begin{bmatrix} sC_1 & 1 \\ 1 & 0 \end{bmatrix} \tag{7.9}$$

$$S_{11}(s) = \frac{F(s)/\varepsilon_R}{E(s)} \qquad S_{22}(s) = \frac{F_{22}(s)/\varepsilon_R}{E(s)} \qquad S_{21}(s) = S_{12}(s) = \frac{P(s)/\varepsilon}{E(s)} \qquad \begin{array}{l}\text{Transfer and reflection}\\ \text{polynomials}\end{array}$$

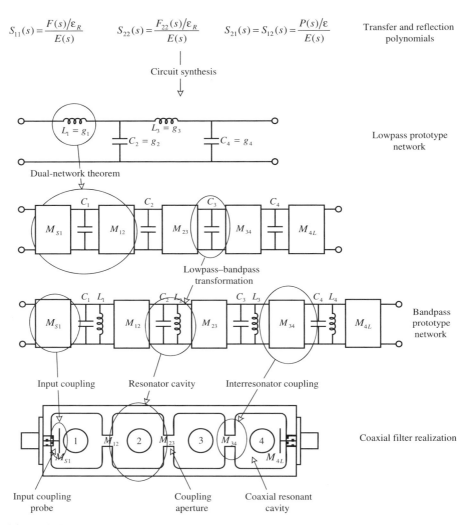

Figure 7.5   Steps in the synthesis process for a fourth-order coaxial resonator bandpass filter.

Cascade the second inverter M_{12}:

$$\begin{bmatrix} sC_1 & 1 \\ 1 & 0 \end{bmatrix} \cdot j \begin{bmatrix} 0 & 1 \\ 1 & 0 \end{bmatrix} = -\begin{bmatrix} 1 & sC_1 \\ 0 & 1 \end{bmatrix} \tag{7.10}$$

Cascade the second shunt capacitor C_2:

$$-\begin{bmatrix} 1 & sC_1 \\ 0 & 1 \end{bmatrix}\begin{bmatrix} 1 & 0 \\ sC_2 & 1 \end{bmatrix} = -\begin{bmatrix} 1 + s^2C_1C_2 & sC_1 \\ sC_2 & 1 \end{bmatrix} \tag{7.11}$$

Cascade the third inverter M_{23}:

$$-\begin{bmatrix} 1 + s^2 C_1 C_2 & s C_1 \\ s C_2 & 1 \end{bmatrix} \cdot j \begin{bmatrix} 0 & 1 \\ 1 & 0 \end{bmatrix} = -j \begin{bmatrix} s C_1 & 1 + s^2 C_1 C_2 \\ 1 & s C_2 \end{bmatrix} \qquad (7.12)$$

At this stage, we have included two frequency-variant elements C_1 and C_2 and have, in effect, a second-degree filter network ($N = 2$). Note the form of the polynomials:

$A(s)$ and $D(s)$ odd polynomial degree $N - 1$
$B(s)$ even polynomial degree N
$C(s)$ even polynomial degree $N - 2$

Then cascade the third shunt capacitor C_3:

$$\begin{bmatrix} s C_1 & 1 + s^2 C_1 C_2 \\ 1 & s C_2 \end{bmatrix} \cdot \begin{bmatrix} 1 & 0 \\ s C_3 & 1 \end{bmatrix} = -j \begin{bmatrix} s(C_1 + C_3) + s^3 C_1 C_2 C_3 & 1 + s^2 C_1 C_2 \\ 1 + s^2 C_2 C_3 & s C_2 \end{bmatrix}$$
$$(7.13)$$

Cascade the fourth inverter M_{3L}:

$$-j \begin{bmatrix} s(C_1 + C_3) + s^3 C_1 C_2 C_3 & 1 + s^2 C_1 C_2 \\ 1 + s^2 C_2 C_3 & s C_2 \end{bmatrix} \cdot j \begin{bmatrix} 0 & 1 \\ 1 & 0 \end{bmatrix}$$
$$= \begin{bmatrix} 1 + s^2 C_1 C_2 & s(C_1 + C_3) + s^3 C_1 C_2 C_3 \\ s C_2 & 1 + s^2 C_2 C_3 \end{bmatrix} \qquad (7.14)$$

Note again the forms of the polynomials for this third degree circuit ($N = 3$):

$A(s)$ and $D(s)$ even polynomial degree $N - 1$
$B(s)$ odd polynomial degree N
$C(s)$ odd polynomial degree $N - 2$

These cascade operations reveal some important features of the $A(s)$, $B(s)$, $C(s)$, and $D(s)$ polynomials that represent the filter ladder network. All are polynomials in variable s with real coefficients and are even or odd, depending on whether N is even or odd. Independently of whether N is even or odd, $B(s)$ is of degree N and is the highest degree. $A(s)$ and $D(s)$ are of degree $N - 1$, and $C(s)$ is the lowest degree $N - 2$.

Next, we relate the $A(s)$, $B(s)$, $C(s)$, and $D(s)$ polynomials with the $S_{21}(s)$ and $S_{11}(s)$ transfer and reflection polynomials ($E(s)$, $P(s)/\varepsilon$ and $F(s)/\varepsilon_R$). For the general case, the coefficients of $E(s)$ and $F(s)$ can have complex coefficients (see Section 3.10). To account for the general cases, it is necessary to add a frequency-independent reactance (FIR) in parallel with each shunt frequency-dependent element (the capacitors C_i) such that the value of the admittance at each node, instead of being simply sC_i, now becomes $sC_i + jB_i$.

The effect of adding the FIRs at each node is to convert the $A(s)$, $B(s)$, $C(s)$, and $D(s)$ polynomials for the ladder network into complex-even and complex-odd forms, that is, their coefficients alternate between purely real and purely imaginary as the power of s increases. This is demonstrated by taking, for example, the $B(s)$ polynomial for the 2nd-degree $[ABCD]$ matrix (7.12) with no FIR added as follows:

$$B(s) = -j(1 + s^2 C_1 C_2)$$

Replacing sC_1 with $sC_1 + jB_1$ and sC_2 with $sC_2 + jB_2$, we obtain

$$B(s) = -j[(1 - B_1 B_2) + j(B_1 C_2 + B_2 C_1)s + C_1 C_2 s^2]$$
$$= j[b_0 + jb_1 s + b_2 s^2] \tag{7.15}$$

where the coefficients b_i are purely real. Similarly for $D(s)$, we obtain

$$
\begin{aligned}
D(s) &= -j[sC_2] && \text{without a FIR} \\
D(s) &= -j[jB_2 + sC_2] && \text{with the FIR } jB_2 \\
&= j[jd_0 + d_1 s] && \tag{7.16}
\end{aligned}
$$

where the coefficients d_i are the coefficients of the $D(s)$ polynomial and are purely real.

In a similar fashion, the coefficients of $A(s)$ and $C(s)$ are derived. In general, for the even-degree cases:

$$
\begin{aligned}
A(s) &= j[ja_0 + a_1 s + ja_2 s^2 + a_3 s^3 + \cdots + a_{N-1} s^{N-1}] \\
B(s) &= j[b_0 + jb_1 s + b_2 s^2 + jb_3 s^3 + \cdots + jb_{N-1} s^{N-1} + b_N s^N] \\
C(s) &= j[c_0 + jc_1 s + c_2 s^2 + jc_3 s^3 + \cdots + c_{N-2} s^{N-2}] \\
D(s) &= j[jd_0 + d_1 s + jd_2 s^2 + d_3 s^3 + \cdots + d_{N-1} s^{N-1}]
\end{aligned} \tag{7.17}
$$

Note that all the coefficients a_i, b_i, c_i, d_i, $i = 0, 1, 2, \ldots, N$, are real. In a similar way it can be shown that for the odd-degree case [see the 3rd-degree matrix in equation (7.14)]

$$
\begin{aligned}
A(s) &= a_0 + ja_1 s + a_2 s^2 \\
B(s) &= jb_0 + b_1 s + jb_2 s^2 + b_3 s^3 \\
C(s) &= jc_0 + c_1 s \\
D(s) &= d_0 + jd_1 s + d_2 s^2
\end{aligned} \tag{7.18}
$$

Typically, for N odd, we obtain

$$
\begin{aligned}
A(s) &= a_0 + ja_1 s + a_2 s^2 + ja_3 s^3 + \cdots + a_{N-1} s^{N-1} \\
B(s) &= jb_0 + b_1 s + jb_2 s^2 + b_3 s^3 + \cdots + jb_{N-1} s^{N-1} + b_N s^N \\
C(s) &= jc_0 + c_1 s + jc_2 s^2 + c_3 s^3 + \cdots + c_{N-2} s^{N-2} \\
D(s) &= d_0 + jd_1 s + d_2 s^2 + jd_3 s^3 + \cdots + d_{N-1} s^{N-1}
\end{aligned} \tag{7.19}
$$

Note that the j outside the brackets in equation (7.17) is a consequence of using inverters as coupling elements, each with its associated frequency-invariant 90° phase shift.

For the even-degree case, there are an odd number of inverters along the main path of the signal, and therefore, an odd number of 90° phase shifts; thus the multiplication by j. For the purpose of the synthesis method described next, the j can be safely ignored.

The reason for adding the FIRs is to introduce the purely imaginary coefficients ja_i, jb_i, into the $A(s)$, $B(s)$, $C(s)$, and $D(s)$ polynomials in addition to the purely real coefficients. These purely imaginary coefficients need to be present in the $A(s)$, $B(s)$, $C(s)$, and $D(s)$ polynomials to ensure that they are properly matched with the corresponding terms in the $E(s)$ and $F(s)$ polynomials when the latter are representing an asymmetric transfer function. In these cases, $E(s)$ and $F(s)$ have some complex coefficients, and with the inclusion of the FIRs, it is possible to relate network polynomials $A(s)$, $B(s)$, $C(s)$, and $D(s)$ with the transfer and reflection polynomials $E(s)$ $F(s)$ and $P(s)$ as follows.

Knowing now the form of the $A(s)$, $B(s)$, $C(s)$, and $D(s)$ polynomials, of the polynomials constituting $S_{11}(s)$ and $S_{21}(s)$ ($E(s)$, $F(s)/\varepsilon_R$ and $P(s)/\varepsilon$), and also the relationship between them [equation set (7.1)], the coefficients of $A(s)$, $B(s)$, $C(s)$, and $D(s)$ may be directly expressed in terms of the coefficients of $E(s)$ and $F(s)/\varepsilon_R$.

From equation set (7.1),

$$[ABCD] = \frac{1}{P(s)/\varepsilon} \begin{bmatrix} A(s) & B(s) \\ C(s) & D(s) \end{bmatrix}$$

where, for N even, we have

$$A(s) = j\,\mathrm{Im}(e_0 + f_0) + \mathrm{Re}(e_1 + f_1)s + j\,\mathrm{Im}(e_2 + f_2)s^2 + \cdots + j\,\mathrm{Im}(e_N + f_N)s^N$$

$$B(s) = \mathrm{Re}(e_0 + f_0) + j\,\mathrm{Im}(e_1 + f_1)s + \mathrm{Re}(e_2 + f_2)s^2 + \cdots + \mathrm{Re}(e_N + f_N)s^N$$

$$C(s) = \mathrm{Re}(e_0 - f_0) + j\,\mathrm{Im}(e_1 - f_1)s + \mathrm{Re}(e_2 - f_2)s^2 + \cdots + \mathrm{Re}(e_N - f_N)s^N$$

$$D(s) = j\,\mathrm{Im}(e_0 - f_0) + \mathrm{Re}(e_1 - f_1)s + j\,\mathrm{Im}(e_2 - f_2)s^2 + \cdots + j\,\mathrm{Im}(e_N - f_N)s^N$$

$$(7.20a)$$

and for N odd

$$A(s) = \mathrm{Re}(e_0 + f_0) + j\,\mathrm{Im}(e_1 + f_1)s + \mathrm{Re}(e_2 + f_2)s^2 + \cdots + j\,\mathrm{Im}(e_N + f_N)s^N$$

$$B(s) = j\,\mathrm{Im}(e_0 + f_0) + \mathrm{Re}(e_1 + f_1)s + j\,\mathrm{Im}(e_2 + f_2)s^2 + \cdots + \mathrm{Re}(e_N + f_N)s^N$$

$$C(s) = j\,\mathrm{Im}(e_0 - f_0) + \mathrm{Re}(e_1 - f_1)s + j\,\mathrm{Im}(e_2 - f_2)s^2 + \cdots + \mathrm{Re}(e_N - f_N)s^N$$

$$D(s) = \mathrm{Re}(e_0 - f_0) + j\,\mathrm{Im}(e_1 - f_1)s + \mathrm{Re}(e_2 - f_2)s^2 + \cdots + j\,\mathrm{Im}(e_N - f_N)s^N$$

$$(7.20b)$$

Here, e_i and f_i, $i = 0, 1, 2, \ldots, N$, are the complex coefficients of $E(s)$ and $F(s)/\varepsilon_R$, respectively.

It is easily demonstrated that these polynomials satisfy the equation set (7.1), and are the correct form for the representation of the electrical performance of a passive, reciprocal, lossless microwave circuit.

Form

- The coefficients of $A(s)$, $B(s)$, $C(s)$, and $D(s)$ polynomials alternate between purely real and purely imaginary (or vice versa) as the power of s increases.
- For noncanonical networks, $\varepsilon_R = 1$ and so the highest-degree coefficients of $E(s)$ and $F(s)/\varepsilon_R$, e_N, and f_N, respectively, also equal unity. Therefore, the highest degree coefficient of $B(s)$, $b_N = \mathrm{Re}(e_N + f_N) = 2$, and the highest-degree coefficients of the $A(s)$, $C(s)$, and $D(s)$ polynomials, $a_N = j\,\mathrm{Im}(e_N + f_N)$, $c_N = \mathrm{Re}(e_N - f_N)$, and $d_N = j\,\mathrm{Im}(e_N - f_N)$, respectively, are all zero (e_N and f_N are real numbers equal to unity). In addition, the process for building the $E(s)$ polynomial from the $F(s)$ and $P(s)$ polynomials results in $\mathrm{Im}(e_{N-1}) = \mathrm{Im}(f_{N-1})$ for $n_{fz} < N$; therefore, $c_{N-1} = j\,\mathrm{Im}(e_{N-1} - f_{N-1}) = 0$ as well. Thus $B(s)$ is of degree N, $A(s)$ and $D(s)$ are of degree $N - 1$, and $C(s)$ is of degree $N - 2$, in accordance with equations (7.17) and (7.19). The exception to this is the canonical transfer functions, where $n_{fz} = N$, that is, where $\varepsilon_R \neq 1$. Here, the polynomial $C(s)$ is also of degree N.

Equations

- It is easily shown by simple addition and subtraction in equation set (7.20) that the following relationships hold:

$$A(s) + B(s) + C(s) + D(s) = 2E(s)$$
$$A(s) + B(s) - C(s) - D(s) = 2F(s)$$
$$-A(s) + B(s) - C(s) + D(s) = 2F(s)^* = 2F_{22}(s) \quad (N \text{ even})$$
$$-A(s) + B(s) - C(s) + D(s) = -2F^*(-s) = -2F(s)^* = 2F_{22}(s) \quad (N \text{ odd}) \qquad (7.21)$$

which clearly satisfy equation set (7.1), repeated here for convenience as follows:

$$[ABCD] = \frac{1}{jP(s)/\varepsilon}\begin{bmatrix} A(s) & B(s) \\ C(s) & D(s) \end{bmatrix}$$

where

$$S_{12}(s) = S_{21}(s) = \frac{P(s)/\varepsilon}{E(s)} = \frac{2P(s)/\varepsilon}{A(s) + B(s) + C(s) + D(s)}$$

$$S_{11}(s) = \frac{F(s)/\varepsilon}{E(s)} = \frac{A(s) + B(s) - C(s) - A(s)}{A(s) + B(s) + C(s) + D(s)}$$

$$S_{22}(s) = \frac{(-1)^N F(s)^*/\varepsilon_R}{E(s)} = \frac{D(s) + B(s) - C(s) - A(s)}{A(s) + B(s) + C(s) + D(s)}$$

If the $A(s)$, $B(s)$, $C(s)$, and $D(s)$ polynomials are derived from equation (7.20), the constant ε is used directly, as determined previously. If the polynomials are normalized to b_N, the highest-degree coefficient of $B(s)$, ε is rederived by using the property of the [ABCD] matrix that at any frequency, its determinant equals unity:

$$A(s)D(s) - B(s)C(s) = \left[\frac{P(s)}{\varepsilon}\right]^2 \tag{7.22}$$

If we evaluate equation (7.22) at zero frequency, we need to consider only the constant coefficients of the polynomials

$$a_0d_0 - b_0c_0 = \left[\frac{p_0}{\varepsilon}\right]^2$$

or

$$\varepsilon = \left|\frac{p_0}{\sqrt{a_0d_0 - b_0c_0}}\right| \tag{7.23}$$

As described in Chapter 8, short-circuit admittance parameters (y parameters) are used for the direct synthesis of coupling matrices for filter networks. At this stage, the y-parameter matrix $[y]$ is found by using the $[ABCD] \rightarrow [y]$ parameter conversion formula [8] as follows

$$\frac{1}{P(s)/\varepsilon}\begin{bmatrix} A(s) & B(s) \\ C(s) & D(s) \end{bmatrix} \Longrightarrow \begin{bmatrix} y_{11}(s) & y_{12}(s) \\ y_{21}(s) & y_{22}(s) \end{bmatrix}$$

$$\begin{bmatrix} y_{11}(s) & y_{12}(s) \\ y_{21}(s) & y_{22}(s) \end{bmatrix} = \frac{1}{y_d(s)}\begin{bmatrix} y_{11n}(s) & y_{12n}(s) \\ y_{21n}(s) & y_{22n}(s) \end{bmatrix} \tag{7.24}$$

$$= \frac{1}{B(s)}\begin{bmatrix} D(s) & \dfrac{-\Delta_{ABCD}P(s)}{\varepsilon} \\ \dfrac{-P(s)}{\varepsilon} & A(s) \end{bmatrix}$$

where $y_{ijn}(s)$, $i,j = 1,2$, are the numerator polynomials of $y_{ij}(s)$, $y_d(s)$ is their common denominator polynomial, and Δ_{ABCD} is the determinant of the [ABCD] matrix, which for a reciprocal network, $\Delta_{ABCD} = 1$. From this, it becomes clear that

$$y_d(s) = B(s)$$
$$y_{11n}(s) = D(s)$$
$$y_{22n}(s) = A(s) \tag{7.25}$$
$$y_{21n}(s) = y_{12n}(s) = \frac{-P(s)}{\varepsilon}$$

demonstrating that the short-circuit admittance parameters can be built up from the coefficients of $S_{11}(s)$ and $S_{21}(s)$, in the same way as the [ABCD] polynomials are by using equation set (7.20). The open-circuit impedance parameters z_{ij} are built up in a similar way.

The $y(s)$ polynomials have the same form as the [ABCD] polynomials; namely, their coefficients alternate between purely real and purely imaginary as the power of s increases, $y_d(s)$ is of degree N, $y_{11n}(s)$ and $y_{22n}(s)$ are of degree $N - 1$, and $y_{21n}(s)$ and $y_{12n}(s)$ are of degree n_{fz} (the number of finite-position transmission zeros). The short-circuit admittance parameters are used for the direct coupling matrix synthesis method described in the next chapter.

7.2.2 Synthesis of the [*ABCD*] Polynomials for the Singly Terminated Filter Prototype

Singly terminated filter networks are designed to operate from very high- or very low- impedance sources. They were once used to couple directly from the outputs of thermionic valve amplifiers and certain transistor amplifiers with very high internal impedances (effectively, current sources). Singly terminated filters are considered here because the admittance looking in at the input part has a characteristic which is particularly useful for the design of contiguous-channel manifold multiplexers, described in Chapter 18.

Figure 7.6 shows the development of a network with a zero-impedance source to the Thévenin equivalent network with a load impedance Z_L attached. The voltage v_L at the load can be written [17–18] as follows:

$$v_L = \frac{-y_{12}v_S}{y_{22}} \cdot \frac{Z_L}{(Z_L + 1/y_{22})} \tag{7.26}$$

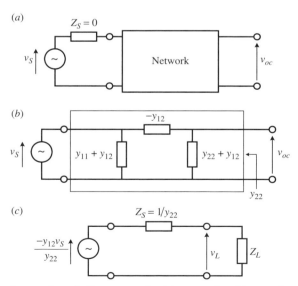

Figure 7.6 Singly terminated filter network: (*a*) network with zero-impedance voltage source; (*b*) network replaced with equivalent π network; (*c*) Thévenin equivalent circuit with load impedance Z_L attached.

Thus, the voltage gain of this network is

$$S_{21}(s) = \frac{P(s)/\varepsilon}{E(s)} = \frac{v_L}{v_S} = \frac{-y_{12}Z_L}{1 + Z_L y_{22}} \tag{7.27}$$

Setting the terminating impedance $Z_L = 1 \ \Omega$ and replacing y_{12} and y_{22} with $y_{12n}(s)/y_d(s)$ and $y_{22n}(s)/y_d(s)$, respectively, we write

$$\frac{P(s)/\varepsilon}{E(s)} = \frac{-y_{12n}(s)}{y_d(s) + y_{22n}(s)} \tag{7.28}$$

By splitting $E(s)$ into its complex-even and complex-odd components, we obtain

$$E(s) = m_1 + n_1$$

where

$$m_1 = \mathrm{Re}(e_0) + j\,\mathrm{Im}(e_1)s + \mathrm{Re}(e_2)s^2 + \cdots + \mathrm{Re}(e_N)s^N$$

$$n_1 = j\,\mathrm{Im}(e_0) + \mathrm{Re}(e_1)s + j\,\mathrm{Im}(e_2)s^2 + \cdots + j\mathrm{Im}(e_N)s^N \tag{7.29}$$

Substituting equation (7.29) into equation (7.28) yields

$$\frac{P(s)/\varepsilon}{m_1 + n_1} = \frac{-y_{12n}(s)}{y_d(s) + y_{22n}(s)} \tag{7.30}$$

$E(s)$ is normalized such that its highest-degree coefficient e^N equals unity. Also, we know that the coefficients of $y_{12n}(s)$, $y_{22n}(s)$, and $y_d(s)$ polynomials must alternate between purely real and purely imaginary as the power of s increases, and that the $y_d(s)$ polynomial must be one degree greater than $y_{22n}(s)$. As a result

For N even:	$y_d(s) = m_1,$	$y_{22n}(s) = n_1$
For N odd:	$y_d(s) = n_1,$	$y_{22n}(s) = m_1$
For N even or odd:	$y_{12n}(s) = y_{21n}(s) = \dfrac{-P(s)}{\varepsilon}$	(7.31)

Now the polynomials $y_{12n}(s)$, $y_{22n}(s)$, and $y_d(s)$ for the singly ended filter network are known in terms of the desired transfer function. In addition, this gives the $A(s)$ and $B(s)$ polynomials of the [ABCD] matrix, since $A(s) = y_{22n}(s)$ and $B(s) = y_d(s)$.

Thus, the $A(s)$ and $B(s)$ polynomials are easily found from the coefficients of the known $E(s)$ polynomial in a fashion similar to that used for the doubly terminated case. However, it remains to find the $C(s)$ and $D(s)$ polynomials to commence network synthesis. This is easily accomplished by a method due to Levy [12]. This method takes advantage of the property of an [ABCD] matrix for a reciprocal network, for which, the determinant of the matrix is unity at any value of the frequency variable s:

$$A(s)D(s) - B(s)C(s) = \left(\frac{P(s)}{\varepsilon}\right)^2 \tag{7.32}$$

Taking a 4th degree example and this time writing the relationship (7.32) out in matrix form:

4th degree case, $N = 4$

degree of $P(s)/\varepsilon$ $= n_{fz}$ = number of finite-position transmission zeros

degree of $A(s)$ and $D(s)$ $= N - 1 = 3$

degree of $B(s)$ $= N = 4$

degree of $C(s)$ $= N - 2 = 2$

                            $= N = 4$ for fully canonical cases ($n_{fz} = N$):

$$
\begin{bmatrix}
a_0 & 0 & 0 & 0 & b_0 & 0 & 0 & 0 & 0 \\
a_1 & a_0 & 0 & 0 & b_1 & b_0 & 0 & 0 & 0 \\
a_2 & a_1 & a_0 & 0 & b_2 & b_1 & b_0 & 0 & 0 \\
a_3 & a_2 & a_1 & a_0 & b_3 & b_2 & b_1 & b_0 & 0 \\
0 & a_3 & a_2 & a_1 & b_4 & b_3 & b_2 & b_1 & b_0 \\
0 & 0 & a_3 & a_2 & 0 & b_4 & b_3 & b_2 & b_1 \\
0 & 0 & 0 & a_3 & 0 & 0 & b_4 & b_3 & b_2 \\
0 & 0 & 0 & 0 & 0 & 0 & 0 & b_4 & b_3 \\
0 & 0 & 0 & 0 & 0 & 0 & 0 & 0 & b_4
\end{bmatrix}
\cdot
\begin{bmatrix}
d_0 \\ d_1 \\ d_2 \\ d_3 \\ -c_0 \\ -c_1 \\ -c_2 \\ -c_3 \\ -c_4
\end{bmatrix}
= \frac{1}{\varepsilon^2}
\begin{bmatrix}
p_0 \\ p_1 \\ p_2 \\ p_3 \\ \cdots \\ \cdots \\ p_{2nfz} \\ 0 \\ 0
\end{bmatrix}
\tag{7.33}
$$

where p_i, $i = 0, 1, 2, \ldots, 2n_{fz}$ are the $2n_{fz} + 1$ coefficients of $P(s)^2$. By inverting the left-hand square matrix (7.33) and postmultiplying it by the column matrix on the right-hand side (RHS), containing the $2n_{fz} + 1$ coefficients of $P(s)^2$, the coefficients of the $C(s)$ and $D(s)$ polynomials are found. Because $y_{11n}(s) = D(s)$, and $y_{21n}(s)$, $y_{22n}(s)$, and $y_d(s)$ are already known from equation (7.25), the admittance matrix $[y]$ for the filter network is also fully defined.

In practice, it is very rare that a singly terminated fully canonical filter design is needed. For the noncanonical cases where $n_{fz} < N$, the last two rows and columns of the square matrix and the last two rows of the two column matrices can be omitted before matrix (7.33) is evaluated.

7.3 LADDER NETWORK SYNTHESIS

The preceding sections describe the evaluation of the overall $[ABCD]$ and the equivalent admittance $[y]$ matrix from the characteristic polynomials of the desired filter function. From this matrix, we can synthesize the network given its topology. The synthesis process is similar to that described in Section 7.1.

From equation (7.1), the $[ABCD]$ matrix for the transfer functions is represented by

$$
[ABCD] = \frac{1}{jP(s)/\varepsilon} \cdot \begin{bmatrix} A(s) & B(s) \\ C(s) & D(s) \end{bmatrix}
$$

where $A(s)$, $B(s)$, $C(s)$, $D(s)$, and $P(s)$ are polynomials in the frequency variable s with coefficients that alternate between purely real or purely imaginary (or vice versa), as the power of s increases. For an Nth-degree characteristic with n_{fz} finite-position

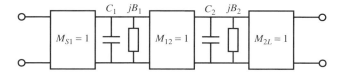

Figure 7.7 Second-order ladder network with FIRs and inverters.

transmission zeros, $A(s)$ and $D(s)$ are of degree $N - 1$, $B(s)$ is of degree N, $C(s)$ is of degree $N - 2$, [except for fully canonical cases where $C(s)$ is of degree N as well as $B(s)$], and $P(s)$ is of degree n_{fz}. When $n_{fz} = 0$ (an all-pole network), then $P(s)/\varepsilon$ is of zero degree (i.e., a constant $= 1/\varepsilon$) and the network can be synthesized as a regular ladder network without cross-couplings. If $n_{fz} > 0$, indicating that there are finite-position transmission zeros, then the network must be synthesized including the cross-couplings (couplings between nonadjacent resonator nodes), or by other means such as extracted pole.

To illustrate the synthesis procedure, we begin with the synthesis of a simple second-degree network without cross couplings but including the FIRs (see Fig. 7.7). Usually, FIRs are associated with cross-coupled networks that are realizing asymmetric filter characteristics. However, in this non-cross-coupled case, the network is of the asynchronously-tuned or conjugately-tuned symmetric type, introduced in Chapter 6 (predistorted filter with off-axis reflection zeros). The process for the extraction of parallel-coupled inverters is outlined, leading the way to the synthesis of more advanced networks, capable of realizing asymmetric and linear phase filters.

The [ABCD] matrix for the network in Figure 7.7 is constructed from equations (7.15) and (7.16):

$$[ABCD] = -j \begin{bmatrix} (jB_1 + sC_1) & (1 - B_1B_2) + j(B_1C_2 + B_2C_1)s + C_1C_2s^2 \\ 1 & (jB_2 + sC_2) \end{bmatrix} \quad (7.34)$$

The procedure for extracting the elements is the reverse of building up the [ABCD] matrix from the individual elements in the ladder network. Knowledge of the type of element and the order in which it appears in the network is required beforehand. Each element is extracted by premultiplying the overall [ABCD] matrix with the inverse [ABCD] matrix of the elemental component to be extracted, leaving a remainder matrix in series with a unit matrix, which can be ignored. The procedure for the extraction of series- and shunt-connected frequency-variant components is given in Section 7.1.

In this case, the first extracted element is a frequency-invariant inverter of unity characteristic admittance $J = M_{S1} = 1$ (see Fig. 7.8):

$$[ABCD] = \begin{bmatrix} 0 & j/J \\ jJ & 0 \end{bmatrix}$$

$$\text{Inverse} = \begin{bmatrix} 0 & -j/J \\ -jJ & 0 \end{bmatrix} = -j \begin{bmatrix} 0 & 1/J \\ J & 0 \end{bmatrix} \quad (7.35)$$

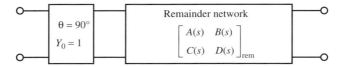

Figure 7.8 Extraction of the admittance inverter.

Putting $J = 1$ (unit inverter) and premultiplying the overall $[ABCD]$ matrix (7.34) yields

$$
\begin{bmatrix} A(s) & B(s) \\ C(s) & D(s) \end{bmatrix}_{\text{rem}} = -j \begin{bmatrix} 0 & 1 \\ 1 & 0 \end{bmatrix} \cdot (-j) \begin{bmatrix} A(s) & B(s) \\ C(s) & D(s) \end{bmatrix} = - \begin{bmatrix} C(s) & D(s) \\ A(s) & B(s) \end{bmatrix}
$$
$$
= - \begin{bmatrix} 1 & (jB_2 + sC_2) \\ (jB_1 + sC_1) & (1 - B_1 B_2) + j(B_1 C_2 + B_2 C_1)s + C_1 C_2 s^2 \end{bmatrix}
$$

(7.36)

The effect of extracting the unit inverter is to exchange the top and bottom rows of the $[ABCD]$ matrix. Since the admittance, looking into the network, is given by (assuming a unity load termination)

$$
Y_{\text{in}} = \frac{C(s) + D(s)}{A(s) + B(s)}
$$

(7.37)

extracting the inverter, in effect, inverts the input admittance:

$$
Y_{\text{in(rem)}} = \frac{1}{Y_{\text{in}}}
$$

(7.38)

The next element in the chain to be extracted is the shunt capacitor C_1. We select the frequency-variant component first, instead of the frequency-invariant reactance next to it. At this stage in the synthesis, the degree of the $D(s)$ polynomial is one degree greater than the $B(s)$ polynomial, and $C(s)$ is one degree greater than $A(s)$.

The short-circuit admittance parameter y_{11} is calculated by

$$
y_{11} = \frac{D(s)}{B(s)} = \frac{(1 - B_1 B_2) + j(B_1 C_2 + B_2 C_1)s + C_1 C_2 s^2}{(jB_2 + sC_2)}
$$

(7.39)

The numerator polynomial is one degree higher than the denominator, indicating that the next extracted component must be frequency-variant. Before the value of the component can be determined, we need to divide both sides of equation (7.39) by s and evaluate it as $s \to j\infty$:

$$
\left. \frac{y_{11}}{s} \right|_{s \to j\infty} = \left. \frac{D(s)}{sB(s)} \right|_{s \to j\infty} = C_1
$$

(7.40a)

Therefore, $y_{11} = sC_1$, indicating that C_1 is the value of a frequency-variant admittance. Alternatively, the open-circuit admittance z_{11} may be used:

$$z_{11} = \frac{A(s)}{C(s)} = \frac{1}{(jB_1 + sC_1)}$$

Multiplying both sides by s and evaluating at $s = j\infty$ yields

$$sz_{11}\bigg|_{s \to j\infty} = \frac{sA(s)}{C(s)}\bigg|_{s \to j\infty} = \frac{1}{C_1} \tag{7.40b}$$

or $z_{11} = 1/sC_1$.

Having determined the value of the shunt capacitor C_1, we now extract it from matrix (7.36) by

$$\begin{bmatrix} A(s) & B(s) \\ C(s) & D(s) \end{bmatrix}_{\text{rem}} = -\begin{bmatrix} 0 & 1 \\ -sC_1 & 0 \end{bmatrix}$$

$$\cdot \begin{bmatrix} 1 & (jB_2 + sC_2) \\ (jB_1 + sC_1) & (1 - B_1B_2) + j(B_1C_2 + B_2C_1)s + C_1C_2s^2 \end{bmatrix}$$

$$= -\begin{bmatrix} 1 & (jB_2 + sC_2) \\ jB_1 & (1 - B_1B_2) + jsB_1C_2 \end{bmatrix} \tag{7.41}$$

Note that in the remainder matrix, both $C(s)$ and $D(s)$ decrease in degree by one, and are now the same degree as $A(s)$ and $B(s)$, respectively. This indicates that the next component to be extracted should be frequency-invariant. Also, note that there is no C_1 component in the remainder matrix, which indicates that the extraction of the parallel capacitor C_1 has been successful.

Again by evaluating at $s = j\infty$, but this time, without multiplying or dividing by s, we obtain

$$y_{11}\big|_{s \to j\infty} = \frac{D(s)}{B(s)}\bigg|_{s \to j\infty} = \frac{(1 - B_1B_2) + jsB_1C_2}{(jB_2 + sC_2)}\bigg|_{s \to j\infty} = jB_1$$

$$z_{11}\big|_{s \to j\infty} = \frac{A(s)}{C(s)}\bigg|_{s \to j\infty} = \frac{1}{jB_1} \tag{7.42}$$

Now this shunt frequency-invariant reactance can be extracted from the matrix in (7.40) by

$$\begin{bmatrix} A(s) & B(s) \\ C(s) & D(s) \end{bmatrix}_{\text{rem}} = -\begin{bmatrix} 0 & 1 \\ -jB_1 & 0 \end{bmatrix} \cdot \begin{bmatrix} 1 & (jB_2 + sC_2) \\ jB_1 & (1 - B_1B_2) + jsB_1C_2 \end{bmatrix}$$

$$= -\begin{bmatrix} 1 & (jB_2 + sC_2) \\ 0 & 1 \end{bmatrix} \tag{7.43}$$

Again, it is evident that both $C(s)$ and $D(s)$ have dropped by one degree, and that jB_1 no longer appears in the remainder matrix, indicating a successful extraction.

The next component in the cascade is another unity admittance inverter that is extracted, as before, leaving a remainder matrix

$$
\begin{bmatrix} A(s) & B(s) \\ C(s) & D(s) \end{bmatrix}_{\text{rem}} = j \begin{bmatrix} 0 & 1 \\ 1 & (jB_2 + sC_2) \end{bmatrix}
\tag{7.44}
$$

C_2 and jB_2 can now be extracted in the same way as for C_1 and jB_1. Because the $C(s)$ and $D(s)$ polynomials are one degree greater than $A(s)$ and $B(s)$, respectively, the next component to be extracted is required to be frequency-variant (C_2), followed by the frequency-invariant reactance (jB_2). Following the same procedure as before, the remainder matrix after extracting C_2 and jB_2 is

$$
\begin{bmatrix} A(s) & B(s) \\ C(s) & D(s) \end{bmatrix}_{\text{rem}} = j \begin{bmatrix} 1 & 0 \\ -jB_2 - sC_2 & 1 \end{bmatrix} \begin{bmatrix} 0 & 1 \\ 1 & (jB_2 + sC_2) \end{bmatrix}
$$

$$
= j \begin{bmatrix} 0 & 1 \\ 1 & 0 \end{bmatrix}
\tag{7.45}
$$

which is simply the final inverter.

Now the circuit is fully synthesized and the $A(s)$, $B(s)$, $C(s)$, and $D(s)$ polynomials are reduced to constants or zero. During the synthesis process, the polynomials are reduced degree by degree, while the ladder network is built up, element by element.

However, all this has not yet affected the $P(s)$ polynomial, which forms the denominator of the $[ABCD]$ matrix. The $P(s)$ polynomial is simply a constant ($=1/\varepsilon$) for the all-pole transfer functions; that is, no finite-position transmission zeros (TZs) are present in the complex s plane. As a result, no cross-couplings (couplings between nonadjacent resonator nodes) are required. When TZs do appear in finite positions in the complex plane (on the imaginary axis or in pairs symmetric about the imaginary axis), then the $P(s)$ polynomial is finite in degree, the degree corresponding to the number of finite-position zeros n_{fz}. To reduce the $P(s)$ polynomial to a constant, along with the $A(s)$, $B(s)$, $C(s)$, and $D(s)$ polynomials, *parallel coupled inverters* (PCIs) need to be extracted at appropriate stages in the synthesis procedure. The theory behind the extraction of a PCI is presented. It is followed by an example of the synthesis process for the fourth-degree, asymmetric (4–2) filter, used as the example in Chapter 6 for the synthesis of the $S_{21}(s)$ and $S_{11}(s)$ polynomials for the Chebyshev type of transfer function. The order in which the circuit elements are extracted from the polynomials is important, and so are the rules governing the extraction sequence.

Parallel-Coupled Inverters—Extraction Process

The inverter is extracted in parallel from between the input and output terminals of the two-port network, leaving a remainder matrix from which further parallel-coupled inverters can be extracted if necessary (see Fig. 7.9).

The initial $[ABCD]$ matrix is in the form, described by equation (7.1), where the $P(s)$ polynomial is finite in degree (i.e., where TZs are present). If $P(s)$ is less in degree than

(a)

(b)

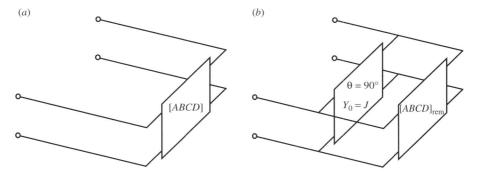

$\theta = 90°$

$Y_0 = J$

[ABCD]

[ABCD]$_{\text{rem}}$

Figure 7.9 Extraction of a parallel-coupled inverter (PCI): (*a*) overall network as represented by the [*ABCD*] matrix; (*b*) with an inverter of characteristic admittance *J* in parallel with the remainder matrix [*ABCD*]$_{\text{rem.}}$

the initial or remainder network that it is being extracted from, then the value of the characteristic admittance, *J*, of the PCI, is zero. Note also that the polynomial $P(s)$ is multiplied by *j*; this allows the cross-couplings to be extracted as 90° inverters (or transformers). This is in addition to the multiplication of $P(s)$ by *j*, when $N + n_{f_z}$ is even.

Extraction of Parallel Admittance Inverters The overall [*ABCD*] matrix is

$$[ABCD] = \frac{1}{jP(s)/\varepsilon} \cdot \begin{bmatrix} A(s) & B(s) \\ C(s) & D(s) \end{bmatrix} \tag{7.46}$$

The equivalent *y* matrix is

$$[y] = \begin{bmatrix} y_{11} & y_{12} \\ y_{21} & y_{22} \end{bmatrix} = \frac{1}{B(s)} \cdot \begin{bmatrix} D(s) & \dfrac{-jP(s)}{\varepsilon} \\ \dfrac{-jP(s)}{\varepsilon} & A(s) \end{bmatrix} \tag{7.47}$$

The [*ABCD*] matrix for the parallel admittance inverter *J* is

$$[ABCD]_{\text{inv}} = \begin{bmatrix} 0 & \dfrac{j}{J} \\ jJ & 0 \end{bmatrix} \tag{7.48}$$

and its equivalent *y* matrix is

$$[y]_{\text{inv}} = \begin{bmatrix} 0 & jJ \\ jJ & 0 \end{bmatrix} \tag{7.49}$$

The overall *y* matrix [*y*] is the sum of the admittance matrix for the inverter and the admittance matrix [*y*]$_{\text{rem}}$, the remaining matrix of the network after the inverter is extracted:

$$[y] = [y]_{\text{inv}} + [y]_{\text{rem}} \tag{7.50}$$

Therefore

$$[y]_{\text{rem}} = [y] - [y]_{\text{inv}} = \frac{1}{B(s)} \cdot \begin{bmatrix} D(s) & \dfrac{-jP(s)}{\varepsilon} \\ \dfrac{-jP(s)}{\varepsilon} & A(s) \end{bmatrix} - \begin{bmatrix} 0 & jJ \\ jJ & 0 \end{bmatrix}$$

$$= \frac{1}{B(s)} \cdot \begin{bmatrix} D(s) & -j\left(\dfrac{P(s)}{\varepsilon} + JB(s)\right) \\ -j\left(\dfrac{P(s)}{\varepsilon} + JB(s)\right) & A(s) \end{bmatrix} \qquad (7.51)$$

Reconverting to the [ABCD] matrix yields

$$[ABCD]_{\text{rem}} = \frac{1}{jP_{\text{rem}}(s)} \cdot \begin{bmatrix} A_{\text{rem}}(s) & B_{\text{rem}}(s) \\ C_{\text{rem}}(s) & D_{\text{rem}}(s) \end{bmatrix} = \frac{-1}{y_{21\text{rem}}} \cdot \begin{bmatrix} y_{22\text{rem}} & 1 \\ \Delta_{y\text{rem}} & y_{11\text{rem}} \end{bmatrix} \qquad (7.52)$$

$$A_{\text{rem}}(s) = \frac{-y_{22\text{rem}}}{y_{21\text{rem}}} = \frac{A(s)}{-j(P(s)/\varepsilon + JB(s))} \qquad (7.53\text{a})$$

$$B_{\text{rem}}(s) = \frac{-1}{y_{21\text{rem}}} = \frac{B(s)}{j(P(s)/\varepsilon + JB(s))} \qquad (7.53\text{b})$$

$$C_{\text{rem}}(s) = \frac{-\Delta_{y\text{rem}}}{y_{21\text{rem}}} = \frac{1}{B(s)} \cdot \frac{\left(A(s)D(s) + (P(s)/\varepsilon + JB(s))^2\right)}{j(P(s)/\varepsilon + JB(s))}$$

$$= \frac{A(s)D(s) + (P(s)/\varepsilon)^2 + 2JB(s)P(s)/\varepsilon + (JB(s))^2}{jB(s)(P(s)/\varepsilon + JB(s))}$$

Since $A(s)D(s) - B(s)C(s) = -(P(s)/\varepsilon)^2$, it follows that

$$C_{\text{rem}}(s) = \frac{B(s)C(s) + 2JB(s)P(s)/\varepsilon + (JB(s))^2}{jB(s)(P(s)/\varepsilon + JB(s))} = \frac{C(s) + 2JP(s)/\varepsilon + J^2B(s)}{j(P(s)/\varepsilon + JB(s))}$$

$$(7.53\text{c})$$

and finally

$$D_{\text{rem}}(s) = \frac{-y_{11\text{rem}}}{y_{21\text{rem}}} = \frac{D(s)}{j(P(s)/\varepsilon + JB(s))} \qquad (7.53\text{d})$$

Thus, the remainder [ABCD] matrix has the following form:

$$[ABCD]_{\text{rem}} = \frac{1}{jP_{\text{rem}}(s)/\varepsilon} \cdot \begin{bmatrix} A_{\text{rem}}(s) & B_{\text{rem}}(s) \\ C_{\text{rem}}(s) & D_{\text{rem}}(s) \end{bmatrix}$$

$$= \frac{1}{j(P(s)/\varepsilon + JB(s))} \cdot \begin{bmatrix} A(s) & B(s) \\ C(s) + 2JP(s)/\varepsilon + J^2B(s) & D(s) \end{bmatrix} \qquad (7.54)$$

Now the elements of the remainder $[ABCD]$ matrix, $[ABCD]_{rem}$, are known in terms of the elements of the original matrix and the value of parallel inverter J. The value J is found by knowing that after the extraction of the parallel inverter, the degree of $P_{rem}(s)$, the denominator polynomial of $[ABCD]_{rem}$, must be one degree less than that of $P(s)/\varepsilon$ polynomial of the original $[ABCD]$ matrix.

In order for the inverter to be extracted with a nonzero value, the $P(s)$ and $B(s)$ polynomials are both nth degree, and are written out in terms of their coefficients as follows:

$$\frac{P(s)}{\varepsilon} = p_0 + p_1 s + p_2 s^2 + \cdots + p_n s^n \tag{7.55a}$$

and

$$B(s) = b_0 + b_1 s + b_2 s^2 + \cdots + b_n s^n \tag{7.55b}$$

Consequently, $jP_{rem}(s)/\varepsilon = j(P(s)/\varepsilon + JB(s))$, and therefore, the highest-degree coefficient of $P_{rem}(s)/\varepsilon$ is computed as follows:

$$jp_{nrem} = j(p_n + Jb_n)$$

$$= 0 \quad \text{if} \quad (p_n + Jb_n) = 0, \text{ i.e., } \quad J = \frac{-p_n}{b_n} \tag{7.56}$$

Since the value of J is now known, the polynomials of the remainder matrix may be determined:

$$A_{rem}(s) = A(s) \qquad C_{rem}(s) = C(s) + \frac{2JP(s)}{\varepsilon} + J^2 B(s)$$

$$B_{rem}(s) = B(s) \qquad D_{rem}(s) = D(s) \tag{7.57}$$

$$\frac{P_{rem}(s)}{\varepsilon} = \frac{P(s)}{\varepsilon} + JB(s)$$

It is evident that the only polynomials affected by the extraction of a parallel-connected inverter are $P(s)$ (which drops by one degree) and $C(s)$. At the end of the synthesis procedure, the $P(s)$ polynomial will have reduced to zero degree, a constant.

Summary of the Principal Components of the Prototype Network

The main elements employed to synthesize the parallel-resonator cross-coupled network are summarized in Figure 7.10. Where it is appropriate, the formulas used to determine the elements' values from the $A(s)$, $B(s)$, $C(s)$, $D(s)$, and $P(s)$ polynomials, and the operations on the polynomials used to obtain the remainder polynomials, are included.

(a)

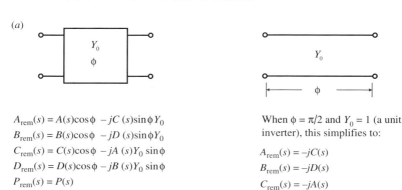

$$A_{\text{rem}}(s) = A(s)\cos\phi - jC(s)\sin\phi\, Y_0$$
$$B_{\text{rem}}(s) = B(s)\cos\phi - jD(s)\sin\phi\, Y_0$$
$$C_{\text{rem}}(s) = C(s)\cos\phi - jA(s)Y_0\sin\phi$$
$$D_{\text{rem}}(s) = D(s)\cos\phi - jB(s)Y_0\sin\phi$$
$$P_{\text{rem}}(s) = P(s)$$

When $\phi = \pi/2$ and $Y_0 = 1$ (a unit inverter), this simplifies to:

$$A_{\text{rem}}(s) = -jC(s)$$
$$B_{\text{rem}}(s) = -jD(s)$$
$$C_{\text{rem}}(s) = -jA(s)$$
$$D_{\text{rem}}(s) = -jB(s)$$
$$P_{\text{rem}}(s) = P(s)$$

(b)   (c)

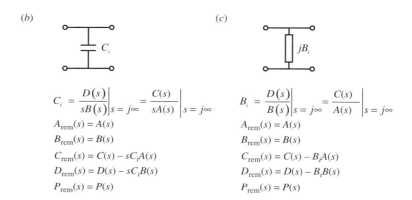

$$C_i = \left.\frac{D(s)}{sB(s)}\right|_{s=j\infty} = \left.\frac{C(s)}{sA(s)}\right|_{s=j\infty}$$
$$A_{\text{rem}}(s) = A(s)$$
$$B_{\text{rem}}(s) = B(s)$$
$$C_{\text{rem}}(s) = C(s) - sC_iA(s)$$
$$D_{\text{rem}}(s) = D(s) - sC_iB(s)$$
$$P_{\text{rem}}(s) = P(s)$$

$$B_i = \left.\frac{D(s)}{B(s)}\right|_{s=j\infty} = \left.\frac{C(s)}{A(s)}\right|_{s=j\infty}$$
$$A_{\text{rem}}(s) = A(s)$$
$$B_{\text{rem}}(s) = B(s)$$
$$C_{\text{rem}}(s) = C(s) - B_iA(s)$$
$$D_{\text{rem}}(s) = D(s) - B_iB(s)$$
$$P_{\text{rem}}(s) = P(s)$$

(d)

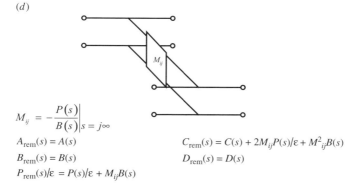

$$M_{ij} = -\left.\frac{P(s)}{B(s)}\right|_{s=j\infty}$$
$$A_{\text{rem}}(s) = A(s)$$
$$B_{\text{rem}}(s) = B(s)$$
$$P_{\text{rem}}(s)/\varepsilon = P(s)/\varepsilon + M_{ij}B(s)$$

$$C_{\text{rem}}(s) = C(s) + 2M_{ij}P(s)/\varepsilon + M^2{}_{ij}B(s)$$
$$D_{\text{rem}}(s) = D(s)$$

Figure 7.10 Cross-coupled array components and their extraction formulas: (a) transmission line, and special case where $\phi = \pi/2$ and $Y_0 = 1$ (unit inverter); (b) shunt-connected frequency-variant capacitor; (c) shunt-connected frequency-invariant reactance; (d) parallel cross-coupling inverter.

7.4 SYNTHESIS EXAMPLE OF AN ASYMMETRIC (4–2) FILTER NETWORK

This example is intended to clarify the rules to be observed when synthesizing a network [19]:

- Extractions are carried out at the source or load terminals only.
- Extractions are carried out on the [*ABCD*] matrix remaining from the previous step, starting with the full matrix.
- For the shunt $sC + jB$ pairs, C is extracted first.
- For parallel (cross-coupling) inverters, $sC + jB$ are extracted at each end first.
- For fully canonical networks, a parallel inverter between the source and load terminals is extracted first.
- The last element to be extracted must be a parallel inverter.

At the end of extractions, the $A(s)$, $B(s)$, $C(s)$, $D(s)$, and $P(s)$ polynomials should all be zeros, apart from the constant in $B(s)$.

To demonstrate the circuit synthesis approach, the (4–2) asymmetric prototype (see Fig. 7.11), for which the $S_{21}(s)$ and $S_{11}(s)$ functions are developed in Chapter 6, is chosen. The network is constructed as a folded array, with two cross-couplings, one diagonal and one straight. Because the original prototype is asymmetric, the shunt-connected FIRs in the network will be nonzero.

The polynomials corresponding to the zeros of the numerators and denominators of the $S_{21}(s)$ and $S_{11}(s)$ rational functions, $E(s)$, $F(s)$, and $P(s)$, derived in Chapter 6 for the asymmetric (4–2) prototype, are summarized as follows:

$s^i, i =$	$E(s)$	$F(s)/\varepsilon_R$	$P(s)$
0	$-0.1268 - j2.0658$	$+0.0208$	$+2.3899$
1	$+2.4874 - j3.6255$	$-j0.5432$	$+j3.1299$
2	$+3.6706 - j2.1950$	$+0.7869$	-1.0
3	$+2.4015 - j0.7591$	$-j0.7591$	—
4	$+1.0$	$+1.0$	—
		$\varepsilon_R = 1.0$	$\varepsilon = 1.1548$

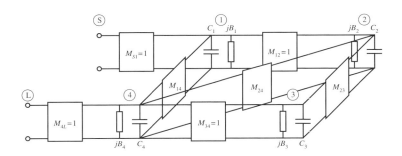

Figure 7.11 Folded cross-coupled network for the (4–2) asymmetric prototype.

Note that $P(s)$ has been multiplied by j twice here, once because $N - n_{fz}$ is an even number, and once to allow the cross-coupling elements to be extracted as inverters. With knowledge of the coefficients of $E(s)$, $F(s)$, and $P(s)$, we can build up the $A(s)$, $B(s)$, $C(s)$, and $D(s)$ polynomials according to equations (7.20), (7.29), and (7.31), summarized as follows:

s^i, $i =$	$A(s)$	$B(s)$	$C(s)$	$D(s)$	$P(s)/\varepsilon$
0	$-j2.0658$	-0.1059	-0.1476	$-j2.0658$	$+2.0696$
1	$+2.4874$	$-j4.1687$	$-j3.0823$	$+2.4874$	$+j2.7104$
2	$-j2.1950$	$+4.4575$	$+2.8836$	$-j2.1950$	-0.8660
3	$+2.4015$	$-j1.5183$	—	$+2.4015$	—
4	—	$+2.0$	—	—	—

Network synthesis begins with the extraction of the input coupling inverter at the source end of the network:

Extraction Number	Extracted Element	Working Node
1	Series unit inverter M_{S1}	S

The effect of extracting a unit inverter is to exchange the $B(s)$ and $D(s)$ polynomials, with the $A(s)$ and $C(s)$ polynomials, and to multiply them all by $-j$ (see Fig. 7.10a):

s^i, $i =$	$A(s)$	$B(s)$	$C(s)$	$D(s)$	$P(s)/\varepsilon$
0	$+j0.1476$	-2.0658	-2.0658	$+j0.1059$	$+2.0696$
1	-3.0823	$-j2.4874$	$-j2.4874$	-4.1687	$+j2.7104$
2	$-j2.8836$	-2.1950	-2.1950	$-j4.4575$	-0.8660
3	—	$-j2.4015$	$-j2.4015$	-1.5183	—
4	—	—	—	$-j2.0$	—

Now we are at working node 1, and are able to extract a frequency-variant capacitor C_1 and a FIR jB_1, in that order:

Extraction Number	Extracted Element	Working Node
2	Capacitor C_1	1
3	Reactance jB_1	1

Evaluate C_1 from

$$C_1 = \left. \frac{D(s)}{sB(s)} \right|_{s=j\infty} \quad \text{or} \quad C_1 = \left. \frac{C(s)}{sA(s)} \right|_{s=j\infty} = 0.8328$$

Extract C_1:

$s^i, i =$	$A(s)$	$B(s)$	$C(s)$	$D(s)$	$P(s)/\varepsilon$
0	$+j0.1476$	-2.0658	-2.0658	$+j0.1059$	$+2.0696$
1	-3.0823	$-j2.4874$	$-j2.6103$	-2.4482	$+j2.7104$
2	$-j2.8836$	-2.1950	$+0.3720$	$-j2.3860$	-0.8660
3	—	$-j2.4015$	—	$+0.3098$	—
4	—	—	—	—	—

Evaluate jB_1 from

$$B_1 = \left.\frac{D(s)}{B(s)}\right|_{s=j\infty} \quad \text{or} \quad B_1 = \left.\frac{C(s)}{A(s)}\right|_{s=j\infty} = j0.1290$$

Extract jB_1:

$s^i, i =$	$A(s)$	$B(s)$	$C(s)$	$D(s)$	$P(s)/\varepsilon$
0	$+j0.1476$	-2.0658	-2.0658	$+0.3724$	$+2.0696$
1	-3.0823	$-j2.4874$	$-j2.2127$	-2.7691	$+j2.7104$
2	$-j2.8836$	-2.1950	—	$-j2.1028$	-0.8660
3	—	$-j2.4015$	—	—	—
4	—	—	—	—	—

After C_1 and jB_1 are extracted at working node 1 (see Fig. 7.10), one of the terminals of the cross-coupling inverter M_{14} is encountered. To access the second of the two terminals, we need to turn the network [which is simply the exchange of the $A(s)$ and $D(s)$ polynomials] and extract the series unit inverter M_{4L}, followed by the capacitor–reactance pair $C_4 + jB_4$. This is accomplished according to steps 1–3 (in the preceding tabular list), yielding $C_4 = 0.8328$ and $B_4 = 0.1290$ (the same values as C_1 and B_1).

Extraction Number	Extracted Element	Working Node
Turn network	—	—
4	Series unit invertor M_{4L}	L
5	Capacitor C_4	L
6	Reactance jB_4	L

Having extracted these components, the remainder polynomials are

$s^i, i =$	$A(s)$	$B(s)$	$C(s)$	$D(s)$	$P(s)/\varepsilon$
0	$+j2.0468$	$+0.1476$	$+0.6364$	$+j2.0468$	$+2.0696$
1	-2.2127	$+j3.0823$	$+j1.3499$	-2.2127	$+j2.7104$
2	—	-2.8836	—	—	-0.8660
3	—	—	—	—	—
4	—	—	—	—	—

Now it may be seen that the polynomial $P(s)$ is the same degree as $B(s)$, indicating that the network is ready for the extraction of the cross-coupling inverter M_{14}:

Extraction Number	Extracted Element	Working Node
7	Parallel inverter M_{14}	1,4

Evaluate

$$M_{14} = \left. \frac{-P(s)}{\varepsilon B(s)} \right|_{s=j\infty} = -0.30003$$

from equation (7.56), and then extract by using equation (7.57). The remainder polynomials are given in the next tabular list. Note that only $C(s)$ and $P(s)$ have been affected by the extraction of the cross coupling inverter:

s^i, $i =$	$A(s)$	$B(s)$	$C(s)$	$D(s)$	$P(s)/\varepsilon$
0	$+j2.0468$	$+0.1476$	-0.5933	$+j2.0468$	$+2.0253$
1	-2.2127	$+j3.0823$	—	-2.2127	$+j1.7848$
2	—	-2.8836	—	—	—
3	—	—	—	—	—
4	—	—	—	—	—

At this stage we have encountered one of the terminals of the diagonal cross-coupling inverter M_{24}, and need to turn the network, extract the series unit inverter M_{12}, and then the capacitor/reactance pair $C_2 + jB_2$ at node 2 before we can access its other terminal at node 2. Now M_{24} may be extracted in the same way as M_{14} above:

Extraction Number	Extracted Element	Working Node
Turn network	—	—
8	Series unit inverter M_{12}	1
9	Capacitor C_2	2
10	Reactance jB_2	3
11	Parallel inverter M_{24}	2,4

Thus, $C_2 = 1.3032$ and $B_2 = -0.1875$, followed by $M_{24} = -0.8066$. The polynomials, after they are extracted, are listed as follows:

s^i, $i =$	$A(s)$	$B(s)$	$C(s)$	$D(s)$	$P(s)/\varepsilon$
0	$+j0.5933$	$+2.0468$	—	$+j0.2362$	$+0.3744$
1	—	$+j2.2127$	—	—	—
2	—	—	—	—	—
3	—	—	—	—	—
4	—	—	—	—	—

Finally, one of the terminals of the last cross-coupling inverter M_{23} (which is actually a mainline inverter, but since it is the last inverter, it is treated as a cross-coupling inverter) is encountered at node 2. In a similar way to steps 8–11, the network is turned, and the series unit inverter M_{34} and the capacitor/reactance pair $C_3 + jB_3$ at node 3 are extracted. Only then can we access the other terminal at node 3 and extract it:

Extraction Number	Extracted Element	Working Node
Turn network	—	—
12	Series unit inverter M_{34}	4
13	Capacitor C_3	3
14	Reactance jB_3	3
15	Parallel inverter M_{23}	2,3

This gives $C_3 = 3.7296$ and $B_2 = -3.4499$, followed by $M_{23} = -0.6310$. After these elements are extracted, the remaining polynomials are all zeros except for the constant of $B(s) = 0.5933$. The synthesis is now complete. The values of the 15 elements that have been extracted are summarized as follows:

$$C_1 + jB_1 = 0.8328 + j0.1290 \qquad M_{s1} = 1.0000 \qquad M_{14} = -0.3003$$
$$C_2 + jB_2 = 1.3032 - j0.1875 \qquad M_{12} = 1.0000 \qquad M_{24} = -0.8066$$
$$C_3 + jB_3 = 3.7296 - j3.4499 \qquad M_{34} = 1.0000 \qquad M_{23} = -0.6310$$
$$C_4 + jB_4 = 0.8328 + j0.1290 \qquad M_{4L} = 1.0000$$

Scaling at Resonator Nodes The general canonical lowpass prototype network, synthesized using this circuit element extraction approach, consists of an array of N shunt-connected capacitors. Each is in parallel with a frequency-invariant reactance (FIR), and the sequentially numbered capacitor/FIR nodes are coupled directly by 90° inverters (mainline couplings). If the filter transfer function contains finite-position transmission zeros, the nonsequential nodes are coupled again by inverters (cross-couplings). In addition, there are two more inverters to couple the input (source) to the first node, and the last node to the output terminal (load). In some cases, other internal nodes, apart from the first and last, are also connected to the source and load terminals. For fully canonical transfer functions, the source and load terminals are directly coupled with an inverter. In the real bandpass filter structure, the capacitor/FIR pairs become resonators, offset (or asynchronously tuned) with respect to the center frequency of the bandpass filter. If the FIR is not equal to zero in value; the inverters become input, output, or interresonator coupling devices, such as apertures, probes, and loops.

When each node is surrounded by inverters, it is possible to scale the impedance levels at each node to arbitrary values by adjusting the characteristic immitances of the inverters. Essentially, this is equivalent to adjusting the transformer ratios of the entry and exit coupling transformers such that the total energy transferring through the node remains the same.

For a network of nodes and couplings such as in Figure 7.10, the amount of energy passing through each inverter is represented by the scalar coupling coefficient k_{ij}, computed by

$$k_{ij} = \frac{M_{ij}}{\sqrt{C_i \cdot C_j}} \tag{7.58}$$

where M_{ij} is the characteristic immitance of the inverter coupling between the ith and jth nodes. For a given transfer function, all k_{ij} in the circuit must be kept constant. Therefore, M_{ij}, C_i, and C_j can be scaled to new values M'_{ij}, C'_i, and C'_j, provided that the following relationship is satisfied:

$$k_{ij} = \frac{M_{ij}}{\sqrt{C_i \cdot C_j}} = \frac{M'_{ij}}{\sqrt{C'_i \cdot C'_j}} \tag{7.59}$$

Thus, for a given network topology, there are an infinite number of combinations of M'_{ij}, C'_i, and C'_j, that can be chosen. The different combinations result in the same electrical performance provided that the proper relationship with the starting values M_{ij}, C_i, and C_j is maintained by using equation (7.59).

In this chapter, we use two typical schemes.

- All mainline coupling inverters M_{ij} are scaled to unity value, giving the shunt capacitors C_i (and in most cases, the terminating admittance G_L) nonunity values. This form leads directly to the ladder network lowpass prototype forms often used for classical circuit-based filter design.
- All shunt capacitors at the resonator nodes are scaled to unity, which gives the coupling inverters nonunity values. This form is adopted for the construction of the coupling matrix, and is dealt with in the next chapter.

Scaling of Inverters to Unity   Figure 7.12 depicts a typical lowpass prototype network (without cross-couplings) with source and load admittances, shunt capacitors C_i, and mainline coupling inverters M_{ij}. Typically, these components have nonunity values. To scale the inverters to unity, we progress through the network, starting at the source end. The first inverter at this end, M_{S1}, is scaled to unity by applying equation (7.59) with $M'_{S1} = 1$. Then

$$\frac{M_{S1}}{\sqrt{G_S C_1}} = \frac{1}{\sqrt{G'_S C'_1}} \tag{7.60}$$

Therefore, the new value for $C_1 (= C'_1)$ is

$$C'_1 = \frac{G_S C_1}{G'_S M^2_{S1}} \tag{7.61}$$

If we must retain the original terminating admittance value $(G'_S = G_S)$, then

$$C'_1 = \frac{C_1}{M^2_{S1}} \tag{7.62}$$

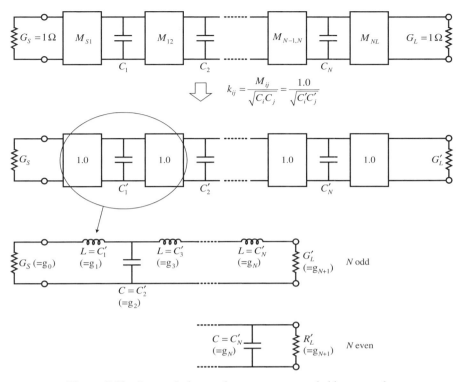

Figure 7.12 Lumped-element lowpass prototype ladder networks.

By applying (7.59) to the second inverter M_{12}, we obtain

$$\frac{M_{12}}{\sqrt{C_1 C_2}} = \frac{M'_{12}}{\sqrt{C'_1 C'_2}}$$

If $M'_{12} \to 1$, then

$$C'_2 = \frac{C_1 C_2}{C'_1 M^2_{12}} \tag{7.63}$$

This is repeated throughout the network until the last inverter M_{NL} is reached as follows:

$$\frac{M_{NL}}{\sqrt{C'_N G_L}} = \frac{1}{\sqrt{C'_N G'_L}} \quad \text{or} \quad G'_L = \frac{G_L}{M^2_{NL}} \tag{7.64}$$

In general this means that the new terminating admittance G'_L is nonunity in value. However, for filter functions with a reflection zero (the point of perfect transmission) at zero frequency, G'_L is unity. The second network shown in Figure 7.12 shows the scaled network, with all inverters = 1.

Once the mainline couplings are scaled to unity, and values C_i' are known, any cross-coupling inverters M_{ik} must be scaled to account for the new capacitor values at each end as follows:

$$M_{ik}' = M_{ik}\sqrt{\frac{C_i' C_k'}{C_i C_k}} \qquad (7.65)$$

If there are nonzero FIRs (jB_i) at the resonator nodes, they are scaled in the same proportion as the capacitor at that node by

$$B_i' = B_i \frac{C_i'}{C_i} \qquad (7.66)$$

Finally, if it is required that both terminations are unity (which is the usual case), scaling can be applied from both ends of the network, progressing toward the center, which leaves an internal inverter with a nonunity value at the meeting point. This is the natural form of the networks synthesized by the element extraction procedure described earlier in this chapter. For the asymmetric (4–2) filter example in Section 7.4, all the mainline inverter and termination values are unity except for the central mainline inverter M_{23}, which has the value 0.6310.

Figure 7.12 further demonstrates how the dual network theorem is applied to alternate C_i', and unit inverter pairs to create series-connected inductances of values C_i'. The network is now in the classic form [8], where the C_i' is equivalent to g_i, as well as $G_S = g_0$ and G_L (or R_L) $= g_{N+1}$. Note that $G_L'(R_L')$ equals unity for symmetric odd-degree functions. For the general case, $G_L'(R_L') \neq 1$.

SUMMARY

In the previous chapters, we introduced methods to generate the transfer and the reflection polynomials for a variety of filter functions. The next step in the design process is to translate these polynomials into a prototype electrical circuit from which a practical microwave filter can be developed. There are two ways to accomplish it: the classical circuit synthesis method and the direct-coupling matrix synthesis approach. In this chapter, the circuit synthesis approach, based on the $[ABCD]$ transfer matrix or the chain matrix, as it is sometimes known, is presented.

Circuit elements used for the classical filter synthesis include the lossless inductors and capacitors, and the terminating resistors. The lowpass prototype filters derived from these elements inherently lead to symmetric response filters. In Sections 3.10 and 6.1, such prototype networks are generalized to include asymmetric response filters by including the hypothetical frequency-invariant reactance, referred to by the acronym FIR. The inclusion of such elements gives rise to transfer and reflection polynomials with complex coefficients, rendering the lowpass prototype filter response to be asymmetric. In practical bandpass or bandstop filters, FIR elements appear as frequency offsets to the resonant circuits. Another element that is required in the synthesis of filters at microwave frequencies is the frequency-invariant impedance or admittance inverter, commonly referred to as the *immitance inverter*. Such inverters significantly simplify the physical realization of the distributed microwave filter structure. In a microwave circuit, immitance inverters can be realized approximately over narrow bandwidths by a variety of microwave structures, such as inductive irises and coupling probes.

The chapter begins by establishing the relationship between the *ABCD* parameters and the scattering parameters, related to the transfer and reflection polynomials of the filter function. It goes on to show the buildup of the overall *ABCD* matrix of a network by cascading the *ABCD* matrices of the individual elements. The reverse process—derivation of the individual elements from the overall *ABCD* matrix—provides the basis of circuit synthesis technique. The procedure described in this chapter includes the synthesis of both the symmetric, and asymmetric lowpass prototype filters.

The first step in the synthesis procedure is to compute the *ABCD* matrix from the optimized lowpass prototype filter polynomials. The next step is to follow the procedure reverse of that used for cascading a number of *ABCD* matrices corresponding to individual elements. In other words, the individual elements in the filter are extracted from the overall *ABCD* matrix in the reverse order, from the highest to the lowest values. The synthesis procedure is broken down into two stages. The first stage of the synthesis process involves lumped-element lossless inductors, capacitors, and FIR elements. The second stage of the synthesis process includes the immitance inverters. Use of such inverters allows for the prototype electrical circuit in a form suitable for realization with intercoupled microwave resonators. The technique is applicable for synthesizing lowpass prototype filters with a symmetric or asymmetric response in ladder form, as well as cross-coupled topologies. A further generalization is introduced to allow the synthesis of singly terminated filters. Such filters are required in the design of contiguous multiplexers, described in Chapter 18. The well-known g_k parameters are readily derived by the synthesis procedure provided in this chapter. An example of an asymmetric filter is included to demonstrate the synthesis procedure.

REFERENCES

1. S. Darlington, Synthesis of reactance 4-poles which produce insertion loss characteristics, *J. Math. Phys.* **18**, 257–353 (1939).

2. W. Cauer, *Synthesis of Linear Communication Networks*, McGraw-Hill, New York, 1958.

3. E. A. Guillemin, *Synthesis of Passive Networks*, Wiley, New York, 1957.

4. H. W. Bode, *Network Analysis and Feedback Amplifier Design*, Van Nostrand, Princeton, NJ, 1945.

5. M. E. van Valkenburg, *Network Analysis*, Prentice-Hall, Englewood Cliffs, NJ, 1955.

6. J. D. Rhodes, *Theory of Electrical Filters*, Wiley, New York, 1976.

7. R. F. Baum, Design of unsymmetrical band-pass filters, *IRE Trans. Circuit Theory* **CT-4**, 33–40 (June 1957).

8. G. Matthaei, L. Young, and E. M. T. Jones, *Microwave Filters, Impedance Matching Networks and Coupling Structures*, Artech House, Norwood, MA, 1980.

9. H. J. Carlin, The scattering matrix in network theory, *IRE Trans. Circuit Theory* **CT-3**, 88–96 (June 1956).

10. H. J. Orchard and G. C. Temes, Filter design using transformed variables, *IEEE Trans. Circuit Theory* **CT-15**, 385–408 (Dec. 1968).

11. H. C. Bell, Transformed-variable synthesis of narrow-bandpass filters, *IEEE Trans. Circuits Syst.* **CAS-26**, 389–394 (June 1979).

12. R. Levy, Synthesis of general asymmetric singly- and doubly-terminated cross-coupled filters, *IEEE Trans. Microwave Theory Tech.* **42**, 2468–2471 (Dec. 1994).

13. N. Marcuvitz, *Waveguide Handbook*, Electromagnetic Waves Series 21, IEE, London, 1986.

14. S. B. Cohn, Direct coupled cavity filters, *Proc. IRE*, **45**, 187–196 (Feb. 1957).

15. L. Young, Direct coupled cavity filters for wide and narrow bandwidths, *IEEE Trans. Microwave Theory Tech.* **MTT-11**, 162–178 (May 1963).

16. R. Levy, Theory of direct coupled cavity filters, *IEEE Trans. Microwave Theory Tech.* **MTT-11**, 162–178 (May 1963).

17. M. E. van Valkenburg, *Introduction to Modern Network Synthesis*, Wiley, New York, 1960.

18. M. H. Chen, Singly terminated pseudo-elliptic function filter, *COMSAT Tech. Rev.* **7**, 527–541 (Fall 1977).

19. R. J. Cameron, General prototype network synthesis methods for microwave filters, *ESA J.* **6**, 193–206 (1982).

CHAPTER 8

COUPLING MATRIX SYNTHESIS OF FILTER NETWORKS

In this chapter, we examine the *coupling matrix* representation of microwave filter circuits. Modeling the circuit in matrix form is particularly useful because matrix operations can then be applied, such as inversion, similarity transformation, and partitioning. Such operations simplify the synthesis, reconfiguration of the topology, and performance simulation of complex circuits. Moreover, the coupling matrix is able to include some of the real-world properties of the elements of the filter. Each element in the matrix can be identified uniquely with an element in the finished microwave device. This enables us to account for the attributions of electrical characteristics of each element, such as the Q_u values for each resonator cavity, different dispersion characteristics for the various types of mainline coupling and cross-coupling within the filter. This is difficult or impossible to achieve with a polynomial representation of the filter's characteristics.

The basic circuit that the coupling matrix represents is reviewed, and the method used to construct the matrix directly from a lowpass prototype circuit, synthesized in Chapter 7, is outlined. This is followed by the presentation of two methods for synthesis of the coupling matrix directly from the filter's transfer and the reflection polynomials, the $N \times N$ and the $N + 2$ matrices.

8.1 COUPLING MATRIX

In the early 1970s, Atia and Williams introduced the concept of the coupling matrix as applied to dual-mode symmetric waveguide filters [1–4]. The circuit model they

Microwave Filters for Communication Systems: Fundamentals, Design, and Applications,
by Richard J. Cameron, Chandra M. Kudsia, and Raafat R. Mansour
Copyright © 2007 John Wiley & Sons, Inc.

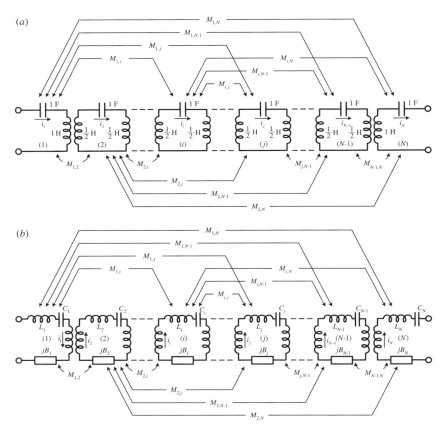

Figure 8.1 Multicoupled series-resonator *bandpass prototype* network: (*a*) classical representation (courtesy A. E. Atia); (*b*) modified to include FIR elements and separate self-inductors.

investigated was a *bandpass prototype*, which has the form shown in Figure 8.1*a*. The circuit is comprised of a cascade of lumped-element series resonators intercoupled by transformers; each resonator consists of a capacitor of 1 farad (F) in series with the self inductances of the mainline transformers, which total 1 henry (H) within each loop. This gives a center frequency of 1 radian per second (rad/s), and the couplings are normalized for a bandwidth of 1 rad/s. In addition, each loop is theoretically coupled to every other loop through cross-mutual couplings between the mainline transformers. This circuit model supports symmetric characteristics.

In the synthesis procedure outlined in this chapter, we have included the hypothetical FIR elements (Sections 3.10 and 6.1) and have modified the prototype circuit. This is illustrated in Figure 8.1*b*. In the modified circuit, self-inductances of the mainline transformers are separated out and represented as separate inductors within each loop. The inclusion of FIR elements enables the circuit to represent asymmetric characteristics.

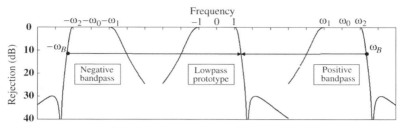

Figure 8.2   Bandpass–lowpass frequency mapping.

8.1.1   Bandpass and Lowpass Prototypes

The positive frequency characteristics of the bandpass prototype (BPP) circuit can be related to the lowpass prototype (LPP) characteristics that we considered in Section 3.8.1 through the lumped-element bandpass-to-lowpass (bandpass–lowpass) mapping as follows:

$$s = j \frac{\omega_0}{\omega_2 - \omega_1} \left[\frac{\omega_B}{\omega_0} - \frac{\omega_0}{\omega_B} \right] \tag{8.1}$$

Here, $\omega_0 = \sqrt{\omega_1 \omega_2}$ is the center frequency of the bandpass prototype ($= 1$ rad/s), ω_2 and ω_1 are the upper and lower band-edge frequencies, respectively (which are easily shown to be $\omega_2 = 1.618$ and $\omega_1 = 0.618$ rad/s for a positive BPP, i.e., a *bandwidth*[1] of $\omega_2 - \omega_1 = 1$ rad/s), and ω_B is the bandpass frequency variable. The mapping formula is written in this way to ensure that the negative frequency BPP characteristic correctly maps to the same point on the lowpass characteristic as the corresponding positive BPP frequency, even for the asymmetric cases (see Fig. 8.2). For the negative frequency BPP, all the frequency terms in equation (8.1) become negative so that an attenuation at $-\omega_B$ on the negative frequency BPP characteristic maps to the same point on the LP characteristic in the lowpass domain as an attenuation at $+\omega_B$ on the positive BPP characteristic in the positive BP domain.

Because the coupling elements are frequency-invariant, the series resonator circuit itself can be transformed to the lowpass domain by the following steps:

1. Replace all the mutual inductive couplings, provided by transformers, with inverters with the same values as the mutual couplings of the transformers. The inverters then provides the same amount of coupling energy between the resonator nodes as the transformers, and with the same $90°$ phase change (Fig. 8.3).
2. Transform the bandpass network to a lowpass prototype network with the band edges at $\omega = \pm 1$ by letting the value of the series capacitance go to infinity (zero series impedance).

---

[1]The word *bandwidth* is in italics because the negative frequency BPP, $\omega_2 - \omega_1$ becomes $(-\omega_2) - (-\omega_1) = -1$ rad/s, such that its *bandwidth* is negative.

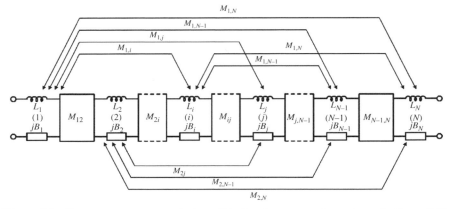

Figure 8.3 Lowpass prototype equivalent of the bandpass network, as shown in Figure 8.1*b*, with inverter coupling elements.

This is now in the form of the lowpass prototype circuits that were synthesized from the filter $S_{21}(s)$ and $S_{11}(s)$ polynomials in Chapter 7. Because the coupling elements are assumed frequency-invariant, the synthesis in the lowpass or bandpass domains yields the same values for the circuit elements, and under analysis the insertion loss and rejection amplitudes are the same; the frequency variables in the lowpass and bandpass domains are connected by the lumped-element frequency mapping formula (8.1).

8.1.2 Formation of the General $N \times N$ Coupling Matrix and its Analysis

The two-port network of Figure 8.1*b* (in either its BPP or LPP form) operates between a voltage source generating e_g volts and an internal impedance of R_S ohms (Ω) and a load impedance of R_L ohms. As a series resonator circuit with currents circulating in the loops, the overall circuit including the source and load terminations are represented with the impedance matrix $[z']$ in Figure 8.4.

Kirchhoff's nodal law (stating that the vector sum of all the currents entering a node is equal to zero) is applied to the currents circulating in the series resonators of the circuit shown in Figure 8.1*a*, leading to a series of equations that may be represented with the matrix equation [1–4]

$$[e_g] = [z'] \quad [i] \tag{8.2}$$

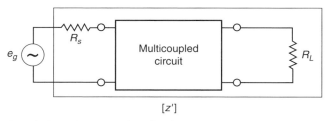

Figure 8.4 Overall impedance matrix $[z']$ of the series resonator circuits of Figure 8.3 operating between a source impedance R_S and a load impedance R_L.

where $[z']$ is the impedance matrix of the N-loop network plus its terminations. Equation (8.2) is expanded as follows:

$$e_g[1,0,0,\ldots,0]^t = [j\mathbf{M} + s\mathbf{I} + \mathbf{R}] \cdot [i_1, i_2, i_3, \ldots, i_N]^t \tag{8.3}$$

where $[\cdot]^t$ denotes matrix transpose and $\mathbf{I}$ is the unit matrix, e_g is the source voltage, and $i_1, i_2, \ldots, i_N$ are the currents in each of the N loops of the network.

It is evident that the impedance matrix $[z']$ is itself the sum of three $N \times N$ matrices:

Main Coupling Matrix $j\mathbf{M}$ This is the $N \times N$ matrix containing the values of the mutual couplings between the nodes of the network (provided by the transformers in Fig. 8.1, and equal in value to the immittance inverters in the lowpass prototype of Fig. 8.3). If the coupling is between sequentially numbered nodes, $M_{i,i+1}$, it is referred to as a *mainline coupling*. The entries on the main diagonal $M_{i,i}$ ($\equiv B_i$, the FIR at each node) are the *self-couplings*, whereas all the other couplings between the nonsequentially numbered nodes are known as *cross-couplings*:

$$j\mathbf{M} = j\begin{bmatrix} B_1 & M_{12} & M_{13} & \cdots & & M_{1N} \\ M_{12} & B_2 & & & & \\ M_{13} & & \ddots & & & \\ \vdots & & & \ddots & & M_{N-1,N} \\ M_{1N} & & & M_{N-1,N} & & B_N \end{bmatrix} \tag{8.4}$$

Because of the reciprocity of the passive network, $M_{ij} = M_{ji}$, and generally, all the entries are nonzero. In the RF domain, any variation of the coupling values with the frequency (dispersion) may be included at this stage.

Frequency Variable Matrix $s\mathbf{I}$ This diagonal matrix contains the frequency-variable portion (either the lowpass prototype or the bandpass prototype) of the impedance in each loop, giving rise to an $N \times N$ matrix with all entries at zero except for the diagonal filled with $s = j\omega$ as follows:

$$s\mathbf{I} = \begin{bmatrix} s & 0 & 0 & \cdots & 0 \\ 0 & s & & & \\ 0 & & \ddots & & \\ \vdots & & & \ddots & 0 \\ 0 & & & 0 & s \end{bmatrix} \tag{8.5}$$

For bandpass filters, the effects of a noninfinite resonator unloaded quality factor Q_u are included by offsetting s by a positive real factor δ (i.e., $s \rightarrow \delta + s$), where $\delta = f_0/(\text{BW} \cdot Q_u)$, f_0 is the center frequency of the passband, and BW is its design bandwidth.

***Termination Impedance Matrix* R**   This $N \times N$ matrix contains the values of the source and load impedances in the R_{11} and R_{NN} positions; all the other entries are zero:

$$\mathbf{R} = \begin{bmatrix} R_S & 0 & 0 & \cdots & & 0 \\ 0 & 0 & & & & \\ 0 & & \ddots & & & \\ \vdots & & & \ddots & & 0 \\ 0 & & & & 0 & R_L \end{bmatrix} \tag{8.6}$$

\raster="rg_u01"

N* × *N* and *N* + 2 *Coupling Matrices   The $N \times N$ impedance matrix for the series resonator network are separated out into the matrix's purely resistive and purely reactive parts [equations (8.2) and (8.3)] by

$$[z'] = \mathbf{R} + [\, jM + s\mathbf{I}] = \mathbf{R} + [z] \tag{8.7}$$

Now, the impedance matrix $[z]$ represents the circuit in Figure 8.5*a*, a purely reactive network operating between a voltage source with internal impedance R_S and a load R_L.

Typically, the source and load terminations are nonzero, and can be normalized to unity impedance by insertion of impedance inverters M_{S1} and M_{NL} of impedance values $\sqrt{R_S}$ and $\sqrt{R_L}$, respectively, on the source and load side of the network (Fig. 8.5*a*,*b*). In both cases in Figure 8.5*a*,*b*, the impedance as seen looking out from the network on the input side is R_S and on the output side is R_L.

The action of placing the two inverters on either side of the $N \times N$ impedance matrix has two effects:

1. The terminating impedances become terminating conductances $G_S = 1/R_S$ and $G_L = 1/R_L$, respectively, (and also the voltage source e_g is transformed into a current source $i_g = e_g/R_S$).[2]

[2]Terminating conductances G_S at the source end of the network are usually associated with zero conductance current sources, and resistances R_S with zero impedance voltage sources. If it is convenient to do so, they can be interchanged by using the Norton or Thévenin equivalent circuits, explained as follows:

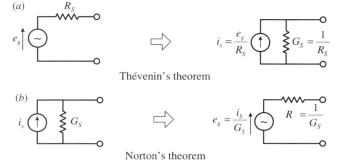

2. The $[z]$ matrix surrounded by two inverters can be replaced by the dual network, which is an admittance matrix $[y]$. The values of the input/output inverters themselves are then absorbed in the $[y]$ matrix by surrounding the $N \times N$ matrix by an extra row top and bottom and an extra column on each side, creating an $N + 2 \times N + 2$ matrix, commonly known as the $N + 2$ matrix (Fig. 8.5c). The dual of this network, which has series resonators and impedance inverter coupling elements, is illustrated in Figure 8.5d. Whether an impedance or admittance type, the $N + 2$ matrices will have the same values for their mainline and cross-coupling inverters.

The full $N + 2$ network and the corresponding coupling matrix is depicted in Figures 8.6 and 8.7. We can see that in addition to the main-line input/output couplings M_{S1} and M_{NL}, it is now possible to include other couplings between the source and/or load terminations, and the internal resonator nodes within the core $N \times N$ matrix. Also, it is possible to accommodate the direct source–load coupling M_{SL} in order

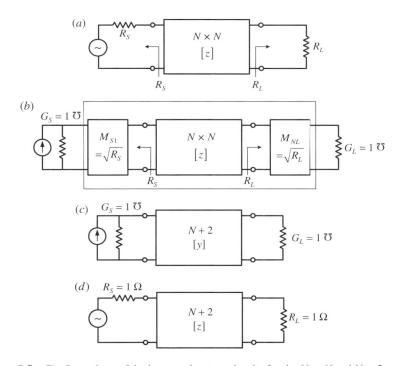

Figure 8.5 Configurations of the input and output circuits for the $N \times N$ and $N + 2$ coupling matrices: (*a*) the series resonator circuit in Figure 8.4 represented as an $N \times N$ impedance coupling matrix between terminations R_S and R_L; (*b*) circuit in (*a*) with inverters to normalize the terminations to unity; (*c*) $N + 2$ matrix (parallel resonators) and normalized terminating conductances G_S and G_L; (*d*) $N + 2$ impedance matrix with series resonators and normalized terminating resistances R_S and R_L, the dual network of (*c*).

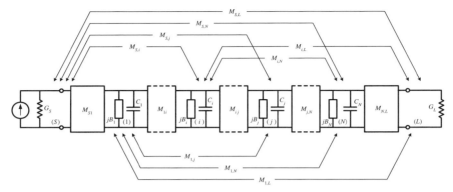

Figure 8.6 $N + 2$ multicoupled network with parallel lowpass resonators.

	S	1	2	3	4	L
S	M_{SS}	M_{S1}	M_{S2}	M_{S3}	M_{S4}	M_{SL}
1	M_{S1}	M_{11}	M_{12}	M_{13}	M_{14}	M_{1L}
2	M_{S2}	M_{12}	M_{22}	M_{23}	M_{24}	M_{2L}
3	M_{S3}	M_{13}	M_{23}	M_{33}	M_{34}	M_{3L}
4	M_{S4}	M_{14}	M_{24}	M_{34}	M_{44}	M_{4L}
L	M_{SL}	M_{1L}	M_{2L}	M_{3L}	M_{4L}	M_{LL}

Figure 8.7   Fourth-degree $N + 2$ coupling matrix with all possible cross-couplings. The core $N \times N$ matrix is indicated within the double lines. The matrix is symmetric about the principal diagonal: i.e. $M_{ij} = M_{ji}$.

to realize the fully canonical filter functions. As an admittance matrix, it has parallel resonators and admittance inverters for the coupling elements.

8.1.3   Formation of the Coupling Matrix from the Lowpass Prototype Circuit Elements

In the previous chapter, the methods for building up the cross-coupled folded array network from the polynomials representing the desired filter response were introduced. The resultant network is an array of parallel nodes (shunt-connected capacitors C_i and FIRs jB_i), interconnected by mainline and cross-coupling inverters. A coupling matrix can be built up directly from the value of these components, but first the frequency-variant components in the circuit (the capacitors C_i at the nodes) need to be normalized to unity such that the resultant coupling matrix properly corresponds with the circuits in Figure 8.1.

The scaling can be done directly on the coupling matrix once its nonzero elements are filled with the values of the components from the prototype circuit. Figure 8.7 exhibits a general $N + 2$ coupling matrix for a fourth-degree network that has all the possible couplings, symmetric about the principal diagonal. Because the circuit synthesis procedure has resulted in parallel lowpass resonators (the shunt $sC_i + jB_i$ at the nodes), the matrix is shown as an admittance matrix with conductances $G_S = 1/R_S$ and $G_L = 1/R_L$ at its source and load terminals, respectively. Now the standard scaling process can be applied to the matrix. To scale the capacitor C_i at the node $M_{i,i}$ to unity, multiply row i and column i by $\left[\sqrt{C_i}\right]^{-1}$ for $i = 1, 2, 3, \ldots, N$:

$$
\begin{array}{cccccc}
\times(\sqrt{G_S})^{-1} & \times(\sqrt{C_1})^{-1} & \times(\sqrt{C_2})^{-1} & \times(\sqrt{C_3})^{-1} & \times(\sqrt{C_4})^{-1} & \times(\sqrt{G_L})^{-1} \\
\downarrow & \downarrow & \downarrow & \downarrow & \downarrow & \downarrow
\end{array}
$$

$$
\begin{array}{c}
\times(\sqrt{G_S})^{-1} \to \\
\times(\sqrt{C_1})^{-1} \to \\
\times(\sqrt{C_2})^{-1} \to \\
\times(\sqrt{C_3})^{-1} \to \\
\times(\sqrt{C_4})^{-1} \to \\
\times(\sqrt{G_L})^{-1} \to
\end{array}
\begin{bmatrix}
G_S + jB_S & jM_{S1} & jM_{S2} & jM_{S3} & jM_{S4} & jM_{SL} \\
M_{S1} & sC_1 + jB_1 & jM_{12} & jM_{13} & jM_{14} & jM_{1L} \\
jM_{S2} & jM_{12} & sC_2 + jB_2 & jM_{23} & jM_{24} & jM_{2L} \\
jM_{S3} & jM_{13} & jM_{23} & sC_3 + jB_3 & jM_{34} & jM_{3L} \\
jM_{S4} & jM_{14} & jM_{24} & jM_{34} & sC_4 + jB_4 & jM_{4L} \\
jM_{SL} & jM_{1L} & jM_{2L} & jM_{3L} & jM_{4L} & G_L + jB_L
\end{bmatrix}
$$

$$(8.8)$$

By applying the scaling operation, the elements on the diagonal are multiplied by C_i^{-1}, which scales the frequency dependent capacitors and the FIRs to values $sC_i \to s$ and $jB_i \ (\equiv M_{ii}) \to jB_i/C_i$ respectively. The off-diagonal coupling elements M_{ij} are also scaled by the row/column multiplications:

$$
M_{ij} \to \frac{M_{ij}}{\sqrt{C_i \cdot C_j}} \qquad i, \ j = 1, 2, \ldots, N, \qquad i \neq j \qquad (8.9a)
$$

In addition, the input and output terminations can be scaled to unity by multiplying the first row and column by $1/\sqrt{G_S}(=\sqrt{R_S})$. Then the M_{SS} element scales to $M_{SS} = G_S + jB_S \to 1 + jB_S/G_S$, and the mainline coupling from the source to the first resonator node M_{S1} scales as follows:

$$
M_{S1} \to \frac{M_{S1}}{\sqrt{G_S C_1}} = M_{S1}\sqrt{\frac{R_S}{C_1}} \qquad (8.9b)
$$

Similarly, multiplying the last row and column by $1/\sqrt{G_L} \left(= \sqrt{R_L}\right)$ scales the M_{LL} element to $M_{LL} = G_L + jB_L \to 1 + jB_L/G_L$, and the mainline coupling from the last resonator node to the load termination M_{NL} scales by

$$M_{NL} \rightarrow \frac{M_{NL}}{\sqrt{C_N G_L}} = M_{NL} \sqrt{\frac{R_L}{C_N}}. \tag{8.9c}$$

The $(4-2)$ asymmetric filter function that was employed for the demonstration of the synthesis of the prototype electrical network in Section 7.4 used again as an example of the scaling for the construction of a coupling matrix. The values of the extracted C_i, B_i, and M_{ij} are summarized as follows:

$C_1 = 0.8328$	$B_1 = j0.1290$	$M_{12} = 1.0$	$G_S = 1.0$
$C_2 = 1.3032$	$B_2 = -j0.1875$	$M_{23} = -0.6310$	$G_L = 1.0$
$C_3 = 3.7296$	$B_3 = -j3.4499$	$M_{34} = 1.0$	
$C_4 = 0.8328$	$B_4 = j0.1290$	$M_{14} = -0.3003$	$M_{S1} = 1.0$
		$M_{24} = -0.8066$	$M_{4L} = 1.0$

Equation set (8.9) is now applied to normalize C_i to unity and derive the new values for B_i and M_{ij} (B_i' and M_{ij}', respectively):

$C_1' = 1.0$	$B_1' = j0.1549$	$M_{12}' = 0.9599$	$G_S' = 1.0$
$C_2' = 1.0$	$B_2' = -j0.1439$	$M_{23}' = -0.2862$	$G_L' = 1.0$
$C_3' = 1.0$	$B_3' = -j0.9250$	$M_{34}' = 0.5674$	
$C_4' = 1.0$	$B_4' = j0.1549$	$M_{14}' = -0.3606$	$M_{S1}' = 1.0958$
		$M_{24}' = -0.7742$	$M_{4L}' = 1.0958$

The coupling matrix for the $(4-2)$ network can now be constructed. The same scaling process, this time by -1.0, is then used to change the signs of any negative main line couplings to positive if required. Therefore

$$
\mathbf{M} =
\begin{array}{c}
\\ S \\ 1 \\ 2 \\ 3 \\ 4 \\ L
\end{array}
\begin{array}{c}
\begin{array}{cccccc}
S & 1 & 2 & 3 & 4 & L
\end{array} \\
\left[
\begin{array}{cccccc}
0.0 & 1.0958 & 0 & 0 & 0 & 0 \\
1.0958 & 0.1549 & 0.9599 & 0 & 0.3606 & 0 \\
0 & 0.9599 & -0.1439 & 0.2862 & 0.7742 & 0 \\
0 & 0 & 0.2862 & -0.9250 & 0.5674 & 0 \\
0 & 0.3606 & 0.7742 & 0.5674 & 0.1549 & 1.0958 \\
0 & 0 & 0 & 0 & 1.0958 & 0.0
\end{array}
\right]
\end{array}
\tag{8.10}
$$

Because all the capacitors C_i have been scaled to unity, the diagonal frequency matrix $s\mathbf{I}$ [matrix (8.5)] is now defined, and the termination matrix $\mathbf{R} = 0$, except for $R_{SS} = G_S = 1.0$, and $R_{LL} = G_L = 1.0$ [matrix (8.6)]. Now the $N + 2$ immittance matrix for this asymmetric $(4-2)$ network can be constructed as follows:

$$[y'] \quad \text{or} \quad [z'] = j\mathbf{M} + s\mathbf{I} + \mathbf{R}. \tag{8.11}$$

8.1.4   Analysis of the Network Represented by the Coupling Matrix

The network as represented by the coupling matrix can be analyzed in two ways:

1. As an [ABCD] matrix that can then be cascaded with the [ABCD] matrices of the other components to form a compound network, for example, in a

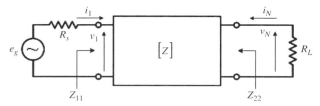

Figure 8.8   Network between source e_g of impedance R_S and load of impedance R_L.

manifold multiplexer where the matrices of the different channel filters are interconnected via lengths of transmission line, waveguide, coaxial cables, and so on.

2. As a standalone network, incorporating the terminating impedances R_S and R_L.

The network relevant to the first case is illustrated in Figure 8.8, where R_S and R_L have been separated out from the main network. By writing out equation (8.3) in full for this case ($N \times N$ matrix):[3]

$$
\begin{bmatrix} e_g \\ 0 \\ \vdots \\ \vdots \\ 0 \end{bmatrix} = \begin{bmatrix} R_S & 0 & 0 & \cdots & 0 \\ 0 & 0 & & & \\ 0 & & \ddots & & \\ \vdots & & & \ddots & 0 \\ 0 & & & 0 & R_L \end{bmatrix} \begin{bmatrix} i_1 \\ i_2 \\ \vdots \\ \\ i_N \end{bmatrix}
$$

$$
+ \begin{bmatrix} s + jM_{11} & jM_{12} & jM_{13} & \cdots & & jM_{1N} \\ jM_{12} & s + jM_{22} & & & & \\ jM_{13} & & \ddots & & & \\ \vdots & & & \ddots & & jM_{N-1,N} \\ jM_{1N} & & & jM_{N-1,N} & s + jM_{NN} \end{bmatrix} \begin{bmatrix} i_1 \\ i_2 \\ \vdots \\ \\ i_N \end{bmatrix} \quad (8.12)
$$

$$
\begin{bmatrix} e_g - R_S i_1 \\ 0 \\ \vdots \\ \\ -R_L i_N \end{bmatrix} = \begin{bmatrix} s + jM_{11} & jM_{12} & jM_{13} & \cdots & & jM_{1N} \\ jM_{12} & s + jM_{22} & & & & \\ jM_{13} & & \ddots & & & \\ \vdots & & & \ddots & & jM_{N-1,N} \\ jM_{1N} & & & jM_{N-1,N} & s + jM_{NN} \end{bmatrix} \begin{bmatrix} i_1 \\ i_2 \\ \vdots \\ \\ i_N \end{bmatrix}.
$$

$$ (8.13) $$

[3]For the $N + 2$ matrix, the frequency variable s is not entered into the top left and bottom right corners of the matrix, because these will be nonresonant nodes.

Since $e_g - R_S i_1 = v_1$, and $-R_L i_N = v_N$ (see Fig. 8.8), it is seen that the coupling matrix here is in the form of an open-circuit impedance matrix $[z]$ [5]:

$$
\begin{bmatrix} v_1 \\ 0 \\ \vdots \\ v_N \end{bmatrix} = \begin{bmatrix} & & \\ & [z] & \\ & & \end{bmatrix} \begin{bmatrix} i_1 \\ i_2 \\ \vdots \\ i_N \end{bmatrix}. \tag{8.14}
$$

Inverting the $[z]$ matrix will give the currents in terms of the short-circuit admittance matrix $[y]$,

$$
\begin{bmatrix} i_1 \\ i_2 \\ i_3 \\ \vdots \\ i_N \end{bmatrix} = \begin{bmatrix} & & \\ & [y] & \\ & (=[z]^{-1}) & \\ & & \end{bmatrix} \begin{bmatrix} v_1 \\ 0 \\ 0 \\ \vdots \\ v_N \end{bmatrix}. \tag{8.15}
$$

Since we are interested, at the moment, only in the voltages and currents at the terminals for this analysis, we rewrite matrix (8.15) as

$$
\begin{bmatrix} i_1 \\ i_N \end{bmatrix} = \begin{bmatrix} [y]_{11} & [y]_{1N} \\ [y]_{N1} & [y]_{NN} \end{bmatrix} \cdot \begin{bmatrix} v_1 \\ v_N \end{bmatrix} \tag{8.16}
$$

where $[y]_{11}$, $[y]_{1N}$, and so on are the corner elements of the admittance matrix $[y]$. This is converted into $[ABCD]$ parameters using the standard $[y] \rightarrow [ABCD]$ parameter transform [5], including the normalization to the source and load impedance values

$$
\begin{bmatrix} A & B \\ C & D \end{bmatrix} = \frac{-1}{[y]_{N1}} \begin{bmatrix} \sqrt{\dfrac{R_L}{R_S}}[y]_{NN} & \dfrac{1}{\sqrt{R_S R_L}} \\ \Delta_{[y]}\sqrt{R_S R_L} & \sqrt{\dfrac{R_S}{R_L}}[y]_{11} \end{bmatrix} \tag{8.17}
$$

where $\Delta_{[y]}$ is the determinant of the submatrix in (8.16), given by $\Delta_{[y]} = [y]_{11}$ $[y]_{NN} - [y]_{1N}[y]_{N1}$. If a rapid analysis is required, for example, during a real-time parameter deembedding process when a coupling matrix is being optimized to correspond to a measured characteristic, it is more efficient to solve matrix (8.14) by a Gaussian elimination procedure [6], rather than a full inversion on the $[z]$ matrix. In this case, the short circuit z parameters $[z]_{11}$, $[z]_{1N}$, $[z]_{N1}$, and $[z]_{NN}$, and the equivalent $[ABCD]$ matrix are given by

$$
\begin{bmatrix} A & B \\ C & D \end{bmatrix} = \frac{1}{[z]_{N1}} \begin{bmatrix} \sqrt{\dfrac{R_L}{R_S}}[z]_{11} & \dfrac{\Delta_{[z]}}{\sqrt{R_S R_L}} \\ \sqrt{R_S R_L} & \sqrt{\dfrac{R_S}{R_L}}[z]_{NN} \end{bmatrix} \tag{8.18}
$$

where $\Delta_{[z]} = [z]_{11}[z]_{NN} - [z]_{1N}[z]_{N1}$. If the values of the input/output coupling inverters are employed instead, $\left(M_{S1} = \sqrt{R_s}, M_{NL} = \sqrt{R_L}\right)$, then

$$\begin{bmatrix} A & B \\ C & D \end{bmatrix} = \frac{1}{[z]_{N1}} \begin{bmatrix} \dfrac{M_{NL}}{M_{S1}}[z]_{11} & \dfrac{\Delta_{[z]}}{M_{S1}M_{NL}} \\ M_{S1}M_{NL} & \dfrac{M_{S1}}{M_{NL}}[z]_{NN} \end{bmatrix} \tag{8.19}$$

Now the matrix may be cascaded with the $[ABCD]$ matrices of other networks such as transmission lines.

8.1.5 Direct Analysis

For the second analysis case, the full coupling matrix, including the source–load terminations is used, as in equations (8.2) and (8.3)

$$[e_g] = [z']\cdot[i] \quad \text{or} \quad [i] = [z']^{-1}[e_g] = [y'][e_g] \tag{8.20}$$

where $[z']$ and $[y']$ are the network open-circuit impedance and short-circuit admittance matrices, respectively, with the source and load impedances included. Referring to Figure 8.8 and equation (8.20), we easily see that

$$i_1 = [y']_{11}e_g \tag{8.21a}$$

$$i_N = [y']_{N1}e_g = \frac{v_N}{R_L} \tag{8.21b}$$

By substituting equation (8.21b) into the transmission coefficient definition for S_{21} [5], we obtain

$$S_{21} = 2\sqrt{\frac{R_S}{R_L}} \cdot \frac{v_N}{e_g} = 2\sqrt{\frac{R_S}{R_L}} \cdot R_L[y']_{N1}$$

$$= 2\sqrt{R_S R_L} \cdot [y']_{N1} \tag{8.22}$$

The reflection coefficient at the input port is computed by

$$S_{11} = \frac{Z_{11} - R_S}{Z_{11} + R_S}$$

$$= \frac{Z_{11} + R_S - 2R_S}{Z_{11} + R_S}$$

$$= 1 - \frac{2R_S}{Z_{11} + R_S} \tag{8.23}$$

where $Z_{11} = v_1/i_1$ is the impedance looking in at the input port (see Fig. 8.8). The potential divider at the input port gives v_1 and equation (8.21a) yields i_1, and so, Z_{11} is expressed as

$$Z_{11} = \frac{v_1}{i_1} = \frac{e_g Z_{11}}{Z_{11} + R_S} \cdot \frac{1}{e_g [y']_{11}} \quad \text{or} \quad \frac{1}{Z_{11} + R_S} = [y']_{11} \tag{8.24}$$

Substituting equation (8.24) into equation (8.23), we find the reflection coefficient at the input port of the network:

$$S_{11} = 1 - 2R_S[y']_{11} \tag{8.25a}$$

Similarly, for the output port, we obtain

$$S_{22} = 1 - 2R_L[y']_{NN} \tag{8.25b}$$

Then, the voltage v_1 incident at the network input (Fig. 8.8) is related to the generator voltage e_g by making use of equations (8.25a) and (8.21a):

$$v_1 = (e_g - R_S i_1) = \frac{e_g(1 + S_{11})}{2} \tag{8.26a}$$

This shows that at a point of perfect transmission ($S_{11} = 0$), $v_1 = e_g/2$, meaning that the impedance Z_{11} looking into the network at the input side is R_S. At this point, the maximum power available from the generator is being transferred to the load, and if e_g is set at 2 V (volts) and the source impedance $R_S = 1\ \Omega$ (ohm), the power P_i incident at the input is 1 W (watt). Also, if e_g in equation (8.26a) is substituted by using (8.22), v_N is then directly related to v_1 through the network S parameters by

$$\frac{v_N}{v_1} = \sqrt{\frac{R_L}{R_S}} \cdot \frac{S_{21}}{(1 + S_{11})} \tag{8.26b}$$

8.2 DIRECT SYNTHESIS OF THE COUPLING MATRIX

In this section, two methods for the direct synthesis of the coupling matrix are presented, the first for the $N \times N$ matrix and the second for the $N + 2$ matrix. In both cases, the approach is the same, namely, to formulate the two-port short-circuit admittance parameters in two ways: (1), from the coefficients of the polynomials $F(s)/\varepsilon_R$, $P(s)/\varepsilon$, and $E(s)$ that make up the desired transfer and reflection characteristics $S_{21}(s)$ and $S_{11}(s)$; and (2), from the elements of the coupling matrix itself. By equating the two formulations, the coupling values of the matrix are related to the coefficients of the transfer and reflection polynomials.

Although the $N + 2$ matrix is more flexible than the $N \times N$ matrix and also happens to be a little easier to synthesize, it is informative to explore the principles behind the $N \times N$ matrix synthesis technique. We will examine this method before proceeding to the synthesis of the $N + 2$ matrix.

8.2.1   Direct Synthesis of the $N \times N$ Coupling Matrix

The two-port short-circuit admittance matrix for the overall network, defined by equation (8.16) [4], is given by

$$\begin{bmatrix} i_1 \\ i_N \end{bmatrix} = \begin{bmatrix} [y]_{11} & [y]_{1N} \\ [y]_{N1} & [y]_{NN} \end{bmatrix} \cdot \begin{bmatrix} v_1 \\ v_N \end{bmatrix}$$

where $[y]_{11}$, $[y]_{1N}$ and so on are the corner elements of the admittance matrix $[y]$, itself the inverse of the impedance matrix $[z] = j\mathbf{M} + s\mathbf{I}$ [equation (8.7)]. Taking the standard definitions for the $[y]$ matrix elements, the two-port y parameters can be written in terms of the coupling matrix $\mathbf{M}$ and the frequency variable $s = j\omega$ as follows:

$$y_{11}(s) = [z]_{11}^{-1} = \frac{i_1}{v_1}\bigg|_{v_N = 0} = [j\mathbf{M} + s\mathbf{I}]_{11}^{-1} = j[-\mathbf{M} - \omega\mathbf{I}]_{11}^{-1} \tag{8.27a}$$

$$y_{22}(s) = [z]_{NN}^{-1} = \frac{i_N}{v_N}\bigg|_{v_1 = 0} = [j\mathbf{M} + s\mathbf{I}]_{NN}^{-1} = j[-\mathbf{M} - \omega\mathbf{I}]_{NN}^{-1} \tag{8.27b}$$

$$y_{12}(s) = y_{21}(s) = [z]_{N1}^{-1} = \frac{i_N}{v_1}\bigg|_{v_N = 0} = [j\mathbf{M} + s\mathbf{I}]_{N1}^{-1} = j[-\mathbf{M} - \omega\mathbf{I}]_{N1}^{-1} \tag{8.27c}$$

This is an essential step in the network synthesis procedure that relates the transfer function, expressed in purely mathematical terms to the real world of the coupling matrix, each element of which corresponds uniquely to a physical coupling element in the realized filter.

Since $\mathbf{M}$ is real and symmetric about its principal diagonal, all of its eigenvalues are real [6]. Thus, an $N \times N$ matrix $\mathbf{T}$ with rows of orthogonal vectors exists and satisfies the equation

$$-\mathbf{M} = \mathbf{T} \cdot \Lambda \cdot \mathbf{T}^t \tag{8.28}$$

where $\Lambda = \mathrm{diag}[\lambda_1, \lambda_2, \lambda_3, \ldots, \lambda_N]$, λ_i are the eigenvalues of $-\mathbf{M}$, and $\mathbf{T}^t$ is the transpose of $\mathbf{T}$ such that $\mathbf{T} \cdot \mathbf{T}^t = \mathbf{I}$. The $y_{ij}(s)$ polynomials for the singly and doubly terminated cases have already been derived [equation sets (7.20) and (7.25)] from the coefficients of the $E(s)$, $F(s)/\varepsilon_R$, and $P(s)/\varepsilon$ polynomials. In fact, only two of the y parameters are required, and we choose $y_{21}(s)$ and $y_{22}(s)$ to avoid a sign ambiguity, and because $y_{22}(s)$ results directly from the $A(s)$ and $B(s)$ polynomials for the singly terminated network [the $C(s)$ and $D(s)$ polynomials are not needed]. Substituting equation (8.28) into equations (8.27b) and (8.27c) yields

$$y_{21}(s) = j\left[\mathbf{T} \cdot \Lambda \cdot \mathbf{T}^t - \omega\mathbf{I}\right]_{N1}^{-1} \tag{8.29a}$$

$$y_{22}(s) = j\left[\mathbf{T} \cdot \Lambda \cdot \mathbf{T}^t - \omega\mathbf{I}\right]_{NN}^{-1} \tag{8.29b}$$

The general solution for an element i, j of an inverse eigenmatrix problem such as the right-hand side (RHS) of equation (8.29) is

$$\left[\mathbf{T} \cdot \mathbf{\Lambda} \cdot \mathbf{T}^t - \omega \, \mathbf{I} \right]_{ij}^{-1} = \sum_{k=1}^{N} \frac{T_{ik} T_{jk}}{\omega - \lambda_k}, \qquad i, j = 1, 2, \ldots, N \tag{8.30}$$

Therefore, from equation (8.29), we obtain

$$y_{21}(s) = \frac{y_{21n}(s)}{y_d(s)} = j \sum_{k=1}^{N} \frac{T_{Nk} T_{1k}}{\omega - \lambda_k} \tag{8.31a}$$

and

$$y_{22}(s) = \frac{y_{22n}(s)}{y_d(s)} = j \sum_{k=1}^{N} \frac{T_{Nk}^2}{\omega - \lambda_k} \tag{8.31b}$$

Equations (8.31) reveal that the eigenvalues λ_k of $-\mathbf{M}$, multiplied by j, are also the roots of the denominator polynomial $y_d(s)$ which is common to the rational polynomial admittance functions $y_{21}(s)$ and $y_{22}(s)$. So now we can develop the first and last rows T_{1k} and T_{Nk} of the orthogonal matrix $\mathbf{T}$ by equating the residues of $y_{21}(s)$ and $y_{22}(s)$ with $T_{1k} \, T_{Nk}$ and T_{Nk}^2, respectively, at corresponding eigenvalue poles λ_k. Knowing the numerator and denominator polynomials of $y_{21}(s)$ and $y_{22}(s)$ [equations (7.20) and (7.25) for doubly terminated networks, and (7.29) and (7.31) for singly terminated networks], we determine their residues r_{21k} and r_{22k} from the partial fraction expansions[4]

$$y_{21}(s) = \frac{y_{21n}(s)}{y_d(s)} = -j \sum_{k=1}^{N} \frac{r_{21k}}{\omega - \lambda_k}, \qquad y_{22}(s) = \frac{y_{22n}(s)}{y_d(s)} = -j \sum_{k=1}^{N} \frac{r_{22k}}{\omega - \lambda_k}$$

giving

$$T_{Nk} = \sqrt{r_{22k}}$$

$$T_{1k} = \frac{r_{21k}}{T_{Nk}} = \frac{r_{21k}}{\sqrt{r_{22k}}}, \qquad k = 1, 2, \ldots, N \tag{8.32}$$

With equation (8.27a) and by following the same procedure, it is demonstrated that $T_{1k} = \sqrt{r_{11k}}$ where r_{11k} are the residues of $y_{11}(s)$. From this and equation (8.32), it is seen that $T_{1k} = \sqrt{r_{11k}} = r_{21k} / \sqrt{r_{22k}}$, confirming the relationship $r_{21k}^2 = r_{11k} r_{22k}$ for a

---

[4]The residues of rational polynomials can be calculated as follows [6]:

$$r_{21k} = \left. \frac{y_{21n}(s)}{y_d'(s)} \right|_{s=j\lambda_k}, \qquad r_{22k} = \left. \frac{y_{22n}(s)}{y_d'(s)} \right|_{s=j\lambda_k}, \qquad k = 1, 2, \ldots, N$$

where $j\lambda_k$ are the roots of $y_d(s)$, and $y_d'(s)$ denotes the differentiation of the polynomial $y_d(s)$ with respect to s.

realizable network [20]. Having derived T_{1k} from $y_{21}(s)$ and $y_{22}(s)$ [equation (8.32)], the numerator polynomial of $y_{11}(s) = y_{11n}(s)/y_d(s)$ can be constructed.

The network can be directly connected between the terminating resistances R_S and R_L which, in general are nonunity in value. To scale the terminating impedances to 1 Ω, input/output inverter values M_{S1} and M_{NL} are found by scaling the magnitudes of the row vectors T_{1k} and T_{Nk} to unity for the "inner" network of Figure 8.5b by

$$M_{S1}^2 = R_S = \sum_{k=1}^{N} T_{1k}^2 \qquad M_{NL}^2 = R_L = \sum_{k=1}^{N} T_{Nk}^2 \qquad (8.33)$$

Then $T_{1k} \to T_{1k}/M_{S1}$ and $T_{Nk} \to T_{Nk}/M_{NL}$, where M_{S1} and M_{NL} are equivalent to the turns ratios n_1 and n_2 of the two transformers at the source and load ends of the network respectively, matching the terminating impedances with the internal network [4].

With the first and last rows of **T** now determined, the remaining orthogonal rows are constructed by the Gram–Schmitt orthonormalization process or similar methods [7,8,11]. Finally, the coupling matrix **M** is synthesized by using equation (8.28).

8.3 COUPLING MATRIX REDUCTION

The elements of the coupling matrix **M** that emerge from the synthesis procedure, explained in Section 8.2, all have the nonzero values observed in Figure 8.7. The nonzero values, occurring in the diagonal elements of the coupling matrices for electrically asymmetric networks, represent the offsets from the center frequency of each resonance (asynchronously tuned). The nonzero entries everywhere else mean that, in the network that **M** represents, couplings exist between each resonator or termination node, and every other resonator or termination node. Since this is clearly impractical, it is usual to annihilate the couplings with a sequence of similarity transforms (sometimes called "rotations") [9,10], until a more convenient form with a minimal number of couplings is obtained. The use of similarity transforms ensures that the eigenvalues and eigenvectors of the matrix **M** are preserved such that under analysis, the transformed matrix yields exactly the same transfer and reflection characteristics as the original matrix.

There are several practical canonical forms for the transformed coupling matrix **M**. Two of the better-known forms are the *arrow* form [12] and the more useful *folded* form [13,14], illustrated in Figure 8.9. Either of these canonical forms can be used directly if it is convenient to realize the couplings. Alternatively, the form can be adopted as a starting point for the application of further transforms to create topologies better suited for the physical and electrical constraints of the technology with which the filter is eventually realized [15,16]. The method for the reduction of the coupling matrix to the folded form is described here. The arrow form can be derived using a very similar method.

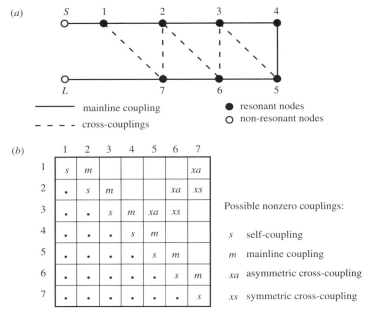

Possible nonzero couplings:

s self-coupling

m mainline coupling

xa asymmetric cross-coupling

xs symmetric cross-coupling

Couplings are symmetric about the principal diagonal

Figure 8.9 $N \times N$ folded canonical network coupling matrix form, seventh-degree example; s and xa couplings are zero for symmetric characteristics.

8.3.1 Similarity Transformation and Annihilation of Matrix Elements

A similarity transform (or rotation) on an $N \times N$ coupling matrix $\mathbf{M}_0$ is carried out by pre- and postmultiplying $\mathbf{M}_0$ by an $N \times N$ rotation matrix $\mathbf{R}$ and its transpose $\mathbf{R}^t$ [9,10]

$$\mathbf{M}_1 = \mathbf{R}_1 \cdot \mathbf{M}_0 \cdot \mathbf{R}_1^t \tag{8.34}$$

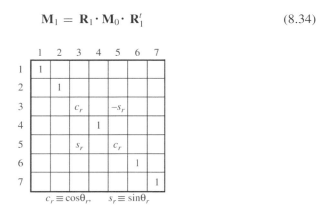

Figure 8.10 Example of seventh-degree rotation matrix $\mathbf{R}_r$, pivot [3,5], and angle θ_r.

where $\mathbf{M}_0$ is the original matrix, $\mathbf{M}_1$ is the matrix after the transform operation, and the rotation matrix $\mathbf{R}$ is as defined in Figure 8.10. The pivot $[i, j]$ $(i \neq j)$ of $\mathbf{R}_r$ indicates that elements $R_{ii} = R_{jj} = \cos \theta_r, R_{ji} = -R_{ij} = \sin \theta_r$, $(i, j \neq 1$ or $N)$, and θ_r are the angle of the rotation. The other principal diagonal entries are equal to one, and all other off-diagonal entries are zero.

After the transform, the eigenvalues of the matrix $\mathbf{M}_1$ are the same as those of the original matrix $\mathbf{M}_0$, which indicates that an arbitrarily long series of transforms with arbitrarily defined pivots and angles can be applied, starting with $\mathbf{M}_0$. Each transform in the series takes the following form

$$\mathbf{M}_r = \mathbf{R}_r \cdot \mathbf{M}_{r-1} \cdot \mathbf{R}_r^t \qquad r = 1, 2, 3, \dots R \qquad (8.35)$$

and under analysis, the resultant matrix $\mathbf{M}_R$ at the end of the series of transforms yields the same performance as that of the original matrix $\mathbf{M}_0$.

When a similarity transform of pivot $[i, j]$ and angle θ_r $(\neq 0)$ is applied to coupling matrix $\mathbf{M}_{r-1}$, the elements in rows i and j and columns i and j of the resultant matrix $\mathbf{M}_r$ change, in value, from the corresponding element values in $\mathbf{M}_{r-1}$. For the kth element in the row or column i or j of $\mathbf{M}_r$, and not on the cross-points of the pivot (i.e., $k \neq i, j$), the value changes according to the following formulas:

$$M'_{ik} = c_r M_{ik} + s_r M_{jk} \quad \text{for an element in row } i$$
$$M'_{jk} = s_r M_{ik} + c_r M_{jk} \quad \text{for an element in row } j$$
$$M'_{ki} = c_r M_{ki} - s_r M_{kj} \quad \text{for an element in column } i$$
$$M'_{kj} = s_r M_{ki} + c_r M_{kj} \quad \text{for an element in column } j \qquad (8.36a)$$

where k $(\neq i, j) = 1, 2, 3, \dots N$, $c_r = \cos \theta_r, s_r = \sin \theta_r$; the undashed matrix elements belong to the matrix $\mathbf{M}_{r-1}$; and the dashed to $\mathbf{M}_r$. For the elements on the cross-points of the pivot $(M_{ii}, M_{jj}, M_{ij} (= M_{ji}))$:

$$M'_{ii} = c_r^2 M_{ii} - 2 s_r c_r M_{ij} + s_r^2 M_{jj}$$
$$M'_{jj} = s_r^2 M_{ii} + 2 s_r c_r M_{ij} + c_r^2 M_{jj}$$
$$M'_{ij} = M_{ij}(c_r^2 - s_r^2) + s_r c_r(M_{ii} - M_{jj}) \qquad (8.36b)$$

Two properties of the similarity transform are exploited for the matrix reduction process:

1. Only those elements in the rows and columns i and j of the pivot $[i,j]$ of a transform are affected by the transform (provided its angle $\theta_r \neq 0$). All the other elements retain their previous values.

2. If two elements, facing each other across the rows and columns of the pivot of a transform, are both zero before the application of the transform, they are still zero after the transform. For example, if the elements M_{23} and M_{25} in

	1	2	3	4	5	6	7
1	s	m	④	③	②	①	xa
2	.	s	m	⑨	⑧	xa	xs
3	.	.	s	m	xa	xs	⑤
4	.	.	.	s	m	⑩	⑥
5	.	.	.	.	s	m	⑦
6	.	.	.	.	.	s	m
7	.	.	.	.	.	.	s

Figure 8.11 Seventh-degree coupling matrix reduction sequence for folded canonical form. The shaded elements are those that can be affected by a similarity transform at pivot [3,5] and angle θ_r ($\neq 0$). All the others remain unchanged.

Figure 8.11 happen to be both zero before the transform with pivot [3,5], they will still be zero after the transform, regardless of the transform angle θ_r.

The equations in (8.36) can be applied to annihilate (zero) specific elements in the coupling matrix. For example, to annihilate the nonzero element M_{15} (and simultaneously, M_{51}) in the seventh-degree coupling matrix of Figure 8.11, a transform of pivot [3,5] and angle $\theta_1 = -\tan^{-1}(M_{15}/M_{13})$ is applied to the coupling matrix [see the last formula in (8.36a) with $k = 1$, $i = 3$, and $j = 5$]. In the transformed matrix, M'_{15} and M'_{51} are zero and all the values in rows and columns three and five (shown shaded in Fig. 8.11) are changed. Equation set (8.37) summarizes the angle formulas for annihilating specific elements in the coupling matrix with a rotation at pivot $[i, j]$:

$$\theta_r = \tan^{-1}\left(M_{ik}/M_{jk}\right) \qquad \text{for the } k\text{th element in row } i \ (M_{ik}) \tag{8.37a}$$

$$\theta_r = -\tan^{-1}\left(M_{jk}/M_{ik}\right) \qquad \text{for the } k\text{th element in row } j \ (M_{ik}) \tag{8.37b}$$

$$\theta_r = \tan^{-1}\left(M_{ki}/M_{kj}\right) \qquad \text{for the } k\text{th element in column } i \ (M_{ik}) \tag{8.37c}$$

$$\theta_r = -\tan^{-1}\left(M_{kj}/M_{ki}\right) \qquad \text{for the } k\text{th element in column } j \ (M_{ik}) \tag{8.37d}$$

$$\theta_r = \tan^{-1}\left(\frac{-M_{ij} \ \pm \ \sqrt{M_{ij}^2 - M_{ii}M_{jj}}}{M_{jj}}\right) \qquad \text{for cross-pivot element } (M_{ii}) \tag{8.37e}$$

$$\theta_r = \tan^{-1}\left(\frac{M_{ij} \ \pm \ \sqrt{M_{ij}^2 - M_{ii}M_{jj}}}{M_{ii}}\right) \qquad \text{for cross-pivot element } (M_{jj}) \tag{8.37f}$$

$$\theta_r = \frac{1}{2}\tan^{-1}\left(\frac{2M_{ij}}{(M_{jj} - M_{ii})}\right) \qquad \text{for cross-pivot element } (M_{ij}) \tag{8.37g}$$

The method for reducing the full coupling matrix $\mathbf{M}_0$, resultant from the synthesis procedure of Section 8.2 to the folded form of Figure 8.9, involves applying a series of similarity transforms to the matrix that progressively annihilates the unrealizable or inconvenient elements one by one. The transforms are applied in a certain order and pattern that takes advantage of the two effects, mentioned previously, ensuring that once annihilated, an element is not regenerated by a subsequent transform in the sequence.

Reduction Procedure of a Full Coupling Matrix to Folded Canonical Form

A number of transform sequences are followed to reduce the full coupling matrix to the folded form. The sequence here involves alternately annihilating the elements right to left along the rows, and top to bottom down the columns, as shown in the seventh-order example in Figure 8.11, starting with the element in the first row and the $(N-1)$th column (M_{16}).

Element M_{16} is annihilated with a transform of pivot [5,6] and angle $\theta_1 = -\tan^{-1}(M_{16}/M_{15})$ [see equation (8.37d)]. This is followed by a second transform, pivot [4,5], angle $\theta_2 = -\tan^{-1}(M_{15}/M_{14})$ to annihilate element M_{15}. The previously annihilated element M_{16} is unaffected by this transform because it is lying outside the rows and columns of the latest pivot, and remains at zero. Now third and fourth transforms at pivots [3,4] and [2,3], and angles $\theta_3 = -\tan^{-1}(M_{14}/M_{13})$ and $\theta_4 = -\tan^{-1}(M_{13}/M_{12})$ annihilate M_{14} and M_{13}, respectively, again without disturbing the previously annihilated elements.

After these four transforms, the elements in the first row of the matrix between the main line coupling M_{12} and the element in the last (seventh) column are zero. Because of the symmetry about the principal diagonal, the elements between M_{21} and M_{71} in the first column will also be zero.

Next, the three elements in column 7, M_{37}, M_{47}, and M_{57}, are annihilated with transforms at pivots [3,4], [4,5], and [5,6] and angles $\tan^{-1}(M_{37}/M_{47})$,

TABLE 8.1 Seventh-Degree Example of Similarity Transform (Rotation) Sequence for Reduction of Full Coupling Matrix to Folded Form[a]

Transform Number r	Element to be Annihilated	Pivot $[i,j]$	$\theta_r = \tan^{-1}(c\,M_{kl}/M_{mn})$				
			k	l	m	n	c
1	M_{16} In row 2	[5,6]	1	6	1	5	−1
2	M_{15} In row 2	[4,5]	1	5	1	4	−1
3	M_{14} In row 2	[3,4]	1	4	1	3	−1
4	M_{13} In row 2	[2,3]	1	3	1	2	−1
5	M_{37} In column 7	[3,4]	3	7	4	7	+1
6	M_{47} In column 7	[4,5]	4	7	5	7	+1
7	M_{57} In column 7	[5,6]	5	7	6	7	+1
8	M_{25} In row 2	[4,5]	2	5	2	4	−1
9	M_{24} In row 2	[3,4]	2	4	2	3	−1
10	M_{46} In column 6	[4,5]	4	6	5	6	+1

[a]Total number of transforms $R = \sum_{n=1}^{N-3} n = 10$.

$\tan^{-1}(M_{47}/M_{57})$, and $\tan^{-1}(M_{57}/M_{67})$, respectively [see equation (8.37a)]. As with the rows, the columns are cleared down to the first mainline coupling encountered in that column. The couplings M_{13}, M_{14}, M_{15}, and M_{16} annihilated in the first sweep remain at zero, because they face each other across the pivot columns of the transforms of the second sweep, and are, therefore, unaffected.

A third sweep along row 2 annihilates M_{25} and M_{24} in that order, and the final sweep annihilates M_{46} in column 6. At this point, it is evident that the form of the folded canonical coupling matrix is achieved (Fig. 8.9) with its two cross-diagonals, containing the symmetric and asymmetric cross-couplings. Table 8.1 summarizes the entire annihilation procedure.

The final values and positions of the elements in the cross-diagonals are automatically determined; no specific action to annihilate the couplings within them needs to be taken. As the number of finite-position prescribed transmission zeros that the transfer function is realizing grows from one to the maximum permitted for the $N \times N$ matrix $(N - 2)$, the entries in the cross-diagonals progressively become nonzero, beginning with the asymmetric entry nearest to the principal diagonals (M_{35} in the 7th-degree example). If the original filter function is symmetric, then the asymmetric cross-couplings M_{35}, M_{26}, and M_{17} are zero (and in most cases the self-couplings in the principal diagonal M_{11} to M_{77}).

The regular pattern and order of the annihilation procedure makes it very amenable to computer programming, for any degree of coupling matrix.

Example of Usage We now illustrate the reduction procedure by choosing an example of a seventh-degree 23 dB return loss singly terminated asymmetric filter. A complex pair of transmission zeros at $\pm 0.9218 - j0.1546$ in the s plane are positioned to give group delay equalization over approximately 60% of the bandwidth, and a single zero is placed at $+j1.2576$ on the imaginary axis to give a rejection lobe level of 30 dB on the upper side of the passband.

TABLE 8.2 **(7–1–2) Asymmetric Singly Terminated Filter with Coefficients of Transfer and Reflection Polynomials**

s^n	Polynomial Coefficients		
$n =$	$P(s)$	$F(s)$	$E(s)$
0	-1.0987	$-j0.0081$	$+0.1378 - j0.1197$
1	$-j0.4847$	$+0.0793$	$+0.8102 - j0.5922$
2	$+0.9483$	$-j0.1861$	$+2.2507 - j1.3346$
3	$+j1.0$	$+0.7435$	$+3.9742 - j1.7853$
4		$-j0.5566$	$+4.6752 - j1.6517$
5		$+1.6401$	$+4.1387 - j0.9326$
6		$-j0.3961$	$+2.2354 - j0.3961$
7		$+1.0$	$+1.0$
	$\varepsilon = 6.0251$	$\varepsilon_R = 1.0$	

	1	2	3	4	5	6	7
1	0.0586	−0.0147	−0.2374	−0.0578	0.4314	−0.4385	0
2	−0.0147	−0.0810	0.4825	0.3890	0.6585	−0.0952	−1.3957
3	−0.2374	0.4825	0.2431	−0.0022	0.3243	−0.2075	0.1484
4	−0.0578	0.3890	−0.0022	−0.0584	−0.3047	0.4034	−0.0953
5	0.4314	0.6585	0.3243	−0.3047	0.0053	−0.5498	−0.1628
6	−0.4385	0.0952	−0.2075	0.4034	−0.5498	−0.5848	−0.1813
7	0	−1.3957	0.1484	−0.0953	−0.1628	−0.1813	0.0211

Figure 8.12 $N \times N$ coupling matrix of the (7−1−2) asymmetric, singly terminated filter, before application of reduction process. Element values are symmetric about the principal diagonal, and $R_1 = 0.7220$, $R_N = 2.2354$.

With knowledge of the positions of the three Tx (transmission) zeros, the numerator polynomial $F(s)$ of $S_{11}(s)$ is constructed by the recursive technique of Section 6.4. Then, by using the given return loss and the constant ε, we determine the denominator polynomial $E(s)$ common to $S_{11}(s)$ and $S_{21}(s)$. The s-plane coefficients of these polynomials are given in Table 8.2. Note that the coefficients of the polynomial $P(s)$ are multiplied by j, because $N - n_{fz} = 4$ is an even number.

From these coefficients, the numerators and denominators of y_{21} and y_{22} can be built up with equation sets (7.20) and (7.25). Now the residues, resultant from the partial-fraction expansions of y_{21} and y_{22} yield the first and last rows of the orthogonal matrix **T** (8.32). After scaling (8.33), the remaining rows of **T** are found by using an orthogonalization process [11], and finally the coupling matrix **M** is formed by equation (8.28). The element values of **M** are listed in Figure 8.12.

To reduce this full coupling matrix to the folded form, a series of 10 similarity transforms can now be applied to **M**, according to Table 8.2 and equation (8.34) in Section 8.3.1. Each transform is applied to the coupling matrix resultant from the previous transform, starting with **M** [$=\mathbf{M}_0$ in equation (8.35)]. After the last of the transforms in the series, the nonzero couplings in the matrix $\mathbf{M}_{10}$ topologically correspond with the couplings between the filter resonators arranged in a folded

	1	2	3	4	5	6	7
1	0.0586	0.6621	0	0	0	0	0
2	0.6621	0.0750	0.5977	0	0	0.1382	0
3	0	0.5977	0.0900	0.4890	0.2420	0.0866	0
4	0	0	0.4890	−0.6120	0.5038	0	0
5	0	0	0.2420	0.5038	−0.0518	0.7793	0
6	0	0.1382	0.0866	0	0.7793	0.0229	1.4278
7	0	0	0	0	0	1.4278	0.0211

Figure 8.13 (7−1−2) asymmetric single-terminated filter for the $N \times N$ coupling matrix after reduction to the folded form ($\mathbf{M}_{10}$).

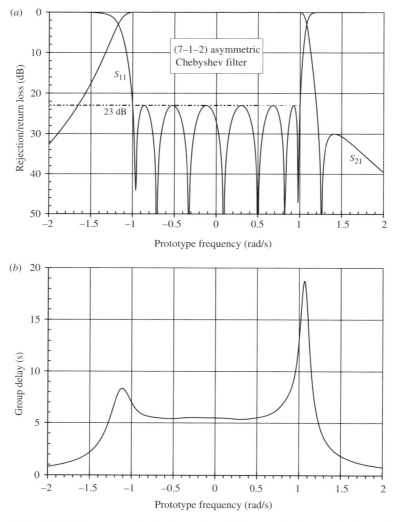

Figure 8.14 $(7-1-2)$ asymmetric single-terminated filter synthesis example for analysis of the folded coupling matrix: (a) rejection and return loss; (b) group delay.

pattern, ready for a direct realization in a suitable technology (Fig. 8.13). Note that the couplings M_{17} and M_{27} in the cross-diagonals, which are not needed to realize this particular transfer function, are automatically at zero. No specific action to annihilate them needs to be taken.

The results of analyzing this coupling matrix are plotted in Figure 8.14a (rejection/return loss) and Figure 8.14b (group delay). The return loss, as shown here, is indicative only, added to demonstrate that the in-band insertion loss for this single-terminated filter is equiripple. It can be seen that the 30 dB lobe level and equalized in-band group delay have not been affected by the transformation process.

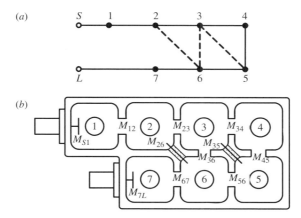

Figure 8.15 Realization in folded configuration: (*a*) folded network coupling and routing schematic; (*b*) corresponding realization in coaxial resonator technology.

Figure 8.15*a* illustrates the topology of the folded network, corresponding to the coupling matrix of Figure 8.13, and Figure 8.15*b* shows a possible realization for the filter in coupled coaxial resonator cavities. In this particular case, all the cross-couplings happen to be the same sign as the mainline couplings, but in general they may be mixed in sign.

8.4 SYNTHESIS OF THE $N+2$ COUPLING MATRIX

In this section, a method is presented for synthesis of the fully canonical or $N+2$ folded coupling matrix, which overcomes some of the shortcomings of the conventional $N \times N$ coupling matrix [17]. The $N+2$ folded coupling matrix is actually easier to synthesize, not needing the Gram–Schmitt orthonormalization step. The $N+2$ or extended coupling matrix has an extra pair of rows, top and bottom, and an extra pair of columns, left and right, surrounding the core $N \times N$ coupling matrix, which carry the input and output couplings from the source and load terminations to the resonator nodes in the core matrix. The $N+2$ matrix has some advantages as compared with the conventional coupling matrix:

- Multiple input/output couplings can be accommodated; that is, couplings can be made directly from the source and/or to the load to the internal resonators, in addition to the main input/output couplings to the first and last resonators in the filter circuit.
- Fully canonical filter functions can be synthesized.
- During certain synthesis procedures that employ a sequence of similarity transforms (rotations), it is sometimes convenient to temporarily park couplings in the outer rows or columns while the other rotations are carried out elsewhere in the matrix.

The $N + 2$ coupling matrix for a filter function is directly created by first synthesizing the coupling matrix for an Nth-degree "transversal" circuit, and then reducing this to the folded form by the same reduction technique as the one to reduce the "full" $N \times N$ coupling matrix to the folded form described in Section 8.3.

8.4.1   Synthesis of the Transversal Coupling Matrix

To synthesize the $N + 2$ transversal coupling matrix, we need to construct the two-port short-circuit admittance parameter matrix $[Y_N]$ for the overall network in two ways. First, the matrix is constructed from the coefficients of the rational polynomials of the transfer and reflection scattering parameters $S_{21}(s)$ and $S_{11}(s)$, which represent the characteristics of the filter to be realized, and the second from the circuit elements of the transversal array network. By equating the $[Y_N]$ matrices, derived by these two methods, the elements of the coupling matrix, associated with the transversal array network, are related to the coefficients of the $S_{21}(s)$ and $S_{11}(s)$ polynomials.

Synthesis of Admittance Function [Y_N] from the Transfer and Reflection Polynomials   The transfer and reflection polynomials that are generated in Section 6.4 for the general Chebyshev filter function have the form [equations (6.5) and (6.6)]

$$S_{21}(s) = \frac{P(s)/\varepsilon}{E(s)}, \qquad S_{11}(s) = \frac{F(s)/\varepsilon_R}{E(s)} \tag{8.38}$$

where

$$\varepsilon = \frac{1}{\sqrt{10^{RL/10} - 1}} \cdot \left| \frac{P(s)}{F(s)} \right|_{s = \pm j}$$

RL is the prescribed return loss in dB, and the polynomials $E(s)$, $F(s)$, and $P(s)$ are assumed to have been normalized to their respective highest-degree coefficients. Both $E(s)$ and $F(s)$ are Nth-degree polynomials; N is the degree of the filtering function; and $P(s)$, which contains the finite-position prescribed transmission zeros, is of degree n_{fz}, where n_{fz} is the number of finite-position transmission zeros (TZs) that have been prescribed. For a realizable network, n_{fz} must be $\leq N$.

The value of ε_R is unity for all the cases except for the fully canonical filtering functions where all the TZs are prescribed at finite frequencies (i.e., $n_{fz} = N$). In this case, the value of $S_{21}(s)$ (in dB) is finite at infinite frequency and if the highest-degree coefficient of the polynomials $E(s)$, $F(s)$, and $P(s)$ are each normalized to unity, ε_R has a value slightly greater than unity [see equation (6.26)] as follows:

$$\varepsilon_R = \frac{\varepsilon}{\sqrt{\varepsilon^2 - 1}} \tag{8.39}$$

The numerator and denominator polynomials for the $y_{21}(s)$ and $y_{22}(s)$ elements of $[Y_N]$ are built up directly from the transfer and reflection polynomials for $S_{21}(s)$ and $S_{11}(s)$ [equation sets (7.20) and (7.25)]. In a doubly terminated network with source and load terminations of $1\,\Omega$ [18], we obtain

$$\text{For } N \text{ even:} \quad y_{21}(s) = \frac{y_{21n}(s)}{y_d(s)} = \frac{(P(s)/\varepsilon)}{m_1(s)}$$

$$y_{22}(s) = \frac{y_{22n}(s)}{y_d(s)} = \frac{n_1(s)}{m_1(s)}$$

$$\text{For } N \text{ odd:} \quad y_{21}(s) = \frac{y_{21n}(s)}{y_d(s)} = \frac{(P(s)/\varepsilon)}{n_1(s)}$$

$$y_{22}(s) = \frac{y_{22n}(s)}{y_d(s)} = \frac{m_1(s)}{n_1(s)}$$

where

$$m_1(s) = \text{Re}(e_0 + f_0) + j\,\text{Im}(e_1 + f_1)s + \text{Re}(e_2 + f_2)s^2 + \cdots$$

$$n_1(s) = j\,\text{Im}(e_0 + f_0) + \text{Re}(e_1 + f_1)s + j\,\text{Im}(e_2 + f_2)s^2 + \cdots \quad (8.40)$$

and e_i and f_i, $i = 0, 1, 2, 3, \ldots, N$, are the complex coefficients of $E(s)$ and $F(s)/\varepsilon_R$, respectively. Also, $y_{11}(s)$ can be found here, but, like the $N \times N$ matrix, it is not needed for synthesis of the $N + 2$ matrix. In a similar way, the $y_{21}(s)$ and $y_{22}(s)$ polynomials for singly terminated networks may be found from equations (7.29) and (7.31).

Knowing the denominator and numerator polynomials for $y_{21}(s)$ and $y_{22}(s)$, we may find their residues r_{21k} and r_{22k}, $k = 1, 2, \ldots, N$, with partial fraction expansions; and the purely real eigenvalues λ_k of the network found by rooting the denominator polynomial $y_d(s)$, common to both $y_{21}(s)$ and $y_{22}(s)$. The Nth degree polynomial $y_d(s)$ has purely imaginary roots $= j\lambda_k$ [see equation (8.31)]. Expressing the residues in matrix form yields the following equation for the admittance matrix $[Y_N]$ for the overall network:

$$[Y_N] = \begin{bmatrix} y_{11}(s) & y_{12}(s) \\ y_{21}(s) & y_{22}(s) \end{bmatrix} = \frac{1}{y_d(s)} \begin{bmatrix} y_{11n}(s) & y_{12n}(s) \\ y_{21n}(s) & y_{22n}(s) \end{bmatrix}$$

$$= j \begin{bmatrix} 0 & K_\infty \\ K_\infty & 0 \end{bmatrix} + \sum_{k=1}^{N} \frac{1}{(s - j\lambda_k)} \cdot \begin{bmatrix} r_{11k} & r_{12k} \\ r_{21k} & r_{22k} \end{bmatrix} \quad (8.41)$$

Here, the real constant $K_\infty = 0$, except for the fully canonical case where the number of finite-position transmission zeros n_{fz} in the filtering function is equal to the filter degree N. In this case, the degree of the numerator of $y_{21}(s)$ ($y_{21n}(s) = jP(s)/\varepsilon$)) is equal to its denominator $y_d(s)$, and K_∞ needs to be extracted from $y_{21}(s)$ first, to reduce the degree of its numerator polynomial $y_{21n}(s)$ by one before its residues

r_{21k} can be found. Note that in the fully canonical case, where the integer quantity $N - n_{fz} = 0$ is even, it is necessary to multiply $P(s)$ by j to ensure that the unitary conditions for the scattering matrix are satisfied.

Independent of s, K_∞ is evaluated at $s = j\infty$ as follows:

$$jK_\infty = \left. \frac{y_{21n}(s)}{y_d(s)} \right|_{s = j\infty} = \left. \frac{jP(s)/\varepsilon}{y_d(s)} \right|_{s = j\infty} \tag{8.42}$$

The process for building up y_d from equation (8.40) results in its highest-degree coefficient with a value of $1 + 1/\varepsilon_R$; and since the highest-degree coefficient of $P(s) = 1$, the value of K_∞ is found as follows:

$$K_\infty = \frac{1}{\varepsilon} \cdot \frac{1}{(1 + 1/\varepsilon_R)} = \frac{\varepsilon_R}{\varepsilon} \frac{1}{(\varepsilon_R + 1)} \tag{8.43a}$$

With equation (8.39), an alternative form for K_∞ is derived as follows:

$$K_\infty = \frac{\varepsilon}{\varepsilon_R} (\varepsilon_R - 1) \tag{8.43b}$$

The new numerator polynomial $y'_{21n}(s)$ is now determined as

$$y'_{21n}(s) = y_{21n}(s) - j K_\infty y_d(s) \tag{8.44}$$

which is the degree $N - 1$, and the residues r_{21k} of $y'_{21}(s) = y'_{21n}(s)/y_d(s)$ may now be found as normal.

Synthesis of the Admittance Function [Y_N] by the Circuit Approach   In

addition, the two-port short-circuit admittance parameter matrix [Y_N] for the overall network can be synthesized directly from the fully canonical transversal network. The general form is depicted in Figure 8.16a. The matrix consists of a series of N individual 1st-degree lowpass sections, connected in parallel between the source and load terminations, but not to each other. The direct source–load coupling inverter M_{SL} is included to allow the fully canonical transfer functions to be realized, according to the minimum path rule (i.e., n_{fzmax}), the maximum number of finite position TZs that can be realized by the network $= N - n_{min}$, where n_{min} is the number of resonator nodes in the shortest route through the network between the source and load terminations. In fully canonical networks, $n_{min} = 0$, and so $n_{fzmax} = N$, the degree of the network.

Each of the N lowpass sections consists of one parallel-connected capacitor C_k and one frequency invariant susceptance B_k, connected through admittance inverters of characteristic admittances M_{Sk} and M_{Lk} to the source and load terminations, respectively. The circuit of the kth lowpass section is exhibited in Figure 8.16(b).

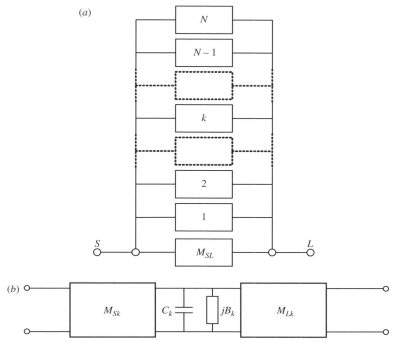

Figure 8.16   Canonical transversal array: (*a*) N-resonator transversal array including direct source–load coupling M_{SL}; (*b*) equivalent circuit of the kth lowpass resonator in the transversal array.

Fully Canonical Filter Functions   The direct source–load inverter M_{SL} in Figure 8.16*a* is zero except for the fully canonical filtering functions, where the number of finite-position zeros equals the degree of the filter. At infinite frequency ($s = \pm j\infty$), all the capacitors C_k in the lowpass sections become parallel short-circuits. They appear as open- circuits at the source–load ports through the inverters M_{Sk} and M_{Lk}. So, the only path between the source and load is via the frequency-invariant admittance inverter M_{SL}.

If the load impedance is 1 Ω, the driving point admittance $Y_{11\infty}$ looking in at the input port, as depicted in Figure 8.17, is

$$Y_{11\infty} = M_{SL}^2 \qquad (8.45)$$

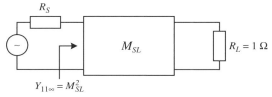

Figure 8.17   Equivalent circuit of transversal array at $s = \pm j\infty$.

Therefore the input reflection coefficient $S_{11}(s)$ at $s = j\infty$ is

$$S_{11}(s)\bigg|_{s=j\infty} \equiv |S_{11\infty}| = \left|\frac{(1 - Y_{11\infty})}{(1 + Y_{11\infty})}\right| \tag{8.46}$$

Making use of the conservation of energy principle and substituting for $|S_{11\infty}|$ in equation (8.46), we obtain

$$|S_{21\infty}| = \sqrt{1 - |S_{11\infty}|^2} = \frac{2\sqrt{Y_{11\infty}}}{(1 + Y_{11\infty})}$$

$$= \frac{2M_{SL}}{(1 + M_{SL}^2)} \tag{8.47}$$

Solving for M_{SL}, we now obtain

$$M_{SL} = \frac{1 \pm \sqrt{1 - |S_{21\infty}|^2}}{|S_{21\infty}|} = \frac{1 \pm |S_{11\infty}|}{|S_{21\infty}|} \tag{8.48}$$

At infinite frequency $|S_{21}(j\infty)| = |(P(j\infty)/\varepsilon)/E(j\infty)| = 1/\varepsilon$, because for a fully canonical filtering function, $P(s)$ and $E(s)$ are both Nth degree polynomials with their highest-degree coefficients normalized to unity. Similarly, $|S_{11}(j\infty)| = |(F(j\infty)/\varepsilon_R)/E(j\infty)| = 1/\varepsilon_R$. Therefore

$$M_{SL} = \frac{\varepsilon(\varepsilon_R \pm 1)}{\varepsilon_R} \tag{8.49}$$

Since ε_R is slightly greater than unity for a fully canonical network, we select the negative sign to give a relatively small value for M_{SL}

$$M_{SL} = \frac{\varepsilon(\varepsilon_R - 1)}{\varepsilon_R} \tag{8.50a}$$

which correctly gives $M_{SL} = 0$ for the noncanonical filters where $\varepsilon_R = 1$. Furthermore, it can be shown that the positive sign yields a second solution $M'_{SL} = 1/M_{SL}$ [see equation (8.43)], but since this is a large number, it is never used in practice [19]. An alternative form of equation (8.50a) involving ε_R can be found with the use of equation (8.39)

$$M_{SL} = \sqrt{\frac{\varepsilon_R - 1}{\varepsilon_R + 1}} \tag{8.50b}$$

where, again, it is evident that $M_{SL} = 0$ when $\varepsilon_R = 1$.

Synthesis of the Two-Port Admittance Matrix [Y_N] Cascading the elements in Figure 8.16b gives an [ABCD] transfer matrix for the k-h lowpass resonator as follows:

$$[ABCD]_k = - \begin{bmatrix} \dfrac{M_{Lk}}{M_{Sk}} & \dfrac{(sC_k + jB_k)}{M_{Sk}M_{Lk}} \\ 0 & \dfrac{M_{Sk}}{M_{Lk}} \end{bmatrix} \tag{8.51}$$

This is then directly converted into the following equivalent short-circuit y-parameter matrix:

$$
\begin{aligned}
[y_k] &= \begin{bmatrix} y_{11k}(s) & y_{12k}(s) \\ y_{21k}(s) & y_{22k}(s) \end{bmatrix} = \frac{M_{Sk}\,M_{Lk}}{(sC_k + jB_k)} \cdot \begin{bmatrix} \dfrac{M_{Sk}}{M_{Lk}} & 1 \\ 1 & \dfrac{M_{Lk}}{M_{Sk}} \end{bmatrix} \\
&= \frac{1}{(sC_k + jB_k)} \cdot \begin{bmatrix} M_{Sk}^2 & M_{Sk}\,M_{Lk} \\ M_{Sk}\,M_{Lk} & M_{Lk}^2 \end{bmatrix}
\end{aligned}
\tag{8.52}
$$

The two-port short-circuit admittance matrix $[Y_N]$ for the parallel-connected transversal array is the sum of the y-parameter matrices for N individual sections, plus the y-parameter matrix $[y_{SL}]$ for the direct source–load coupling inverter M_{SL} such that

$$
\begin{aligned}
[Y_N] &= \begin{bmatrix} y_{11}(s) & y_{12}(s) \\ y_{21}(s) & y_{22}(s) \end{bmatrix} = [y_{SL}] + \sum_{k=1}^{N} \begin{bmatrix} y_{11k}(s) & y_{12k}(s) \\ y_{21k}(s) & y_{22k}(s) \end{bmatrix} \\
&= j \begin{bmatrix} 0 & M_{SL} \\ M_{SL} & 0 \end{bmatrix} + \sum_{k=1}^{N} \frac{1}{(sC_k + jB_k)} \begin{bmatrix} M_{Sk}^2 & M_{Sk}\,M_{Lk} \\ M_{Sk}\,M_{Lk} & M_{Lk}^2 \end{bmatrix}
\end{aligned}
\tag{8.53}
$$

Synthesis of the $N+2$ Transversal Matrix Now the two expressions for $[Y_N]$, the first in terms of the residues of the transfer function matrix (8.41) and the second in terms of the circuit elements of the transversal array matrix (8.53), can be equated. It is obvious that $M_{SL} = K_\infty$, and for the elements with subscripts 21 and 22 in the matrices in the right-hand sides (RHSs) of equations (8.41) and (8.53), we obtain

$$\frac{r_{21k}}{(s - j\lambda_k)} = \frac{M_{Sk}M_{Lk}}{(sC_k + jB_k)} \tag{8.54a}$$

$$\frac{r_{22k}}{(s - j\lambda_k)} = \frac{M_{Lk}^2}{(sC_k + jB_k)} \tag{8.54b}$$

The residues r_{21k} and r_{22k} and the eigenvalues λ_k have already been derived from the S_{21} and S_{22} polynomials of the desired filtering function [see equation (8.41)], and so by equating the real and imaginary parts in equations (8.54a) and (8.54b), it is possible to relate them directly to the circuit parameters as

$$C_k = 1, \qquad B_k(\equiv M_{kk}) = -\lambda_k$$

$$M_{Lk}^2 = r_{22k}, \qquad M_{Sk}M_{Lk} = r_{21k}$$

$$M_{Lk} = \sqrt{r_{22k}} = T_{Nk}$$

$$M_{Sk} = r_{21k}/\sqrt{r_{22k}} = T_{1k}, \quad k = 1, 2, \ldots, N \qquad (8.55)$$

It may be recognized at this stage that M_{Sk} and M_{Lk} constitute the unscaled row vectors T_{1k} and T_{Nk} of the orthogonal matrix $\mathbf{T}$ as defined in Section 8.2.1. The capacitors C_k of the parallel networks are all unity; and the frequency-invariant susceptances B_k ($= -\lambda_k$, representing the self-couplings $M_{11} \rightarrow M_{NN}$), the input couplings M_{Sk}, the output couplings M_{Lk}, and the direct source–load coupling M_{SL} are all now defined. As a result, the reciprocal $N+2$ transversal coupling matrix $\mathbf{M}$, representing the network in Figure 8.16a, can now be constructed. M_{Sk} ($= T_{1k}$) are the N input couplings and occupy the first row and column of the matrix from positions 1 to N, as in Figure 8.18. Similarly, M_{Lk} ($= T_{Nk}$) are the N output couplings and they occupy the last row and column of $\mathbf{M}$ from positions

	S	1	2	3	..	k	..	$N-1$	N	L
S		M_{S1}	M_{S2}	M_{S3}	..	M_{Sk}	..	$M_{S,N-1}$	M_{SN}	M_{SL}
1	M_{1S}	M_{11}								M_{1L}
2	M_{2S}		M_{22}							M_{2L}
3	M_{3S}			M_{33}						M_{3L}
:	:				.		.			:
k	M_{kS}					M_{kk}				M_{kL}
:	:						.			:
$N-1$	$M_{N-1,S}$							$M_{N-1,N-1}$		$M_{N-1,L}$
N	M_{NS}								M_{NN}	M_{NL}
L	M_{LS}	M_{L1}	M_{L2}	M_{L3}	..	M_{Lk}	..	$M_{L,N-1}$	M_{LN}	

Figure 8.18 $N+2$ fully-canonical coupling matrix $\mathbf{M}$ for the transversal array. The "core" $N \times N$ matrix is indicated within the double lines. The matrix is symmetric about the principal diagonal, i.e. $M_{ij} = M_{ji}$.

1 to N. All the other entries are zero. Finally, M_{S1}^2 and M_{NL}^2 are equivalent to the terminating impedances R_S and R_L.

8.4.2 Reduction of the $N+2$ Transversal Matrix to the Folded Canonical Form

With N input and output couplings, the transversal topology is clearly impractical to realize for most cases and must be transformed to a more suitable topology. A more convenient form is the folded configuration [13], realized either directly or as the starting point for further transformations to other topologies that are more appropriate for the technology it is intended to use for the construction of the filter.

To reduce the transversal matrix to the folded form, the formal procedure, described in Section 8.3, is applied for the $N+2$ matrix instead of the $N \times N$ coupling matrix. This procedure involves a series of similarity transforms (rotations), which eliminate the unwanted coupling matrix entries, alternately, right to left along the rows, and top to bottom down the columns, starting with the outermost rows and columns, and working inward toward the center of the matrix, until the only remaining couplings are those that can be realized by filter resonators in a folded structure, as conveyed in Figure 8.19.

As with the $N \times N$ matrix, no special action needs to be taken to eliminate the unneeded xa and xs couplings in the cross-diagonals, since they automatically become zero if they are not necessary to realize the particular filter characteristic under consideration.

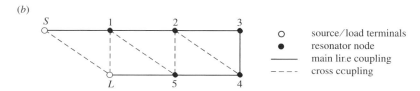

(a)

	S	1	2	3	4	5	L
S		m					xa
1	.	s	m			xa	xs
2	.	.	s	m	xa	xs	
3	.	.	.	s	m		
4	.	.	.	.	s	m	
5	.	.	.	.	.	s	m
L	.	.	.	.	.	.	

Possible non-zero couplings:

s — self coupling
m — main line coupling
xa — asymmetric cross-coupling
xs — symmetric cross-coupling
. — couplings are symmetric about the principal diagonal

All unspecified matrix entries are zero.

(b)

o — source/load terminals
● — resonator node
—— main line coupling
---- cross coupling

Figure 8.19 Folded $N+2$ canonical network coupling matrix form, fifth-degree example: (a) folded coupling matrix form—s and xa couplings are, in general, zero for the symmetric characteristics; couplings are symmetric about the principal diagonal (all unspecified matrix entries are zero); (b) coupling and routing schematic.

8.4.3 Illustrative Example

To illustrate the $N + 2$ matrix synthesis procedure, the following example is considered: a fully canonical fourth degree asymmetric filtering function with a 22 dB return loss and four transmission zeros, two at $-j3.7431$ and $-j1.8051$ that produce two attenuation lobes of 30 dB each on the lower side of the passband, and two at $+j1.5699$ and $+j6.1910$, producing a lobe of 20 dB on the upper side.

Application of the recursive technique of Section 6.4 yields the coefficients for the numerator and denominator polynomials of $S_{11}(s)$ and $S_{21}(s)$, repeated here for convenience as follows:

$$S_{21}(s) = \frac{P(s)/\varepsilon}{E(s)} \qquad S_{11}(s) = \frac{F(s)/\varepsilon_R}{E(s)} \tag{8.56}$$

Computed values of the coefficients are shown in Table 8.3. The value of ε_R is derived from equation (8.39). Note that because $N - n_{fz} = 0$, and is, therefore, an even number, the coefficients of $P(s)$ have been multiplied by j in Table 8.3.

Now the numerator and denominator polynomials of $y_{21}(s)(= y_{21n}(s)/y_d(s))$ and $y_{22}(s)(= y_{22n}(s)/y_d(s))$ can be constructed by using equation (8.40). The coefficients of $y_d(s)$, $y_{22n}(s)$ and $y_{21n}(s)$, normalized to the highest-degree coefficient of $y_d(s)$, are summarized in Table 8.4.

The next step is to find the residues of $y_{21}(s)$ and $y_{22}(s)$ with partial fraction expansions. Because the numerator of $y_{22}(s)$ $(y_{22n}(s))$ is one less in degree than its denominator $y_d(s)$, finding the associated residues r_{22k} is straightforward. However, the degree of the numerator of $y_{21}(s)$ $(y_{21n}(s))$ is the same as its denominator $y_d(s)$, and the factor K_∞ $(= M_{SL})$ needs to be extracted first, to reduce $y_{21n}(s)$ by one degree.

This is easily accomplished by first finding M_{SL} by evaluating $y_{21}(s)$ at $s = j\infty$ that is, M_{SL} equals the ratio of the highest degree coefficients in the numerator and denominator polynomials of $y_{21}(s)$ [see equation (8.42)]

$$jM_{SL} = y_{21}(s)\Big|_{s=j\infty} = \frac{y_{21n}(s)}{y_d(s)}\Big|_{s=j\infty} = j0.01509 \tag{8.57}$$

TABLE 8.3   Coefficients of Polynomials $E(s)$, $F(s)$, and $P(s)$ for the Fourth-Degree (4–4) Filter Function

s^i i	Coefficients of S_{11} and S_{21} Denominator Polynomial $E(s)$ (e_i)	Coefficients of S_{11} Numerator Polynomial $F(s)$ (f_i)	Coefficients of S_{21} Numerator Polynomial $P(s)$ (p_i)
0	$1.9877 - j0.0025$	0.1580	$j65.6671$
1	$+3.2898 - j0.0489$	$-j0.0009$	$+1.4870$
2	$+3.6063 - j0.0031$	$+1.0615$	$+j26.5826$
3	$+2.2467 - j0.0047$	$-j0.0026$	$+2.2128$
4	$+1.0$	$+1.0$	$+j1.0$
		$\varepsilon_R = 1.000456$	$\varepsilon = 33.140652$

TABLE 8.4 (4–4) Filtering Function — Coefficients of Numerator and Denominator Polynomials of $y_{21}(s)$ and $y_{22}(s)$, and $y'_{21n}(s)$

s^i i	Coefficients of Denominator Polynomial of $y_{22}(s)$ and $y_{21}(s)$ $(y_d(s))$	Coefficients of Numerator Polynomial of $y_{22}(s)$ $(y_{22n}(s))$	Coefficients cf Numerator Polynomial of $y_{21}(s)$ $(y_{21n}(s))$	Coefficients of Numerator Polynomial of $y_{21}(s)$ After Extraction of M_{SL} $(y'_{21n}(s))$
0	1.0730	$-j0.0012$	$j0.9910$	$j0.9748$
1	$-j0.0249$	$+1.6453$	$+0.0224$	$+0.0221$
2	$+2.3342$	$-j0.0016$	$+j0.4012$	$+j0.3659$
3	$-j0.0036$	$+1.1236$	$+0.0334$	$+0.0333$
4	$+1.0$	—	$+j0.0151$	—

which is seen to be the highest-degree coefficient of $y_{21n}(s)$ in Table 8.4. Alternatively, M_{SL} can be derived from equation (8.50).

From equation (8.44), M_{SL} is now extracted from the numerator of $y_{21}(s)$ by

$$y'_{21n}(s) = y_{21n}(s) \ - j M_{SL} \ y_d(s) \tag{8.58}$$

At this stage, $y'_{21n}(s)$ is one degree less than $y_d(s)$ (see Table 8.4) and the residues r_{21k} may be found as normal. The residues, the eigenvalues λ_k [where $j\lambda_k$ are the roots of $y_d(s)$], and the associated eigenvectors T_{1k} and T_{Nk} [see equation (8.55)] are listed in Table 8.5.

Note that for doubly terminated lossless networks with equal source and load terminations, r_{22k} is positive real for a realizable network, and $|r_{21k}| = |r_{22k}|$ [20].

Knowing the values of the eigenvalues λ_k, the eigenvectors T_{1k} and T_{Nk}, and M_{SL}, we now complete the $N+2$ transversal coupling matrix (Fig. 8.18) as shown in Figure 8.20.

Using the same reduction process as described in Section 8.3.1 but now operating on the $N+2$ matrix, we reduce the transversal matrix to the folded form with a series of six rotations, annihilating the elements M_{S4}, M_{S3}, M_{S2}, M_{2L}, M_{3L}, and finally, M_{13} in sequence (see Table 8.6). The resulting folded configurat:on coupling matrix is shown in Figure 8.21a, and its corresponding coupling and routing schematic in Figure 8.21b.

TABLE 8.5 Residues, Eigenvalues, and Eigenvectors for the Fourth-Degree (4–4) Filter Function

k	Eigenvalues λ_k	Residues		Eigenvectors	
		r_{22k}	r_{21k}	$T_{Nk} = \sqrt{r_{22k}}$	$T_{1k} = r_{21k}/\sqrt{r_{22k}}$
1	-1.3142	0.1326	0.1326	0.3641	0.3641
2	-0.7831	0.4273	-0.4247	0.6537	-0.6537
3	0.8041	0.4459	0.4459	0.6677	0.6677
4	1.2968	0.1178	-0.1178	0.3433	-0.3433

	S	1	2	3	4	L
S	0	0.3641	−0.6537	0.6677	−0.3433	0.0151
1	0.3641	1.3142	0	0	0	0.3641
2	−0.6537	0	0.7831	0	0	0.6537
3	0.6677	0	0	−0.8041	0	0.6677
4	−0.3433	0	0	0	−1.2968	0.3433
L	0.0151	0.3641	0.6537	0.6677	0.3433	0

Figure 8.20 Transversal coupling matrix for the fully canonical (4−4) filter function. Matrix is symmetric about the principal diagonal.

Analysis of this coupling matrix is plotted in Figure 8.22. It is evident that the return loss and rejection characteristics are unchanged from those obtained from the analysis of the original S_{11} and S_{21} polynomials.

TABLE 8.6 Pivots and Angles of the Similarity Transform Sequence for Reduction of Transversal Matrix to Folded Configuration for Fourth-Degree Filter[a]

Transform Number r	Pivot $[i, j]$	Element to be Annihilated	Figure 8.20	$\theta_r = \tan^{-1}(cM_{kl}/M_{mn})$				
				k	l	m	n	c
1	[3,4]	M_{S4}	In row S	S	4	S	3	−1
2	[2,3]	M_{S3}	In row S	S	3	S	2	−1
3	[1,2]	M_{S2}	In row S	S	2	S	1	−1
4	[2,3]	M_{2L}	In column L	2	L	3	L	+1
5	[3,4]	M_{3L}	In column L	3	L	4	L	+1
6	[2,3]	M_{13}	In row 1	1	3	1	2	−1

[a]Total number of transforms $R = \sum_{n=1}^{N-1} n = 6$.

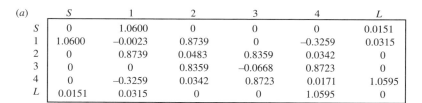

(a)

	S	1	2	3	4	L
S	0	1.0600	0	0	0	0.0151
1	1.0600	−0.0023	0.8739	0	−0.3259	0.0315
2	0	0.8739	0.0483	0.8359	0.0342	0
3	0	0	0.8359	−0.0668	0.8723	0
4	0	−0.3259	0.0342	0.8723	0.0171	1.0595
L	0.0151	0.0315	0	0	1.0595	0

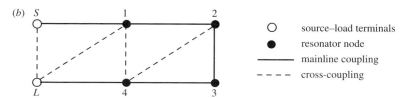

(b)

○ source–load terminals
● resonator node
—— mainline coupling
- - - - cross-coupling

Figure 8.21 The fully canonical synthesis example with the folded coupling matrix for the (4−4) filter function: (a) coupling matrix, symmetric about the principal diagonal; (b) coupling and routing schematic.

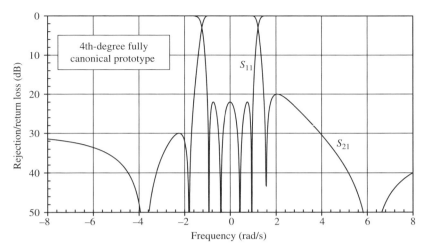

Figure 8.22   (4−4) fully canonical synthesis example: analysis of folded coupling matrix. Rejection as $s \rightarrow \pm j\infty = 20 \log_{10}(\varepsilon) = 30.407$ dB.

SUMMARY

In the 1970s, the synthesis technique, based on the coupling matrix of a filter, was introduced for the design of dual-mode bandpass filters. The technique was based on a *bandpass prototype* circuit, consisting of lossless lumped-element resonators inter-coupled via lossless transformers. Such a filter is capable of realizing a symmetric filter response. There are two principal advantages of this synthesis technique. Once the general coupling matrix with all the permissible couplings has been synthesized, it allows matrix operations on the coupling matrix to realize a variety of filter topologies. The second advantage is that the coupling matrix represents the practical bandpass filter topology. Therefore, it is possible to identify each element of the practical filter uniquely, including its Q value, dispersion characteristics, and sensitivity. This permits a more accurate determination of the practical filter characteristics and an insight into ways to optimize filter performance.

At the beginning of this chapter, the concept of a $N \times N$ coupling matrix for synthesizing bandpass filters is introduced. The concept is then extended to include the hypothetical frequency invariant reactance (FIR) element, as described in Sections 3.10 and 6.1. The FIRs make the synthesis process more general, and allows inclusion of the asymmetric filter response as well. A second generalization is introduced by establishing the equivalent lowpass prototype circuit of the coupling matrix. Such a prototype circuit allows the application of the synthesis techniques discussed in Chapter 7.

The procedure for synthesis of the $N \times N$ coupling matrix is described in depth. The concept is extended to the $N + 2$ coupling matrix by separating out the purely resistive and purely reactive portions of the $N \times N$ matrix. The synthesis process yields the general coupling matrix with finite entries for all the couplings. A filter that is based on such a coupling structure is clearly impractical. The next step in the process is to derive topologies with a minimum number of couplings, referred to as *canonical forms*. This is achieved by applying similarity transformations to the coupling matrix. Such transformations preserve the eigenvalues and eigenvectors of the matrix, thus ensuring that the desired filter response remains unaltered.

There are several practical canonical forms for the transformed coupling matrix **M**. Two of the better-known forms are the *arrow* form and the more generally useful *folded* form. Either of these canonical forms can be used directly, if it is convenient to realize the couplings. Alternatively, either form may be used as a starting point for the application of further transforms to create topologies better suited for the technology with which the filter is eventually realized. The method for the reduction of the general coupling matrix to the folded form is described in depth. The arrow form can be derived using a very similar method. Examples are included to illustrate the synthesis process for both the $N \times N$ and $N + 2$ coupling matrices.

REFERENCES

1. A. E. Atia and A. E. Williams, New types of bandpass filters for satellite transponders, *COMSAT Tech. Rev.* **1**, 21–43 (Fall 1971).

2. A. E. Atia and A. E. Williams, Narrow-bandpass waveguide filters, *IEEE Trans. Microwave Theory Tech.* **MTT-20**, 258–265 (April 1972).

3. A. E. Atia and A. E. Williams, Nonminimum-phase optimum-amplitude bandpass waveguide filters, *IEEE Trans. Microwave Theory Tech.* **MTT-22**, 425–431 (April 1974).

4. A. E. Atia, A. E. Williams, and R. W. Newcomb, Narrow-band multiple-coupled cavity synthesis, *IEEE Trans. Circuits Syst.* **CAS-21**, 649–655 (Sept. 1974).

5. G. Matthaei, L. Young, and E. M. T. Jones, *Microwave Filters, Impedance Matching Networks and Coupling Structures*, Artech House, Norwood, MA, 1980.

6. E. Kreyszig, *Advanced Engineering Mathematics*, 3rd ed., Wiley, New York, 1972.

7. J. S. Frame, Matrix functions and applications, Part IV: Matrix functions and constituent matrices, *IEEE Spectrum* **1** (March–July 1964), (series of 5 articles).

8. H. L. Hamburger and M. E. Grimshaw, *Linear Transformations in n-Dimensional Space*, Cambridge Univ. Press, London, 1951.

9. F. R. Gantmacher, *The Theory of Matrices*, Vol. 1, Chelsea Publishing, New York, 1959.

10. C. E. Fröberg, *Introduction to Numerical Analysis*, Addison-Wesley, Reading, MA, 1965, Chapter 6.

11. G. H. Golub and C. F. van Loan, *Matrix Computations*, 2nd ed., John Hopkins Univ. Press, Baltimore, 1989.

12. H. C. Bell, Canonical asymmetric coupled-resonator filters, *IEEE Trans. Microwave Theory Tech.* **MTT-30**, 1335–1340 (Sept. 1982).

13. J. D. Rhodes, A lowpass prototype network for microwave linear phase filters, *IEEE Trans. Microwave Theory Tech.* **MTT-18**, 290–300 (June 1970).

14. J. D. Rhodes and A. S. Alseyab, The generalized Chebyshev low pass prototype filter, *Int. J. Circuit Theory Appl.* **8**, 113–125 (1980).

15. R. J. Cameron and J. D. Rhodes, Asymmetric realizations for dual-mode bandpass filters, *IEEE Trans. Microwave Theory Tech.* **MTT-29**, 51–58 (Jan. 1981).

16. R. J. Cameron, A novel realization for microwave bandpass filters, *ESA J.* **3**, 281–287 (1979).

17. R. J. Cameron, Advanced coupling matrix synthesis techniques for microwave filters, *IEEE Trans. Microwave Theory Tech.* **MTT-51**, 1–10 (Jan. 2003).

18. O. Brune, Synthesis of a finite two-terminal network whose driving point impedance is a prescribed function of frequency, *J. Math. Phys.* **10**(3), 191–236 (1931).

19. S. Amari, Direct synthesis of folded symmetric resonator filters with source-load coupling, *IEEE Microwave Wireless Compon. Lett.* **MTT-11**, 264–266 (June 2001).

20. M. E. van Valkenburg, *Network Analysis*, Prentice-Hall, Englewood Cliffs, NJ, 1955.

CHAPTER 9

RECONFIGURATION OF THE FOLDED COUPLING MATRIX

In Chapters 7 and 8 we described two different techniques for synthesizing the canonical folded-array coupling matrix from the transfer and the reflection polynomials of the desired filter function. The folded array has several advantages for the filter designer:

- Layout possibilities are relatively simple.
- Positive and negative couplings can be implemented to realize advanced filter characteristics.
- The maximum $(N-2)$ number of finite-position transmission zeros can be realized if required.
- The diagonal couplings, necessary for asymmetric characteristics, can be implemented in some cases.

However, if the filter is to be realized in dual-mode technology (with two orthogonal resonant modes supported in the same physical resonator), a disadvantage becomes apparent. For a folded-array realization, the input and output of the filter are in the same physical cavity. This imposes a limit on the input/output isolation that can be achieved with bandpass filters, typically 25 dB for the dual-TE_{11n} (Transverse Electric) mode in cylindrical resonator cavities or dual-TE_{10n} mode in square resonator cavities. Consequently, for dual-mode filters, the input and output couplings must occur in separate resonators. This chapter is devoted to the methods for the similarity transformations of the folded coupling array to realize a wide range of topologies suitable for dual-mode filter networks.

Microwave Filters for Communication Systems: Fundamentals, Design, and Applications,
by Richard J. Cameron, Chandra M. Kudsia, and Raafat R. Mansour
Copyright © 2007 John Wiley & Sons, Inc.

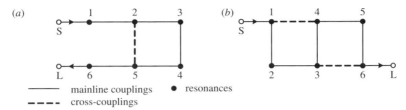

Figure 9.1 Sixth-degree network (*a*) cross-coupled folded configuration; (*b*) after conversion to inline topology.

9.1 SYMMETRIC REALIZATIONS FOR DUAL-MODE FILTERS

In the early 1970s, the folded cross-coupled array was first introduced by Rhodes in a classical series of papers [1–3]. For symmetric even-degree characteristics, a method for synthesizing dual-mode longitudinal structure from the folded coupling matrix was also introduced [4]. The input to the filter is at one end of the structure and the output is at the other end, avoiding the low isolation problems associated with the folded realization.

Figure 9.1 illustrates the conversion of a (6–2) folded network to an inline or propagating topology. With the inline topology we can realize fewer finite-position transmission zeros than with the folded topology (minimum path rule). So the prototype function must be designed accordingly: two TZs for the 6th order, four TZs for the 8th and 10th orders, and six TZs for the 12th order.

For the inline topology, Rhodes and Zabalawi adopted the even-mode coupling matrix as a base [4]. For even degree symmetric networks, the even mode network is formed by reflecting the quadrants of the $N \times N$ folded coupling matrix about the horizontal and vertical lines of symmetry, superimposed into the upper left quadrant. An alternative method for generating even-mode and odd-mode submatrices is outlined in Section 9.5.1. Equation (9.1) demonstrates the process for sixth-degree examples: (a) a 6×6 symmetric coupling matrix with the lines of symmetry; (b) the 3×3 even-mode matrix $\mathbf{M_e}$ and the associated K matrix, which is simply the $\mathbf{M_e}$ matrix with the subscripts to the elements along the diagonal adjusted to reflect their real position within the matrix, ($K_{22} = M_{25}$); (c) even-mode matrix after the rotation sequence; and (d) after unfolding to the full matrix:

$$\mathbf{M} = \left[\begin{array}{ccc|ccc} 0 & M_{12} & 0 & 0 & 0 & 0 \\ M_{12} & 0 & M_{23} & 0 & M_{25} & 0 \\ 0 & M_{23} & 0 & M_{34} & 0 & 0 \\ \hline 0 & 0 & M_{34} & 0 & M_{23} & 0 \\ 0 & M_{25} & 0 & M_{23} & 0 & M_{12} \\ 0 & 0 & 0 & 0 & M_{12} & 0 \end{array}\right] \qquad (9.1a)$$

$$\mathbf{M_e} = \begin{bmatrix} 0 & M_{12} & 0 \\ M_{12} & M_{25} & M_{23} \\ 0 & M_{23} & M_{34} \end{bmatrix} = \begin{bmatrix} 0 & K_{12} & 0 \\ K_{12} & K_{22} & K_{23} \\ 0 & K_{23} & K_{33} \end{bmatrix} \tag{9.1b}$$

$$\mathbf{M'_e} = \begin{bmatrix} 0 & K'_{12} & K'_{13} \\ K'_{12} & 0 & K'_{23} \\ K'_{13} & K'_{23} & K'_{33} \end{bmatrix} \tag{9.1c}$$

$$\mathbf{M'} = \left[\begin{array}{ccc|ccc} 0 & K'_{12} & 0 & K'_{13} & 0 & 0 \\ K'_{12} & 0 & K'_{23} & 0 & 0 & 0 \\ 0 & K'_{23} & 0 & K'_{33} & 0 & K'_{13} \\ \hline K'_{13} & 0 & K'_{33} & 0 & K'_{23} & 0 \\ 0 & 0 & 0 & K'_{23} & 0 & K'_{12} \\ 0 & 0 & K'_{13} & 0 & K'_{12} & 0 \end{array} \right] \tag{9.1d}$$

Now the matrix is in the correct form for the sixth-degree inline filter (Figure 9.1b). Next, a series of rotations are applied to the even-mode matrix, with angles derived from the elements from the original folded coupling matrix, according to Table 9.1. However, the sequence of pivots and angles for each degree of the matrix does not seem to follow any particular pattern, and needs to be determined individually for each degree. Table 9.1 lists the pivots and the order that the rotations need to be applied to the even mode matrix for even degrees 6–12 inclusive. The

TABLE 9.1 Symmetric Even-Degree Prototypes: Pivot and Angle Definitions for Rotation Sequences to Reconfigure the Folded Coupling Matrix to Inline Topology

Degree N	Rotation No. r	Rotation Pivot $[i, j]$	Angle θ_r	Angle Equation	Annihilated Elements in $[\mathbf{M_e}]$
6	1	[2,3]	θ_1	(9.2)	K_{22}
8	1	[3,4]	θ_1	(9.3a)	—
	2	[2,4]	θ_2	(9.3b)	$K_{22}\ K_{24}$
10	1	[4,5]	θ_1	(9.4a)	—
	2	[3,5]	θ_2	(9.4b)	—
	3	[2,4]	θ_3	(9.4c)	$K_{44}\ K_{33}\ K_{25}$
12	1	[4,5]	θ_1	(9.5a)	—
	2	[5,6]	θ_2	(9.5b)	—
	3	[4,6]	θ_3	$\theta_3 = \tan^{-1}(K_{46}/K_{66})$	—
	4	[3,5]	θ_4	$\theta_4 = \tan^{-1}(K_{35}/K_{55})$	$K_{33}\ K_{35}\ K_{44}\ K_{46}$
	5	[2,4]	θ_5	$\theta_5 = \tan^{-1}(K_{25}/K_{45})$	K_{25}

(a)
$$M = \begin{bmatrix} 0 & 0.8867 & 0 & 0 & 0 & 0 \\ 0.8867 & 0 & 0.6050 & 0 & -0.1337 & 0 \\ 0 & 0.6050 & 0 & 0.7007 & 0 & 0 \\ 0 & 0 & 0.7007 & 0 & 0.6050 & 0 \\ 0 & -0.1337 & 0 & 0.6050 & 0 & 0.8867 \\ 0 & 0 & 0 & 0 & 0.8867 & 0 \end{bmatrix}$$

(b)
$$M_e = \begin{bmatrix} 0 & 0.8867 & 0 \\ 0.8867 & -0.1337 & 0.6050 \\ 0 & 0.6050 & 0.7007 \end{bmatrix}$$

(c)
$$M'_e = \begin{bmatrix} 0 & 0.8820 & -0.0919 \\ 0.8820 & 0 & 0.6780 \\ -0.0919 & 0.6780 & 0.5670 \end{bmatrix}$$

(d)
$$M' = \begin{bmatrix} 0 & 0.8820 & 0 & -0.0919 & 0 & 0 \\ 0.8820 & 0 & 0.6780 & 0 & 0 & 0 \\ 0 & 0.6780 & 0 & 0.5670 & 0 & -0.0919 \\ -0.0919 & 0 & 0.5670 & 0 & 0.6780 & 0 \\ 0 & 0 & 0 & 0.6780 & 0 & 0.8820 \\ 0 & 0 & -0.0919 & 0 & 0.8820 & 0 \end{bmatrix}$$

Figure 9.2 Stages in the transformation of a sixth-degree folded coupling matrix to the symmetric inline form.

corresponding formulas for the angles for each rotation are given by equations (9.2)–(9.5). The coupling elements in the angle formulas are taken from the original $N \times N$ folded coupling matrix.

9.1.1 Sixth-Degree Filter

Only a single rotation is required to transform the folded sixth-degree network to the inline form in Figure 9.1, with pivot [2,3] and angle θ_1 given by

$$\theta_1 = \tan^{-1}\left[\frac{M_{23} \pm \sqrt{M_{23}^2 - M_{25}M_{34}}}{M_{34}}\right] \tag{9.2}$$

A symmetric sixth-degree Chebyshev filter function with a 23 dB return loss and two TZs at $\pm j1.5522$ that produce two 40-dB rejection lobes is used as an example to show the realization. The $N \times N$ folded coupling matrix **M** for this filter is depicted in Figure 9.2a and the corresponding even mode matrix M_e, in Figure 9.2b. The application of equation (9.2) yields two solutions for θ_1: $-5.948°$ and $61.359°$. Taking the first of these solutions and applying a rotation to M_e at pivot [2,3], we annihilate M_{22} to attain the modified even mode matrix M'_e, (Fig. 9.2c). Finally, the unfolding yields the propagating form of the coupling matrix with symmetrically valued coupling elements, as presented in Figure 9.2d.

9.1.2 Eighth-Degree Filter

Refer to Figure 9.3.

The angles θ_1 and θ_2 required for the two rotations for the eighth-degree case are given by

$$\theta_1 = \tan^{-1}\left[\frac{M_{27}M_{34} \pm \sqrt{M_{27}^2 M_{34}^2 + M_{27}M_{45}\left(M_{23}^2 - M_{27}M_{36}\right)}}{M_{23}^2 - M_{27}M_{36}}\right] \tag{9.3a}$$

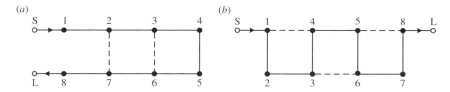

Figure 9.3 Eighth-degree network: (*a*) cross-coupled folded configuration; (*b*) after conversion to inline topology.

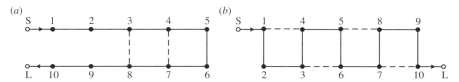

Figure 9.4 10th-degree network (*a*) cross-coupled folded configuration; (*b*) after conversion to inline topology.

and

$$\theta_2 = \tan^{-1}\left[\frac{M_{27}}{M_{23}\sin\theta_1}\right] \tag{9.3b}$$

9.1.3 10th-Degree Filter

Refer to Figure 9.4.

The 10th degree case needs three rotations with angles, resulting from

$$\theta_1 = \tan^{-1}\left[\frac{M_{45} \pm \sqrt{M_{45}^2 - M_{47}M_{56}}}{M_{56}}\right] \tag{9.4a}$$

$$\theta_2 = \tan^{-1}\left[\frac{s_1 M_{34} \pm \sqrt{s_1^2 M_{34}^2 - M_{38}(M_{47} + M_{56})}}{M_{47} + M_{56}}\right] \tag{9.4b}$$

$$\theta_3 = \tan^{-1}\left[\frac{t_2 M_{23}}{M_{45} - t_1 M_{56} + c_1 t_2 M_{34}}\right] \tag{9.4c}$$

where $c_1 \equiv \cos\theta_1$, $t_2 \equiv \tan\theta_2$, and so on.

9.1.4 12th-Degree Filter

Refer to Figure 9.5.

The 12th-degree filter case begins with the solution of a quartic equation, which is achieved analytically as follows [5]

$$t_1^4 + d_3 t_1^3 + d_2 t_1^2 + d_1 t_1 + d_0 = 0 \tag{9.5a}$$

(a)

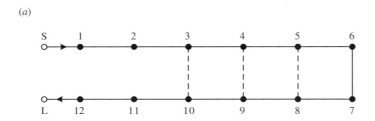

(b)

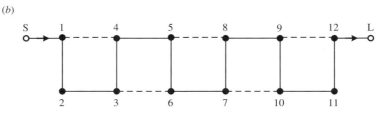

Figure 9.5 12th-degree network: (a) cross-coupled folded configuration; (b) after conversion to inline topology.

where $t_1 \equiv \tan\theta_1$, and

$$d = a_2 c_3^2 + a_3 c_2^2$$

$$d_0 = \frac{a_0 + a_3 c_0^2 + a_4 c_0}{d}$$

$$d_1 = \frac{a_1 + 2a_0 c_3 + 2a_3 c_0 c_1 + a_4(c_1 + c_0 c_3)}{d}$$

$$d_2 = \frac{a_2 + 2a_1 c_3 + a_0 c_3^2 + a_3(c_1^2 + 2c_0 c_2) + a_4(c_2 + c_1 c_3)}{d}$$

$$d_3 = \frac{2a_2 c_3 + a_1 c_3^2 + 2a_3 c_1 c_2 + a_4 c_2 c_3}{d}$$

$$a_0 = M_{58} \qquad\qquad b_0 = M_{49}M_{67} \qquad\qquad c = a_3 b_5 - a_4 b_4$$

$$a_1 = 2M_{45} \qquad\qquad b_1 = -2M_{45}M_{67} \qquad c_0 = \frac{a_0 b_4 - a_3 b_0}{c}$$

$$a_2 = M_{49} - \frac{M_{34}^2}{M_{3,10}} \quad b_2 = M_{58}M_{67} - M_{56}^2 \quad c_1 = \frac{a_1 b_4 - a_3 b_1}{c}$$

$$a_3 = M_{67} \qquad\qquad b_3 = -2M_{56}M_{45} \qquad c_2 = \frac{a_2 b_4 - a_3 b_2}{c}$$

$$a_4 = -2M_{56} \qquad\quad b_4 = M_{49}M_{58} - M_{45}^2 \quad c_3 = \frac{a_3 b_3}{c}$$

$$b_5 = 2M_{56}M_{49}$$

Any one of the previous four solutions of t_1 from the quartic is then used to find θ_2, as follows:

$$\theta_2 = \tan^{-1}\left[\frac{c_0 + c_1 t_1 + c_2 t_1^2}{(1 + c_3 t_1)\sqrt{1 + t_1^2}}\right] \tag{9.5b}$$

Three further rotations are required, according to Table 9.1, by using the elements from the matrix resultant from the previous rotation in each case. After the final rotation and unfolding, the inline configuration in Figure 9.5*b* is obtained.

Conditions for Symmetric Realizations The angle formulas for the rotations in reconfiguring the folded coupling matrix to the symmetric inline form involve the solutions of quadratic equations. Inevitably, there are certain patterns of transmission zeros in the original prototype filter functions that produce negative square roots and, therefore, cannot be realized with a symmetric structure. These conditions are discussed in more detail in the literature [4]. One useful pattern that does violate the "positive square-root condition" is the case of a real-axis TZ pair and an imaginary-axis TZ pair in an eighth order prototype (8−2−2). This particular problem is resolved by using the CQ method, or the asymmetric realization method, discussed next.

9.2 ASYMMETRIC REALIZATIONS FOR SYMMETRIC CHARACTERISTICS

The originating folded coupling matrix, and the resultant inline topology of the asymmetric inline realization are exactly the same as those for the symmetric equivalent. However, the values of the inline coupling matrix and, thus, the dimensions of the physical elements that realize them, are not equivalued about the physical center of the structure [6]. Although this means a greater design effort is needed to develop and manufacture a working filter, there are no restrictions on the pattern of TZs that the prototype can incorporate (apart from the minimum path rule, and the symmetry of the pattern of TZs about the imaginary axis (unitary condition), and about the real axis (symmetric characteristics)). Moreover, the computations required to produce the inline configuration are significantly less complex.

Like the symmetric configurations, there does not appear to be a rule governing the sequence of rotations or angles of rotation for the different degrees of filters. Each (even) degree has to be considered individually. For $N = 4$, the folded and inline matrices are of the same form and no further transformation is necessary. The first nontrivial case is $N = 6$, and solutions for $N = 6$, 8, 10, 12, 14 have been derived. Table 9.2 is a summary of the pivots and angles θ_r that are applied to obtain the inline configuration for these degrees. For the rth rotation, the elements M_{l_1,l_2} and M_{m_1,m_2} are again taken from the coupling matrix of the previous rotation,

TABLE 9.2   Pivot Positions and Rotation Angles for the General Asymmetric Inline Realization, for Degrees 6, 8, 10, 12, and 14

Degree N	Rotation Number r	Pivot $[i,j]$	$\theta_r = \tan^{-1}[c\,M_{l_1,l_2}/M_{m_1,m_2}]$				
			l_1	l_2	m_1	m_2	c
6	1	[2,4]	2	5	4	5	+1
8	1	[4,6]	3	6	3	4	−1
	2	[2,4]	2	7	4	7	+1
	3	[3,5]	2	5	2	3	−1
	4	[5,7]	4	7	4	5	−1
10	1	[4,6]	4	7	6	7	+1
	2	[6,8]	3	8	3	6	−1
	3	[7,9]	6	9	6	7	−1
12	1	[5,9]	4	9	4	5	−1
	2	[3,5]	3	10	5	10	+1
	3	[2,4]	2	5	4	5	+1
	4	[6,8]	3	8	3	6	−1
	5	[7,9]	6	9	6	7	−1
	6	[8,10]	5	10	5	8	−1
	7	[9,11]	8	11	8	9	−1
14	1	[6,10]	5	10	5	6	−1
	2	[4,6]	4	11	6	11	+1
	3	[7,9]	4	9	4	7	−1
	4	[8,10]	7	10	7	8	−1
	5	[9,11]	6	11	6	9	−1
	6	[10,12]	9	12	9	10	−1
	7	[5,7]	4	7	4	5	−1
	8	[7,9]	6	9	6	7	−1
	9	[9,11]	8	11	8	9	−1
	10	[11,13]	10	13	10	11	−1

and then applied to that matrix, beginning with the original folded matrix for $r = 1$. The resultant inline topologies are exactly the same as those for the symmetric inline structure except for the 6th-, 10th-, and 14th-degree cases. Here, the cross-coupling closest to the output of the filter is zero.

9.3   "PFITZENMAIER" CONFIGURATIONS

In 1977, G. Pfitzenmaier introduced a configuration to avoid the input/output isolation problem associated with the folded configuration in a dual-mode structure for sixth-degree symmetric filter characteristics [7]. Pfitzenmaier demonstrated that the synthesized sixth-degree circuit can be transformed (not by using coupling matrix methods) to a topology where the input and output resonances (1 and 6) are in adjacent

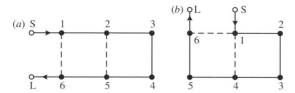

Figure 9.6 Pfitzenmaier configuration for (6–4) symmetric filtering characteristic: (*a*) original folded configuration; (*b*) after transformation to the Pfitzenmaier configuration.

cavities of the dual-mode structure, avoiding the isolation problem. Furthermore, because it is possible to directly cross couple resonances 1 and 6, the signal has only two resonances to pass through between the input and output. As a result, by the minimum path rule, the Pfitzenmaier configuration is able to realize $N - 2$ transmission zeros, the same as the folded structure. The coupling and routing diagram for a sixth degree Pfitzenmaier configuration is illustrated in Figure 9.6.

The Pfitzenmaier configuration may be easily obtained for the sixth-degree filter, and any even-degree symmetric characteristic above 6, by using a sequence of coupling matrix rotations [8]. Unlike the asymmetric inline realization, the pivots and angles of the rotations in the sequence may be defined with simple equations.

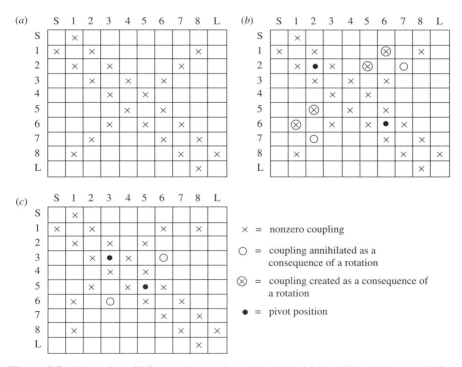

Figure 9.7 Formation of Pfitzenmaier topology: (*a*) original folded CM; (*b*) pivot at [2,6] to annihilate M_{27}, creating M_{16} and M_{25}; (*c*) pivot at [3,5] to annihilate M_{36}.

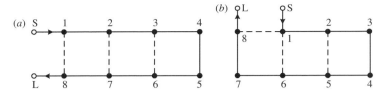

Figure 9.8 Pfitzenmaier configuration for 8th-degree symmetric filtering characteristic: (*a*) original folded configuration; (*b*) after transformation to Pfitzenmaier configuration.

Starting with the folded matrix, a series of $R = (N - 4)/2$ rotations is applied according to equation (9.6), after which the Pfitzenmaier configuration is obtained.

For the *k*th rotation, pivot $= [i, j]$ and the associated angle $= \theta_r$, where

$$\left. \begin{array}{l} i = r + 1 \\ j = N - i \\ \theta_r = \tan^{-1} \dfrac{-M_{i,N-r}}{M_{j,N-r}} \end{array} \right\} r = 1, 2, 3, \ldots, R \qquad (9.6)$$

and N is the degree of the filter ($N =$ even integer ≥ 6).

The process is also demonstrated for an eighth-degree filter with six transmission zeros, where the total number of required rotations is $R = (N - 4)/2 = 2$, tabulated below.

The transformation of the coupling matrix from folded to Pfitzenmaier topology is illustrated in Figure 9.7.

					$\theta_r = \tan^{-1}(cM_{i,N-r}/M_{j,N-r})$			
Filter Degree N	Rotation Number r	Pivot $[i,j]$	Element to be Annihilated	i	$N - r$	j	$N - r$	c
8	1	[2,6]	M_{27}	2	7	6	7	-1
	2	[3,5]	M_{36}	3	6	5	6	-1

It is evident from Figure 9.8 that the Pfitzenmaier topology is easily accommodated by a dual-mode structure, for example, the cylindrical TE_{113}-mode waveguide or the dual-$TE_{01\delta}$-mode dielectric resonators. The M_{18} coupling is unaffected by this process and has the same (small) value as that in the original folded coupling matrix, benefiting the far-out-of-band rejection performance. Also, there are no restrictions on the pattern of the TZs, apart from those previously mentioned for the asymmetric inline configuration.

9.4 CASCADED QUARTETS (CQs)—TWO QUARTETS IN CASCADE FOR DEGREES 8 AND ABOVE

In the next chapter, a method is presented for creating a cascade of resonator node quartets by using trisections. However, a more direct method is available where the

TABLE 9.3 Rotation Sequence for CQ Configuration

Filter Degree N	Rotation Number r	Pivot $[i,j]$	Element to be Annihilated	$\theta_r = \tan^{-1}\left[c\, M_{l_1,l_2}/M_{m_1,m_2}\right]$				
				l_1	l_2	m_1	m_2	c
8	1	[3,5]	—					
	2	[4,6]	M_{36}	3	6	3	4	−1
	3	[5,7]	M_{27}, M_{47}	4	7	4	5	−1
	4	[2,4]	M_{25}	2	5	4	5	+1

degree of the filter is 8 or higher, and there are two pairs of transmission zeros (each pair realized by one of the two quartets in the cascade [6]). Because the angle of the first rotation is the solution of a quadratic equation, there are restrictions on the pattern of transmission zeros that can be realized; however, these restrictions are different from those for the symmetric inline configuration. For the CQ case, the two zero pairs can lie on the real axis or on the imaginary axis or one pair on each axis. Each pair must be symmetrically positioned with respect to both axes, which means that no off-axis pairs are allowed. The CQ configuration is able to realize the useful (8−2−2) case where there are two TZ pairs: one on the real axis and one on the imaginary axis. This pattern, though, contravenes the symmetric inline realizability conditions.

The procedure for synthesizing the pair of quartets necessitates a series of four rotations to an eighth-degree coupling matrix, starting with the folded coupling matrix. The first rotation angle is obtained by solving a quadratic equation (two solutions), expressed in terms of the coupling elements taken from the original folded coupling matrix

$$t_1^2(M_{27}M_{34}M_{45} - M_{23}M_{56}M_{67} + M_{27}M_{36}M_{56})$$
$$+ t_1(M_{23}M_{36}M_{67} - M_{27}(M_{34}^2 - M_{45}^2 - M_{56}^2 + M_{36}^2)$$
$$- M_{27}(M_{36}M_{56} + M_{34}M_{45}) = 0, \tag{9.7}$$

where $t_1 \equiv \tan\theta_1$. Then, three further rotations are applied according to Table 9.3, after which the CQ topology for the 8th-degree filter is obtained as seen in Figure 9.9.

If the minimum path rule is applied to the topology in Figure 9.9b, it is obvious that the CQ solution realizes only four TZs in total. Accordingly, the original

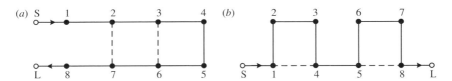

Figure 9.9 Cascade quartet (CQ) configuration with the eighth-order symmetric filtering characteristic: (a) original folded configuration; (b) after the transformation to form two CQs.

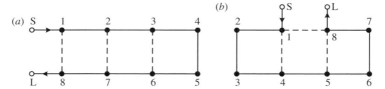

Figure 9.10 Quasi-Pfitzenmaier topology for canonical 8th-degree symmetric filtering characteristic: (*a*) original folded configuration; (*b*) after transformation to form two CQs and retaining the M_{18} coupling.

prototype characteristic should also incorporate only four TZs. Consequently, the M_{18} coupling in the folded coupling matrix, synthesized from the prototype polynomials, is zero. However, if there are six TZs in the prototype and the M_{18} coupling is present, it is not affected by the CQ rotations and needs to be realized within the CQ topology. This leads to the Pfitzenmaier-like topology, as depicted in Figure 9.10, which may have advantages, if unusual layout restrictions apply.

For higher-degree characteristics, the folded matrix is synthesized as normal; then the angle formulas and rotation pivots are applied to the central 8×8 submatrix with all the coupling value subscripts and pivot numbers increased by $(N - 8)/2$. For example, for a 10–2–2 prototype, $(N - 8)/2 = 1$. Coupling M_{67} in equation (9.7) becomes M_{78} and so on, and pivot [3,5] becomes pivot [4,6] and so forth. The quartets are shifted to different positions on the diagonal of the coupling matrix by the application of a single rotation according to Table 9.4 (where $i =$ the number of the first resonance node of the quartet). For example, to shift the first quartet ($i = 1$) in Figure 9.9*b* down the diagonal of the coupling matrix by one position (to become conjoined with the second quartet), we apply a rotation at pivot [2,4] and angle $\theta = \tan^{-1}[-M_{14}/M_{12}]$ to annihilate coupling M_{14} and to create coupling M_{25}. Shifting the quartet up the diagonal creates a dual-input filter, with both couplings M_{S1} and M_{S3} present.

If the original prototype is asymmetric, each CQ section includes a diagonal cross-coupling, and this synthesis method is not applicable. For the asymmetric cases, the circuit must first be synthesized by using trisections, followed by rotations (hybrid synthesis), detailed in Chapter 10.

The CQ structure has some practical advantages. One is that the topology can be easily implemented in a dual-mode structure of some kind, and another is that each CQ is responsible for the production of a specific TZ pair, facilitating the

TABLE 9.4 **Rotations to Shift a Quartet Up or Down the Diagonal of the Coupling Matrix by One Position ($i =$ Number of First Resonance Node of Quartet)**

		Element to be	$\theta = \tan^{-1}\left[c\, M_{l_1,l_2}/M_{m_1,m_2}\right]$				
Quartet Shift	Pivot	Annihilated	l_1	l_2	m_1	m_2	c
Up diagonal	$[i, i+2]$	$M_{i,i+3}$	i	$i+3$	$i+2$	$i+3$	$+1$
Down diagonal	$[i+1, i+3]$	$M_{i,i+3}$	i	$i+3$	i	$i+1$	-1

development and tuning processes. The cross coupling in the CQ is positive if the TZ pair that it is realizing is on the real axis, and negative if the pair is on the imaginary axis.

In the next two sections, two novel realizations are considered: the parallel-connected two-port networks and the cul-de-sac configuration [9]. The first is based on the $N + 2$ transverse matrix, and is derived by grouping residues and forming separate two-port subnetworks, which are then connected in parallel between the source and load terminations. The second is formed by a series of similarity transforms, operating on the folded coupling matrix.

9.5 PARALLEL-CONNECTED TWO-PORT NETWORKS

Closely related to the short-circuit admittance parameters, the eigenvalues and corresponding residues (see Chapter 8) of the filter function are separated into groups and subnetworks by the same matrix rotation procedures as those for the full matrix. Then, the subnetworks are connected in parallel between the source and load terminations to recover the original filtering characteristics. The transverse array itself is regarded as a parallel connection of N single-resonator groups.

Although the choice of residue groupings is arbitrary, difficult-to-realize couplings are created within the subnetworks, and between the internal nodes of the subnetworks and the source/load terminations. The choice of the filter function and of the residue groupings must not be restricted. The restrictions are summarized next:

- Filter functions may be fully canonical but must be symmetric and of even degree.
- There must be complementary pairs of eigenvalues and their associated residues within each group; that is, if an eigenvalue λ_i and its associated residues r_{21i} and r_{22i} are selected to be part of one group, there must also be the eigenvalue $\lambda_j = -\lambda_i$ and its associated residues $r_{21j}(= -r_{21i})$ and $r_{22j}(= r_{22i})$ in the same group. This implies that only doubly terminated between equal value source and load terminations can be synthesized with this simplified configuration.

If these restrictions are observed, the overall network consists of a number of two-port networks. The number corresponds to the number of groups that the residues are divided into, each connected in parallel between the source and the load terminals. If the filter function is fully canonical, the direct source–load coupling M_{SL} is also present.

After the residues are divided into groups, the synthesis of the submatrices and their reduction to the folded form follows the same process as that for a single network, described for the transversal array synthesis in Chapter 8. To illustrate the process, we examine a 23-dB return loss sixth-degree filter characteristic, with two symmetrically placed transmission zeros at $\pm j1.3958$ and a pair of real-axis

TABLE 9.5 (6–2–2) Symmetric Filter Function with Residues, Eigenvalues and Eigenvectors

k	Eigenvalues λ_k	Residues		Eigenvectors	
		r_{22k}	r_{21k}	$T_{Nk} = \sqrt{r_{22k}}$	$T_{1k} = r_{21k}/\sqrt{r_{22k}}$
1	−1.2225	0.0975	−0.0975	0.3122	−0.3122
2	−1.0648	0.2365	0.2365	0.4863	0.4863
3	−0.3719	0.2262	−0.2262	0.4756	−0.4756
4	0.3719	0.2262	0.2262	0.4756	0.4756
5	1.0648	0.2365	−0.2365	0.4863	−0.4863
6	1.2225	0.0975	0.0975	0.3122	0.3122

zeros at ±1.0749. This filter is synthesized as two subnetworks, one of degree 2 and one of degree 4. The residues and eigenvalues are calculated based on the procedure outlined in Chapter 8, and the values are summarized in Table 9.5.

By grouping the eigenvalues and associated residues $k = 1$ and 6, we find the following folded matrix for the second-degree subnetwork, depicted in Figure 9.11.

Now, the grouping of eigenvalues/residues $k = 2,3,4,5$ yields the folded coupling matrix for the fourth-degree subnetwork of Figure 9.12.

Superimposing the two matrices yields the overall matrix, as shown in Figure 9.13.

The results of analyzing the overall coupling matrix are revealed in Figure 9.14a (rejection/return loss) and Figure 9.14b (group delay), indicating that the 25 dB lobe level and equalized inband group delay are preserved.

Other solutions for this topology are available, depending on the combinations of the residues that are chosen for the subnetworks. However, whatever combination is chosen, at least one of the input/output couplings is negative. Of course, the number of topology options increases as the order of the filter function increases; for example, a 10th-degree filter can be realized as two parallel-connected two-port networks, one 4th-degree and one 6th-degree, or as three networks, one 2nd-degree

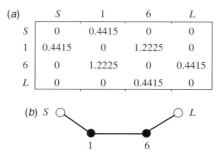

Figure 9.11 Coupling submatrix (a) and coupling–routing diagram (b) for residues $k = 1$ and 6.

(a)

	S	2	3	4	5	L
S	0	0.9619	0	0	0	0
2	0.9619	0	0.7182	0	0.3624	0
3	0	0.7182	0	0.3305	0	0
4	0	0	0.3305	0	0.7182	0
5	0	0.3624	0	0.7182	0	-0.9619
L	0	0	0	0	-0.9619	0

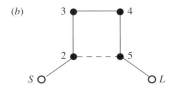

Figure 9.12 Coupling submatrix (*a*) and coupling–routing diagram (*b*) for residue group $k = 2,3,4,5$.

and two 4th-degree, all connected in parallel between the source and load terminations. Also, each subnetwork itself can be reconfigured to other two-port topologies with further transformations.

If the network is synthesized as $N/2$ parallel-coupled pairs, as signified in Figure 9.15*b*, a more direct synthesis route exists. Beginning with the transversal matrix, it is necessary to apply only a series of rotations to annihilate half the couplings in the top row from positions $M_{S,N}$ back to the midpoint $M_{S,N/2+1}$, implying $N/2$ rotations (see Fig. 9.15*a*). Because of the symmetry of the values in the outer

(a)

	S	1	2	3	4	5	6	L
S	0	0.4415	0.9619	0	0	0	0	0
1	0.4415	0	0	0	0	0	1.2225	0
2	0.9619	0	0	0.7182	0	0.3624	0	0
3	0	0	0.7182	0	0.3305	0	0	0
4	0	0	0	0.3305	0	0.7182	0	0
5	0	0	0.3624	0	0.7182	0	0	-0.9619
6	0	1.2225	0	0	0	0	0	0.4415
L	0	0	0	0	0	-0.9619	0.4415	0

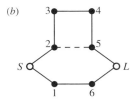

Figure 9.13 Superimposed 2nd and 4th-degree submatrices: (*a*) coupling matrix; (*b*) coupling–routing diagram.

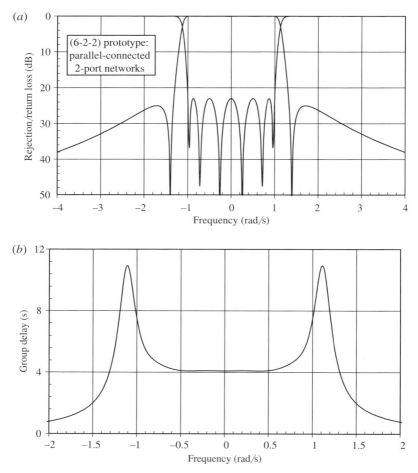

Figure 9.14 Analysis of parallel-connected two-port coupling matrix: (a) rejection and return loss; (b) group delay.

rows and columns of the transversal matrix, the corresponding entries M_{1L} to $M_{N/2,L}$ in the last column are annihilated simultaneously.

The pivots of the rotations to annihilate these couplings start at position $[1,N]$ and progress toward the center of the matrix to position $[N/2, N/2 + 1]$. For the sixth degree example, this is a sequence of $N/2 = 3$ rotations, according to Table 9.6, and is applied to the transversal matrix.

After the series of rotations, the matrix as shown in Figure 9.15a is obtained, which corresponds to the coupling and routing diagram in Figure 9.15b. In each case, at least one of the input/output couplings is negative. An interesting example of a fourth-degree implementation of this topology in dielectric resonator technology is provided in Ref. 10.

(a)	S	1	2	3	4	5	6	L
S	0	0.4415	0.6877	0.6726	0	0	0	0
1	0.4415	0	0	0	0	0	1.2225	0
2	0.6877	0	0	0	0	1.0648	0	0
3	0.6726	0	0	0	0.3720	0	0	0
4	0	0	0	0.3720	0	0	0	0.6726
5	0	0	1.0648	0	0	0	0	−0.6877
6	0	1.2225	0	0	0	0	0	0.4415
L	0	0	0	0	0.6726	−0.6877	0.4415	0

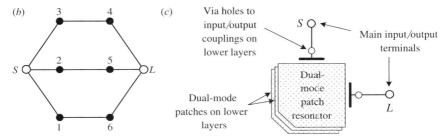

Figure 9.15 Symmetric (6−2−2) filter configured as parallel-coupled pairs: (a) coupling matrix; (b) coupling−routing diagram; (c) multilayer dual-mode patch resonator realization.

TABLE 9.6 Symmetric (6−2−2) Filter Example: Rotation Sequence for Reduction of the Transversal Matrix to the Parallel-Coupled Pair Form

Transform Number	Pivot $[i,j]$	Elements to be Annihilated	$\theta_r = \tan^{-1}(cM_{kl}/M_{mn})$				
			k	l	m	n	c
1	[1,6]	M_{S6} (and M_{1L})	S	6	S	1	−1
2	[2,5]	M_{S5} (and M_{2L})	S	5	S	2	−1
3	[3,4]	M_{S4} (and M_{3L})	S	4	S	3	−1

A possible realization is with three planar dual-mode resonant patches, arranged in a triple-substrate stack, as shown in Figure 9.15c. Low-temperature cofired ceramic (LTCC) multilayer substrates technology is well suited for building such filters.

9.5.1 Even-Mode and Odd-Mode Coupling Submatrices

The even-mode and odd-mode submatrices for filter functions can be synthesized by the eigenvalue grouping process. If the eigenvalues of the function are arranged in the order of descending value, then the odd-numbered group represents the odd-mode coupling matrix, and the even-numbered group, the even-mode matrix. The two coupling matrices are synthesized by the same methods as those for the full $N + 2$ matrix, by working with the two subgroups of eigenvalues and their associated residues. Once the two transversal submatrices are obtained, the rotation operations are carried out on the submatrices to annihilate the unwanted couplings.

The even-mode coupling matrices (Section 9.1), representing symmetric inline filter topology, can be generated by this method of residue grouping. For symmetric sixth degree filter that was used to demonstrate the reconfiguration of the folded coupling matrix to the inline form, the eigenvalues are computed as follows: $\lambda_1 = -\lambda_6 = 1.2185$, $\lambda_2 = -\lambda_5 = 1.0729$, and $\lambda_3 = -\lambda_4 = 0.4214$. Using the even-numbered group of eigenvalues, and their associated residues and eigenvectors, we can synthesize the corresponding $N + 2$ even-mode matrix $\mathbf{M_e}$, as denoted in matrix (9.8a). If the odd-numbered eigenvalue group is used instead, the odd-mode matrix $\mathbf{M_o}$ is produced [also shown in (9.8a)]:

$$
\mathbf{M_e} =
\begin{bmatrix}
0 & 0.7472 & 0 & 0 & 0 \\
0.7472 & 0 & 0.7586 & -0.4592 & 0.7472 \\
0 & 0.7586 & -0.4459 & 0.0891 & 0 \\
0 & -0.4592 & 0.0891 & 1.0129 & 0 \\
0 & 0.7472 & 0 & 0 & 0
\end{bmatrix},
$$

$$
\mathbf{M_o} =
\begin{bmatrix}
0 & 0.7472 & 0 & 0 & 0 \\
0.7472 & 0 & 0.6151 & 0.6387 & -0.7472 \\
0 & 0.6151 & 0.3053 & 0.4397 & 0 \\
0 & 0.6387 & 0.4397 & -0.8723 & 0 \\
0 & -0.7472 & 0 & 0 & 0
\end{bmatrix}
\tag{9.8a}
$$

In this case, one rotation is necessary to transform the $\mathbf{M_e}$ and $\mathbf{M_o}$ matrices to the folded form that is used in the reconfiguration example in Section 9.1. Rotation is carried out at pivot [2,3] to annihilate M_{13} as portrayed in (9.8b). Now it can be seen that the core $N \times N$ matrix of $\mathbf{M_e}$ is the same as the even-mode matrix in Figure 9.2b, and as expected, $\mathbf{M_o}$ is its conjugate. These matrices can also be produced by applying the cul-de-sac synthesis process as outlined in Section 9.6.1:

$$
\mathbf{M_e} =
\begin{bmatrix}
0 & 0.7472 & 0 & 0 & 0 \\
0.7472 & 0 & 0.8867 & 0 & 0.7472 \\
0 & 0.8867 & -0.1337 & 0.6050 & 0 \\
0 & 0 & 0.6050 & 0.7007 & 0 \\
0 & 0.7472 & 0 & 0 & 0
\end{bmatrix},
$$

$$\mathbf{M_o} = \begin{bmatrix} 0 & 0.7472 & 0 & 0 & 0 \\ 0.7472 & 0 & 0.8867 & 0 & -0.7472 \\ 0 & 0.8867 & 0.1337 & 0.6050 & 0 \\ 0 & 0 & 0.6050 & -0.7007 & 0 \\ 0 & -0.7472 & 0 & 0 & 0 \end{bmatrix} \qquad (9.8b)$$

9.6 CUL-DE-SAC CONFIGURATION

The cul-de-sac configuration [9] in its basic form is restricted to doubly terminated networks and realizes a maximum of $N - 3$ transmission zeros. Otherwise, the configuration can accommodate any even- or odd-degree symmetric or asymmetric prototype. Moreover, such a form lends itself to a certain amount of flexibility in the physical layout of its resonators.

A typical cul-de-sac configuration is illustrated in Figure 9.16a for a 10th-degree prototype with the maximum allowable seven TZs (in this case three imaginary axes and two complex pairs). There is a central core of a quartet of resonators in a square formation (1, 2, 9, and 10 in Fig. 9.16a), straight-coupled to each other (no diagonal cross-couplings). One of these couplings is consistently negative; the choice of which one is arbitrary. The entry to and exit from the core quartet are from the opposite corners of square 1 and 10, respectively, in Figure 9.16a.

Some or all of the other resonators are strung out in cascade from the other two corners of the core quartet in equal numbers (even-degree prototypes) or one more than the other (odd degree prototypes). The last resonator in each of the two chains has no output coupling; hence the nomenclature "cul-de-sac" for this configuration. Another possible configuration for an odd-degree (seventh-degree) characteristic is shown in Figure 9.16b. If it is practical to include a diagonal cross-coupling between the input and output of the core quartet, an extra transmission zero can be realized. This coupling will have the same value as in the original folded coupling matrix.

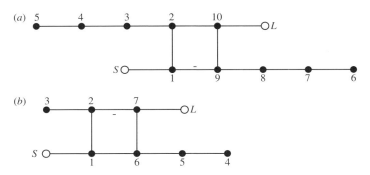

Figure 9.16 Cul-de-sac network configurations: (a) (10–3–4) filter network; (b) (7–1–2) filter network.

The procedure for the synthesis of the cul-de-sac network takes advantage from the symmetry of the ladder lattice network for both the symmetric or the asymmetric responses of a doubly terminated filter network [11]. This leads to a particularly simple and regular series of similarity rotations. Starting with the folded coupling matrix, elements are annihilated by a series of regular similarity transforms (for odd-degree filters), and cross-pivot transforms (for even-degree filters), beginning with a mainline coupling near the center of the matrix, and working outwards along or parallel to the cross-diagonal.

For odd-degree filters, the angle formula takes the conventional form [see equation (8.37a)]:

$$\theta_r = \tan^{-1} \frac{M_{ik}}{M_{jk}} \tag{9.9}$$

where $[i,j]$ is the pivot coordinate and $k = j - 1$. For the even-degree characteristics, the cross-pivot transform is adopted [see equation (8.37g)]:

$$\theta_r = \frac{1}{2}\tan^{-1}\left(\frac{2M_{ij}}{(M_{jj} - M_{ii})}\right) \tag{9.10}$$

Table 9.7 lists the pivot coordinates and angle formula for the sequence of similarity transforms to be applied to the folded coupling matrix for degrees 4–9, and a general formula for the pivot coordinates for any degree ≥ 4.

TABLE 9.7   Pivot Coordinates for Reduction of the Folded Matrix to the Cul-de-Sac Configuration

| | Pivot Position $[i,j]$ and Element to be Annihilated | | | | |
					Transform Angle Formula θ_r
		Similarity Transform Number r			
	$r = 1, 2, 3, \ldots R$		$R = (N-2)/2$ (N even) $= (N-3)/2$ (N odd)		
Degree N	$r = 1$	2	3	r	
4	$[2,3]\ M_{23}$				(9.10)
5	$[2,4]\ M_{23}$				(9.9)
6	$[3,4]\ M_{34}$	$[2,5]\ M_{25}$			(9.10)
7	$[3,5]\ M_{34}$	$[2,6]\ M_{25}$			(9.9)
8	$[4,5]\ M_{45}$	$[3,6]\ M_{36}$	$[2,7]\ M_{27}$		(9.10)
9	$[4,6]\ M_{45}$	$[3,7]\ M_{36}$	$[2,8]\ M_{27}$		(9.9)
—	—	—	—	—	—
N (even)	$[i,j]\ M_{i,j}$ $i = (N+2)/2 - 1$ $j = N/2 + 1$	—	—	$[i,j]\ M_{i,j}$ $i = (N+2)/2 - r$ $j = N/2 + r$	(9.10)
N (odd)	$[i,j]\ M_{i,j-1}$ $i = (N+1)/2 - 1$ $j = (N+1)/2 + 1$	—	—	$[i,j]\ M_{i,j-1}$ $i = (N+1)/2 - r$ $j = (N+1)/2 + r$	(9.9)

The folded coupling matrix for the doubly-terminated version of the example, which is employed to demonstrate the coupling matrix reduction process in Section 8.3, is shown in Figure 9.17*a*. This is a seventh degree, 23-dB return loss, asymmetric filter, with a complex pair of transmission zeros to give group delay equalization over approximately 60% of the bandwidth, and a single zero on the imaginary axis to give a rejection lobe level of 30dB on the upper side of the passband.

To transform the folded network coupling matrix to the cul-de-sac configuration, two transformations, according to Table 9.7, are applied (pivots [3,5] and [2,6]) with the angles calculated from equation (9.9). Then the coupling matrix in Figure 9.17*b* is obtained, corresponding to the coupling–routing diagram of Figure 9.16*b*. The results of analyzing this coupling matrix are plotted in Figure 9.18*a* (rejection/return loss) and Figure 9.18*b* (group delay). It is evident that the 30 dB lobe level and equalized in-band group delay are not affected by the transformation process. Note that when the source and load are directly connected to the corners of the core quartet and the number of finite-position zeros of the prototype is less than $N-3$, the values of the four couplings in the core quartet have the same absolute value.

The topologies of the filters in coaxial resonator technology, corresponding to the two coupling matrices of Figure 9.17, are displayed in Figure 9.19*a* (folded) and Figure 9.19*b* (cul-de-sac). This demonstrates the rather simple construction of the cul-de-sac form, compared with that of the folded form. There are no diagonal cross-couplings and only one negative coupling in the cul-de-sac network.

(a)

	S	1	2	3	4	5	6	7	L
S	0	1.0572	0	0	0	0	0	0	0
1	1.0572	0.0211	0.8884	0	0	0	0	0	0
2	0	0.8884	0.0258	0.6159	0	0	0.0941	0	0
3	0	0	0.6159	0.0193	0.5101	0.1878	0.0700	0	0
4	0	0	0	0.5101	−0.4856	0.4551	0	0	0
5	0	0	0	0.1878	0.4551	−0.0237	0.6119	0	0
6	0	0	0.0941	0.0700	0	0.6119	0.0258	0.8884	0
7	0	0	0	0	0	0	0.8884	0.0211	1.0572
L	0	0	0	0	0	0	0	1.0572	0

(b)

	S	1	2	3	4	5	6	7	L
S	0	1.0572	0	0	0	0	0	0	0
1	1.0572	0.0211	0.6282	0	0	0	0.6282	0	0
2	0	0.6282	−0.0683	0.5798	0	0	0	−0.6282	0
3	0	0	0.5798	−0.1912	0	0	0	0	0
4	0	0	0	0	−0.4856	0.6836	0	0	0
5	0	0	0	0	0.6836	0.1869	0.6499	0	0
6	0	0.6282	0	0	0	0.6499	0.1199	0.6282	0
7	0	0	−0.6282	0	0	0	0.6282	0.0211	1.0572
L	0	0	0	0	0	0	0	1.0572	0

Figure 9.17 Example of the seventh-degree filter in cul-de-sac configuration: (*a*) original folded coupling matrix; (*b*) after transformation to cul-de-sac configuration.

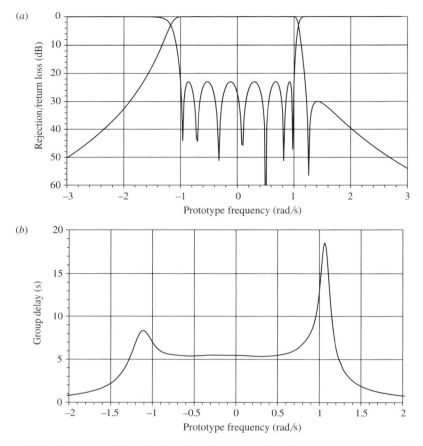

Figure 9.18 Example of the simulated performance of a $(7-1-2)$ asymmetric filter: (a) rejection and return loss; (b) group delay.

9.6.1 Further Cul-de-Sac Forms

The cul-de-sac filter is defined in Figure 9.16 for an even-degree example and an odd-degree example. This basic form consists of a core quartet of resonator nodes with two opposite corners directly connected to the source and load terminations; the rest of the resonator nodes form two chains connected to the other two corners. The form progressively evolves as a series of R rotations is applied to an originating folded coupling matrix, with pivots and angles as listed in Table 9.7. The number of rotations to be applied is given by $R = (N - 2)/2$ (N even) or $R = (N - 3)/2$ (N odd).

Two further forms can be obtained by stopping the series of rotations at the penultimate (i.e., $R - 1$ rotations), or continuing on with one extra rotation

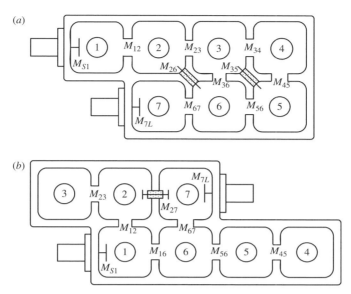

Figure 9.19 Example of a $(7-1-2)$ asymmetric filter in coaxial cavity configurations: (*a*) folded network configuration; (*b*) cul-de-sac configuration.

$(R + 1$ rotations). The first form leaves the core quartet not directly connected to the source and load terminations, as shown in Figure 9.20*a* for an eighth-degree example. The minimum path rule dictates that this form realizes fewer finite-position transmission zeros: $n_{fz} = N-5$ for these cases, compared with $N-3$ for the basic form. However, the advantage is that there is a greater isolation between input and output, and the fact that the topology is suitable for realization in dual-mode cavities.

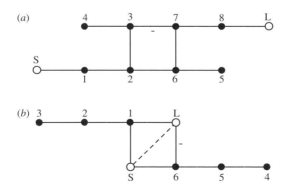

Figure 9.20 Two alternative forms for the cul-de-sac configuration: (*a*) the indirect-coupled; (*b*) fully canonical form.

	S	1	2	3	4	5	6	L
S	0	0.7472	0	0	0	0	0.7472	0
1	0.7472	0	0.8867	0	0	0	0	0.7472
2	0	0.8867	−0.1337	0.6050	0	0	0	0
3	0	0	0.6050	0.7007	0	0	0	0
4	0	0	0	0	−0.7007	0.6050	0	0
5	0	0	0	0	0.6050	0.1337	0.8867	0
6	0.7472	0	0	0	0	0.8867	0	−0.7472
L	0	0.7472	0	0	0	0	−0.7472	0

Figure 9.21   Canonical cul-de-sac coupling matrix for symmetric (6−2) characteristic.

By continuing on with one extra rotation, the canonical cul-de-sac form is obtained, where the source and load terminations form part of the core quartet as in Figure 9.20b. This form realizes $N-1$ transmission zeros, and, by including the direct source–load coupling M_{SL} (shown by the dashed line in Fig. 9.2b), the form becomes fully canonical ($n_{fz} = N$). It can be demonstrated that the two branches form the even-mode and odd-mode subnetworks of the filtering function, and so, become very similar in principle of operation to the hybrid reflection mode bandpass and bandstop filters, described by Hunter and others [12,13].

To demonstrate the synthesis of the canonical form of the cul-de-sac network, (Fig. 9.20b), the sixth-degree characteristic that is an example for the symmetric inline synthesis (Section 9.1.1) is used again. We take the $N+2$ folded coupling matrix for this filter, and apply three cross-pivot rotations at [3,4], [2,5], and [1,6] to annihilate M_{34}, M_{25}, and M_{16}, respectively. Figure 9.21 shows the coupling matrix. It can be seen that the two branches of the canonical cul-de-sac network form the even-mode and odd-mode networks, as derived in Sections 9.1.1 and 9.5.1.

Proof-of-Concept Model   To illustrate the previous discussion, a proof-of-concept S-band model, realizing an asymmetric eighth-degree filter function, is synthesised in the indirect-coupled cul-de-sac configuration in Figure 9.20a, realized as a coaxial cavity filter, and measured. The prototype is an eighth-degree Chebyshev filter, with a 23 dB return loss and three prescribed transmission zeros at $s = -j1.3553$, $+j1.1093$, and $+j1.2180$, producing one rejection lobe level of 40 dB on the lower side of the passband and two of 40 dB on the upper side. Starting with the folded coupling matrix for this prototype (Fig. 9.22a) and applying a series of $R-1 = 2$ similarity transforms at pivots [4,5] and [3,6] of Table 9.7, we achieve the coupling matrix in Figure 9.22b, which corresponds to the configuration in Figure 9.20a.

The simulated and measured rejection and return loss performances of this filter are shown in Figure 9.23, which clearly shows the three transmission zeros. The somewhat large bandwidth of this filter (110 MHz) means that the return loss performance has degraded slightly because of the dispersion effects (the coupling values varying with frequency); however, these effects seem to be well predicted by the simulation model.

(a)

	S	1	2	3	4	5	6	7	8	L
S	0	1.0428	0	0	0	0	0	0	0	0
1	1.0428	0.0107	0.8623	0	0	0	0	0	0	0
2	0	0.8623	0.0115	0.5994	0	0	0	0	0	0
3	0	0	0.5994	0.0133	0.5356	0	-0.0457	-0.1316	0	0
4	0	0	0	0.5356	0.0898	0.3361	0.5673	0	0	0
5	0	0	0	0	0.3361	-0.8513	0.3191	0	0	0
6	0	0	0	-0.0457	0.5673	0.3191	-0.0073	0.5848	0	0
7	0	0	0	-0.1316	0	0	0.5848	0.0115	0.8623	0
8	0	0	0	0	0	0	0	0.8623	0.0107	1.0428
L	0	0	0	0	0	0	0	0	1.0428	0

(b)

	S	1	2	3	4	5	6	7	8	L
S	0	1.0428	0	0	0	0	0	0	0	0
1	1.0428	0.0107	0.8623	0	0	0	0	0	0	0
2	0	0.8623	0.0115	0.3744	0	0	-0.4681	0	0	0
3	0	0	0.3744	-0.0439	0.8166	0	0	0.3744	0	0
4	0	0	0	0.8166	0.1976	0	0	0	0	0
5	0	0	0	0	0	-0.9590	0.2093	0	0	0
6	0	0	-0.4681	0	0	0.2093	0.0499	0.4681	0	0
7	0	0	0	0.3744	0	0	0.4681	0.0115	0.8623	0
8	0	0	0	0	0	0	0	0.8623	0.0107	1.0428
L	0	0	0	0	0	0	0	0	1.0428	0

Figure 9.22 $(8-3)$ asymmetric filter example: (*a*) folded coupling matrix; (*b*) same matrix after transformation to the indirect cul-de-sac configuration.

A sketch of the filter resonator layout is provided in Figure 9.24, showing the very uncomplicated coupling arrangement. The omission of the third rotation has left the core quartet (2, 3, 6, and 7) not directly connected to the input/output terminations, which should benefit far out-of-band rejection. This same construction can support 8th-degree prototypes with the number of transmission zeros ranging from 0 to 3, and with symmetric or asymmetric characteristics.

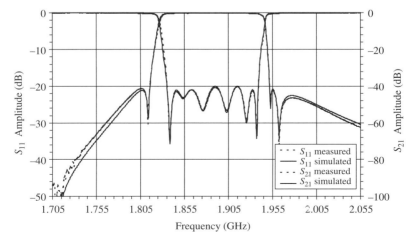

Figure 9.23 Simulated and measured performance of the $(8-3)$ cul-de-sac filter.

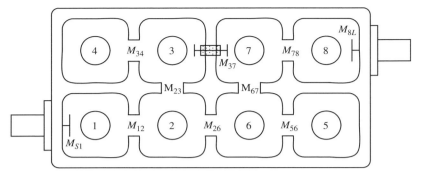

Figure 9.24 Indirect-coupled cul-de-sac configuration for the (8–3) asymmetric filter.

As noted, all the couplings are positive except for one in the core quartet. It can be moved to any one of the four couplings for the greatest convenience, and implemented as a capacitive probe. For example, the filter can be realized in coaxial resonator technology and the other couplings as inductive apertures or inductive loops. Moreover, there are no diagonal couplings, even though the original prototype is asymmetric.

Some of the more important features of the cul-de-sac configuration are now summarized:

1. The cul-de-sac topology may be used to realize a completely general class of electrical filter functions. It can be symmetric or asymmetric, even or odd degree, provided that the filter function is designed to operate between the equivalue source and load terminations (double-terminated), and that the number of transmission zeros embodied by the filtering function is less than or equal to $N - 3$, where N is the degree of the function.

2. A filter function, incorporating transmission zeros, is realized with an absolute minimum number of coupling elements and their associated adjustment mechanisms. For a 10th-degree function with seven transmission zeros, 10 internal couplings are required. The equivalent folded structure requires 16 internal couplings, some of which are diagonal couplings.

3. All the couplings are straight, that is, they are either horizontal or vertical between the resonators, if these are arranged in a regular grid pattern. This significantly simplifies the design and eliminates the need for diagonal couplings, which are typically difficult to manufacture, assemble, and adjust and are quite sensitive to vibration and temperature variations. The absence of diagonal couplings enables the realization of asymmetric filter characteristics with dual-mode resonators, such as in dielectric or waveguide, and also usage of the high-Q_u TE_{011} cylindrical waveguide resonant mode.

4. The same structure can be used to embody any doubly terminated symmetric or asymmetric filter function of the same degree (provided that $n_{fz} \leq N - 3$). Therefore, the same basic housing, incorporating the coupling elements can be

used for a variety of filter characteristics. If the changes in the coupling values from the original design are small, for example, to correct for a dispersion distortion, the changes are achieved with tuning and iris adjustment screws without the need to alter the mechanical dimensions of the interresonator coupling apertures or to add further cross-couplings.

5. Regardless of the transfer characteristic, all the internal couplings have the same sign, with the exception of the one within the central quartet of resonators. The range over which the values of the internal interresonator couplings are spread is comparatively small, allowing the use of a common design for all the couplings with the same sign.

6. The values of the interresonator couplings are relatively large, compared with the levels of stray and spurious couplings typically found within complex filter structures. Not only is the design and tuning process eased; the values of all interresonator couplings are also externally adjustable with screws, reducing the need for tight tolerances during manufacture and enabling optimal performance to be obtained during production.

7. The Q_u factor of a microwave resonant cavity decreases as the number of coupling elements is augmented, increasing the insertion loss of the filter of which the cavity forms a part. Owing to the minimal number of interresonator couplings in the cul-de-sac filter, with only one of those being a negative coupling (usually realized as a probe in coaxial and dielectric filters and can be relatively lossy, particularly if it is a diagonal coupling), the insertion loss is less than that of the equivalent filter with more coupling elements.

8. By short-circuiting certain cavities, the filter is split into smaller sections, easing the development and production tuning processes.

9. The cul-de-sac filter can be manufactured as a simple two-dimensional two-part housing and lid assembly with all the tuning screws on top. The result is a filter that is appropriate for mass production and tuning methods. The form is amenable to volume manufacturing processes such as die casting. Moreover, a certain amount of flexibility is available in the layout of the filter cavities, enabling compact integration with other subsystems and components in the same system.

9.6.2 Sensitivity Considerations

Having the minimum number of internal couplings, it is not surprising that the cul-de-sac filter is more sensitive to variations in its coupling element values and tuning state than is the equivalent filter realized in the folded configuration, or with trisections and so on. To assess the nature of this sensitivity and compare the alternative configurations for the same filter function, a brief qualitative simulation study was made.

There are two principal types of variations in the coupling and resonator tuning states: *random*, due to component and manufacturing/assembly tolerances, and *uni-* or *bidirectional*, caused by temperature fluctuations during operation. The first

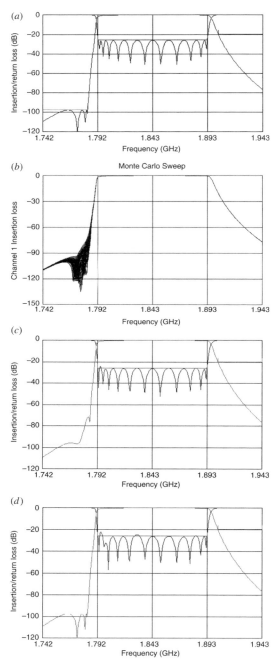

Figure 9.25   Sensitivity to variations in the coupling value: (*a*) nominal 11−3 characteristic; (*b*) cul-de-sac—random variations to all the couplings; (*c*) cul-de-sac—0.2% increase to the self-couplings, 0.2% decrease to all others; (*d*) box filter—0.2% increase to self-couplings, 0.2% decrease to all others.

category is assessed by a Monte Carlo sensitivity analysis package on an 11th-degree characteristic. There are three three transmission zeros on the lower side, providing 97 dB rejection lobe levels, realized in folded, box section, trisection (see Chapter 10) and cul-de-sac configurations. Here, the box and trisection were relatively little affected by 1% standard deviation variations in coupling values. However, the rejection performances of the folded and cul-de-sac configurations are substantially affected, as plotted in Figure 9.25b. Although the variations in the values in the coupling elements are adjustable with tuning screws during production, counteracting the effect of the manufacturing and assembly tolerances and restoring the rejection performance, there is an impact on volume production where more accurate tuning is required.

The second category of variation is investigated by applying a 0.2% increase to the self-couplings and a 0.2% decrease to the other coupling values. Here, the cul-de-sac fares the worst overall with the rejection performance suffering significantly as traced in Figure 9.25c. By comparison, the box filter is relatively unaffected (Fig. 9.25d). This indicates that the thermal stability of the coupling values and resonant frequencies needs to be improved, or that the mechanical design gives shifts in the same direction under the influence of temperature changes.

SUMMARY

In Chapters 7 and 8, two different techniques for synthesizing the canonical folded array coupling matrix from the transfer and reflection polynomials of the desired filter function are described. For certain applications, this canonical form is used directly, if it is convenient to realize the couplings. Alternatively, the form can be used as a starting point for the application of further transforms to create topologies better suited for the technology with which the filter is eventually realized. One such technology that has had a major impact on filter design is the dual-mode filter realization, which utilizes two orthogonally polarized degenerate modes, supported in a single physical resonator, either a cavity, a dielectric disk, or a planar structure. The technology allows a significant reduction in the size of filters with a minor penalty in the unloaded Q and spurious response. Owing to the large reduction in size and mass, dual-mode filters are widely used in satellite and wireless communication systems. This chapter is devoted to the methods of similarity transformations to realize a wide range of topologies, appropriate for dual-mode filter networks. Besides longitudinal and folded structures, the chapter also includes structures referred to as *cascade quartets* and *cul-de-sac filters*. The chapter concludes with examples and a discussion of the sensitivity of some dual-mode filter topologies.

REFERENCES

1. J. D. Rhodes, A lowpass prototype network for microwave linear phase filters, *IEEE Trans. Microwave Theory Tech.* **MTT-18**, 290–301 (June 1970).
2. J. D. Rhodes, The generalized interdigital linear phase filter, *IEEE Trans. Microwave Theory Tech.* **MTT-18**, 301–307 (June 1970).

3. J. D. Rhodes, The generalized direct-coupled cavity linear phase filter, *IEEE Trans. Microwave Theory Tech.* **MTT-18**, 308–313 (June 1970).

4. J. D. Rhodes and I. H. Zabalawi, Synthesis of symmetrical dual mode in-line prototype networks, *Circuit Theory and Application,* Vol. 8, Wiley, New York, 1980, pp. 145–160.

5. M. Abramowitz and I. A. Stegun, eds. *Handbook of Mathematical Functions*, Dover Publications, New York, Nov. 1970.

6. R. J. Cameron and J. D. Rhodes, Asymmetric realizations for dual-mode bandpass filters, *IEEE Trans. Microwave Theory Tech.* **MTT-29**, 51–58 (Jan. 1981).

7. G. Pfitzenmaier, An exact solution for a six-cavity dual-mode elliptic bandpass filter, *IEEE MTT-S Int. Microwave Symp. Digest*, San Diego, 1977, pp. 400–403.

8. R. J. Cameron, Novel realization of microwave bandpass filters, *ESA J.* **3**, 281–287 (1979).

9. R. J. Cameron, A. R. Harish, and C. J. Radcliffe, Synthesis of advanced microwave filters without diagonal cross-couplings, *IEEE Trans. Microwave Theory Tech.* **MTT-50**, 2862–2872 (Dec. 2002).

10. V. Pommier, D. Cros, P. Guillon, A. Carlier, and E. Rogeaux, Transversal filter using whispering gallery quarter-cut resonators, *IEEE MTT-S Int. Symp. Digest*, Boston, 2000, pp. 1779–1782.

11. H. C. Bell, Canonical asymmetric coupled-resonator filters, *IEEE Trans. Microwave Theory Tech.* **MTT-30**, 1335–1340 (Sept. 1982).

12. I. C. Hunter, J. D. Rhodes, and V. Dassonville, Triple mode dielectric resonator hybrid reflection filters, *IEE Proc. Microwaves Anten. Propag.* **145**, 337–343 (1998).

13. I. C. Hunter, *Theory and Design of Microwave Filters*, Electromagnetic Waves Series 48, IEE, London, 2001.

CHAPTER 10

SYNTHESIS AND APPLICATION OF EXTRACTED POLE AND TRISECTION ELEMENTS

Although the folded and transversal array coupling matrices form bases from which a wide variety of useful filter configurations may be derived, there are certain sections that are difficult or impossible to synthesize by coupling matrix operations alone. In this Chapter, we shall follow the circuit-based method for synthesizing the "extracted pole" section, and its direct application in bandpass and bandstop filter design. Then the rôle of the extracted pole section in the formation of another useful section able to realize a single transmission zero, the "trisection", will be examined. The trisection may be realized directly or used as part of a cascade with other sections. However it also finds application in the synthesis of more complex networks, where firstly a cascade of trisections is synthesized and then further reconfiguration is carried out using coupling matrix transforms. The techniques used to transform a cascade of trisections to form cascaded quartets, quintets and sextets are outlined. Finally the synthesis of the "box section" from a trisection is described, and its derivative, the "extended box" configuration.

10.1 EXTRACTED POLE FILTER SYNTHESIS

The low-loss requirement for high-power filters usually means that the designs are low-degree, with no in-band equalization provision (group delay equalization increases the insertion loss). Other requirements are a relatively steep out-of-band isolation to minimize the effect of spectrum spreading (due to the nonlinear nature of the HPAs), and to provide adequate channel-to-channel isolation.

Microwave Filters for Communication Systems: Fundamentals, Design, and Applications, by Richard J. Cameron, Chandra M. Kudsia, and Raafat R. Mansour
Copyright © 2007 John Wiley & Sons, Inc.

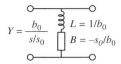

Figure 10.1   Shunt-connected series pair producing a short circuit at $s = s_0$.

Although it is simple to synthesise the prototype network for a low-degree filter with symmetric or asymmetric transmission zeros, its realization can present some difficulties if negative cross-couplings are required.[1] In a low-power environment, the negative coupling is often realized as a capacitive probe; however, in high-power conditions, this probe can overheat (it is usually thermally decoupled from an effective heat sink) and cause arcing problems and an overall reduction in the Q_u of the cavity.

The need for negative couplings can be avoided by using the extracted pole principle. In effect, the transmission zeros are removed from the network before the synthesis of the main body, and realized as bandstop cavities. After the transmission zeros are extracted, the remaining network does not require cross-couplings, either straight or diagonal, unless there are further transmission zeros that have not been extracted. If the remaining zeros are a real-axis pair for the group-delay equalization of a symmetric filter, the cross-couplings needed for their realization are the same sign as the mainline couplings.

10.1.1   Synthesis of the Extracted Pole Element

In the following section, the circuit-based procedure for extracting single (imaginary axis) transmission zeros is detailed. We follow the synthesis procedure based on the $(4-2)$ asymmetric prototype filter example in Section 7.3. The procedure outlined here gives more flexibility than the method given in Ref. 1, which is for conjugate pole pairs in symmetric networks.

The prototype circuit that produces the transmission zero (or pole) is shown in Figure 10.1. It consists of a FIR of value $B = -s_0/b_0$ in series with an inductive element $L = 1/b_0$, shunt connected to the transmission line; b_0 is known as the *residue* of the pole. It is evident that when $s = s_0$, the impedance of the series pair is zero, short-circuiting the transmission line.

The pole is extracted from the $[ABCD]$ matrix that represents the overall filter network in a three-stage operation:

1. A frequency-invariant phase shifter is first removed to prepare the circuit for extraction of the pole at $s = s_0$ (partial removal).
2. The shunt-connected resonant pair is removed at $s = s_0$.
3. A further phase length is extracted to prepare the remaining network for the extraction of another pole if required, or for the extraction of a capacitor/ FIR pair $(C + jB)$ at the input to the main body of the filter.

---

[1]"Negative" here means negative with respect to the other couplings in the network, and inductive if the other couplings are capacitive, and vice versa.

Removal of Phase Length The $[ABCD]$ matrix, representing the overall circuit, is considered as a cascade of a length of transmission line in cascade with a remainder matrix $[A'B'C'D']$ [the (s) has been omitted from the polynomials here for reasons of clarity, and so $A \equiv A(s)$, $B \equiv B(s)$, etc. and P ($\equiv P(s)/\varepsilon$) is assumed to have absorbed the constant ε]:

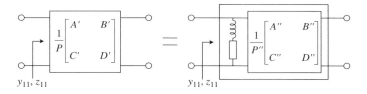

$$\frac{1}{P}\begin{bmatrix} A & B \\ C & D \end{bmatrix} = \begin{bmatrix} \cos\varphi & j\sin\varphi/J \\ jJ\sin\varphi & \cos\varphi \end{bmatrix} \cdot \frac{1}{P}\begin{bmatrix} A' & B' \\ C' & D' \end{bmatrix}$$

$$\therefore \frac{1}{P}\begin{bmatrix} A' & B' \\ C' & D' \end{bmatrix} = \frac{1}{P}\begin{bmatrix} A\cos\varphi - jC\sin\varphi/J & B\cos\varphi - jD\sin\varphi/J \\ C\cos\varphi - jAJ\sin\varphi & D\cos\varphi - jBJ\sin\varphi \end{bmatrix} \quad (10.1)$$

The remainder matrix contains the shunt-connected resonant pair as the leading element:

The open-circuit admittance z_{11} at the input to the remainder matrix approaches zero and the short-circuit admittance y_{11} approaches infinity as s approaches the pole frequency s_0, expressed as

$$z_{11} = \frac{A'}{C'} = 0 \quad \text{and} \quad y_{11} = \frac{D'}{B'} = j\infty \quad \text{at} \quad s = s_0 \quad (10.2)$$

Therefore, $A' = 0$ or $B' = 0$ at $s = s_0$. Taking the A' and B' elements from matrix (10.1), we obtain

$$A' = A\cos\varphi - \frac{jC\sin\varphi}{J} = 0,$$

$$\therefore \tan\varphi_0 = \frac{AJ}{jC}\bigg|_{s=s_0} \quad \text{or} \quad J_0 = \frac{jC\tan\varphi}{A}\bigg|_{s=s_0} \quad (10.3a)$$

$$B' = B\cos\varphi - \frac{jD\sin\varphi}{J} = 0,$$

$$\therefore \tan\varphi_0 = \frac{BJ}{jD}\bigg|_{s=s_0} \quad \text{or} \quad J_0 = \frac{jD\tan\varphi}{B}\bigg|_{s=s_0} \quad (10.3b)$$

Equations (10.3) show that either the admittance J of the inverter can be prescribed, giving a phase length φ_0 at which A' and B' are both equal zero, or the phase length φ can be prescribed ($\neq 90°$), which gives a characteristic admittance J_0 by which A' and B' go to zero. In practice, it is more common to prescribe $J = 1$ so that the interconnecting phase lengths are the same characteristic admittance as those of the interfacing transmission-line medium. If J is prescribed to be unity, the phase shift φ_0 is evaluated from equation (10.3), resulting in

$$\varphi_0 = \tan^{-1}\frac{A}{jC}\bigg|_{s=s_0} \tag{10.4a}$$

$$\varphi_0 = \tan^{-1}\frac{B}{jD}\bigg|_{s=s_0} \tag{10.4b}$$

Because the polynomials A' and B' go to zero when $s = s_0$, they are both divisible by the same factor $(s - s_0)$. This factor can be divided out to obtain the intermediate polynomials A^x and B^x:

$$A^x = \frac{A'}{(s - s_0)}, \qquad B^x = \frac{B'}{(s - s_0)} \tag{10.5}$$

Removal of Resonant Pair   After extraction of the phase length, the residue b_0 of the resonant pair is evaluated. Referring to Figure 10.2, we observe that as s approaches s_0, the impedance or admittance of the network is dominated by the shunt resonant pair.

At $s = s_0$

$$y_{11} = \frac{D'}{B'} = \frac{b_0}{(s - s_0)}\bigg|_{s=s_0}, \quad \therefore b_0 = \frac{(s - s_0)D'}{B'}\bigg|_{s=s_0} = \frac{D'}{B^x}\bigg|_{s=s_0} \tag{10.6a}$$

$$z_{11} = \frac{A'}{C'} = \frac{(s - s_0)}{b_0}\bigg|_{s=s_0}, \quad \therefore b_0 = \frac{(s - s_0)C'}{A'}\bigg|_{s=s_0} = \frac{C'}{A^x}\bigg|_{s=s_0} \tag{10.6b}$$

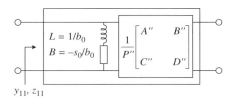

Figure 10.2   Input impedance/admittance as s approaches s_0.

Now we extract the resonant pair

$$
\frac{1}{P''}\begin{bmatrix} A'' & B'' \\ C'' & D'' \end{bmatrix} = \begin{bmatrix} 1 & 0 \\ -b_0/(s-s_0) & 1 \end{bmatrix} \cdot \frac{1}{P}\begin{bmatrix} A' & B' \\ C' & D' \end{bmatrix}
$$

$$
= \frac{1}{P}\begin{bmatrix} A' & B' \\ C' - \dfrac{b_0}{(s-s_0)}A' & D' - \dfrac{b_0}{(s-s_0)}B' \end{bmatrix}
$$

$$
= \frac{1}{P}\begin{bmatrix} A' & B' \\ C' - b_0 A^x & D' - b_0 B^x \end{bmatrix}, \tag{10.7a}
$$

where $[A''B''C''D'']$ is the remainder matrix. By multiplying the right-hand matrix in equation (10.7a) top and bottom by the factor $(s - s_0)$, we obtain

$$
\frac{1}{P''}\begin{bmatrix} A'' & B'' \\ C'' & D'' \end{bmatrix} = \frac{(s-s_0)}{P}\begin{bmatrix} \dfrac{A'}{(s-s_0)} & \dfrac{B'}{(s-s_0)} \\ \dfrac{C' - b_0 A^x}{(s-s_0)} & \dfrac{D' - b_0 B^x}{(s-s_0)} \end{bmatrix} \tag{10.7b}
$$

Thus it is easily seen that $A'' = A'/(s - s_0) = A^x$ and $B'' = B'/(s - s_0) = B^x$, and that the numerators of C'' and D'' equal $C' - b_0 A^x$ and $D' - b_0 B^x$, respectively. Since $b_0 = C'/A^x = D'/B^x$ [equation (10.6)], the numerators of C'' and D'' in equation (10.7b) go to zero at $s = s_0$. So, by extracting the resonant pair with residue b_0, numerators of C'' and D'' are conditioned to be divisible by the factor $(s - s_0)$, in the same way as the extraction of the phase length conditioned the numerators of A' and B' to be exactly divisible by $(s - s_0)$. By definition, the P polynomial of the original matrix is also divisible by $(s - s_0)$.

The process for extracting a single transmission zero at $s = s_0$ and obtaining the remainder matrix $[A''B''C''D'']$ is summarized as follows:

1. Calculate the phase length φ_0 or characteristic admittance J_0 of the length of transmission line to be extracted from (10.3).
2. Extract the transmission line to obtain the matrix $[A'B'C'D']$ [equation (10.1)].
3. Divide A' and B' by the factor $(s - s_0)$ to obtain the intermediate polynomials $A^x = A''$ and $B^x = B''$ from equation (10.5).
4. Calculate the value of the residue b_0 from equation (10.6).
5. Compute $C' - b_0 A^x$ and $D' - b_0 B^x$.
6. Divide these two polynomials by the factor $(s - s_0)$ to obtain C'' and D''.
7. Finally, divide the P polynomial by the factor $(s - s_0)$ to obtain P''.

Thus, the extraction of the pole has left all the polynomials in the remainder matrix one degree less than the corresponding polynomials in the original matrix. From the remainder matrix, a second pole may now be extracted by removing another length of transmission line at the new pole frequency s_{02}. If the next component to be extracted is an inverter or a shunt-connected $C + jB$ pair to start the synthesis of the main body of the filter, the phase length is removed at $s_0 = j\infty$. The main body of the filter itself is synthesized as a cross-coupled network to produce further transmission zeros (usually real-axis pairs for group delay equalization), and if required, the filter's topology can be reconfigured with rotations (subject to the usual restrictions), regardless of whether extracted poles are present. Also, the procedure can be applied to create extracted poles within the main body of the filter, separating the mainline resonators into groups (see Fig. 10.6b). This has certain advantages in tuning the filter, but does tend to result in rather extreme values for the coupling elements.

10.1.2 Example of Synthesis of Extracted Pole Network

To illustrate the extracted pole procedure, the (4–2) filter for which we have already derived the prototype characteristics and [ABCD] polynomials is used again to synthesize an extracted pole circuit. The method for single pole extraction described in the previous section allows for the extraction of a pole at any stage of the synthesis of the network, even between the main-body resonators. For a two-pole network, synthesizing the two poles to be physically on either side of the network gives the most practical coupling values; however, for the purposes of this demonstration, we choose to synthesize both poles to be physically on the input side of the filter. The procedure starts with the [ABCD] polynomials for the (4–2) characteristic (see Section 7.3):

$s^i, i =$	$A(s)$	$B(s)$	$C(s)$	$D(s)$	$P(s)/\varepsilon$
0	$-j2.0658$	-0.1059	-0.1476	$-j2.0658$	$+2.0696$
1	$+2.4874$	$-j4.1687$	$-j3.0823$	$+2.4874$	$+j2.7104$
2	$-j2.1950$	$+4.4575$	$+2.8836$	$-j2.1950$	-0.8656
3	$+2.4015$	$-j1.5183$	—	$+2.4015$	—
4	—	$+2.0$	—	—	—

To prepare the network for the removal of the first pole, extract a transmission line of phase length θ_{S1} and unity characteristic admittance at $s_{01} = j1.8082$ as follows:

$$\theta_{S1} = \tan^{-1}\frac{B(s)}{jD(s)}\bigg|_{s=s_{01}} = 48.9062°$$

The polynomials after extracting this phase length are listed next:

$s^i, i =$	$A'(s)$	$B'(s)$	$C'(s)$	$D'(s)$	$P(s)/\varepsilon$
0	$-j1.2466$	-1.6265	-1.6539	$-j1.2780$	$+2.0696$
1	-0.6880	$-j4.6146$	$-j3.9006$	$+1.5067$	$+j2.7104$
2	$-j3.6160$	$+0.2411$	$+0.2411$	$-j4.8021$	-0.8656
3	$+1.5785$	$-j2.8078$	$-j1.8099$	$+0.4343$	—
4	—	$+1.3146$	—	$-j1.5073$	—

Divide the $A'(s)$ and $B'(s)$ polynomials by $(s - s_{01})$ to form the intermediate polynomials $A^x(s)$ and $B^x(s)$:

$s^i, i =$	$A^x(s)$	$B^x(s)$	$C'(s)$	$D'(s)$	$P(s)/\varepsilon$
0	+0.6894	−j0.8995	−1.6539	−j1.2780	+2.0696
1	−j0.7618	+2.0546	−j3.9006	+1.5067	+j2.7104
2	+1.5785	−j0.4308	+0.2411	−j4.8021	−0.8656
3	—	+1.3146	−j1.8099	+0.4343	—
4	—	—	—	−j1.5073	—

Calculate the residue of the shunt-connected resonant pair b_{01} from

$$b_{01} = \frac{D'(s)}{B^x(s)}\Big|_{s=s_{01}} = 1.9680$$

Compute $C'(s) - b_{01}A^x(s)$ and $D'(s) - b_{01}B^x(s)$, then divide each by $(s - s_{01})$ to obtain $C''(s)$ and $D''(s)$. Finally divide $P(s)$ by $(s - s_{01})$ to obtain $P''(s)$:

$s^i, i =$	$A''(s) (= A^x(s))$	$B''(s) (= B^x(s))$	$C''(s)$	$D''(s)$	$P''(s)/\varepsilon$
0	+0.6894	−j0.8995	−j1.6650	−0.2722	+j1.1446
1	−j0.7618	+2.0546	+0.4073	−j2.9189	−0.8660
2	+1.5785	−j0.4308	−j1.8099	+0.5726	—
3	—	+1.3146	—	−j1.5073	—
4	—	—	—	—	—

Now the previous steps can be repeated to extract the second pole at $s_{02} = j1.3217$, giving $\theta_{12} = 27.5430°$ and $b_{02} = 3.6800$, and leaving the remainder polynomials:

$s^i, i =$	$A''(s)$	$B''(s)$	$C''(s)$	$D''(s)$	$P''(s)/\varepsilon$
0	−j0.1200	+0.5082	+1.0240	−j1.9122	−0.8660
1	+0.5627	−j0.0274	−j2.3347	+1.1540	—
2	—	+0.4686	—	−j1.9443	—
3	—	—	—	—	—
4	—	—	—	—	—

Prepare the network to remove $C_3 + jB_3$, which is done by extracting a phase length at $s = j\infty$ such that

$$\theta_{23} = \tan^{-1} \frac{B(s)}{jD(s)}\Big|_{s=j\infty} = 13.5508°$$

Extracting this phase length leaves the following polynomials:

$s^i, i =$	$A(s)$	$B(s)$	$C(s)$	$D(s)$	$P(s)/\varepsilon$
0	$-j0.3566$	$+0.0460$	$+0.9673$	$-j1.9781$	-0.8660
1	—	$-j0.2970$	$-j2.4015$	$+1.1540$	—
2	—	—	—	$-j2.0000$	—
3	—	—	—	—	—
4	—	—	—	—	—

Synthesis of the main body can now proceed conventionally as in Section 7.4:

1. Extract $C_3 = 6.7343$ and $B_3 = 2.7126$.
2. Turn the network. If there is a pole present on the output side, a phase length θ_{5L} must be extracted now to prepare the network for the removal of the resonant pair, but in this case, both poles are on the input side, therefore, $\theta_{5L} = 0$.
3. Extract transmission line $90°$ and $Y_0 = M_{45} = 1$ (output coupling inverter).
4. Extract $C_4 = 0.8328$ and $B_4 = 0.1290$.
5. Finally extract a parallel-connected inverter $M_{34} = 2.4283$.

It is convenient to introduce two unity admittance line lengths of $-90°$ and $+90°$, which have zero net effect on the transfer function, between the phase length θ_{23} and $C_3 + jB_3$. The first length is absorbed into θ_{23} ($\theta_{23} \rightarrow \theta_{23} - 90° = -79.4492°$), and the second forms the input coupling inverter to the main body of the filter M_{23}.

The extracted pole design in Figure 10.3 can be transformed through the dual-network theorem into a parallel resonant pair, connected through an inverter in parallel with the mainline (Fig. 10.4a), or a series-connected parallel resonant pair between the two unit inverters in the mainline (Fig. 10.4b). These circuits are now in the right form for realization in rectangular waveguide H-plane and E-plane open T junctions, respectively. Note Figure 10.4a, where the equivalent circuit for the H-plane junction includes an inverter [7].

By using the dual-network theorem, the shunt resonant pairs in the circuits in Figure 10.5 can also be developed into an inverter plus a length of transmission line that is $180°$, or an integral number of half-wavelengths at the resonant frequency of the transmission zero s_0 and short-circuited at one end. Eventually, this length is

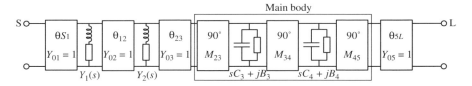

Figure 10.3 (4−2) extracted pole filter, synthesized network with both poles on input side.

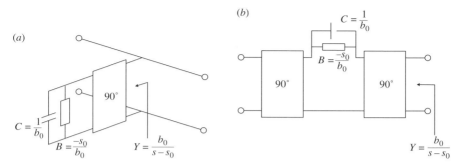

Figure 10.4 Possible realizations for the extracted pole: (*a*) parallel-connected (waveguide *H* plane); (*b*) series-connected (waveguide *E* plane).

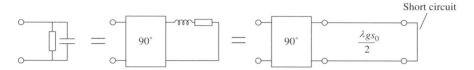

Figure 10.5 Development of the shunt resonant pair into an inverter and short-circuited half-wavelength transmission line.

realized as a resonant cavity with the inverter becoming its input coupling. In practice, the inverter can be equated to its equivalent circuit, a shunt inductor at the center of a short length of negative transmission line. This is usually realized with an inductive aperture in waveguide.

Figure 10.6*a* depicts the extracted pole filter with TE_{011}-mode cylindrical resonant cavities with both poles on the input side of the main body. Although the structure incorporates two transmission zeros, there are no negative couplings required (which would be difficult to implement with TE_{011} cavities). The couplings into and out of each mainline cavity are at right angles to each other to suppress the TM_{111} resonant mode, which is degenerate with the TE_{011} mode in cylindrical cavities. Figure 10.6*b* displays an alternative arrangement, where one of the extracted pole cavities is realized between the two mainline cavities.

10.1.3 Analysis of the Extracted Pole Filter Network

The extracted pole network can be analyzed by using the classical cascaded [*ABCD*] matrix method. A more useful method is based on the admittance matrix, which includes the coupling matrix [*M*] (see equation 8.7), and which may be analyzed in the same way as for a coupling matrix.

The overall admittance matrix [*Y*] is built up by the superimposition of the individual 2×2 [*y*] submatrices. These submatrices themselves may make up more than one element, and are formed by building the [*ABCD*] matrix of the element or

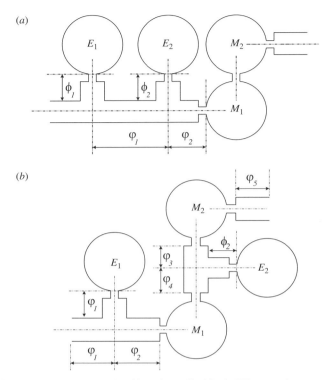

Figure 10.6 (4–2) extracted pole filter in cylindrical TE_{011}-mode resonator cavities: (*a*) both poles cavities on the input side; (*b*) one pole cavity synthesized within the main body.

"subcascade," and then converting it to [*y*] parameters. Figure 10.7 presents four such submatrices. The first is for a length of transmission line of phase θ_i and characteristic admittance Y_{0i}, followed by an extracted pole of admittance $Y_i(s)$; the second is for a length of transmission line in cascade with an inverter of characteristic admittance $Y_0 = M_{jk}$, representing the phase length between the last extracted pole and the input coupling inverter of the main body of the filter (M_{S1}). The third is for a shunt capacitor and FIR in cascade with a mainline coupling inverter, and the fourth is the same as the third but with an additional phase length. Pairing the components in this way reduces the number of nodes, reducing the size of the overall admittance matrix that eventually is be inverted. Of course, the $[y]^{(i)}$ matrices can be reversed (for working at the output side of the filter) by simply exchanging the y_{11} and y_{22} elements in the submatrix.

The procedure for superimposing the individual 2×2 $[y]^{(i)}$ submatrices for a (4–2) filter with two extracted poles on the input side is shown in Figure 10.8, where the submatrix $[y]^{(1)}$ represents the input phase length + extracted pole 1, $[y]^{(2)}$ represents the interpole phase length + the second extracted pole, $[y]^{(3)}$ is

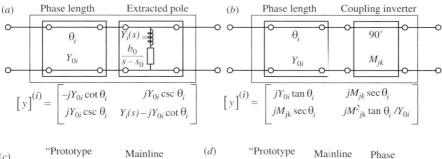

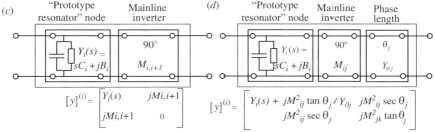

Figure 10.7 Admittance submatrices: (a) phase length + extracted pole; (b) phase length + coupling inverter; (c) prototype resonator node + inverter; (d) prototype resonator node + inverter + phase length.

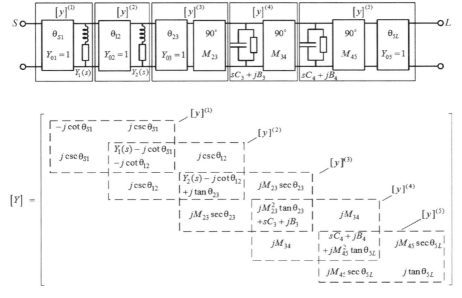

Figure 10.8 Superimposition of the admittance submatrices to form an overall admittance matrix.

the last phase length + input coupling inverter of the main body of the filter, and $[y]^{(4)}$ and $[y]^{(5)}$ represent the main body of the filter (which can be represented as an $N \times N$ cross-coupled matrix). The input phase length θ_{S1} and the final phase length θ_{5L} at the unity impedance source and load terminations define the input and output phase planes; they have no effect on the amplitudes of the transfer and reflection performance. It is in fact better to physically put the two extracted poles on either side of the main body of the filter if it is possible. This tends to result in more practical coupling values.

Adding the termination impedances (1 Ω) to the y_{11} and y_{66} positions of the matrix, inverting it, and applying the transfer and reflection formulas [(8.22) and (8.25)] yields the transfer and reflection performance of this network in the same way as for a regular coupling matrix.

10.1.4 Direct-Coupled Extracted Pole Filters

An alternative to the interpole phase lengths, and the phase lengths that couple between the last extracted pole and the main body of the filter, may be created through the use of the circuit equivalence as shown in Figure 10.9. This equivalence relates a length of transmission line of frequency-invariant phase length θ_i and characteristic admittance Y_0 to an inverter of admittance $Y_0 \operatorname{cosec}\theta_i$ surrounded by 2 FIRs each of value $-jY_0 \cot\theta_i$.

The phase length between an extracted pole and the first $C + jB$ of the main body of the filter may be directly replaced with the inverter and the two FIRs as shown in Fig. 10.9 below. The inverter itself forms a convenient input coupling inverter to the first resonator of the main body of the filter (avoiding the need to add the $-90°$ and $+90°$ line lengths as was done in the extracted pole synthesis example above), whilst the right-hand FIR may be added to the admittance $(sC + jB)$ of the first resonator, and realized as an additional frequency offset of that resonator. The FIR on the left-hand side (LHS) may be realized as a small capacitive or inductive obstacle or screw at the input port of the filter, or a nonresonant node (NRN) if convenient.

The process for developing a direct-coupled extracted pole filter from a phase-coupled type is most easily demonstrated through an example. Figure 10.10 below shows the network for extracted pole synthesis of the $(4-2)$ asymmetric

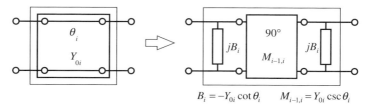

$$B_i = -Y_{0i}\cot\theta_i \qquad M_{i-1,i} = Y_{0i}\csc\theta_i$$

Figure 10.9 Equivalence of phase length with inverter surrounded by two FIRs.

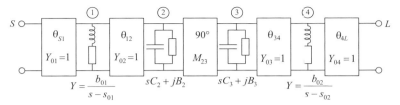

Figure 10.10 (4–2) asymmetric filter—phase-coupled extracted pole network.

filter; however, this time, with the two pole cavities realized on either side of the main body. The order of synthesis was to extract:

1. Phase length θ_{S1}
2. Resonant pair 1
3. Phase length θ_{12}
4. $C_1 + jB_1$ and turn network
5. Phase length θ_{4L}
6. Resonant pair 2
7. Phase length θ
8. $C_2 + jB_2$, and finally
9. Parallel coupled inverter M_{23}

The values obtained for these components are summarized in Table 10.1.

Figure 10.11 shows the configuration of the network after the phase lengths θ_{12} and θ_{34} have been replaced with the equivalent FIRs and inverters. Also the series resonant pairs Y_1 and Y_2 have transformed through the unit inverters M_{S1} and M_{4L}

TABLE 10.1   4–2 Extracted-Pole Filter—Prototype Network Element Values

TZ No. i	TZ Frequency s_{0i}	Extracted Pole Residue b_{0i}	$Y_i(s) = sC_i + jB_i$	Phase Length $\theta_{i-1,i}$	Characteristic Admittance Y_{0i}
1	$s_{01} = j1.3217$	$b_{01} = 0.5767$		$\theta_{s1} = +23.738°$	$Y_{01} = 1.0$
				$\theta_{12} = +66.262°$	$Y_{02} = 1.0$
			$C_2 = 1.3937$		
			$B_2 = 0.8101$		
				$90.0°$	$M_{23} = 1.7216$
			$C_3 = 2.8529$		
			$B_3 = 1.7180$		
				$\theta_{34} = +41.094°$	$Y_{03} = 1.0$
2	$s_{02} = j1.8082$	$b_{02} = 1.9680$		$\theta_{4L} = +48.906°$	$Y_{04} = 1.0$

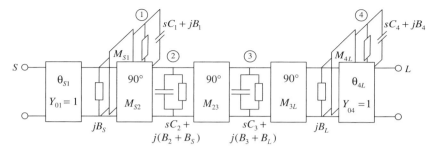

Figure 10.11   (4–2) asymmetric filter—direct-coupled extracted pole realization.

to form parallel resonators (see Fig. 10.4). The values of the new components are:

$$M_{S2} = \csc\theta_{12} = 1.0924 \qquad M_{3L} = \csc\theta_{34} = 1.5214$$
$$B_S = -\cot\theta_{12} = -0.4398 \qquad B_L = -\cot\theta_{34} = -1.1466$$
$$C_1 = \frac{1}{b_{01}} = 1.7340 \qquad C_4 = \frac{1}{b_{02}} = 0.5081$$
$$B_1 = -\frac{s_{01}}{b_{01}} = -2.2918 \qquad B_4 = -\frac{s_{02}}{b_{02}} = -0.9188$$
$$B_2 \rightarrow B_2 + B_S = 0.3703 \qquad B_3 \rightarrow B_3 + B_L = 0.5714 \qquad (10.8a)$$

Now scaling at the nodes to set all the C_i to unity:

$$M_{S1} = \frac{1}{\sqrt{C_1}} = 0.7594 \qquad\qquad M_{4L} = \frac{1}{\sqrt{C_4}} = 1.4029$$
$$B_1 \rightarrow \frac{B_1}{C_1} = -s_{01} = -1.3217 \equiv M_{11} \quad B_4 \rightarrow \frac{B_4}{C_4} = -s_{02} = -1.8082 \equiv M_{44}$$
$$M_{S2} \rightarrow \frac{M_{S2}}{\sqrt{C_2}} = 0.9254 \qquad\qquad M_{3L} \rightarrow \frac{M_{3L}}{\sqrt{C_3}} = 0.9007$$
$$B_2 \rightarrow \frac{B_2}{C_2} = 0.2657 \equiv M_{22} \qquad\qquad B_3 \rightarrow \frac{B_3}{C_3} = 0.2003 \equiv M_{33}$$
$$M_{23} \rightarrow \frac{M_{23}}{\sqrt{C_2 C_3}} = 0.8634$$
$$M_{SS} \equiv B_S = -0.4398 \qquad\qquad M_{LL} \equiv B_L = -1.1466$$
$$C_1, C_2, \ C_3, \ C_4 \rightarrow 1 \qquad\qquad (10.8b)$$

With all the C_i set to unity, the coupling matrix for the direct-coupled extracted pole filter may be constructed. Each FIR may be directly realized as a small obstacle as in Figure 10.11, eg a screw near the input/output coupling devices, or as a

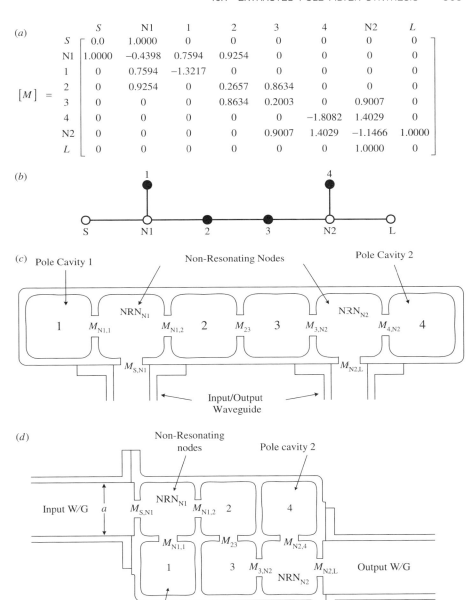

Figure 10.12 (4–2) asymmetric filter, the direct-coupled extracted pole realization: (*a*) coupling matrix. Nodes N1 and N2 (as well as the S and L nodes) are NRNs (*b*) coupling and routing diagram. (*c*) realization in coaxial cavities. (*d*) realization with waveguide resonant cavities.

'non-resonant node' (NRN), in effect a detuned resonator, coupling the pole cavity and main body cavity simultaneously [3]. In the coupling matrix and the corresponding coupling and routing diagram shown in Figures 10.12a and b respectively, unity value inverters have been added to provide the input and output couplings to non-resonant cavities at the source and load ends of the network respectively. Note that the values of the M_{11} and M_{NN} elements are the negatives of the prototype transmission zero positions (s_{01} and s_{02}, respectively). Also the values of B_S and B_L appear in the $M_{N1,N1}$ and $M_{N2,N2}$ positions, representing the parallel-connected FIRs at the input and output terminals. Figure 10.12c gives a possible coaxial-cavity realization and Fig. 10.12d a waveguide realization of the direct-coupled extracted pole network.

10.2 SYNTHESIS OF BANDSTOP FILTERS USING THE EXTRACTED POLE TECHNIQUE

Bandstop filters are necessary when a specific narrow band of frequencies needs to be attenuated, for example, the second-harmonic frequency at the output of a high-power saturated amplifier. Although the same job can be accomplished by a band-pass filter, it is often difficult to optimize the rejection performance over the stop-band, and at the same time maintain a low insertion loss and good return loss performance over the main signal band. On the other hand, the bandstop filter needs only to be optimized over its stopband, and the bandstop cavities designed to be spurious-free over the passband.

Bandstop filters are composed of a series of bandstop resonant cavities, placed at intervals along the main transmission-line medium carrying the signal, but not in the signal path, such that the power handling problems are mitigated. Figure 10.13 shows an example of a fourth-degree bandstop filter realized in a rectangular waveguide structure [2,4].

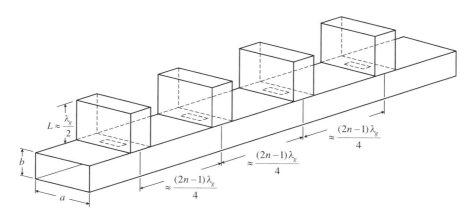

Figure 10.13 Fourth-degree bandstop filter realized in a rectangular waveguide.

Because the bandstop cavities of bandstop filters have the same equivalent circuit as the bandstop resonant cavities that realize the TZs of extracted pole bandpass filters, the same technique, used to synthesize the extracted poles, may be used for the bandstop filters. This approach was first noted [5] and then explained in greater detail in [6]. The difference now is that the $S_{21}(s)$ and $S_{11}(s)$ characteristics are exchanged, so that what used to be return loss becomes the transfer characteristic and vice versa. If the characteristic is a Chebyshev, the prescribed return loss level now becomes the equiripple stopband attenuation, whereas the former rejection characteristics become the out-of-stopband return loss performance.

Thus, the numerator $F(s)/\varepsilon_R$ of the reflection function $S_{11}(s)$ becomes the numerator of the transfer function $S_{21}(s)$, and $P(s)/\varepsilon$ becomes the new numerator of $S_{11}(s)$. Therefore, for an Nth-degree bandstop characteristic, there are N transmission zeros to extract at the former reflection zero positions, in any order. Having extracted all the poles and their associated phase lengths, we need to extract a parallel-connected inverter. This can be located anywhere in the cascade, but is equal to unity in all cases except for the canonical asymmetric and canonical even-degree symmetric filtering functions. This nonunity inverter represents a small reactive frequency-independent element, giving the finite return loss value at an infinite frequency that is inherent in the reflection polynomials for these cases.

The (4−4) asymmetric canonical prototype that is used for the $N+2$ synthesis example in Section 8.4 is also adopted to demonstrate the bandstop synthesis process:

1. Exchange the $F(s)/\varepsilon_R$ and $P(s)/\varepsilon$ polynomials of the (4−4) lowpass prototype transfer and reflection functions $S_{21}(s)$ and $S_{11}(s)$. The j that multiplies $P(s)$ because $N - n_{fz}$ is even, and the j that multiplies $P(s)$ to allow cross-couplings to be realized as inverters, should now multiply $F(s)/\varepsilon_R$, because this is now the new numerator of $S_{21}(s)$.

2. Extract the first phase length θ_{S1} at $s_{01} = -j0.9384$, the lowest of the $F(s)$ zeros, followed by the extraction of the resonant pair Y_1 at s_{01}.

3. Similarly, extract θ_{12} and Y_2 at $s_{02} = -j0.4228$.

4. Extract the phase length $\theta_{23}^{(1)}$ at $s_0 = j\infty$.

5. Turn the network and extract θ_{4L} and Y_4 at $s_{04} = j0.9405$ [the highest-frequency zero of $F(s)$].

6. Similarly extract θ_{34} and Y_3 at $s_{03} = j0.4234$.

7. Extract the phase length $\theta_{23}^{(3)}$ at $s_0 = j\infty$. Also, it is usually necessary to extract an extra $90°$ to attain the polynomials in the right form for the extraction of the final parallel-coupled inverter.

8. Extract the parallel-coupled inverter M_{23}. M_{23} is unity for all cases except when the original lowpass prototype is a symmetric even-degree canonical or any asymmetric canonical. In these cases, M_{23} has a value $\varepsilon_R - \sqrt{\varepsilon_R^2 - 1}$, and is approximately realized with a small reactive element, ideally a FIR, to provide the finite return loss that these prototypes have at infinite frequency.

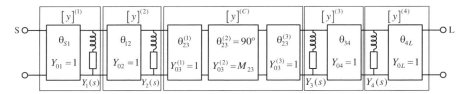

Figure 10.14 (4−4) bandstop filter network.

TABLE 10.2 Element Values of the (4−4) Asymmetric Canonical Bandstop Filter

TZ Number i	TZ Frequency s_{0i}	Phase Length $\theta_{i-1,i}$	Extracted Pole Residue b_{0i}	Characteristic Admittance Y_{0i}
1	$s_{01} = -j0.9384$	$\theta_{S1} = +34.859°$	$b_{01} = 0.6243$	$Y_{01} = 1.0$
2	$s_{02} = -j0.4228$	$\theta_{12} = +75.104°$	$b_{02} = 1.6994$	$Y_{02} = 1.0$
		$\theta_{23}^{(1)} = -19.962°$	—	$Y_{03}^{(1)} = 1.0$
		$\theta_{23}^{(2)} = +90.0°$	—	$Y_{03}^{(2)} = M_{23} = 0.9703$
3	$s_{03} = +j0.4234$	$\theta_{23}^{(3)} = 18.772°$	$b_{03} = 1.6024$	$Y_{03}^{(3)} = 1.0$
4	$s_{04} = +j0.9405$	$\theta_{34} = -72.334°$	$b_{04} = 0.6328$	$Y_{04} = 1.0$
		$\theta_{4L} = -36.438°$	—	$Y_{0L} = 1.0$
		$\sum \theta = k \cdot 90°$		

The form of the synthesized (4−4) bandstop network is shown in Figure 10.14, and the corresponding element values are given in Table 10.2. Half-wavelengths (180°) can be added to the interresonator phase lengths, if these turn out to be negative or small for a more practical physical layout for the bandstop filter; however, the lengths should not be too long, because, ideally, they should be dispersionless (phase lengths frequency invariant), which in practice they are not.

For this example, the transmission zeros are taken in ascending order, but there is no restriction on this, and they can be taken in any order. In addition, there is no restriction on where the final inverter is realized. It can be placed at the end of the network. In this case, the four poles are extracted in sequence, then a phase length at infinity, and finally the parallel inverter.

10.2.1 Direct-Coupled Bandstop Filters

The network synthesis methods presented in Chapter 7 can be used once the $S_{21}(s)$ and $S_{11}(s)$ functions are exchanged, to create a bandpass-like filter configuration with bandstop filter characteristics. The resonant cavities are direct-coupled; therefore, the wideband performance is potentially better, and because the cavities are tuned to frequencies within the *stopband*, the main signal power routes mainly through the direct input/output coupling, bypassing the resonators, and giving a minimal insertion loss and relatively high power handling.

As with the extracted pole types, to generate a bandstop characteristic from the regular lowpass prototype polynomials, it is necessary only to exchange the reflection and transfer functions (including the constants):

$$S_{11}(s) = \frac{P(s)/\varepsilon}{E(s)}, \qquad S_{21}(s) = \frac{F(s)/\varepsilon_R}{E(s)} \qquad (10.9)$$

Since $S_{21}(s)$ and $S_{11}(s)$ share a common denominator polynomial $E(s)$, the unitary conditions for a passive lossless network are preserved. If the characteristics are Chebyshev, then the original prescribed equiripple *return loss* characteristic becomes the transfer response, with a minimum reject level equal to the original prescribed return loss level. Because the degree of the new numerator polynomial for $S_{21}(s)$ ($=F(s)/\varepsilon_R$) is now the same as its denominator $E(s)$, the network that is synthesized is fully canonical. The new numerator of $S_{11}(s)$ is the original transfer function numerator polynomial $P(s)/\varepsilon$, and can have any number n_{fz} of prescribed transmission zeros provided $n_{fz} \leq N$, the degree of the characteristic. If $n_{fz} < N$, then the constant $\varepsilon_R = 1$.

The synthesis of the coupling matrix for the bandstop network is very similar to the synthesis of fully canonical lowpass prototypes for bandpass filters, as discussed in Section 8.4. First, we exchange the $P(s)/\varepsilon$ (degree n_{fz}) and $F(s)/\varepsilon_R$ (degree N) polynomials, and then derive the polynomials for the rational short-circuit admittance parameters $y_{21}(s)$ and $y_{22}(s)$ of the network. For an even-degree doubly terminated network with source and load terminations of 1 Ω, we obtain

$$y_{21}(s) = \frac{y_{21n}(s)}{y_d(s)} = \frac{(F(s)/\varepsilon_R)}{m_1(s)}$$

$$y_{22}(s) = \frac{y_{22n}(s)}{y_d(s)} = \frac{n_1(s)}{m_1(s)} \qquad (10.10a)$$

For an odd-degree network

$$y_{21}(s) = \frac{y_{21n}(s)}{y_d(s)} = \frac{(F(s)/\varepsilon_R)}{n_1(s)}$$

$$y_{22}(s) = \frac{y_{22n}(s)}{y_d(s)} = \frac{m_1(s)}{n_1(s)} \qquad (10.10b)$$

where

$$m_1(s) = \mathrm{Re}(e_0 + p_0) + j\mathrm{Im}(e_1 + p_1)s + \mathrm{Re}(e_2 + p_2)s^2 + \cdots$$

$$n_1(s) = j\mathrm{Im}(e_0 + p_0) + \mathrm{Re}(e_1 + p_1)s + j\mathrm{Im}(e_2 + p_2)s^2 + \cdots$$

and e_i and p_i, $i = 0,1,2,3, \ldots , N$, are the complex coefficients of $E(s)$ and $P(s)/\varepsilon$ respectively. If n_{fz} is less than N (which is the usual case), then the highest-degree coefficient of $P(s)$ ($=p_N$) is zero.

Having built up the numerator polynomials $y_{21n}(s)$, $y_{22n}(s)$, and the denominator polynomial $y_d(s)$, the coupling matrix synthesis proceeds in exactly the same way as that in Section 8.4. Because the numerator of $S_{21}(s)$ is now the same degree as its denominator, the characteristic is now fully canonical and the coupling matrix representing it needs to incorporate a direct source–load coupling M_{SL}. M_{SL} is calculated as follows [see equation (8.42)]:

$$jM_{SL} = \frac{y_{21}(s)}{y_d(s)}\bigg|_{s=j\infty} = \frac{jF(s)/\varepsilon_R}{y_d(s)}\bigg|_{s=j\infty} \tag{10.11}$$

If the original bandpass characteristic is noncanonical, that is, the degree n_{fz} of $P(s)$ is less than N, and therefore $p_N = 0$, it can be seen from equation (10.10) that the leading coefficient of $y_d(s) = 1$. For a Chebyshev characteristic, the leading coefficient of $F(s)$ always equals unity, as does ε_R for noncanonical cases, and so from equation (10.11), it is evident that $M_{SL} = 1$. In other words, the direct source–load coupling inverter is the same characteristic impedance as the interfacing transmission lines from the source and to the load, and it can be simply constructed from a $90°$ section of that line. For the fully canonical characteristics, M_{SL} is slightly less than unity, and provides the finite return loss at infinite frequency that the fully canonical prototype requires.

The rest of the synthesis of the coupling matrix proceeds as for a folded lowpass prototype network for a bandpass filter. Diagonal cross-couplings appear if the characteristic is asymmetric. Figure 10.15 gives an example of a symmetric fourth-degree bandstop filter in folded configuration, showing that the main signal path is through the direct input/output coupling M_{SL}.

In general it is not desirable to include complex couplings in direct-coupled bandstop filters in folded form. It is best to restrict the applications to symmetric prototypes. Let us consider an example of a symmetric fourth-degree filter with 22 dB return loss (which becomes the stopband reject level), and two TZs (reflection zeros now) at $\pm j2.0107$ to give an out-of-band (return loss) lobe level of 30 dB. After synthesizing the folded ladder network and scaling all the C_i to unity, the following coupling matrix is obtained:

$$
M = \begin{array}{c}
\begin{array}{c} S \\ 1 \\ 2 \\ 3 \\ 4 \\ L \end{array}
\left[
\begin{array}{cccccc}
S & 1 & 2 & 3 & 4 & L \\
0.0 & 1.5109 & 0 & 0 & 0 & 1.0000 \\
1.5109 & 0.0 & 0.9118 & 0 & 1.3363 & 0 \\
0 & 0.9118 & 0.0 & -0.7985 & 0 & 0 \\
0 & 0 & -0.7985 & 0.0 & 0.9118 & 0 \\
0 & 1.3363 & 0 & 0.9118 & 0.0 & 1.5109 \\
1.0000 & 0 & 0 & 0 & 1.5109 & 0.0
\end{array}
\right]
\end{array} \tag{10.12a}
$$

Note that the direct input/output coupling M_{SL} is unity for this noncanonical case, meaning that this coupling inverter is the same characteristic admittance as the

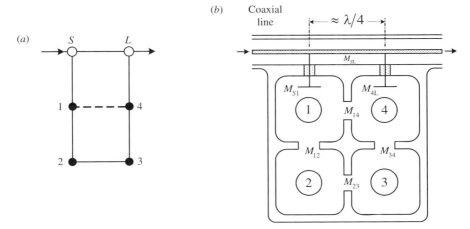

Figure 10.15 Direct-coupled (4–2) bandstop filter: (*a*) coupling–routing diagram; (*b*) possible realization with coaxial cavities.

input and output lines. It can be formed from a 90° ($\approx\lambda/4$) part of that line carrying the main signal power, between the input and output tap points to the main body of the filter. Figure 10.15*b* shows a possible realization with coaxial cavities; however, coupling M_{23} is negative in equation (10.12a), and needs to be realized as a probe.

It is interesting to note that with bandstop transfer and reflection characteristics, when the number of reflection zeros (formerly *transmission* zeros) is less than the degree of filter N (i.e., $n_{fz} < N$), a second solution is possible by working with the dual of the network. The dual network is obtained simply by multiplying the coefficients of the reflection numerator polynomials $F(s)$ and $P(s)$ by -1, and is equivalent to placing unit inverters at the input and output of the network. If this is done for the (4–2) symmetric example and the bandstop matrix is resynthesized, the coupling matrix (10.12b) is obtained with all-positive couplings. In general, this uniformity of coupling sign does not occur.

$$M = \begin{matrix} & \begin{matrix} S & 1 & 2 & 3 & 4 & L \end{matrix} \\ \begin{matrix} S \\ 1 \\ 2 \\ 3 \\ 4 \\ L \end{matrix} & \left[\begin{matrix} 0.0 & 1.5109 & 0 & 0 & 0 & 1.0000 \\ 1.5109 & 0.0 & 0.9118 & 0 & 0.9465 & 0 \\ 0 & 0.9118 & 0.0 & 0.7985 & 0 & 0 \\ 0 & 0 & 0.7985 & 0.0 & 0.9118 & 0 \\ 0 & 0.9465 & 0 & 0.9118 & 0.0 & 1.5109 \\ 1.0000 & 0 & 0 & 0 & 1.5109 & 0.0 \end{matrix} \right] \end{matrix} \qquad (10.12b)$$

Cul-de-Sac Forms for the Direct-Coupled Bandstop Matrix   If the number of reflection zeros of the bandstop characteristic is less than the degree of the

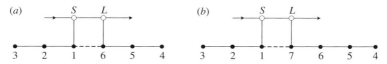

Figure 10.16 Cul-de-sac forms for direct-coupled bandstop filters: (*a*) sixth degree; (*b*) seventh degree.

network ($n_{fz} < N$), and the network is doubly terminated between equal source and load terminations, then a cul-de-sac form for the bandstop network, similar to that for bandpass filters (see Section 9.6), can be obtained by introducing two unity impedance 45° phase lengths at either end of the network. This is equivalent to multiplying the $F(s)$, $F_{22}(s)$, and $P(s)$ polynomials by j, which has no effect on the overall transfer and reflection responses of the network apart from the 90° phase changes.

Synthesizing the network by using either the circuit approach or the direct coupling matrix approach yields networks as shown in Figure 10.16. These networks are characterized by the square-shaped *core* quartet of couplings, with the source and load terminals at adjacent corners at the input/output end, the other resonators are strung out in two chains from the other two corners in equal numbers, if N is even, and one more than the other if N is odd. There are no diagonal couplings even for asymmetric characteristics, and all the couplings are of the same sign. For these characteristics, where $n_{fz} < N$, the direct source–load coupling M_{SL} is always unity in value.

Let us take the example of the (4–2) asymmetric characteristic that is used as the synthesis example in Chapters 7 and 8. Exchanging the $F(s)$ and $P(s)$ polynomials, multiplying them by j, and synthesizing the folded coupling matrix (Chapter 8) yields the coupling matrix (10.13). Figure 10.17 shows the corresponding coupling and routing diagram and a possible realization with rectangular waveguide resonators. The section of the waveguide transmission line between the input and output connections forms the direct input/output coupling M_{SL}, and can be an odd number of quarter wavelengths, the less, the better.

$$
M = \begin{array}{c} \\ S \\ 1 \\ 2 \\ 3 \\ 4 \\ L \end{array}
\begin{bmatrix}
\begin{array}{cccccc}
S & 1 & 2 & 3 & 4 & L \\
0.0 & 1.5497 & 0 & 0 & 0 & 1.0000 \\
1.5497 & 0.5155 & 1.2902 & 0 & 1.2008 & 0 \\
0 & 1.2902 & -0.0503 & 0.0 & 0 & 0 \\
0 & 0 & 0.0 & -1.0187 & 0.4222 & 0 \\
0 & 1.2008 & 0 & 0.4222 & -0.2057 & 1.5497 \\
1.0000 & 0 & 0 & 0 & 1.5497 & 0.0
\end{array}
\end{bmatrix} \quad (10.13)
$$

The resonators of the direct-coupled bandstop filter can be realized with coaxial, dielectric, or planar resonators. A possible application is in high-power signal

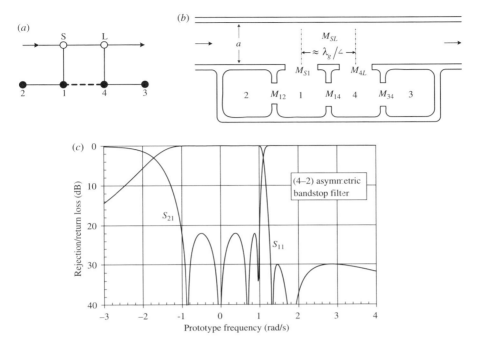

Figure 10.17 (4–2) Direct-coupled cul-de-sac bandstop filter: (*a*) coupling and routing diagram; (*b*) possible realization with waveguide cavities; (*c*) rejection and return loss performance.

diplexing, when ultralow loss is important but where out-of-band rejection is not a serious problem (or is provided by a low-loss wideband cover filter), and large levels of isolation are not required. For a diplexer application, each of the two bandstop filters is tuned to reject the frequency of the channel on the opposite arm of the combination point.

10.3 TRISECTIONS

Trisections are another section that, like the extracted pole section, are able to realize one transmission zero each. A trisection consists of three couplings between three nodes, the first and third of which can be source or load terminals, or $C_i + jB_i$ *lowpass resonators*. The middle node is usually a lowpass resonator [5, 7–8]. Figure 10.18 shows four possible configurations. Figure 10.18*a* is an internal trisection, and Figures 10.18*b* and 10.18*c* show the input and output trisections, respectively, where one node is the source or load termination. When the first and third nodes are the source and load terminations, respectively (Fig. 10.18*d*), we have a canonical network of degree 1 with the direct source–load coupling M_{SL}

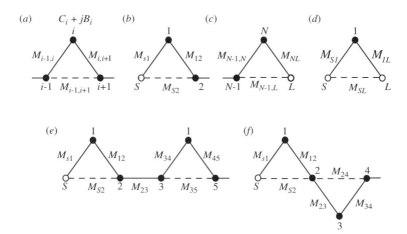

Figure 10.18 Coupling–routing diagram for trisections: (*a*) internal; (*b*) source-connected; (*c*) load-connected; (*d*) canonical; (*e*) nonconjoined cascaded; (*f*) conjoined cascaded.

providing the single transmission zero. Moreover, trisections can be cascaded with other trisections, either separately or conjoined (Figs 10.18*e*, *f*).

In general, if the transmission zero that the trisection is realizing is above the main passband, then the cross-coupling (e.g., M_{S2} in Fig. 10.18*b*) is positive, and if the zero is below the passband, it is negative. The cross-coupling can be realized by a diagonal capacitive probe for a negative coupling and by an inductive loop for positive coupling.

Trisections have some properties that are very useful for the synthesis of advanced cross coupled network topologies:

- A particular transmission zero may be associated with the cross-coupling in a particular trisection.
- The trisection may be represented in a coupling matrix.

The freedom to be able to synthesize the trisections in more or less unrestricted positions within the overall filter network, and then to apply rotations to the associated coupling matrix to obtain a final topology, opens up some configuration possibilities that cannot be easily obtained by using the folded or transversal matrix as the starting point. Two approaches for the synthesis of an individual trisection or a cascade of trisections are described: a hybrid method based on the extraction of circuit elements followed by conversion to a coupling matrix, and an alternative method based purely on the coupling matrix. The methods used to transform the trisection-configured coupling matrix into cascaded sections (*n-tuplets* and *box* sections) are outlined.

10.3.1 Synthesis of the Trisection—Circuit Approach

The synthesis of the trisection follows a route similar to that for the synthesis of the extracted pole circuit, involving the extraction of only parallel-connected elements and lengths of transmission line [5]. For the extracted pole, a phase length is first extracted from the overall [ABCD] matrix that represents the filtering characteristic desired in order to prepare the network for the extraction of the parallel-connected series resonant pair to realize the transmission zero at $s = s_0$. In the case of a tri-section, a shunt-connected FIR of admittance J_{13}, evaluated at $s = s_0$, is extracted first, followed by a unity admittance inverter to prepare the network for the extraction of the resonant pair by the same procedure as with the extracted pole. Then a further FIR of value $-1/J_{13}$ is extracted, another unity admittance inverter, finally a parallel FIR of value J_{13}, the same value as the first FIR. The network is then transformed into the trisection network using the dual-network theorem and a circuit transform.

The procedure is illustrated in Table 10.3. Step 1 shows the initial network, consisting of two inverters, three parallel-connected FIRs and the resonant pair, that is synthesized from the $A(s)$, $B(s)$, $C(s)$, $D(s)$, and $P(s)/\varepsilon$ polynomials. After these components have been extracted, all the remainder polynomials are one degree less than the originals, in the same way as with the extracted pole design. Steps 2 and 3 show the transformation of the network to the "π" network shown in step 3.

Now the well-known transform can be applied to the components in the series arm of the π network to form the new π network as shown in step 4. The transform is readily proved by equating the short-circuit input admittances y_{11} of the networks in the series arms of the π networks at $s = 0$ and $s = s_0$. At this stage, the π network formed by the two shunt admittances J_{13} at either side, and the admittance $-J_{13}$ that bridges them, are replaced by the equivalent admittance inverter of characteristic admittance $-J_{13}$ [2]. Lastly, the dual-network theorem is applied to the remaining series components in the series arm to form the shunt-connected pair $C_{S2} + B_{S2}$, the central *lowpass resonator* of the trisection.

In Table 10.3, the input and output ends of the trisection (nodes 1 and 3, respectively) are connected to nodes that are themselves lowpass resonators, but these nodes can be the source or load terminals. In this case, the filter is a dual-input type, with two couplings from the source terminal to internal resonators 1 and 2, and/or $N-1$ and N to the load terminal. Of course, the trisections can be synthesized anywhere within larger-degree networks by simply indexing the subscripts of C_S, B_S, and J appropriately. If the central shunt-connected pair of the trisection is at node position k within the overall network, then k is called the *index* of the trisection, and the subscripts of the cross-coupling inverter become $-J_{k-1,k+1}$ ($k = 2$ for the trisection in Table 10.3). The value of this and the values of the trisection's central resonant pair $C_{Sk} + jB_{Sk}$ are entered into a coupling matrix along with the values of the other elements that have been extracted, and scaled along with these other elements such that the values of all the capacitors are unity. Then, rotations may be applied to reconfigure the network.

TABLE 10.3 Trisection Synthesis Procedure

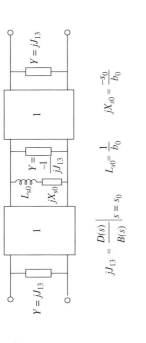

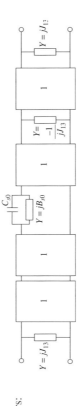

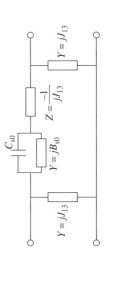

1. Extract:
   a. Parallel frequency-invariant
      admittance jJ_{13}, to prepare network
      for extraction of pole at $s = s_0$
   b. Unit inverter
   c. Parallel-connected series inductor
      L_{s0} ($= 1/b_0$) + frequency-invariant reactance
      jX_{s0} ($= -s_0/b_0$) (extracted pole; see Section 10.1.1)
   d. Parallel frequency-invariant admittance $-1/jJ_{13}$
   e. Unit inverter
   f. Parallel frequency-invariant admittance jJ_{13}

2. Dual-network theorem
   a. Convert parallel-connected $L_{s0} + jX_{s0}$
      to series-connected parallel pair between two unit invertors:
      $$B_{S0} = X_{S0}$$
      $$C_{S0} = L_{S0}$$

   b. Convert parallel-connected admittance $-1/jJ_{13}$
      between two unit inverters to series reactance $-1/jJ_{13}$

$$jJ_{13} = \left.\frac{D(s)}{B(s)}\right|_{s = s_0} \qquad L_{s0} = \frac{1}{b_0} \qquad jX_{s0} = \frac{-s_0}{b_0}$$

374

4. Circuit transform:

$$X_{S2} = \frac{J_{13} - B_{s0}}{-J_{13}^2} = \frac{J_{13} + \omega_0/b_0}{-J_{13}^2}$$

$$L_{S2} = \frac{-jB_{S0}}{s_0 J_{13}^2} = \frac{1}{b_0 J_{13}^2}$$

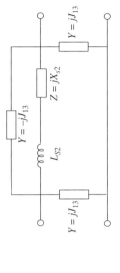

5. a. Inverter of characteristic admittance $-J_{13}$ formed by π network of J_{13} susceptances.

b. Dual-network theorem: Convert series-connected L_{S2} and jX_{S2} to parallel pair C_{S2} and jB_{S2} between two unit inverters:

$$C_{S2} = L_{S2}$$
$$B_{S2} = X_{S2}$$

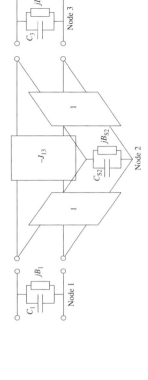

For the elements of the trisection at index k within an Nth-degree network, the values of the scaled entries to the coupling matrix are

$$M_{k-1,k} = \frac{1}{\sqrt{C_{k-1}C_{Sk}}}$$

$$M_{k,k+1} = \frac{1}{\sqrt{C_{Sk}C_{k+1}}}$$

$$M_{k-1,k+1} = \frac{-J_{k-1,k+1}}{C_{k-1}C_{k+1}} \qquad (10.14)$$

$$B_{Sk} \ (\equiv M_{k,k}) \to \frac{B_{Sk}}{C_{Sk}}$$

$$C_{Sk} \to 1$$

where C_{Sk} and B_{Sk} are the values of the central shunt elements of the trisection and C_{k-1} and C_{k+1} are the values of the shunt capacitors extracted immediately before and immediately after the trisection. If $k = 1$, $C_{k-1} = 1$ or if $k = N$, $C_{k+1} = 1$.

After the values of the trisection, scaled or unscaled, are obtained, the frequency of its transmission zero is checked by application of the following formulas:

Unscaled values: $\omega_0 = -\dfrac{1}{C_{Sk}} \left[\dfrac{1}{J_{k-1,k+1}} + B_{Sk} \right]$ \qquad (10.15a)

If the values are taken from the coupling matrix; then

$$\omega_0 = \frac{M_{k-1,k} \cdot M_{k,k+1}}{M_{k-1,k+1}} - M_{k,k} \qquad (10.15b)$$

which is the solution to the trisection determinant evaluated at $s = j\omega_0$:

$$\det \begin{vmatrix} M_{k-1,k} & M_{k-1,k+1} \\ \omega_0 + M_{k,k} & M_{k,k+1} \end{vmatrix} = 0 \qquad (10.15c)$$

In Figure 10.19, the $(4-2)$ asymmetric prototype described in Section 10.1.2 is synthesized in a form similar to that in Figure 10.18e.

The [$ABCD$] matrix polynomials that are synthesized for the $(4-2)$ characteristic are provided in Section 7.4. The synthesis of the first trisection at index $k = 1$ with TZ located at $s_0 = j1.8082$ proceeds as follows (see item 1 in Table 10.3):

1. Calculate the value of the parallel-connected FIR J_{S2} from

$$jJ_{S2} = \frac{D(s)}{B(s)}\bigg|_{s=s_0}, \qquad J_{S2} = -0.8722$$

2. Extract the FIR J_{S2} and the unit inverter from the $A(s)$, $B(s)$, $C(s)$, and $D(s)$ polynomials to obtain the remainder polynomials $A'(s)$, $B'(s)$, $C'(s)$, and $D'(s)$.

Now the procedure is similar to that used to extract the shunt-connected resonant pair (extracted pole), described in Section 10.1.

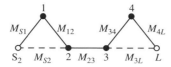

Figure 10.19 Trisection realization for the (4–2) asymmetric filter characteristic.

3. Divide $A'(s)$ and $B'(s)$ by the factor $(s - s_0)$ to obtain the polynomials A'' and B''.
4. Evaluate the value of the residue b_0 from

$$b_0 = \left.\frac{D'(s)}{B''(s)}\right|_{s=s_0}, \quad b_0 = 1.1177$$

Calculate B_{S1} and C_{S1} from

$$B_{S1} = -\frac{J_{S2} + \omega_0/b_0}{J_{S2}^2}, \quad C_{S1} = \frac{1}{b_0 J_{S2}^2}$$

and then from $B_{S1} = -0.9801$ and $C_{S1} = 1.1761$.

5. Compute $C'(s) - b_0 A''(s)$ and $D'(s) - b_0 B''(s)$.
6. Divide these two polynomials by the factor $(s - s_0)$ to obtain $C''(s)$ and $D''(s)$.
7. Divide the $P(s)$ polynomial by the factor $(s - s_0)$ to obtain $P''(s)$. At this stage the coefficients of the polynomials have the the values, which are listed next:

$s^i, i =$	$A''(s)$	$B''(s)$	$C''(s)$	$D''(s)$	$P''(s)/\varepsilon$
0	+0.9148	$-j1.1936$	$-j1.7080$	-0.7964	$+j1.1446$
1	$-j1.0108$	+2.7263	-0.1938	$-j3.5503$	-0.8660
2	+2.0945	$-j0.5716$	$-j2.4015$	+0.1484	—
3	—	+1.7443	—	$-j2.0000$	—
4	—	—	—	—	—

8. Extract a parallel-connected FIR of value $-1/J_{S2}$ from $[A''B''C''D'']$ matrix, followed by a unit inverter, and finally, a FIR of value J_{S2}.

This completes the extraction of the first trisection, leaving the coefficients of the [*ABCD*] polynomials with the following values:

$s^i, i =$	$A(s)$	$B(s)$	$C(s)$	$D(s)$	$P(s)/\varepsilon$
0	-0.6591	$-j0.5721$	$-j1.4896$	-0.6946	$+j1.1446$
1	$-j0.9652$	-0.4244	-0.1690	$-j3.0964$	-0.8660
2	—	$-j0.8038$	$-j2.0945$	+0.1294	—
3	—	—	—	$-j1.7443$	—
4	—	—	—	—	—

TABLE 10.4 Element Values of the (4–2) Asymmetric Filter with Two Trisections

Frequency-Variant Capacitors at Nodes	Frequency-Invariant Reactances at Nodes	Sequential Coupling Inverters	Nonsequential Coupling Inverters
$C_{S1} = 1.1761$	$B_{S1} = -0.9801$	$J_{S1} = 1.0$	$-J_{S2} = 0.8722$
$C_2 = 2.1701$	$B_2 = 1.3068$	$J_{12} = 1.0$	$-J_{3L} = 2.2739$
$C_3 = 7.2065$	$B_3 = 4.1886$	$J_{23} = 3.4143$	—
$C_{S4} = 2.0694$	$B_{S4} = -2.2954$	$J_{34} = 1.0$	—
—	—	$J_{4L} = 1.0$	—

It is noted that at the end of this cycle, the $D(s)$ polynomial has the highest degree, instead of $B(s)$, which is typical. This is the consequence of not extracting an inverter at the beginning of the procedure, but has no effect on the rest of the synthesis. Since the trisection is removed, the shunt-connected pair $C_2 + jB_2$ may be extracted, where $C_2 = 2.1701$ and $B_2 = 1.3068$.

The second trisection at index $k = 4$ with the central shunt-connected pair $C_{S4} + jB_{S4}$, realizing the transmission zero $s_0 = j1.3217$, is now synthesized by turning the network [exchanging $A(s)$ and $D(s)$] and applying the same eight steps. The corresponding parameter values are $J_{3L} = -2.2739$, $b_0 = 0.0935$, $C_{S4} = 2.0694$, and $B_{S4} = -2.2954$. The following shunt-connected pair $C_3 + jB_3$ is extracted, where $C_3 = 7.2064$ and $B_3 = 4.1886$; and finally the parallel-connected inverter J_{23} is extracted as 3.4143 (see Table 10.4).

Scaling all the capacitors at the nodes to unity gives the $N + 2$ coupling matrix, corresponding to that in Figure 10.20a. Analyzing this matrix provides exactly the same prototype performance characteristics as those of the original folded

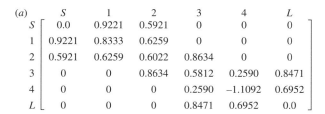

$$(a) \quad \begin{array}{c c c c c c c} & S & 1 & 2 & 3 & 4 & L \\ S & 0.0 & 0.9221 & 0.5921 & 0 & 0 & 0 \\ 1 & 0.9221 & 0.8333 & 0.6259 & 0 & 0 & 0 \\ 2 & 0.5921 & 0.6259 & 0.6022 & 0.8634 & 0 & 0 \\ 3 & 0 & 0 & 0.8634 & 0.5812 & 0.2590 & 0.8471 \\ 4 & 0 & 0 & 0 & 0.2590 & -1.1092 & 0.6952 \\ L & 0 & 0 & 0 & 0.8471 & 0.6952 & 0.0 \end{array}$$

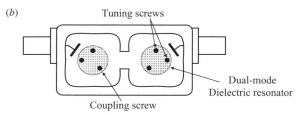

(b)

Tuning screws

Coupling screw

Dual-mode Dielectric resonator

Figure 10.20 $N + 2$ coupling matrix for the (4–2) asymmetric characteristic synthesized with trisections: (a) coupling matrix; (b) possible realization with two dual-mode dielectric resonator cavities.

configuration. Figure 10.20*b* shows a possible realization by employing two dual-mode dielectric resonator cavities. The angled input and output probes provide the dual input/output couplings M_{S1} and M_{S2}, and M_{4L} and M_{3L}, respectively.

10.3.2 Cascade Trisections—Coupling Matrix Approach

A method based purely on the coupling matrix for the generation of a cascade of trisections has been introduced by Tamiazzo and Macchiarella [9]. This method takes as its basis the *Bell's wheel* or *arrow* form for the canonical coupling matrix. This has the advantage that only the similarity transform operations are required, starting with any canonical form for the coupling matrix (e.g., full matrix, transverse matrix), which itself may be synthesized directly from the transfer and reflection polynomials of the filter function, discussed in Chapter 8. There is no need to first synthesize a circuit and then convert it to an equivalent coupling matrix. The pure coupling matrix approach allows a seamless sequence of rotations to be applied to the starting coupling matrix: to generate the *arrow* configuration for the coupling matrix, cascading of trisections, and then to reconfigure to another form if it is required (e.g., cascade quartets).

Synthesis of the Arrow Canonical Coupling Matrix   The folded cross-coupled circuit and its corresponding coupling matrix were introduced in Chapters 7 and 8 as one of the basic canonical forms of the coupling matrix, capable of realizing *N* transmission zeros in an *N*th-degree network. A second form was introduced by Bell [10] in 1982. The configuration later became known as the *wheel* or *arrow* form. As with the folded form, all the mainline couplings are present, and in addition, the source terminal and each resonator node is cross-coupled to the load terminal.

Figure 10.21*a* is a coupling–routing diagram for a fifth-degree fully canonical filter circuit, showing clearly why this configuration is referred to as the *wheel*

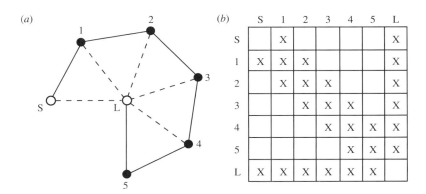

(a)	(b)	S	1	2	3	4	5	L
	S		X					X
	1	X	X	X				X
	2		X	X	X			X
	3			X	X	X		X
	4				X	X	X	X
	5					X	X	X
	L	X	X	X	X	X	X	

Figure 10.21 Fifth-degree wheel or arrow canonical circuit: (*a*) coupling–routing diagram (wheel); (*b*) $N + 2$ coupling matrix (arrow).

canonical form with the mainline couplings forming the (partially incomplete) "rim" and the cross-couplings and input/output coupling forming the "spokes." Figure 10.21b displays the corresponding coupling matrix, where the cross-coupling elements are all in the last row and column, and along with the mainline and self-couplings on the main diagonals, give the matrix the appearance of an arrow pointing downward toward the lower right corner of the matrix. The network reversed, where all the cross-couplings are now in the first row and column of the matrix, gives the upward-pointing arrow form. If there are no transmission zeros at all in the transfer function (all-pole function), the only entry in the last row and column of the arrow matrix is the output coupling M_{NL}. With one TZ, the entry $M_{N-1,L}$ appears in the last row/column, with two TZs, $M_{N-2,L}$ appears, and so on until finally there are N TZs (fully canonical) and the entry M_{SL} becomes nonzero (as in Figure 10.21b). Thus, as the number of TZs in the filter function increases from zero to N, the head of the arrow grows from the lower right corner of the matrix until the last row and column are completely filled.

Synthesis of the Arrow Coupling Matrix Form   The arrow matrix can be formed from the transversal or any other matrix form by a series of rotations very similar to those needed to form the folded coupling matrix (Section 8.3). The total number of rotations R in the sequence needed to generate an $N + 2$ arrow matrix is given by $R = \sum_{r=1}^{N-1} r$, which equals 10 for the fifth-degree example in Figure 10.21. Table 10.5 gives the sequence of pivots and angle formula for the fifth-degree case, reducing the initial $N + 2$ coupling matrix (transversal, folded, or other) to the downward-pointing arrow form. As with the folded matrix, the final values and positions of the elements in the last row and column are automatically determined. No specific action to the annihilate unneeded elements needs to be taken. The pattern of pivots and angle computations in the sequence is regular and is easily programmed.

TABLE 10.5   Fifth-Degree Example: Similarity Transform (Rotation) Sequence for Reduction of the $N + 2$ Coupling Matrix to the Arrow Form

Transform Number r	Pivot $[i,j]$	Element to Be Annihilated		$\theta_r = -\tan^{-1} (M_{k1}/M_{mn})$			
				k	l	m	n
1	[1,2]	M_{S2}	In row 1	S	2	S	1
2	[1,3]	M_{S3}	—	S	3	S	1
3	[1,4]	M_{S4}	—	S	4	S	1
4	[1,5]	M_{S5}	—	S	5	S	1
5	[2,3]	M_{13}	In row 2	1	3	1	2
6	[2,4]	M_{14}	—	1	4	1	2
7	[2,5]	M_{15}	—	1	5	1	2
8	[3,4]	M_{24}	In row 3	2	4	2	3
9	[3,5]	M_{25}	—	2	5	2	3
10	[4,5]	M_{35}	In row 4	3	5	3	4

***Creation and Repositioning of Trisections within the Arrow Coupling
Matrix***   The procedure followed to create a trisection, associated with the ith transmission zero $s_{0i} = j\omega_{0i}$ within the arrow-configured coupling matrix, involves applying a single rotation with pivot $[N - 1, N]$ near the tip of the arrow, at an angle
calculated to satisfy the following determinant for the first trisection associated
with the first TZ $s_{01} = j\omega_{01}$ embedded in the transformed matrix, [see equation
(10.15c)]:

$$\det \begin{vmatrix} M^{(1)}_{N-2,N-1} & M^{(1)}_{N-2,N} \\ \omega_{01} + M^{(1)}_{N-1,N-1} & M^{(1)}_{N-1,N} \end{vmatrix} = 0 \qquad (10.16)$$

where $M^{(1)}_{ij}$ are elements in the transformed matrix $\mathbf{M}^{(1)}$. By following through the
algebra associated with the application of the rotation at pivot $[N - 1, N]$ to the original arrow matrix $\mathbf{M}^{(0)}$ [see equation set (8.34)], the angle θ_{01}, required to satisfy
equation (10.16), is determined by

$$\theta_{01} = \tan^{-1}\left[\frac{M^{(0)}_{N-1,N}}{\omega_{01} + M^{(0)}_{N,N}} \right] \qquad (10.17)$$

where $M^{(0)}_{N-1,N}$ and $M^{(0)}_{N,N}$ are taken from the original arrow matrix $\mathbf{M}^{(0)}$.
    After application of this rotation at pivot $[N - 1, N]$ and angle θ_{01}, coupling
$M^{(1)}_{N-2,N}$ is created in the transformed matrix $\mathbf{M}^{(1)}$. A trisection is now embedded
in the arrow structure at index position $k = N - 1$. This trisection can be pulled
up the diagonal to index position $k = N-2$ with the application of a rotation at
pivot $[N - 2, N - 1]$ and angle θ_{12}, after which element $M^{(2)}_{N-2,N}$ is zero and
$M^{(2)}_{N-3,N-1}$ is created in the new matrix $\mathbf{M}^{(2)}$ [see equation (8.34a)]:

$$\theta_{12} = \tan^{-1} \frac{M^{(1)}_{N-2,N}}{M^{(1)}_{N-1,N}} \qquad (10.18)$$

The trisection determinant (10.16) for the new position of the trisection yields
$M^{(1)}_{N-2,N-1} \cdot M^{(1)}_{N-1,N} = M^{(1)}_{N-2,N} \cdot (\omega_{01} + M^{(1)}_{N-1,N-1})$, which is rearranged and
substituted into equation (10.18) such that

$$\theta_{12} = \tan^{-1}\left[\frac{M^{(1)}_{N-2,N}}{M^{(1)}_{N-1,N}} \right] = \tan^{-1}\left[\frac{M^{(1)}_{N-2,N-1}}{\omega_{01} + M^{(1)}_{N-1,N-1}} \right] \qquad (10.19)$$

Thus, from equations (10.17), (10.18), and (10.19) it is evident that application of the
rotation at pivot $[N - 2, N - 1]$ and angle θ_{12}, to pull the trisection up the diagonal
by one position, automatically conditions it in its new position to again give a value
of zero for its determinant. This applies for a shift of the trisection wherever it is on
the diagonal of the arrow matrix.

The trisection is then pulled up the diagonal by a further index position ($k = N - 3$) by another rotation at pivot $[N - 3, \ N - 2]$ and angle $\theta_{23} = \tan^{-1}\left(M^{(2)}_{N-3,N-1}/M^{(2)}_{N-2,N-1}\right)$. The result is that element $M^{(3)}_{N-3,N-1}$ is zero, and $M^{(3)}_{N-4,N-2}$ is created in the new matrix $\mathbf{M}^{(3)}$, and so on until the trisection is in its desired position on the diagonal. Once the trisection has been pulled clear of the arrow structure of the matrix, the outermost cross-coupling elements in the last row and column disappear; in other words, the coupling diagram wheel (Fig. 10.21a) "loses a spoke."

If the transfer function incorporates a second TZ at $s_{02} = j\omega_{02}$, this procedure is repeated to produce a second trisection. This time it is associated with s_{02}, and so on, until a cascade of trisections realizing $s_{01}, s_{02}, \ldots$ is formed. Then further rotations are applied to the cascade to form quartets and quintets, introduced in Section 10.3.3.

Illustrative Example To demonstrate the application of the procedure, let us consider an example of a 23 dB return loss (8–2–2) asymmetric filter function with two TZs at $-j1.2520$ and $-j1.1243$ on the lower side of the passband producing two rejection lobes of 40 dB each, and a complex zero pair at $\pm 0.8120 + j0.1969$ to equalize the group delay over approximately 50% of the passband. Two trisections realizing the two imaginary-axis zeros are synthesized and then merged to form a quartet.

Generating the $N+2$ coupling matrix for this transfer function and converting it to the arrow form using the reduction procedure outlined above (28 rotations) results in the matrix shown in Figure 10.22a. Note that because there are four transmission zeros in the transfer function, there are four cross-coupling elements in the last row and column of the matrix, in addition to the mainline coupling M_{8L}.

For the first imaginary axis zero $s_{01} = -j1.2520$, the first rotation angle is evaluated from equation (10.17) as $\theta_{01} = -j1.2747°$. Applying the first rotation at pivot $[7,8]$ to create the trisection (Fig. 10.22c) and then five further rotations to pull the trisection up the diagonal from index position 7 to index position 2 (Fig. 10.22e) completes the formation and positioning of the first trisection realizing the first TZ at s_{01}. The second cycle begins by evaluating θ_{67} with $s_{02} = -j1.1243$ in equation (10.17), taking the coupling values from the matrix resultant at the end of the first cycle. Applying the first rotation of the new cycle again at pivot $[7,8]$ and then pulling the newly created trisection up the diagonal with three rotations to index position 4 completes the creation and positioning of the second trisection. The whole process is summarized in Table 10.6.

After the two trisections are created and repositioned, the coupling matrix and its corresponding coupling and routing diagram take the form as depicted in Figure 10.23a,b. Note that because two TZs are extracted from the arrow structure and realized as trisections, the cross-couplings M_{4L} and M_{5L} at the extremes of the arrowhead in the last row and column now have zero values. This automatically occurs during the trisection shifting process. No specific action to annihilate them is needed.

The two trisections can now be a merged into a quartet by pulling the second trisection up the diagonal by one further position with a rotation at pivot $[3,4]$

(a)

	S	1	2	3	4	5	6	7	8	L
S	0	1.0516	0	0	0	0	0	0	0	0
1	1.0516	−0.0276	0.8784	0	0	0	0	0	0	0
2	0	0.8784	−0.0324	0.6147	0	0	0	0	0	0
3	0	0	0.6147	−0.0464	0.5813	0	0	0	0	0
4	0	0	0	0.5813	−0.1049	0.6073	0	0	0	−0.1789
5	0	0	0	0	0.6073	−0.2413	0.6171	0	0	0.2370
6	0	0	0	0	0	0.6171	−0.0477	0.5511	0	−0.3621
7	0	0	0	0	0	0	0.5511	0.3380	0.0027	0.8164
8	0	0	0	0	0	0	0	0.0027	1.1293	0.4691
L	0	0	0	0	−0.1789	0.2370	−0.3621	0.8164	0.4691	0

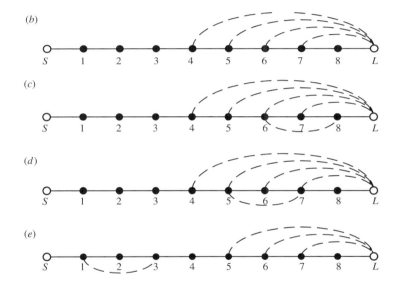

Figure 10.22 (8−2−2) filter synthesis example: (a) initial arrow coupling matrix; (b) corresponding coupling–routing diagram; (c) rotation 1 creates first trisection at index position 7; (d) rotation 2 pulls the trisection to index position 6; (e) rotation 6 positions the first trisection at index position 2.

annihilating M_{35}, and then one rotation at [2,3] annihilating either M_{13} or M_{24}. A second quartet (realizing the complex TZ pair) already exists as the remainder of the arrow formation (see Fig. 10.23a); this can be pulled up the diagonal by one position by applying two rotations, one at pivot [7,8] to annihilate M_{7L}, and then one at [6,8] to annihilate M_{6L}. The final form of the (8−2−2) cascade quartet coupling/ routing diagram is shown in Figure 10.23c.

Trisections Realizing Complex Transmission Zeros By applying the method described above, each trisection is created individually within the arrow

TABLE 10.6 (8–2–2) Filter Example: Rotation Sequence for the Creation and Positioning of Two Trisections

Rotation Number r	Pivot $[i,j]$	Element to Be Annihilated	$\theta_{r-1,r} = \tan^{-1}(M_{kl}^{(r-1)}/M_{mn}^{(r-1)})$ [Equation (8.37a)]				Comments[a]
			k	l	m	n	
1	[7,8]		$\theta_{01} = -1.2747°$				Create first trisection at IP7 with $\omega_{0i} = \omega_{01} = -1.2520$ [see equation (10.17)]; coupling M_{68} is created
2	[6,7]	M_{68}	6	8	7	8	Rotations 2–6: to shift trisection 1 from IP7 to IP2
3	[5,6]	M_{57}	5	7	6	7	
4	[4,5]	M_{46}, M_{4L}	4	6	5	6	
5	[3,4]	M_{35}	3	5	4	5	
6	[2,3]	M_{24}	2	4	3	4	
7	[7,8]		$\theta_{67} = 79.3478°$				Create second trisection at IP7 with $\omega_{0i} = \omega_{02} = -1.1243$ [see equation (10.17)]; coupling M_{68} is created again
8	[6,7]	M_{68}	6	8	7	8	Rotations 8–10: to shift trisection 2 from IP7 to IP4
9	[5,6]	M_{57}, M_{5L}	5	7	6	7	
10	[4,5]	M_{46}	4	6	5	6	

[a]IP = index position.

384

(a)

	S	1	2	3	4	5	6	7	8	L
S	0	1.0516	0	0	0	0	0	0	0	0
1	1.0516	−0.0276	0.6953	−0.5368	0	0	0	0	0	0
2	0	0.6953	0.6920	0.4324	0	0	0	0	0	0
3	0	−0.5368	0.4324	−0.1272	0.4045	−0.4041	0	0	0	0
4	0	0	0	0.4045	0.7072	0.4167	0	0	0	0.0000
5	0	0	0	−0.4041	0.4167	−0.0435	0.6252	0	0	0.0000
6	0	0	0	0	0	0.6252	0.0058	0.7449	0	0.4142
7	0	0	0	0	0	0	0.7449	−0.0436	0.5911	0.0858
8	0	0	0	0	0	0	0	0.5911	−0.1961	0.9628
L	0	0	0	0	0.0000	0.0000	0.4142	0.0858	0.9628	0

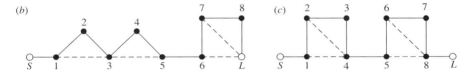

Figure 10.23 $(8-2-2)$ asymmetric filter function: (a) coupling matrix after synthesis of two trisections; (b) coupling–routing diagram; (c) coupling–routing diagram after formation of two quartets.

structure and then moved to a new position on the diagonal. Each trisection has one transmission zero uniquely associated with it. By forming the trisection and moving it from the arrow formation, effectively that single TZ is removed from the arrow formation too. When the TZ s_{0i} is pure imaginary, equation (10.17) gives a real value for the rotation angle θ_{ti}, and the rotations may proceed conventionally.

However, only imaginary-axis TZs can exist individually. For off-axis TZs (to give group delay equalization), these must exist in mirror-image pairs about the imaginary axis for a realizable network. Because each trisection is capable of realizing only one of the zeros from the complex or real-axis pair, equation (10.17) yields a complex value $\theta_{ti} = a + jb$ for the initial rotation angle to be applied to the matrix (rotation 1 in Table 10.6). In such cases, the complex values for $\sin\theta_{ti}$ and $\cos\theta_{ti}$, which are needed for the rotation (see Section 8.3.1), are calculated [11] as follows:

$$\theta_{ti} = \tan^{-1}\left[\frac{M_{N-1,N}}{\omega_{0i} + M_{N,N}}\right] = \tan^{-1}(x + jy) = a + jb \qquad (10.20)$$

where

$$a = \frac{1}{2}\tan^{-1}\left[\frac{2x}{1 - x^2 - y^2}\right], \quad b = \frac{1}{4}\log_e\left[\frac{x^2 + (y+1)^2}{x^2 + (y-1)^2}\right]$$

Now knowing a and b, we can calculate $\sin\theta_{ti}$ and $\cos\theta_{ti}$ from

$$\sin\theta_{ti} = \sin a \cosh b + j\cos a \sinh b$$
$$\cos\theta_{ti} = \cos a \cosh b - j\sin a \sinh b \tag{10.21}$$

The application of this first rotation by a complex angle gives complex values to some of the coupling elements of the matrix, and when the trisection is to be pulled up the diagonal, the rotation angles required to annihilate the necessary elements (e.g., rotations 2–6 in Table 10.6) also acquire complex values. For the rth rotation

$$\theta_r = \tan^{-1}\frac{M_{ij}}{M_{kl}} = \tan^{-1}(x_r + jy_r) = a_r + jb_r \tag{10.22}$$

and $\sin\theta_r$ and $\cos\theta_r$ are calculated in the same way as the initial rotation by using (10.20) and (10.21).

Consequently, after the first series of rotations has been applied to create and reposition the first complex TZ s_{01}, the coupling matrix has complex values, which of course are unrealizable. However, if the next trisection is created for the second TZ of the mirror-image pair ($s_{02} = -s_{01}^*$), and is pulled up the diagonal and merged with the first trisection, (as in the eighth-degree example) to form a quartet (Fig. 10.23c), then all the imaginary parts of the complex coupling values cancel out, leaving pure real coupling values. This procedure is necessary if cascade quartets are synthesized for more than one complex pair, as in Figure 10.24.

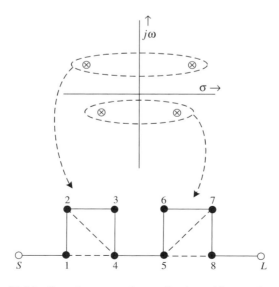

Figure 10.24   Complex zero pairs: realization with cascade quartets.

10.3.3 Techniques Based on the Trisection for Synthesis of Advanced Circuits

In this section, the synthesis of two useful network configurations is explained: the cascaded *n*-tuplet (such as quartets, quintets, sextuplets) and the *box* section, including the method for its development to *extended box* sections. Both are valuable for microwave filter designers, offering advantages in design and performance, and in the manufacturing and tuning processes. Typically, the procedure for the synthesis of these networks starts with the generation of a cascade of trisections, which is achieved by the circuit methods described earlier in this section, or by using the arrow matrix method.

Synthesis Procedure for Cascade Quartets, Quintets, and Sextets   The method for synthesis of cascaded quartets is summarized first, followed by the extension of the method to cascaded quintets and sextuplets. It is possible to extend the method beyond sextuplets, but no practical application is foreseen for such configurations.

Cascaded Quartets   The procedure begins with the synthesis of a pair of conjoined trisections at s_{01} and s_{02}, as shown for the asymmetric $(8-2-2)$ filter (two imaginary-axis TZs, one complex pair) example in Figure 10.23a. The rest of the network may then be synthesized conventionally as a folded network, if it is appropriate [automatically forming the second quartet of the $(8-2-2)$ example]. The next step is to apply two rotations, the first at pivot [3,4] to annihilate M_{35}. This rotation creates two new couplings M_{14} and M_{24}. Now it can be seen that a quartet with both diagonal couplings M_{13} (which was originally present in the first trisection) and M_{24} has been formed. Either of these couplings can be eliminated with a second rotation pivot [2,3] to annihilate either M_{13} or M_{24}.

Formal Procedure for Cascade Quartets, Quintets, and Sextets   When we follow the procedure for the synthesis of cascade quartets, it is noticeable that the sequence of rotations for the annihilation of the couplings to finally obtain the quartet has two stages:

1. Rotations are carried out to annihilate any trisection cross-coupling that is outside the 4×4 submatrix in the main matrix that the quartet eventually occupies [M_{35} in the $(8-2-2)$ example]. Additional couplings are created during this process, but the new couplings are within the 4×4 submatrix. This stage in the process is known as "gathering in."

2. A sequence of rotations is applied within the 4×4 matrix to reduce the couplings (in general, they are nonzero) to the folded form. This sequence of rotations is the same as that used to reduce the $N \times N$ or $N + 2$ matrix in their raw forms to the folded form. The process is summarized in Table 10.7, and illustrated graphically in Figure 10.25. The $(8-2-2)$ example is used for this illustration, where the first trisection is synthesized at index $k = 2$ and

TABLE 10.7 (8–2–2) Example of the Rotation Procedure to Transform the Two Trisections to the Cascaded Quartet Form

Rotation Number r	Pivot $[i, j]$	Element to Be Annihilated	$\theta_r = \tan^{-1}(cM_{kl}/M_{mn})$				
			k	l	m	n	c
		"Gather in" to 4×4 Submatrix					
1	[3,4]	M_{35}	3	5	4	5	+1
		Reduction to Folded Form within 4×4 Submatrix					
2	[2,3]	M_{13} In row 1	1	3	1	2	−1
or 2	[2,3]	M_{24} In row 1	2	4	3	4	+1

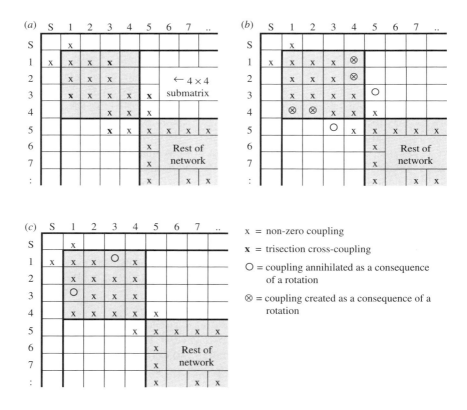

Figure 10.25 Formation of cascade quartet: (*a*) realization of TZs with two trisections (trisection cross-couplings shown in bold); (*b*) annihilation of M_{35}, creating M_{14} and M_{24} (Fig. 10.26*b*); (*c*) rotation at [2,3] to annihilate M_{13}, and thus create first quartet in Figure 10.26*c*.

the second at $k = 4$, creating M_{13} and M_{35}. However, the process can be applied to the trisection group, wherever it occurs in the main matrix, indexing the subscripts of the couplings and the rotation coordinates appropriately.

Quartets are easily shifted by one index position within the network by applying a couple of rotations to the coupling matrix. For example, to move the first quartet in Figure 10.26*b* down the diagonal by one position (to become conjoined with the second quartet at node 5), two rotations are applied, the first at pivot [2,4] to annihilate M_{14} (creating M_{25}), and the second at pivot [3,4] to annihilate M_{24} (if required).

Cascaded Quintets   A quintet (see Fig. 10.27 and Table 10.8) is capable of realizing three transmission zeros and is created by first generating a cascade of three conjoined trisections from the three TZs (Fig. 10.28*a*).

Cascaded Sextets   A sextet (see Fig. 10.29 and Table 10.9) is generated from a cascade of four conjoined trisections (Fig. 10.30a), and is capable of realizing four TZs.

Multiple Cascaded Sections   Although this formal procedure is able to form cascaded *n*-tet sections relatively easily, it does actually need more "room" along the diagonal of the matrix as the number of sections or their order increases. This is because after the formation of the first trisection, which realizes the first transmission zero, each new conjoined trisection that is synthesized to realize one additional TZ takes up two additional circuit nodes. This may be seen in Figure 10.30*a*, for example, where four trisections take up the source termination plus eight resonator nodes for the production of four TZs, where the eventual sextet realizing those four zeros only occupies the first six nodes not including the

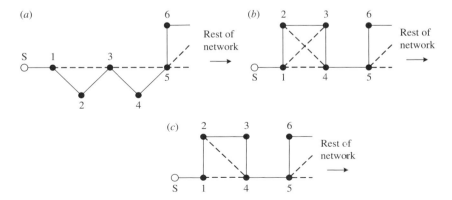

Figure 10.26   Transformation of two conjoined trisections to form a quartet section.

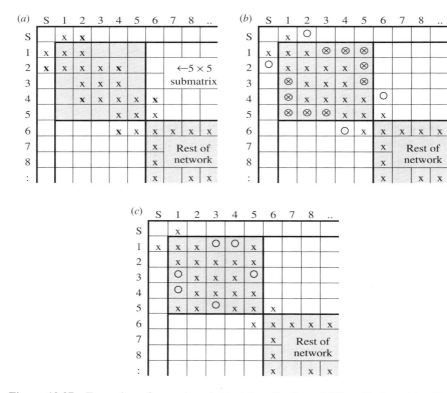

Figure 10.27   Formation of cascade quintet: (*a*) realization of TZs with three trisections (trisection cross-couplings shown in bold); (*b*) annihilation of M_{S2} and M_{46}; (*c*) after series of three rotations to create the folded network in 5×5 submatrix.

TABLE 10.8   Rotation Procedure to Transform Three Conjoined Trisections to the Cascaded Quintet Form

Rotation Number r	Pivot $[i, j]$	Element to Be Annihilated	$\theta_r = \tan^{-1}(cM_{kl}/M_{mn})$				
			k	l	m	n	c
		"Gather in" to 5×5 Submatrix					
1	[1,2]	M_{S2}	S	2	S	1	-1
2	[4,5]	M_{46}	4	6	5	6	$+1$
		Reduction to Folded Form within 5×5 Submatrix					
3	[3,4]	M_{14} In row 1	1	4	1	3	-1
4	[2,3]	M_{13} —	1	3	1	2	-1
5	[3,4]	M_{35} In column 5	3	5	4	5	$+1$

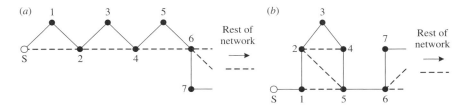

Figure 10.28 Transformation of three conjoined trisections to form a quintet section.

source node (Fig. 10.30*b*). This of course, leaves fewer nodes available in the coupling matrix for the synthesis of further trisections and *n*-tet sections.

The problem may be overcome if the direct-coupling matrix method to create the basic cascaded trisections is used (Section 10.3.2). Because we are working purely with rotations on coupling matrices, the *n*-tet sections may be individually created as the appropriate number of trisections are pulled out of the arrow formation. For example, two trisections may be first extracted from the arrow formation, and

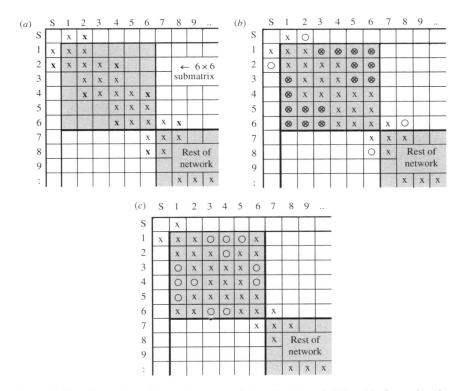

Figure 10.29 Formation of cascade sextet: (*a*) realization of TZs with four trisections (trisection cross-couplings are shown in bold); (*b*) annihilation of M_{S2} and M_{68}; (*c*) after series of six rotations to create a folded network within the 6 × 6 submatrix.

TABLE 10.9 Rotation Procedure to Transform Four Trisections to Cascaded Sextet Form

Rotation Number r	Pivot $[i, j]$	Element to Be Annihilated		$\theta_r = \tan^{-1}(cM_{kl}/M_{mn})$				
				k	l	m	n	c
		"Gather in" to 6×6 submatrix						
1	[1,2]	M_{S2}		S	2	S	1	-1
2	[6,7]	M_{68}		6	8	7	8	$+1$
3	[5,6]	M_{57}		5	7	6	7	$+1$
4	[4,6]	M_{47}		4	7	6	7	$+1$
		Reduction to Folded Form within 6×6 Submatrix						
5	[4,5]	M_{15}	In row 1	1	5	1	4	-1
6	[3,4]	M_{14}	—	1	4	1	3	-1
7	[2,3]	M_{13}	—	1	3	1	2	-1
8	[3,4]	M_{36}	In column 6	3	6	4	6	$+1$
9	[4,5]	M_{46}	—	4	6	5	6	$+1$
10	[3,4]	M_{24}	In row 2	2	4	2	3	-1

then merged using the methods above to create a quartet section in its final position in the coupling matrix. Then further trisections may be extracted and merged, to create a second n-tet section in cascade with the first. By forming the n-tet sections as the appropriate number of basic trisections are pulled out of the arrow formation, less "room" is needed along the diagonal of the coupling matrix than if all the trisections are produced first in a cascade, and then the n-tet sections formed as a second step in the overall process.

Alternative circuit based methods for the formation of trisections quartet and quintet sections may be found in references [12–15].

10.4 BOX SECTION AND EXTENDED BOX CONFIGURATIONS

For certain applications, specifications for channelizing filters require asymmetric rejection characteristics. This is particularly true for the front-end transmit/receive diplexers in the base stations of mobile communications systems, and requires the use of filter characteristics optimally tailored to the rejection

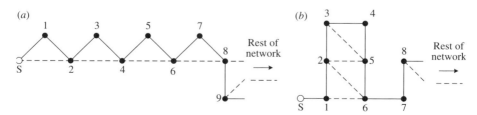

Figure 10.30 Transformation of four conjoined trisections to form a sextet section.

requirements; at the same time the maximum in-band amplitude and group delay linearity and lowest insertion loss must be maintained. In the course of synthesizing the networks for such characteristics, and reconfiguring the interresonator mainline and cross-couplings, it is frequently found that diagonal couplings are required. In this section, a synthesis method is presented for the realization of symmetric or asymmetric filtering characteristics without the need for diagonal cross-couplings: the *box* configuration and a derivation known as the *extended box* configuration [16].

10.4.1 Box Sections

The box section is similar to the cascade quartet section, with four resonator nodes arranged in a square formation, but with the input and the output of the quartet at opposite corners of the square. Figure 10.31*a* illustrates the conventional quartet arrangement for a fourth-degree filter with a single transmission zero, realized with a trisection. Figure 10.31*b* displays the equivalent box section realizing the same transmission zero, but without the need for the diagonal coupling. The application of the minimum path rule indicates that the box section can realize only a single transmission zero.

The box section is created by the application of a cross-pivot similarity transform [equation (8.37g)] to a trisection that has been synthesized in the coupling matrix for the filter. A cross-pivot similarity transform is the case where the coordinates of the coupling matrix element to be annihilated are the same as the pivot of the transform itself, that is, the element to be annihilated lies on the cross-points of the pivot [see equation (9.10)]. Arbitrary numbers of 90° are added to the cross-pivot rotation angle to create alternative solutions given by

$$\theta_r = \frac{1}{2}\tan^{-1}\left(\frac{2M_{ij}}{(M_{jj} - M_{ii})}\right) \pm k\frac{\pi}{2} \qquad (10.23)$$

For the box section, the pivot is set to annihilate the second mainline coupling of the trisection in the coupling matrix; thus, pivot = [2,3] annihilates element M_{23} (and M_{32}) in the fourth-degree example of Figure 10.31*a*. The equivalent coupling and routing schematic is given in Figure 10.32*a*. In the process of annihilating the

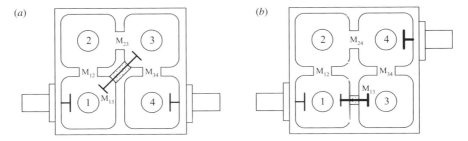

Figure 10.31 (4−1) asymmetric filter function: (*a*) realized with the conventional diagonal cross-coupling (M_{13}); and (*b*) realized with the box configuration.

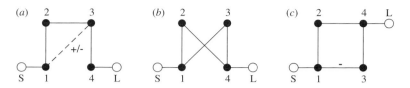

Figure 10.32 Formation of the box section for the $(4-1)$ filter: (*a*) the trisection, (*b*) annihilation of M_{23} and creation of M_{24}; (*c*) untwisting to obtain box section.

mainline coupling M_{23}, the coupling M_{24} is created, as depicted in (Figure 10.32*b*). Then, by untwisting this section, the box section in (Figure 10.32*c*) is formed. In the resultant box section, one of the couplings is always negative, irrespective of the sign of the cross-coupling (M_{13}) in the original trisection.

To illustrate the procedure, we examine a fourth-degree 25-dB return loss Chebyshev filter with a single transmission zero at $s = +j2.3940$, providing a lobe level of 41 dB on the upper side of the stopband.

(*a*)

	S	1	2	3	4	L
S	0	1.1506	0	0	0	0
1	1.1506	0.0530	0.9777	0.3530	0	0
2	0	0.9777	−0.4198	0.7128	0	0
3	0	0.3530	0.7128	0.0949	1.0394	0
4	0	0	0	1.0394	0.0530	1.1506
L	0	0	0	0	1.1506	0

(*b*)

	S	1	2	3	4	L
S	0	1.1506	0	0	0	0
1	1.1506	0.0530	0.5973	−0.8507	0	0
2	0	0.5973	−0.9203	0	0.5973	0
3	0	−0.8507	0	0.5954	0.8507	0
4	0	0	0.5973	0.8507	0.0530	1.1506
L	0	0	0	0	1.1506	0

(*c*)

	S	1	2	3	4	L
S	0	1.1506	0	0	0	0
1	1.1506	−0.0530	0.5973	−0.8507	0	0
2	0	0.5973	0.9203	0	0.5973	0
3	0	−0.8507	0	−0.5954	0.8507	0
4	0	0	0.5973	0.8507	−0.0530	1.1506
L	0	0	0	0	1.1506	0

Figure 10.33 $(4-1)$ filter coupling matrices: (*a*) the trisection; (*b*) after transformation to box section (TZ on upper side of passband); (*c*) TZ on lower side of passband.

Figure 10.33a gives the $N+2$ coupling matrix for the (4−1) filter showing the M_{13} trisection cross-coupling, corresponding to the coupling diagram of Figure 10.32a. Figure 10.33b shows the coupling matrix after transformation to the box configuration. Figure 10.34a plots the measured results of a coaxial resonator (4−1) filter configured as in Figure 10.31b. Good correlation is obtained between the measured and simulated results.

If the transmission zero is placed at $-j2.3940$ below the passband instead of above, then the transformation to the box section results in the same values for the interresonator couplings, but complementary values for the self-couplings ($M_{11}, M_{22}, \ldots$ along the principal diagonal of the coupling matrix, in Fig. 10.33c). Since the self-couplings represent offsets from the center frequency and are adjustable by tuning screws, the same filter structure can be used for the complementary filters of a transmit/receive diplexer, for example. Figure 10.34b shows the

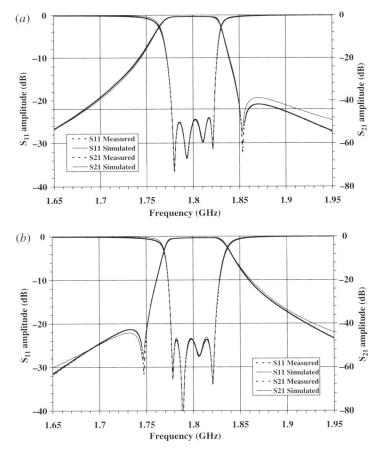

Figure 10.34 Measured results of the (4−1) filter: (a) transmission zero on upper side; (b) transmission zero on lower side.

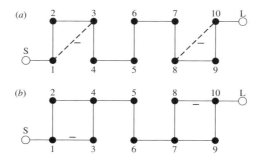

Figure 10.35 Coupling and routing diagrams of the (10–2) asymmetric filter: (*a*) synthesized with two trisections; (*b*) after transformation of trisections to two box sections by application of two cross-pivot rotations at pivots [2,3] and [8,9]. This form is suitable for realization in a dual-mode technology.

measured results of the same structure that gave the single TZ on the upper side of the passband (Fig. 10.34*a*), now retuned to place the TZ below the passband.

In addition, box sections can be cascaded within higher-degree filters, indexing the coordinates of the cross pivot rotation appropriately to correspond correctly with the position of each trisection. Figure 10.35 represents the coupling and routing diagrams for a tenth order filter with two transmission zeros on the lower side of the passband. Figure 10.36 shows the measured and simulated return loss and rejection characteristics of a test filter.

It is evident in Figure 10.35*b* that we can realize this asymmetric characteristic in dual-mode cavities (there are no diagonal cross-couplings).

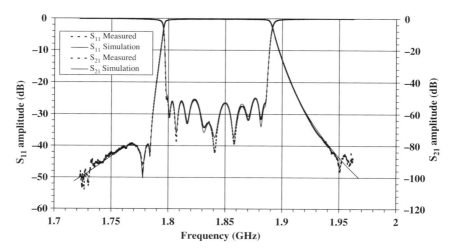

Figure 10.36 Simulated and measured return loss and rejection of the (10–2) asymmetric filter.

10.4.2 Extended Box Sections

A series of box sections may be formed in higher degree networks by cascading tri-sections and applying the single cross-pivot rotation to eliminate the mainline coupling within each one, and then untwisting. However the closest that box sections may be created by this method is where two sections share a common resonator, for example, the 8th-degree network of Figure 10.37. This leads to a rather awkward and restricted layout and heavily loads the common resonator (resonator 4 of Fig. 10.37) with four couplings. Moreover, only two transmission zeros may be realized by this network.

A rather more convenient arrangement in terms of both physical layout and the number of transmission zeros that may be realized is shown in Figure 10.38. Here the basic 4th-degree box section is shown and then the addition of pairs of resonators to form 6th-, 8th-, and 10th-degree networks. Application of the minimum path rule indicates that a maximum of $1, 2, 3, 4, \ldots, (N-2)/2$ transmission zeros may be realized by the 4th-, 6th-, 8th-, 10th-, $\ldots$, Nth-degree networks, respectively. The resonators are arranged in two parallel rows with half the total number of resonators in each row; input is at the corner at one end, and output is from the diagonally opposite corner at the other end. Even though asymmetric characteristics may be prescribed, there are no diagonal cross-couplings.

There appears to be no regular pattern for determining the sequence of rotations to synthesize the coupling matrix for the extended box sections from the folded network or any other canonical network. The sequences in Table 10.10 are derived by first determining the sequence of rotations to reduce the extended box section network to the corresponding folded network coupling matrix, as described in Section 8.3. This sequence of rotations is then reversed to transform the folded coupling matrix (which is readily derived from the S_{21} and S_{11} polynomials by using the methods of Chapters 7 and 8), back to the box section matrix.

Reversing the sequence means that some of the rotation angles θ_r are unknown a priori, and can be determined only by relating the coupling elements of the final matrix with those of the initial folded matrix and then solving to find θ_r (similar to the procedure for synthesizing cascade quadruplets in an 8th-degree network [17]). Deriving the equations to solve for θ_r analytically involves a formidable amount of algebra, and the task is most easily performed by setting up a solver procedure.

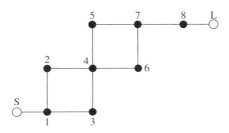

Figure 10.37 8th-degree network — two box sections conjoined at one corner.

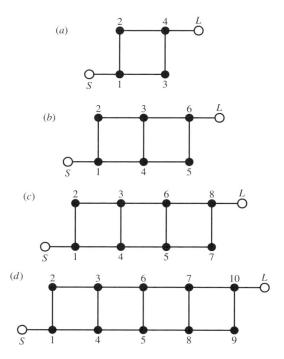

Figure 10.38 Coupling and routing diagrams for extended box section networks (*a*) 4th degree (basic box section); (*b*) 6th degree; (*c*) 8th degree; (*d*) 10th degree

The procedure starts by setting the initial values for the unknown rotation angles (one for the 6th degree, two for the 8th degree, and five for the 10th degree), and then running through the sequence of rotations as shown in Table 10.10 to obtain the transformed coupling matrix. With these initial angles, some of the unwanted coupling matrix elements will have nonzero values. An error function can be evaluated by forming, for example, the root sum square of the values of all the elements in the coupling matrix that should be zero. The initial angles are then adjusted by the solver routine until the cost function is zero. By varying the value of the integer k in the cross-pivot angle formula [equation (10.23)] when it is used in the rotation sequences, several unique solutions can be found in most cases. This allows the choice of the coupling matrix that gives the most convenient coupling values for the technology to be used for the realization.

Let us consider an 8th-degree filter with 23 dB return loss and three prescribed transmission zeros at $s = -j1.3553$, $+j1.1093$, and $+j1.2180$, which produce one rejection lobe level of 40 dB on the lower side of the passband and two at 40 dB on the upper side. The folded $N + 2$ coupling matrix is shown in Figure 10.39a. By finding the two unknown angles θ_1 and θ_2 of one of the possible solutions ($+63.881°$ and $+35.865°$, respectively), and applying the sequence of seven rotations (Table 10.10), the coupling matrix for the extended box section is obtained

TABLE 10.10 Pivot Coordinates for Reduction of Folded Matrix to Extended Box Section Configuration

Degree N	Rotation Number r	Rotation Privot $[i,j]$	Angle θ_r	Annihilating Element(s)
6	1	[3,4]	θ_1	—
	2	[2,4]		M_{24}
	3	[3,5]		M_{25}, M_{35}
8	1	[4,5]	θ_1	—
	2	[5,6]	θ_2	—
	3	[3,5]		M_{37}
	4	[3,4]		M_{46}
	5	[3,5]		M_{35}
	6	[2,4]		M_{24}, M_{25}
	7	[6,7]		M_{67}, M_{37}
10	1	[5,6]	θ_1	—
	2	[5,7]	θ_2	—
	3	[6,7]	θ_3	—
	4	[4,7]	θ_4	—
	5	[4,6]	θ_5	—
	6	[4,5]		M_{47}
	7	[6,8]		M_{48}
	8	[3,5]		M_{37}, M_{38}
	9	[7,8]		M_{57}, M_{68}
	10	[3,4]		M_{46}, M_{35}
	11	[7,9]		M_{79}, M_{69}

(Fig. 10.39*b*). It corresponds to the coupling and routing diagram in Figure 10.38*c*. The analyzed transfer and reflection characteristics for this coupling matrix are shown in Figure 10.40, demonstrating that the performance has been preserved intact.

Another optimization technique, due to Seyfert [18], is a synthesis method based on computer algebra, which captures a much larger number of solutions. The process actually finds all solutions, but some are unrealizable because of complex coupling values, and the number of real solutions depends on the pattern of transmission zeros in the prototype. Table 10.11 summarizes the approximate number of real solutions that are be expected from prototypes of degrees 6–12 with the maximum number of TZs allowed by the minimum path rule.

It can be seen from Figure 10.38 that because there are no diagonal cross-couplings, extended box sections are also suitable for realization with dual-mode resonator cavities. A practical advantage arises from all the straight couplings being present, in that any spurious stray couplings become less significant. Also 7th-, 9th-, 11th-, and 13th- odd-degree filters can be synthesized, by using the solutions for the 6th-, 8th-, 10th-, and 12th-degree filters, respectively.

By working on the subsections of the main coupling matrix, hybrid forms may also be created. An example is an 11th-degree filter with 3 TZs, initially synthesized

(a)

	S	1	2	3	4	5	6	7	8	L
S	0	1.0428	0	0	0	0	0	0	0	0
1	1.0428	0.0107	0.8623	0	0	0	0	0	0	0
2	0	0.8623	0.0115	0.5994	0	0	0	0	0	0
3	0	0	0.5994	0.0133	0.5356	0	−0.0457	−0.1316	0	0
4	0	0	0	0.5356	0.0898	0.3361	0.5673	0	0	0
5	0	0	0	0	0.3361	−0.8513	0.3191	0	0	0
6	0	0	0	−0.0457	0.5673	0.3191	−0.0073	0.5848	0	0
7	0	0	0	−0.1316	0	0	0.5848	0.0115	0.8623	0
8	0	0	0	0	0	0	0	0.8623	0.0107	1.0428
L	0	0	0	0	0	0	0	0	1.0428	0

(b)

	S	1	2	3	4	5	6	7	8	L
S	0	1.0428	0	0	0	0	0	0	0	0
1	1.0428	0.0107	0.2187	0	−0.8341	0	0	0	0	0
2	0	0.2187	−1.0053	0.0428	0	0	0	0	0	0
3	0	0	0.0428	−0.7873	0.2541	0	−0.2686	0	0	0
4	0	−0.8341	0	0.2541	0.0814	0.4991	0	0	0	0
5	0	0	0	0	0.4991	0.2955	0.4162	0.1937	0	0
6	0	0	0	−0.2686	0	0.4162	−0.2360	0	−0.7644	0
7	0	0	0	0	0	0.1937	0	0.9192	0.3991	0
8	0	0	0	0	0	0	−0.7644	0.3991	0.0107	1.0428
L	0	0	0	0	0	0	0	0	1.0428	0

Figure 10.39 Extended box section configuration of an 8th-degree example: (*a*) original folded coupling matrix; (*b*) after transformation to extended box section configuration.

with three trisections (Fig. 10.41*a*). The first two trisections are then transformed into an asymmetric cascade quad section with two similarity transforms at pivots [4,5] and [3,4] to eliminate couplings M_{46} and M_{24}, respectively (Section 8.3.3). Then the iterative procedure described above for the 6th-degree case is applied to

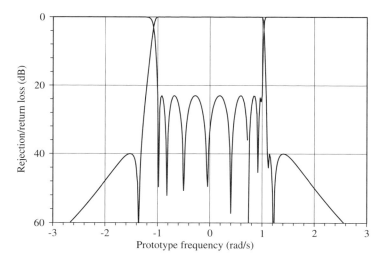

Figure 10.40 (8−3) asymmetric filter in an extended box section configuration — simulated rejection and return loss performance.

TABLE 10.11   Number of Real Solutions for the Extended Box Configuration

Extended Box Degree	Maximum Number of Finite-Position Transmission Zeros n_{fz}	Number of Real Solutions (Approximate)
6	2	6
8	3	16
10	4	58
12	5	>2000

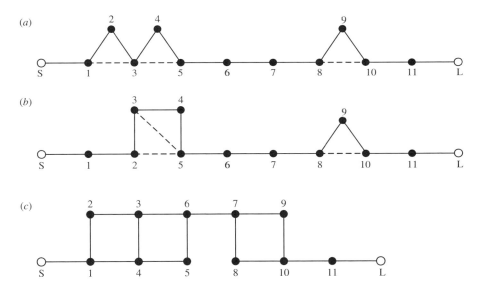

Figure 10.41   Stages in the synthesis of an $(11-3)$ network with cascaded box sections: (a) initial synthesis with trisections; (b) formation of cascaded quad section; (c) formation of the 6th-degree extended box section in cascade with a basic box section.

the upper left 6×6 submatrix of the main matrix, to form the extended box section at the left-hand side of the network (Fig. 10.41c). Finally, the basic box section at the right is formed with a cross-pivot transform at pivot [8,9] to eliminate coupling M_{89}. Alternatively, applying the cross-pivot transform at pivot [9,10] forms the basic box section at the extreme right of the network.

SUMMARY

Synthesis techniques described in Chapters 7, 8, and 9 are based on lossless lumped-element inductors, capacitors, frequency-invariant reactance (FIR), and frequency-invariant K and J inverters. This leads to the folded and transversal array coupling matrices from which a wide variety of filter topologies are derived.

In this chapter, two advanced circuit sections are introduced, the extracted pole section and the trisection. These sections are capable of realizing one transmission zero each. Also, they can be cascaded with other circuit elements in the filter network. Application of these sections extends the range of filter topologies for realizing microwave filters.

For high-power applications, it is preferable to avoid negative cross-couplings. Such couplings are often realized as capacitive probes, susceptible to overheating under high-power conditions. Negative couplings are effectively removed by using the extracted pole principle, presented in this chapter. It allows removal of transmission zeros (realized as bandstop cavities) before the synthesis of the remainder network. Since the transmission zeros are extracted, the remaining network does not require cross-couplings, either straight or diagonal, unless there are further transmission zeros that have not been extracted. The synthesis procedure based on the extracted pole section is described. Many examples are included to illustrate the procedure, and the role of the extracted pole section in the formation of a trisection is explained in depth.

Trisections have some properties that are advantageous for the synthesis of more advanced cross-coupled network topologies. A particular transmission zero can be associated with the cross-coupling in a particular trisection, and a trisection can be represented in a coupling matrix. The freedom to be able to synthesize trisections in more or less unrestricted positions in the filter network, and apply rotations to the coupling matrix to obtain a final topology, opens up some configuration possibilities that cannot be easily obtained by using the folded or transversal matrix as the starting point. The trisection is realized directly or as part of a cascade with other sections. It allows the synthesis of more complex networks, where a cascade of trisections is synthesized first and then further reconfiguration is carried out by using coupling matrix transforms. The power of this technique is demonstrated by applying it to transform a cascade of trisections to form cascaded quartet, quintet, and sextet filter topologies. Finally, the synthesis of the box section and its derivative, the extended box configuration, is explained. Examples are included to illustrate the intricacies of this synthesis procedure.

REFERENCES

1. J. D. Rhodes and R. J. Cameron, General extracted pole synthesis technique with applications to low-loss TE_{011} mode filters, *IEEE Trans. Microwave Theory Tech.* **MTT-28**, 1018–1028 (Sept. 1980).

2. G. Matthaei, L. Young, and E. M. T. Jones, *Microwave Filters, Impedance Matching Networks and Coupling Structures*, Artech House, Norwood, MA, 1980.

3. S. Amari, U. Rosenberg, and J. Bornemann, Singlets, cascaded singlets, and the nonresonating node model for advanced modular design of elliptic filters, *IEEE Microwave Wireless Compon. Lett.* **14**, 237–239 (May 2004).

4. J. D. Rhodes, Waveguide bandstop elliptic filters, *IEEE Trans. Microwave Theory Tech.* **MTT-20**, 715–718 (Nov. 1972).

5. R. J. Cameron, General prototype network synthesis methods for microwave filters, *ESA J.* **6**, 193–206 (1982).

6. S. Amari and U. Rosenberg, Direct synthesis of a new class of bandstop filters, *IEEE Trans. Microwave Theory Tech.* **MTT-52**, 607–616 (Feb. 2004).

7. R. Levy, Filters with single transmission zeros at real or imaginary frequencies, *IEEE Trans. Microwave Theory Tech.* **MTT-24**, 172–181 (April 1976).

8. R. Levy and P. Petre, Design of CT and CQ filters using approximation and optimization, *IEEE Trans. Microwave Theory Tech.* **MTT-49**, 2350–2356 (Dec. 2001).

9. S. Tamiazzo and G. Macchiarella, An analytical technique for the synthesis of cascaded *N*-tuplets cross-coupled resonators microwave filters using matrix rotations, *IEEE Trans. Microwave Theory Tech.* **MTT-53**, 1693–1698 (May 2005).

10. H. C. Bell, Canonical asymmetric coupled-resonator filters, *IEEE Trans. Microwave Theory Tech.* **MTT-30**, 1335–1340 (Sept. 1982).

11. I. A. Stegun and M. Abramowitz, eds., *Handbook of Mathematical Functions*, Dover Publications, New York, Nov. 1970.

12. R. Levy, Direct synthesis of cascaded quadruplet (CQ) filters, *IEEE Trans. Microwave Theory Tech.* **MTT-43**, 2939–2944 (Dec. 1995).

13. N. Yildirim, O. A. Sen, Y. Sen, M. Karaaslan, and D. Pelz, A revision of cascade synthesis theory covering cross-coupled filters, *IEEE Trans. Microwave Theory Tech.* **MTT-50**, 1536–1543 (June 2002).

14. T. Reeves, N. Van Stigt, and C. Rossiter, A method for the direct synthesis of general sections, *IEEE MTT-S Int. Microwave Symp. Digest*, Phoenix, 2001, pp. 1471–1474.

15. T. Reeves and N. Van Stigt, A method for the direct synthesis of cascaded quintuplets, *IEEE MTT-S Int. Microwave Symp. Digest*, Seattle, 2002, pp. 1441–1444.

16. R. J. Cameron, A. R. Harish, and C. J. Radcliffe, Synthesis of advanced microwave filters without diagonal cross-couplings, *IEEE Trans. Microwave Theory Tech.* **MTT-50**, 2862–2872 (Dec. 2002).

17. R. J. Cameron and J. D. Rhodes, Asymmetric realizations for dual-mode bandpass filters, *IEEE Trans. Microwave Theory Tech.* **MTT-29**, 51–58 (Jan. 1981).

18. F. Seyfert et al., Design of microwave filters: Extracting low pass coupling parameters from measured scattering data, *Int. Workshop Microwave Filters*, Toulouse, France, June 24–26 2002.

CHAPTER 11

MICROWAVE RESONATORS

Resonators are the basic building blocks of any bandpass filter. A resonator is an element that is capable of storing both frequency-dependent electric energy and frequency-dependent magnetic energy. A simple example is an LC resonator, where the magnetic energy is stored in the inductance L and the electric energy is stored in the capacitance C. The resonant frequency of a resonator is the frequency at which the energy stored in electric field equals the energy stored in the magnetic field. At microwave frequencies, resonators can take various shapes and forms. The shape of the microwave structure affects the field distribution and hence the stored electric and magnetic energies. Potentially, any microwave structure should be capable of constructing a resonator whose resonant frequency is determined by the structure's physical characteristics and dimensions. In this chapter, we examine various resonator configurations highlighting their main applications and limitations. We also review the available techniques to calculate and measure the resonant frequency and unloaded Q factors of microwave resonators.

11.1 MICROWAVE RESONATOR CONFIGURATIONS

The main design considerations of microwave resonators are the resonator size, unloaded Q, spurious performance, and power handling capability. The unloaded Q represents the inherent losses in the resonator. The higher the losses are, the lower is the Q value. It is therefore desirable to use resonators with high Q values

Microwave Filters for Communication Systems: Fundamentals, Design, and Applications,
by Richard J. Cameron, Chandra M. Kudsia, and Raafat R. Mansour
Copyright © 2007 John Wiley & Sons, Inc.

TABLE 11.1   **Comparison Between Various Modes of Operation**

Parameter	Single-Mode	Dual-Mode	Triple-Mode
Size	Large	Medium	Small
Spurious response	Good	Fair	Fair
Unloaded Q	High	Medium	Medium
Power handling capability	High	Medium	Medium
Design complexity	Low	Medium	High

since this reduces the insertion loss of the filter and improves its selectivity performance. In contrast to LC resonators, which have only one resonant frequency, microwave resonators can support an infinite number of electromagnetic field configurations or resonant modes. The spurious performance of a resonator is determined by how close the neighboring resonant modes are to the operating mode. The neighboring resonant modes act as spurious modes interfering with the filter performance. It is therefore desirable to increase the spurious free window of the resonator in order to improve the filter out-of-band rejection performance.

The resonant modes in microwave resonators exist in the form of a single mode representing one electric resonator or in the form of degenerate modes (i.e., modes having the same resonance frequency with different field distributions). These degenerate modes allow the realization of two electric resonators within the same physical resonator (dual-mode resonators) or three electric resonators within the same physical resonator (triple-mode resonators). Examples of dual modes are TE_{11} modes, which exist in circular waveguide cavities [1–4]; HE_{11} modes, which exist in dielectric resonators [5,6]; or TM_{11} modes, which exist in circular or square patch microstrip resonators [7]. Cubic waveguide cavities and cubic dielectric resonators can support triples modes [8]. The key advantage of operating in dual-mode or triple-mode configurations is size reduction. However, these modes do have an impact on the unloaded Q, spurious performance, and power handling capability of the cavity resonator. A summary of the features of each mode of operation is given in Table 11.1.

Microwave resonators are grouped into three categories: lumped-element LC resonators, planar resonators, and three-dimensional (3D) cavity-type resonators. Figure 11.1a shows a lumped-element resonator constructed by using a chip

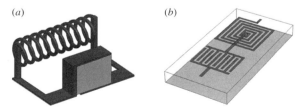

(a)   *(b)*

Figure 11.1   Lumped-element resonators realized by (*a*) a chip inductor and a chip capacitor; (*b*) spiral inductor and interdigital capacitor.

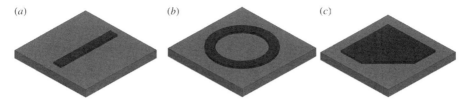

Figure 11.2 Examples of microstrip resonator configurations: (*a*) half-wavelength resonator; (*b*) ring resonator; (*c*) rectangular patch resonator.

inductor and a chip capacitor. The lumped-element resonator can be printed on a dielectric substrate in the form of a spiral inductor and an interdigital capacitor, as shown in Figure 11.1*b*. Lumped-element resonators are very small in size and offer a wide spurious free window, however, they have a relatively low Q value.

Examples of planar resonators are depicted in Figure 11.2. Planar resonators can take the form of a length of a microstrip transmission line, terminated in a short circuit or open circuit, or it can take the form of meander line, folded line, ring resonator, patch resonator or any other configuration [9]–[11]. Any printed structure effectively acts as a resonator whose resonant frequency is determined by the resonator's dimensions, substrate dielectric constant and substrate height. Figure 11.3 illustrates examples of three-dimensional (3D) coaxial, rectangular waveguide, circular waveguide, and dielectric resonators. The coaxial resonator consists of a length of a coaxial line, terminated in a short circuit at both ends. The waveguide resonators are rectangular and circular waveguides terminated also in a short circuit at both ends. The dielectric resonator cavity consists of a dielectric resonator with a high

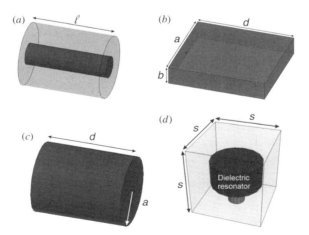

Figure 11.3 Examples of three-dimensional cavity resonators: (*a*) coaxial resonator; (*b*) rectangular waveguide resonator; (*c*) circular waveguide resonator; (*d*) dielectric resonator.

dielectric constant mounted on a support structure with a low dielectric constant inside a below-cutoff metallic housing. 3D resonators are bulky in size; however, they offer very high Q values; in addition, they are capable of handling high RF power levels.

The selection of a resonator or filter configuration for a particular application involves several tradeoffs among filter insertion loss (i.e., resonator Q), filter size, cost, power handling requirements etc. A comparison between these resonators is given Figures 11.4 and 11.5. Typically, lumped-element filters are employed in low-frequency applications. The filters are suitable for integration in microwave monolithic integrated circuits (MMICs) or RFIC circuits. A typical Q value for lumped-element resonators is between 10 and 50 at 1 GHz. Planar resonators are usually employed in wideband, compact, and low-cost applications. The typical Q value for planar filters is in the range of 50–300 at 1 GHz. If planar filters are implemented in a superconductor [9], they can offer Q values ranging from 20,000 to 50,000 at 1 GHz. However, in this case the filters need to be cooled down to very low temperatures, below 90 K. Coaxial, waveguide, and dielectric resonators offer a Q value ranging from 3000 to 30,000 at 1 GHz. 3D resonators are widely employed to construct filters for low loss wireless and space communication applications [10,11]. The dielectric resonator technology, in particular, is emerging as the technology of choice for miniature high Q filters, Chapter 16 is dedicated to dielectric resonator filters.

Figures 11.1–11.3 illustrate only a few of the most common resonators; hundreds of other resonators with different configurations have been reported in the literature. The availability of commercial electromagnetic (EM) software simulators in the early 1990s has helped researchers in the field to devise and propose different planar and 3D resonator configurations. In particular, there have been tremendous innovations in planar resonator configurations. Simple and complex planar resonator configurations can be handled by the same low-cost

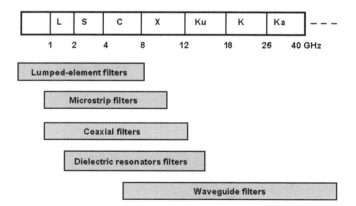

Figure 11.4 Application of the various resonator configurations.

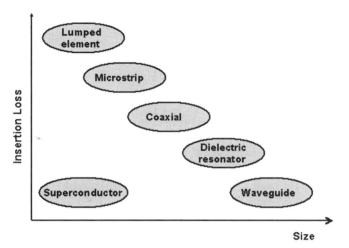

Figure 11.5 Typical relative size and insertion loss of various resonators.

photolithographic fabrication process, enabling researchers to readily build prototype units to verify their designs.

11.2 CALCULATION OF RESONANT FREQUENCY

11.2.1 Resonance Frequency of Conventional Transmission-Line Resonators

For resonators that are constructed from a section of waveguide, coaxial, or microstrip line, terminated in a short circuit or open circuit, the resonant frequency is calculated from the knowledge of the guide wavelength λ_g. Consider the transmission
line shown in Figure 11.6, which is terminated in short circuit at both ends. By
assuming a lossless structure, the resonant frequency is determined by standing at
any point along the transmission line and calculating the impedance or the admittance seen from both sides. At resonance, we have

$$Z_1 + Z_2 = 0 \quad \text{or} \quad Y_1 + Y_2 = 0 \tag{11.1}$$

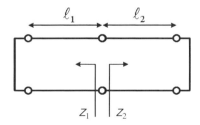

Figure 11.6 A transmission line terminated in short circuit at both ends.

Z_1 and Z_2 are the impedances seen at the input of transmission lines terminated in short circuit and are given by

$$Z_1 = j \tan \beta l_1 \quad \text{and} \quad Z_2 = j \tan \beta l_2 \tag{11.2}$$

Then, $Z_1 + Z_2 = 0$ gives $\tan \beta l_1 = -\tan \beta l_2$, which leads to

$$\beta l_1 = -\beta l_2 + q\pi, \quad \beta(l_1 + l_2) = \beta l = \frac{2\pi}{\lambda_g} l = q\pi, \quad q = 1, 2, 3, \ldots \tag{11.3}$$

$$l = q\frac{\lambda_g}{2}, \quad q = 1, 2, 3, \ldots \tag{11.4}$$

Using equation (11.4), we now consider calculating the resonant frequency of transmission-line coaxial, microstrip, and waveguide resonators:

1. *Coaxial Line Resonators (Fig. 11.3a).* Coaxial lines are nondispersive media with λ_g given by

$$\lambda_g = \frac{c_0}{f\sqrt{\varepsilon_r}}, \quad f_{0q} = q\frac{c_0}{2l\sqrt{\varepsilon_r}} \tag{11.5}$$

where ε_r is the relative permittivity of the dielectric material that completely fills the space between the inner and outer connectors, c_0 is the free-space speed of light, $c_0 = 2.998 \times 10^8$ m/s.

2. *Microstrip Line Resonators (Fig. 11.2a).* In microstrip lines, the electromagnetic fields are not completely confined in the dielectric. The propagation is described by defining an effective dielectric constant ε_{eff} [10]. The guide wavelength λ_g is then given by

$$\lambda_g = \frac{c_0}{f\sqrt{\varepsilon_{\text{eff}}}}, \quad f_{0q} = q\frac{c_0}{2l\sqrt{\varepsilon_{\text{eff}}}} \tag{11.6}$$

Microstrip lines are dispersive media, and ε_{eff} is frequency-dependent. However, at low frequencies ε_{eff} can be approximated as [2]

$$\varepsilon_{\text{eff}} = \frac{\varepsilon_r + 1}{2} + \frac{\varepsilon_r - 1}{2}\frac{1}{\sqrt{1 + 12\,h/W}} \tag{11.7}$$

where h is the substrate height and W is the microstrip width. More accurate expressions for ε_{eff} are found in [10] and [11].

3. *Rectangular Waveguide Resonators (Fig. 11.3b).* The propagation constant β in rectangular waveguides, which is given by [1]

$$\beta = \sqrt{\left(\frac{2\pi f}{c}\right)^2 - \left(\frac{n\pi}{a}\right)^2 - \left(\frac{m\pi}{b}\right)^2} \tag{11.8}$$

where $c = 1/\sqrt{\mu\varepsilon}$, applying equation (11.3), gives:

$$f_{omnq} = \frac{c}{2\pi}\sqrt{\left(\frac{n\pi}{a}\right)^2 + \left(\frac{m\pi}{b}\right)^2 + \left(\frac{q\pi}{d}\right)^2} \quad \text{for TE}_{nmq} \text{ and TM}_{nmq} \text{ modes} \quad (11.9)$$

4. *Circular Waveguide Resonators (Fig. 11.7c)*. The propagation constants of TE and TM modes in circular waveguides are given by [1].

$$\beta = \sqrt{\left(\frac{2\pi f}{c}\right)^2 - \left(\frac{\rho'_{nm}}{a}\right)^2} \quad \text{for TE}_{nm} \text{ modes} \quad (11.10)$$

$$\beta = \sqrt{\left(\frac{2\pi f}{c}\right)^2 - \left(\frac{\rho_{nm}}{a}\right)^2} \quad \text{for TM}_{nm} \text{ modes} \quad (11.11)$$

where ρ_{nm} and ρ'_{nm} are mth roots of the Bessel functions $J_n(x)$ and $J'_n(x)$, respectively [1]. The resonant frequencies of TE$_{nmq}$ and TM$_{nmq}$ modes are given

$$f_{omnq} = \frac{c}{2\pi}\sqrt{\left(\frac{p'_{nm}}{a}\right)^2 + \left(\frac{q\pi}{d}\right)^2} \quad \text{for TE}_{nmq} \text{ modes} \quad (11.12)$$

$$f_{omnq} = \frac{c}{2\pi}\sqrt{\left(\frac{\rho_{nm}}{a}\right)^2 + \left(\frac{q\pi}{d}\right)^2} \quad \text{for TM}_{nmq} \text{ modes} \quad (11.13)$$

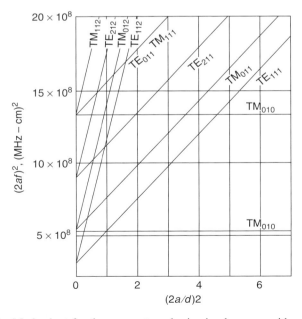

Figure 11.7 Mode chart for the resonant modes in circular waveguide resonators [1].

A mode chart for the circular cavities is depicted in Figure 11.7. The mode chart helps designers to select the proper resonator dimensions (a and d) to obtain the desired resonant frequency and to design for a wide spurious free window.

Detailed analyses of coaxial, rectangular, and circular waveguides are well covered in several textbooks on microwaves [1,2], providing information on the field distribution of the various modes. Filter designers need to be familiar with the field distribution of these modes to identify the optimum location for the input/output coupling as well to determine the optimum interresonator coupling configuration.

11.2.2 Resonance Frequency Calculation Using the Transverse Resonance Technique

Loading the transmission-line resonator with a capacitive/inductive element or with any type of discontinuity such as a screw alters the resonant frequency of the resonator. Figure 11.8 shows a microstrip transmission-line resonator loaded with a rectangular strip along the resonator's length. The rectangular strip represents a discontinuity in the transmission line. The field distribution in the resonator must change from that of the transmission line in order to satisfy the new boundary conditions caused by the discontinuity. This, in turn, changes the stored electric and magnetic energy, causing a shift in the resonant frequency. The new resonant frequency can be calculated with the use of the transverse resonance technique [12]. The discontinuity is represented by a two-port network, whose S parameters can be calculated using a commercial EM simulator such as HFSS [13] or sonnet em [14], or any other software tool. The resonator can be represented as a two-port network connected to two open-circuited transmission lines of lengths l_1 and l_2 as shown in Figure 11.9. Γ_L is given by

$$\Gamma_L = e^{-2j\beta l_2} \tag{11.14}$$

With knowledge of the S parameters of the discontinuity and in view of Chapter 5 (Section 5.1.5), we can express the input reflection coefficient Γ_{in} as follows:

$$\Gamma_{in} = \frac{V_1^-}{V_1^+} = S_{11} + \frac{S_{12}\Gamma_L S_{21}}{1 - \Gamma_L S_{22}} \tag{11.15a}$$

Figure 11.8   A microstrip transmission line loaded with a rectangular strip.

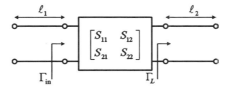

Figure 11.9   A two-port representation of the discontinuity shown in Figure 11.8.

However, the incident and reflected voltages V_1^+ and V_1^- at the input of the two-port network are related as follows:

$$V_1^+ = e^{-2j\beta l_1} V_1^- \qquad (11.15b)$$

Combining equations (11.15a) and (11.15b) gives

$$[1 - e^{-2j\beta l_1} \Gamma_{in}]V_1^- = 0 \qquad (11.16)$$

Equation (11.16) is known as the "characteristic equation." At resonance, the coefficient of V_1^+ must be equal to zero. Then, the resonant frequency of the resonator is obtained by solving the following equation:

$$\left[1 - e^{-2j\beta l_1}\left(S_{11} + \frac{S_{12}e^{-2j\beta l_2}S_{21}}{1 - e^{-2j\beta l_2}S_{22}}\right)\right] = 0 \qquad (11.17)$$

Note that both the propagation constant β and the scattering parameters of the discontinuity are functions of frequency. For a first approximation, the S parameters are calculated at the resonant frequency of the unperturbed resonator, and are assumed to be frequency-independent. Solving equation (11.17) provides β, which in turn gives an approximate value for the new resonant frequency. For a more accurate evaluation, equation (11.17) needs to be solved numerically. Alternatively, the adaptive frequency sampling technique, described in Chapter 15, can be used to find frequency-dependent polynomial expressions as an approximation for the scattering parameters.

11.2.3   Resonance Frequency of Arbitrarily Shaped Resonators

For resonators with arbitrarily shaped structures, an EM simulation tool is required for accurate calculation of the resonant frequency. This is achieved by using either an eigenmode analysis [15] or an S parameter analysis. Both techniques are capable of calculating the resonant frequency of the fundamental resonant mode, as well as the resonant frequencies of the higher-order modes.

Eigenmode Analysis   Several EM packages such as HFSS [13] and CST [16] have eigenmode solution tools that are capable of analyzing any microwave resonator providing information on the resonant frequency, unloaded Q factor, and field distribution. The user specifies the frequency range, and the eigenmode tool identifies all the resonant modes that exist within this frequency range. Figure 11.10

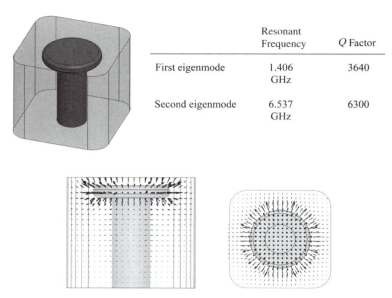

	Resonant Frequency	Q Factor
First eigenmode	1.406 GHz	3640
Second eigenmode	6.537 GHz	6300

Figure 11.10   The field distribution of the first eigenmode solution of a top-loaded coaxial resonator.

shows the eigenmode solution of a top-loaded coaxial cavity obtained by HFSS [13]. The figure gives the first two resonant frequencies and the unloaded Q and the field distribution of the first resonant TEM mode.

The results show that the resonant frequency of the fundamental mode is 1.406 GHz, whereas that of the second mode is 6.537 GHz; thus, the resonator has a spurious free window of 5.131 GHz.

S-Parameters Analysis   This method is useful in cases where the EM software tool available to the designer has no eigenmode analysis tool, but has the capability to calculate the scattering parameters. Figure 11.11 shows the same resonator given in Figure 11.10 fed by a probe. The input reflection coefficient S_{11} seen at the probe

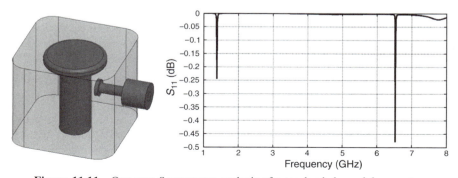

Figure 11.11   One-port S-parameter analysis of a top-loaded coaxial resonator.

versus frequency is also shown in Figure 11.11. For most of the frequency range, the RF energy is reflected back. The location of the dips indicate the frequencies where the RF energy is stored in the resonator, specifically, the resonant frequencies of the various modes. The S_{11} response shows only the modes that the probe can excite inside the resonator. Thus, we need to have a rough idea about the field distribution of the resonant modes, in order to ensure the proper probe excitation.

The width of the dips reflects the losses in the cavity. For lossless cavities this width is infinitesimally small and the dips may not show up in simulation. Therefore, in using the one-port S-parameter analysis, we need to account for the losses. Having losses widens the dips, allowing them to be shown in the S_{11} response. It should also be mentioned that for an accurate calculation of the resonant frequency, we need to use very loose coupling between the probe and the resonator to minimize the loading on the resonator. This explains the slight shift in the resonant frequency usually obtained between the results of the eigenmode analysis and those of the one-port S-parameter analysis.

Alternatively, one can use a two-port analysis where the resonator is fed by an input/output port. However, the accuracy of the two-port analysis deteriorates when the input/output coupling is strong. Figure 11.12 compares the results of a

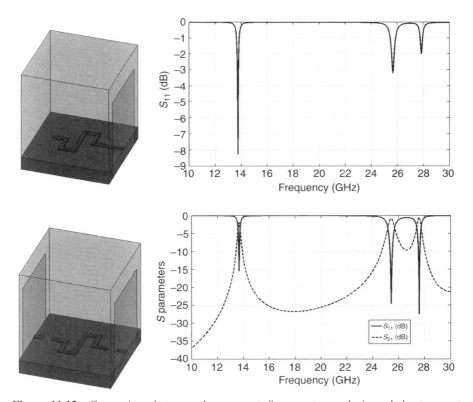

Figure 11.12 Comparison between the one-port S-parameter analysis and the two-port S-parameter analysis for a microstrip resonator as calculated by HFSS.

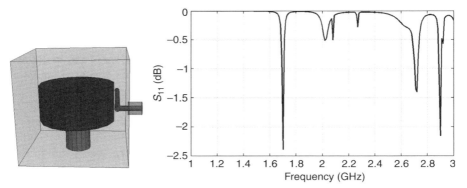

Figure 11.13 One-port analysis of a dielectric resonator illustrating the resonant frequencies of single and dual modes.

microstrip resonator fed by one port and by two-ports. It can be seen that the two-port analysis provides results that are slightly different from those of the one-port analysis, which is attributed to loading caused by the output probe.

The one-port S-parameter analysis is very useful in measuring the resonant frequency. A simple S_{11} measurement directly can provide information on the resonant frequencies of the resonator and its spurious performance.

Dual-mode and triple-mode of resonant frequencies can also be calculated by using either eigenmode analysis or the S-parameter analysis. The eigenmode solution would display almost identical resonant frequencies with two different field distributions; while in the one-port S-parameter analysis, the dual-mode shows up in the S_{11} simulation as two dips very close to each other. Figure 11.13 illustrates S_{11} as simulated by HFSS for a dielectric resonator fed by an input probe. The first dip represents the resonant frequency of the single TE_{01} mode, while the two dips represent the resonant frequency of the hybrid HE dual-mode [5]. As shown in Chapter 14, the spacing between the two dips represents the coupling between the two dual modes. The coupling, in this case, is caused by the input probe.

11.3 RESONATOR UNLOADED Q FACTOR

The unloaded Q_u factor of a resonator describes its frequency-selective properties, namely, the sharpness of the resonance curve. It is defined as

$$Q_u = \omega_0 \frac{W_T}{P_{\text{loss}}} \tag{11.18}$$

where ω_0 is the radian resonance frequency, W_T is the average energy stored, and P_{loss} is the power dissipated in the resonator. The energy stored is the sum of the stored electric energy W_e and stored magnetic energy W_m:

$$W_T = W_e + W_m = 2W_e = 2W_m \quad \text{(since } W_e = W_m \text{ at resonance)} \tag{11.19}$$

$$W_e = \frac{\varepsilon}{4} \int_v E \cdot E^* dv \tag{11.20}$$

$$W_m = \frac{\mu}{4} \int_v H \cdot H^* dv \tag{11.21}$$

The total power dissipated in the resonator is the sum of the power dissipated in resonator conducting walls P_c and the power dissipated in dielectric P_d which can be written as:

$$P_c = \frac{R_s}{2} \int_{\text{walls}} |H_t|^2 ds \tag{11.22}$$

$$P_d = \frac{\omega_0 \varepsilon''}{2} \int_v E \cdot E^* dv \tag{11.23}$$

Here, H_t is the tangential magnetic field at the surface; R_s is the surface resistance of the conducting walls, $R_s = \sqrt{\omega \mu_0 / 2\sigma}$, where σ is the conductivity of the resonator metallic walls; and ε'' is the imaginary part of the permittivity, which is related to the loss tangent $\tan \delta$ by

$$\varepsilon = \varepsilon' - j\varepsilon'' = \varepsilon_r \varepsilon_0 (1 - j \tan \delta) \tag{11.24}$$

By factoring in the conducting and dielectric losses, we can express the unloaded Q_u as follows:

$$Q_u = \omega_0 \frac{W_T}{(P_c + P_d)} \tag{11.25}$$

By defining Q_c as $Q_c = \omega_0 W_T / P_c$ and Q_d as $Q_d = \omega_0 W_T / P_d$, we can further express the unloaded Q_u by

$$\frac{1}{Q_u} = \frac{1}{Q_c} + \frac{1}{Q_d} \tag{11.26}$$

In view of equations (11.20) and (11.23), Q_d is written as follows:

$$Q_d = \omega_0 \frac{2W_e}{P_d} = \frac{\varepsilon'}{\varepsilon''} = \frac{1}{\tan \delta} \tag{11.27}$$

From equations (11.26) and (11.27), it is concluded that for resonators that have dielectric loading, the overall unloaded Q will be less than $1/(\tan \delta)$.

11.3.1 Unloaded *Q* Factor of Conventional Resonators

A Coaxial λ/2 Resonator Terminated in a Short Circuit For coaxial resonators we adopt equations (11.18)–(11.23) and the field distribution inside the coaxial line [1,2] to determine the exact analytical expression for Q_c. However, an approximate expression for Q_c can be obtained by considering the resonator as a transmission line terminated in a short circuit as shown in Figure 11.14*a*. The input impedance near resonance is approximated as a series *RLC* resonant circuit:

$$Z_{\mathrm{in}} = Z_0 \tanh(\alpha l + j\beta l) = Z_0 \frac{\tanh \alpha l + j \tan \beta l}{1 + j \tan \beta l \tanh \alpha l} \tag{11.28}$$

Around resonance, $\omega = \omega_0 + \Delta\omega$, and $\beta l = \pi + \pi\Delta\omega/\omega_0$. For small losses we can approximate $\tanh \alpha l$ as $\tanh \alpha l \approx \alpha l$; Z_{in} can then be approximated by

$$Z_{\mathrm{in}} \approx Z_0 \left(\alpha l + j\frac{\pi\Delta\omega}{\omega_0} \right)$$

which is in the form of the input impedance of a series *RLC* resonant circuit with R, L, and C given by [1,2].

$$R = Z_0 \alpha l, \quad L = \frac{Z_0 \pi}{2\omega_0}, \quad C = \frac{1}{\omega_0^2 L} \tag{11.29}$$

where Z_0 is the characteristic impedance of the coaxial transmission line, $Z_0 = \eta/2\pi \ln b/a$, and α is the attenuation constant. The unloaded Q of the equivalent series resonant circuit is represented by

$$Q = \frac{\omega_0 L}{R} = \frac{\pi}{2\alpha l} = \frac{\beta}{2\alpha} \tag{11.30}$$

Thus Q_c in coaxial lines is thus approximated as

$$Q_c \approx \frac{\beta}{2\alpha_c} \tag{11.31}$$

Assuming a coaxial line of an inner radius a and an outer radius b, α_c is given by

$$\alpha_c = \frac{R_s}{2\eta \ln b/a} \left(\frac{1}{a} + \frac{1}{b} \right) \tag{11.32}$$

$$Q_c \approx \frac{ab\beta Z_0 \ln b/a}{(a+b)R_s} \tag{11.33}$$

η is the free space impedance, $\eta = 376.7\ \Omega$. Note that the expression of Q_c given in equation (11.33) does not account for the losses in the short-circuit end plates.

A $\lambda/2$ Microstrip Resonator   Since microstrip lines are inhomogeneous structures, there are no accurate closed-form expressions for the field distribution in microstrip lines. Therefore, it is difficult to use equations (11.18)–(11.23) to obtain an exact analytical expression for Q_c in microstrip lines. The same concept applied for coaxial line resonators can also be applied for microstrip line resonators to obtain an approximate value for Q_c. By considering the resonator as $\lambda/2$ transmission line terminated in an open circuit (see e.g., Fig. 11.14b), the input impedance around the resonant frequency can be approximated as a parallel RLC resonant circuit. The input impedance Z_{in} is given by

$$Z_{in} = Z_0 \coth(\alpha l + j\beta l) = Z_0 \frac{1 + j\tan\beta l \tanh\alpha l}{\tanh\alpha l + j\tan\beta l} \tag{11.34}$$

For small losses and near resonance we have $\omega = \omega_0 + \Delta\omega$, $\beta l = \pi + \pi\Delta\omega/\omega_0$, and $\tanh\alpha l \approx \alpha l$. We can then approximate Z_{in} as

$$Z_{in} \approx \frac{Z_0}{\alpha l + j\dfrac{\pi\Delta\omega}{\omega_0}}$$

which is in the form of the input of a parallel RLC resonant circuit with R, L, and C given by [1,2].

$$R = \frac{Z_0}{\alpha l}, \quad C = \frac{\pi}{2\omega_0 Z_0}, \quad L = \frac{1}{\omega_0^2 C} \tag{11.35}$$

The unloaded Q of the equivalent parallel resonant circuit is given by

$$Q = \omega_0 RC = \frac{\pi}{2\alpha l} = \frac{\beta}{2\alpha} \tag{11.36}$$

Thus Q_c and Q_d of a $\lambda/2$ microstrip resonator can be approximated as

$$Q_c \approx \frac{\beta}{2\alpha_c}, \quad Q_d \approx \frac{\beta}{2\alpha_d} \tag{11.37}$$

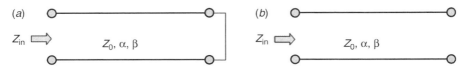

Figure 11.14   (*a*) A $\lambda/2$ short-circuit transmission line; (*b*) $\lambda/2$ open-circuit transmission line.

There are no accurate closed-form expression for α_c and α_d in microstrip lines. They can be approximated as [2]

$$\alpha_c \approx \frac{R_s}{Z_0 W} \tag{11.38}$$

$$\alpha_d = \frac{k_0 \varepsilon_r (\varepsilon_{\text{eff}} - 1) \tan \delta}{2\sqrt{\varepsilon_{\text{eff}}}(\varepsilon_r - 1)} \tag{11.39}$$

R_s is the surface resistivity $R_s = \sqrt{\omega \mu_0 / 2\delta}$. This expression for Q_c does not factor in the effects of the fringing field and the losses of the metallic enclosure that houses the resonator.

Rectangular Waveguide Resonators   In rectangular waveguide resonators (Fig. 11.3b) there exist analytical expressions for the field distribution inside the cavity. Equations (11.18)–(11.23) can then be employed to derive an analytical expression for the unloaded Q_c. For a rectangular cavity of dimensions a, b, d operating in TE_{10q} modes, Q_c is given by [1,2].

$$Q_c = \omega_0 \frac{W_T}{P_c} = \frac{(k_{10q} a d)^3 b \eta}{2\pi^2 R_s} \frac{1}{[2q^2 a^3 b + 2bd^3 + q^2 a^3 d + ad^3]} \tag{11.40a}$$

where

$$k_{10q} = \left[\left(\frac{\pi}{a}\right)^2 + \left(\frac{q\pi}{b}\right)^2 \right]^{1/2}, \quad \eta = \sqrt{\frac{\mu}{\varepsilon}} \tag{11.40b}$$

Circular Cavities   Similarly, analytical expressions for the field distribution are employed in equations (11.18)–(11.23) to obtain an exact expression for Q_c. The unloaded Q_c of a circular cavity of radius a and length d, operating in TE_{nmq} modes, is given by [1]

$$Q_c = \omega_0 \frac{W_T}{P_c} = \frac{\lambda_0}{\delta_s} \frac{\left[1 - \left(\frac{n}{p'_{nm}}\right)^2\right]\left[(p'_{nm})^2 + \left(\frac{q\pi a}{d}\right)^2\right]^{3/2}}{2\pi \left[(p'_{nm})^2 + \frac{2a}{d}\left(\frac{q\pi a}{d}\right)^2 + \left(1 - \frac{2a}{d}\right)\left(\frac{nq\pi a}{p'_{nm}d}\right)^2\right]} \tag{11.41}$$

where δ_s is the skin depth and $\delta_s = 1/\sqrt{\pi f \mu_0 \sigma}$. Note that Q_c varies as λ_0/δ_s and thus decreases as $\sqrt{f}$.

Figure 11.15 displays a plot of Q for various modes for circular waveguide cavities (Fig. 11.3c) [1]. Note that mode TE_{011} has a Q value higher than that of the TE_{111} resonant mode. It is maximum around $a = 2d$. However, as seen from the mode chart given in Figure 11.7, the TE_{011} mode is degenerate with the TM_{111} mode; therefore, one needs to be careful in selecting the coupling scheme that does not excite the TM_{111} mode [17].

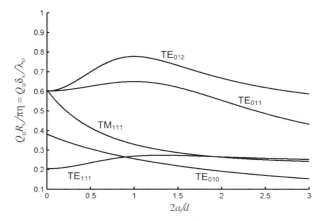

Figure 11.15 A plot of Q for various modes for circular waveguide cavities [1].

11.3.2 Unloaded Q of Arbitrarily Shaped Resonators

The eigenmode analysis is used to calculate both the resonant frequency and the unloaded Q of the resonator. With the eigenvalue (resonant frequency) determined, the eigenvector (field distribution) can be calculated, which in turn can be used to calculate the Q values. The process has been automated in most commercial EM software packages directly offering the frequency, unloaded Q, and field distribution. As shown in Figure 11.10, the eigenmode analysis for the top-loaded coaxial resonator also provides the unloaded Q values.

11.4 MEASUREMENT OF LOADED AND UNLOADED Q FACTOR

An accurate calculation of the unloaded Q factor of microwave resonators requires knowledge of the conductivity of the metal walls as well as the dielectric loss tangent for resonators that are loaded with dielectric materials. Typically, high-conductivity metals such as gold, copper, and silver are used to achieve high Q values. Such metals are usually deposited or electroplated on metal surfaces, or dielectric surfaces in the case of planar filters. The achievable conductivity can vary from sample to sample, depending on the deposition technique used and the surface roughness achieved.

In addition, there is a considerable uncertainty regarding the loss tangent value of dielectric materials. Often, the loss tangent of dielectric substrates and dielectric resonators is specified by the manufacturer at a particular frequency that may be different from the operating frequency. Most designers assume the product [frequency $\cdot$ ($1/\tan\delta$)] constant and scale the loss tangent according to the operating frequency. Although this scaling yields a good approximation, the scaling does not give the exact value of the loss tangent at the operating frequency. Moreover, like

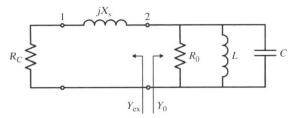

Figure 11.16   A resonator inductively coupled to an input port.

the dielectric constant ε_r, the loss tangent $(1/\tan \delta)$ also varies from batch to batch. Furthermore, in actual filter implementations, resonators are usually loaded with tuning elements that have a significant impact on the Q value. Because of these factors, measurement is the only approach that can provide an accurate evaluation of the unloaded Q factor of resonators. The measurement is achieved through one- or two-port measurement [18–28]. In this chapter we address the one-port measurement, where reflection-type measurements are taken at an input port, coupled capacitively or inductively to the resonator.

To illustrate the concept, let us consider the circuit shown in Figure 11.16, which represents the equivalent circuit of a resonator, coupled to an input transmission line by a reactance X_s. The unloaded resonator has a resonant frequency of $f_0 = 1/2\pi\sqrt{LC}$ and an unloaded Q_0 of $Q_0 = \frac{2\pi f_0 C}{G_0}$. The unloaded resonator at port 2 is loaded by an external admittance Y_{ex} as depicted in Figure 11.17 and is given by

$$Y_{ex} = G_{ex} + jB_{ex} = \frac{1}{R_c + jX_s} \tag{11.42}$$

The effect of the loading is to change the resonant frequency to a new resonant frequency f_L due to the addition of B_{ex}, and to reduce the Q factor to a loaded value of Q_L due to the addition of G_{ex}. Near resonance, the admittance $Y_L = (Y_0 + Y_{ex})$ of the loaded resonator can be approximated as

$$Y_L \approx G_{ex} + G_0 + j\left(2G_0Q_0\frac{f - f_0}{f_0} + B_{ex}\right) \tag{11.43}$$

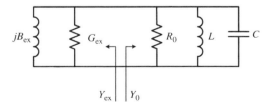

Figure 11.17   Resonator loading caused by input coupling.

The loaded resonant frequency f_L is then given by equating the imaginary part of Y_L to zero as follows:

$$f_L = f_0 \left(1 - \frac{B_{ex}}{2Q_0 G_0}\right) \tag{11.44}$$

The loaded $Q_L = \dfrac{2\pi f_0 C}{G_0 + G_{ex}}$ is written in terms of Q_0 as

$$Q_L = Q_0 \frac{G_0}{G_0 + G_{ex}} = Q_0 \frac{1}{1 + \dfrac{G_{ex}}{G_0}} \tag{11.45}$$

$$Q_L = \frac{Q_0}{1 + k} \tag{11.46}$$

where k is the coupling coefficient, representing the ratio of the power dissipated in the external load to the power dissipated in the resonator:

$$k = \frac{P_{loss-ex}}{P_{loss-res}} = \frac{V^2 G_{ex}}{V^2 G_0} = \frac{G_{ex}}{G_0} \tag{11.47}$$

When the amount of power dissipated in the external circuit equals that dissipated in the resonator, the coupling is said to be critical and the coupling coefficient $k = 1$. Undercoupling ($k < 1$) corresponds to the case where the power dissipated in the external circuit is less than that dissipated in the resonator, whereas overcoupling ($k > 1$) corresponds to the case where the power dissipated in the external circuit is more than that dissipated in the resonator.

In view of equation (11.42), the coupling coefficient is expressed in terms of the coupling reactance element X_c as

$$k = \frac{\dfrac{R_0}{R_c}}{1 + \dfrac{X_s^2}{R_c^2}} \tag{11.48}$$

The resonant frequency of the loaded resonator is written in terms of k, the unloaded Q_0 factor, and the equivalent circuit elements X_s and R_c as

$$f_L = f_0 \left(1 + \frac{k X_s}{2 Q_0 R_c}\right) \tag{11.49}$$

The measurement of the reflection coefficient S_{11} versus the frequency at port 1 in Figure 11.16 provides the required information to calculate Q_L and k and Q_u. In view of Figure 11.6, the input reflection coefficient Γ at port 1 is written as

$$S_{11} = \Gamma = \frac{jX_s + \dfrac{1}{Y_0} - R_c}{jX_s + \dfrac{1}{Y_0} + R_c} \tag{11.50}$$

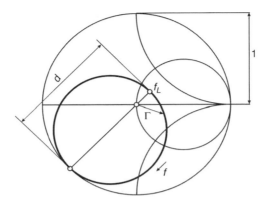

Figure 11.18 The Q circle on the Smith chart.

Kajfez [20] has shown that equation (11.50) can be reduced to

$$\Gamma = \Gamma_d + \frac{2k}{1+k} \frac{e^{-j2\delta}}{1 + 2jQ_L \dfrac{f - f_L}{f_0}} \tag{11.51}$$

where

$$\Gamma_d = \frac{jX_s - R_c}{jX_s + R_c} = -\frac{Y_{\text{ex}}}{Y_{\text{ex}}^*} = -e^{-j2\delta}, \quad \delta = \tan^{-1}\frac{X_s}{R_c} \tag{11.52}$$

The vector $\Gamma - \Gamma_d$ describes a circle on the Smith chart, as shown in Figure 11.18. The circle is known as the Q circle [20]. For frequencies far from resonance (i.e., detuned $f > f_L$), the second term in equation (11.51) is small and we can approximate the reflection coefficient as $\Gamma_{\text{untuned}} \approx \Gamma_d$. At resonance $f = f_L$, the reflection coefficient is given by

$$\Gamma_{\text{res}} = \Gamma_d + \frac{2k}{1+k} e^{-j2\delta} = \frac{k-1}{k+1} e^{-j2\delta} \tag{11.53}$$

The following observations are based on equations (11.51)–(11.53):

1. The magnitude of the reflection coefficient for $f > f_L$ is almost 1. The reflection coefficient corresponding to these frequencies lies on the circumference of the Smith chart.
2. The reflection coefficient at resonance has the angle $(-j2\delta)$. Depending on the k value, it is either in phase or $180°$ out of phase with Γ_{untuned} or Γ_d.
3. The reflection coefficient represents a circle on the Smith chart with Γ_{res} corresponding to the point on the circle with the shortest distance to the center of the Smith chart. As the frequency increases, δ also increases, and the input reflection coefficient moves along the Q circle in a clockwise direction.

4. The difference between $\Gamma_d - \Gamma_{res} = [2k/(1+k)]$, which represents the diameter d of the Q circle. Thus the diameter d is written as

$$d = \frac{2k}{1+k} \tag{11.54}$$

With measurement of the diameter d on the Smith chart, the coupling coefficient k is calculated as follows:

$$k = \frac{d}{2-d} \tag{11.55}$$

A glance at the Q circle on the Smith chart will reveal whether the coupling is under- or overcoupled. A small Q circle indicates undercoupling ($k < 1$); a large Q circle signifies overcoupling ($k > 1$). A Q circle of diameter 1 denotes critical coupling ($k = 1$).

For example, Figure 11.19 illustrates a $\lambda/2$ resonator capacitively coupled to an input port. The resonator is a $\lambda/2$ transmission line, terminated in an open circuit. The transmission line has a length of 53 mm, and a dielectric constant of 2 resonating at 2 GHz. The transmission line is assumed to have an attenuation of 1 dB/m. A simple calculation shows that the structure is critically coupled ($k = 1$) when the capacitance C is 0.125 pF. The reflection coefficient is simulated using $\theta = 0$ with $C = 0.1$ pF (undercoupling), $C = 0.2$ pF (overcoupling), and $C = 0.125$ pF (critical coupling). The Q charts of these three cases are shown in Figures 11.20a, 11.20b, and 11.20c, respectively. It is clear that the diameter of the Q circle of the reflection coefficient depends on the coupling coefficient k. It is also noted that the circles are shifted from the $0°$ line on the Smith chart by -2δ. The circuit is then simulated by assuming a capacitance of $C = 0.125$ pF (critical coupling) and $\theta = 35°$. The corresponding Q chart is shown in Figure 11.20d. The Q circle now is shifted from $0°$ by $-(2\delta + 2\theta)$. This case illustrates the fact that the Q circle can be located anywhere in the Smith chart depending on the reference plane, where measurements of the reflection coefficient are taken.

To determine the loaded Q factor, it is necessary to identify the resonant frequency of the loaded resonator f_L and the phase values of the reflection coefficient at two frequency points around the resonant frequency. The frequency on the Q circle that provides the shortest length for Γ (i.e., minimum reflection) is the resonant frequency. By using equation (11.51) and in view of Figure 11.21, we can

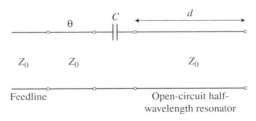

Figure 11.19 A $\lambda/2$ transmission-line resonator capacitively coupled to an input port.

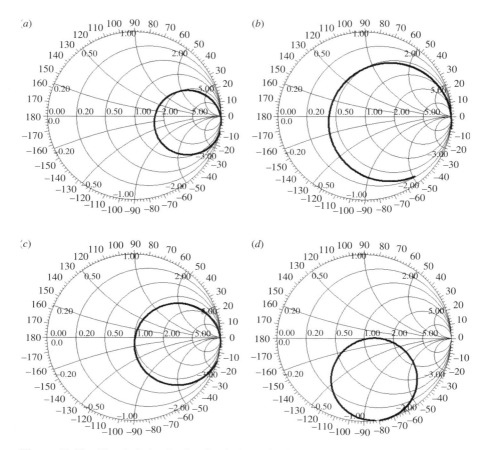

Figure 11.20 The Q circles for the circuit shown in Figure 11.19: (*a*) $\theta = 0$, undercoupled; (*b*) $\theta = 0$, overcoupled; (*c*) $\theta = 0$, critical coupled; (*d*) $\theta = 35$, critical coupled.

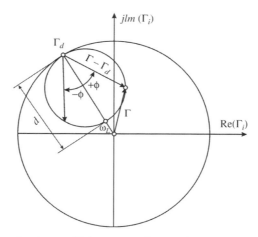

Figure 11.21 The reflection coefficient at two frequencies around f_L on the Q circle [20].

readily show that by moving the marker to frequencies f_1 and f_2 respectively below and above the resonant frequency f_L, the corresponding phases ϕ_1 and ϕ_2 become

$$\tan \phi_1 = -2Q_L \frac{f_1 - f_L}{f_0} \tag{11.56}$$

$$\tan \phi_2 = -2Q_L \frac{f_2 - f_L}{f_0} \tag{11.57}$$

Subtracting equations (11.56) and (11.57) gives

$$Q_L = \frac{1}{2} \frac{f_L}{f_2 - f_1} (\tan \phi_1 - \tan \phi_2) \tag{11.58}$$

If we select f_1 and f_2 such that $\phi_1 = 45°$ and $\phi_2 = -45°$, we get

$$Q_L = \frac{f_L}{f_2 - f_1} \tag{11.59}$$

Note that the angle ϕ cannot be read directly from the Smith chart display, but it can be calculated indirectly from the angle of the reflection coefficient. Therefore, for the polar display it may be easier to use equation (11.58) rather than equation (11.59).

In view of the analysis above, the measurement of unloaded Q is achieved by using either the polar display (Smith chart) or the linear display (dB measurements) of the reflection coefficient. The steps are now summarized:

Measurement of Unloaded Q Using Polar Display of Reflection Coefficient

Step 1. Find f_L by identifying the frequency on the Q circle that corresponds to the minimum reflection coefficient.

Step 2. Calculate the coupling coefficient by measuring the diameter of the Q circle $[k = d/(2 - d)]$.

Step 3. Identify two frequencies f_1 and f_2 around f_L, and calculate Q_L by using either equation (11.58) or equation (11.59).

Step 4. Calculate Q_u by using $Q_u = Q_L(1 + k)$.

Measurement of Unloaded Q Using Linear Display of Reflection Coefficient

The reflection coefficient on the linear scale is depicted in Figure 11.22. In view of Figure 11.21, it can be easily shown that when $\phi = 45°$, we have

$$|\Gamma|^2 = \frac{1 + |\Gamma_{\text{res}}|^2}{2} \tag{11.60}$$

Step 1. Find f_L by locating the frequency marker at the dip of the reflection coefficient.

Step 2. With the marker located at the dip, read $S_{11}^{\min}$ in dB. Note that $S_{11}^{\min}$ is the reflection coefficient Γ_{res} given in equation (11.53).

$$S_{11}^{\min} = 20 \log |\Gamma_{\text{res}}| = 20 \log \left| \frac{k - 1}{k + 1} \right| \tag{11.61}$$

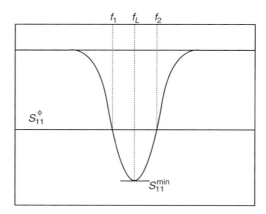

Figure 11.22 The reflection coefficient on a linear scale.

Thus, the coupling coefficient k can be calculated in terms of $S_{11}^{\min}$ as follows:

$$k = \frac{1 - 10^{-S_{11}^{\min}/20}}{1 + 10^{-S_{11}^{\min}/20}} \quad \text{if the resonator is undercoupled} \tag{11.62}$$

$$k = \frac{1 + 10^{-S_{11}^{\min}/20}}{1 - 10^{-S_{11}^{\min}/20}} \quad \text{if the resonator is overcoupled} \tag{11.63}$$

To determine whether the coupling is undercoupled or overcoupled, one needs to switch to the polar display to see the radius of the Q circle. Note that no calculation is done in the polar display. All the calculations are obtained by using the linear display.

Step 3. Determine S_{11}^{ϕ} in dB at $\phi = 45°$ using the formula

$$S_{11}^{\phi} = 10 \log \frac{1 + 10^{-S_{11}^{\min}/10}}{2} \tag{11.64}$$

Corresponding to $S_{11} = S_{11}^{\phi}$, determine the frequencies f_1 and f_2 as shown in Figure 11.22. Calculate Q_L as

$$Q_L = \frac{f_L}{f_2 - f_1} \tag{11.65}$$

Step 4. Calculate Q_u by using $Q_u = Q_L(1 + k)$.

SUMMARY

This chapter provides an overview of the theoretical and experimental techniques for evaluating the resonant frequency and unloaded Q-factor of microwave resonators. Closed form expressions are included for the resonant frequency and unloaded Q of standard rectangular and circular waveguide resonators. The use of the transverse resonance technique for the evaluation of the resonant frequency of transmission line resonators is described by considering a microstrip resonator loaded with a capacitive discontinuity. The chapter also illustrates two approaches for calculating the resonant frequency of arbitrarily-shaped resonators: the eigen-mode analysis and the *S*-parameter analysis. Examples are given illustrating the implementation of these two techniques using EM-based commercial software tools such as HFSS.

The computation of the Q value is affected by the uncertainties in the properties of materials used in the construction of microwave filters. For example, in dielectric resonators filters, there is usually an uncertainty in the value of the loss tangent of the dielectric materials. Moreover, tuning screws can degrade the Q-value by a large amount. For such resonator structures, Q-measurement is perhaps the only approach that can provide reliable results. A step by step procedure is outlined for measuring the loaded and unloaded Q values using either the polar display of a vector network analyzer or the linear display of a scalar network analyzer.

REFERENCES

1. R. E. Collin, *Foundation for Microwave Engineering*, McGraw-Hill, New York, 1966.

2. D. Pozar, *Microwave Engineering*, 2nd ed., Wiley, New York, 1998.

3. A. E. Atia and A. E. Williams, Narrow bandpass waveguide filters, *IEEE Trans. Microwave Theory Tech.* **MTT-20**, 238–265 (1972).

4. A. E. Atia and A. E. Williams, New types of waveguide bandpass filters for satellite transponders, *COMSAT Tech. Rev.* **1**(1), 21–43 (1971).

5. S. J. Fiedziusko, Dual-mode dielectric resonator loaded cavity filter, *IEEE Trans. Microwave Theory Tech.* **MTT-30**, 1311–1316 (1982).

6. D. Kajfez and P. Guilon, *Dielectric Resonators*, Artech House, Norwood, MA, 1986.

7. R. R. Mansour, Design of superconductive multiplexes using single-mode and dual-mode filters, *IEEE Trans. Microwave Theory Tech.* **MTT-42**, 1411–1418 (1994).

8. V. Walker and I. C. Hunter, Design of triple mode TE_{01} resonator transmission filters, *IEEE Microwave Wireless Compon. Lett.* **12**, 215–217 (June 2002).

9. R. R. Mansour, Microwave superconductivity, *IEEE Trans. Microwave Theory Tech.* **MTT-50**, 750–759 (1972).

10. E. O. Hammerstad and O. Jensen, Accurate models for microstrip computer-aided design, *IEEE-MTT-S Digest*, 1980, pp. 407–409.

11. I. Bahl and P. Bhartia, *Microwave Solid State Circuit Design*, 2nd ed., Wiley, New York, April 2003.

12. T. Itoh, *Numerical Techniques for Microwave and Millimeter-Wave Passive Structures*, Wiley, New York, April 1989.

13. High frequency system simulator (HFSS), www.Ansoft.com.

14. em sonnet, www.sonnetusa.com.

15. R. E. Collin, *Field Theory of Guided Waves*, McGraw-Hill, New York, 1960.

16. www.cst.com, CST Microwave Studio.

17. A. E. Atia and A. E. Williams, General TE01 mode waveguide bandpass filters, *IEEE Trans. Microwave Theory Tech.* **MTT-24**, 640–648 (Oct. 1976).

18. E. L. Ginzton, *Microwave Measurements*, McGraw-Hill, New York, 1957.

19. M. Sucher and J. Fox, eds., *Handbook of Microwave Measurements*, Polytechnic Press, New York, 1963.

20. D. Kajfez, *Q Factor*, Vector Forum, Oxford, MS, 1994.

21. D. Kajfez and E. J. Hwan, Q-factor measurement with network analyzer, *IEEE Trans. Microwave Theory Tech.* **MTT-32**, 666–670 (July 1984).

22. W. P. Wheless and D. Kajfez, Microwave resonator circuit model from measured data fitting, *1986 IEEE MTT-S Symp. Digest*, Baltimore, June 1986, pp. 681–684.

23. A. Asija and A. Gundavajhala, Quick measurement of unloaded *Q* using a network analyzer, *Rf Design* 48–52 (Oct. 1994).

24. D. Kajfez, *Q*-factor measurement with a scalar network analyzer, *IEE Proc. Microwave Anten. Propag.* **142** (Oct. 1995).

25. M. C. Sanchez, E. Martin, and J. M. Zamarro, Unified and simplified treatment of techniques for characterizing transmission, reflection, or absorption resonators, *IEE Proc.* **137**(4) (Pt. H). (Aug. 1990).

26. T. Miura, T. Takahashi, and M. Kobayashi, Accurate Q-factor evaluation by resonance curve area method and it's application to the cavity perturbation, *IEICE Trans. Electron.* **E77-C**(6), 900–907 (June 1994).

27. K. Leong and J. Mazierska, Accurate measurements of surface resistance of HTS films using a novel transmission mode Q-factor technique, *J. Superconduct.* **14**(1), 93–103 (2001).

28. K. Leong and J. Mazierska, Accounting for lossy coupling in measurements of Q-factors in transmission mode dielectric resonators, *Proc. Asia Pacific Microwave conf.*, APMC'98, Yokohama, Dec. 8–11, 1998, pp. 209–212.

CHAPTER 12

WAVEGUIDE AND COAXIAL LOWPASS FILTERS

Synthesis methods presented in Chapters 7, 8, and 9 are based on lossless, lumped-element prototype filters. Such a model is suitable for realizing practical narrowband (typically <2% bandwidth) filters at microwave frequencies. The bulk of filter applications in a communication system is for bandpass filters with bandwidths of less than 2%. The lumped-element model serves well to satisfy this need. However, communication systems also require lowpass filters. The bandwidth requirement for lowpass filters is much higher (typically in the GHz range), and therefore, prototype models based on lumped elements are not suitable for realization at microwave frequencies. It requires use of distributed elements for the prototype filters and, as a consequence, modification of the synthesis techniques.

In this chapter, we study the methods for synthesizing the transfer and reflection polynomials, suitable for two important classes of lowpass filter (LPF): the distributed *stepped impedance* (SI) filter and the *mixed lumped distributed* filter. The theory behind the synthesis and realization of lowpass filters is due mainly to Levy [1–5], who produced a series of papers in the 1960s and 1970s. He introduced a class of prototype polynomials that is suitable for representing the LPF transfer and reflection performance, and from which a lowpass prototype circuit may then be synthesized. These include the all-pole Chebyshev function, the Chebyshev function of the second kind, and the Achieser–Zolotarev function [2,6]. Once a circuit is synthesized from these functions, a microwave LPF can be realized in rectangular waveguide, coaxial transmission-line, or planar (TEM) technology, for example, stripline.

Microwave Filters for Communication Systems: Fundamentals, Design, and Applications, by Richard J. Cameron, Chandra M. Kudsia, and Raafat R. Mansour
Copyright © 2007 John Wiley & Sons, Inc.

Following on now is a description of the types of prototype polynomial suitable for realization of lowpass filters in a waveguide, coaxial or planar technology, and some methods for their generation. Then we shall examine the commensurate line elements that go into the lowpass prototype network.

Firstly the transforms that are used for the Stepped Impedance lowpass filter will be studied, and the resultant transfer polynomials matched with those derived from the structure itself, which leads to the network synthesis method. A useful by-product of this synthesis method, the short-step impedance transformer, which has a filter-like transfer performance, will also be presented. The slightly more complex synthesis procedure for the lumped/distributed lowpass filter will then be detailed, and the methods for dimensioning the structure outlined.

12.1 COMMENSURATE-LINE BUILDING ELEMENTS

An essential component in the realization of microwave filters is the *commensurate-line elements*. These are short-lengths transmission line, all of the same electrical length θ_c. With these elements, the distributed equivalents of lumped capacitors and inductors may be created.

Commensurate elements have the same sign and value as those of their lumped equivalents, but a frequency dependence that varies as $t = j \tan \theta$ instead of $s = j\omega$, where $\theta = \omega l/v_p = 2\pi l/\lambda = \beta l$ is the electrical length of the element at frequency ω, l is the physical length of the commensurate line, and v_p is the velocity of propagation in the transmission-line medium. Assuming that v_p is constant at all frequencies, $\theta_0 = \omega_0 l/v_p$, where θ_0 is the electrical length at a reference frequency ω_0. This yields $\theta = (\omega/\omega_0)\theta_0$, the frequency variable θ, in terms of ω. $t = j \tan \theta$ is known as *Richard's transform* [7], and is used extensively as the frequency variable in commensurate-line networks.

The lumped components and their commensurate-line equivalents shown in Figure 12.1 have the same value of impedance or admittance at the frequency f_c at which the length of line is θ_c radians, but deviate from each other as the frequency moves away from f_c. Notably, the lumped element's reactance changes monotonically as the frequency moves toward infinity, whereas the distributed component's reactance is cyclic, repeating every π radians.

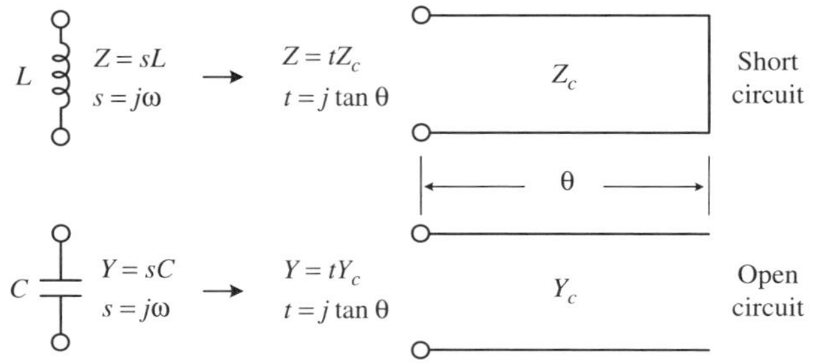

Figure 12.1 Commensurate-line equivalents for (*a*) lumped inductor and (*b*) lumped capacitor.

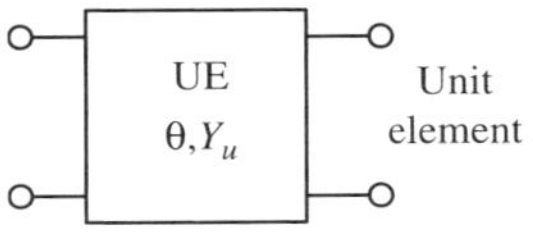

$$\begin{bmatrix} A & B \\ C & D \end{bmatrix} = \begin{bmatrix} \cos\theta & j\sin\theta/Y_u \\ jY_u\sin\theta & \cos\theta \end{bmatrix}$$

$$\begin{bmatrix} A & B \\ C & D \end{bmatrix} = \frac{1}{\sqrt{(1-t^2)}} \cdot \begin{bmatrix} 1 & t/Y_u \\ Y_u t & 1 \end{bmatrix}$$

$$\begin{bmatrix} y_{11} & y_{12} \\ y_{21} & y_{22} \end{bmatrix} = \begin{bmatrix} D/B & -\Delta_{ABCD}/B \\ -1/B & A/B \end{bmatrix} = \frac{1}{t} \cdot \begin{bmatrix} Y_u & -Y_u\sqrt{(1-t^2)} \\ -Y_u\sqrt{(1-t^2)} & Y_u \end{bmatrix}$$

Figure 12.2 Unit element (UE) and associated [ABCD] transfer and admittance matrix.

Another transmission-line component that is used extensively is the unit element (UE), which has no direct lumped-element equivalent. Figure 12.2 shows a unit element of phase length θ and characteristic admittance Y_u, and gives the associated [ABCD] transfer and admittance matrices in terms of $t = j\tan\theta$, obtained with the use of the identities $\cos\theta = 1/\sqrt{1+\tan^2\theta}$ and $\sin\theta = \tan\theta/\sqrt{1+\tan^2\theta}$.

These elements are also widely used for the synthesis and modeling of combline, interdigital bandpass filters, couplers, impedance transformers, and matching devices.

12.2 LOWPASS PROTOTYPE TRANSFER POLYNOMIALS

For the synthesis of stepped impedance and lumped/distributed lowpass filters, the microwave structures can support only certain types of filter functions.

Stepped impedance

- All-pole functions (no transmission zeros)
- Even- or odd-degree functions
- Equiripple all-pole Chebyshev, the most common class, although others can be used
- All-pole Zolotarev functions

Lumped/distributed

- Only odd-degree polynomials can be realized
- One pair of "half-zeros" necessary in the transfer polynomials
- Chebyshev functions of the second kind with a "half-zero" pair
- Zolotarev functions, also with a half-zero pair

12.2.1 Chebyshev Polynomials of the Second Kind

The equiripple all-pole Chebyshev polynomial of the first kind is the regular polynomial that is used for the design of bandpass filters. The methods for the synthesis of these polynomials are detailed in Chapter 6. The transfer and reflection functions are given by

$$S_{21}(\omega) = \frac{1}{\varepsilon E(\omega)}, \qquad S_{11}(\omega) = \frac{F(\omega)}{E(\omega)} \qquad (12.1)$$

where $E(\omega)$ is an Nth-degree monic Chebyshev polynomial of the first kind. The lumped/distributed lowpass filter functions require Chebyshev polynomials of the second kind, with one symmetric half-zero pair of transmission zeros. The transfer and reflection functions have the form

$$S_{21}(\omega) = \frac{\sqrt{\omega^2 - a^2}}{\varepsilon E(\omega)}, \qquad S_{11}(\omega) = \frac{F(\omega)}{E(\omega)} \qquad (12.2)$$

where $\pm a$ are the positions of the half-zeros, and $E(\omega)$ is now an Nth-degree Chebyshev polynomial of the second kind. This is compared with the more familiar Chebyshev function of the first kind with one symmetric "full-zero" pair, for which the synthesis is established in Chapter 6, represented by

$$S_{21}(\omega) = \frac{(\omega^2 - a^2)}{\varepsilon E(\omega)}, \qquad S_{11}(\omega) = \frac{F(\omega)}{E(\omega)} \qquad (12.3)$$

Figure 12.3 compares the performances of the two kinds of Chebyshev function, showing that the transmission zeros of the second kind are "thinner" than the first kind, although the far-out-of-band rejection is better. In these seventh-degree cases, the positions of the full-zero position of the first kind is adjusted until the rejection lobe level is 50 dB. Then the positions of the half-zeros of the second kind are set in the same positions. It is evident that the lobe levels of the second kind are lower, and close-to-band selectivity is less. The in-band return loss performance levels of both kinds are is almost identical.

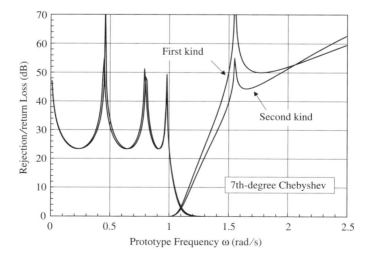

Figure 12.3   Comparison of Chebyshev functions of the first and second kinds, with a full-zero pair (first kind), and a half-zero pair (second kind).

The Nth-degree numerator and denominator polynomials for the Chebyshev function of the second kind for up to N half-zero pairs is generated by a zero-mapping formula and recursive method, similar to that used in Chapter 6 for generation of the $F(\omega)$ polynomial of the first kind from the given $P(\omega)$ polynomial. Using the same symbols as previously [8], we obtain

$$C_N(\omega) = \frac{F(\omega)}{P(\omega)} = \cosh\left[\sum_{n=1}^{N} \cosh^{-1}(x_n)\right] \tag{12.4}$$

where

$$x_n = \omega\sqrt{\frac{(1 - 1/\omega_n^2)}{(1 - \omega^2/\omega_n^2)}}$$

The factors a_n and b_n are calculated by

$$a_n = x_n = \omega\sqrt{\frac{\left(1 - 1/\omega_n^2\right)}{\left(1 - \omega^2/\omega_n^2\right)}}$$

$$b_n = \sqrt{x_n^2 - 1} = \frac{\omega'}{\sqrt{\left(1 - \omega^2/\omega_n^2\right)}} \tag{12.5}$$

where $\omega' = \sqrt{\omega^2 - 1}$, the transformed frequency variable.

Finally, the C_N polynomial is formed:

$$C_N(\omega) = \frac{\displaystyle\prod_{n=1}^{N}\left[\omega\sqrt{\left(1 - 1/\omega_n^2\right)} + \omega'\right] + \prod_{n=1}^{N}\left[\omega\sqrt{\left(1 - 1/\omega_n^2\right)} - \omega'\right]}{2\displaystyle\prod_{n=1}^{N}\sqrt{\left(1 - \omega^2/\omega_n^2\right)}}$$

$$= \frac{\displaystyle\prod_{n=1}^{N}[c_n + d_n] + \prod_{n=1}^{N}[c_n - d_n]}{2\displaystyle\prod_{n=1}^{N}\sqrt{\left(1 - \omega^2/\omega_n^2\right)}} \tag{12.6}$$

where $c_n = \omega\sqrt{\left(1 - 1/\omega_n^2\right)}$ and $d_n = \omega'$.

Using the same recursion method as that in Section 6.3, the $F(\omega)$ polynomial is formed, and knowing the prescribed half-zero-position polynomial $P(\omega)$, one can determine the denominator polynomial $E(\omega)$.

The routine to generate the $F(\omega)$ polynomial is similar to the one in Chapter 6. Again, the complex array XP contains the prescribed half-zero positions, including those at $\omega = \infty$:

```
X=1.0/XP(1)**2          initialize with the first
                        prescribed half-zero ω₁

Y=CDSQRT(1.0-X)
U(1)=0.0
```

```
        U(2)=CDSQRT(1.0D0-X)
        V(1)=1.0
        V(2)=0.0
C
        DO 10 K=3, N+1                    multiply by the second and
                                         subsequent prescribed half-zeros
        X=1.0 / XP(K-1)**2
        Y=CDSQRT(1.0-X)
        DO 11 J=1, K-1                   multiply by the constant terms
        U2(J)=-V(J)
11      V2(J)=U(J)
        DO 12 J=2, K                     multiply by the terms in ω
        U2(J)=U2(J)+Y * U(J-1)
12      V2(J)=V2(J)+Y * V(J-1)
        DO 13 J=3, K                     multiply by the terms in ω²
13      U2(J)=U2(J)+V(J-2)
        DO 14 J=1, K                     update Uₙ and vₙ
        U(J)=U2(J)
14      V(J)=V2(J)
10      CONTINUE                         reiterate for 3rd, 4th, ..., Nth
                                         prescribed zero
```

As before, the array U now contains the coefficients of the $F(\omega)$ polynomial, and computing its roots yield the N singularities of the numerator of the reflection function $S_{11}(\omega)$. If it is required, the $N-1$ in-band reflection maxima can be found by determining the roots of polynomial V. The alternating-pole principle (see Section 6.3) cannot be used here directly to create the coefficients of $E(\omega)$ for odd numbers of half-zero pairs, because of the square root in the numerator of $S_{21}(\omega)$. However, it is shown in the section on lumped-distributed LPFs that after transformation to the t-plane, $S_{21}(t)$ will always have an even number of half-zeros. Thus, the square root sign in the numerator of $S_{21}(\omega)$ disappears, the $P(t)$ polynomial formed, and the alternating principle can be used as before to find $E(t)$.

12.2.2   Achieser–Zolotarev Functions

Achieser–Zolotarev [2] functions are similar to the Chebyshev functions of the second kind in that they have an equiripple in-band amplitude characteristic. However, they possess an extra design parameter that allows the peak nearest to the origin to exceed the preset equiripple level. Figure 12.4 illustrates the in-band amplitude behavior of an even-degree (8th) and an odd-degree (9th) Achieser–Zolotarev function.

Figure 12.5 compares the in-band and rejection performances of the seventh-degree Chebyshev, second kind, and Zolotarev functions. In Figure 12.5a, the higher level first ripple level of the Zoloterev function is clearly seen, whereas the

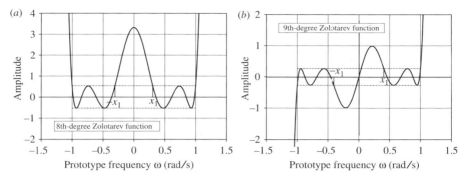

Figure 12.4   In-band (voltage) amplitude of (*a*) an 8th-degree function and (*b*) a 9th-degree Zolotarev function.

out-of-band rejection of Figure 12.5*b* indicates that the rejection performances are quite similar, with Zolotarev somewhat superior. However, it is not for this reason that the Zolotarev function is valuable in the design of LPFs. The function tends to yield better element values with less abrupt transitions, and greater internal gap dimensions that help with high power design.

Synthesis of Achieser–Zolotarev Functions   The all-pole even-degree Achieser–Zolotarev function (used for the stepped impedance LPF but not for the lumped/distributed types) is easily generated from the zeros of the regular Chebyshev function of the same degree and prescribed return loss level through the following mapping formula

$$s_k' = \pm\sqrt{s_k^2(1 - x_1^2) - x_1^2}\qquad(12.7)$$

where s_k is the original position of the singularity in the complex s plane, s_k' is the transformed position, and x_1 ($|x_1| < 1$) is the frequency point in the band at which the equiripple behavior starts (see Fig. 12.4*a*). For an *N*th degree all-pole Chebyshev

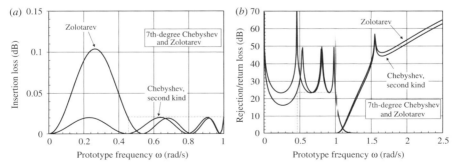

Figure 12.5   Comparison of seventh-degree Chebyshev and Zolotarev functions: (*a*) in-band amplitude; (*b*) out-of-band rejection.

function (N even or odd) with a prescribed return loss level RL (dB), the poles of $S_{21}(s)$ [the roots s_{pk} of $E(s)$] and the zeros of $S_{11}(s)$ [the roots s_{zk} of $F(s)$] are easily determined from [9]

$$s_{pk} = j \cosh(\eta + j\theta_k)$$
$$s_{zk} = j \cos \theta_k$$

where

$$\eta = \frac{1}{N} \ln(\varepsilon_1 + \sqrt{\varepsilon_1^2 + 1}), \quad \varepsilon_1 = \sqrt{10^{R/10-1}}, \quad \theta_k = \frac{(2k-1)\pi}{2N},$$
$$k = 1, 2, \ldots, N \tag{12.8}$$

When $x_1 = 0$, the even-degree Zolotarev function degenerates to the pure all-pole Chebyshev function. These same even-degree Zolotarev functions are sometimes used for the design of dual-passband filter prototypes.

In contrast to the ease with which the polynomials of the even-degree Zolotarev function are determined, those for the odd-degree Zolotarev function, with or without half-zero pairs, are significantly more difficult to generate. The polynomial synthesis procedure is described as clearly as it can be by Levy [2], and tables of element values are also included in Ref. 6.

12.3 SYNTHESIS AND REALIZATION OF THE DISTRIBUTED STEPPED IMPEDANCE LOWPASS FILTER

The prototype stepped impedance LPF is realized with a cascade of commensurate-line elements, all of the same electrical length θ_c, but with adjacent elements having different impedances in the direction of propagation. The objective of this section is to demonstrate that the polynomials that represent the transfer and reflection characteristics of such a cascade are in the same form as the polynomials that represent the design transfer and reflection characteristics, transformed by the ω-plane to θ-plane mapping function

$$\omega = \frac{\sin \theta}{\sin \theta_c} = a \sin \theta \tag{12.9}$$

where θ_c is the commensurate-line length and $a = 1/\sin \theta_c$.

The effect of applying this mapping function to the ω-plane transfer function is shown in Figure 12.6a, where, the transfer and reflection characteristics of an all-pole Zolotarev function are seen in the ω plane with its cutoff frequency at $\omega = \pm 1$ and return loss in the equiripple region of 30 dB. Figure 12.6b exhibits the characteristics after mapping to the θ plane by the function equation (12.9) with a commensurate length $\theta_c = 20°$, that is, $a = 2.9238$.

As θ increases from zero, the corresponding frequency variable in the ω plane ω increases, reaching band edge at $\omega = +1$ when $\theta = \theta_c$ (20°). The range $\theta_c \le \theta \le 90°$ maps into the range $1 \le \omega \le a$. As θ increases beyond 90° towards 180°, ω retraces its path back to zero and then on to $-a$ as θ reaches 270°. Thus, the repeating pattern in Figure 12.6b is mapped from the portion of the ω-plane characteristic between $\omega = \pm a$.

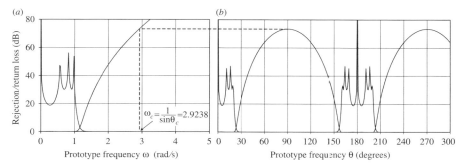

Figure 12.6 Stepped impedance LPF mapping of an all-pole transfer and reflection function from the ω plane to the θ plane through the mapping function $\omega = a \sin \theta$.

12.3.1 Mapping the Transfer Function S_{21} from the ω Plane to the θ Plane

By using equation (12.9) (mapping function) and applying the identity $\sin \theta = \tan \theta / \sqrt{1 + \tan^2 \theta}$, we obtain

$$\omega = \frac{a \tan \theta}{\sqrt{1 + \tan^2 \theta}} \tag{12.10}$$

Putting $t = j \tan \theta$, it follows that

$$s = j\omega = \frac{at}{\sqrt{1 - t^2}} \tag{12.11a}$$

or

$$t = \frac{\pm s}{\sqrt{a^2 + s^2}} = \frac{\pm s \sin \theta_c}{\sqrt{1 + (s \sin \theta_c)^2}} \tag{12.11b}$$

The s-plane all-pole transfer function has the form $S_{21}(s) = 1/(\varepsilon E(s))$, where $E(s)$ is the Nth-degree denominator polynomial of the all-pole Chebyshev [equation (12.8)] or Zolotarev transfer function. By changing the frequency variable from s to t using equation (12.11a), we obtain

$$S_{21}(t) = \frac{1}{\varepsilon E\left(\dfrac{at}{\sqrt{1 - t^2}}\right)} = \frac{\left[\sqrt{1 - t^2}\right]^N}{\varepsilon' E'(t)} \tag{12.12}$$

where $E'(t)$ is another Nth-degree polynomial in variable $t = j \tan \theta$. The most accurate method for determining $E'(t)$ is to transform the N s-plane singularities of $E(s)$ to the t plane by using equation (12.11b), and then factorizing them to form the new polynomial in t. Then, normalizing constant ε' is found by evaluating it at $\theta = \theta_c$ (corresponding to $\omega = 1$), where the return loss level is known. Note that the N transmission half-zeros of $S_{21}(t)$ occur at $t = \pm 1$, equivalent to $s = \pm j\infty$ in the s plane.

Form of $S_{21}(t)$ for the Circuit Approach   As mentioned, the Nth-degree stepped impedance LPF consists of a cascade of N commensurate-line lengths

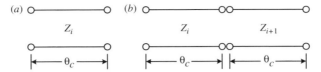

Figure 12.7   Commensurate-line-lengths (UEs): (*a*) single UE; (*b*) two UEs in cascade.

with different characteristic impedances Z_i, $i = 1, 2, \ldots, N$. One such line length is shown in Figure 12.7*a*, and its [*ABCD*] transfer matrix is given by

$$[ABCD] = \begin{bmatrix} \cos\theta & jZ_i \sin\theta \\ \dfrac{jZ_i}{\sin\theta} & \cos\theta \end{bmatrix} = \frac{1}{\sqrt{1-t^2}} \begin{bmatrix} 1 & Z_i t \\ \dfrac{t}{Z_i} & 1 \end{bmatrix} \tag{12.13}$$

since $t = j\tan\theta$, $\cos\theta = 1/\sqrt{1-t^2}$ and $j\sin\theta = t/\sqrt{1-t^2}$.

Now, by cascading two UEs of impedance Z_i and Z_{i+1} (Fig. 12.7*b*), we obtain

$$[ABCD] = \left[\frac{1}{\sqrt{1-t^2}}\right] \cdot \begin{bmatrix} 1 + t^2 \dfrac{Z_l}{Z_{l+1}} & t(Z_l + Z_{l+1}) \\ t\left(\dfrac{1}{z_l} + \dfrac{1}{z_{l+1}}\right) & 1 + t^2 \dfrac{Z_{l+1}}{Z_l} \end{bmatrix}$$

$$= \left[\frac{1}{\sqrt{1-t^2}}\right]^2 \cdot \begin{bmatrix} A_2(t) & B_1(t) \\ C_1(t) & D_2(t) \end{bmatrix} \tag{12.14}$$

For *N* lines

$$[ABCD] = \left[\frac{1}{\sqrt{1-t^2}}\right]^N \cdot \begin{bmatrix} A_N(t) & B_{N-1}(t) \\ C_{N-1}(t) & D_N(t) \end{bmatrix} \quad (N \text{ even})$$

$$= \left[\frac{1}{\sqrt{1-t^2}}\right]^N \cdot \begin{bmatrix} A_{N-1}(t) & B_N(t) \\ C_N(t) & D_{N-1}(t) \end{bmatrix} \quad (N \text{ odd}) \tag{12.15}$$

where the subscripts indicate the degree of the polynomials in the variable *t*. The transfer function associated with this [*ABCD*] matrix is derived from

$$S_{21}(t) = \frac{2[1-t^2]^{N/2}}{A(t) + B(t) + C(t) + D(t)} = \frac{[1-t^2]^{N/2}}{\varepsilon_t E'(t)} \tag{12.16a}$$

$$= \frac{[(1-t)(1+t)]^{N/2}}{\varepsilon_t E'(t)} = \frac{\sqrt{P(t)/\varepsilon_t}}{E'(t)} \tag{12.16b}$$

where $P(t)$ is the *N*th-degree numerator polynomial of $S_{21}(t)$ and constants are consolidated into ε_t. It is obvious at this stage that equation (12.16a) is in the same form as equation (12.12), derived by directly transforming the all-pole Chebyshev or Zolotarev function to the *t* plane with the mapping function (12.11). Note that the numerator polynomial $P(t)$ is of degree N, with N transmission half-zeros at $t = \pm 1$, which means that the function is fully canonical. The corresponding reflection function is $S_{11}(t) = \frac{F(t)/\varepsilon_{Rt}}{E'(t)}$, and because the conservation of energy condition must still be satisfied, ε_{Rt} is nonunity (although very close to unity for the higher degrees).

Although it is not possible to form the polynomial $\sqrt{P(t)}$ for odd-degree functions, it is actually not needed for the synthesis procedure. ε_t is evaluated at a convenient frequency point in the filter function where $S_{21}(t)$ and $S_{11}(t)$ are known, such as at the cutoff frequency, where $\theta = \theta_c$, the cutoff angle. Then

$$\varepsilon_t = \frac{[1 - t^2]^{N/2}}{\left(\sqrt{1 - 10^{-RL/10}}\right) \cdot |E'(t)|}\Bigg|_{t=t_c} \quad \text{and} \quad \varepsilon_{Rt} = \frac{\varepsilon_t}{\sqrt{\varepsilon_t^2 - 1}} \tag{12.17}$$

where $t_c = j \tan \theta_c$.

12.3.2 Synthesis of the Stepped Impedance Lowpass Prototype Circuit

The synthesis of the stepped impedance LPF cascade follows lines very similar to those used for the LP prototype networks for bandpass filters, with a knowledge of the *form* of the $A(t)$, $B(t)$, $C(t)$, and $D(t)$ polynomials [see equation (12.15)]. As a symmetric characteristic about $t = 0$, the $F(t)$ polynomial is even or odd, depending on whether N is even or odd, and both $E'(t)$ and $F(t)$ have pure real coefficients.

The impedance Z_{in}, looking in at the input of the circuit, is

$$Z_{\text{in}} = \frac{A(t)Z_L + B(t)}{C(t)Z_L + D(t)} = \frac{1 + S_{11}(t)}{1 - S_{11}(t)} = \frac{E'(t) + F(t)/\varepsilon_{Rt}}{E'(t) - F(t)/\varepsilon_{Rt}} \tag{12.18}$$

where Z_L is the load impedance terminating the output of the network, and it is assumed that the source impedance $Z_S = 1$. In the synthesis procedure, we shall be taking ratios such as $A(t)/C(t)$ and $B(t)/D(t)$ to evaluate the Z_i, the characteristic impedances of the UE cascade. Therefore, it is unimportant whether or not Z_L is included in equation (12.18). For reasons of clarity, we omit Z_L for the synthesis procedure described next.

Knowing that $A(t)$ and $D(t)$ are even polynomials, and $B(t)$ and $C(t)$ are odd for N even, and vice versa for N odd [see equation (12.15)], the $A(t)$, $B(t)$, $C(t)$, and $D(t)$ polynomials are constructed from the coefficients of the $E'(t)$ and $F(t)/\varepsilon_{Rt}$ polynomials as follows

$$A(t) = (e_0 + f_0) + (e_2 + f_2)t^2 + (e_4 + f_4)t^4 + \cdots$$

$$B(t) = (e_1 + f_1)t + (e_3 + f_3)t^3 + (e_5 + f_5)t^5 + \cdots$$

$$C(t) = (e_1 - f_1)t + (e_3 - f_3)t^3 + (e_5 - f_5)t^5 + \cdots$$

$$D(t) = (e_0 - f_0) + (e_2 - f_2)t^2 + (e_4 - f_4)t^4 + \cdots \tag{12.19}$$

where e_i and f_i, $i = 0, 1, 2, \ldots, N$ are the coefficients of the $E'(t)$ and $F(t)/\varepsilon_{Rt}$ poly-nomials, respectively. Writing the polynomials in this way ensures that the form is correct, and that the transfer and reflection functions are satisfied [see equation (12.16a)]:

$$S_{12}(t) = S_{21}(t) = \frac{\sqrt{P(t)}/\varepsilon_t}{E(t)} = \frac{2\left[1 - t^2\right]^{\frac{N}{2}}/\varepsilon_t}{A(t) + B(t) + C(t) + D(t)} \tag{12.20a}$$

$$S_{11}(t) = S_{22}(t) = \frac{F(t)/\varepsilon_{Rt}}{E(t)} = \frac{A(t) + B(t) - C(t) - D(t)}{A(t) + B(t) + C(t) + D(t)} \tag{12.20b}$$

Network Synthesis   Following the methods described in Section 7.2.1 to extract inverters, UEs are removed in sequence from the [$ABCD$] matrix, leaving a remainder matrix after each extraction. First, the overall [$ABCD$] matrix is decomposed into the first UE and a remainder matrix expressed as

$$\frac{\varepsilon_t}{[1 - t^2]^{N/2}} \cdot \begin{bmatrix} A(t) & B(t) \\ C(t) & D(t) \end{bmatrix} = \frac{1}{[1 - t^2]^{1/2}} \cdot \begin{bmatrix} 1 & Z_1 t \\ t/Z_1 & 1 \end{bmatrix} \frac{\varepsilon_t}{[1 - t^2]^{(N-1)/2}} \begin{bmatrix} A_{\text{rem}}(t) & B_{\text{rem}}(t) \\ C_{\text{rem}}(t) & D_{\text{rem}}(t) \end{bmatrix}$$

$$= \frac{\varepsilon_t}{[1 - t^2]^{N/2}} \begin{bmatrix} A_{\text{rem}}(t) + tZ_1 C_{\text{rem}}(t) & B_{\text{rem}}(t) + tZ_1 D_{\text{rem}}(t) \\ C_{\text{rem}}(t) + \dfrac{tA_{\text{rem}}(t)}{Z_1} & D_{\text{rem}}(t) + \dfrac{tB_{\text{rem}}(t)}{Z_1} \end{bmatrix} \tag{12.21}$$

The open-circuit impedance of this circuit may be calculated in two ways:

$$z_{11} = \left. \frac{A(t)}{C(t)} \right|_{t=1} = \left. \frac{A_{\text{rem}}(t) + tZ_1 C_{\text{rem}}(t)}{C_{\text{rem}}(t) + tA_{\text{rem}}(t)/Z_1} \right|_{t=1} = Z_1 \tag{12.22a}$$

$$z_{11} = \left. \frac{B(t)}{D(t)} \right|_{t=1} = \left. \frac{B_{\text{rem}}(t) + tZ_1 D_{\text{rem}}(t)}{D_{\text{rem}}(t) + tB_{\text{rem}}(t)/Z_1} \right|_{t=1} = Z_1 \tag{12.22b}$$

Having evaluated Z_1 from either of the formulas in equation (12.22), the first UE is extracted from the initial [$ABCD$] matrix, leaving the remainder matrix [$ABCD$]$_{\text{rem}}$:

$$\frac{\varepsilon_t}{[1 - t^2]^{(N-1/2)}} \begin{bmatrix} A_{\text{rem}}(t) & B_{\text{rem}}(t) \\ C_{\text{rem}}(t) & D_{\text{rem}}(t) \end{bmatrix} = \frac{1}{[1 - t^2]^{1/2}} \cdot \begin{bmatrix} 1 & -Z_1 t \\ -t & 1 \\ Z_1 & \end{bmatrix} \cdot \frac{\varepsilon_t}{[1 - t^2]^{N/2}} \begin{bmatrix} A(t) & B(t) \\ C(t) & D(t) \end{bmatrix}$$

$$= \frac{\varepsilon_t}{[1 - t^2]^{(N+1)/2}} \begin{bmatrix} A(t) - tZ_1 C(t) & B(t) - tZ_1 D(t) \\ C(t) - \dfrac{tA(t)}{Z_1} & D(t) - \dfrac{tB(t)}{Z_1} \end{bmatrix}$$

$$\tag{12.23}$$

Finally, the right-hand matrix in equation (12.23) must be divided top and bottom by $(1 - t^2)$ to reduce the denominator term to the same power as that of the remainder matrix in the left-hand side (LHS) of equation (12.21). After the extraction of Z_1, the numerator polynomials are divisible by $(1 - t^2)$, leaving the polynomials $A_{rem}(t)$, $B_{rem}(t)$, $C_{rem}(t)$, $D_{rem}(t)$, which will now each be one degree less than the original $A(t)$, $B(t)$, $C(t)$, and $D(t)$ polynomials. The process is repeated on the remainder matrix to extract the second UE Z_2, and so on until all N of the UEs are extracted.

The load termination is calculated by recognizing that at zero frequency, the cascade of UEs is effectively transparent, and $Z_{in} = Z_L$. Thus, we evaluate equation (12.18) at $t = 0$ by

$$Z_{in}\Big|_{t=0} = \frac{E'(t) + F(t)/\varepsilon_{Rt}}{E'(t) - F(t)/\varepsilon_{Rt}}\Big|_{t=0} = \frac{e_0 + f_0}{e_0 - f_0} = Z_L \tag{12.24}$$

For odd-degree characteristics, $f_0 = 0$ (point of perfect transmission at $t = 0$), and so Z_L is equal to unity and equal to the source impedance Z_S. For even-degree characteristics, $f_0 \neq 0$ and $Z_L > 1$.

It can also be shown [3] that the UE values of odd-degree networks are equal about the center of the network, whereas those of even-degree networks are antimetric:

$$Z_k Z_{N-k+1} = 1 \ (N \ \text{odd}) \quad \text{and} \quad Z_k Z_{N-k+1} = Z_L \ (N \ \text{even}), \ k = 1, 2, \ldots, N \tag{12.25}$$

This useful feature, where the second half of UE values are calculated from the first half, is used to improve or check the accuracy of the synthesis.

Either of the expressions for z_{11} in equation (12.22) is used to carry out the network synthesis. However, to avoid the accumulation of errors during the synthesis process, it is best to use only one or the other throughout the synthesis, not to mix them. Best results are usually obtained by carrying out parallel independent synthesis by using both formulas, and then using symmetry (odd degree) or antimetry (even degree) to determine which synthesis route has diverged the least.

12.3.3 Realization

The cascade of UEs that form the prototype stepped impedance lowpass filter can, in theory, be realized directly in waveguide/coaxial, or planar structures as a series of transmission lines, all of the same length. The characteristic impedances are set according to the Z_i, obtained during the synthesis procedure. For a rectangular waveguide realization, this means that the waveguide height dimension (b) varies in proportion to the value of Z_i. When Z_i is high, then b is high, and vice versa. b is normalized to the incoming waveguide dimension, either full-height or transformed down. A sixth-degree example is given in Figure 12.8.

However, this gives some quite abrupt impedance changes associated with stray capacitances at the junctions, and performance can be quite severely degraded from the ideal. Better results are obtained if redundant impedance inverters are introduced

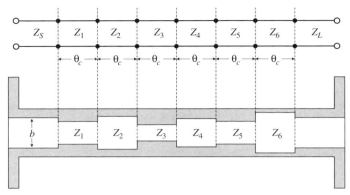

Figure 12.8   Sixth-order stepped impedance LPF prototype circuit and direct rectangular waveguide realization.

at the junctions by using the dual-network theorem on alternate UEs. This tends to equalize the line impedances and gives a degree of freedom, allowing the lines impedances to be prescribed.

Once the redundant inverters are introduced, their impedances are scaled (see Fig. 12.9), keeping the coupling coefficient $k_{i,i+1}$ constant:

$$k_{i,i+1} = \frac{K'_{i,i+1}}{\sqrt{Z'_i \cdot Z'_{i+1}}} = \frac{K''_{i,i+1}}{\sqrt{Z''_i \cdot Z''_{i+1}}}$$

(12.26)

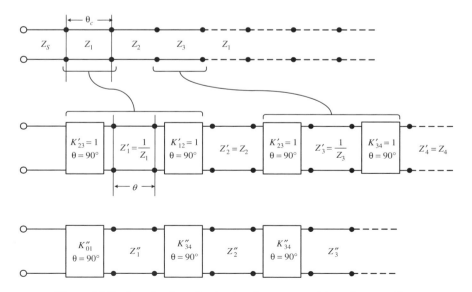

Figure 12.9   Introduction of redundant inverters, and impedance scaling.

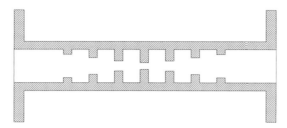

Figure 12.10 Sixth-degree stepped impedance LPF in a rectangular waveguide.

For example, if all Z_i'' are set at unity (a common case), the impedance values of the inverters are given by

$$K_{i,i+1}'' = \frac{1}{\sqrt{Z_i' \cdot Z_{i+1}'}}, \qquad i = 0, 1, 2, \cdots, N \tag{12.27}$$

where $Z_0 \equiv Z_S = 1$ and $Z_{N+1} \equiv Z_L$. Then, $Z_1'' = Z_2'' = \cdots = Z_N'' = Z_L'' = 1$. The inverters now have nonunity values, and are realized as capacitive irises in a rectangular waveguide structure. If the network is antimetric (even degree) to start with, it becomes symmetric, operating between the unity terminations as in the odd-degree cases. Figure 12.10 illustrates a possible realization in the waveguide for a sixth-degree filter.

To illustrate the design procedure, the design of the prototype circuit for a sixth degree Zolotarev stepped impedance LPF, with a return loss of 26 dB, start of the equiripple region θ_1 set at $8°$, and the cutoff angle $\theta_c = 22°$, is selected.

Initially, equation (12.8) is applied to generate the all-pole Chebyshev singularities [roots of $E(\omega)$ and $F(\omega)$], and then by using the mapping function (12.7), the ω-plane singularities are transformed to those for the even-degree Zolotarev function, as shown in the first two columns of Table 12.1. These singularities are then transformed to the t plane using equation (12.11), and are listed in the second two columns of Table 12.1.

Reforming the t-plane polynomials, creating the $[ABCD]$ polynomials, and synthesizing the cascade of six UEs by using the evaluation and extraction method gives the values for the Hi-Lo cascade of UEs in the first two columns of

TABLE 12.1 Poles and Zeros of Sixth-Degree Zolotarev Function Singularities in ω and t Planes

ω Plane		t Plane	
Poles [Roots of $E(\omega)$]	Zeros [Roots of $F(\omega)$]	Mapped Poles	Mapped Zeros
$-0.1486 \pm j1.1328$	$\pm j0.9707$	$-0.0747 \pm j0.4654$	$\pm j0.3904$
$-0.3952 \pm j0.8516$	$\pm j0.7543$	$-0.1707 \pm j0.3234$	$\pm j0.2946$
$-0.4907 \pm j0.3429$	$\pm j0.4425$	$-0.1851 \pm j0.1230$	$\pm j0.1681$
		$\varepsilon = 704.58$	$\varepsilon_R = 1.000001$

TABLE 12.2 **Element Values of Stepped Impedance Lowpass Filter**

	Hi-Lo Cascade		After Inclusion of Inverters $K'_{i,i+1}$		
	Unit Element Impedances		UE Impedances ($K'_{i,i+1} = 1.0$)	Unit Elements and Z_L Scaled to Unity	
Z_1	2.62571		2.62571	K''_{S1}	0.61713
Z_2	0.32087		3.11657	K''_{12}	0.34957
Z_3	5.22472		5.22472	K''_{23}	0.24782
Z_4	0.32208		3.10479	K''_{34}	0.24829
Z_5	5.24455		5.24455	K''_{45}	0.24782
Z_6	0.64089		1.56033	K''_{56}	0.34957
Z_L	1.68279			K''_{6L}	0.61713

Table 12.2. Now, inserting unity impedance inverters $K_{i,i+1}$ between the elements (and one at the beginning of the cascade) has the effect of reciprocating every second UE (dual-network theorem), as shown in the third column. Finally, the impedances of the UEs and the terminating impedances are scaled to unity by equation (12.26) (see the final column in Table 12.2). It is evident that the values of the network are now symmetric about the center.

A circuit analysis of this network yields the lowpass return loss and rejection characteristics shown in Figure 12.11. It clearly shows the reduced level of the first ripple of the return loss characteristic up to $\theta = 8°$, but maintaining 26 dB between $\theta = 8°$ and $22°$.

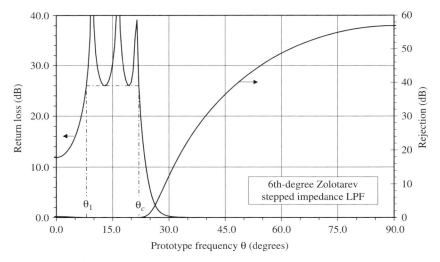

Figure 12.11 Rejection and return loss of the sixth degree 26 dB return loss stepped impedance lowpass filter, with $\theta_1 = 8°$ and $\theta_c = 22°$.

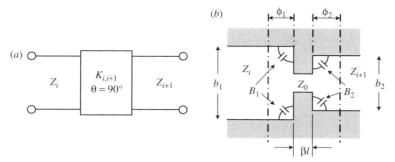

Figure 12.12   Equivalent circuits: (a) impedance inverter; (b) capacitive iris in a rectangular waveguide.

Realization of the Impedance Inverters   The impedance inverters of the prototype network are equated to a capacitive iris in a rectangular waveguide, situated between two phase planes (see Fig. 12.12). Then, the characteristic impedance of the prototype inverter, which is proportional to the height of the iris b_0, can be related to the impedance Z_0 of the iris according to the following procedure:

$$
\begin{bmatrix} a'' & jb'' \\ jc'' & d'' \end{bmatrix} \qquad \begin{bmatrix} a' & jb' \\ jc' & d' \end{bmatrix}
$$

$$a'' = d'' = 0 \qquad a' = Z_{n1}(c - sB_2Z_0)$$

$$b'' = \frac{K_{i,i+1}}{Z_{n2}} \qquad b' = s\frac{Z_0}{Z_{n2}}$$

$$c'' = \frac{Z_{n2}}{K_{i,i+1}} \qquad c' = Z_{n2}\left[c(B_1 + B_2) + s\left[\frac{1}{Z_0} - B_1B_2Z_0 \right] \right]$$

$$d' = \frac{c - sB_1Z_0}{Z_{n1}} \tag{12.28}$$

where

$$Z_{n1} = \sqrt{Z_{i+1}/Z_i} \text{ and } Z_{n2} = \sqrt{Z_iZ_{i+1}}$$

B_1, B_2 = the fringing capacitances of the capacitive steps [10]

l = thickness of the iris (usually preset)

$\beta = 2\pi/\lambda_{gi}$, λ_{gi} wavelength in iris section

$c = \cos\beta l$, $s = \sin\beta l$

Z_i, Z_{i+1} and $K_{i,i+1}$ = values of characteristic impedances of UEs and inverters from prototype

The value of the unknown, the impedance Z_0 of the capacitive step, may be found by equating the matrix elements in the following equation, and iteratively solving

$$1 + \frac{1}{4}(a' - d')^2 + \frac{1}{4}(b' - c')^2 = 1 + \frac{1}{4}(a'' - d'')^2 + \frac{1}{4}(b'' - c'')^2 \tag{12.29}$$

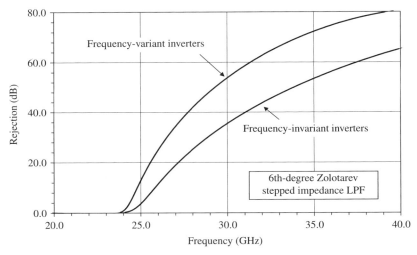

Figure 12.13 Comparison of the rejection performance of a sixth-degree stepped impedance LPF with and without frequency variation in the UEs.

The reference phase planes are then calculated by [11]

$$\tan 2\phi_1 = \frac{2(b'd' - a'c')}{(a'^2 - d'^2) + (b'^2 - c'^2)}$$

$$\tan 2\phi_2 = \frac{2(a'b' - c'd')}{(d'^2 - a'^2) + (b'^2 - c'^2)} \qquad (12.30)$$

It should be noted that inclusion of the inverters in the prototype network has no effect on the performance of the prototype network, because the inverters are frequency-independent. However, the inverters are converted into capacitive irises to allow realization of the lowpass characteristic in a waveguide structure. These irises are frequency-dependent, and therefore behave as additional elements of filter design. Thus, the structure in Figure 12.10, although derived from a 6th-degree prototype network, exhibits the performance of a 13th-degree $(6 + 7)$ filter. Figure 12.13 illustrates the effect for an LPF with a cutoff frequency of 23.5 GHz.

12.4 SHORT-STEP TRANSFORMERS

The short-step transformer is introduced here because of the similarity of its design procedure to that for a stepped impedance LPF, using even-degree Zolotarev functions and the same commensurate-line synthesis method. The short-step transformer is a useful matching element between transmission lines of different impedance levels, from a full-height to reduced-height waveguide. A special case of the short-step transformer is noted in which it degenerates to the more familiar quarter-wavelength transformer.

It was noted above that when the synthesis procedure for stepped impedance LPFs is applied to an even-degree LPF filter function, the resulting UE characteristic impedances Z_i are antimetric in value about the center of the filter. Consequently, the load impedance Z_L is unequal in value to the source Z_S. This is so, because at zero frequency, the even-degree filtering function has a nonzero insertion loss, and the cascade of UEs realizing it is transparent; that is, the source of impedance $Z_S = 1 \, \Omega$ is directly connected to the load Z_L. Thus, it is the nonunity Z_L and the resultant mismatch between the source and load that is responsible for the finite return loss at $\omega = 0$, according to the relationship

$$S_{11}(\omega)|_{\omega=0} = \frac{f_0}{e_0} = 10^{-RL_0/10}$$

$$Z_L = \frac{1 + S_{11}(\omega)}{1 - S_{11}(\omega)}\bigg|_{\omega=0} = \frac{e_0 + f_0}{e_0 - f_0} \qquad (12.31)$$

where RL_0 is the return loss of the LP filter function at zero frequency and e_0 and f_0 are the constant coefficients of the polynomials $E(\omega)$ and $F(\omega)$, respectively.

The large and adjustable first ripple of the even-degree Zolotarev function is ideal for providing the necessary return loss RL_0 at zero frequency for a given degree, this value can be adjusted by altering the equiripple return loss level or the bandwidth of the Zolotarev function [the equiripple region between $\omega = x_1$ and $\omega = 1$ in the ω plane (see Fig. 12.4), or the equivalents $\theta = \theta_1$ and $\theta = \theta_c$ in the θ-plane]. Usually, the return loss level is chosen as the variable, and a simple single-variable solving routine is set up to converge at the value of return loss that gives the desired Z_L through equation (12.31), fixing the degree and bandwidth. If the resultant value for RL_0 is unacceptably low or inappropriately high, the degree or bandwidth may then be adjusted and the process repeated.

The design process for the short-step transformer allows for a tradeoff between wide equiripple bandwidth and out-of-band rejection levels on the upper side (the rejection level on the lower side is defined by the required transformer ratio).

Defining $\theta_b = \theta_c - \theta_1$ as the bandwidth and $\theta_0 = (\theta_c + \theta_1)/2$ as the center frequency of the transformer, setting θ_0 below $45°$ tends to increase the rejection on the upper side to levels above those on the lower side, and vice versa for θ_0 above $45°$. Widening the equiripple bandwidth θ_b also tends to lower the rejection levels on the upper side, but, in practice, all of these considerations of rejection are usually secondary to those of a wide bandwidth of operation, good return loss, and ease of realization.

We illustrate the procedure by taking an example of a sixth-degree transformer that is required to transform to an impedance level of $Z_L = 4Z_S$, and $\theta_1 = 20°$, and $\theta_c = 50°$, that is, $\theta_0 = 35°$. The iterative procedure results in an equiripple return loss level of 21.57 dB between $\theta = 20°$ and $50°$. At this return loss level, the sixth-degree Zolotarev function gives the desired insertion loss of 1.94 dB (4.44 dB return loss) at zero frequency, corresponding to the specified $1:4$ source:load impedance ratio. The insertion loss and return loss characteristics are shown in Figure 12.14a, and the six synthesized line impedance values are given in Table 12.3.

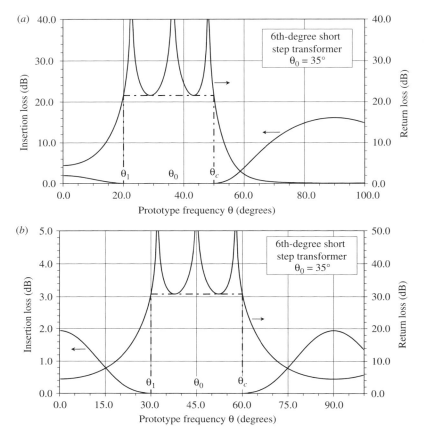

Figure 12.14 Insertion loss and return loss characteristics of the symmetric $1:4$ short-step transformer: (a) $\theta_0 = 35°$ and (b) $\theta_0 = 45°$.

TABLE 12.3 Short-Step Transformer Line Impedances for Center Frequencies $\theta_0 = 35°$ and $\theta_0 = 45°$

	Line Impedances; $\theta_0 = 35°$	Line Impedances; $\theta_0 = 45°$
Z_S	1.0	1.0
Z_1	1.6427	1.2417
Z_2	0.9661	1.2417
Z_3	2.9717	2.0000
Z_4	1.3460	2.0000
Z_5	4.1403	3.2215
Z_6	2.4350	3.2215
Z_L	4.0	4.0

An interesting feature of the short-step transformer is that if the center frequency is set at 45°, the impedance values of the adjacent pairs of the UEs along the length of the transformer become equal in value. Let us now take the same example, where $\theta_1 = 30°$ and $\theta_c = 60°$, that is $\theta_0 = 45°$. For this case, the equiripple return loss level is 30.8 dB, and the insertion loss and return loss characteristics are shown in Figure 12.14b. Note that they have become symmetric about 45°. The synthesized line impedance values for this symmetric case are listed in Table 12.3. It should be noted that the adjacent internal steps have become equal-valued. Now, the short-step transformer has become, in effect, a regular quarter-wave transformer, with three steps, each $2 \times 45° = 90°$ at the center frequency. The impedance levels are still antimetric about the center, with the value of the middle step equal to $\sqrt{Z_L}$, if $N/2$ is odd.

12.5 SYNTHESIS AND REALIZATION OF MIXED LUMPED/ DISTRIBUTED LOWPASS FILTER

Like the stepped impedance circuit, the lumped/distributed (L/D) lowpass filter circuit is composed of a series of UEs, but is distinct from the stepped impedance LPF circuit in that the UEs are in pairs, each pair having the same characteristic impedance Z_i, and at the junction of each pair, a distributed shunt capacitor is located. Figure 12.15 shows a cascade of one such UE pair with an open-circuited UE connected in shunt at each end, forming the distributed capacitors (see Fig. 12.15b). As such, this circuit represents a third-degree prototype of a tapered–corrugated lowpass filter.

At a frequency where the lengths θ_c of the UEs become 90°, it can be seen that the open circuits at the ends of the shunt UEs transform to short circuits across the main-line, and the double-length UEs become 180°. Thus, this third-degree circuit gives a transmission zero at the frequency at which θ_c of the UEs become 90° (or 90° plus an integer multiple of 180°).

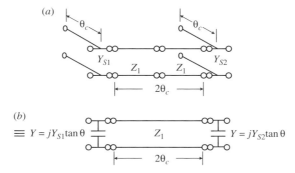

Figure 12.15 Basic prototype elements of the lumped/distributed filter: (*a*) double-length unit element with a shunt-connected open-circuited UE at each end (distributed capacitors); (*b*) with the distributed capacitors represented by their lumped equivalents.

Like the stepped impedance LPF, we demonstrate that the class of circuit as shown in Figure 12.15 has the same form of transfer function as a Chebyshev or Zolotarev function of the second type, with a pair of transmission half-zeros at a frequency of $\omega = 1/(\sin \theta_c)$. Here, θ_c is the chosen cutoff frequency for the distributed prototype, and is transformed with the same mapping formula as the one for the stepped impedance LPF.

12.5.1   Formation of the Transfer and Reflection Polynomials

The mapping formula for the stepped impedance LPF is repeated here

$$\omega = \frac{\sin \theta}{\sin \theta_c} = a \sin \theta \tag{12.32}$$

where θ_c is the commensurate-line length and $a = 1/\sin \theta_c$. In this case, a represents the position of the half-zero pair in the ω plane, which maps to $90°$ in the θ plane (see Fig. 12.16). Again, it can be seen that the repeating pattern in Figure 12.16b is mapped from the portion of the ω-plane characteristic between $\omega = \pm a$. The transfer function for the Chebyshev function of the second kind or Zolotarev function with a single half-zero pair has the following form in the ω plane:

$$S_{21}(\omega) = \frac{\sqrt{\omega^2 - a^2}}{\varepsilon E(\omega)} \tag{12.33}$$

Applying the transformation to $S_{21}(\omega)$, we obtain

$$S_{21}(\theta) = \frac{\sqrt{a^2 - a^2 \sin^2 \theta}}{\varepsilon E(a \sin \theta)} = \frac{\cos \theta}{\varepsilon_t E(a \sin \theta)} \tag{12.34}$$

where the constants are consolidated into ε_t.

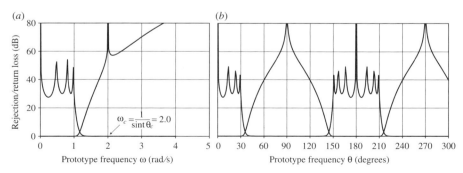

Figure 12.16   Mapping of the seventh-degree Zolotarev transfer and reflection functions from the ω plane (a) to the θ plane (b) through the mapping function $\omega = a \sin \theta$, where $a = 1/\sin(\theta_c)$. In this case $\theta_1 = 10°$, $\theta_c = 30°$, and the return loss = 30 dB.

Applying the identities $\sin\theta = \tan\theta/\sqrt{1 + \tan^2\theta}$ and $\cos\theta = 1/\sqrt{1 + \tan^2\theta}$, and letting $t = j\tan\theta$, we obtain

$$S_{21}(t) = \frac{1}{\sqrt{1 - t^2}} \cdot \frac{1}{\varepsilon_t E\left(-jta/\sqrt{1 - t^2}\right)} = \frac{\left[\sqrt{1 - t^2}\right]^{N-1}}{\varepsilon_t' E'(t)} \qquad (12.35)$$

where ε_t' is another constant and $E'(t)$ is an odd-degree polynomial in the variable t. The numerator is a polynomial $\left[\sqrt{1 - t^2}\right]^{N-1}$, which, because $N - 1$ is an even number, can be formed into the $(N - 1)$th-degree even polynomial $P(t)$, the numerator of $S_{21}(t)$. From this, we observe that $S_{21}(t)$ has $N - 1$ zeros at $t = \pm 1$.

Knowing the form of the $S_{21}(t)$ function from the mapped prototype polynomials that represent the lowpass transfer function, we can check whether the transfer function of the commensurate-line circuit, as shown in Figure 12.15, has the same form. Once this is established, the $[ABCD]$ polynomials are formed from the prototype polynomials and the synthesis of the circuit can begin.

If we take a cascade of two UEs of commensurate length θ_c and both of the same characteristic impedance Z_1 (Fig. 12.17a), then the transfer $[ABCD]$ matrix for the UE pair is

$$[ABCD] = \begin{bmatrix} \cos\theta & jZ_1\sin\theta \\ \dfrac{j\sin\theta}{Z_1} & \cos\theta \end{bmatrix}^2 = \cos^2\theta \begin{bmatrix} 1 & jZ_1\tan\theta \\ \dfrac{j\tan\theta}{Z_1} & 1 \end{bmatrix}^2$$

$$= \frac{1}{1 - t^2} \cdot \begin{bmatrix} 1 + t^2 & 2Z_1 t \\ \dfrac{2t}{Z_1} & 1 + t^2 \end{bmatrix} \qquad (12.36)$$

As before, $t = j\tan\theta$. Now we pre- and postmultiply by the distributed capacitive stubs (Fig. 12.17b) and obtain

$$\frac{1}{1 - t^2} \cdot \begin{bmatrix} 1 & 0 \\ Y_{S1}t & 1 \end{bmatrix} \cdot \begin{bmatrix} 1 + t^2 & 2Z_1 t \\ \dfrac{2t}{Z_1} & 1 + t^2 \end{bmatrix} \cdot \begin{bmatrix} 1 & 0 \\ Y_{S2}t & 1 \end{bmatrix} = \frac{1}{1 - t^2} \cdot \begin{bmatrix} A_2(t) & B_1(t) \\ C_3(t) & D_2(t) \end{bmatrix} \qquad (12.37)$$

where again the subscripts represent the degree of the polynomial in the variable t. This is now the $[ABCD]$ matrix for a third-degree lumped/distributed lowpass circuit as shown in Figure 12.15.

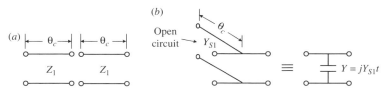

Figure 12.17 Components of the tapered–corrugated LPF: (a) double unit element; (b) distributed shunt capacitor.

With the addition of more UE pairs and shunt capacitors, the general form for an Nth-degree $[ABCD]$ matrix becomes

$$[ABCD] = \frac{1}{[1-t^2]^{(N-1)/2}} \cdot \begin{bmatrix} A_{N-1}(t) & B_{N-2}(t) \\ C_N(t) & D_{N-1}(t) \end{bmatrix} \tag{12.38}$$

from which the transfer function $S_{21}(t)$ is derived as

$$S_{21}(t) = \frac{2[1-t^2]^{(N-1)/2}}{A_{N-1}(t) + B_{N-2}(t) + C_N(t) + D_{N-1}(t)} = \frac{[1-t^2]^{(N-1)/2}}{\varepsilon_t E_N(t)} \tag{12.39}$$

This is in the same form as equation (12.35), which results from the transform of the transfer function polynomials.

Figure 12.15 confirms that the class of circuit has the same transfer characteristic (and because it is lossless, it also has the same reflection characteristic) as the Chebyshev or Zolotarev polynomials. At this point, we can generate the $E(t)$ and $F(t)$ polynomials to satisfy a particular specification, relate them to the $[ABCD]$ polynomials of the L/D prototype circuit, and then synthesize the values of the circuit elements.

12.5.2   Synthesis of the Tapered–Corrugated Lowpass Prototype Circuit

Using the methods presented in Section 12.1, the reflection function numerator polynomial $F(\omega)$, corresponding to the transfer function $S_{21}(\omega)$ with a single half-zero pair $P(\omega) = \sqrt{\omega^2 - a^2}$ as the numerator, can be found. The next task is to find $E(\omega)$, the denominator polynomial common to both $S_{11}(\omega)$ and $S_{21}(\omega)$, so that the network synthesis for the lumped/distributed prototype network can begin. However, because the numerator polynomial of $S_{21}(\omega)$ is in the form $P(\omega) = \sqrt{\omega^2 - a^2}$, the alternating pole principle cannot be used directly to find $E(\omega)$.

However, we now know that with transformation of the frequency variable from ω to $j \tan \theta$, a rational function for $S_{21}(t)$ in the form of equation (12.35) is formed. This has $P(t) = [1-t^2]^{(N-1)/2}$ as its denominator, which, because N is an odd number, represents a polynomial in the variable t.

So the method used to find the polynomials of the reflection function $S_{11}(t) = F(t)/E(t)$ would proceeds as follows:

1. With the desired lowpass cutoff angle θ_c, which determines the width of the reject band before the second harmonic passband cuts in (see Fig. 12.16), calculate the Tx half-zero position $a = 1/\sin \theta_c$. Then, with the return loss equiripple level that is desired over the passband, form the polynomial $F(\omega)$ by using the method of Section 12.1 for the Chebyshev (second kind) case or [2] for the odd-degree Zolotarev function, both with the single half-zero pair at $\omega = \pm a$ in the numerator.

2. Root the polynomial $F(\omega)$ and transform its singularities to the t plane by the transform (12.11b). Form the polynomial $F(t)$ from the t-plane singularities, and the polynomial $P(t)$ from $P(t) = [1-t^2]^{(N-1)/2}$.

3. Derive ε_t at a convenient frequency where the return loss level is known, such as at the cutoff frequency where $t = t_c = j\tan\theta_c$:

$$\varepsilon_t = \frac{1}{|F(t_c)|} \cdot \frac{[1 - t_c^2]^{(N-1)/2}}{\sqrt{10^{RL/10} - 1}} \tag{12.40}$$

4. Form the polynomial $E(t) = P(t) \pm \varepsilon_t F(t)$, root it, and reflect any singularities that are in the right half-plane back to the left half-plane to preserve the Hurwitz condition. Now the Hurwitz polynomial $E(t)$ can be formed.

With the $E(t)$ and $F(t)$ polynomials, we begin the synthesis of the network, employing the same methods as for the stepped impedance lowpass filter. Noting the form of the polynomials in the cascade matrix [equation (12.38)], we can form the $A(t)$, $B(t)$, $C(t)$, and $D(t)$ polynomials of the t plane $[ABCD]$ matrix as follows:

$$A(t) = (e_0 - f_0) + (e_2 - f_2)t^2 + (e_4 - f_4)t^4 + \cdots$$

$$B(t) = (e_1 - f_1)t + (e_3 - f_3)t^3 + (e_5 - f_5)t^5 + \cdots$$

$$C(t) = (e_1 + f_1)t + (e_3 + f_3)t^3 + (e_5 + f_5)t^5 + \cdots$$

$$D(t) = (e_0 + f_0) + (e_2 + f_2)t^2 + (e_4 + f_4)t^4 + \cdots \tag{12.41}$$

where e_i and f_i, $i = 0, 1, 2, \ldots, N$, are the coefficients of the $E(t)$ and $F(t)/\varepsilon_{Rt}$ polynomials, respectively. ε_{Rt} is unity here because the numerator polynomial is one less in degree than the denominator (i.e., the filtering function is noncanonical).

Network Synthesis   For the first synthesis cycle, the overall ladder network is assumed to consist of a shunt capacitor, followed by a double-length UE and the remainder matrix $[ABCD]^{(2)}$.

The first component to be extracted is the shunt capacitor Y_{S1}, to prepare the network for the extraction of the double-length unit element of characteristic impedance Z_1. The overall network is assumed at first to be composed of the shunt capacitor in cascade with the remainder matrix $[ABCD]^{(1)}$ (see Fig. 12.18):

$$\frac{1}{[1 - t^2]^{(N-1)/2}} \cdot \begin{bmatrix} A(t) & B(t) \\ C(t) & D(t) \end{bmatrix} = \begin{bmatrix} 1 & 0 \\ tY_{S1} & 1 \end{bmatrix} \cdot \frac{1}{[1 - t^2]^{(N-1)/2}} \cdot \begin{bmatrix} A^{(1)} & B^{(1)} \\ C^{(1)} & D^{(1)} \end{bmatrix}$$

$$= \frac{1}{[1 - t^2]^{(N-1)/2}} \cdot \begin{bmatrix} A^{(1)} & B^{(1)} \\ C^{(1)} + tY_{S1}A^{(1)} & D^{(1)} + tY_{S1}B^{(1)} \end{bmatrix} \tag{12.42}$$

The short-circuit admittance y_{11}, looking into this network, is

$$y_{11} = \frac{D(t)}{B(t)} = y_{11R} + tY_{S1} = \frac{D^{(1)}}{B^{(1)}} + tY_{S1} \tag{12.43}$$

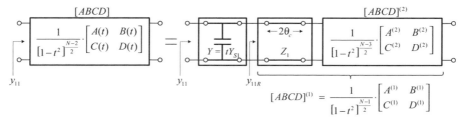

Figure 12.18 Initial synthesis cycle for the lumped/distributed LPF ladder network; the original matrix comprises a shunt capacitor + double UE + remainder matrix $[ABCD]^{(2)}$.

The problem now is to evaluate Y_{S1}. Unlike most other extractions, this one cannot be done by simply evaluating at a convenient value of t, for example, at $t = 0, 1, t_c$, or ∞ etc. However, we can exploit a feature of a network that consists of a double UE as the leading component (e.g., $[ABCD]^{(1)}$ in Fig. 12.18), that the *differential* of its input admittance with respect to t is zero at $t = \pm 1$. This is easily proved by differentiating the input admittance y_{11R} of the cascaded double-length line plus remainder network ($[ABCD]^{(1)}$ in Fig. 12.18) with respect to t, when it is evaluated at $t = 1$.

The differential y'_{11} of the input admittance y_{11} of the overall network is derived from equation (12.43) as

$$y'_{11} = \frac{dy_{11}}{dt} = \left[\frac{D(t)}{B(t)}\right]' = \left[\frac{D^{(1)}}{B^{(1)}}\right]' + Y_{S1} \qquad (12.44)$$

where the prime here denotes differentiation with respect to t. Since the first term on the right-hand side (RHS) of equation (12.44) is equal to zero when it is evaluated at $t = 1$, we have

$$y'_{11}\Big|_{t = \pm 1} = \left[\frac{D(t)}{B(t)}\right]'\Bigg|_{t = \pm 1} = Y_{S1} \qquad (12.45)$$

Now, with the value of Y_{S1} known, the synthesis continues rather more conventionally:

1. Extract the shunt admittance Y_{S1} from the overall matrix;

$$\begin{bmatrix} 1 & 0 \\ -tY_S & 1 \end{bmatrix} \cdot \frac{1}{[1-t^2]^{(N-1)/2}} \cdot \begin{bmatrix} A & B \\ C & D \end{bmatrix}$$

$$= \frac{1}{[1-t^2]^{(N-1)/2}} \cdot \begin{bmatrix} A & B \\ C - tY_S A & D - tY_S B \end{bmatrix}$$

$$= \frac{1}{[1-t^2]^{(N-1)/2}} \cdot \begin{bmatrix} A^{(1)} & B^{(1)} \\ C^{(1)} & D^{(1)} \end{bmatrix} \qquad (12.46)$$

2. The remainder matrix $[ABCD]^{(1)}$ contains a double UE plus a remainder matrix $[ABCD]^{(2)}$ (see Fig. 12.18). Evaluate the characteristic impedance Z_1 of the first double UE by

$$
\frac{1}{[1-t^2]^{(N-1)}} \cdot \begin{bmatrix} A^{(1)} & B^{(1)} \\ C^{(1)} & D^{(1)} \end{bmatrix} = \frac{1}{(1-t^2)} \cdot \begin{bmatrix} 1+t^2 & 2t\,Z_1 \\ \dfrac{2t}{Z_1} & 1+t^2 \end{bmatrix}
$$

$$
\times \frac{1}{[1-t^2]^{(N-3)/2}} \cdot \begin{bmatrix} A^{(2)} & B^{(2)} \\ C^{(2)} & D^{(2)} \end{bmatrix}
$$

$$
= \frac{1}{[1-t^2]^{(N-1)/2}} \cdot \begin{bmatrix} (1+t^2)A^{(2)} + 2t\,C^{(2)}Z_1 & (1+t^2)B^{(2)} + 2t\,D^{(2)}Z_1 \\ (1+t^2)C^{(2)} + \dfrac{2t\,A^{(2)}}{Z_1} & (1+t^2)D^{(2)} + \dfrac{2t\,B^{(2)}}{Z_1} \end{bmatrix}
$$

$$\tag{12.47}$$

The admittance y_{11R}, looking in at the input of $[ABCD]^{(1)}$, is

$$
y_{11R} = \frac{D^{(1)}}{B^{(1)}} = \frac{(1+t^2)D^{(2)} + 2t\,B^{(2)}/Z_1}{(1+t^2)B^{(2)} + 2t\,D^{(2)}Z_1} \tag{12.48}
$$

By evaluating at $t = 1$, we obtain

$$
y_{11R}\bigg|_{t=1} = \frac{D^{(1)}}{B^{(1)}}\bigg|_{t=1} = \frac{2B^{(2)}/Z_1 + 2D^{(2)}}{2B^{(2)} + 2D^{(2)}Z_1} = \frac{1}{Z_1} \tag{12.49}
$$

3. Knowing the value of Z_1, we can now extract the double-length line from $[ABCD]^{(1)}$ such that

$$
\frac{1}{(1-t^2)} \cdot \begin{bmatrix} 1+t^2 & -2Z_1 t \\ \dfrac{-2t}{Z_1} & 1+t^2 \end{bmatrix} \cdot \frac{1}{[1-t^2]^{(N-1)/2}} \cdot \begin{bmatrix} A^{(1)} & B^{(1)} \\ C^{(1)} & D^{(1)} \end{bmatrix}
$$

$$
= \frac{1}{[1-t^2]^{(N+1)/2}} \cdot \begin{bmatrix} A^{(2)} & B^{(2)} \\ C^{(2)} & D^{(2)} \end{bmatrix} \tag{12.50}
$$

4. Finally, divide the top and bottom of the remainder matrix $[ABCD]^{(2)}$ by $(1-t^2)^2$ to obtain the correct degree $(N-3)/2$ for the denominator polynomial (see $[ABCD]^{(2)}$ in Fig. 12.18). Having extracted the first shunt capacitor Y_{S1}, the numerator polynomials $A^{(2)}$, $B^{(2)}$, $C^{(2)}$, and $D^{(2)}$ are exactly divisible by $(1-t^2)^2$.

After the first shunt capacitor and first double-length line are extracted, the cycle is repeated by using the remainder matrix $[ABCD]^{(2)}$, and so on until the last shunt capacitor, which is extracted at $t = 1$, as usual.

12.5.3 Realization

Figure 12.19 shows the network for a seventh-degree prototype network for a lumped/distributed LPF, with shunt distributed capacitors Y_{Si} and the double line lengths with characteristic impedances Z_i. The shunt capacitors (Fig. 12.20a) in the prototype network translate into capacitive irises (Fig. 12.20b) in the waveguide or coaxial structure, and the matrix matching procedure to determine the value of the iris impedance, which is very similar to the method used for the SI LPF, is outlined in Figure 12.11 and equation (12.28) [11].

$$\begin{bmatrix} a'' & jb'' \\ jc'' & d'' \end{bmatrix} \qquad \begin{bmatrix} a' & jb' \\ jc' & d' \end{bmatrix}$$

$$a'' = Z_{n1} \qquad\qquad a' = Z_{n1}(c - sB_2 Z_0)$$

$$b'' = 0 \qquad\qquad b' = s\frac{Z_0}{Z_{n2}}$$

$$c'' = t_c Y_S Z_{n2} \qquad c' = Z_{n2}\left[c(B_1 + B_2) + s\left[\frac{1}{Z_0} - B_1 B_2 Z_0\right]\right]$$

$$d'' = \frac{1}{Z_{n1}} \qquad\qquad d' = (c - sB_1 Z_0)/Z_{n1} \tag{12.51}$$

where

$$Z_{n1} = \sqrt{Z_{i+1}/Z_i}, \ Z_{n2} = \sqrt{Z_i Z_{i+1}}$$

$B_1, B_2 =$ fringing capacitances of capacitive steps [10]

$l =$ thickness of iris (usually preset)

$\beta = 2\pi/\lambda_{gi}, \lambda_{gi} =$ wavelength in iris section

$c = \cos\beta l, \ s = \sin\beta l, \ t_c = \tan\theta_c$

$Z_i, Z_{i+1}, Y_{Si} =$ values of double unit element impedances and shunt capacitors from prototype

In a method similar to that used for the stepped impedance LPF, the value of the unknown, the impedance Z_0 of the capacitive step, is found by equating the matrix elements in the following equation and iteratively solving:

$$1 + \frac{1}{4}(a' - d')^2 + \frac{1}{4}(b' - c')^2 = 1 + \frac{1}{4}(a'' - d'')^2 + \frac{1}{4}(b'' - c'')^2 \tag{12.52}$$

A first estimate for Z_0 may be obtained for this iterative process if the capacitive step fringing capacitors are ignored. Solving equation set (12.51) by using equation (12.52) with B_1 and $B_2 = 0$ yields

$$Z_0 = \frac{Z_{n2}}{\sqrt{2}}\left[\sqrt{x-1} - \sqrt{x+1}\right]$$

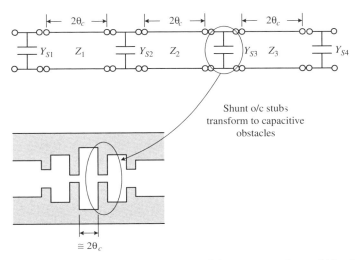

Figure 12.19 Relationships of the components of the prototype lumped/distributed LPF network to a waveguide structure (o/c = open/closed).

where

$$x = \frac{1}{2}\left[\frac{Z_2}{Z_1} + \frac{Z_1}{Z_2}\right] + \frac{Z_1 Z_2\, t_c Y_{si}}{2\sin^2 \beta l} \qquad (12.53)$$

The reference phase planes, which tend to shorten the interiris phase lengths, are calculated from equation (12.30). Small adjustments can be applied to Z_0 to compensate for proximity effects, which are the capacitive couplings that occur between adjacent irises [13,14]. A very similar method is used to design the capacitive sections for coaxial LPFs.

Synthesis Accuracy Considerations   As described in the previous section, the synthesis of the prototype network for a lumped/distributed or stepped

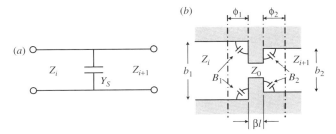

Figure 12.20 Equivalent circuits: (*a*) shunt distributed capacitor; (*b*) capacitive iris in rectangular waveguide.

impedance LPF is a progressive process: extracting the electrical components one by one in sequence from the defining polynomials, building up the ladder network, and finishing when the final component has been extracted and the polynomials are of zero degree. Accompanying such processes, computation errors accumulate, which sooner or later cause the synthesis to become unstable, usually in a spectacular and sudden fashion. More accurate input data and double-precision computing help the situation, but other simple measures can also improve the accuracy:

1. *Parallel Synthesis.* The synthesis procedure made use of the short-circuit parameter y_{11}, calculated from the B and D polynomials of the t-plane $[ABCD]$ matrix. y_{11} can also be calculated from the A and C polynomials ($y_{11} = C(t)/A(t)$) [equation (12.42)] and the element values of the network may be determined using the same procedure. This gives two sets of values for Y_{Si} and Z_i. Since we know that the values should be symmetric about the center of the filter, the set that gives the better symmetry is selected. After the set that is the most accurate is determined, the values of the components in the second half of the network can be made to be symmetric with those in the first half. It is not advisable to average the values from the two sets, particularly after each step of the synthesis process.

2. *Higher-Degree LPFs.* By using the symmetry property, an accurate synthesis may be carried out for LPFs up to degrees as high as 25 before synthesis failures start to occur. By using an adapted z-plane synthesis technique [15–17], filters with degrees up to 31 can be synthesized with reasonable accuracy.

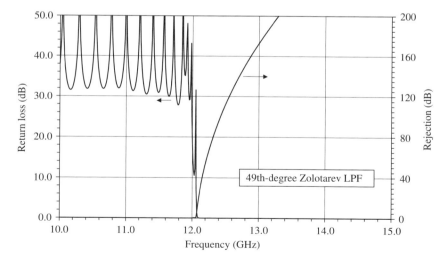

Figure 12.21   Rejection and return loss performance of a degree 49 Zolotarev LPF, designed by extrapolation from a degree 31 prototype.

To design LPFs of degrees greater than 31, we take advantage of the observation that the dimensions of the central regions of higher-degree LPFs tend to be uniform; that is, the heights of the high and low impedance sections are not significantly different from each other in the central regions of the LPF. Thus, for the high degree LPF (degree >31) it is feasible to design a degree 31 LPF (or whatever degree the limit of accuracy allows), and then "copy and paste" the dimensions of its central section repeatedly until the desired degree of LPF is obtained. Analysis indicates that this usually achieves good results with only slight degradation in return loss level near the cutoff regions. Care should be taken to ensure that after "copying and pasting," the network is still physically symmetric about the center.

Figure 12.21 shows the results of extending the central dimensions of a degree 31 Zolotarev LPF, with a design return loss of 30 dB and cutoff frequency of 12 GHz, to a degree 49 LPF. We can see that the 30 dB level is largely preserved, except for the regions near to the cutoff frequency.

Spurious Responses Since the main function of a lowpass filter is to reject noise and interfering signals up to frequencies that can be several times the cutoff frequency, preservation of the integrity of the reject bands is critical. Spurious responses occur for three main reasons:

1. The repeat passbands of the commensurate line network
2. Propagation of the higher-order TE_{n0} modes, $n = 2, 3, \ldots$
3. Mode conversion within the filter structure

Repeat Passbands The repeat passbands of a commensurate-line ladder network are an inherent feature of this class of microwave circuits and cannot be altogether avoided. Because there exist nonideal commensurate-line elements within the network such as the capacitive irises in rectangular waveguide that are realizing the distributed capacitors in lumped/distributed LPFs, the repeat passbands tend to occur at lower frequencies than the prototype predicts. This effect may be counteracted to a certain extent by choosing a lower cutoff frequency for the prototype, reducing the passband width, but this tends to make the variations in the values of the synthesized elements more extreme.

Higher-Order TE_{n0} Modes in Waveguide As the frequency of the signal increases, the width of the waveguide will become progressively cut-on for the higher ordered modes, first the TE_{20} mode, then the TE_{30} and so on. If present at the input to the LPF, these modes will propagate through the filter at the higher frequencies with theoretically the same transfer characteristics as the fundamental TE_{10} mode, although with reduced bandwidths. The spurious bands are approximately predicted as follows [5].

In the lower passband edge f_{0n}, $f_{0n} = nf_{01}$, where $f_{01} = c/2a \approx 15/a$ (GHz) is the fundamental waveguide cutoff frequency, c is the velocity of light, a (cm) is the waveguide broad dimension, and f_{0n} (GHz) is the cutoff frequency of the waveguide of the TE_{n0} mode passband. The upper band edge f_{cn} is given by

$$f_{cn} = \sqrt{f_{c1}^2 + (n^2 - 1)f_{01}^2} \qquad (12.54)$$

where f_{c1} and f_{cn} are the upper edges of the fundamental and the TE_{n0}-mode pass-bands, respectively. For waveguide WR42, which has a broad dimension of $a = 1.0668$ cm, the higher-order passbands for an LPF characteristic with a cutoff frequency of 23.5 GHz are calculated as shown in Table 12.4.

These higher-order passbands are depicted in Figure 12.22 for the TE_{20} and TE_{30} modes. If these modes are incident at the input of the LPF, in theory, they should propagate through without attenuation within their respective passbands. Ideally, there should be no such incident modes, but they can be generated by certain coaxial-to-waveguide launch methods, waveguide bends, and misalignments between sections of waveguides. The effects can be reduced to a certain extent by tapering the inner sections of the filter to scatter the higher passband responses,

TABLE 12.4   Bands of Propagation for Higher-Order Waveguide Modes

n	Mode TE_{n0}	Lower Band Edge f_{0n} (GHz)	Upper Band Edge f_{cn} (GHz)
1	TE_{10}	14.06	23.50
2	TE_{20}	28.12	33.84
3	TE_{30}	42.18	46.19
4	TE_{40}	56.24	59.31

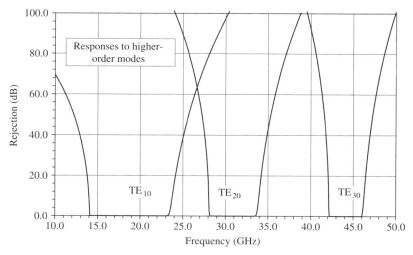

Figure 12.22   Theoretical passbands of the lowpass filter to higher-order TE modes. The TE_{10} mode forms the primary passband.

but the amount of tapering is limited by the need to avoid creating sections within the filter that are cut off to the fundamental mode.

Mode Conversion The internal generation and propagation of higher-order modes can severely limit the performance of lowpass filters in coaxial or waveguide, if the dimensions of the inner regions of the LPF allow them to exist. The higher ordered modes are generated by conversion of the fundamental mode at discontinuities within the filter, and carry energy through the filter before reconverting to the fundamental mode nearer to the output. In effect the rejection properties of the LPF are being bypassed.

These higher-order modes can be substantially reduced by ensuring that the widths and heights of all sections within the structure are no greater than those of the incoming waveguide. Usually, this necessitates the introduction of step transformers at either end to reduce the height of the waveguide at the input. Because all of the internal heights are referred to the height of this waveguide feeding the first LPF iris, this dimension is chosen to ensure that all the internal heights are less than those of the main waveguide from the RF source.

Scaling the internal heights means that the minimum heights within the filter, usually those of capacitive irises near the center, are also reduced, compromising power handling capability. This is where the advantages of using a Zolotarev LPF function becomes apparent; for the same degree of filter function, the gap sizes tend to be greater than for the Chebyshev, as well as rather less variation in the heights of the corrugations.

Example of Synthesis To illustrate the procedure, the design of the prototype circuit for a degree 17 Chebyshev lumped/distributed LPF with a return loss level of 26 dB and the cutoff angle $\theta_c = 22°$ is traced. The LPF is realized in waveguide WR42 (1.0668×0.4318 cm) with iris lengths of 0.1 cm and a design cutoff frequency of 23.5 GHz. The input/output to the LPF is in reduced-height (0.2 cm) waveguide.

In this case, the ω-plane position of the half-zero is at $\omega = a = 1/\sin \theta_c = 2.6695$. By using the procedure given in Section 12.2.1, the zeros of the reflection function $S_{11}(\omega)$ are generated and are listed in the second column of Table 12.5. The third column gives the zeros after mapping to the t plane by using equation (12.11b). The final column provides the reflection poles [denominator of $S_{11}(t)$], derived by using the alternating singularity principle.

With the t-plane poles and zeros of the reflection function, polynomials $E(t)$ and $F(t)$ are computed by using equation (12.41), and the synthesis procedure for the lumped/distributed ladder network is applied. The resultant prototype values for the admittances of the shunt capacitors and double UEs are given in columns 2 and 3 of Table 12.6. Column 4 lists the characteristic admittances of the capacitive irises, calculated from the shunt capacitance values Y_{Si}, but ignoring for the moment, the fringing capacitances B_1 and B_2. The corresponding electrical lengths of the elements are given in columns 5 and 6. Note that the lengths of the double UEs are less than $2\theta_c = 44°$, due to the difference in the phase reference planes of the shunt capacitors and the capacitive irises.

TABLE 12.5 Poles and Zeros of Degree 17 Chebyshev Reflection Function $S_{11}(\omega)$ in ω and t Planes

	ω Plane	t Plane	
	Zeros [Roots of $F(\omega)$]	Zeros [Roots of $F(t)$]	Poles [Roots of $E(t)$]
1	± 0.9958	$\pm j0.4020$	$-0.0095 \pm j0.4130$
2	± 0.9622	$\pm j0.3864$	$-0.0276 \pm j0.3965$
3	± 0.8960	$\pm j0.3563$	$-0.0437 \pm j0.3649$
4	± 0.7993	$\pm j0.3138$	$-0.0568 \pm j0.3206$
5	± 0.6753	$\pm j0.2615$	$-0.0669 \pm j0.2665$
6	± 0.5281	$\pm j0.2018$	$-0.0739 \pm j0.2052$
7	± 0.3626	$\pm j0.1371$	$-0.0786 \pm j0.1392$
8	± 0.1845	$\pm j0.0693$	$-0.0811 \pm j0.0703$
9	0.0000	0.0000	$-0.0819 \pm j0.0000$

Table 12.7 summarizes the characteristic admittances of the irises after correction for the fringing capacitances and proximity effects, which also means adjustments to the lengths of the double UEs. The final three columns give the actual dimensions of the elements, where the heights are referred to the incoming waveguide b dimension (0.2 cm), and the lengths are calculated from the wavelength in WR42 at 23.5 GHz. The widths here are constant, but can be tapered to scatter the higher-frequency spurious responses.

Analyzing these dimensions with EM mode matching software, we obtain the return loss and rejection characteristics as shown in Figure 12.24. It may be seen that the return loss is at or above the design 26 dB for about 6 GHz within the passband. An outline sketch of the 17th-degree Chebyshev lumped/distributed LPF with transformers is shown in Figure 12.23.

TABLE 12.6 Prototype Values for Degree 17 Chebyshev Lumped/Distributed Lowpass Filter Network

	Synthesized Admittances			Element Lengths	
Element i	Shunt Capacitor Y_{Si}	Double UE $1/Z_i$	Capacitive Iris Admittance $1/Z_{0i}$	Capacitive Iris	Double UE
1, 17	1.8532	—	2.2329	22.6190°	—
2, 16	—	0.4948	—	—	40.2709°
3, 15	4.0907	—	4.3459	22.6190°	—
4, 14	—	0.4219	—	—	41.8787°
5, 13	4.5219	—	4.7863	22.6190°	—
6, 12	—	0.4086	—	—	42.0657°
7, 11	4.6289	—	4.8965	22.6190°	—
8, 10	—	0.4051	—	—	42.1092°
9	4.6533	—	4.9217	22.6190°	—

TABLE 12.7   Lumped/Distributed Lowpass Filter Element Values and Electrical Lengths Corrected for the Fringing Capacitance and Proximity Effects, and Dimensions

Element	Capacitive Irises		Double UE		Heights	Lengths	Widths
i	Y_{Si}	θ_i (degs)	Y_i	θ_i (degs)	(cm)	(cm)	(cm)
1, 17	1.7021	22.6190°	—	—	0.1175	0.1000	1.0668
2, 16	—	—	0.4948	39.2847°	0.4042	0.1737	1.0668
3, 15	2.7701	22.6190°	—	—	0.0722	0.1000	1.0668
4, 14	—	—	0.4219	40.5960°	0.4741	0.1795	1.0668
5, 13	3.0036	22.6190°	—	—	0.0665	0.1000	1.0668
6, 12	—	—	0.4086	40.8383°	0.4894	0.1805	1.0668
7, 11	3.0658	22.6190°	—	—	0.0652	0.1000	1.0668
8, 10	—	—	0.4051	40.9016°	0.4937	0.1808	1.0668
9	3.0804	22.6190°	—	—	0.0649	0.1000	1.0668

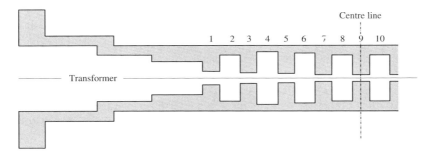

Figure 12.23   17th-degree lumped/distributed lowpass filter with transformers.

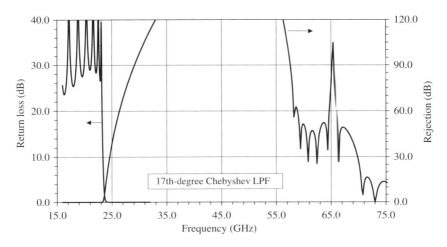

Figure 12.24   Mode-matching analysis of the 17th-degree Chebyshev lumped/distributed LPF.

SUMMARY

Synthesis methods in Chapters 7, 8, and 9 are based on lumped-element parameters. The prototype models derived from these techniques are suitable for realizing narrowband (typically <2% bandwidth), bandpass, and bandstop filters at microwave frequencies. In the microwave region, say, between 1 and 12 GHz, a 2% bandwidth translates to 20 and 240 MHz, respectively. The lowpass filters, which are required to suppress the harmonics generated by the high-power amplifiers, require bandwidths in the GHz range. As a consequence, the lumped-element prototype model is not applicable for the design of such lowpass filters. It is necessary to introduce distributed elements in the prototype filter, leading to the so-called distributed lowpass prototype model.

This chapter is devoted to the synthesis techniques for the realization of lowpass filters at microwave frequencies. The chapter begins with a review of the properties of distributed circuit elements, basically lengths of transmission line called a *unit element* (UE). It should be noted that there is no lumped element equivalent to a UE. At the prototype design stage, these UEs can differ in length, but more commonly, they are chosen to be the same length, the "commensurate line" length. The class of transfer and reflection polynomials appropriate for realization in the commensurate line circuit are then introduced.

Chebyshev functions of the second kind and the Achieser–Zolotarev function represent the prime candidates for derivation of the characteristic polynomials for a lowpass filter design. A method for evaluation of such functions is described. This is followed by the detailed procedure for the synthesis of the distributed stepped impedance lowpass filter and the mixed lumped–distributed lowpass filter. Examples of stepped impedance and lumped–distributed (tapered corrugated design) filters in waveguide structures are included to illustrate the design procedure. Brief discussions of the spurious response of such filters and methods to minimize it are included.

REFERENCES

1. R. Levy and S. B. Cohn, A history of microwave filter research, design and development, *IEEE Trans. Microwave Theory Tech.* **MTT-32**, 1055–1066 (Sept. 1984).

2. R. Levy, Generalized rational function approximation in finite intervals using Zolotarev functions, *IEEE Trans. Microwave Theory Tech.* **MTT-18**, 1052–1064 (Dec. 1970).

3. R. Levy, Tables of element values for the distributed low-pass prototype filter, *IEEE Trans. Microwave Theory Tech.* **MTT-13**, 514–536 (Sept. 1965).

4. R. Levy, A new class of distributed prototype filters with application to mixed lumped/distributed component design, *IEEE Trans. Microwave Theory Tech.* **MTT-18**, 1064–1071 (Dec. 1970).

5. R. Levy, Tapered corrugated waveguide low-pass filters, *IEEE Trans. Microwave Theory Tech.* **MTT-21**, 526–532 (Aug. 1973).

6. R. Levy, Characteristics and element values of equally terminated Achieser–Zolotarev quasi-lowpass filters, *IEEE Trans. Circuit Theory* **CT-18**, 538–544 (Sept. 1971).

7. P. I. Richards, Resistor-transmission-line circuits, *Proc IRE.* **36**, 217–220 (Feb. 1948).

8. J. D. Rhodes and S. A. Alseyab, The generalized Chebyshev low-pass prototype filter, *IEEE Trans. Circuit Theory* **8**, 113–125 (1980).

9. I. C. Hunter, *Theory and Design of Microwave Filters*, Electromagnetic Waves Series 48, IEE, London, 2001.

10. N. Marcuvitz, *Waveguide Handbook*, Electromagnetic Waves Series 21, IEE, London, 1986.

11. R. Levy, A generalized design technique for practical distributed reciprocal ladder networks, *IEEE Trans. Microwave Theory Tech.* **MTT-21**, 519–525 (Aug. 1973).

12. G. L. Matthaei, Short-step Chebyshev impedance transformers, *IEEE Trans. Microwave Theory Tech.* **MTT-14**, 372–383 (Aug. 1966).

13. P. I. Somlo, The computation of coaxial line step capacitances, *IEEE Trans. Microwave Theory Tech.* **MTT-15**, 48–53 (Jan. 1967).

14. H. E. Green, The numerical solution of some important transmission line problems, *IEEE Trans. Microwave Theory Tech.* **MTT-13**, 676–692 (Sept. 1965).

15. R. Saal and E. Ulbrich, On the design of filters by synthesis, *IRE Trans. Circuit Theory* **CT-5**, 284–327 (Dec. 1958).

16. J. A. C. Bingham, A new method of solving the accuracy problem in filter design, *IEEE Trans. Circuit Theory* **CT-11**, 327–341 (Sept. 1964).

17. H. J. Orchard and G. C. Temes, Filter design using transformed variables, *IEEE Trans. Circuit Theory* **CT-15**, 385–407 (Dec. 1968).

CHAPTER 13

WAVEGUIDE REALIZATION OF SINGLE- AND DUAL-MODE RESONATOR FILTERS

In the preceding chapters, we have studied methods for generation of the polynomials for filtering functions that have the widest practical application in today's microwave systems for terrestrial and airborne telecommunications, radar, and earth observation and scientific systems. In the great majority of cases the Chebyshev equiripple filtering function is employed, as it has the optimal balance between in-band linearity, close-to-band selectivity, and far out-of-band rejection. Specialist variants are available where high performance is required—asymmetric, group delay equalized, prescribed transmission zeros, and combinations of the foregoing. Other classes of filtering function were briefly touched on—including elliptic Butterworth, where very linear group delay performance is required, and predistorted for linear amplitude and compact size.

From the generation of transfer and reflection polynomials, attention was then focused on exact synthesis techniques for the electrical networks corresponding to the original filtering function polynomials, concentrating on those configurations that may be realized with a microwave structure of some kind, such as the coaxial resonator or waveguide cavity resonator. From the electrical network, or directly from the transfer and electrical polynomials themselves, a coupling matrix may be generated for most configurations. Having a direct correspondence with individual elements in the filter structure, the coupling matrix (CM) is a very useful representation of the filter's electrical components, bridging the gap between the pure mathematics of the transfer/ reflection polynomials and the real world of mainline couplings, resonator tuning offsets, and so on.

Microwave Filters for Communication Systems: Fundamentals, Design, and Applications,
by Richard J. Cameron, Chandra M. Kudsia, and Raafat R. Mansour
Copyright © 2007 John Wiley & Sons, Inc.

The prototype CM may be generated using the methods of Chapter 8 or by optimization methods. Once the CM has been generated, purely mathematical operations may be applied to it [e.g., similarity transforms (plane rotations) or "scaling at nodes"] to create, annihilate, or change coupling element values or tuning states. These operations preserve the CM's eigenvalues, and so an arbitrarily large series of them may be applied to the CM without destroying the unique one-to-one relationship between the CM elements in the eventually realized hardware.

In this chapter we shall study some of the more practical aspects of the filter design process, making use of the methods that have been developed to create useful filtering functions, then their realization in a microwave structure, and where possible to give design information relating the electrical values in the CM to actual dimensions in the filter structure.

In particular, dual-mode waveguide and dielectric realizations will be addressed. Dual-mode filters have found very wide application in both (1) the low-power channel filtering stages of a telecommunications system, where high selectivity and in-band linearity are required, and (2) the high-power output stages, where low loss is a priority. For satellite systems in particular, the advanced electrical performance has to be combined with lowest possible mass and volume, high stability of the electrical performance in the presence of environmental temperature fluctuations, and (for output filters) high power handling capability (conduction of dissipated RF energy to a cooling plate, ability to withstand multipactor breakdown in vacuum, corona in partial pressure conditions). And finally, the filters and multiplexer assemblies need to be robust enough to survive the rigors of the launch (vibration, shock).

Since their introduction to practical use in the early 1970s by Atia and Williams [1–4], waveguide dual-mode filters have been developed by the space industry [5–8] and refined to deliver advanced performance in the harshest of environmental conditions. More recently the methods for realizing asymmetric filtering characteristics without an accompanying increase in complexity have been introduced [9]. Asymmetric responses are useful for correcting the inherent group delay slopes that appear in wideband filters, or some of the distortions encountered in the channels at the edge of a contiguous group in an output multiplexer, or providing an efficient "tailor-made" rejection characteristic where rejection specifications are themselves asymmetric.

13.1   SYNTHESIS PROCESS

In Chapter 7, a typical process for the design and realization of a microwave Chebyshev bandpass filter in coaxial resonator technology was described (see Fig. 7.5). The procedure can be broadly broken up into three stages:

1. Design and optimization of the filter function (polynomials, singularities) by using a suitable simulator, to ensure compliance with the required specifications.

2. Synthesis of the prototype electrical network, usually in the form of a coupling matrix. The elements of the coupling matrix correspond to individual physical elements or resonator tuning states in the filter structure—diagonal elements represent tuning offsets (if any); off-diagonal elements correspond to coupling apertures, probes, loops or screws, or input/output couplings, and so on.

3. Lowpass prototype to RF bandpass mapping, dimensioning of resonant cavities, coupling apertures, and inclusion of dispersions (frequency dependencies) specific to each individual coupling element, and the unloaded Q factor (Q_u) specific to each resonator.

13.2 DESIGN OF THE FILTER FUNCTION

Design of the filter function and tradeoffs are described in Sections 3.7, 4.4, and 6.2. In Chapter 6, the design method for the general Chebyshev filter function is illustrated by including an example. The general Chebyshev function, in conjunction with transmission zeros, is most often used in the design of filters for a wide range of applications. The prescribed transmission zeros can be on the imaginary axis of the complex plane (creating transmission zeros and rejection lobes in the filter's transfer characteristic), or on the real axis (group delay equalization of symmetric characteristics) or complex (group delay equalization of symmetric or asymmetric characteristics), or combinations of the above. The recursion method described in Chapter 6 for the generation of Chebyshev polynomials will deal with TZ numbers up to the maximum possible (N zeros in an Nth-degree function, i.e., fully canonical), and by bypassing the recursion routine for the generation of the reflection function numerator polynomial $F(s)$, and putting all the reflection zeros at $s = 0$ [i.e., $F(s) = s^N$], Butterworth filtering functions may also be generated.[1]

When transmission zeros are required, it is necessary to optimize the location of these zeros with respect to the specification requirement of the amplitude or group delay or both.

13.2.1 Amplitude Optimization

In Chapter 3, we described the concept of unified design charts (UDCs) to design filters on the basis of their amplitude response. For the Butterworth, Chebyshev, and elliptic function filters, UDCs can be readily generated and are included as Appendix 3A. Use of these charts provides a good first approximation in optimizing the amplitude response of a filter. In Chapter 4, we introduced a computer-aided optimization procedure to generate filter functions with arbitrary amplitude response. A subset of this procedure is the ability to optimize the amplitude response

---

[1]Pure Butterworth functions are sometimes used when high far-out-of-band rejection slopes are required. Butterworth filters have the highest asymptotic rejection slopes of any filtering function—$6N$ dB per octave. Also, because of the relatively slow cutoff rates near to the bandedges, Butterworth filters have relatively linear in-band group delay characteristics, and are sometimes used where exceptionally flat group delay performance over a greater part of the passband is required.

of a filter function, given its degree and the number of transmission zeros. This implies optimization of the location of the transmission and reflection zeros to achieve the desired filter response.

The flow of the process to generate characteristics with prescribed rejection lobe levels and/or in-band group delay equalization is not always routine, depending sometimes on the particular filtering function under consideration. The process is often a matter of personal preference for the individual designer, but one method for optimizing rejection lobe levels and for group delay equalization is suggested here.

13.2.2   Rejection Lobe Optimization

For rejection lobes a cost function may be built up from the difference between the target lobe levels that have been set and those actually analyzed on the prototype. Because $E(s)$ is a complex polynomial with its zeros in the complex plane, its turning points (which indicate the frequency positions of the rejection lobes) cannot be directly derived by differentiating the function $S_{21}(s) = P(s)/\varepsilon E(s)$. However all the turning points may be found by differentiating the scalar power function:

$$S_{21}(s)S_{21}(s)^* = \frac{P(s)P(s)^*}{P(s)P(s)^* \pm \varepsilon^2 F(s)F(s)^*/\varepsilon_R^2} \qquad (13.1)$$

If it is assumed that all the zeros of $P(s)$ and $F(s)$ (the transmission zeros and reflection zeros, respectively) are either symmetric about the imaginary axis or lie on the imaginary axis itself, it may be shown that the numerator of this function after differentiating is:

$$\left(\frac{\varepsilon}{\varepsilon_R}\right)^2 P(s)F(s)[P'(s)F(s) - F'(s)P(s)] \qquad (13.2)$$

where the prime indicates differentiation with respect to s. By equating (13.2) to zero and rooting it, all the $2(N + n_{fz}) - 1$ turning points of the power transfer function may be found.

However, the computational effort may be reduced by recognizing that the polynomial $P(s)$ contains the n_{fz} turning points associated with the positions of the n_{fz} transmission zeros; likewise, $F(s)$ contains the N turning points associated with the positions of the N reflection zeros (points of perfect transmission), and so the remaining $(N + n_{fz}) - 1$ turning points may be found by solving the polynomial equation

$$F'(s)P(s) - P'(s)F(s) = 0 \qquad (13.3)$$

which is the same as evaluating the following differential equation:

$$\frac{d}{ds}\left(\frac{F(s)}{P(s)}\right) = 0 \qquad (13.4)$$

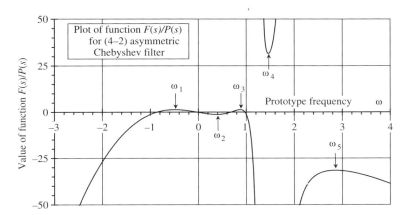

Figure 13.1 Plot of function $F(s)/P(s)$ for a (4–2) asymmetric filtering function.

Thus it is necessary only to construct the polynomial function (13.3) and root it to find the $(N + n_{fz}) - 1$ turning points, and then to select the out-of-band real frequencies that will correspond to the rejection lobe positions. Evaluating $-20 \log (S_{21}(s))$ at these out-of-band turning frequencies will give the levels of the associated rejection lobes in dB. For a Chebyshev function there will be $N-1$ in-band real frequency turning points, corresponding to the points of reflection maxima within the passband. Figure 13.1 gives a plot of the function $F(s)/P(s)$ for the (4–2) asymmetric filtering function that was designed in Chapter 6, indicating the three in-band turning points and the two rejection lobe positions.

This method of finding out-of-band rejection lobe levels involves rooting polynomials of degree $(N + n_{fz}) - 1$ instead of $2(N + n_{fz}) - 1$, and so will be faster and more accurate. For fully canonical functions ($n_{fz} = N$), the highest degree coefficients will cancel in (13.3), and there will be one less root in total ($2N - 2$). This means that if all N transmission zeros of the fully canonical filtering function are on the imaginary axis, the rejection characteristic will have $N - 1$ lobes produced by the N transmission zeros, in contrast to the n_{fz} lobes that are produced by n_{fz} transmission zeros in a noncanonical filtering function.

The method depends on the polynomials $F(s)$ and $P(s)$ having symmetry about the imaginary axis. For $P(s)$ this will always be the case, but for the $F(s)$ polynomial, the restriction must be applied that all its zeros must lie on the imaginary axis itself. It may be recalled that the same constraint applied when using the alternating pole principle (Chapter 6).

By associating the calculated lobe levels with their associated target levels, the cost function may be evaluated. The optimizer may then move the positions of the transmission zeros, the new $F(s)$ polynomial determined, and then the new lobe levels and cost function evaluated. If there is an improvement from the previous iteration, the zero positions may be moved again in the same direction, and so on until the cost function is arbitrarily close to zero.

Constraints must be placed on the optimizer to prevent it from stepping the TZs (transmission zeros) too close to the band edge, or within the passband itself ($|s| > j$). Otherwise the process seems to be stable and converges rapidly even for fully canonical functions. The $(4-2)$ asymmetric filtering function with 30 dB lobe levels that has been used as an example in previous chapters was designed using this method.

If the computer optimization routines being used to optimize the amplitude function require gradients with respect to the TZ locations, these may be computed analytically as described in Section 4.6. The relationships are summarized as follows:

$$\frac{\partial T}{\partial b_i} = -40 \log e \cdot |\rho|^2 \cdot \frac{b_i}{(s^2 \pm b_i^2)}$$

$$\frac{\partial R}{\partial b_i} = 40 \log e \cdot |t|^2 \cdot \frac{b_i}{(s^2 \pm b_i^2)}$$

where the zero locations are at $s^2 = \pm b_i^2$. The positive sign applies to the zeros on the imaginary axis, and the negative sign applies to zeros located along the real axis. T and R represent the transmission and reflection loss in dB, respectively. In terms of the scattering parameters: $T = -10 \log |S_{21}(s)|^2$ dB and $R = -10 \log |S_{11}(s)|^2$ dB. The complex zeros, located symmetrically about the imaginary axis, are defined as

$$Q(s) = (s + \sigma_1 \pm j\omega_1)(s - \sigma_1 \pm j\omega_1)$$

$$= s^2 \pm j2s\omega_1 - (\sigma_1^2 + \omega_1^2)$$

The gradients with respect to the complex zero locations are given by

$$\frac{\partial T}{\partial \sigma_1} = -40 \log e \cdot |\rho|^2 \sigma_1 \frac{(s^2 - \sigma_1^2 - \omega_1^2)}{|Q(s)|^2}$$

and

$$\frac{\partial R}{\partial \omega_1} = -40 \log e |t|^2 \omega_1 \frac{(3s^2 - \sigma_1^2 - \omega_1^2)}{|Q(s)|^2}$$

The inclusion of these analytic gradients improves the efficiency of the optimization process.

13.2.3 Group Delay Optimization

Phase and group delay responses of minimum phase filters are uniquely related to their amplitude response [10]. Minimum phase filters are characterized by the distribution of the zeros of the transfer function along the imaginary axis (elliptic and quasielliptic filters) or at infinity (maximally flat and Chebyshev filters). For such filters, once the amplitude response is optimized, we must accept the group delay response that is unique to that amplitude response. It is possible to improve the group delay without affecting the amplitude by the use of allpass equalizers that are discussed in Chapter 17.

An alternative way to improve the group delay response is by locating some of the zeros of the transfer function [zeros of $P(s)$], along the real axis or as pairs distributed symmetrically about the imaginary axis. Zeros of $P(s)$ must have symmetry about the imaginary axis to satisfy the realizability conditions, as described in Sections 3.11 and 6.1. Such filters are referred to as *nonminimum phase, linear phase,* or *self-equalized filters.* Amplitude and phase of this class of filters is no longer uniquely related. Linear phase filters can be optimized to provide an improved group delay performance at the expense of a poorer amplitude response. This implies that we need to resort to a higher-degree filter to obtain the desired amplitude response and improved group delay. As always, there is a tradeoff between the filter complexity and performance. In order to identify different types of zeros of $P(s)$, often referred to as *transmission zeros* (TZ), a notation that is widely used indicates the degree of the filter, TZs along the imaginary axis followed by TZs elsewhere. As an example, $(10-2-4)$ represents a 10th-degree filter with two TZs along the $j\omega$ axis (not necessarily symmetrically located) and four TZs distributed symmetrically about the $j\omega$ axis. The last index in this notation is always an even integer.

Next, the problem is to determine the location of the zeros to minimize the group delay variations. Because they exist only in pairs symmetrically located about the imaginary axis, the zeros themselves do not contribute to the overall phase characteristic of the filter as s varies along the imaginary axis (except for integer multiples of π radians), and therefore have no effect on group delay. However, the introduction of zeros changes the characteristic polynomials and thus, the transfer and reflection performance of the filter. Polynomials $F(s)$ and $E(s)$ need to be regenerated in order to restore the original features of the filter function (e.g., equiripple passband in the case of the Chebyshev family). As described in Section 3.4, the phase and group delay characteristics are determined entirely by the $E(s)$ polynomial.

Group Delay and Attenuation Slope Calculation

The group delay and attenuation slope characteristics of a filter function can be conveniently calculated from the transfer polynomials as follows.

If $S_{21}(s) = P(s)/\varepsilon E(s)$ is a transfer function and $S'_{21}(s)$ is its first derivative with respect to the frequency variable s, then the real part of the quantity $-S'_{21}(s)/S_{21}(s)$ is the group delay function $\tau(s)$ and the imaginary part is the gain slope of the transfer characteristic. The proof follows directly from the fact that

$$-\frac{S'_{21}(s)}{S_{21}(s)} = -\frac{d}{ds}[\ln S_{21}(s)] = -\frac{d}{ds}\Big[\ln\big(|S_{21}(s)|e^{j\phi}\big)\Big]$$

$$= -\frac{d}{ds}[\ln|S_{21}(s)| + j\phi]$$

where ϕ is the phase of $S_{21}(s)$. Because $s = j\omega$ and therefore $ds = jd\omega$, it follows that

$$-\frac{S'_{21}(s)}{S_{21}(s)} = j\frac{d}{d\omega}[\ln|S_{21}(s)| + j\phi] = -\frac{d\phi}{d\omega} + j\frac{dA}{d\omega} \qquad (13.5)$$

where $A = \ln|S_{21}(s)|$ is the gain in nepers. Thus $\mathrm{Re}[-S'_{21}(s)/S_{21}(s)]$ is the group delay in seconds and $-8.6859\ \mathrm{Im}[-S'_{21}(s)/S_{21}(s)]$ is the attenuation slope in dB per rad/s (radians per second). It can be easily shown that

$$-\frac{S'_{21}(s)}{S_{21}(s)} = \frac{E'(s)}{E(s)} - \frac{P'(s)}{P(s)} \qquad (13.6)$$

where $E'(s)$ and $P'(s)$ are the differentials of the polynomials $E(s)$ and $P(s)$ with respect to s. Because any transmission zeros are either on the imaginary axis or are in mirror-image pairs symmetrically positioned about the imaginary axis, the second term in (13.6) will always be purely imaginary and need not be calculated to obtain group delay values.

Group Delay Optimization Procedure   Although methods exist to analytically design certain types of delay-equalized filters (also known as *linear phase filters*) [10,11], the most generally useful procedure is with optimization. With the speed and power of today's PCs, the optimization process converges within a few seconds, even for the most complex cases. Moreover, with the optimization method, it is possible to simultaneously optimize transmission zeros for prescribed rejection lobe levels, and to deal with asymmetric cases. A flow diagram showing one possible organization of the group delay optimization process is given in Figure 13.2.

The procedure may be summarized as follows:

1. Check that the total number of TZs (for group delay equalization and rejection lobes) complies with the minimum path rule for the intended filter topology. The next step is to select a fraction of the bandwidth over which group delay equalization is to be carried out; 50% for a real-axis pair, and 65–70% for a quad is usually realistic. A certain amount of trading off may be done between equalization bandwidth and the amplitude of the group delay ripple within that bandwidth.

2. Postulate initial positions for the complex TZs: $\pm 0.8 + j0.0$ for real-axis pairs and $\pm 0.8 \pm j0.5$ for quads are generally successful for symmetric characteristics.

3. With the prescribed TZ positions, form the $P(s)$ polynomial, and generate the $F(s)$ and $E(s)$ polynomials. This is accomplished by optimizing the amplitude response using the analytic gradients and optimization procedure as explained in Chapter 4. For Chebyshev-type filters, it is more expedient to generate $F(s)$ and $E(s)$ polynomials by using the method described in Chapter 6.

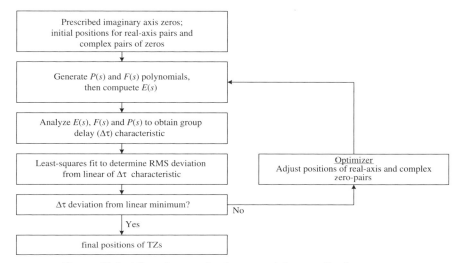

Figure 13.2 Flow diagram for the group delay equalization process.

4. Calculate the group delay τ_i at a number N_S sample points s_i across the equalization bandwidth (or half the bandwidth from the center to the edge for symmetric characteristics). With polynomial $E(s)$, the group delay at a sample frequency s_i is calculated from equation (13.6):

$$\tau_i = \text{Re}\left[\frac{E'(s_i)}{E(s_i)}\right] \tag{13.7}$$

where the prime denotes differentiation.

5. Calculate the least-squares best-fit straight line (BFSL) to the data:

$$\tau(\omega) = m\omega + c$$

where $m = \dfrac{\sum_{i=1}^{N_S} \tau_i(\omega_i - \omega_{\text{av}})}{\sum_{i=1}^{N_S} (\omega_i - \omega_{\text{av}})^2}$

$c = \tau_{\text{av}} - m\omega_{\text{av}}$

ω_{av} = average of the frequency sampling points = $\left[\sum_{i=1}^{N_S} \omega_i\right]/N_S$

τ_{av} = average delay = $\left[\sum_{i=1}^{N_S} \tau_i\right]/N_S$

$\omega_i = -js_i$ = frequency of the ith sample point

m = slope of the BFSL, c is its constant and N_S is the number of sample points (13.8)

Once the slope and the constant of the BFSL have been computed, the root-mean-square (rms) deviation of the group delay data from the BFSL is calculated by

$$\Delta\tau_{\text{rms}} = \left[\sum_{i=1}^{N_S} [\tau_i - (m\omega_i + c)]^2\right]^{1/2} \tag{13.9}$$

Both $\Delta\tau_{rms}$ and the slope of the BFSL m can be built into the error function to be minimized by the optimization routine to be as close to zero as possible. In addition, the slope bias can be built in to the error function, to counter any dispersion effects introduced by the technology that the filter is realized in, for example, a positive slope to counter the natural negative group delay slope in an iris-coupled waveguide bandpass filter. With the appropriate weightings, the imaginary-axis transmission zeros can also be built into the error function along with the complex and real-axis equalization pairs, to form a symmetric or asymmetric quasielliptic self-equalized filter function. As an example, the response of an eighth-degree filter with 23 dB return loss is displayed in Figure 13.3. Here, there is a group delay slope of 1 second over the prototype bandwidth, and two symmetric amplitude lobes of

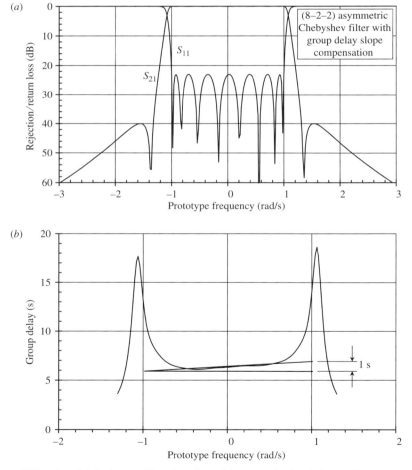

Figure 13.3   An eighth-degree filter function with two rejection lobes and a group delay compensation slope: (*a*) rejection and return loss; (*b*) group delay.

40 dB each on either side of the passband. This is achieved with two TZs at $+j1.3527$ and $-j1.3692$, and a group delay self-equalization pair at $\pm 0.8592 \pm j0.1091$.

In some cases, it is possible to optimize the characteristic with the coupling matrix model of the filter, rather than the poles and zeros of the filter function. Because of the direct correspondence between the coupling matrix entries and the actual coupling devices in the filter, individual dispersion behavior can be assigned to each coupling, and the frequency variation characteristics of each resonator can be modeled more precisely. Although the overhead of synthesizing and analyzing the matrix at each iteration adds to the computation time, a closer fit with the real performance is achieved.

13.3 REALIZATION AND ANALYSIS OF THE MICROWAVE FILTER NETWORK

Once the prototype filter function has been designed, a prototype network may be synthesized. If the filter function is an all-pole type, such as a pure Chebyshev or Butterworth with all TZs at $s = \pm j\infty$, then there are no cross-couplings, and the lowpass prototype circuit is a simple ladder structure. Typically, such all-pole networks are synchronously tuned, meaning that the FIRs at the nodes are zero in value. For filter functions with finite-position TZs, it is necessary to have cross-couplings between nonsequentially numbered resonator nodes.

In the majority of practical cases, a coupling matrix is formed, either directly from the filter function polynomials, or indirectly via a prototype circuit synthesis. The circuit route is slightly more flexible, allowing the synthesis of multiple cascaded trisections and extracted poles. The configurations of both the trisection and the direct-coupled extracted pole are amenable to a representation by a coupling matrix, but only the main body of the phase-coupled extracted pole type [12] is represented by a coupling matrix.

The next step in the realization process is to apply similarity transforms to the prototype coupling matrices to reconfigure them to the desired topology. Measurable electrical values for each RF coupling element of the filter can be calculated, using as a basis the corresponding entry from the prototype coupling matrix. Figure 13.4 illustrates the procedure for a sixth-degree symmetric filter function with two transmission zeros on either side of the passband creating two rejection lobes, and realized in rectangular waveguide "inline" configuration.

For the rectangular waveguide example in Figure 13.4, the mainline couplings are realized as inductive apertures and the cross-couplings (some of which can be negative in the coupling matrix, depending on the positions and type of transmission zeros) as inductive apertures or loops (mainly in coaxial resonator filters), or capacitive probes if negative. The next step in the realization process for the couplings is to assign RF electrical values to them, which are then dimensioned either by measurements on the test sample pieces or by electromagnetic techniques discussed in later chapters.

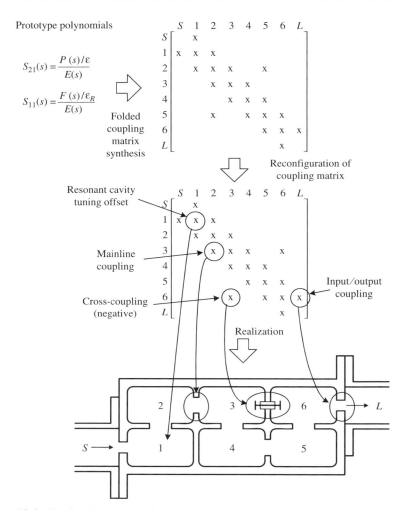

Figure 13.4   Realization process, from prototype filtering function via a coupling matrix to the realization in a rectangular waveguide.

For a waveguide coupling in the direction of propagation of the waveguide, the shunt inductance is a useful equivalent to the impedance or admittance inverter taken from the coupling matrix. Matthaei et al. [13] show, using matrix matching techniques, that the electrical equivalent of an impedance inverter is a shunt inductor X symmetrically placed in a length φ of transmission line of characteristic impedance Z_0 (see Fig. 13.5). (The values of all the impedances or admittances of the inverters or lumped components are normalized to the characteristic impedance Z_0 or admittance Y_0, respectively.)

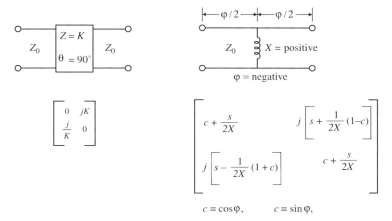

$$c \equiv \cos\varphi, \qquad c \equiv \sin\varphi,$$

Figure 13.5   Equivalence of impedance inverter and shunt reactance symmetrically placed in a length of transmission line.

The phase length φ is found by equating the A elements:

$$\varphi = -\tan^{-1} 2X = -\cot^{-1}\frac{B}{2} \qquad (13.10a)$$

and the susceptance B, by equating and then subtracting the B and C elements:

$$B = \frac{1}{|X|} = K - \frac{1}{K} \qquad (13.10b)$$

This equivalent circuit for waveguide coupling is particularly useful because the lengths of line can be absorbed into the adjacent line lengths (the lengths of waveguide that form the resonant cavities), shortening them slightly. Moreover, the RF susceptance B of a waveguide aperture or iris is a directly measurable parameter, and is directly related to the prototype coupling values taken from the coupling matrix.

For input and output couplings [13]

$$K_{S1} = \frac{M_{S1}}{\sqrt{\alpha}}, \qquad K_{LN} = \frac{M_{LN}}{\sqrt{\alpha}} \qquad (13.11a)$$

and for the interresonator couplings

$$K_{ij} = \frac{M_{ij}}{\alpha} \qquad (13.11b)$$

where M_{ij} are the coupling values taken from the prototype coupling matrix; K_{ij} are the RF inverter values; $\alpha = [(\lambda_{g1} + \lambda_{g2})/n\pi(\lambda_{g1} - \lambda_{g2})]$, λ_{g1}, and λ_{g2} are the guide wavelengths at the lower and upper band-edge frequencies respectively; and n is the number of half-wavelengths of the waveguide resonator cavity. For example, For a

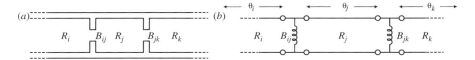

Figure 13.6 Rectangular waveguide resonators where the endwall is coupled: (a) rectangular waveguide resonator R_j; (b) open-wire electrical equivalent circuit.

TE_{10} mode, resonating in a rectangular waveguide cavity that is three half-wavelengths long, the resonant mode is TE_{103} and so $n = 3$. α is the lowpass to waveguide bandpass scaling factor[2] and is usually > 1.

Equation (13.11) can now be applied to find the susceptances of the waveguide coupling elements, normalized to the waveguide admittances. For endwall couplings between resonator cavities R_i and R_j, and between cavities R_j and R_k in the direction of propagation in the waveguide (see Fig. 13.6), the mainline susceptances B_{ij} and B_{jk} are given by [13]

$$B_{ij} = K_{ij} - \frac{1}{K_{ij}}, \qquad B_{jk} = K_{jk} - \frac{1}{K_{jk}} \tag{13.12}$$

Because α is usually > 1, the RF coupling susceptances B_{ij} have negative values when the corresponding prototype coupling M_{ij} is positive (they can be represented electrically as shunt inductors or physically as inductive apertures or irises). The total electrical length of resonator cavity R_j is then given by

$$\theta_j = n\pi + \frac{1}{2}\left(\cot^{-1}\frac{B_{ij}}{2} + \cot^{-1}\frac{B_{jk}}{2}\right) \quad \text{radians} \tag{13.13}$$

Sidewall couplings between resonator cavities can be represented by a series coupling from approximately the midpoint of the two resonators (or the voltage maximum points; see Fig. 13.6). The effect of the series coupling is to insert a shunt FIR approximately halfway along the length of the cavities R_j and R_l [14]. Equation (13.13) for the electrical length of the cavity can be modified to account for the sidewall coupling, as well as for the endwall couplings. Also, it is convenient now to include any tuning offset for the resonance, represented by the diagonal elements of the coupling matrix (asynchronous tuning):

$$
\begin{aligned}
\theta_{1j} &= \frac{n\pi}{2} + \frac{1}{2}\left(\cot^{-1}\frac{B_{ij}}{2} - \sin^{-1}(B_{jl} - B_{jj})\right) \\
\theta_{2j} &= \frac{n\pi}{2} + \frac{1}{2}\left(\cot^{-1}\frac{B_{jk}}{2} - \sin^{-1}(B_{jl} - B_{jj})\right)
\end{aligned} \tag{13.14}
$$

[2]For coaxial bandpass filters, α is the reciprocal of the bandwidth ratio: $\alpha = f_0/\text{BW}$.

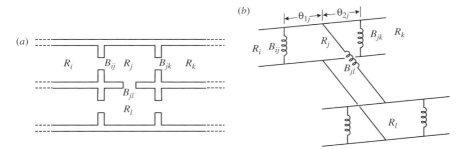

Figure 13.7 Rectangular waveguide resonators with sidewall coupling: (*a*) sidewall coupling between rectangular waveguide resonators R_j and R_l; (*b*) open-wire electrical equivalent circuit.

where $B_{jl} = M_{jl}/\alpha$ is the sidewall coupling between resonant cavities R_j and R_l (see Fig. 13.7), $B_{jj} = M_{jj}/\alpha$ is the tuning offset factor, and $\theta_{1j} + \theta_{2j}$ is the total electrical length of the resonator cavity R_j.

The sidewall coupling component B_{jl} is shown as an inductor in Figure 13.7*b*, indicating that this coupling has the same sign as that of the endwall couplings, such as B_{ij}, and so it can also be realized as an inductive iris (Fig. 13.7*a*). If the sign of B_{jl} is opposite to the signs of the aperture couplings, it can be represented electrically as a capacitor and realized as a capacitive probe. It will be explained in the next section how resonances R_j and R_l can be accommodated by two orthogonally polarized waveguide modes within a single cylindrical or square cross-sectional dual-mode cavity. In a dual-mode cavity, the sidewall coupling B_{jl} is realized as a 45° offset screw on the side of the cavity, and the intercavity couplings B_{ij}, B_{jk}, and so on are realized as slots in the cavity endwalls.

It should be remembered that when the prototype circuit is mapped to the waveguide circuit, a transform from the *parallel* resonator of the prototype circuit (shunt-connected $C_j + jB_j$) to the *series* resonator of the waveguide circuit, ($\approx n\lambda/2$ lengths of transmission line) needs to be included. This is done by applying the dual-network theorem; add an inverter at both ends of the prototype network to form the input/output irises B_{S1} and B_{NL}, as depicted in Figure 13.9.

To demonstrate that this open-wire model, representing a waveguide bandpass filter, offers a reasonably accurate analysis, an example is taken of a sixth-degree asymmetric filter realized in extended box topology. Electrically, the filtering function is a (6–2) asymmetric Chebyshev with 23 dB return loss and two positive transmission zeros, producing two rejection lobes of 50 dB each on the upper side of the passband. At RF, the passband width is 40 MHz centered at 6.65 GHz. Physically, the filter is realized in rectangular TE_{103}-mode cavities of broad dimension 3.485 cm (WR137).

Following the procedures laid out in earlier chapters, the prototype polynomials are determined and the folded coupling matrix synthesized. Then the folded matrix can be reconfigured to the extended box form by using the methods outlined in

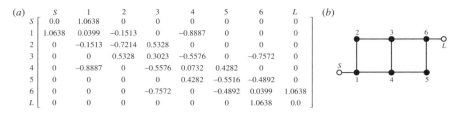

Figure 13.8 Extended box (6–2) filter prototype: (a) prototype coupling matrix; (b) coupling–routing diagram.

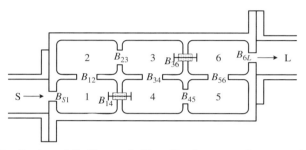

Figure 13.9 Realization of (6–2) extended box filter in rectangular waveguide TE_{103}-mode cavities.

Chapter 10. The final $N + 2$ coupling matrix and the corresponding coupling and routing diagram are shown in Figure 13.8. Note that in the coupling matrix the signs of the sequential (mainline) couplings have been organized such that after the application of equation (13.12), the RF values B_{ij} are all negative so that they

TABLE 13.1 (6–2) Asymmetric Extended Box Filter with RF Values of Coupling Susceptances[a] and Cavity Lengths

	Couplings		Cavity Lengths (degrees)		
	Endwall	Sidewall	j	θ_{1j}	θ_{2j}
Mainline Couplings	$B_{S1} = -4.0234$	—	1	257.05	272.75
	—	$B_{12} = -0.0074$	2	269.20	267.72
	$B_{23} = -38.4940$	—	3	269.71	273.31
	—	$B_{34} = -0.0272$	4	273.36	269.69
	$B_{45} = -47.9005$	—	5	268.72	269.91
	—	$B_{56} = -0.0238$	6	272.85	257.52
	$B_{6L} = -4.0234$	—	—	—	—
Cross Couplings	$B_{14} = +23.0492$	—	—	—	—
	$B_{36} = +27.0672$	—	—	—	—

[a]Normalized to the rectangular waveguide characteristic admittance Y_0.

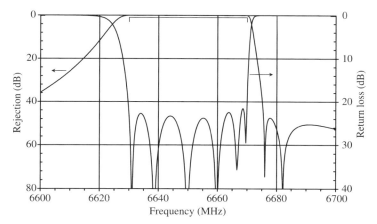

Figure 13.10 RF analysis of open-wire circuit model of (6−2) asymmetric filter in extended box configuration.

may be realized as apertures. This leaves the cross couplings B_{14} and B_{36} positive in value, and they are realized with capacitive probes (see Fig. 13.9).

Applying equations (13.11), (13.12), and (13.14), $\alpha = 2C.5219$, and then the coupling values and cavity lengths are calculated as shown in Table 13.1. An analysis of the open-wire circuit with these values is given in Figure 13.10, which shows that the return loss level and rejection lobe levels are reasonably well preserved. The return loss characteristic is not perfectly equiripple, due to the dispersion effects that are included in the analysis.

13.4 DUAL-MODE FILTERS

Dual-mode filters are based on a polarized mode degeneracy within the same cavity, rather than two *different* modes within the cavity [15]. *Mode degeneracy* means that two modes with the same mode indices can exist independently, that is, orthogonally, until the modes are coupled by some means such as a coupling screw that perturbs the fields. Both TE and TM modes can be degenerate, and dual TE_{11n} mode in cylindrical cavities (see Fig. 13.11) is used most often. Dual-mode filters became popular in the early 1970s for satellite transponders because of their compact size and ability to realize advanced filtering functions [1]–[4].

Each cylindrical cavity supports two orthogonally polarized TE_{11n} mode resonances. They are tuned by the two adjustable screws set 90° apart around the circumference of the cavity, and which define and stabilize the polarization directions. The two resonances exist independently; that is, if vertical resonance 1 is excited by the horizontal input iris, then resonance 2 (the horizontal) does not exist until the circular symmetry of the cavity is perturbed, for example, by an adjustable screw at 45°. This asymmetry distorts the field of the first resonance, producing field components in the direction of the second resonance's field lines such that it is sustained. In

(a)

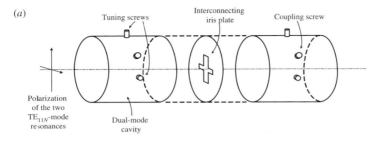

(b)

(c)

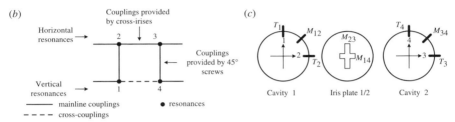

Figure 13.11 Dual-mode resonator cavities: (*a*) two cylindrical waveguide dual-mode cavities and interconnecting iris plate; (*b*) corresponding coupling–routing diagram; (*c*) relations between the mode polarization vectors and tuning screws, coupling screws, and coupling irises.

effect, the coupling screw is realizing the coupling M_{12} from the coupling matrix, and the depth of its penetration into the cavity determines the strength of that coupling.

The intercavity couplings are provided by the cross-iris plates between the cavities. Again, if we assume that resonance 1 in cavity 1 is vertically polarized, vertical resonance 4 in cavity 2 is generated through the horizontal iris slot. Thus, this iris slot realizes the coupling M_{14} from the coupling matrix; the strength of the coupling depends on the length, width, and thickness of the slot. Similarly, the vertical iris slot realizes the coupling M_{23} between the horizontal resonance 2 in cavity 1, and horizontal resonance 3 in cavity 2. Because the iris slots are usually longer than they are wide, the cross-polarization couplings are negligibly small. For example, M_{14}, which couples resonances 1 and 4, will not couple strongly between resonances 2 and 3. There are other iris configuration possibilities, various input/output coupling schemes, and many useful design formulas discussed in the literature [15].

13.4.1 Virtual Negative Couplings

When we synthesize and reconfigure the coupling matrix for a filter function that includes finite-position transmission zeros, we find that certain patterns of those TZs unavoidably cause some of the couplings to be negative. An example is a

(4−2) filter characteristic with two symmetrically placed TZs located on the imaginary axis. In the coupling matrix for this characteristic, the cross-coupling M_{14} is negative with respect to M_{12}, M_{23}, and M_{34}. In a filter realized with single-mode cavities, M_{14} would be realized with a capacitive probe, while the others are inductive loops or apertures.

For filters with dual-mode cavities, the negative coupling is achieved by the appropriate positioning of the 45° coupling screws relative to each other, obviating the need to provide special capacitive coupling devices that increase the complexity considerably, and reduce Q_u (unloaded Q), the reliability, and the stability.

To explain how this "virtual" negative coupling is formed, a convention is adopted where the positive polarization direction (i.e., the heads of the vector arrows) of the horizontally and vertically polarized modes in a dual-mode cavity, remain adjacent to the 45° screw that couples them in the same cavity. With this convention, it is evident from Figure 13.12a, where the coupling screws M_{12} and M_{34} are in the same axial position around the peripheries of cavities 1 and 2, that resonances 1 and 4 are polarized in the same direction as seen through their intercavity coupling iris M_{14}. However, if the coupling screw M_{34} in the second cavity is moved around the periphery of the cavity to be axially displaced 90° relative to coupling screw M_{12}, as shown in Figure 13.12b, then resonance 4 is polarized opposite to resonance 1, as seen through the coupling iris M_{14}. Thus, the phase of resonance 4 is 180° different from that of resonance 1. The overall effect is that iris M_{14} behaves as a virtual negative coupling, even though it is electrically inductive.

When we analyze such couplings with an open-wire model, the negative sign should be allocated to the shunt inductor B_{14} that represents the coupling M_{14} (its susceptance B_{14} is positive), but retains the frequency dependency of an inductive coupling (λ_g/λ_{g0}).

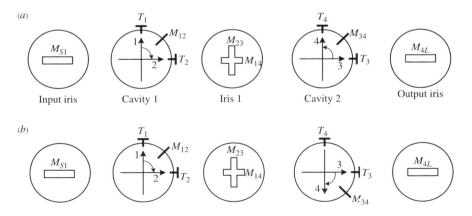

Figure 13.12 Creation of a virtual negative coupling between dual-mode cavities: (a) M_{14} positive; (b) M_{14} negative.

13.5 COUPLING SIGN CORRECTION

The ability to realize both positive and negative couplings within a structure with only inductive couplings is one of the main attractions of the dual-mode resonant cavity. It not only offers mass and volume reductions over the single-mode equivalent but is also able to realize the couplings of advanced filter functions which, in general, require some negative couplings, without the need for capacitive probe structures.

However, before embarking on the design of a dual-mode filter, the coupling matrix that embodies the filter characteristic to be realized by the dual-mode structure needs to be conditioned to ensure that the negative couplings are assigned to one of the two arms of the coupling iris, that is, the horizontal couplings in coupling/ routing diagrams such as those shown in Figure 13.11. After synthesizing and reconfiguring a coupling matrix for a particular filter function and topology, negative couplings tend to appear rather randomly, sometimes corresponding to 45° coupling screws in the dual-mode cavities. These couplings need to be shifted to a coupling to be realized by an intercavity iris, where they can be realized "virtually" by 90° coupling screw axial offsets as explained in 13.4.1.

The sign of any coupling M_{ij} in the coupling matrix can be "shifted" (i.e., changed to positive) by applying a scaling of -1 to all the entries in row and column i, or row and column j of the CM. The effect of doing this is to shift the negative sign from M_{ij} to the other couplings in the row and column that the scaling is applied to. For example, if the coupling M_{12} (which is accomplished with a 45° screw in a dual-mode structure) is negative in the CM of Figure 13.8, then multiplying all the entries in row 2 and column 2 by -1 shifts the negative sign to mainline coupling M_{23} and any other cross-coupling in row/column 2. Coupling M_{23} is realized as one of the slots in the iris plate between cavities 1 and 2, and so can be "virtually" realized as a 90° axial offset in the position of the coupling screws in cavities 1 and 2. Note that the diagonal entry $M_{22,}$ representing the frequency offset of resonance 2, is unaffected by this "scaling" because it is multiplied by -1 twice.

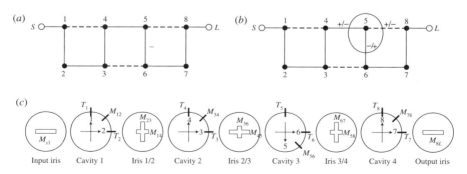

Figure 13.13 Coupling sign correction by the method of enclosures: (*a*) negative sign on M_{56}; (*b*) moving the negative sign to iris couplings—every coupling the enclosure cuts changes the sign; (*c*) dual-mode realization.

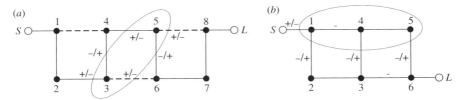

Figure 13.14 Two examples of multiple sign corrections on coupling–routing diagrams by the method of enclosures: (*a*) M_{34} and M_{56} negative; (*b*) M_{12}, M_{34}, and M_{56} negative.

Because the scaling has no effect on the electrical performance of the coupling matrix, the -1 scaling can be applied as many times as is necessary to shift the negative couplings to the right positions in the CM, although multiple scalings become quite cumbersome and prone to error. However, an easier graphical method, working with the coupling–routing diagram, may be used instead.

The method involves drawing arbitrarily shaped "enclosures" on the coupling-routing diagram. Each coupling line that the boundary of the enclosure crosses changes the sign of that coupling. Figure 13.13 gives an example of an eighth-degree longitudinal dual-mode filter, where the synthesis/reconfiguration process leaves a negative sign on M_{56} (a screw coupling); all others are positive. The enclosure drawn in Figure 13.13*b* changes the sign of M_{56} to positive, and M_{45} and M_{58}, both of which correspond to an iris coupling, to negative. Now the filter is realized as a dual-mode structure, as shown in Figure 13.13*c*. This example is equivalent to multiplying row/column 5 in the CM by -1.

Multiple enclosures or a single enclosure enveloping more than one resonator node and correcting several negative signs at once, can be drawn. Also, the boundary of an enclosure can cut single or multiple input/output couplings, changing their signs, if required. Figure 13.14 gives two further examples, the second of which is the correction of the sixth-degree extended box coupling matrix shown in Figure 13.8 into a form suitable for realization with dual-mode cavities.

13.6 DUAL-MODE REALIZATIONS FOR SOME TYPICAL COUPLING MATRIX CONFIGURATIONS

In previous chapters, the synthesis and topology reorganization procedures for several types of filter configuration were described. Typically, the canonical form that emerges from the direct coupling matrix synthesis procedure is the folded array (sometimes known as the *reflex array*). Similarity transforms (rotations) can be applied to the matrix to form other topologies (minimum path rule permitting), many of which can be realized as a dual-mode structure. These configurations include the following:

- Propagating or inline configuration
    Symmetric structure
    Asymmetric structure

- Pfitzenmaier configuration
- Cascade quartet
- Extended box

Each of these configurations has its own advantages and disadvantages, which are outlined along with the corresponding diagrammatic arrangement of dual-mode cavities, and intercoupling iris plates and screws.

13.6.1   Folded Array

The canonical folded coupling matrix can be directly realized with a dual-mode structure, as portrayed in Figure 13.15. The polarized mode enters the filter in one polarization that propagates through the resonators in that polarization towards

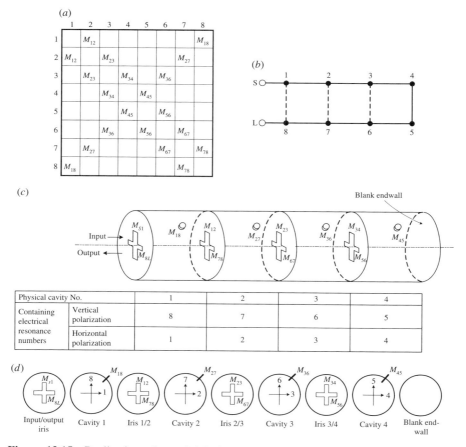

Figure 13.15 Realization of an eighth-degree folded configuration coupling matrix in dual-mode filter structure: (*a*) folded $N \times N$ coupling matrix; (*b*) corresponding coupling–routing diagram; (*c*) dual-mode cavity arrangement; (*d*) cavity–iris arrangement.

the cavity at the other end of the structure. The 45° coupling screw in this cavity generates the orthogonal polarization of resonance in this end resonator, then the signal propagates back through the cavities toward the front end of the filter in this orthogonal polarization, and emerges from the same cavity as it entered. The 45° coupling screws in the cavities provide the cross-couplings.

The advantages of the reflex dual-mode filter are that it is capable of realizing up to $N-2$ finite-position transmission zeros ($n_{fz} = 6$ in the eight-pole filter example), and that the couplings values are symmetric about the cross-diagonal of the CM. This means that the values of the two coupling slots in each iris plate are equal (and therefore, it can be circular hole). However, the big disadvantage is the existence of the input/output couplings in the same cavity, leading to poor input/output isolation. Furthermore, an orthomode transducer or sidewall coupling to the first cavity is needed to separate the input and output signals.

13.6.2 Pfitzenmaier Configuration

The Pfitzenmaier configuration (Fig. 13.16) overcomes the poor isolation problem of the folded structure by having the input/output couplings located in adjacent cavities, but can realize the same maximum numbers of finite-position TZs.

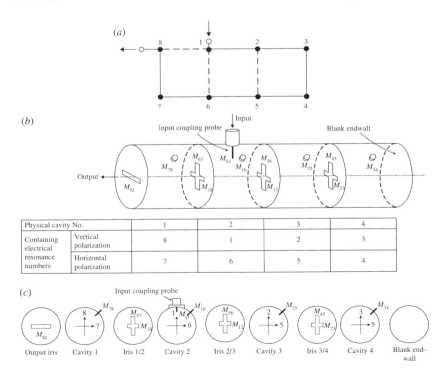

Physical cavity No.		1	2	3	4
Containing electrical resonance numbers	Vertical polarization	8	1	2	3
	Horizontal polarization	7	6	5	4

Figure 13.16 Pfitzenmaier dual-mode filter arrangement—eighth-degree example: (*a*) coupling–routing diagram; (*b*) dual-mode structure; and (*c*) cavity–iris relationships.

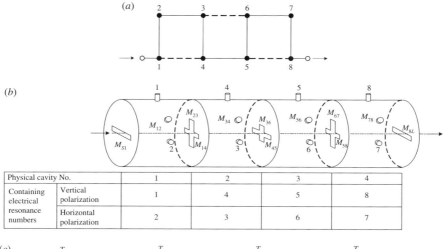

Physical cavity No.		1	2	3	4
Containing electrical resonance numbers	Vertical polarization	1	4	5	8
	Horizontal polarization	2	3	6	7

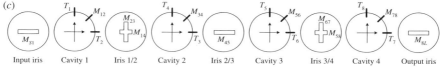

Figure 13.17 "Propagating" dual-mode filter arrangement for an eighth-degree filter; (*a*) coupling–routing diagram; (*b*) dual-mode structure; (*c*) cavity–iris relationships.

13.6.3 Propagating Forms

Both the symmetric and asymmetric propagating forms can be realized with the dual-mode arrangement, indicated in Figure 13.17, as an example of an eighth-degree filter.

The asymmetric propagating structure can realize any symmetric filter function (minimum path rule permitting), whereas the symmetric form has some restrictions on the type of filter function that it can realize. With the input and output at the opposite ends of the filter, a very good input/output isolation is attained, but the structure is able to realize fewer finite-position TZs as compared to the folded structure (e.g., $n_{fz} = 4$ for the above eighth-degree example). Note that if $N/2$ is odd, the output coupling slot is $90°$ axially rotated with respect to the input slot.

13.6.4 Cascade Quartet

The cascade quartet is similar to the propagating form, but simplified in that some of the internal irises have single slots. Tuning is also easier, because each quartet is uniquely associated with the production of a symmetric pair of TZs, either on the imaginary axis or the real axis (see Fig. 13.18).

13.6.5 Extended Box

Extended box filters are unique in their ability to realize asymmetric filtering functions, and in retaining the simple construction of the regular propagating dual-mode

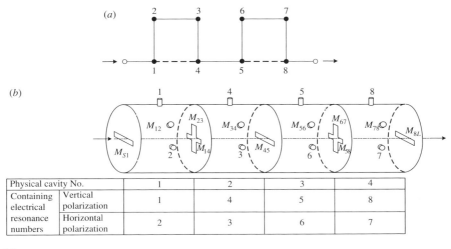

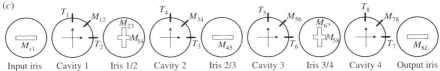

Physical cavity No.		1	2	3	4
Containing electrical resonance numbers	Vertical polarization	1	4	5	8
	Horizontal polarization	2	3	6	7

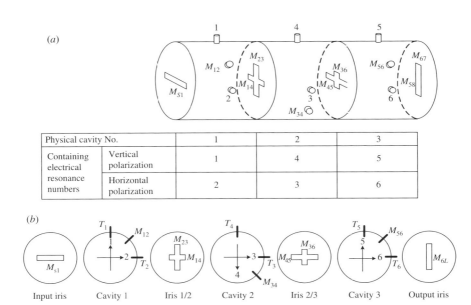

Figure 13.18 Cascade quartet dual-mode filter arrangement for an eighth-degree filter: (*a*) coupling–routing diagram; (*b*) dual-mode structure; (*c*) cavity–iris relationships.

Physical cavity No.		1	2	3
Containing electrical resonance numbers	Vertical polarization	1	4	5
	Horizontal polarization	2	3	6

Figure 13.19 Extended box dual-mode filter arrangement for a sixth-degree filter: (*a*) dual-mode structure; (*b*) cavity–iris relationships.

structure for symmetric filtering characteristics. The output coupling slot is always $90°$ axially rotated with respect to the input slot for any (even) degree of filter. Also, the resonances are asynchronously tuned, which can complicate the design of the dual-mode cavities in some cases. The example in Figure 13.19 is for the $(6-2)$ asymmetric filter function for which the coupling–routing diagram is given in Figure 13.8. Here, M_{14} and M_{36} are both negative, and it may be seen in Figure 13.19 that to implement these negative couplings, the coupling screw M_{34} is $90°$ rotated with respect to coupling screws M_{12} and M_{56}.

13.7 PHASE- AND DIRECT-COUPLED EXTRACTED POLE FILTERS

In Chapter 10, the method for extracting the transmission zeros from a network and realizing them as bandstop cavities was detailed. Once the TZs are extracted, the main body of the filter can be synthesized and reconfigured as required, in the same way as for a nonextracted pole filter. With the TZs extracted, the cross-couplings in the main body, if any, are positive, which is useful for single-mode cavities, such as the TE_{011}-mode resonant cavity, where negative couplings must be formed as capacitive probes with all the associated consequences of poor power handling and reliability.

Figure 13.20a is an outline sketch of a rectangular waveguide phase-coupled extracted pole filter for a $(6-2)$ filter function, where the TZs are realized with two extracted pole cavities, situated at either end of the main body and connected to the body through lengths of waveguide. Because there are only the two TZs in the transfer function (i.e., no real axis or complex pairs), the main body of the filter has no cross-couplings and is realized as a straight cascade of cavities. Figure 13.20b illustrates the development of one of the shunt-connected series resonant pairs from the extracted pole prototype circuit to the rectangular waveguide equivalent circuit for a bandstop cavity, connected to the main waveguide through an E-plane (series-connected) junction [9]. Equations (13.9), (13.10), and (13.12) are adapted to determine the electrical values by treating the extracted pole cavity as the first resonator of a regular filter but with no output coupling and no cross coupling, and resonant at the frequency of the transmission zero being considered.

Pole Cavity and Interconnecting Stub   Using equations (13.11) and (13.12), the susceptance of the input iris to the pole cavity is given by

$$B_{SP1} = \frac{1}{\sqrt{\alpha C_{P1}}} - \sqrt{\alpha C_{P1}} \tag{13.15}$$

where $C_{P1} = 1/b_{01}$ and b_{01} is the residue of extracted pole 1. The phase length of the pole cavity is given by

$$\beta_{P1} = n\pi + \frac{1}{2}\cot^{-1}\frac{B_{SP1}}{2} - \sin^{-1}(B_{P1}) \tag{13.16}$$

where $B_{P1} = \omega_{01}/\alpha$ and $s_{01} = j\omega_{01}$ is the prototype position of the relevant transmission zero on the imaginary axis. This ensures that the pole cavity resonates at the RF frequency of the transmission zero. The shortening effect of the inductive iris means that the interconnecting length of line α_{P1}, between the pole cavity and the wall of the waveguide, must be negative for proper electrical operation. Consequently, a length equivalent to $180°$ phase is added to bring it to a positive value. Since this length is loaded by the same inductive iris as the main pole cavity and is normalized to the pole frequency, equation (13.16) applies and $\alpha_{P1} = \beta_{P1}$.

Phase Lengths Between the Filter Body and Pole Cavities   The unity admittance length of waveguide between the filter body and the T junction φ_{P1} is required to accommodate three factors: (1) phase length ψ_{P1} in the prototype circuit between the first shunt $C_1 + jB_1$ of the main body of the filter and the adjacent extracted pole; (2) the short negative length associated with the input coupling iris of the main body of the filter; and (3) if the pole cavity is connected to the main waveguide run via an E-plane junction, a $90°$ inverter associated with the equivalent circuit needs to be added, as exhibited in Figure 13.20. Thus, the prototype phase length between the body of the filter and the nearest extracted pole is modified by the following equation

$$\varphi_{P1} = \psi_{P1} + \frac{\pi}{2} - \frac{1}{2}\left(\cot^{-1}\frac{B_{S1}}{2}\right) \tag{13.17}$$

where ψ_{P1} is the phase length between the extracted pole and the main body of the filter obtained from the prototype circuit, and B_{S1} is the susceptance of the input iris of the main body of the filter.

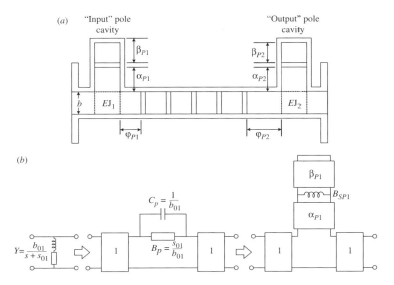

Figure 13.20   Sixth-degree rectangular waveguide filter with two E-plane extracted poles: (*a*) waveguide realization; (*b*) development of one of the prototype extracted poles to the waveguide equivalent circuit.

Phase Lengths Between Adjacent Pole Cavities For the phase lengths between the adjacent pole stubs, adjacent admittance inverters (if any) cancel, and the effect of the first body iris is nil. The phase lengths need only be modified by an arbitrary number of half-wavelengths, which should be added or subtracted for a convenient physical layout:

$$\varphi_P \to \varphi_P \pm k\pi \quad \text{radians,} \quad k = \text{integer.} \tag{13.18}$$

✸13.8 THE "FULL INDUCTIVE" DUAL-MODE FILTER

The dual-mode resonant cavity operates on the basis of two orthogonally polarized modes of the same kind in each cavity, resonating independently until intercoupled by a 45° screw. A novel "full inductive" (all inductive couplings) dual-mode form, introduced by Guglielmi [16], is based on specially dimensioned rectangular waveguide resonator cavities. Two different rectangular waveguide modes exist independently within each cavity, launched by a single critically dimensioned and positioned aperture at the input end. At the other end of the cavity, the two modes are coupled through one or possibly two apertures to the same two modes in the second cavity.[3] The couplings between the same modes through the intercavity aperture are referred to as "straight" couplings and those between the unlike modes are "cross-mode" couplings.

Figure 13.21 is a sketch of a rectangular waveguide resonator cavity, showing the cross-sectional broad and narrow dimensions a and b, respectively, and the length l in the direction of propagation.

For a mode pair TE_{m0n} and TE_{p0q} to resonate simultaneously at frequency f_0 in a cavity of width a and length l

$$a = \frac{c}{2f_0} \sqrt{\frac{(mq)^2 - (np)^2}{q^2 - n^2}} \tag{13.19a}$$

$$l = \frac{c}{2f_0} \sqrt{\frac{(mq)^2 - (np)^2}{m^2 - p^2}} \tag{13.19b}$$

$$\frac{a}{l} = \frac{c}{2f_0} \sqrt{\frac{m^2 - p^2}{q^2 - n^2}} \tag{13.19c}$$

where c is the velocity of propagation.

For the proper operation of the true dual-mode cavity, the first and last mode indices must be different between the two modes: $m \neq p$ and $n \neq q$. If the middle index of the modes is zero, then the b dimension of the waveguide is arbitrarily

---

[3]It is not strictly necessary to have the same modes in the second cavity as in the first; the principles still apply if the modes are different, provided that they can couple through the aperture(s) provided.

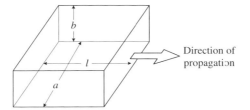

Figure 13.21   Dual-mode rectangular waveguide resonant cavity.

chosen for the best Q_u value and spurious mode rejection. The resonant frequencies of the two modes track each other, if there is a linear expansion of the cavity dimensions caused by a rise in the temperature of cavities.

Figure 13.22 shows a possible configuration for a $(4-2)$ dual-mode filter with two resonator cavities, each supporting one TE_{102}-mode N resonance and one TE_{201}-mode resonance. The tuning screws are placed to be symmetric with respect to each mode to prevent an internal cross-mode coupling, and at points which are at an electric field maximum for one mode, and minimal or zero for the other mode. The dimensional and positional variables of the input and output apertures and the intercavity apertures are as indicated. These dimensions can be used by an EM-based optimization routine to give the required straight and cross-mode coupling values simultaneously. We can also employ the same routine to fine-tune the dimension the cavities. However, in practice, tuning screws tend to perturb the fields, and upset the balance of the multiple coupling values provided by each aperture.

There is a certain small amount of coupling between the two modes in the same cavity at the iris interfaces, which can also be brought into the aperture optimization process. Also, extra degrees of freedom can be brought into play by introducing a second aperture and possibly a third aperture, for example, to couple exclusively between like modes in the two cavities. Here, the first aperture is designed to provide the correct cross-mode couplings, which will unavoidably also introduce some straight coupling as well. The second and third iris is then used to fill in and bring the straight couplings to their required values.

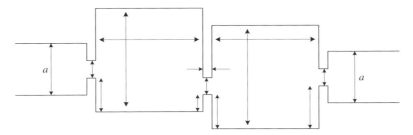

Figure 13.22   An example of a fourth-degree filter with two full-inductive dual-mode resonator cavities.

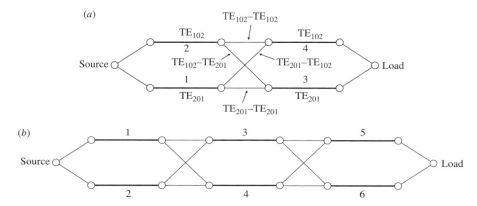

Figure 13.23 Full-inductive dual-mode filter equivalent circuits: (*a*) two cascaded cavities (4th degree) with TE_{102}/TE_{201} mode dual resonances; (*b*) three cascaded cavities (6th degree) with TE_{m0n}/TE_{p0q}-mode dual resonances.

13.8.1 Synthesis of the Equivalent Circuit

Figure 13.23 shows the equivalent circuit of a fourth-degree dual-mode filter, and also a three-cavity cascade forming a sixth-degree dual-mode filter.

The dual-input/output couplings from the interfacing rectangular waveguide are indicated, and the straight and cross-mode couplings that are provided by the inter-cavity iris aperture(s). This topology can be created from any of the inline coupling matrices, including the asymmetric dual-mode, the cascade quartet, or even the asymmetric extended box section form.

In a true dual-mode cavity, there is no direct coupling between the two modes that it is supporting. This is in contrast to the regular dual-mode configuration where intracavity mode coupling is provided by the $45°$ coupling screw. Therefore, the synthesis procedure for the true dual-mode coupling matrix involves starting with an inline form, and applying a series of $N/2$ cross-pivot rotations [see equation (8.37 g)] at pivots [1, 2], [3, 4], . . . , [N−1, N], at each stage, annihilating the main-line coupling M_{12}, M_{34}, . . . , $M_{N-1,N}$, where N is the degree of the filtering characteristic. In the process of eliminating M_{12}, a second input coupling M_{S2} is created; as well as M_{13} and M_{24}. The second rotation annihilates M_{34} but creates M_{35} and M_{46}, and so on until all the intracavity couplings are annihilated, and the second output coupling $M_{N-1,L}$ is created.

The process is illustrated for the (6−2) extended box configuration that was used as an example for the network analysis, earlier in this chapter. Figure 13.8 gives the original extended box coupling matrix for the (6−2) prototype, and Figure 13.24*a* gives the matrix after the application of three cross-pivot rotations at pivots [1,2], [3,4], and [5,6], annihilating M_{12}, M_{34}, and M_{56} in sequence and some sign correction. Figure 13.24*b* presents the analysis of a open-wire equivalent circuit of this true dual-mode filter, by assuming a TE_{102} and a TE_{201} mode in each cavity.

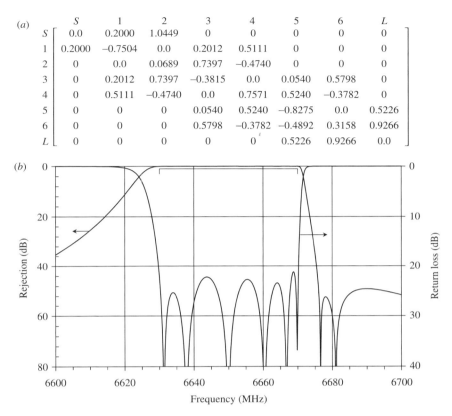

(a)

	S	1	2	3	4	5	6	L
S	0.0	0.2000	1.0449	0	0	0	0	0
1	0.2000	−0.7504	0.0	0.2012	0.5111	0	0	0
2	0	0.0	0.0689	0.7397	−0.4740	0	0	0
3	0	0.2012	0.7397	−0.3815	0.0	0.0540	0.5798	0
4	0	0.5111	−0.4740	0.0	0.7571	0.5240	−0.3782	0
5	0	0	0	0.0540	0.5240	−0.8275	0.0	0.5226
6	0	0	0	0.5798	−0.3782	−0.4892	0.3158	0.9266
L	0	0	0	0	0	0.5226	0.9266	0.0

Figure 13.24 Full-inductive dual-mode realization for asymmetric (6–2) filter characteristic: (a) the coupling matrix; (b) analysis of an equivalent circuit at RF.

SUMMARY

Dual-mode filters employ two orthogonally polarized modes in the physical realization of the filter structure. This implies that each physical cavity or resonator can support two electrical resonances, independent of each other. This leads to a size reduction of almost a factor of 2 and a concomitant reduction in mass as well. However, these advantages come at the expense of reduced spurious free frequency bands and some degradation in the unloaded Q. Owing to the large savings in mass and volume, dual-mode filters are widely used in communication systems.

In addition, dual-mode filters can be configured in waveguide, dielectric-loaded, and planar structures in a variety of topologies. This chapter describes the process of practical filter design by using dual-mode resonators that operate in the dominant mode, as well as in the higher-degree propagation modes. A variety of examples are included to illustrate the design procedure: longitudinal and canonical configurations, the extended box design, the extracted pole filter, and filters with all inductive couplings. The examples also include symmetric and asymmetric response filters. The steps involved in the simultaneous optimization of amplitude and group delay response are described for the design of a filter that may be realized in a variety of technologies. Examples in this chapter span the analysis and synthesis techniques described in Chapters 3–10.

REFERENCES

1. A. E. Atia and A. E. Williams, New types of bandpass filters for satellite transponders, *COMSAT Tech. Rev.*, **1**, 21–43 (Fall 1971).

2. A. E. Williams, A four-cavity elliptic waveguide filter, *IEEE Trans. Microwave Theory Tech.* **MTT-18**, 1109–1114 (Dec. 1970).

3. A. E. Atia and A. E. Williams, Narrow bandpass waveguide filters, *IEEE Trans. Microwave Theory Tech.* **MTT-20**, 258–265 (April 1972).

4. A. E. Atia and A. E. Williams, Non-minimum phase optimum-amplitude bandpass waveguide filters, *IEEE Trans. Microwave Theory Tech.* **MTT-22**, 425–431 (April 1974).

5. C. M. Kudsia, K. R. Ainsworth, and M. V. O'Donovan, Status of filter and multiplexer technology in the 12 GHz frequency band for space applications, in *IEEE Selected Reprint Series in Satellite Communications*, H. L. Van Trees, ed., 1979, pp. 540–548.

6. S. J. Fiedziuszko, Dual mode dielectric resonator loaded cavity filter, *IEEE Trans. Microwave Theory Tech.* **MTT-30**, 1311–1316 (Sept. 1982).

7. R. J. Cameron, W. C. Tang, and C. M. Kudsia, Advances in dielectric loaded filters and multiplexers for communications satellites, *Proc. AIAA 13th Int. Satellite System Conf.*, March 1990.

8. C. Kudsia, R. Cameron, and W. C. Tang, Innovations in microwave filters and multiplexing networks for communications satellite systems, *IEEE Trans. Microwave Theory Tech.* **40**, 1133–1149 (June 1992).

9. R. J. Cameron, A. R. Harish, and C. J. Radcliffe, Synthesis of advanced microwave filters without diagonal cross-couplings, *IEEE Trans. Microwave Theory Tech.* **MTT-50**, 2862–2872 (Dec. 2002).

10. J. D. Rhodes, *Theory of Electrical Filters*, Wiley, New York, 1976.

11. J. D. Rhodes, Filters approximating ideal amplitude and arbitrary phase characteristics, *IEEE Trans. Circuit Theory* **CT-20**, 150–153 (March 1973).

12. J. D. Rhodes and R. J. Cameron, General extracted pole synthesis technique with applications to low-loss TE_{011} mode filters, *IEEE Trans. Microwave Theory Tech.* **MTT-28**, 1018–1028 (Sept. 1980).

13. G. Matthaei, L. Young, and E. M. T. Jones, *Microwave Filters, Impedance Matching Networks and Coupling Structures*, Artech House, Norwood, MA, 1980.

14. J. D. Rhodes, The generalized direct-coupled cavity linear phase filter, *IEEE Trans. Microwave Theory Tech.* **MTT-18**, 308–313 (June 1970).

15. J. Uher, J. Bornemann, and U. Rosenberg, *Waveguide Components for Antenna Feed Systems—Theory and CAD*, Artech House, Norwood, MA, 1993.

16. M. Guglielmi *et al.*, Low-cost dual-mode asymmetric filters in rectangular waveguide, *IEEE MTT-S Int. Symp. Digest*, Phoenix, 2001, pp. 1787–1790.

CHAPTER 14

DESIGN AND PHYSICAL REALIZATION OF COUPLED RESONATOR FILTERS

This chapter addresses the physical realization of microwave filters illustrating how one can combine filter circuit models with electromagnetic (EM) simulation tools to determine the filter physical dimensions. The techniques illustrate a direct approach to synthesis filter physical dimensions from the elements of the coupling matrix model, the K-impedance inverter model, or the J-admittance inverter model. The approach is fast, yet it provides reasonably accurate results. Numerical examples are given in this chapter to demonstrate, step by step, the application of the approach to the design of dielectric resonator, waveguide, and microstrip filters.

Bandpass filters typically consist of resonators coupled to each other's by inductive or capacitive coupling elements. Figure 14.1 shows some of the common microwave filter structures. In waveguide, coaxial, and dielectric resonator filters, interresonator coupling is achieved with the use of an iris, whereas input/output coupling is realized with the use of a probe or an iris. In microstrip filters, interresonator and input/output coupling can take various forms depending on the circuit layout. The objective is to determine the physical dimensions of the individual resonators, interresonator couplings, and input/output couplings. The filter design process usually involves the following four main steps:

Step 1. Identification of the filter order and filter function according to specification requirements

Step 2. Synthesis of the coupling matrix [M], K, or J inverter circuit models that can realize the required filter function

Microwave Filters for Communication Systems: Fundamentals, Design, and Applications,
by Richard J. Cameron, Chandra M. Kudsia, and Raafat R. Mansour
Copyright © 2007 John Wiley & Sons, Inc.

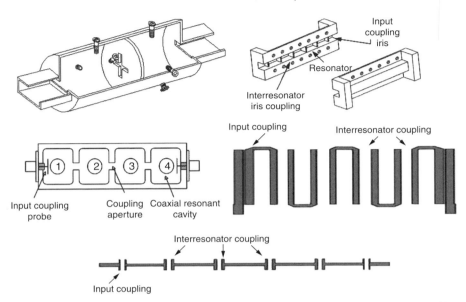

Figure 14.1   Several waveguide, dielectric resonator, and coaxial and microstrip filter components illustrating interresonator coupling and input/output coupling.

Step 3. Identification of the filter type (waveguide, coaxial dielectric, microstrip, etc.), on the basis of size requirements, Q value, and power handling capability

Step 4. Identification of the filter physical dimensions

Chapters 3, 4, 6–10 dealt with steps 1 and 2, whereas Chapter 11 addressed some aspects of step 3. This chapter and Chapter 15 deal with step 4. In Chapter 15 several advanced EM-based techniques for filter design are presented. The techniques given in Chapter 15 are very accurate, yet they are very computationally intensive. In this chapter (Chapter 14), we employ an approach that combines filter circuit models with an EM simulator to get a fast yet reasonably accurate design. The approach employs a "divide and conquer strategy": to divide the filter structure into the design of interresonator couplings and input/output couplings. The accuracy of this approach is limited by the inherent approximation of the circuit models and by the fact that elements of the circuit model are assumed to be frequency-independent. While the solution obtained using this approach is not expected to yield an exact design, it usually suffices, as an initial solution. The exact design can be obtained by fine-tuning of the initial design or by using any of the advanced EM-based techniques presented in Chapter 15.

14.1   CIRCUIT MODELS FOR CHEBYSHEV BANDPASS FILTERS

We use Chebyshev filters in this chapter to demonstrate the concept of using filter circuit models and an EM simulator to synthesize the physical dimensions of the

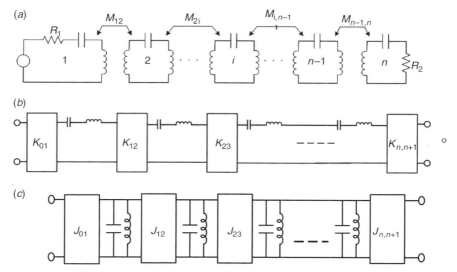

Figure 14.2   (*a*) Coupling matrix model for Chebyshev filters; (*b*) impedance inverter model for Chebyshev filters; (*c*) admittance inverter model for Chebyshev filters.

filter. This concept is not limited to Chebyshev filters; it can be applied to other filter functions as well. Figure 14.2 illustrates three circuit models for Chebyshev filters: the coupling matrix model, the K-impedance inverter model, and the J-admittance inverter model. Chapters 7 and 8 provide details of these models. The three filter circuit models are equivalent to each other. The coupling elements $M_{n,n+1}$, the impedance inverters $K_{n,n+1}$, and the admittance inverters $J_{n,n+1}$ are related to the lumped-element lowpass prototype elements $g_0, g_1, g_2, \ldots, g_{N+1}$ as [1] follows:

Coupling Matrix Model:

$$M_{j,j+1} = \frac{1}{\sqrt{g_j g_{j+1}}}, \qquad j = 1, 2, \ldots, N-1 \tag{14.1}$$

$$R_1 = \frac{1}{g_0 g_1}, \qquad R_N = \frac{1}{g_N g_{N+1}} \tag{14.2}$$

K-Impedance Inverter Model:

$$\frac{K_{j,j+1}}{Z_0} = \frac{\pi \Delta}{2\sqrt{g_j g_{j+1}}}, \qquad j = 1, 2, \ldots, N-1 \tag{14.3}$$

$$\frac{K_{01}}{Z_0} = \sqrt{\frac{\pi \Delta}{2 \, g_0 g_1}}, \qquad \frac{K_{N,N+1}}{Z_0} = \sqrt{\frac{\pi \Delta}{2 \, g_N g_{N+1}}} \tag{14.4}$$

J-Admittance Inverter Model:

$$\frac{J_{j,j+1}}{Y_0} = \frac{\pi\Delta}{2\sqrt{g_j g_{j+1}}}, \qquad j = 1, 2, \ldots, N-1 \tag{14.5}$$

$$\frac{J_{01}}{Y_0} = \sqrt{\frac{\pi\Delta}{2\,g_0\,g_1}}, \qquad \frac{J_{N,N+1}}{Y_0} = \sqrt{\frac{\pi\Delta}{2\,g_N g_{N+1}}} \tag{14.6}$$

The parameter Δ in equations (14.3)–(14.6) is the fraction bandwidth. For a dispersive medium, Δ is given by [1] $\Delta = (\lambda_{g1} - \lambda_{g2})/\lambda_{g0}$, where λ_{g0} is the guide wavelength at the center frequency and λ_{g1} and λ_{g2} are the guide wavelengths at the band-edge frequencies. The lowpass g values for maximally flat response and Chebyshev response are determined using equations (14.7) and (14.8), respectively [1, 2]:

Maximally Flat Filters:

$$g_0 = 1$$

$$g_k = 2\sin\left[\frac{(2k-1)\pi}{2n}\right], \qquad k = 1, 2, \ldots, n \tag{14.7}$$

$$g_{n+1} = 1$$

Chebyshev Filters:

$$g_0 = 1$$

$$\beta = \ln\left(\coth\left(\frac{L_{AR}}{17.37}\right)\right)$$

$$\gamma = \sinh\left(\frac{\beta}{2n}\right)$$

$$a_k = \sin\left[\frac{(2k-1)\pi}{2n}\right], \qquad k = 1, 2, \ldots, n$$

$$b_k = \gamma^2 + \sin^2\left(\frac{k\pi}{n}\right), \qquad k = 1, 2, \ldots, n$$

$$g_1 = \frac{2a_1}{\gamma}$$

$$g_k = \frac{4a_{k-1}a_k}{b_{k-1}g_{k-1}} \qquad k = 2, 3, \ldots, n$$

$$g_{n+1} = \begin{cases} 1 & n \text{ odd} \\ \coth^2\left(\frac{\beta}{4}\right) & n \text{ even} \end{cases} \tag{14.8}$$

where n is the filter order and L_{AR} is the ripple level in the passband that can be calculated from the filter return loss. The g values for Chebyshev filters with various ripple levels are given in Table 14.1. As illustrated in Chapter 7, the impedance and admittance inverters are basically impedance and admittance transformers,

TABLE 14.1 Element Values for Chebyshev Filters for Various Ripple Levels
$(g_0 = 1, \omega_c = 1)$

	g_1	g_2	g_3	g_4	g_5	g_6	g_7	g_8	g_9	g_{10}	g_{11}
					0.01 dB Ripple						
1	0.0960	1.0000									
2	0.4488	0.4077	1.1007								
3	0.6291	0.9702	0.6291	1.0000							
4	0.7128	1.2003	1.3212	0.6476	1.1007						
5	0.7563	1.3049	1.5773	1.3049	0.7563	1.0000					
6	0.7813	1.3600	1.6896	1.5350	1.4970	0.7098	1.1007				
7	0.7969	1.3924	1.7481	1.6331	1.7481	1.3924	0.7969	1.0000			
8	0.8072	1.4130	1.7824	1.6833	1.8529	1.6193	1.5554	0.7333	1.1007		
9	0.8144	1.4270	1.8043	1.7125	1.9057	1.7125	1.8043	1.4270	0.8144	1.0000	
10	0.8196	1.4369	1.8192	1.7311	1.9362	1.7590	1.9055	1.6527	1.5817	0.7446	1.1007
					0.0138 dB Ripple						
1	0.1128	1.0000									
2	0.4886	0.4365	1.1194								
3	0.6708	1.0030	0.6708	1.0000							
4	0.7537	1.2254	1.3717	0.6734	1.1194						
5	0.7965	1.3249	1.6211	1.3249	0.7965	1.0000					
6	0.8210	1.3770	1.7289	1.5445	1.5414	0.7334	1.1194				
7	0.8362	1.4075	1.7846	1.6368	1.7846	1.4075	0.8362	1.0000			
8	0.8463	1.4269	1.8172	1.6837	1.8847	1.6234	1.5973	0.7560	1.1194		
9	0.8533	1.4400	1.8380	1.7109	1.9348	1.7109	1.8380	1.4400	0.8533	1.0000	
10	0.8583	1.4493	1.8521	1.7281	1.9636	1.7542	1.9344	1.6546	1.6223	0.7668	1.1194
					0.1 dB Ripple						
1	0.3052	1.0000									
2	0.8430	0.6220	1.3554								
3	1.0315	1.1474	1.0315	1.0000							
4	1.1088	1.3061	1.7703	0.8180	1.3554						
5	1.1468	1.3712	1.9750	1.3712	1.1468	1.0000					
6	1.1681	1.4039	2.0562	1.5170	1.9029	0.8618	1.3554				
7	1.1811	1.4228	2.0966	1.5733	2.0966	1.4228	1.1811	1.0000			
8	1.1897	1.4346	2.1199	1.6010	2.1699	1.5640	1.9444	0.8778	1.3554		
9	1.1956	1.4425	2.1345	1.6167	2.2053	1.6167	2.1345	1.4425	1.1956	1.0000	
10	1.1999	1.4481	2.1444	1.6265	2.2253	1.6418	2.2046	1.5821	1.9628	0.8853	1.3554
					0.2 dB Ripple						
1	0.4342	1.0000									
2	1.0378	0.6745	1.5386								
3	1.2275	1.1525	1.2275	1.0000							
4	1.3028	1.2844	1.9761	0.8468	1.5386						
5	1.3394	1.3370	2.1660	1.3370	1.3394	1.0000					
6	1.3598	1.3632	2.2394	1.4555	2.0974	0.8838	1.5386				
7	1.3722	1.3781	2.2756	1.5001	2.2756	1.3781	1.3722	1.0000			
8	1.3804	1.3875	2.2963	1.5217	2.3413	1.4925	2.1349	0.8972	1.5386		
9	1.3860	1.3938	2.3093	1.5340	2.3728	1.5340	2.3093	1.3938	1.3860	1.0000	
10	1.3901	1.3983	2.3181	1.5417	2.3904	1.5536	2.3720	1.5066	2.1514	0.9034	1.5386

respectively. These inverters can be realized using any of the lumped-element circuits shown in Table 14.2 [3]. Equations (14.9)–(14.16) provide the relations between the values of the circuit elements and their equivalent inverter values.

For the impedance inverter circuits shown in Table 14.2c and 14.2d we have

$$K = Z_0 \tan\left|\frac{\phi}{2}\right|, \qquad \phi = -\tan^{-1}\frac{2X}{Z_0} \tag{14.9}$$

$$\left|\frac{X}{Z_0}\right| = \frac{\dfrac{K}{Z_0}}{1 - \left(\dfrac{K}{Z_0}\right)^2} \tag{14.10}$$

TABLE 14.2   Equivalent Circuits for Impedance and Admittance Inverters

Impedance Inverter   Admittance Inverter

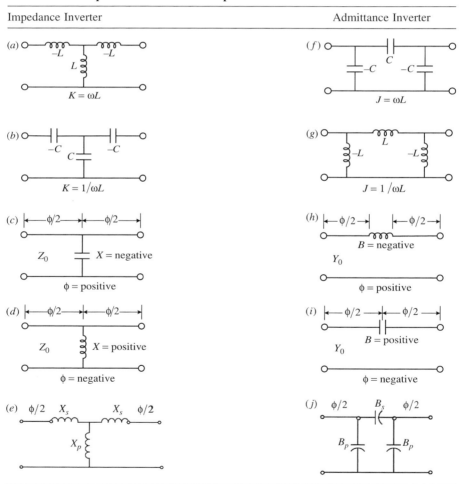

For the admittance inverter circuits shown in Table 14.2h and 14.2i we have

$$J = Y_0 \tan \left| \frac{\phi}{2} \right|, \qquad \phi = -\tan^{-1} \frac{2B}{Y_0} \tag{14.11}$$

$$\left| \frac{B}{Y_0} \right| = \frac{\dfrac{J}{Y_0}}{1 - \left(\dfrac{J}{Y_0} \right)^2} \tag{14.12}$$

For the impedance inverter circuit shown in Table 14.2e we have

$$\frac{K}{Z_0} = \left| \tan \left(\frac{\phi}{2} + \arctan \frac{X_s}{Z_0} \right) \right| \tag{14.13}$$

$$\phi = -\arctan \left(2 \frac{X_p}{Z_0} + \frac{X_s}{Z_0} \right) - \arctan \frac{X_s}{Z_0} \tag{14.14}$$

For the admittance inverter circuit shown in Table 14.2j we have

$$\frac{J}{Y_0} = \left| \tan \left(\frac{\phi}{2} + \arctan \frac{B_p}{Y_0} \right) \right| \tag{14.15}$$

$$\phi = -\arctan \left(2 \frac{B_s}{Y_0} + \frac{B_p}{Y_0} \right) - \arctan \frac{B_p}{Y_0} \tag{14.16}$$

14.2 CALCULATION OF INTERRESONATOR COUPLING

14.2.1 The Use of Electric Wall and Magnetic Wall Symmetry

We consider first the coupling between two resonators, where symmetry is used to divide the problem into single resonators terminated by a magnetic wall and an electric wall. The coupling then is determined from the knowledge of the resonant frequencies of the two individual resonators. The equivalent circuit of two resonators separated by an inductive coupling element is shown in Figure 14.3. The coupling element is represented by a T network consisting of a shunt inductance L_m and two series inductances of value $-L_m$. The network effectively represents the K-impedance inverter given in Table 14.2a. The coupling coefficient M can be computed by evaluating the resonant frequencies of the even- and odd-mode circuits shown in Figure 14.3.

In view of Figure 14.3, the resonant frequencies f_m and f_e of the two circuits shown in Figures 14.3b and 14.3c can be written as

$$f_m = \frac{1}{2\pi \sqrt{(L + L_m)C}} \tag{14.17}$$

$$f_e = \frac{1}{2\pi \sqrt{(L - L_m)C}} \tag{14.18}$$

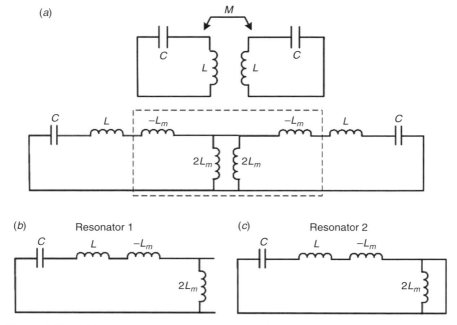

Figure 14.3 (*a*) Impedance inverter—equivalent circuit of two resonators separated by an inductive coupling; (*b*) half of the circuit terminated in a magnetic wall (even mode; open circuit); (*c*) half of the circuit terminated in a electric wall (odd mode; short circuit).

Solving equations (14.17) and (14.18) gives the inductive coupling coefficient k_M as

$$k_M = \frac{L_m}{L} = \frac{f_e^2 - f_m^2}{f_e^2 + f_m^2} \tag{14.19}$$

In view of Chapter 8, the elements of the coupling matrix [*M*] are normalized to the fractional bandwidth. The coupling element *M* between two adjacent resonators is then given by

$$M = \frac{f_0}{\text{BW}} \frac{f_e^2 - f_m^2}{f_e^2 + f_m^2} \tag{14.20}$$

where f_0 is the filter center frequency and BW is the filter bandwidth.

Similarly, for capacitive coupling, the equivalent circuit can be presented as shown in Figure 14.4. The π network shown in this figure effectively represents the admittance inverter *J* shown in Table 14.2*f*.

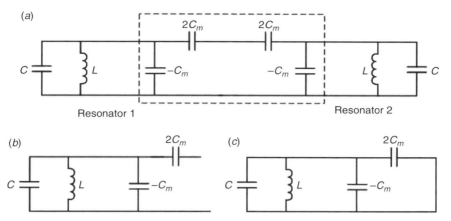

Figure 14.4 (*a*) Admittance inverter—equivalent circuit of two resonators capacitively coupled; (*b*) magnetic wall (even mode); (*c*) electric wall (odd mode).

The resonant frequencies f_m and f_e of the two circuits shown in Figures 14.4*b* and 14.4*c* are given as

$$f_e = \frac{1}{2\pi\sqrt{(C - C_m)L}} \tag{14.21}$$

$$f_m = \frac{1}{2\pi\sqrt{(C + C_m)L}} \tag{14.22}$$

Solving equations (14.21) and (14.22) yields the electric coupling coefficient k_e as

$$k_e = \frac{C_m}{C} = \frac{f_m^2 - f_e^2}{f_m^2 + f_e^2} \tag{14.23}$$

The normalized coupling element M is given by

$$M = \frac{f_0}{\text{BW}} \frac{f_m^2 - f_e^2}{f_m^2 + f_e^2} \tag{14.24}$$

In general, $f_e > f_m$ for inductive coupling and $f_m > f_e$ for capacitive coupling. Only the magnitude of the coupling is needed to calculate the physical dimensions of the coupling iris as illustrated in the example given in Section 14.4. In the analysis above, we assume a synchronous design, where all resonators in the filter have the same resonance frequency. Expressions for the case of nonsynchronous designs are given in [4].

14.2.2 Interresonator Coupling Calculation Using *S* Parameters

The use of electric and magnetic wall symmetry to calculate interresonator coupling is experimentally difficult to implement. It can be only theoretically calculated and

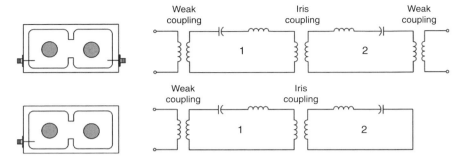

Figure 14.5 Measurement of interresonator coupling using a two-port network and a one-port network.

requires the use of an EM simulator that has an eigenvalue calculation capability such as HFSS [5] or CST [6]. Alternatively, the two resonant frequencies f_e and f_m can be calculated by connecting the two resonators in a two or one-port network configuration as shown in Figure 14.5. The ports need to be weakly coupled to the resonator for this approach to work. A sketch of S_{11} is given in Figure 14.6. The two peaks of S_{11} represent the two frequencies f_e and f_m. The nature of the coupling, whether magnetic or electric, can be determined from the phase information of S_{11} [4].

In waveguide and microstrip filters where the resonators are in the form of transmission lines, the interresonator coupling can be calculated directly from the S parameters by treating the coupling element as a discontinuity between two transmission lines. For example, in waveguide filters where interresonator coupling is realized using an iris, the equivalent circuit of the discontinuity formed by the iris can be represented by a T network as shown in Figure 14.7. The elements of the T network are obtained by converting the two-port S matrix of the waveguide discontinuity to the Z matrix using the formula given in Table 5.2 in Chapter 5.

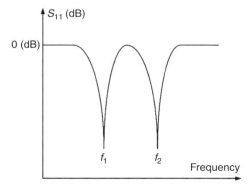

Figure 14.6 The return loss of the circuits shown in Figure 14.5.

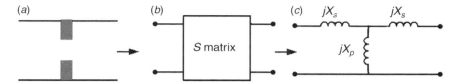

Figure 14.7 Interresonator coupling calculation using S parameters: (*a*) coupling iris in a rectangular waveguide; (*b*) equivalent S matrix; (*c*) equivalent T network that is used to calculate the coupling between two resonators.

The shunt inductance determines the value of the K inverter, while the series inductance represents the loading on the adjacent resonators. Section 14.5 gives a detailed example of the use of this approach in calculating interresonator coupling.

14.3   CALCULATION OF INPUT/OUTPUT COUPLING

14.3.1   Frequency Domain Method

Figure 14.8 shows the equivalent circuit for the first resonator when coupled to the input feed source. The input coupling is represented by the conductance G. The reflection coefficient looking into the single resonator, with respect to a feed line of characteristic admittance G, is given by

$$S_{11} = \frac{G - Y_{in}}{G + Y_{in}} = \frac{1 - Y_{in}/G}{1 + Y_{in}/G} \qquad (14.25)$$

The impedance Y_{in} seen looking into the single resonator is

$$Y_{in} = j\omega C + \frac{1}{j\omega L} = j\omega_0 C \left(\frac{\omega}{\omega_0} - \frac{\omega_0}{\omega} \right) \qquad (14.26)$$

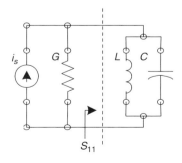

Figure 14.8   Equivalent circuit of the input coupling and first resonator.

where $\omega_0 = 1/\sqrt{LC}$. For frequencies near resonance, $\omega = \omega_0 + \Delta\omega$, where $\Delta\omega \ll \omega_0$, Y_{in} is approximated by

$$Y_{in} \approx j\omega_0 C \frac{2\Delta\omega}{\omega_0} \tag{14.27}$$

Substituting equation (14.27) in the reflection coefficient S_{11} of the resonator and noting that $Q_e = (\omega_0 C/G)$, we get

$$S_{11} = \frac{1 - jQ_e(2\Delta\omega/\omega_0)}{1 + jQ_e(2\Delta\omega/\omega_0)} \tag{14.28}$$

When the frequency offset from resonance is $\Delta\omega_{\mp} = \mp\omega_0/2Q_e$, the phase of S_{11} takes $\pm 90°$. Thus the Q_e is related to the $\pm 90°$ bandwidth in the phase of S_{11} by $\Delta\omega_{\pm 90} = \Delta\omega_+ - \Delta\omega_- = \omega_0/Q_e$:

$$Q_e = \frac{\omega_0}{\Delta\omega_{\pm 90}} \tag{14.29}$$

On the other hand, in view of Chapter 8, the relation between the external quality factor Q_e and normalized input impedance R of the coupling matrix model is given by $Q_e = \omega_0/[R \cdot (\omega_2 - \omega_1)]$, where $(\omega_2 - \omega_1)$ is the filter bandwidth in radians. Thus R can be written as

$$R = \frac{\Delta\omega_{\pm 90}}{(\omega_2 - \omega_1)} \tag{14.30}$$

The normalized impedances R_1 and R_2 of the coupling matrix $[M]$ model can then be extracted from simulation or measurement of the $\pm 90°$ phase of the reflection coefficient of the input and output resonators, respectively.

14.3.2   Group Delay Method

The group delay method for determining the input coupling to a single resonator is based on analysis of the group delay of the reflection coefficient S_{11}. Equation (14.28) can be rewritten as

$$S_{11} = \left| \frac{1 - jQ_e(2\Delta\omega/\omega_0)}{1 + jQ_e(2\Delta\omega/\omega_0)} \right| \angle\varphi \tag{14.31}$$

where

$$\varphi = -2\arctan\left(2Qe\left(2\frac{(\omega - \omega_0)}{\omega_0}\right)\right) \tag{14.32}$$

Using

$$\frac{d}{dx}(\arctan(x)) = \frac{1}{1 + x^2}$$

the group delay $\tau = -(\partial \varphi / \partial \omega)$ is given by

$$\tau = \frac{4Q_e}{\omega_0} \frac{1}{1 + (2Q_e(\omega - \omega_0)/\omega_0)^2} \tag{14.33}$$

Note that the group delay has its maximum value at resonance when $\omega = \omega_0$: $\tau_{\max} = \tau(\omega_0) = (4Q_e/\omega_0)$. Since Q_e is related to the normalized impedance R of the coupling matrix model as $Q_e = \omega_0/[R \cdot (\omega_2 - \omega_1)]$, the normalized impedance R can then be written as

$$R = \frac{4}{(\omega_2 - \omega_1)} \frac{1}{\tau(\omega_0)} \tag{14.34}$$

Thus, the external Q_e values and hence the normalized impedances R_1 and R_2 of the coupling matrix model can be extracted by calculating the group delay of the reflection coefficient of the input and output resonators, respectively, at ω_0.

14.4   DESIGN EXAMPLE OF DIELECTRIC RESONATOR FILTERS USING THE COUPLING MATRIX MODEL

To demonstrate the technique for using the coupling matrix filter circuit model and an EM simulator to synthesize the physical dimensions of the filter, we consider the design of a four-pole dielectric resonator filter. Initially, the design is specified in terms of the Chebyshev bandpass response with a particular center frequency, bandwidth, and return loss. The filter specifications translate into desired coupling matrix elements M_{ij}, R_1, and R_2. The objective is to translate these coupling elements into physical dimensions.

Figure 14.9 shows the four-pole dielectric resonator filter structure. It consists of four dielectric resonators operating in single $TE_{01\delta}$ modes. The coupling

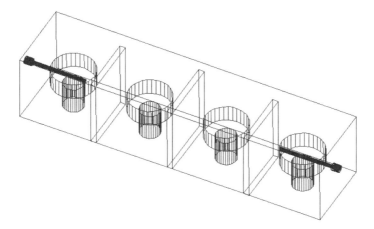

Figure 14.9   The four-pole dielectric resonator filter.

between the resonators is realized using irises, whereas the input/output coupling is achieved with the use of probes. The filter is assumed to have the following specifications:

Filter order $n = 4$

Center frequency $f_0 = 1930$ MHz

Bandwidth   BW $= 15$ MHz (percentage bandwidth $=$ BW$/f_0 = 0.007772$)

Return loss $RL = 20$ dB (equivalent to a passband ripple of 0.0436 dB)

From equation (14.8), the g values corresponding to a return loss of 20 dB are $g_0 = 1$, $g_1 = 0.9332$, $g_2 = 1.2923$, $g_3 = 1.5795$, $g_4 = 0.7636$, $g_5 = 1.2222$.

Using equations (14.1) and (14.2), the elements of the coupling matrix are given by

$$R_1 = \frac{1}{g_0\, g_1} = 1.0715$$

$$R_n = \frac{1}{g_4\, g_5} = 1.0715$$

$$M_{12} = \frac{1}{\sqrt{g_1\, g_2}} = 0.9106$$

$$M_{23} = \frac{1}{\sqrt{g_2\, g_3}} = 0.6999$$

$$M_{34} = \frac{1}{\sqrt{g_3\, g_4}} = 0.9106$$

The diagonal elements $M_{11} = M_{22} = M_{33} = M_{44} = 0$, and $M_{ij} = M_{ji}$.

In view of Chapter 8, with knowledge of the coupling matrix [**M**], the filter S parameters are given by

$$S_{11} = 1 + 2jR_1[\lambda\mathbf{I} - j\mathbf{R} + \mathbf{M}]_{11}^{-1} \tag{14.35}$$

$$S_{21} = -2j\sqrt{R_1 R_n}[\lambda\mathbf{I} - j\mathbf{R} + \mathbf{M}]_{n1}^{-1} \tag{14.36}$$

where **R** is an $n \times n$ matrix with all entries zero except $[\mathbf{R}]_{11} = R_1$ and $[\mathbf{R}]_{nn} = R_n$, **M** is the $n \times n$ symmetric coupling matrix, **I** is the $n \times n$ identity matrix, and

$$\lambda = \frac{f_0}{\text{BW}}\left(\frac{f}{f_0} - \frac{f_0}{f}\right)$$

Figure 14.10 illustrates the simulated ideal response of the filter using the coupling matrix elements of the four-pole filter.

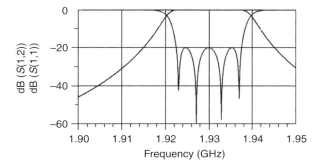

Figure 14.10 The simulated ideal response of the filter based on the coupling matrix model.

14.4.1 Calculation of Dielectric Resonator Cavity Configuration

Figure 14.11 illustrates the cavity configuration and associated dimensions. The dielectric resonator is assumed to have $\varepsilon_r = 34$ and is mounted inside a square cavity of dimensions ($50.8 \times 50.8 \times 51.5$ mm) on a dielectric cylindrical support post with $\varepsilon_r = 10$, a diameter of $D_s = 14.22$ mm, and a height L_s of 20.32 mm. Using HFSS (version 8.5) the EM simulation results show that a dielectric resonator operating in $TE_{01\delta}$ mode, with a diameter of $D = 29.87$ mm, and a height of

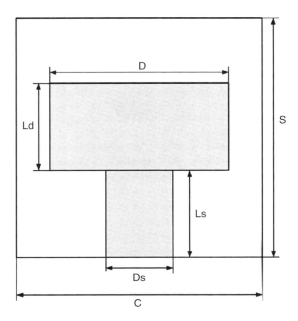

Figure 14.11 Dielectric resonator configuration, $\varepsilon_r = 34$; $D = 29.87$ mm; $L_d = 12.21$ mm; support, $\varepsilon_{rs} = 10$; $D_s = 14.22$ mm; $L_s = 20.32$ mm; cavity, $S = 51.5$ mm; $C = 50.8$ mm; mode 1, $f_0 = 1.931$ GHz (HFSS version 8.5); mode 2, $f_0 = 2.292$ GHz (HFSS version 8.5).

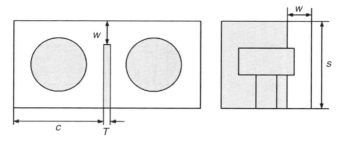

Figure 14.12   The two-coupled dielectric resonator structure.

$L_d = 12.21$ mm, has a resonant frequency of 1.931 GHz. The second mode would have a resonant frequency of 2.929 GHz; thus, the resonator exhibits a spurious-free window of about 300 MHz.

14.4.2   Calculation of Iris Dimensions for Interresonator Coupling

We use the electric wall and magnetic wall symmetry analysis presented in Section 14.2.1 for the calculation of interresonator coupling. The HFSS simulator [5] is employed to calculate the resonant frequency of the even and odd modes. Figure 14.12 illustrates the two-coupled resonator structure used in the HFSS simulations. Note that in calculating the even- and odd-mode resonant frequencies, we assumed the original cavity dimensions (i.e., $C = 50.8$ mm, $S = 51.5$ mm). We assume the same iris thickness of $T = 3.81$ mm for all irises. Table 14.3 shows the calculated values for f_e and f_m and the corresponding coupling factor k.

The physical coupling k_{ij} can be calculated from the normalized coupling M_{ij} as follows:

$$M_{ij} = \frac{f_0}{\text{BW}} \times k_{ij} \qquad (14.37)$$

In view of Figure 14.13, and by using interpolation, we can determine the iris dimensions required to achieve the coupling elements M_{12}, M_{23}, and M_{34}. Table 14.4 gives the iris dimensions and associated coupling values for the structure shown in Figure 14.12.

TABLE 14.3   Calculated Values for f_e and f_m for Various Values of Iris Width W

W(inch)	f_m (GHz)	f_e (GHz)	$k = (f_e^2 - f_m^2)/(f_e^2 + f_m^2)$
0.05	1.9324	1.9355	0.001603
0.07	1.9285	1.9336	0.002641
0.1	1.9261	1.9319	0.003007
0.2	1.9232	1.9319	0.004513
0.25	1.9211	1.9313	0.005295
0.45	1.9170	1.9319	0.007742

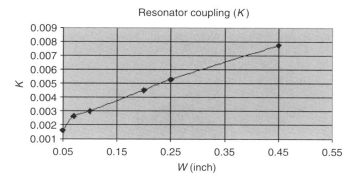

Figure 14.13 Resonator coupling (k) versus iris window (W), calculated from the data given in Table 14.2.

One particular consideration in using this approach is the loading effects of irises and input and output couplings on the adjacent resonators. The result of this loading is a shift in the resonant frequency of the resonator. Adjusting the dimensions of the dielectric resonator to account for the loading is necessary to get good results from this approach.

We can possibly derive a theoretical expression for the shift in resonance frequency due to iris loading of adjacent cavities. We need to adjust the dimensions of the basic resonator configuration shown in Figure 14.11 to account for this frequency shift. Alternatively, as a good approximation, the adjustment of the dielectric resonator dimensions can be obtained by solving the eigenvalue of the structure shown in Figure 14.14, which consists of the dielectric resonator, two irises, and two adjacent cavities with no dielectric resonators. The dielectric resonator dimensions (diameter or height or both) are then adjusted such that the cavity, with the two irises, resonates at the filter center frequency.

The structure shown in Figure 14.14 uses the iris dimensions that were calculated using the original resonator configuration given in Figure 14.11. This assumption is made since the interresonator coupling is not highly sensitive to a small change in resonator resonance frequency. With the use of HFSS [5], the diameter of the resonator is slightly adjusted such that the cavity with the two irises shown in Figure 14.14 resonates at 1.930 GHz. The diameter is found to be 29.66 mm; thus, in comparison with the starting resonator diameter given in Figure 14.11, the resonator diameters of the second and third dielectric resonators need to be reduced by 0.21 mm.

TABLE 14.4 Interresonator Coupling and Iris of the Four-Pole Dielectric Resonator Filter

M_{ij}	k	W (inch)
M_{12} (0.9106)	0.007077	0.396 (W_{01})
M_{23} (0.6999)	0.00544	0.262 (W_{02})
M_{34} (0.9106)	0.007077	0.396 (W_{34})

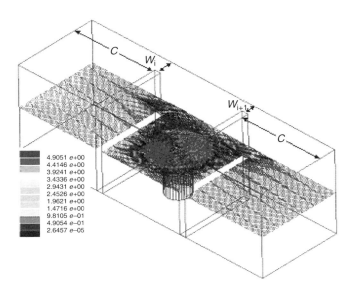

4.9051 e+00	
4.4146 e+00	
3.9241 e+00	
3.4336 e+00	
2.9431 e+00	
2.4526 e+00	
1.9621 e+00	
1.4716 e+00	
9.8105 e−01	
4.9054 e−01	
2.6457 e−05	

Figure 14.14   Calculation of the resonance frequency of the resonator including the effect of the iris. The adjacent cavities are detuned by removing the dielectric resonators.

14.4.3   Calculation of Input/Output Coupling

The input coupling Q_e can be calculated from the values of the normalized input/output coupling resistance R_1 and R_n:

$$Q_e = \frac{f_0}{\text{BW} \times R_1} \qquad (14.38)$$

For $R_1 = 1.07154$, the required $Q_e = 120.08$. We will calculate the Q_e using the group delay method outlined in Section. 14.3.2. With the use of HFSS, the group delay is calculated from the input reflection coefficient S_{11} of the structure shown in Figure 14.15. In order to factor in iris loading effects, we used the structure shown in Figure 14.15 rather than a single-cavity resonator. The first cavity is terminated by an iris W_{12} and the second cavity is detuned by removing the dielectric resonator.

Figure 14.16 shows the details of the probe configuration we used in this filter, as well as the probe's dimensions. We need to vary the probe penetration depth H until we get the proper external Q_e value. However, the probe penetration depth H also affects the resonant frequency of the cavity. Thus, in the presence of the iris W_{12}, we need to adjust both the probe penetration H and dielectric resonator diameter D to achieve the required Q_e and the resonant frequency f_0. This can be achieved by following an iterative approach. First we assume the unloaded original resonator diameter of $D = 29.87$ mm, and then use the group delay calculation and adjust H to get the required Q_e. Assuming this value of H, the dielectric resonator diameter is then adjusted using HFSS such that the cavity with the iris and probe resonates at

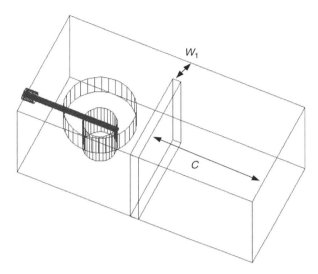

Figure 14.15   Calculation of the Q_e and resonance frequency of the first resonator, factoring in the effect of iris loading.

the filter center frequency f_0. For that new value of D, the probe penetration H is readjusted to get the desired Q_e and the iterative process continues; H and D are adjusted individually to get Q_e and f_0 iteratively until the process converges. Alternatively one may change the probe dimensions and location (H, g, and F shown in Fig. 14.16) until both the required Q_e and resonance frequency f_0 are achieved. The required Q_e is 120.014 and the center frequency is 1.930 GHz. On the basis of HFSS simulations these values are achieved with the use of probe penetration depth H of 7.84 mm and a dielectric resonator diameter D of 29.87 mm.

Table 14.5 summarizes the dimensions of the designed four-pole filter. Note that we used the iris dimensions that we calculated from the model given in Figure 14.12. This is an acceptable approximation since the interresonator coupling is not highly sensitive to the resonance frequency. Figure 14.17 illustrates the HFSS simulated results of the filter based on the physical dimensions given in Table 14.5.

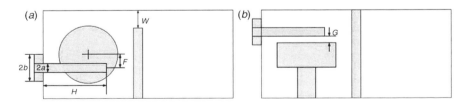

Figure 14.16   Physical dimensions of probe used for the first and last cavities: $a = 0.044$ in.; $b = 0.1$ in.; $G = 0.01$ in., $F = 7.47$ mm; $H =$ length of the probe inside cavity. (*a*) Top-view (*b*) Side view.

TABLE 14.5 Dimensions of Designed Four-Pole Dielectric Resonator Filter (cavity dimensions are given in Figure 14.11)

Parameters	Dimensions
Irises	$W_{01} = W_{34} = 10.058$ mm
	$W_{12} = W_{23} = 6.6548$ mm
DR resonator diameters	$D_1 = D_4 = 29.87$ mm
	$D_2 = D_3 = 29.66$ mm
Input/output probe dimensions	$H = 37.84$ mm
	$G = 0.254$ mm
	$a = 1.117$ mm
	$b = 2.54$ mm
	$F = 7.47$ mm

A comparison between the simulated and the ideal response is given in the same figure. The results obtained are in reasonable agreement with the ideal response and can be considered a reasonable initial solution. A more accurate solution can be obtained by fine-tuning or applying any of the advanced EM modeling techniques described in Chapter 15.

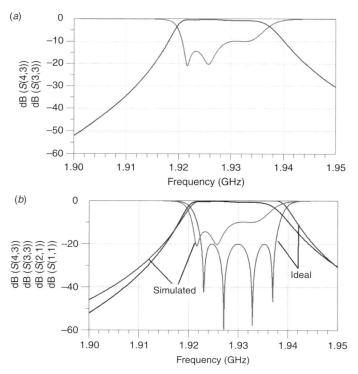

Figure 14.17 (*a*) HFSS simulation results based on physical dimensions given in Table 14.5; (*b*) comparison with ideal response.

In this example, we have changed the dielectric resonator diameters to factor in the effect of iris and input/output probe loading. This leads to a solution where the diameters of resonators 1 and 4 differ from those of resonators 2 and 3. Machining and grinding of dielectric resonators to get the proper length is usually an expensive process. In practical applications, the same diameter and same height are used for the four dielectric resonators and tuning screws are usually used to compensate for design imperfections as well as iris and probe loading effects.

14.5 DESIGN EXAMPLE OF A WAVEGUIDE IRIS FILTER USING THE IMPEDANCE INVERTER MODEL

In this section we consider the design of the four-pole Chebyshev inductive iris-coupled waveguide filter shown in Figure 14.18a. We combine the K-impedance inverter circuit model shown in Figure 14.2b with an EM simulator to calculate the filter physical dimensions. The filter consists of rectangular half-wavelength resonators operating in TE_{101} modes separated by inductive irises. A three-dimensional (3D) illustration of a five-pole version of this filter is shown in Figure 14.1. With the use of an EM simulator, the scattering matrix of the iris and consequently its equivalent T network can be calculated. Each iris is represented by two series inductances denoted by X_s and a shunt inductance denoted by X_p. The waveguide iris filter can then be deduced to the equivalent circuit shown in Figure 14.18b. To arrange the circuit in the form of the K-inverter impedance model shown in Figure 14.2b, we use the impedance inverter circuit shown in Table 14.2e. It consists of an inductive T network and two sections of $\varphi/2$ on each side. The inverter is created by adding a length of $\varphi/2$ and a length of $-\varphi/2$ on each side of the discontinuity as shown in Figure 14.18c. Note that adding the length of $\varphi/2$ and a length of $-\varphi/2$ does not change the original circuit.

The circuit of Figure 14.18c is exactly in the same form as the K-impedance inverter circuit shown in Figure 14.2b. The resonators in this case are waveguide transmission lines of length L_n connected to two transmission lines of artificial lengths $-\varphi_n/2$ and $-\varphi_{n+1}/2$. These lengths represent the loading on the resonator from the adjacent coupling inverters. The filter is assumed to have the following specifications:

Filter center frequency	$f_0 = 11$ GHz
Bandwidth	BW = 300 MHz (percentage bandwidth = 0.02727)
Return loss	$RL = 25$ dB (0.01376 dB ripple)
Waveguide dimensions	WR90: $a = 0.9$ in., $b = 0.4$ in. ($a = 22.86$ mm, $b = 10.16$ mm)
Iris thickness	2 mm (the same for all irises)

Using equation (14.8) or Table 14.1, the lowpass g values corresponding to a return loss of 25 dB are $g_0 = 1$, $g_1 = 0.75331$, $g_2 = 1.22520$, $g_3 = 1.37121$, $g_4 = 0.67310$, $g_5 = 1.11917$.

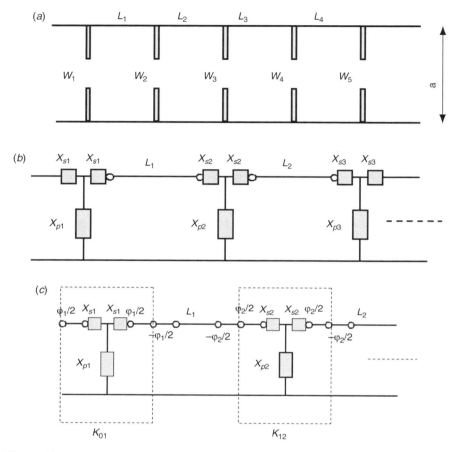

Figure 14.18   Four-pole waveguide iris filter; (*a*) top view of filter; (*b*) equivalent circuit; (*c*) modified equivalent circuit.

The filter center frequency and band-edge frequencies are related by $f_0 = \sqrt{f_1 , f_2}$, BW $= f_1 - f_2$, which gives

$$f_1 = 10.85\,\text{GHz}, \quad f_2 = 11.15\,\text{GHz}$$

$$\Delta = \frac{\lambda_{g1} - \lambda_{g2}}{\lambda_{g0}} = 0.0423$$

$$\frac{K_{01}}{Z_0} = \frac{K_{45}}{Z_0} = \sqrt{\frac{\pi\Delta}{2\,g_0 g_1}} = 0.297$$

$$\frac{K_{12}}{Z_0} = \frac{K_{34}}{Z_0} = \frac{\pi\Delta}{2\sqrt{g_1 g_2}} = 0.0692$$

$$\frac{K_{23}}{Z_0} = \frac{\pi\Delta}{2\sqrt{g_2 g_3}} = 0.0513$$

Using an EM simulator based on the mode matching technique [7], the scattering parameters of the iris can be calculated. From Table 5.2 in Chapter 5, the elements of the T-network shown in Figure 14.7, X_s and X_p are related to the scattering parameters as

$$j\frac{X_s}{Z_0} = \frac{1 - S_{12} + S_{11}}{1 - S_{11} + S_{12}} \tag{14.39}$$

$$j\frac{X_p}{Z_0} = \frac{2S_{12}}{(1 - S_{11})^2 - S_{12}^2} \tag{14.40}$$

where S_{11} and S_{21} are the scattering parameters of the dominant TE_{10} mode. For the K-impedance inverter shown in Table 14.2e, X_s and X_p are related to K and φ as follows:

$$\frac{K}{Z_0} = \left| \tan\left(\frac{\phi}{2} + \arctan\frac{X_s}{Z_0}\right) \right| \tag{14.41}$$

$$\phi = -\arctan\left(2\frac{X_p}{Z_0} + \frac{X_s}{Z_0}\right) - \arctan\frac{X_s}{Z_0} \tag{14.42}$$

Both X_s and X_p and consequently K and φ are functions of the iris width W. Since these functions are not explicit, one needs to run the EM-simulator for a range of iris widths to calculate S parameters and the corresponding X_s, X_p, φ, and K. This will allow the construction of a lookup table or a curve for K versus W. With knowledge of the required K value, we can identify the iris width W and then use the associated φ_i values to calculate the resonator lengths:

$$l_r = \frac{\lambda}{2\pi}\left[\pi + \frac{1}{2}(\phi_r + \phi_{r+1})\right], \qquad r = 1, \dots, N \tag{14.43}$$

Table 14.6 summarizes the physical dimensions of the designed filter. Figure 14.19 shows the mode matching EM simulation results. A comparison obtained using the dimensions given in Table 14.6 with the ideal response that is generated from the coupling matrix is also given in the same figure. The good results obtained are attributed to the fact that the K-impedance inverter model makes it possible to

TABLE 14.6   Synthesized Physical Dimension of Four-Pole Waveguide Filter

Parameters	X_s	X_p	Phase	Width (mm)
$K_{01} = K_{45} = 0.297$	0.358	0.120	-0.816	10.499
$K_{12} = K_{34} = 0.0692$	0.071	0.070	-0.279	6.706
$K_{23} = 0.0513$	0.052	0.063	-0.228	6.147
Cavity length (mm)			$L_1 = L_4 = 15.048$	
			$L_2 = L_3 = 16.329$	

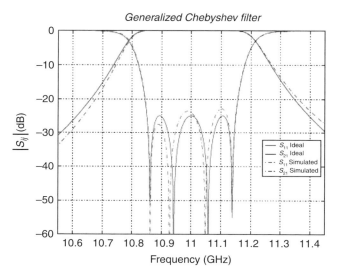

Figure 14.19 A comparison between the ideal response and the mode matching EM simulation results of the designed filter whose dimensions are given in Table 14.6.

accurately include the effect of iris loading on the resonator as seen in equation (14.43). The analysis yields good results for narrowband filters, for wideband filters we need to follow the analysis given in [8] for accurate calculation of the midband guide wavelength λ_g.

14.6 DESIGN EXAMPLE OF A MICROSTRIP FILTER USING THE *J*-ADMITTANCE INVERTER MODEL

In this section we consider the design of the six-pole microstrip filter shown in Figure 14.20a. The filter consists of six microstrip resonators capacitively coupled to each other. We use the *J*-admittance inverter filter model given in Figure 14.2c together with an EM simulator for planar circuits to design this filter. One can easily show that with the use of an EM simulator, the capacitive coupling section can be represented by a π network consisting of a series capacitance C_s and two parallel capacitances of C_p as shown in Figure 14.21. The equivalent circuit of the filter can then be represented as shown in Figure 14.20b. The circuit needs to be modified to conform to the *J*-admittance model given in Figure 14.2c. We use in this example the *J*-admittance inverter model shown in Table 14.2j, which consists of a π network and two transmission-line lengths attached to the sides. In order to modify the circuit to fit the selected model, we add a negative and a positive length of an electrical length φ on each side of the capacitive discontinuity. The resultant circuit is shown in Figure 14.20c. It now has the form of the *J*-admittance inverter filter

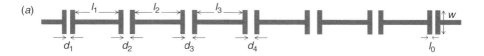

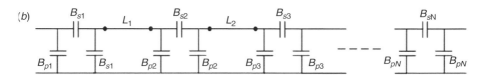

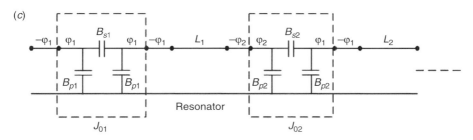

Figure 14.20 The six-pole capacitive coupled microstrip filter: (*a*) top view of filter; (*b*) equivalent circuit; (*c*) modified equivalent circuit.

model, where the resonators consist of microstrip transmission lines of lengths $L_i + \varphi_i + \varphi_{i+1}$. The microstrip filter is assumed to have the following specifications:

Center frequency	$f_0 = 2$ GHz
Bandwidth	BW = 40 MHz (relative bandwidth = 0.02)
Return loss =	20 dB (equivalent to a ripple of 0.0436 dB)
Substrate of	$\varepsilon_r = 10$ and thickness 1.27 mm
Input/output impedance	$Z_0 = 50 \ \Omega$ (microstrip width = 1.234 mm)

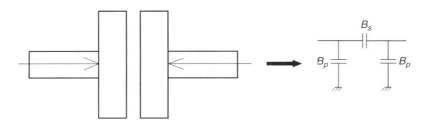

Figure 14.21 The equivalent circuit of the capacitive microstrip discontinuity.

With the knowledge of the lowpass g values, $g_0 = 1$, $g_1 = 0.994$, $g_2 = 1.4131$, $g_3 = 1.8933$, $g_4 = 1.5506$, $g_5 = 1.7253$, $g_6 = 0.8141$, $g_7 = 1.2210$, we can express the J-admittance inverters as

$$J_{01} = J_{67} = Y_0 \sqrt{\frac{\pi\Delta}{2\,g_0\,g_1}} = 3.55 \times 10^{-3}\,\text{S}$$

$$J_{12} = J_{56} = \frac{Y_0\pi\Delta}{2\sqrt{g_1\,g_2}} = 5.3 \times 10^{-4}\,\text{S}$$

$$J_{23} = J_{45} = \frac{Y_0\pi\Delta}{2\sqrt{g_1\,g_2}} = 3.84 \times 10^{-4}\,\text{S}$$

$$J_{34} = \frac{Y_0\pi\Delta}{2\sqrt{g_3\,g_4}} = 3.67 \times 10^{-4}\,\text{S}$$

where Δ is the fraction bandwidth, which is given by

$$\Delta = \frac{f_2 - f_1}{f_0} = 0.02$$

Note that we used the frequency rather than λ_g in defining Δ since the effective dielectric constant of the microstrip line ε_{eff} can be assumed to be constant over such narrow frequency band.

Using em sonnet [9] or Agilent Momentum [10], the scattering parameters of the capacitive discontinuity can be calculated. The elements of the Y-network parameters can be written in terms of the scattering matrix as

$$\frac{B_p}{Y_0} = \frac{1 - S_{12} - S_{11}}{1 + S_{11} + S_{12}} \tag{14.44}$$

$$\frac{B_s}{Y_0} = \frac{2S_{12}}{(1 + S_{11})^2 - S_{12}^2} \tag{14.45}$$

where S_{11} and S_{21} are the scattering parameters of the capacitive microstrip discontinuity. In view of equations (14.15) and (14.16), the J and ϕ in terms of B_s and B_p are given by

$$\frac{J}{Y_0} = \left| \tan\left(\frac{\phi}{2} + \arctan\frac{B_s}{Z_0} \right) \right| \tag{14.46}$$

$$\phi = -\arctan\left(2\frac{B_p}{Y_0} + \frac{B_s}{Y_0} \right) - \arctan\frac{B_s}{Z_0} \tag{14.47}$$

We need to find the dimensions of the capacitive discontinuity that give the required J value. There are three variables, which we can adjust in the microstrip capacitive discontinuity shown in Figure 14.21 to get the proper J: the width l_0, the width W, and the gap d. We assumed the same width W and the same width l_0 for all capacitive discontinuities. The values selected for W and l_0 are $W = 6$ mm and $l_0 = 1.234$ mm.

TABLE 14.7   Lookup Table for Finding Desired Gap Sizes Using Agilent ADS Circuit Simulator

d (mm)*	Y_{11}	Y_{12}	J	ϕ
0.04	$j0.0159294$	$-j0.0057781$	0.0036	-1.296
0.06	—	—	0.0035	—
0.08	—	—	0.0034	—
0.10	—	—	0.0033	—
1.00	—	—	9.4097e-004	—
1.35	—	—	5.6546e-004	—
1.39	$j0.0137513$	$-j0.0007852$	5.3333e-004	-1.2037
1.40	—	—	5.2562e-004	—
1.50	—	—	4.5418e-004	—
1.60	—	—	3.9241e-004	—
1.61	$j0.0137280$	$-j0.0005688$	3.8671e-004	-1.2026
1.63	—	—	3.7558e-004	—
1.65	$j0.0137252$	$-j0.0005365$	3.6479e-004	-1.2024
1.70	—	—	3.3906e-004	—

*The gaps that correspond to the required J values.

TABLE 14.8   Final Design Dimensions (mm) Obtained Using Agilent ADS Circuit Simulator

	1	2	3	4
l	17.312	17.739	17.745	—
d	0.040	1.390	1.610	1.650

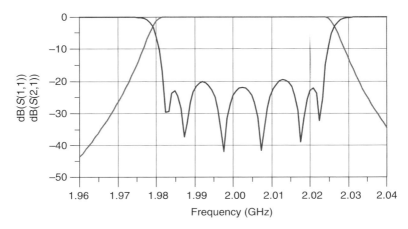

Figure 14.22   Agilent ADS circuit simulation of the design given in Table 14.7.

TABLE 14.9 Lookup Table for Finding Desired Gap Sizes Using Agilent Momentum EM Simulator

d(mm)	Y_{11}	Y_{12}	J	ϕ
0.10	—	—	0.0039	—
0.11	—	—	0.0038	—
0.133	$j0.0165340$	$-j0.0059586$	0.00361	-1.3304
0.138	—	—	0.00357	—
0.14	—	—	0.00356	—
0.15	—	—	0.0035	—
1.56	—	—	5.3439e-004	—
1.565	$j0.0137929$	$-j0.0007837$	5.3128e-004	-1.2065
1.58	—	—	5.2217e-004	—
1.60	—	—	5.1022e-004	—
1.61	—	—	5.0441e-004	—
1.70	—	—	4.5482e-004	—
1.80	—	—	4.0571e-004	—
1.82	—	—	3.9664e-004	—
1.83	—	—	3.9210e-004	—
1.85	$j0.0137662$	$-j0.0005648$	3.8330e-004	-1.2052
1.89	$j0.0137637$	$-j0.0005397$	3.6630e-004	-1.2050
1.90	—	—	3.6224e-004	—

TABLE 14.10 Final Design Dimensions (mm) Obtained Using Agilent Momentum EM Simulator

	1	2	3	4
l	17.188	17.763	17.770	—
d	0.133	1.565	1.850	1.890

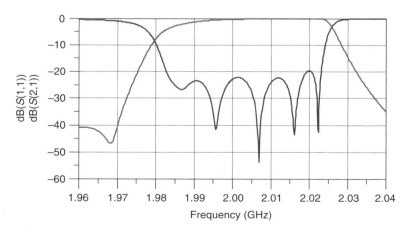

Figure 14.23 Agilent Momentum EM simulation results of the design given in Table 14.10.

We only need to change the capacitive gap d in order to get the required J-inverter value. The discontinuity is simulated for a range of capacitive gaps d, and a lookup table is obtained relating d versus J and φ. The capacitive gaps $d_0, d_1, \ldots, d_N$ are then determined according to the J values given in Table 14.6. With knowledge of φ, one can calculate the resonator lengths as

$$l_r = \frac{\lambda_{g0}}{2\pi} \left[\pi + \frac{1}{2} \left(\phi_r + \phi_{r+1} \right) \right], \qquad r = 1, \ldots, N \qquad (14.48)$$

The lookup table that relates the gap d to J and φ is generated using both the Agilent ADS circuit simulator [11] and the Agilent Momentum EM simulator [10]. Table 14.7 shows the lookup table generated by the ADS circuit simulator. Table 14.8 shows the filter dimensions as calculated based on Table 14.7, whereas Figure 14.22 shows the simulated results using the ADS circuit simulator. A slight deviation from the ideal response is observed and is attributed to the fact the equivalent circuit of the gap capacitance is assumed to be frequency-independent in this approach.

A more accurate analysis can be obtained if the lookup table is generated using an EM simulator. The results for d versus J and φ as calculated by the Agilent Momentum EM simulator are shown in Table 14.9. The corresponding filter design dimensions are given in Table 14.10. The EM simulation performance of the filter using the dimensions given in Table 14.9 is shown in Figure 14.23. The degradation in the roll-off at the lower band edge is attributed to radiation, since the circuit is simulated in Agilent Momentum with no box. It is noted also that the filter response exhibits a transmission zero on the left side. The deviation from the ideal response is due to cross-coupling between resonators, which is not accounted for in the approach presented in this chapter. However, the solution obtained is considered a good initial solution and can be further optimized using the advanced EM-based design techniques described in Chapter 15.

SUMMARY

This chapter illustrates how one can couple the filter circuit models with EM simulation tools to synthesis the physical dimensions of microwave filters. Three coupled resonator filter circuit models are briefly outlined: the coupling matrix model, the impedance inverter model and the admittance inverter model. For Chebychev filters, designers only need to know the low-pass g-values to calculate the M_{ij} elements of the coupling matrix model and the impedance and admittance inverter values of the other two models. For other complicated filter functions. Chapters 7–10 can be used to construct the elements of these circuit models.

With the lack of an EM simulation tool, filter designers may use references such as Microwave Engineer's Handbook or refer to a large pool of papers published in 1970s and early 1980s where microwave discontinuity dimensions were approximated using closed-form expressions. These simple expressions were obtained either empirically or using approximations such as infinitely thin irises in waveguide realizations and quasi-static analyses in TEM structures.

This chapter demonstrates the use of a synthesis approach to translate the coupling matrix elements to physical dimensions using a dielectric resonator filter as an example. The approach is quite general and can be employed to single-mode or dual-mode filters realized using any technology (microstrip, coaxial, waveguide or dielectric). However, it is ideal for dealing with microwave filters where the resonators are not in the form of coupled transmission-lines resonators such as dielectric resonator filters, combline filters, or other filters with arbitrarily shaped resonators. Synthesis approaches that use the impedance and admittance inverter models are ideal for use in filters where the resonators have the form of coupled transmission lines resonators. Examples of such filters are iris-coupled waveguide filters, E-plane septum filters, microstrip capacitively-coupled filters and microstrip parallel-coupled filters. These two approaches are demonstrated by considering the design of a waveguide inductive iris filter and a capacitively-loaded microstrip filter.

The three approaches described in this chapter provide reasonably accurate designs which can be used as initial solutions for further optimization by tuning or by using any of the advanced EM-based techniques presented in Chapter 15. The accuracy of these approaches is limited by the fact that the elements of the filter circuits are assumed to be frequency independent. Therefore, they are more applicable to narrow-band designs. These approaches also do not take into consideration the cross-talk between various filter elements which is a pronounced problem in microstrip filters.

Practical filter designers may find the materials given in this Chapter very useful to come up with a reasonably good initial design for Chebyshev filters. With a minimum tuning effort, this initial design can be easily turned into a perfect design.

REFERENCES

1. G. Matthaei, L. Young, and E. M. T. Jones, *Microwave Filters, Impedance-Matching Networks, and Coupling Structures*, Artech House, Norwood, MA 1980.

2. T. S. Saad et al., eds., *Microwave Engineer's Handbook*, Artech House, Norwood, MA, 1971.

3. R. E. Collin, *Foundation for Microwave Engineering*, McGraw-Hill, New York, 1966.

4. J. Hong and M. J. Lancaster, *Microstrip filters for RF/Microwave Applications*, Wiley, New York, 2001.

5. Ansoft HFSS, http://www.ansoft.com.

6. CST, Microwave Studio, www.cst.com.

7. S. Cogollos, *Design of Iris Filters Using Mode Matching Technique*, technical report, Univ. Waterloo, Dec. 2005.

8. L. Q. Bui, D. Ball, and T. Itoh, Broadband millimeter-wave E-plane bandpass filter, *IEEE Trans., Microwave Theory Tech.*, **32**, 1655–1658 (Dec 1984).

9. em sonnet, http://www.sonnetusa.com.

10. Agilent Momentum, http://eesof.tm.agilent.com.

11. Agilent ADS, http://eesof.tm.agilent.com.

CHAPTER 15

ADVANCED EM-BASED DESIGN TECHNIQUES FOR MICROWAVE FILTERS

A number of software packages are now commercially available for the electromagnetic (EM) simulation of microwave circuits. These EM simulation software tools provide very accurate results, allowing microwave filter designers to innovate and practice on various filter configurations in a very cost-effective approach without the need to build and test the hardware. However, commercial EM software packages are classified as "simulation tools" and not as "design tools." Although they may help users to check the validity of their designs, they cannot be used directly to come up with the design itself. Filter designers must employ other tools and techniques, together with the EM simulator, to carry out the design process. Several EM-based techniques are available for the design of microwave filters; these techniques are divided into three categories:

- EM-based synthesis techniques (EM simulator + circuit models)
- EM-based optimization techniques (EM simulator + optimization tools)
- EM-based advanced design techniques (EM simulator + circuit models + optimization tools)

Chapter 14 addresses the first category of EM-based synthesis techniques. In this chapter (Chapter 15) we address the other two categories with a focus on the third category of EM-based advanced design techniques.

Microwave Filters for Communication Systems: Fundamentals, Design, and Applications,
by Richard J. Cameron, Chandra M. Kudsia, and Raafat R. Mansour

15.1 EM-BASED SYNTHESIS TECHNIQUES

Chapter 14 is dedicated to description of the category of EM-based synthesis design techniques. Three examples are given in Chapter 14, where we demonstrate how EM simulators can be used to synthesize the physical dimensions of dielectric resonator, waveguide, and microstrip filters. In view of the results given in Chapter 14, it is clear that the EM-based synthesis techniques do not provide results that fully meet the design specifications. This is attributed to the fact that the cross-coupling between nonadjacent resonators is neglected in these techniques, and also that the coupling elements are calculated at the filter center frequency. Although these techniques provide reasonable results for waveguide filters, they may fail to provide accurate results for planar filter structures, where the cross-coupling between the resonators is substantial. These techniques also fail to provide accurate results for coaxial and dielectric resonator filters, where it is difficult to accurately account for the effect of iris loading over the filter band. However, EM-based synthesis techniques are useful in obtaining an initial design that can be further optimized by manual tuning or by applying EM-based optimization as described in the following section.

15.2 EM-BASED OPTIMIZATION TECHNIQUES

In these techniques, an accurate EM-based simulator is linked to an optimization software tool that tunes the filter dimensions to achieve the desired response. There are various optimization techniques available in the literature [1–3]. The two main classes are optimization techniques that require an initial solution such as a gradient-type approach [4] and techniques that do not require an initial solution such as a genetic algorithm [5]. Usually a good initial solution reduces the optimization process time and provides a more superior design. In the optimization process, the physical parameters of the filter are iteratively modified by an optimization package until the requirements are satisfied. The flowchart of the design process is shown in Figure 15.1. The initial solution is easily obtained from any synthesis technique such as those described in Chapter 14.

The principal engine of the process is the EM-based simulation tool. The accuracy of the design is determined mainly by the accuracy of this tool. Various forms for the accurate simulator tool are outlined in Figure 15.1. The tool could be a commercial EM simulator or any full-wave analysis tool that is available to the designer. Several variations of this tool can be employed, including the semi-EM simulator or the EM simulator with adaptive frequency sampling, or the use of function interpolations or networks that emulate the EM simulator. Examples of these EM-based interpolations are neural networks, fuzzy logic, and multidimensional Cauchy techniques.

The parameters to be optimized are the filter physical dimensions, which are varied by the optimization routine to minimize an objective function. In determining the error function (objective function) for the optimization process, the filter performance is specified by simulating several frequency points within the passband

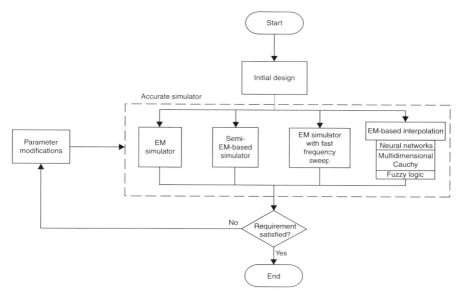

Figure 15.1 EM-based optimization process.

and stopband as depicted in Figure 15.2. The filter response at these frequency points, along with the given filter specification are used to generate two sets of error functions within the passband and stopband. The response can be the magnitude of the S parameters, phase, or group delay. In most cases, however, the

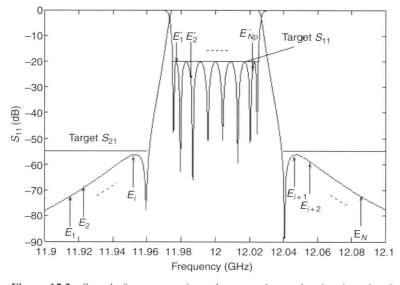

Figure 15.2 Sample frequency points taken over the passband and stopband.

magnitude of the S parameters is adopted at least in the initial optimization stage. A possible definition for the error functions is

$$\text{Passband:} \quad F_j = \left| \frac{S_{11(j)\text{spec.}}}{S_{11(j)\text{resp.}}} \right|, \qquad j = 1, 2, \ldots, N_p \tag{15.1}$$

$$\text{Stopband:} \quad F_j' = \left| \frac{S_{21(j)\text{spec.}}}{S_{21(j)\text{resp.}}} \right|, \qquad j = 1, 2, \ldots, N_s \tag{15.2}$$

where $S_{11(j)\text{resp.}}$ and $S_{21(j)\text{resp.}}$ are the computed S parameters in dB using the accurate simulator at the selected sampling frequency points, while $S_{11(j)\text{spec.}}$ and $S_{21(j)\text{spec.}}$ are the minimum desired S parameters in dB at the same sampling frequency points. For least-square [6] and minimax optimization [7], the error function may be defined as

$$\text{Least-square:} \quad \text{Minimize U} = \left(\sum_{j=1}^{Np} F_j^2 + \sum_{j=1}^{Ns} F_j'^2 \right) \tag{15.3}$$

$$\text{Minimax:} \quad \text{Minimize U} = \text{Max of } (F_1, F_2, \ldots, F_{Np}, F_1', F_2', \ldots, F_{Ns}') \tag{15.4}$$

Several other approaches are available for constructing optimization error functions [1,2].

15.2.1   Optimization Using an EM Simulator

In this case, the EM simulator can be a commercial EM simulator such as HFSS [8], em sonnet [9], Agilent Momentum [10], IE3D [11], CST Microwave Studio [12], Microwave Wizard [13], or any other full-wave simulation analysis software tool. Few commercial EM simulators have built-in optimization tools that allow the designer to specify the parameters to be optimized and offer various ways to build the objective functions. However, at the present time the majority of these commercial software packages still do not offer complete flexibility in selecting the design parameters and typically require an experienced user to exploit the optimization tools.

The use of optimization with a rigorous EM simulation provides the ultimate design accuracy. However, the approach is very computationally intensive. For example, using today's computer, the direct EM-based optimization of an eight-pole dielectric resonator filter can take weeks to complete. For gradient-type optimization, the typical CPU time required for optimization can be approximated by

$$\text{Total time} = N_{\text{freq}} \times N_{\text{var}} \times N_{\text{ite}} \times I_{\text{drev}} \times T_{\text{sim}} \tag{15.5}$$

where
N_{freq} = number of frequency sampling points needed to construct the objective function ($N_{\text{freq}} = N_p + N_s$)
N_{var} = number of variables involved in the optimization process

N_{ite} = number of optimization iterations

I_{drev} = number of function evaluations for numerical gradient calculation

T_{sim} = time taken by the EM simulator to calculate performance of the complete filter circuit at one frequency point

For example, let us consider a symmetric six-pole Chebyshev filter structure, where there are seven variables for optimization: $N_{\text{var}} = 7$. The typical value for the other parameters in equation (15.5) are $N_{\text{freq}} = 20$, $N_{\text{ite}} = 20$, $I_{\text{drev}} = 2$ for a second-order gradient. The total CPU time taken to optimize this filter is then given by

$$\text{Total optimization time} = 20 \times 7 \times 20 \times 2 \times T_{\text{sim}} = 5600\, T_{\text{sim}} \qquad (15.6)$$

The time T_{sim} is dependent on the filter configuration, the EM simulator, and the speed of the computer workstation. With the assumption of one minute per frequency point (i.e., $T_{\text{sim}} = 1$ min), the total CPU time to optimize this filter is 5600 min (close to 4 days). It can be seen that although it only takes 1 minute per frequency point to simulate the filter performance, it takes close to 4 days of CPU time to optimize its performance to the given requirements. Moreover, in some cases, two optimization runs may be required to fine tune the filter to the desired performance; thus close to 8 days of CPU time may be required to design a six-pole filter of this type.

As computer speed continues to increases, the computation time T_{sim} can be significantly reduced. Thus, this brute-force straightforward approach may emerge as the simplest preferred approach for designers in the future. However, at the present time, such approach is useful only for low-order filters. For more complicated filter configurations such as high-order dielectric resonator filters, T_{sim} is very high, prohibiting the use of this technique in the design of such filters. Moreover, with the increasing complexity of filter circuit layouts, the required CPU time increases exponentially, making this approach prohibitively expensive. In addition, there is always a need for advanced techniques to reduce the CPU time needed for the design of more complex structures.

15.2.2 Optimization Using Semi-EM-Based Simulator

The objective of using the semi-EM-based approach is to reduce the simulation time T_{sim} given in equation (15.5) without a major sacrifice in accuracy. In the semi-EM-based simulation approach, the overall filter circuit is divided into subcircuits, whose S parameters are calculated by an EM simulator. The scattering parameters of the overall filter circuit are then computed by using any of the circuit cascading techniques presented in Chapter 5. The reduction in T_{sim} stems from the fact that for most EM simulators the CPU time is roughly proportional to the square of the circuit volume, namely, the CPU $\propto$ (volume^2). Thus, in dividing a circuit into N equal subcircuits, the CPU time taken to simulate these N subcircuits is approximately $N \cdot (T_{\text{overall}}/N^2) \approx (T_{\text{overall}}/N)$, where T_{overall} is the time needed to simulate the overall circuit. For example, the filter circuit in Figure 15.3a is divided into small circuits connected by transmission lines, as shown in Figure 15.3b. In this

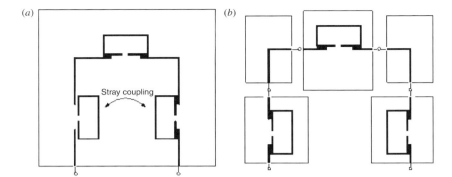

Figure 15.3 (*a*) The complete filter circuit; (*b*) the filter circuit divided into subsections cascaded by transmission lines.

case, the term T_{sim} given in equation (15.5) is the time required to calculate the scattering matrices of the individual three subsections by the EM simulator and the time needed to cascade these subcircuits at one frequency point. This CPU time is considerably less than that required to simulate the overall circuit, which, in turn, reduces the overall optimization time.

However, the improvement in speed is obtained at the price of a less accurate design. The semi-EM simulator approach does not account for the cross-coupling between the nonadjacent resonators. While the problem may not be an issue for waveguide filters, it impacts the accuracy of the scattering parameter results in planar filters, where the cross-resonator coupling between nonadjacent resonators is more pronounced. Figure 15.4 compares the results of the two approaches: EM simulation results of the overall filter circuit and the results obtained from using the cascade of EM-based scattering parameters of the cascaded subcircuits. The transmission zero in Figure 15.4*a* is attributed to the coupling between nonadjacent resonators. It is clear that the cross-coupling effect is not captured by the semi-EM simulator.

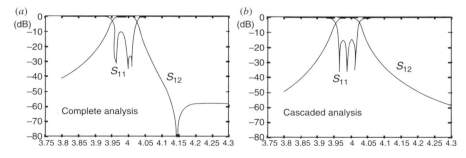

Figure 15.4 (*a*) The EM simulation results of circuit shown in Figure 15.3*a*; (*b*) the simulation results obtained using the semi-EM simulation approach for the circuit shown Figure 15.3*b*.

15.2.3 Optimization Using an EM Simulator with Adaptive Frequency Sampling

This approach targets the reduction of the number of frequency points N_{freq} used in equation (15.5). The idea here is that instead of simulating the filter at N_{freq} points to build the objective functions, we use the EM simulator for only few frequency points (obviously fewer than N_{freq}). These few points are in turn used to create an interpolation polynomial function to describe the filter response over the entire frequency band. This polynomial function is then used to calculate the filter response at the other frequency points. The adaptive frequency sampling is a feature that is available today in the majority of commercial EM simulators (sometimes known as "fast frequency sweep"). The feature allows the designer to get the results over the whole specified band of simulation, whereas the actual simulations are done at only few frequency points.

Using the Padé approximation [14], the filter scattering parameters of passive microwave circuits are accurately described by a rational polynomial function [15–17] in the form of

$$S(f) = \frac{a_0 + a_1 f + a_2 f^2 + \cdots + a_N f^N}{1 + b_1 f + b_2 f^2 + \cdots + b_M f^M} = \frac{a_0 + \sum_{j=1}^{N} a_j f^j}{1 + \sum_{j=1}^{M} b_j f^j} \qquad (15.7)$$

The polynomial given in equation (15.7) is referred to as Cauchy interpolation. The order of the polynomial depends on the sampling simulation points. With the use of K simulation points ($K = 1 + N + M$), a linear system of equations for the polynomial coefficients $a_1, a_2, \ldots, a_N$ and $b_1, b_2, \ldots, b_M$ can be solved by matrix inversion, In order to avoid matrix inversion problems, the Burlisch–Stoer algorithm [18] was proposed in [17] to calculate the polynomial coefficients of equation (15.7).

To demonstrate the efficiency of the fast frequency sweep, the results of the spline and Cauchy interpolation schemes for a four-pole microstrip filter are compared in Figure 15.5 [17]. The response of the filter is sampled at only a few frequency points. It is noted that the Cauchy interpolation, results in almost an exact response.

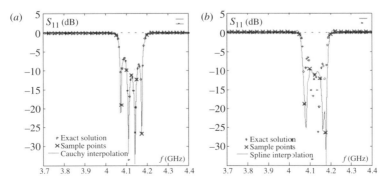

Figure 15.5 Comparison between Cauchy (a) and spline (b) interpolation for four-pole filter.

The number of sampling points can be significantly reduced by the use of adaptive frequency sampling technique [15–17]. In the adaptive approach, a few sample points are first selected uniformly across the band, then more sample points are selected adaptively until convergence is achieved. The adaptive frequency sampling technique minimizes the number of sample points needed to construct the Cauchy polynomial. An example is given in [17], where only nine sampling frequency points are used to build a polynomial that accurately describes the filter. The first five frequency points are selected uniformly across the band of interest. These five sampling points are used to build two different rational polynomials: polynomial 1 with $N = 2, M = 2$; polynomial 2 with $N = 1, M = 3$. These two polynomial approximations are compared over the entire frequency band. Additional samples are selected at the frequency points with the most mismatch. Then new polynomials are calculated and the procedure is repeated until both models agree. The procedure shown in Figure 15.6 illustrates an example where an additional sample frequency point is added at each iteration. It is noted that the two polynomials agree with each other after the adaptive addition of four sample points to the original five sample points.

15.2.4   Optimization Using EM-Based Neural Network Models

This approach targets the reduction of the simulation time T_{sim} in equation (15.5). The EM model of a microwave circuit can be described as a system with an input vector p and an output vector y. The input vector represents the circuit physical dimensions and frequency, whereas the output vector represents the scattering parameters of the circuit. The relationship between p and y is multidimensional and nonlinear. Neural networks [19,20] can be trained by employing EM simulation results to represent the relationship between p and y vectors. Once properly trained, the neural network becomes the model for the circuit and can be used in place of the computationally intensive EM simulator. Typically, the EM-based neural network design approach usually involves the following steps: (1) selecting the appropriate neural network structure, (2) data generation and network training, and (3) network application.

Step 1: Selection of the Appropriate Neural Network Structure   A typical neural network has two basic sets of components; the processing elements and the interconnections among them [19,20]. The processing elements are called *neurons* and the connections between the neurons are known as *links*. Every link has a corresponding weight parameter associated with it. Each neuron receives stimulus from other neurons connected to it, possesses the information, and produces an output. The neurons that receive stimuli from outside the network are called *input neurons*; as neurons whose outputs are externally used are called *output neurons*. Neurons that receive stimuli from other neurons and whose outputs are stimuli for other neurons in the network are known as "hidden neurons." The multilayer perceptions (MLP) format, shown in Figure 15.7, is a popular neural network structure.

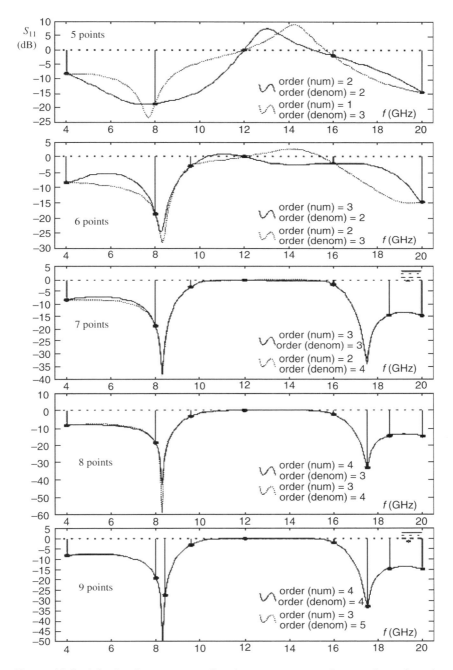

Figure 15.6   Adaptive frequency sampling (num—numerator; denom—denominator).

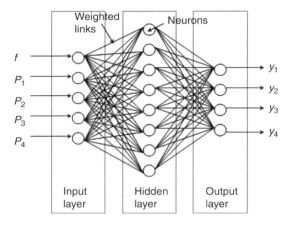

Figure 15.7   A neural network consisting of three-layers.

In the MLP neural network, the neurons are grouped into layers in the form of an input layer, an output layer, and hidden layers. The neurons in the hidden layers are activated by the sigmoid function as

$$a_h = \frac{1}{1 + \exp\left(-\sum_{l=1}^{n} a_l w_{lh} - \theta_h\right)} \tag{15.8}$$

where a_h is the activated neuron, w_{ih} are the adjustable weights between the ith and hth neurons, and θ_h is the threshold. A good starting point for the neural network is to use a three-layer network. The universal approximation theorem [20] states that there always exists a three-layer MLP neural network that can approximate any arbitrary, nonlinear, continuous, and multidimensional functions. The number of hidden neurons must be carefully determined. A small number of neurons cannot model the system precisely; whereas too many neurons tend to overstrain the network. More details on the selection of the neural network structure can be found in [19,20].

Step 2: Data Generation and Network Training   The network must be trained by a number of sample input/output vector data pairs (p, y) in order to construct the network. These data are computed by the EM simulator. The generated data are divided into two sets: training data and test data. The training data are used to guide the training process, to determine the weight factors W_{ih} and threshold θ_h. The test data are used to examine the quality of the trained neural network.

During the training process, the neural network automatically adjusts its weights W_{ik} and threshold values θ_h to minimize the error between the predicted output (by the neural network) and the actual output (by the EM simulator). Typically, the neural training algorithm uses gradient-based training techniques such as backpropagation, conjugate gradient, and quasi-Newton techniques [20]. Global optimization methods such as simulated annealing and genetic algorithms can be used for improving quality of neural network training, but this increases the CPU time associated with the training process.

Step 3: Network Application   After training and testing, the network is ready for use as a substitute for the EM simulator. The network is capable of predicting output y for any given input parameters p in the trained region. The calculation is very fast since only a few basic algebraic operations are required. Once constructed, the neural network can be used also to replace the EM simulator in the optimization process, reducing the design time considerably.

An example [21] illustrating the use of EM-based neural networks in the design of the inset-fed microstrip patch resonator is shown in Figure 15.8. The patch is designed on a substrate of $\varepsilon_r = 2.3$ and a height of $h = 2$ mm. The width of the patch is fixed at 28.7 mm. The objective is to find the dimensions $\vec{P} = (p_1, p_2, p_3)$ such that the microstrip patch resonates at 4 GHz. A total of 500 training data points are obtained by varying the 3 geometric parameters p_1, p_2, p_3 and frequency over the following ranges: p_1, 17–29 mm; p_2, 0–12 mm; p_3, 0–12 mm (f—3.4–4.4 GHz). The MLP network is constructed with input neurons: frequency and 3 geometric parameters, 10 hidden neurons, and 2 output neurons (magnitude and phase of S_{11}).

The network is tested at two points, and the results are shown in Figure 15.9 in comparison with EM simulation. A good agreement is observed. The neural network is then implemented in the optimization process. The optimized dimensions for the patch to resonate at 4 GHz are $\vec{P} = (23.38, 3.5, 1.19 \, \text{mm})$. The S_{11} simulation

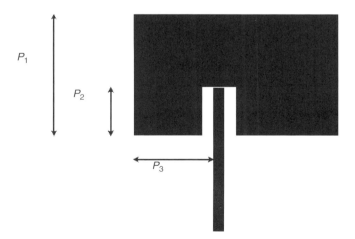

Figure 15.8   An inset-fed microstrip patch resonator.

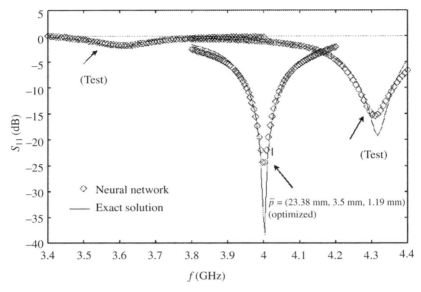

Figure 15.9 Correlation between the S_{11} results obtained by the neural network and EM simulator (exact solution) for the resonator shown Figure 15.8 [21].

results obtained with the EM simulator (exact solution) and the neural network are also compared in Figure 15.9. The optimization process takes only a few seconds. The same optimization process would have taken several hours had the EM simulator been used instead of the neural network model [21].

There are only a few examples reported in literature on the successful use of neural network in the design of low-order (three- or four-pole) microwave filters, where the neural network is used to model the whole filter [21,22]. For high-order filters, the size of vectors p and y can become too large, requiring a large volume of training/test data to properly train the network. The neural network approach can be applied to model subsections of the filters, which can then be cascaded to form the filter. For example, in the semi-EM-based technique presented in Section 15.2.2, the capacitive subsections can be easily modeled using EM-based neural networks. The overall scattering matrix of the filter can then be obtained by cascading these subsections.

A neural network software package is commercially available [23] that has a build-in interface to several commercial EM simulators. Neural network tools are also available from MATLAB [24]. The user, in this case, needs to perform several EM simulations to collect the data pairs. The performance of EM based neural networks can be enhanced by incorporating available knowledge about the circuit into the training process. Several techniques known as knowledge-based neural networks have been reported. More details on EM-based neural network tools for microwave circuit design are given in [19].

15.2.5 Optimization Using EM-Based Multidimensional Cauchy Technique

The multidimensional Cauchy technique is similar to neural networks in that it requires the use of the EM simulator to generate data pairs of inputs (physical parameters and frequency) and output (filter circuit scattering parameters) to build the model. These data pairs are then used to generate a multidimensional Cauchy interpolation that can replace the EM simulator for the analysis and optimization of this particular filter circuit.

The multidimensional Cauchy technique enables the interpolation of the circuit performance with respect to frequency and dimensions. It builds on the one-dimensional (1D) Cauchy interpolation with respect to frequency, described in Section 15.2.3. The multidimensional Cauchy technique, proposed in [17], uses a recursive algorithm, which solves the multidimensional interpolation by using the 1D Cauchy interpolation. The multidimensional technique also allows the application of adaptive sampling, which reduces the required number of sampling data pairs [17]. The technique is applied to the design of the three-pole filter circuit shown in Figure 15.10. Because of symmetry, the filter has four variables: $\vec{p} = (p_1, p_2, p_3, p_4)$. EM simulation training data were obtained by varying these three variables and frequency. The results obtained are shown in Figure 15.11. Excellent results are obtained with only 625 training data points [17].

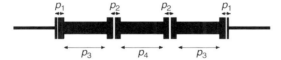

Figure 15.10 Schematic of a three-pole microstrip filter designed by the multidimensional Cauchy technique [17].

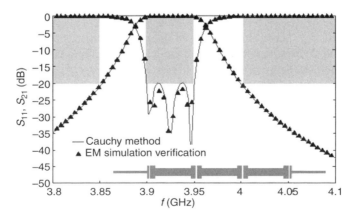

Figure 15.11 Simulation results for the three-pole microstrip filter shown in Figure 15.10 [17].

The multidimensional Cauchy technique is useful to accurately model small circuits, where the number of input variables is relatively low (five or less). As the case for neural networks, the filter circuit can be divided into small subsections. Then, each subsection is represented by a multidimensional Cauchy model to replace the EM simulator. The overall performance of the filter is obtained by cascading these subsections.

15.2.6   Optimization Using EM-Based Fuzzy Logic

The feasibility of using fuzzy logic systems (FLSs) in diagnosis and tuning of microwave filters was demonstrated for the first time, in [27] and is covered in detail in Chapter 19. The FLS is basically a function approximator. It has been shown in [28] that a FLS can approximate any real continuous nonlinear function to arbitrary degree of accuracy. With the use of data generated from an EM simulator, a FLS model can be built to approximate the EM simulator—similar to what can be achieved by neural networks. The FLS model can then replace the accurate simulator in the optimization process shown in Figure 15.1.

The function approximation problem $\vec{y} = f(\vec{x})$ of building a neural network model or a fuzzy logic model, where the input $\vec{x}$ represents the filter physical dimensions and frequency while the output $\vec{y}$ represents the filter scattering parameters, is known as the "forward problem." In [29] the FLS technique was used to deal with the "reverse problem" $\vec{x} = \phi(\vec{y})$, where the data points from the EM simulator were used to build a FLS system that predicts the filter physical dimensions for given specifications or given scattering parameters. This, in turn, eliminates the need to use optimization. It is a direct approach for an approximate design of low order microwave filters. More details are given in [29] and [30].

15.3   EM-BASED ADVANCED DESIGN TECHNIQUES

Several approaches have been reported with the goal of reducing the CPU time of the EM-based design process. The majority of these approaches employ two models: a fine model and a coarse model. The fine model is an accurate but computationally intensive, whereas the coarse model is fast but less accurate. An EM simulator can serve as the fine model, while the coarse model can be a circuit model. The idea in these approaches is to make use of the accuracy of the fine model and the speed of the coarse model to devise a solution that is fast and accurate. These approaches also require the use of an optimization tool. However, in this case the optimization is done using only the coarse model. The fine model is used only a few times during the design process.

Thus the fine model is combined with the coarse model into a hybrid model that is effectively both fast and accurate. While the bulk of computation is done using the fast coarse model, yet such design approaches successfully result in an accurate design. In this chapter, we focus on two techniques: the space mapping (SM) technique and the calibrated coarse model (CCM) technique.

15.3.1 Space Mapping Techniques

The space mapping (SM) technique has been successfully employed in the design of microwave filters and other engineering problems [31–35]. The SM technique establishes a mathematical link (mapping) between the spaces of the design parameters of the two models: the accurate and time-intensive fine model and the less accurate but fast coarse model. The aim is to avoid the computationally expensive calculations encountered in optimizing the filter structure using the time-intensive fine model. SM must use a coarse model that is capable of simulating the filter structure for a wide range of parameters. The coarse model is not updated or changed during the SM process. The SM technique involves two steps: (1) optimizing the design parameters of the coarse model to satisfy the original design specifications and (2) establishing a mapping between parameters spaces of the two models. The space-mapped design is then taken as the mapped image of the optimal coarse model.

We refer to the coarse model parameters as x_c and the fine model parameters as x_f. The optimal coarse model design is denoted as x_c^*. We also denote the response of the coarse model as $R_c(x_c)$ and that of the fine model as $R_f(x_f)$, as shown in Figure 5.12. The mapping between the fine model space and the coarse model space is written as $x_c = P(x_f)$ such that the difference $|R_f(x_f) - R_c(x_c)|$ is minimized. Usually the mapping P is local over a region of space parameters. It is calculated in an iterative way. If we denote $P^{(j)}$ as the mapping calculated after jth iteration, the corresponding fine model design is given by

$$x_f^{(j+1)} = (P^{(j)})^{-1}(X_c^*)$$

If $R_f(x_f^{(j+1)})$ satisfies the specification within certain acceptable accuracy criteria, the solution $x_f^{(j+1)}$ is accepted as the space-mapped design for the coarse solution x_c^*; otherwise, the mapping is updated and a new iteration for P is evaluated.

Bandler et al. introduced the first SM approach in 1994 [31]. In this original SM approach, a linear mapping is assumed between the two parameter spaces of the two models. The mapping is evaluated by a least-square solution of the linear equations, resulting from the corresponding data points in the two spaces. This mapping is

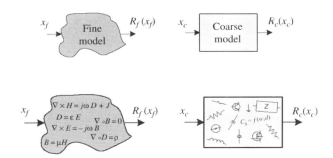

Figure 15.12 Representation of the fine and coarse models [34].

simple but may not converge to a proper solution if a significant misalignment exists between the response of the coarse model and that of the fine model. The misalignment is typically due to the use of very rough coarse model, whose response is far away from the accurate response of the fine model.

A modified version of the original SM technique was also proposed by Bandler et al. [32,33] and is known as *aggressive space mapping* (ASM). ASM does not assume that the mapping is necessarily linear and usually converges to the proper design in a few iterations. Several other SM-based techniques have been also proposed to deal with the nonuniqueness problem of parameter extraction and to minimize the number of iterations. The list includes trust region aggressive space mapping (TRASM), hybrid aggressive space mapping, and neural space mapping. A comprehensive overall review on space mapping techniques is given in [34].

In this chapter, we discuss in details the original space mapping approach and the aggressive space mapping (ASM) approach. We focus particularly on the ASM approach, since it is a more reliable approach. Two examples are given to demonstrate the application of the ASM concept including an example on the design of a six-pole microstrip filter.

The Original Space Mapping Approach   The steps in this approach are summarized as follows:

Step 1. Optimize the coarse model to find optimal solution x_c^*. In this case, x_c^* is the optimum solution based on the coarse model; that is $R_c(x_c^*)$ is the desired performance.

Step 2. Set $x_f^{(1)} = x_c^*$, and select an initial set of points $x_f^{(2)}, x_f^{(3)}, \ldots, x_f^{(j)}$ that are arbitrarily selected in the vicinity of $x_f^{(1)}$. Note that $x_f^{(1)}, x_f^{(2)}, \ldots, x_f^{(j)}$ are vectors representing the filter physical parameters. The set $S_f^{(j)}$ is defined as

$$S_f^{(j)} = \left\{ x_f^{(1)}, x_f^{(2)}, \ldots, x_f^{(m_j)} \right\} \tag{15.9}$$

Step 3. Calculate the fine model response for every point in the set $S_f^{(j)}$.

Step 4. Optimize the coarse model to match the fine model responses calculated in step 3. Use a parameter extraction technique such as optimization for $\min \|R_c(x_c^{(i)}) \simeq R_f(x_f^{(i)})\|$ for $i = 1, 2, \ldots, j$ to extract $x_c^{(1)}, x_c^{(2)}, \ldots, x_c^{(j)}$, which are vectors representing the physical parameters in the coarse model space. A set of coarse model points $S_c^{(j)}$ can be written as

$$S_c^{(j)} = \left\{ x_c^{(1)}, x_c^{(2)}, \ldots, x_c^{(m_j)} \right\} \tag{15.10}$$

Step 5. The two sets of points $S_c^{(j)}$ and $S_f^{(j)}$ are then used to calculate the mapping $x_c = P^j(x_f)$. Bandler [31] assumed that the mapping between the two spaces is linear. Thus

$$x_c = P^{(j)}(x_f) = B^{(j)} x_f + C^{(j)} \tag{15.11}$$

If we define the $A^{(j)}$ as $[C^{(j)}\ \ B^{(j)}]$, it follows that equation (15.11) can be written as

$$
\begin{bmatrix} x_c^{(1)} & x_c^{(2)} & \cdots & x_c^{(m_j)} \end{bmatrix} = A^{(j)} \begin{bmatrix} 1 & 1 & \cdots & 1 \\ x_f^{(1)} & x_f^{(2)} & \cdots & x_f^{(m_j)} \end{bmatrix} \tag{15.12}
$$

A least-squares solution for $A^{(j)}$ is given [35] by

$$
A^{(j)T} = (D^T D)^{-1} D^T Q \tag{15.13}
$$

where

$$
D = \begin{bmatrix} 1 & 1 & \cdots & 1 \\ x_f^{(1)} & x_f^{(2)} & \cdots & x_f^{(m_j)} \end{bmatrix}^T \tag{15.14}
$$

and

$$
Q = \begin{bmatrix} x_c^{(1)} & x_c^{(2)} & \cdots & x_c^{(m_j)} \end{bmatrix}^T \tag{15.15}
$$

Once $A^{(j)}$ is obtained, the space-mapped design is

$$
x_f^{(j+1)} = P^{(j)^{-1}}(x_c^*) = B^{(j)^{-1}}(x_c^* - C^{(j)}). \tag{15.16}
$$

Step 6. Take $x_f^{(j+1)}$ as an approximation to the optimal fine point design x_f, if the condition $\left\| R_f(x_f^{(j+1)}) - R_c(x_c^*) \right\| \le \varepsilon$ is satisfied; the otherwise, the set $S_f^{(j)}$ is augmented to $S_f^{(j+1)}$ by adding point $x_f^{(j+1)}$ and the set $S_c^{(j)}$ is augmented to $S_c^{(j+1)}$ by adding the corresponding point $x_c^{(j+1)}$. A new mapping is then found, and the fine model response of the mapped point is checked. The process is repeated until the fine model response equals the desired response $R_c(x_c^*)$. Figure 15.13 illustrates the original space mapping approach.

This approach assumes a linear mapping between the two models, and this may not be accurate in many problems. Therefore this original space mapping approach is recommended only if the coarse model is very close to the fine model such that a linear mapping can be assumed. The aggressive space mapping technique is preferred, since it is not highly sensitive to the misalignment problems that might exist between the coarse model and the fine model.

Aggressive Space Mapping (ASM) Approach In the aggressive space mapping approach (ASM), the solution x_f^* is obtained by solving the following nonlinear equation:

$$
P(x_f) - x_c^* = 0 \tag{15.17}
$$

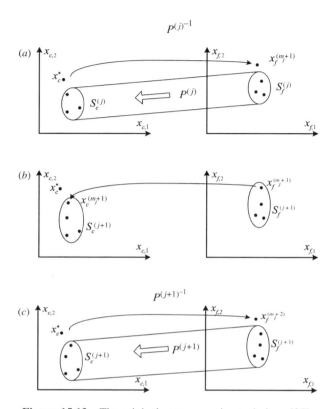

Figure 15.13   The original space mapping technique [35].

By using the quasi-Newton method, the solution is obtained in an iterative manner. If $x_f^{(j)}$ is the jth iterate, the next iteration $x_f^{(j+1)}$ is written as

$$x_f^{(j+1)} = x_f^{(j)} + h^{(j)} \tag{15.18}$$

where $h^{(j)}$ is the solution of $B^{(j)} h^{(j)} = -f^{(j)}$, where

$$f^{(j)} = P(x_f^{(j)}) - x_c^* = x_c^{(j)} - x_c^* \tag{15.19}$$

$$P(x_f^{(j)}) = x_c^{(j)} \quad \text{such that} \quad \|R_c(x_c^{(j)}) - R_f(x_f^{(j)})\| \le \varepsilon \tag{15.20}$$

The vector $x_c^{(j)}$ is obtained by using parameter extraction (optimization) of the coarse model such that the coarse model response equals the fine model response for $x_f^{(j)}$. $B^{(j)}$ is an approximation to the Jacobian of f with respect to x_f at $x_f^{(j)}$.

Using Broyden's rank 1 formula [35], $B^{(j+1)}$ can be written as

$$B^{(j+1)} = B^{(j)} + \frac{f^{(j+1)}h^{(j)^T}}{h^{(j)^T}h^{(j)}} \qquad (15.21)$$

The initial value for $B^{(1)}$ is taken as the identity matrix; $B^{(1)} = U$. The steps of the ASM approach are summarized as follows:

Step 1. Extract x_c^*, such that $R_c(x_c^*) \approx$ desired performance. Initialize $j = 1$, and set $x_f^{(1)} = x_c^*$, $B^{(1)} = U$.
Step 2. Evaluate $R_f(x_f^{(1)})$.
Step 3. Extract $x_c^{(1)}$ such that $R_c(x_c^{(1)}) \approx R_f(x_f^{(1)})$.
Step 4. Evaluate $f^{(1)} = x_c^{(1)} - x_c^*$. Stop if $\left\| f^{(1)} \right\| \leq \eta$.
Step 5. Solve $B^{(j)}h^{(j)} = -f^{(j)}$ for $h^{(j)}$.
Step 6. Set $x_f^{(j+1)} = x_f^{(j)} + h^{(j)}$.
Step 7. Evaluate $R_f(x_f^{(j+1)})$.
Step 8. Extract $x_c^{(j+1)}$ such that $R_c(x_c^{(j+1)}) \approx R_f(x_f^{(j+1)})$.
Step 9. Evaluate $f^{(j+1)} = x_c^{(j+1)} - x_c^*$. Stop if $\left\| f^{(j+1)} \right\| \leq \eta$.

Step 10. Update $B^{(j+1)} = B^{(j)} + \dfrac{f^{(j+1)}h^{(j)^T}}{h^{(j)^T}h^{(j)}}$.

Step 11. Set $j = j + 1$; go to step 4.

The steps are repeated until a solution is found. To illustrate the ASM technique, we will consider the application of the concept to two examples: two Rosenbrock functions and the design of a six-pole microstrip filter.

Example 15.1: An Illustrative Example Demonstrating the Concept of ASM (Two Rosenbrock Functions)   In this example, we apply the ASM technique to find the solution to the Rosenbrock function [36]. The example allows the reader to follow the design process step by step and to duplicate the results by using a calculator. Let us consider that we have the two Roesenbrock functions, $R_c(x)$ and $R_f(x)$ given below.

The coarse model is expressed as $R_c(x) = 100(x_2 - x_1^2)^2 + (1 - x_1)^2$, where

$$x = \begin{bmatrix} x_1 \\ x_2 \end{bmatrix}$$

The fine model is expressed as $R_f(x) = 100(u_2 - u_1^2)^2 + (1 - u_1)^2$, where

$$u = \begin{bmatrix} u_1 \\ u_2 \end{bmatrix} = \begin{bmatrix} 1.1 & -0.2 \\ 0.2 & 0.9 \end{bmatrix} x + \begin{bmatrix} -0.3 \\ 0.3 \end{bmatrix}$$

The problem is to find the values of x_1 and x_2 that is the solution of $R_f(x) = 0$. The two functions are shown in Figure 15.14.

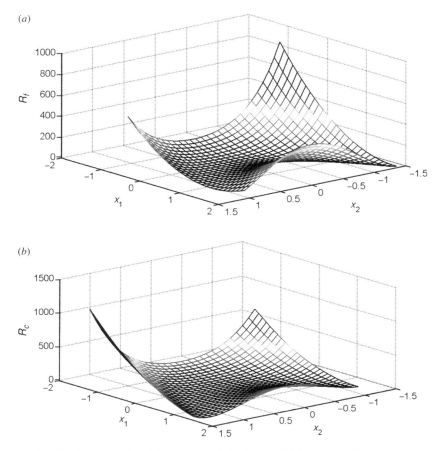

Figure 15.14   The Rosenbrock functions: (*a*) the fine model $R_f(x_1, x_2)$; (*b*) the coarse model $R_c(x_1, x_2)$.

Solution Using the ASM Technique

1. Extract x_c^* such that $R_c^* = R_c(x_c^*) = 0$, $x_c^* = \begin{bmatrix} 1 \\ 1 \end{bmatrix}$, set $x_f^{(1)} = x_c^* = \begin{bmatrix} 1 \\ 1 \end{bmatrix}$,
    $B^{(1)} = U, j = 1$.

2. $R_f(x_f^{(1)}) = 108.32$.

3. Extract $x_c^{(1)}$ for $R_c(x_c^{(1)}) = R_f(x_f^{(1)}) = 108.32$, $x_c^{(1)} = \begin{bmatrix} 0.6 \\ 1.4 \end{bmatrix}$.

4. $f^{(1)} = \begin{bmatrix} 0.6 \\ 1.4 \end{bmatrix} - \begin{bmatrix} 1 \\ 1 \end{bmatrix} = \begin{bmatrix} -0.4 \\ 0.4 \end{bmatrix}$.

5. Since $B^{(1)} = 1$, it follows that $h^{(1)} = \begin{bmatrix} 0.4 \\ -0.4 \end{bmatrix}$.

6. Set $x_f^{(2)} = \begin{bmatrix} 1 \\ 1 \end{bmatrix} + \begin{bmatrix} 0.4 \\ -0.4 \end{bmatrix} = \begin{bmatrix} 1.4 \\ 0.6 \end{bmatrix}$.

7. $R_f(x_f^{(2)}) = 1.8207$.

8. Extract $x_c^{(2)}$ such that $R_c(x_c^{(2)}) = R_f(x_f^{(2)}) = 1.8207, \rightarrow x_c^{(2)} = \begin{bmatrix} 1.12 \\ 1.12 \end{bmatrix}$.

9. $f^{(2)} = \begin{bmatrix} 1.12 \\ 1.12 \end{bmatrix} - \begin{bmatrix} 1 \\ 1 \end{bmatrix} = \begin{bmatrix} 0.12 \\ 0.12 \end{bmatrix}$.

10. $B^{(2)} = B^{(1)} + \dfrac{f^{(2)} h^{(1)^T}}{h^{(1)^T} h^{(1)}} = \begin{bmatrix} 1.15 & -0.15 \\ 0.15 & 0.85 \end{bmatrix}$.

11. $h^{(2)} = -B^{(2)^{-1}} f^{(2)} = \begin{bmatrix} -0.12 \\ -0.12 \end{bmatrix}$.

12. Set $x_f^{(3)} = \begin{bmatrix} 1.4 \\ 0.6 \end{bmatrix} + \begin{bmatrix} -0.12 \\ -0.12 \end{bmatrix} = \begin{bmatrix} 1.28 \\ 0.48 \end{bmatrix}$.

13. $R_f(x_f^{(3)}) = 0.1308$.

14. Extract $x_c^{(3)}$ such that $R_c(x_c^{(3)}) = R_f(x_f^{(3)}), \rightarrow x_c^{(3)} = \begin{bmatrix} 1.012 \\ 0.988 \end{bmatrix}$.

15. $f^{(3)} = \begin{bmatrix} 1.012 \\ 0.988 \end{bmatrix} - \begin{bmatrix} 1 \\ 1 \end{bmatrix} = \begin{bmatrix} 0.012 \\ -0.012 \end{bmatrix}$.

16. $B^{(3)} = B^{(2)} + \dfrac{f^{(3)} h^{(2)^T}}{h^{(2)^T} h^{(2)}} = \begin{bmatrix} 1.1 & -0.2 \\ 0.2 & 0.9 \end{bmatrix}$.

17. $h^{(3)} = -B^{(3)^{-1}} f^{(3)} = \begin{bmatrix} -0.0082 \\ -0.0151 \end{bmatrix}$.

18. $x_f^{(4)} = \begin{bmatrix} 1.28 \\ 0.48 \end{bmatrix} + \begin{bmatrix} -0.0082 \\ 0.0151 \end{bmatrix} = \begin{bmatrix} 1.2718 \\ 0.4951 \end{bmatrix}$.

19. $R_f(x_f^{(4)}) = 9.2 \times 10^{-8}$, then $x_f = x_f^{(4)}$, and we can end the algorithm.

The solution to $R_f(x) = 0$ then is $x_1 = 1.2718$, and $x_2 = 0.4951$.

Example 15.2: Design of a Six-Pole Filter Using Aggressive Space Mapping
Let us examine the design of a six-pole microstrip filter with the layout shown in Figure 15.15. The filter has a center frequency of 2 GHz, a bandwidth of 2%, and a return loss of 20 dB. The filter is realized on an alumina substrate with a height of $h = 0.025$in. (25 mil) and a dielectric constant of $\varepsilon_r = 10.2$. The structure consists of six resonators and seven coupling gaps. All resonators are assumed to have a 50 Ω impedance. The problem is to find the circuit physical dimensions: w, l_0 d_1, d_2, d_3, d_4, l_1, l_2, and l_3.

Figure 15.15 The six-pole filter layout.

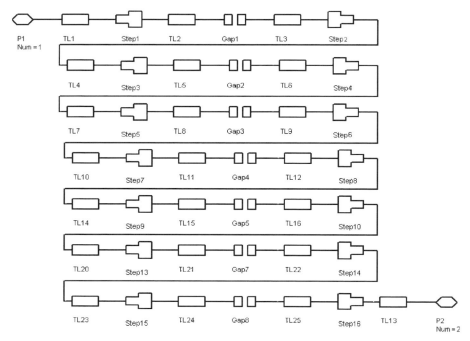

Figure 15.16   The coarse model of the microstrip filter shown in Figure 15.15.

Solution Using the ASM Technique   In this case, the fine model is the EM simulator Agilent Momentum [10], whereas the coarse model is the Agilent ADS circuit simulator ADS [37]. The circuit schematic is shown in Figure 15.16. where the capacitive coupling is presented by two-step discontinuities: two transmission lines and a gap capacitor. The coarse model is fast since ADS uses closed-form empirical models for these discontinuities. We follow the steps of the ASM technique and give the details for the six-pole filter design. The coarse model parameters x_c^* are obtained by optimizing the schematic circuit shown in Figure 15.16. We include all the nine parameters shown in Figure 15.15 in the optimization process. The optimal x_c^* is given in Table 15.1, and Figure 15.17 shows the corresponding coarse model optimal response.

The application of the ASM technique iteratively leads to a set of $x_c^{(j)}$, $x_f^{(j)}$, $f^{(j)}$, $h^{(j)}$. Table 15.2 gives the numerical values for these parameters. The ASM converges in eight iterations. Figure 15.18 shows the filter response for the eight iterations.

TABLE 15.1   The Design Dimensions of the Six-Pole Filter Circuit Shown in Figure 15.16 After ADS Coarse Model Optimization

d_1 (mil)	d_2 (mil)	d_3 (mil)	d_4 (mil)	l_0 (mil)	l_1 (mm)	l_2 (mm)	l_3 (mm)	w (mil)
1.01337	27.9174	32.1179	32.7377	24.9124	17.6166	18.0528	18.061	225.02

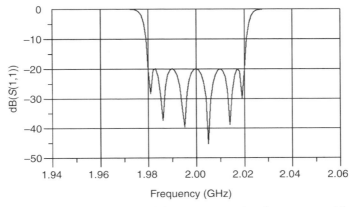

Figure 15.17   The optimized filter response using the coarse model.

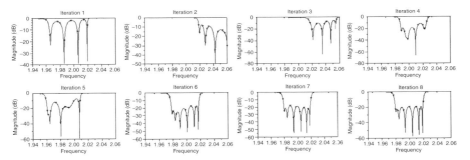

Figure 15.18   The filter responses after each iteration of the ASM technique.

15.3.2   Calibrated Coarse Model (CCM) Techniques

The calibrated coarse model (CCM) technique for filter design was introduced in [38]. A more general version of this technique that makes use of the filter coupling matrix model is presented [39–41]. The CCM technique also employs two models: a coarse model and a fine model. The CCM technique, however, allows the designer to add physical insights into the coarse model. This helps to significantly reduce the required number of EM simulations and gives the designer the flexibility to dynamically change the coarse model to maintain a reasonable alignment with the fine model. For example, in the case of filters, where coupling between nonadjacent resonators usually contributes to transmission zeros, the conventional space mapping technique may not converge, if the coarse model does not include such effects. On the other hand, in the CCM technique, we can begin with the same coarse model. However, it is modified from one iteration to the next to obtain a proper alignment with the fine model.

The most fundamental difference between the conventional SM technique and the CCM technique is that the coarse model in the SM technique is static; the same coarse model is used for all the iterations. The coarse model in the CCM technique is dynamic in the sense that a different coarse model is used in each iteration.

TABLE 15.2 Fine- and Coarse-Model Variables Obtained at Eight Iterations

Aggressive SM Iterations		d_1 (mil)	d_2 (mil)	d_3 (mil)	d_4 (mil)	l_0 (mil)	l_1 (mm)	l_2 (mm)	l_3 (mm)	w (mil)
Iteration 1	$x_f^{(1)}$	1.01337	27.9174	32.1179	32.7377	24.9124	17.6166	18.0528	18.061	225.02
	$x_c^{(1)}$	0.954706	26.8626	29.9675	30.5151	16.8989	18.3516	18.4402	18.4455	267.01
	$f^{(1)}$	−0.0587	−1.0548	−2.1504	−2.2226	−8.0135	0.7350	0.3874	0.3845	41.9900
	$h^{(1)}$	−0.0207	0.0282	0.0014	−0.0258	−0.3206	0.0055	0.0009	0.0006	1.2830
Iteration 2	$x_f^{(2)}$	1.0720	28.9722	34.2683	34.9603	32.9259	16.8816	17.6654	17.6767	183.0280
	$x_c^{(2)}$	1.23176	28.6173	32.9179	33.5163	20.9102	17.0009	17.5378	17.5515	246.608
	$f^{(2)}$	0.2184	0.6999	0.8000	0.7786	−4.0022	−0.6157	−0.5150	−0.5095	21.5880
	$h^{(2)}$	−0.0124	−0.0399	−0.0456	−0.0444	0.2281	0.0351	0.0293	0.0290	−1.2302
Iteration 3	$x_f^{(3)}$	0.6285	27.5506	32.6434	33.3789	41.0547	18.1321	18.7114	18.7117	139.1787
	$x_c^{(3)}$	1.04541	27.7568	30.6978	30.4684	21.5204	18.3551	18.7148	18.6829	212.054
	$f^{(3)}$	0.0320	−0.1606	−1.4201	−2.2693	−3.3920	0.7385	0.6620	0.6219	−12.9660
	$h^{(3)}$	−0.0732	−0.0990	0.5234	0.9662	2.7851	−0.2204	−0.2071	−0.1879	1.0712
Iteration 4	$x_f^{(4)}$	0.6868	27.9074	33.8485	35.1238	41.7942	17.4332	18.0983	18.1265	155.3479
	$x_c^{(4)}$	0.8533	28.1703	32.5873	33.5311	28.4139	17.1644	17.5678	17.5959	217.48
	$f^{(4)}$	−0.1601	0.2529	0.4694	0.7934	3.5015	−0.4522	−0.4850	−0.4651	−7.5400
	$h^{(4)}$	0.0497	−0.2537	−0.4371	−0.6397	−1.3282	0.3691	0.3666	0.3536	−0.1173

Iteration 5									
$x_f^{(5)}$	1.2047	27.6410	32.4441	32.4860	29.4390	18.5305	19.2887	19.2519	178.3964
$x_c^{(5)}$	1.44391	27.6133	31.3112	31.3195	20.1401	18.2778	18.7656	18.7294	239.325
$f^{(5)}$	0.4305	−0.3041	−0.8067	−1.4182	−4.7723	0.6612	0.7128	0.6684	14.3050
$h^{(5)}$	−0.1009	0.1580	0.3879	0.5922	0.6341	−0.2699	−0.2636	−0.2482	−0.01757
Iteration 6									
$x_f^{(6)}$	0.8174	27.7587	33.3458	34.2027	36.9616	17.8706	18.5708	18.5795	163.1318
$x_c^{(6)}$	0.9733	28.0252	32.2081	32.8322	25.1093	17.5888	18.0159	18.0251	225.869
$f^{(6)}$	−0.0401	0.1078	0.0902	0.0945	0.1969	−0.0278	−0.0369	−0.0359	0.8490
$h^{(6)}$	0.0358	−0.0689	0.0084	0.0492	−0.0678	−0.0380	−0.0244	−0.0227	−0.0593
Iteration 7									
$x_f^{(7)}$	0.8584	27.6370	33.2676	34.1409	37.0008	17.8909	18.5998	18.6080	162.0220
$x_c^{(7)}$	1.03403	27.8892	32.1165	32.7635	25.233	17.6111	18.0519	18.0604	223.737
$f^{(7)}$	0.0207	−0.0282	−0.0014	0.0258	0.3206	−0.0055	−0.0009	−0.0006	−1.2830
$h^{(7)}$	−0.0428	0.0286	−0.0038	−0.0201	−0.0686	0.0016	−0.0091	−0.0085	0.0105
Iteration 8									
$x_f^{(8)}$	0.8469	27.6602	33.2659	34.1182	36.7479	17.8937	18.6005	18.6087	162.6640
$x_c^{(8)}$	—	—	—	—	—	—	—	—	—

In fact, the CCM technique is similar to the calibration concept employed in microwave testing. In order to test a device, the measurement setup needs to be calibrated to compensate for losses, phase variations, and mismatch. Typically, a set of adjustments are stored and used to evaluate the performance of the device under test. In the CCM technique, the fine model is used to extract the "standards" needed for calibration.

The circuit, simulated by the coarse model, can be "calibrated" locally so that the coarse model simulations are the same as the fine model simulations. In other words, the CCM technique allows dynamic adjustments to be added to the coarse model to improve its accuracy, so that its solution matches that of the fine model. Since the calibration is done in a local sense, fine readjustments may be needed to the corrected coarse model after each optimization process. However, the end result is that the CCM technique requires only a few EM simulations by the fine model.

The steps of the CCM technique are summarized as follows:

Step 0. Extract x_c^*, such that $R_c(x_c^*) \approx$ desired performance. Set $x_f^{(1)} = x_c^{(1)} = x_c^*$.
Step 1. Evaluate $R_f(x_f^{(1)})$.
Step 2. Extract $Y_c^{(1)}$ such that $R_c(x_c^{(1)}, Y_c^{(1)}) = R_f(x_f^{(1)})$.
Step 3. Extract $x_c^{(2)}$ such that $R_c(x_c^{(2)}, Y_c^{(1)}) =$ desired performance.
Step 4. Set $x_f^{(2)} = x_c^{(2)}$.
Step 5. Evaluate $R_f(x_f^{(2)})$.
Step 6. Extract $Y_c^{(2)}$ such that $R_c(x_c^{(2)}, Y_c^{(2)}) = R_f(x_f^{(2)})$.
Setp 7. Extract $x_c^{(3)}$ such that $R_c(x_c^{(3)}, Y_c^{(2)}) =$ desired performance.
Step 8. Set $x_f^{(3)} = x_c^{(3)}$.
Step 9. Evaluate $R_f(x_f^{(3)})$.

Steps 6–9 are repeated j times, until $R_f(x_f^{(j)})$ meets the desired performance. The solution is then $x_f^{(j)}$.

Note that the flexibility of the CCM technique lies in the ability of adding the term $Y_c^{(j)}$, which allows the calibrated coarse model to provide a solution that matches that of the fine model in a local sense [i.e., around the point $x_c^{(j)}$]. The following two examples demonstrate this concept.

Example 15.3 An Illustrative Example Demonstrating the Concept of the CCM Technique   In this example, we consider the simple functions as initial coarse and fine models to illustrate the CCM concept.

$$R_f(x) = \frac{1}{x^2 + 0.1}$$

The problem is to find x for $R_f(x) = 5.0$:

$$R_c(x) = \frac{1}{x^2} + 1$$

In the CCM technique, we need to add another parameter Y_c to the coarse model that is varied from one iteration to another for a better alignment with the fine model. Figure 15.19 illustrates a plot for the two functions $R_f(x)$ and $R_c(x)$. It is clear that

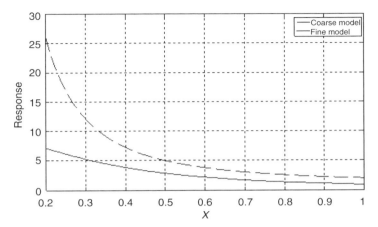

Figure 15.19 Fine and coarse model functions.

the two functions can be close to each other locally (i.e., over a given Δx) by a linear scaling; therefore, the calibrated coarse model is selected as

$$R_c(x, Y_c) = \frac{c}{x^2} + 1$$

Application of the CCM technique then gives:

Step 0. Extract x_c^* such that $R_c(x_c^*) = 5, \rightarrow x_c^* = 0.5$. Then set $x_f^{(1)} = x_c^{(1)} = x_c^* = 0.5$.

Step 1. Evaluate $R_f(0.5) = \dfrac{1}{0.25 + 0.1} = 2.8$.

Step 2. Extract $Y_c^{(1)}$ such that $R_c(x_c^{(1)}, Y_c^{(1)}) = R_f(x_f^{(1)})$, $\dfrac{c}{x^2} + 1 = 2.8 \rightarrow c = 0.45$.

Step 3. Extract $x_c^{(2)}$ such that $R_c(x_c^{(2)}, Y_c^{(1)}) = 5, \dfrac{0.45}{x^2} + 1 = 5, \rightarrow x_c^{(2)} = 0.3354$.

Step 4. Set $x_f^{(2)} = x_c^{(2)} = 0.3354$.

Step 5. Evaluate $R_f(x_f^{(2)}) = \dfrac{1}{0.3354^2 + 0.1} = 4.7060$.

Step 6. Extract $Y_c^{(2)}$ such that

$R_c(x_c^{(2)}, Y_c^{(2)}) = R_f(x_f^{(2)})$, $\dfrac{c}{0.3354^2} + 1 = 4.7060 \rightarrow c = 0.4169$.

Step 7. Extract $x_c^{(3)}$ such that $R_c(x_c^{(3)}, Y_c^{(2)}) = 5, \dfrac{0.4169}{x^2} + 1 = 5, \rightarrow x_c^{(3)} = 0.3228$.

Step 8. Set $x_f^{(3)} = x_c^{(3)} = 0.3228$.

Step 9. Evaluate $R_f(x_f^{(3)}) = \dfrac{1}{0.3228^2 + 0.1} = 4.8972$.

Step 10. Extract $Y_c^{(3)}$ such that

$R_c(x_c^{(3)}, Y_c^{(3)}) = R_f(x_f^{(3)})$, $\dfrac{c}{0.3228^2} + 1 = 4.8972 \rightarrow c = 0.4061$.

Step 11. Extract $x_c^{(4)}$ such that $R_c(x_c^{(4)}, Y_c^{(3)}) = 5, \dfrac{0.4061}{x^2} + 1 = 5 \rightarrow x_c^{(4)} = 0.3186$.

Step 12. Set $x_f^{(4)} = x_c^{(4)} = 0.3186$.

Step 13. Evaluate $R_f(x_f^{(4)}) = \dfrac{1}{0.3186^2 + 0.1} = 4.9626$. (We can end the algorithm here.)

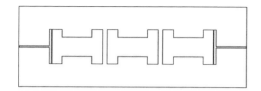

Figure 15.20 The fine EM model of the three-pole microstrip filter.

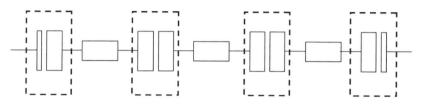

Figure 15.21 The coarse model of the three-pole filter.

After four iterations the CCM technique's solution is $x = 0.3186$. The exact solution for $R_f(x) = 0.5$ is $x = 0.3162$. It is noted that the CCM technique provided a solution that is less than 1% in only four iterations; thus the fine model is used for simulation only 4 times.

Example 15.4: Design of the Three-Pole Filter Using the CCM Technique [38] Consider the three-pole microstrip filter shown in Figure 15.20. The fine model is selected to be the full EM simulation of the overall filter circuit as shown in Figure 15.20, while the coarse model is the semi-EM simulation of the circuit as shown in Figure 15.21. The capacitive parts are simulated by using an EM simulator, and the scattering matrix of the overall filter is obtained by cascading the filter subsections. A comparison between the fine model EM simulation and the coarse model semi-EM simulations is given in Figure 15.22. Although the capacitive parts are calculated by using the same EM simulator, it is evident that there is a significant difference between the results of the two models.

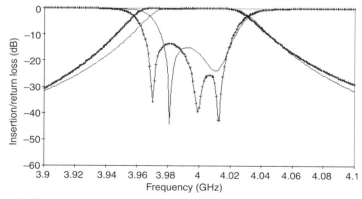

Figure 15.22 Comparison between results of the coarse model and the fine model.

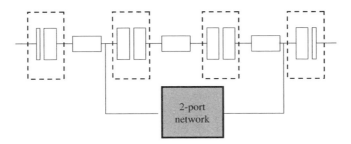

Figure 15.23 The calibrated coarse-model (CCM) technique [38].

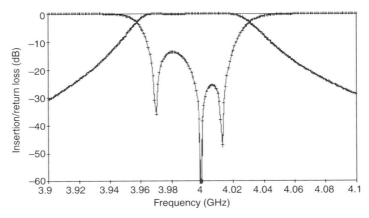

Figure 15.24 A comparison between the fine model (solid line) and the calibrated coarse model (+++).

The coarse model does not account for the effect of cross-coupling between resonators 1 and 3, which is strong in this case. The intuitive addition of a two-port network, as shown in Figure 15.23, can potentially emulate the cross-coupling. This two-port network represents the term Y_c that is added to the coarse model in the CCM technique. This two-port network is varied from iteration to another to get better alignment with the fine model. To illustrate the importance of this element in restoring alignment, Figure 15.24 shows a comparison between the fine model and the calibrated coarse model in one of the iterations. Very good alignment is observed. Details of this example are presented in [38], where it is shown that only two EM simulations were needed to complete the design process.

15.3.3 Generalized Calibrated Coarse Model Technique for Filter Design

The CCM technique, presented in Section 15.3.2, requires some degree of engineering intuition in selecting the configuration of the term Y_c. A proper selection of Y_c may speed up design process. As we have seen in the example of the three-pole filter, only two iterations were needed to complete the design process by properly selecting the location of the two-port network. However, the approach may not be applicable to complicated filter circuits, where it is very difficult to select the proper Y_c that can accurately calibrate the coarse model. A more systematic approach based on the calibrated coarse model technique was introduced in [39] and [40] to circumvent this problem.

The approach uses the EM simulator as the fine model and a circuit simulator as the coarse model. The results of these two models are then used to extract the filter coupling matrix. The calibration is done at the coupling matrix level. The coupling matrix model is selected since it provides direct interpretation of parasitic inter-actions among the filter elements. The technique therefore circumvents the problems in the space mapping approach that could arise when there is a significant misalign-ment between the coarse model and fine model, particularly as a result of the stray coupling between nonadjacent resonators.

For the generalized filter configuration, the filter scattering matrix can be written in terms of the M matrix as

$$S_{11} = 1 + 2jR_1[\lambda I - jR + M]_{11}^{-1} \tag{15.22}$$

$$S_{21} = -2j\sqrt{R_1 R_n}[\lambda I - jR + M]_{n1}^{-1} \tag{15.23}$$

where I is the identity matrix and λ, R, and M are as defined in Chapter 8.

We define $[S]_c$ as the filter response (filter scattering matrix) obtained from the coarse model and $[S]_f$ as the filter response obtained from the fine model. We define $[S^*]_c$ as the scattering matrix of the calibrated coarse model. $[S^*]_c$ is calcu-lated from the calibrated coupling matrix, after modifying the coupling matrix of the coarse model as shown in Figure 15.25.

With knowledge of the filter scattering parameters, we can extract the coupling matrix by optimization or by using any of the parameter extraction techniques (see Chapter 19). Thus, associated with the scattering matrices $[S]_c$, $[S]_f$, and $[S^*]_c$, we define the coupling matrices $[M]_c$, $[M]_f$, and $[M^*]_c$, respectively. The steps of the generalized CCM techniques are

Step 0. Extract x_c^* such that $[S]_c(x_c^*) =$ desired performance. Set $x_f^{(1)} = x_c^*$.
Step 1. Evaluate $[S]_f(x_f^{(1)})$.

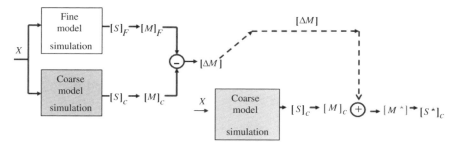

Figure 15.25 The calibrated coupling matrix technique.

Step 2. Extract the coupling matrix $[M]_c(x_c^*)$ from $[S]_c(x_c^*)$.

Step 3. Extract the coupling matrix $[M]_f(x_f^{(1)})$ from $[S]_f(x_f^{(1)})$.

Step 4. Define $[\Delta M]^{(0)} = [M]_f(x_f^{(1)}) - [M]_c(x_c^*)$.

Step 5. Use $[\Delta M]^{(0)}$ to modify the coarse model as illustrated in Figure 15.25. Extract $x_c^{(1)}$ such that $[S^*]_c(x_c^{(1)}) =$ Desired performance.

Step 6. Set $x_f^{(2)} = x_c^{(1)}$. Evaluate $[S]_f(x_f^{(2)})$.

Step 7. Extract the coupling matrix $[M]_c(x_c^{(1)})$ from $[S]_c(x_c^{(1)})$.

Step 8. Extract the coupling matrix $[M]_f(x_f^{(2)})$ from $[S]_f(x_f^{(2)})$.

Step 9. Define $[\Delta M]^{(1)} = [M]_f(x_f^{(2)}) - [M]_c(x_c^{(1)})$.

Step 10. Use $[\Delta M]^{(1)}$ to modify the coarse model as illustrated in Figure 15.27. Extract $x_c^{(2)}$ such that $[S^*]_c(x_c^{(2)}) =$ desired perfromance.

Step 11. Set $x_f^{(3)} = x_c^{(2)}$. Evaluate $[S]_f(x_f^{(3)})$.

And so on until $[S]_f(x_f^{(j)})$ meets the desired performance. The solution is then $x_f^{(j)}$.

Example 15.5: Design of a Six-Pole Filter Using the Generalized Coupling Matrix Technique [40] Figure 15.26 shows an example of a six-pole microstrip filter [40] designed by using the generalized calibrated coarse model technique. The fine model is the EM simulation of the overall filter shown in Figure 15.26. The coarse model is a semi-EM simulation and is obtained by dividing the filter into subsections as shown in Figure 15.27. The scattering matrices of these subsections are calculated by the EM simulator and then cascaded to construct the overall scattering matrix of the whole filter. The two models yield different simulation results due to the stray coupling among the nonadjacent resonators as

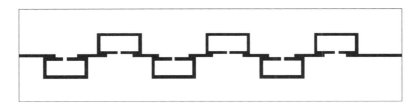

Figure 15.26 A six-pole filter designed using the generalized CCM technique. The EM simulation of the whole filter represents the fine model.

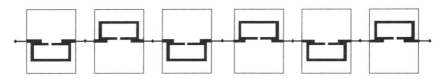

Figure 15.27 The coarse model.

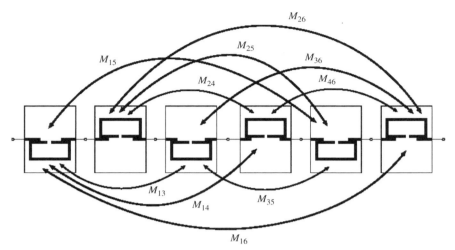

Figure 15.28 Coupling between various resonators of the microstrip filter shown in Figure 15.20.

shown Figure 15.28. This stray coupling is captured by the difference matrix $[\Delta M]$. It should be noted that the difference matrix $[\Delta M]$ does not only capture the stray coupling but also any other deviations between the two models. Figure 15.29 shows a comparison between the results of the EM simulator and the calibrated

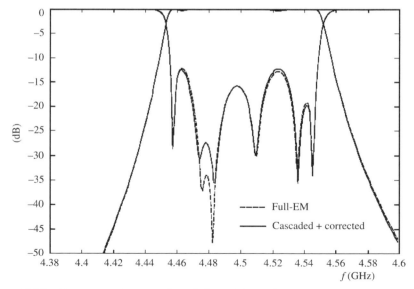

Figure 15.29 Comparison between the solution of the calibrated model (corrected model) and the fine model [40].

Figure 15.30 Photograph of a superconductive filter designed using the CCM technique [40].

coarse model after a few iterations. A good agreement is obtained. A picture of the actual filter is given in Figure 15.30. Details of this design example are given in [39] and [40].

SUMMARY

This chapter presents several techniques for EM-based design of microwave filters. The most direct approach is to combine an accurate EM simulation tool with an optimization software package, and then optimize the physical dimensions of the filter to achieve the desired performance. This is effectively a tuning process where the tuning is done by the optimization package rather than a technologist. A good initial design is needed for the optimization process to converge. Such initial designs can be obtained using the techniques described in Chapter 14. Combining the knowledge of the filter with the optimization procedure is also highly recommended. In other words, one should evaluate the relative sensitivities of the couplings and resonance frequencies of the microwave structure and make use of this information in constraining the optimization procedure. For complex filters and multiplexing networks with a large number of variables, such an approach is deemed essential in reaching a viable solution.

Direct optimization approach, without any simplifying assumptions can be still very computation intensive. The approach becomes impractical, when dealing with large order filters or 3-dimensional filter structures. For example, in view of today's computer workstation speeds, the design of an eight-pole dielectric resonator filter using this approach can take weeks. Several methods are offered to reduce the computation time of this brute force technique. It includes the use of a semi-EM simulator, EM simulators with fast frequency sweep or EM-based interpolation and polynomials to replace the actual EM-simulator. While the use of these approaches may help in reducing the optimization CPU time, it certainly impacts the accuracy of the design.

In this chapter, we also include two advanced EM-based techniques to optimize the physical dimensions of microwave filters: the space mapping (SM) technique, and the calibrated coarse model (CCM) technique. These techniques are described in detail, outlining, step by step, the implementation procedure. To demonstrate the concepts, simple examples are given which should allow the reader to duplicate numerical results using a calculator. Details of a design example of a six pole microstrip filter using the aggressive space mapping (ASM) technique are included to demonstrate the benefits of using such technique in microwave filter design. Two examples for designing filters using the CCM technique are outlined. While the direct optimization process may take 100s of EM simulations to complete the optimization process, only a few EM-simulations are needed when using the ASM or the CCM techniques. Several variations of the space mapping technique have been reported in the literature. This technique is emerging as the most useful approach for accurate EM-based design of microwave filters.

REFERENCES

1. J. W. Bandler and S. H. Chen, Circuit optimization: The state of the art, *IEEE Trans. Microwave Theory Tech.* **36**, 424–443 (Feb. 1988).

2. R. Fletcher, *Practical Methods of Optimization*, 2nd ed., Wiley, New York, 1987.

3. M. B. Steer, J. W. Bandler, and C. M. Snowden, Computer-aided design of RF and microwave circuits and systems, *IEEE Trans. Microwave Theory Tech.* **50**, 996–1005 (March 2002).

4. J. W. Bandler, S. H. Chen, S. Daijavad, and K. Madsen, Efficient optimization with integrated gradient approximations, *IEEE Trans. Microwave Theory Tech.* **36**, 444–455 (Feb. 1988).

5. R. L. Haupt, An introduction to genetic algorithms for electromagnetics, *IEEE Anten. Propag. Mag.* **37**, 7–15 (April 1995).

6. J. Hald and K. Madsen, Combined LP and quasi-Newton methods for nonlinear L1 optimization, *SIAM J. Numer. Anal.* **22**, 68–80 (1985).

7. J. Bandler et al., Microstrip filter design using direct EM field simulation, *IEEE Trans. Microwave Theory Tech.* **42**, 1353–1359 (July 1994).

8. Ansoft HFSS, http://www.ansoft.com.

9. em sonnet, http://www.sonnetusa.com.

10. Agilent Momentum, http://eesof.tm.agilent.com.

11. Zeland Software IE3D, http://www.zeland.com.

12. CST, Microwave Studio, www.cst.com.

13. MiCIAN, Microwave Wizard, http://www.mician.com.

14. C. Brezinski, *Padé-Type Approximation and General Orthogonal Polynomials*, Birkhauser Verlag, Basel, Switzerland, 1980.

15. J. Ureel et al., Adaptive frequency sampling algorithm of scattering parameters obtained by electromagnetic simulation, *IEEE AP-S Symp. Digest*, 1994, pp. 1162–1167.

16. T. Dhaene, J. Ureel, N. Fache, and D. De Zutter, Adaptive frequency sampling algorithm for fast and accurate S-parameter modeling of general planar structures, *IEEE MTT-S Symp. Digest*, 1995, pp. 1427–1431.

17. S. F. Peik, R. R. Mansour, and Y. L. Chow, Multidimensional Cauchy method and adaptive sampling for an accurate microwave circuit modeling, *IEEE Trans. Microwave Theory Tech.* **46**(12) (Dec. 1998).

18. J. Stoer and R. Bulirsch, *Introduction to Numerical Analysis*, Springer-Verlag, Berlin 1980, Sec. 2.2.

19. Q. J. Zhang and K. C. Gupta, *Neural Networks for RF and Microwave Design.* Artech House, Norwood, MA, 2000.

20. Q. J. Zhang, M. Deo, and J. Xu, Neural network for microwave circuits, *Encyclopedia of RF and Microwave Engineering*, Wiley, 2005.

21. S. F. Peik, G. Coutts, and R. R. Mansour, Application of neural networks in microwave circuit modeling, *IEEE CCECE* **2**, 928–931 (May 1998).

22. P. Burrascano, M. Dionigi, C. Fancelli, and M. Mongiardo, A neural network model for CAD and optimization of microwave filters, *IEEE-MTT-S Microwave Symp. Digest*, June 1998, pp. 13–16.

23. Q. J. Zhang, *Neuromodeler Software,* Dept Electronics, Carleton University, Ottawa, Canada.

24. MATLAB software, www.mathworks.com/.

25. P. M. Watson, K. C. Gupta, and R. L. Mahajan, Development of knowledge based artificial neural networks models for microwave components, *IEEE MTT-S Int. Microwave Symp. Digest*, Baltimore, 1998, pp. 9–12.

26. Q. J. Zhang, Knowledge based neuromodels for microwave design, *IEEE Trans. Microwave Theory Tech.* **45**, 2333–2343 (Dec. 1997).

27. V. Miraftab and R. R. Mansour, Computer-aided tuning of microwave filters using fuzzy logic, *IEEE Trans. Microwave Theory Tech.* **50**(12), 2781–2788 (Dec. 2002).

28. L. X. Wang and J. Mendel, Fuzzy basis functions, universal approximation, and orthogonal least-squares learning, *IEEE Trans. Neural Networks* **3**(5) (1992).

29. V. Miraftab and R. R. Mansour, EM-based design tools for microwave circuits using fuzzy logic techniques, *IEEE MTT-S Int. Microwave Symp. Digest*, June 2003, pp. 169–172.

30. V. Miraftab, *Computer-Aided Design and Tuning of Microwave Circuits Using Fuzzy Logic*, Ph.D. thesis, Univ. Waterloo, Ontario, Canada, 2006.

31. J. W. Bandler, R. M. Biernacki, S. H. Chen, P. A. Grobelny, and R. H. Hemmers, Space mapping technique for electromagnetic optimization, *IEEE Trans. Microwave Theory Tech.* **42**, 2536–2544 (Dec. 1994).

32. J. W. Bandler, R. M. Biernacki, S. H. Chen, R. H. Hemmers, and K. Madsen, Electromagnetic optimization exploiting aggressive space mapping, *IEEE Trans. Microwave Theory Tech.* **43**, 2874–2882 (Dec. 1995).

33. J. W. Bandler, R. M. Biernacki, S. H. Chen, and Y. F. Huang, Design optimization of interdigital filters using aggressive space mapping and decomposition, *IEEE Trans. Microwave Theory Tech.* **45**, 761–769 (May 1997).

34. J. W. Bandler et al., Space mapping: the state-of-the-art, *IEEE Trans. Microwave Theory Tech.* **52**, 337–361 (Jan. 2004).

35. M. H. Bakr, J. W. Bandler, K. Madsen, and J. Søndergaard, Review of the space mapping approach to engineering optimization and modeling, *Optimization Eng.* **1**, 241–276 (2000).

36. J. E. Rayas-Sanchez, Space mapping optimization fro engineering design: A tutorial presentation, *Proc. Workshop Next Generation Methodologies for Wireless and Microwave Circuit Design*, McMaster Univ., (June 1999).

37. Agilent ADS, http://eesof.tm.agilent.com.

38. S. Ye and R. R. Mansour, An innovative CAD technique for microstrip filter design, *IEEE Trans. Microwave Theory Tech.* **45**, 780–786 (May 1997).

39. S. F. Peik, *Efficient Design and Optimization Techniques for Planar Microwave Filters*, Ph.D. thesis, Univ. Waterloo, Ontario, Canada, 1999.

40. S. F. Peik and R. R. Mansour, A novel design approach for microwave planar filters, *IEEE MTT-S Int. Microwave Symp. Digest*, June 2002, pp. 1109–1112.

41. S. Bila et al., Direct electromagnetic optimization of microwave filters, *IEEE Microwave Maga.* **2**, 46–51 (March 2001).

CHAPTER 16

DIELECTRIC RESONATOR FILTERS

The term "dielectric resonator" was first reported in 1939 by R. D. Richtmyer [1] of Stanford University, who showed that dielectric objects can function as microwave resonators. It was not, however, before the 1960s that several pioneering works were published on the behavior of dielectrics at microwave frequencies, including the first reports on microwave dielectric resonator filters by Cohn in 1965 [2,3] and by Harrison in 1968 [4]. However, such filters were not implemented in practical applications because of the poor thermal stability of the dielectric resonator materials available at that time. The dielectric constant of these materials exhibited a significant change in their values with temperature variation. This in turn caused the temperature drift of the filter center frequency to be as large as 500 ppm/$^\circ$C.

Advances in the 1970s and 1980s in dielectric materials made it possible to combine high unloaded Q, high dielectric constants, and a small temperature drift in materials suitable for use at microwave frequencies. High-Q dielectric materials with dielectric constants ranging from 20 to 90, and a temperature drift from -6 to $+6$ ppm/$^\circ$C, are now commercially available from various manufacturers. Dielectric resonators with $\varepsilon_r = 29$ are commercially available with quality factor–frequency product $Q \times f$ values of 90,000; thus, an unloaded Q value of more than 50,000 can be achieved at 1.8 GHz. As the dielectric constant increases, the achievable unloaded Q typically decreases. For materials with a dielectric constant of 45, the $Q \times f$ value reduces to 44,000 [5]. Figure 16.1 illustrates various dielectric materials available commercially for dielectric resonator filter applications.

Microwave Filters for Communication Systems: Fundamentals, Design, and Applications, by Richard J. Cameron, Chandra M. Kudsia, and Raafat R. Mansour
Copyright © 2007 John Wiley & Sons, Inc.

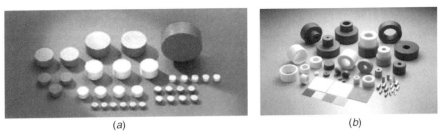

(a)   (b)

Figure 16.1 Various dielectric materials commercially available for dielectric resonator filter applications: (a) resonators; (b) resonators, supports, and substrates.

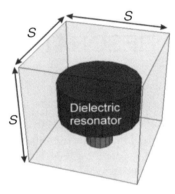

Figure 16.2 A dielectric resonator consisting of a dielectric puck mounted inside a metal enclosure.

Typically, a dielectric resonator consists of a high-dielectric-constant material in a cylindrical form (known as a *dielectric puck*), mounted inside a metal enclosure using a low-dielectric-constant support as shown in Figure 16.2. The dimensions of the metal enclosure are chosen such that metal enclosure (without dielectric) operates in an evanescent mode at the operating frequency. The electromagnetic field is concentrated mainly in the dielectric puck, and the Q factor of the resonator is determined largely by the loss tangent of the dielectric materials.

16.1 RESONANT FREQUENCY CALCULATION IN DIELECTRIC RESONATORS

In contrast to conventional rectangular and circular waveguide resonators, an exact solution of Maxwell's equations in dielectric resonators can be computed only by a numerical EM method such as the mode matching technique [6,7], finite-element analysis [8], or integral equation technique [9]. Commercial software packages such as HFSS [8] and CST Microwave Studio [10] can be readily used to calculate the resonant frequency, field distribution, and resonator Q of dielectric resonators having any arbitrary shape. A software package, based on the mode matching technique, for the analysis of dielectric resonators of cylindrical shapes is also commercially available [11]. The package is capable of accurate calculation of both the resonant frequency and Q at a relatively fast speed.

For $TE_{01\delta}$ modes, one can assume that a magnetic wall surrounds the resonator and deals with the problem as a homogenously filled cylindrical resonator [12]. The magnetic wall assumption is based on the fact that an incident wave from a high-dielectric-constant medium on an air-filled medium sees a reflection coefficient of approximately 1. This assumption simplifies the analysis and provides reasonable results for the resonant frequency and field distribution of $TE_{01\delta}$ modes [12].

Consider the dielectric resonator shown in Figure 16.3. Because of the cylindrical symmetry of the structure, the equations for the radial and azimuthal components of electromagnetic fields with $e^{-j\beta z}$ propagating in the z direction are given by [12,13]

$$E_r = \frac{-j}{k_c^2}\left(\beta\frac{\partial E_z}{\partial r} + \frac{\omega\mu}{r}\frac{\partial H_z}{\partial \phi}\right) \tag{16.1}$$

$$H_r = \frac{j}{k_c^2}\left(\frac{\omega\varepsilon}{r}\frac{\partial E_z}{\partial \phi} - \beta\frac{\partial H_z}{\partial r}\right) \tag{16.2}$$

$$E_\phi = \frac{-j}{k_c^2}\left(\frac{\beta}{r}\frac{\partial E_z}{\partial \phi} - \omega\mu\frac{\partial H_z}{\partial r}\right) \tag{16.3}$$

$$H_\phi = \frac{-j}{k_c^2}\left(\omega\varepsilon\frac{\partial E_z}{\partial r} + \frac{\beta}{r}\frac{\partial H_z}{\partial \phi}\right) \tag{16.4}$$

where $k_c^2 = k^2 - \beta^2$, $k = \sqrt{\varepsilon_r}k_0$. The magnetic field of the TE mode with $E_z = 0$ satisfies the Helmholtz equation, $(\nabla^2 + k^2)H_z = 0$. Expressing the Helmholtz equation in the cylindrical coordinate system, we have

$$\frac{1}{r}\frac{\partial}{\partial r}\left(r\frac{\partial}{\partial r}H_z\right) + \frac{1}{r^2}\frac{\partial^2}{\partial \phi^2}(H_z) + \frac{\partial^2}{\partial z^2}(H_z) + k^2 H_z = 0 \tag{16.5}$$

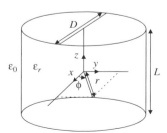

Figure 16.3 Cylindrical dielectric resonator.

Continuing with the assumption of a propagation in the z direction implies, that H_z must vary as $H_z = R(r)\Phi(\phi)e^{-j\beta z}$. Substituting the expression for H_z into equation (16.5) and dividing both sides by $r^2 H_z$ results in

$$\underbrace{\frac{r}{R(r)}\frac{\partial}{\partial r}\left(r\frac{\partial}{\partial r}R(r)\right) + r^2(k^2 - \beta^2)}_{\text{function of } r \text{ only}} + \underbrace{\frac{1}{\Phi(\phi)}\frac{\partial^2}{\partial\phi^2}(\Phi(\phi))}_{\text{function of } \phi \text{ only}} = 0 \qquad (16.6)$$

The first and second terms are functions of r and ϕ, respectively. The sum of these two terms must be a constant number, and hence the terms themselves must have constant values, since the equation must hold for any combination of the two independent degrees of freedom r and ϕ. Letting

$$\frac{1}{\Phi(\phi)}\frac{\partial^2}{\partial\phi^2}(\Phi(\phi)) = -k_\phi^2 \qquad (16.7)$$

with $\Phi(\phi) = \Phi(\phi + 2\pi)$ for self-consistency, we have $\Phi(\phi) = A\cos(k_\phi\phi) + B\sin(k_\phi\phi)$, with $k_\phi = n = 1, 2, 3, \ldots$ with $k_c^2 = k^2 - \beta^2$, the radial equation becomes

$$r^2\frac{\partial^2 R(r)}{\partial r^2} + r\frac{\partial R(r)}{\partial r} + (k_c^2 r^2 - n^2)R(r) = 0 \qquad (16.8)$$

The physically acceptable (finite) solutions to equation (16.8) are the Bessel functions of the first kind [9]:

$$R(r) = CJ_n(k_c r) \qquad (16.9)$$

A plane wave traveling in a dielectric medium with a relative permittivity of $\varepsilon_r \gg 1$, incident normally at a planar dielectric boundary, has very low characteristic impedance, $\eta_D = \eta_0/\sqrt{\varepsilon_r}$, in comparison to the air characteristic impedance. As a result, the propagating field will be reflected according to

$$\vec{E}_{\text{ref}} = \vec{E}_{\text{inc}}\frac{(\eta_0 - \eta_D)}{(\eta_0 + \eta_D)} \approx 1 \qquad (16.10)$$

In analogy with electrical open circuits, the approximation is called the *magnetic wall boundary condition* [12], and simplifies the boundary condition analysis significantly. Imposing the magnetic wall boundary condition further implies that $H_z = 0$ at $r = a$, and quantizes k_c so that

$$J_n(k_c a) = 0, \quad k_c^{(nm)} = \frac{p_{nm}}{a} \qquad (16.11)$$

where p_{nm} is the value of the mth root of the nth Bessel function $J_n(k_c r)$. With this constraint, the total solution for (nm)th mode is

$$H_z = H_0 J_n\left(k_c^{(nm)} r\right)(A \cos(n\phi) + B \sin(n\phi))e^{-j\beta z} \tag{16.12}$$

The derivation of the resonance condition for the $TE_{01\delta}$ mode requires an additional boundary condition on the dielectric$-$air interfaces at $Z = L/2$ and $Z = -L/2$. Setting $n = m = 0$, we have $H_z = AJ_0\left(k_c^{(01)} r\right)e^{-j\beta z}$, which implies $(\partial H_z)/(\partial\phi) = 0$. Using the fact that $E_z = 0$ for TE modes, and $(\partial E_z)/(\partial r) = (\partial E_z)/(\partial\phi) = 0$ for the $TE_{01\delta}$ mode, equations (16.1)–(16.4) reduce to

$$E_\phi = AJ_0'\left(k_c^{(01)} r\right)\cos(\beta z) \tag{16.13}$$

$$H_r = -\frac{jA\beta}{\omega\mu_0}J_0'\left(k_c^{(01)} r\right)\sin(\beta z) \tag{16.14}$$

with $E_r = H_\phi = 0$, for $|Z| \leq L/2$ (i.e., in the dielectric medium). The evanescent decaying solutions in air for $|Z| > L/2$ are therefore [12]

$$E_\phi = BJ_0'\left(k_c^{(01)} r\right)\exp(-\alpha|z|) \tag{16.15}$$

$$H_r = -\frac{jB\alpha}{\omega\mu_0}J_0'\left(k_c^{(01)} r\right)\exp(-\alpha|z|) \tag{16.16}$$

where $\alpha = \sqrt{k_c^2 - k_0^2}$. The dielectric interface boundary condition dictates that the transverse components of the electric field be equal on both sides of the boundary: $E_\phi(L/2^+) = E_\phi(L/2^-)$. Similarly, the radial component of the magnetic flux density must be equal, $H_r(L/2^+) = H_r(L/2^-)$, since both materials are nonmagnetic. Substituting (16.13)–(16.16), we have

$$A\cos\left(\frac{\beta L}{2}\right) = B\exp\left(-\frac{\alpha L}{2}\right) \tag{16.17}$$

$$A\beta\sin\left(\frac{\beta L}{2}\right) = B\alpha\exp\left(-\frac{\alpha L}{2}\right) \tag{16.18}$$

with

$$\beta = \sqrt{\left(\frac{2p_{01}}{D}\right)^2 - \varepsilon_r k_0^2} \tag{16.19}$$

$$\alpha = \sqrt{k_0^2 - \left(\frac{2p_{01}}{D}\right)^2} \tag{16.20}$$

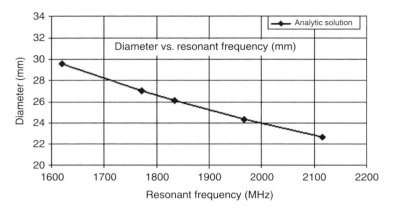

Figure 16.4 Analytic solution of diameter versus resonant frequency. $D/L = 3$ and $\varepsilon_r = 45$.

The solution for diameter D of the dielectric resonator, assuming $D/L = 3$, $\varepsilon_r = 45$, at a particular resonant frequency, is found by numerically solving equations (16.17)–(16.20). A plot of the solutions for several different resonant frequencies is given in Figure 16.4.

The field distribution for the $\mathrm{TE}_{01\delta}$ resonant mode is important when designing the input/ouput coupling. The field equations associated with the $\mathrm{TE}_{01\delta}$ mode are

$$E_\phi = A J_0'\left(k_c^{(01)} r\right) \cos(\beta z) \tag{16.21}$$

$$H_r = -\frac{jA\beta}{\omega\mu_0} J_0'\left(k_c^{(01)} r\right) \sin(\beta z) \tag{16.22}$$

$$H_z = A J_0\left(k_c^{(01)} r\right) e^{-j\beta z} \tag{16.23}$$

$$E_r = E_z = H_\phi = 0 \tag{16.24}$$

In view of these field equations, it is noted that $E_r = 0$ and $E_z = 0$; that is, the E field has only a ϕ component. The magnetic field is a linear combination of vectors pointing in the $\hat{\mathbf{e}}_r$ and $\hat{\mathbf{e}}_z$ directions. Because of the sinusoidal and cosinusoidal variation of $\mathbf{H}_r$ and $\mathbf{H}_z$, respectively, their linear combination produces circular H-field lines perpendicular to the electric field lines.

16.2 RIGOROUS ANALYSES OF DIELECTRIC RESONATORS

Because of the simplification of the boundary condition on the radial and azimuthal fields at the dielectric boundary, the analysis given in Section 16.1 provides only an approximate solution. In addition, dielectric resonators are usually placed inside a metal enclosure to shield the resonator from outside radiation. A dielectric

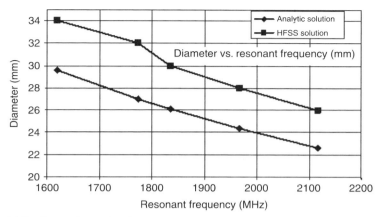

Figure 16.5 Comparison plot of resonant frequency versus diameter, $D/L = 3$ and $\varepsilon_r = 45$.

support of low relative permittivity is used in order to physically suspend the dielectric resonator away from the surfaces of the metal box as shown in Figure 16.2.

These additional parameters are not considered in the approximate analytical solution given in Section 16.1, but they can be included in a full electromagnetic simulation such as the mode matching technique [6,7] or finite-element analyses [8]. The finite-element analysis in particular can deal with any arbitrary shaped dielectric objects inside the enclosure. Figure 16.5 shows a comparison [13] between the results obtained for the resonator structure given in Figure 16.3 using the analytical solution given in Section 16.1, and the finite-element technique HFSS [8]. The resonator is centered inside the cavity. No support structure is assumed in computing the results given in Figure 16.6. The comparison shows that there is about 15% deviation

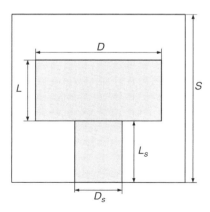

Figure 16.6 Dielectric resonator.

in resonant frequency between the results of the analytical solution and those and the finite-element HFSS solution [8].

16.2.1 Mode Charts for Dielectric Resonators

Consider the cylindrical dielectric resonator shown in Figure 16.16. The parameters of this resonator are given in Table 16.1.

The modes in dielectric resonators can be categorized as TEE, TEH, TME, TMH, HEE, and HEH. The third index was used in [7] and [11] to differentiate between modes with even and odd symmetry along the axis of the resonator at $Z = L/2$. E is used to denote electric wall symmetry (i.e., electric wall at $Z = L/2$), while "H" is used to denote magnetic wall symmetry (i.e., magnetic wall at $Z = L/2$). The dominant TEH mode is basically the $TE_{01\delta}$ mode. Figures 16.7–16.10 show the field distribution of the first four modes TEH, TME, HEH, and HEE and the corresponding resonant frequencies for $D = 1.176$ in. and $L = 0.481$ in. as obtained by HFSS [8]. Both the TEH and TME are single modes, whereas both HEH and HEE are dual hybrid modes. The resonant frequencies of these modes, calculated by assuming a constant diameter D of 1.176 in. for various values of resonator thickness L are given in Table 16.2. The results given in this table translate to the mode chart given in Figure 16.11. Details of theoretical and experimental evaluation of the resonant frequency and unloaded Q factor are given in Chapter 11.

TABLE 16.1   Parameters and Dimensions of the Resonator Given in Figure 16.16

	Parameters and Dimensions
Dielectric resonator	$\varepsilon_r = 34$, $D = 1.176$ in.; $L = 0.481$ in.
Support	$\varepsilon_r = 10$, $D_s = 0.56$ in.; $L_s = 0.8$ in.
Enclosure	$2 \times 2 \times 2.03$ in. ($C = 2$ in., $S = 2.03$ in.)

Single Mode (TEH) = 1.931 GHz

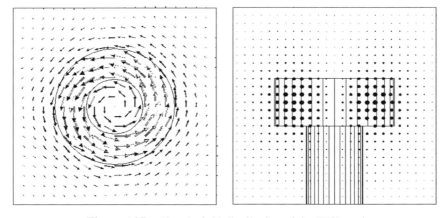

Figure 16.7   Electric field distribution of the TEH mode.

Single Mode (TME) = 2.289 GHz

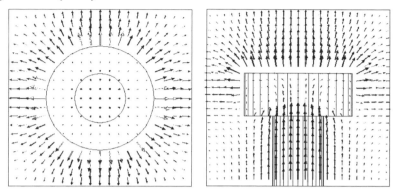

Figure 16.8   Electric field distribution of the TME mode.

Dual Hybrid Mode (HEH) = 2.414 GHz

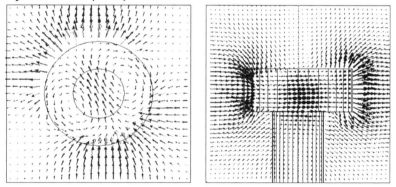

Figure 16.9   Electric field distribution of the HEH mode.

Dual Hybrid Mode (HEE) = 2.483 GHz

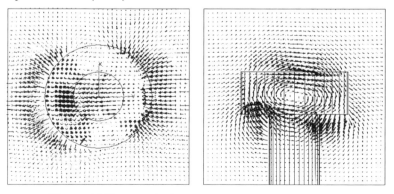

Figure 16.10   Electric field distribution of the HEE mode.

TABLE 16.2   Resonance Frequency of First Four Modes in Dielectric Resonators[a]

Ratio (D/L)	D (in.)	L (in.)	TEH (Single-Mode)	HEH (Dual-Mode)	HEE (Dual-Mode)	TME (Single-Mode)
1	1.176	1.176	1.672	1.909	1.605	1.521
1.5	1.176	0.784	1.739	2.218	1.883	2.027
2	1.176	0.588	1.841	2.331	2.188	2.193
2.445	1.176	0.481	1.931	2.414	2.484	2.289
3	1.176	0.392	2.039	2.518	2.838	2.381
3.5	1.176	0.336	2.131	2.597	3.111	2.437
4	1.176	0.294	2.220	2.669	3.209	2.481

[a]D is fixed while L is varied. Resonator parameters and dimensions are given in Table 16.1.

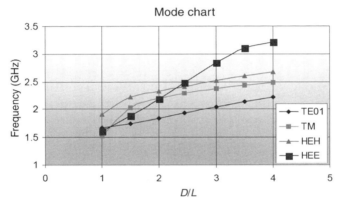

Figure 16.11   The mode chart for first four modes of the dielectric resonator. (Generated from Table 16.2.)

Note that for TME modes, the E-field is in the z-direction and exists in the support. The use of a support of a lower dielectric constant will shift the resonant frequency of the TME mode to a higher frequency range away from that of the TEH mode.

16.3   DIELECTRIC RESONATOR FILTER CONFIGURATIONS

Several dielectric resonator configurations have been reported in literature over the past two decades [14–19]. A typical dielectric resonator filter consists of a number of dielectric resonators mounted inside cavities operating below cutoff. The cavities are separated by irises to provide the necessary coupling between resonators. The dielectric resonators are supported inside the cavity in either planar or axial configuration as shown in Figures 16.12 and 16.13, respectively. While the axial configuration is analogous to cylindrical cavity filters, which makes it easy to select the proper iris shape for coupling, the planar configuration offers simplicity

Figure 16.12   A dielectric resonator filter in a planar configuration.

in designing the resonator support structure. The planar configuration also provides flexibility in folding the filter structure and providing coupling between nonadjacent resonators, which in turn makes it possible to create filters with advanced functions such as those of self-equalized filters.

As shown in Section 16.2, there are several possible operating modes for dielectric resonator filters. These modes have an impact on the filter size, unloaded Q, and spurious performance. Single-mode DR filters that operate in $TE_{01\delta}$ (TEH) modes are the most commonly used designs, particularly in wireless applications. The design offers the highest achievable filter Q in comparison with other modes of operation. For a dielectric constant of 45, the size of a dielectric resonator cavity (DR resonator + enclosure) at 1.8 GHz is about $1.6 \times 1.6 \times 1.7$ in., or a volume of 4.4 in.[3] The volume would be increased by about 5–10% if a cylindrical dielectric resonator with a hole in the center were used rather than a solid disk resonator. The hole in the center would slightly improve the resonator spurious performance, but it

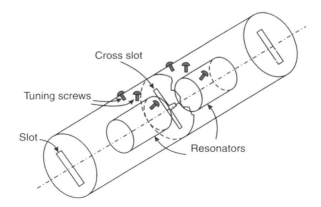

Figure 16.13   A dielectric resonator filter in an axial configuration.

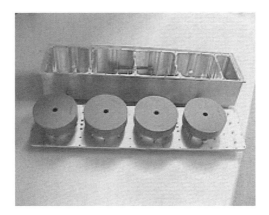

Figure 16.14 An eight-pole dual-mode filter for CDMA applications [20].

is preferred mainly because it facilitates alignment and mounting of the resonator inside the enclosure. Chapter 14 gives a step-by-step numerical example on the design of single-mode dielectric resonator filters.

Dual-mode dielectric resonator filters offer roughly 30% volume saving in comparison with single-mode dielectric resonator filters. Assuming a dielectric constant of 45, the size of a dual-mode DR resonator would be 1.8 × 1.8 × 1.7 in., a volume of 5.5 in.^{3} for two (2) electrical resonators Figure 16.14 illustrates an eight-pole dual-mode filter designed for wireless CDMA (code-division multiple access

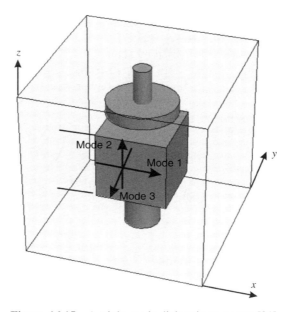

Figure 16.15 A triple-mode dielectric resonator [21].

applications [20]. It consists of four resonators operating in dual HEH modes. The resonators are supported on a plate using low-loss dielectric supports, which in turn are mounted on a housing having four cavities separated by irises.

Triple-mode DR resonators [21] offer roughly 50% volume saving in comparison with single-mode DR filters. Figure 16.15 shows a triple-mode dielectric resonator [21] realized by using a cubic dielectric resonator inserted inside a cubic cavity. The resonator consists of three orthogonal $TE_{01\delta}$ modes that are coupled to each other by tuning screws. Quadruple-mode dielectric resonator filters have been also investigated. Figure 16.16 illustrates a quadruple-mode resonator proposed in [22]. The resonator operates in four orthogonal modes (two TM modes and two TE modes). Details of the coupling between these modes are given in [22]. Both triple-mode and quadruple-mode designs require the use of dielectric resonators of irregular shapes, which are not readily available from dielectric resonator suppliers. Therefore, the cost associated with resonator machining and tuning of such types of filters may overshadow the benefits of the volume reduction.

Table 16.3 summarizes the size and achievable loaded Q values of the above three designs at 1.8 GHz assuming dielectric resonators of $\varepsilon_r = 29, 45$. It is noted

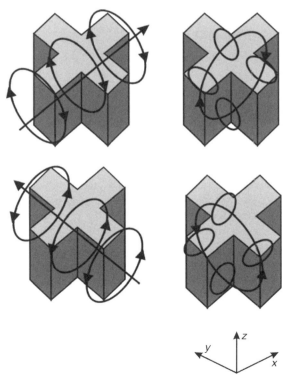

Figure 16.16 The quadruple-mode dielectric resonator illustrating the four orthogonal modes [22].

TABLE 16.3 Size and Achievable Filter Q for Dielectric Resonator Filters Operating in Single Modes, Dual Modes, and Triple Modes at 1.8 GHz

	Unloaded Q $(1/\tan \delta)$	Loaded Filter Q	Volume (in.^{3})	Volume per Electric Resonator
Single-mode (TE$_{01\delta}$) $\varepsilon_r = 29$	50,000	35,000	7.8	7.8 in.^{3}
Dual-mode (HEH) $\varepsilon_r = 29$	50,000	30,000	10.0	5.0 in.^{3}
Dual-mode (HEH) $\varepsilon_r = 29$	50,000	28,000	11.1	3.7 in.^{3}
Single-mode (TE$_{01\delta}$) $\varepsilon_r = 45$	24,000	20,000	4.4	4.4 in.^{3}
Dual-mode (HEH) $\varepsilon_r = 45$	24,000	18,000	5.5	2.8 in.^{3}
Triple-mode (3TE$_{01\delta}$) $\varepsilon_r = 45$	24,000	17,000	6.3	2.1 in.^{3}

that the achievable filter Q value is typically 70–80% that of the unloaded Q value. The reduction in Q is attributed to losses in the metallic enclosure, support structure, and tuning screws.

16.4 DESIGN CONSIDERATIONS FOR DIELECTRIC RESONATOR FILTERS

16.4.1 Achievable Filter Q Value

The Q value quoted by the dielectric resonator suppliers is the unloaded Q and not the actual Q that determines the dielectric resonator filter insertion loss performance. The unloaded Q quoted by the suppliers is typically the inverse of the loss tangent $(1/\tan \delta)$ of the dielectric material. In addition to the loss tangent, the achievable filter loaded Q value is determined by other factors such as size of enclosure, support structure, and tuning screws.

Size of Enclosure Increasing the enclosure size will naturally increase the Q value. The size, however, is usually limited by the bandwidth of the filter. The walls of the enclosure, where the irises are located, must be close enough to the dielectric resonators for the irises to provide the necessary coupling. Typically the enclosure has a square shape with a side width that is 1.5–1.7 times the diameter of the dielectric resonator. The conductivity of the materials used for the enclosure would also have an impact on the achievable filter loaded Q value.

Support Structure The dielectric resonators are typically mounted inside the cavity using a support structure that is made of dielectric material with a low dielectric constant; for example, Trans-Tech support has ε_r of 4.5 with a tan δ of 0.0002. A layer of adhesive material is typically used between the resonator and the dielectric support as shown in Figure 16.17. The combined dielectric resonator–dielectric support is typically mounted inside the cavity using a metallic screw or a plastic screw. The support structure can significantly degrade the Q if it is not properly designed.

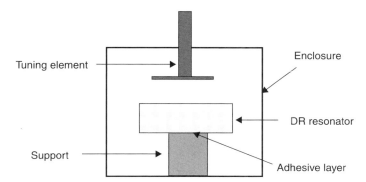

Figure 16.17 The typical support structure of a DR resonator.

Effect of Tuning Elements The tolerances on the dielectric constant are typically ± 1 from batch to batch and ± 0.5 for the same batch. For dielectric resonators with $\varepsilon_r = 29$, a variation in dielectric constant of ± 1 translates to a roughly ± 30 MHz shift in the resonant frequency. Filter suppliers reduced the impact of the tolerance problem by carrying out in-house measurement on each batch. Then a grinding machine is used to adjust the resonator length to compensate for the tolerances in the dielectric constant. However, a full readjustment cannot be made and screws still must be used to tune the resonant frequencies. The design example given in Section 14.4 (Chapter 14) also highlights the need to use tuning screws for the dielectric resonators. In addition, the tolerances on the filter housing are usually large enough to affect the coupling values that are provided by the irises. The coupling values are usually readjusted by using tuning screws mounted inside the irises. The tuning screws for the resonators and irises would certainly degrade the filter Q.

16.4.2 Spurious Performance of Dielectric Resonator Filters

Compared to other microwave filters (waveguide, coaxial, planar, etc.), dielectric resonator filters have the worst spurious performance. At 1.8 GHz, the spurious free window is around 400–500 MHz. Dielectric resonator filter suppliers have solved this problem either by employing the mixed-resonator approach [17], where coaxial cavity resonators are mixed with dielectric resonators as shown in Figure 16.18, or by using a coaxial cavity filter connected in cascade with the dielectric resonator filter as shown in Figure 16.19. The coaxial cavities have a very wide spurious free window at least up to the third harmonic, namely, 3.6 GHz at a center frequency of 1.8 GHz. The addition of such coaxial cavities would certainly solve the spurious problem but would increase the overall filter volume by 15–20% and would increase the filter insertion loss by about 10–15%.

Another approach was proposed [23] to improve the spurious performance of dielectric resonators filters operating in dual modes. The experimental results given in [23] demonstrate that reshaping the resonator structure from the traditional cylindrical shape to the modified resonator structure shown in Figure 16.20

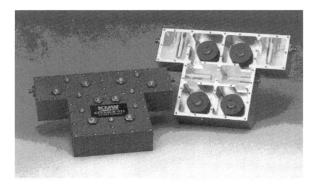

Figure 16.18 Mixing of dielectric resonators with coaxial resonators to improve the filter spurious performance.

Figure 16.19 A coaxial cavity filter is added to improve the filter spurious performance [20].

improves the spurious performance by more than 40%. As can be seen from the mode the charts given in Figure 16.11, the proximity of the HEE mode limits the spurious performance of resonators that operate at HEH dual modes. The electric field of the HEE mode is maximum near the top and bottom surfaces of the resonator, whereas that of HEH mode is almost zeros at the two surfaces. Thus the reduction of the diameter at top and bottom surfaces, as shown in Figure 16.20, should shift the resonant frequency of the HEE mode up with very little impact on that of the HEH mode. More details on the concept are given in [23].

16.4.3 Temperature Drift

The overall temperature drift of the filter is determined by the temperature coefficient of the resonator, the support structure, and the thermal expansion coefficient

Figure 16.20 Modified resonator configuration to improve the spurious performance of resonators operating in dual HEH modes.

of the tuning screws and enclosure. Dielectric resonators are offered with a wide range of temperature coefficients (-6 ppm/°C to $+6$ ppm/°C) to allow designers to compensate for the combined temperature drift that results from the above mentioned factors. Such combined temperature drift could be positive or negative depending on the type of material used. With the proper choice of temperature coefficient of dielectric resonator materials, the overall temperature drift of the filter can be reduced to 1 ppm/°C. (e.g., over a ΔT of 50°C, the frequency drift at a center frequency of 1.8 GHz would be only 90 kHz).

16.4.4 Power Handling Capability

Dielectric resonator filters operating in the $TE_{01\delta}$ mode are the ideal choice for high-peak-power applications because of their field distribution and low-loss performance. The thermal conductivity of the support and the adhesive material are important factors in designing high-power dielectric resonator filters. The support structure is designed to allow the heat generated inside the dielectric resonator to be easily transferred to the filter housing.

16.5 OTHER DIELECTRIC RESONATOR CONFIGURATIONS

Several other novel dielectric resonator configurations have been reported in literature. A half-cut resonator was proposed in [18]. The resonator is almost half the size of the conventional single-mode resonator, with a small Q degradation. The field distribution of the HEH mode in conventional dual-mode dielectric resonators is given in Figure 16.21a. With the use of a metal plate (electric wall) as an image plate [24], half of the resonator can be made, as shown in Figure 16.21b, to resonate at the same resonance frequency of the original full-size resonator. The use of a perfect magnetic wall, as shown in Figure 16.21c [18], should also yield the same result, leading to a resonator with the same resonance frequency f_0. The perfect magnetic wall could be approximated by an dielectric–air interface (nonperfect magnetic wall) leading to a half-cut resonator with a slightly higher resonant frequency $f_0 + \Delta f$ as shown in Figure 16.21d.

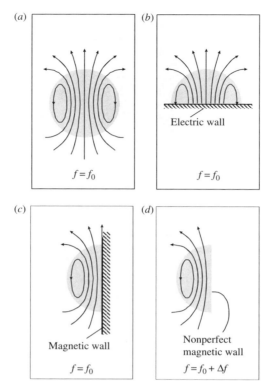

Figure 16.21 Field distribution in dielectric resonators: (*a*) HE dual modes; (*b*) half-cut resonator terminated in a perfect electric wall; (*c*) half-cut resonator terminated in a perfect magnetic wall; (*d*) half-cut resonator, with dielectric–air interface equivalent to termination with a nonperfect magnetic wall [18].

The half-cut resonator was further reshaped to improve its spurious performance [18]. A comparison between a four-pole filter built using a half-cut resonator and a four-pole filter built using a conventional cylindrical single-mode resonator is given in Figure 16.22. It can be seen that the half-cut resonator filter yields 40% reduction in size. The achievable loaded filter Q is 7000, while the achievable loaded Q of the traditional single-mode $TE_{01\delta}$ resonator is 9000 at 4 GHz [18].

The concept of using two single-mode resonators to construct a quasi-dual-mode resonator that has all the features of traditional dual-mode resonators has been proposed [18]. Two half-cut resonators can be combined in a single cavity to form a quasi-dual-mode resonator. A four-pole filter employing four half-cut resonators placed in two cavities is shown in Figure 16.23. The measured results obtained are also shown in the figure.

Other, different filter configurations provide a compromise between the small size advantages of metallic combline resonators filters and the high Q values of dielectric

Figure 16.22 Size comparison between traditional single-mode DR resonator filters and single-mode half-cut DR resonator filters [18].

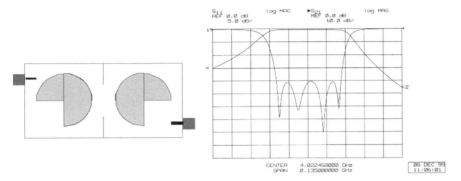

Figure 16.23 A four-pole quasi-dual-mode DR resonator filter constructed by using two half-cut DR resonators in each cavity [18].

resonator filters. Among these filters are the combline dielectric resonator filters [19] and TM dielectric resonator filters. Figure 16.24 illustrates an eight-pole coaxial DR resonator filter [20]. It consists of six DR resonators mounted directly on the housing. The input and output resonators are just normal conductor combline

Figure 16.24 A coaxial TM dielectric resonator filter [20].

Figure 16.25   A TM dielectric resonator filter.

Figure 16.26   Trans-Tech high-K ceramic substrates used to construct low cost DR filters [25].

resonators used to improve the filter spurious performance. In comparison with standard metallic combline resonator filters, the DR combline resonator filter offers 50% increase in Q while maintaining the same size. The TM dielectric resonator filter [20] is shown in Figure 16.25. It consists of dielectric resonators that are in contact with metal housing on both ends. It offers reasonably good Q for the given volume. It is useful in applications where the emphasis is on size reduction.

More recently, a novel configuration of DR filters has been reported in [25]. The filter structure is amenable to low-cost mass production while providing reasonably high Q value. The dielectric resonators are constructed as one piece from a substrate of high-K materials. All of resonators are connected to one another by the same high-K dielectric material, resulting in a whole piece that can be efficiently and accurately cut to the desired shape. The substrate is commercially available from Trans Tech [5] and can be easily cut using low-cost waterjet machining technology. The one piece dielectric resonator filter is assembled inside the filter housing using a support made of low-K substrate such as Teflon. Thus the assembly of an N-pole filter reduces from the assembly of N individual resonators to the assembly of one dielectric substrate. This simplifies the assembly by avoiding the need for alignment of the resonators, resulting in considerable time and cost savings. Figure 16.26 is a photograph of the high-K ceramic substrate used to construct the

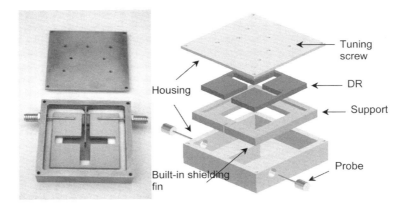

Figure 16.27 An assembled four-pole filter made of high-K ceramic substrates [25].

filter, whereas Figure 16.27 illustrates the assembled four-pole filter. This type of filter promises to be useful in wireless base station filter applications. More details are given in [25].

16.6 CRYOGENIC DIELECTRIC RESONATOR FILTERS

Various dielectric materials exhibit a considerable improvement in loss tangent when cooled to cryogenic temperatures. For example, high-purity sapphire ($\varepsilon_r = 9.4$) at X-band frequencies exhibits a $\tan\delta$ of 10^{-5} at 300 K and a $\tan\delta$ of 10^{-7} at 77 K [21]. Other dielectric material such as magnesium oxides (MgO; $\varepsilon_r = 9.7$), rutile ($\varepsilon_r = 105$), and lanthanum aluminate (LaAlO$_3$; $\varepsilon_r = 23.4$) exhibit a noticeable loss tangent improvement at cryogenic temperatures [26] as compared to room-temperature operation.

Some of the dielectric resonator materials, which were developed for conventional room-temperature filter applications, exhibit low loss at cryogenic temperature while maintaining very small temperature drift. An example is the BaMgTaO (BMT) compound manufactured by Murata [27] with a permittivity of $\sim$22. Figure 16.28 illustrates the measured loss tangent versus temperature [28] of the BMT materials. Other dielectric materials manufactured by Murata and other suppliers exhibit a similar loss tangent improvement at cryogenic temperatures. These results clearly show the potential of improving the loss tangent of dielectric materials filters when operating at cryogenic temperatures.

A fully automated measuring system to measure Q and dielectric constant, over the temperature range 10–310 K, has be been proposed [26]. Figure 16.29 illustrates the DR cavity used in the measurement. It consists of a dielectric resonator supported on a quartz spacer in a copper cavity. Tuning is achieved by using a top plate that is adjustable by means of a screw. The TE$_{01\delta}$ fundamental mode is used

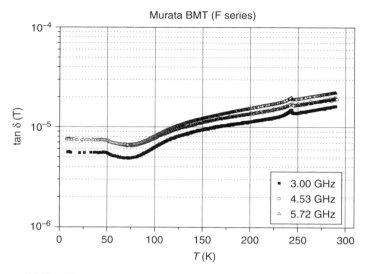

Figure 16.28   The measured loss tangent versus temperature of BMT material.

for measurement. This mode is used because the Q of the mode is easily related to the dielectric loss tangent. Measurements are made using an RF probe that is weakly coupled to the cavity. The RF probe is weakly coupled so that errors in the insertion loss due to uncertainty of the cable loss at different temperatures do not contribute to an error in the unloaded Q. The assembly is placed on the cold head of a cryocooler, which allows for temperature variation between 20 and 300 K. Table 16.4 illustrates the measured Q of four different ceramic compounds at 20, 77, and 300 K [26].

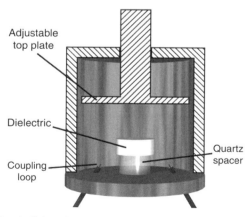

Figure 16.29   A dielectric resonator cavity for measuring unloaded Q [26].

TABLE 16.4 Measured Q Values for Different Ceramic Compounds [26]

Material	Q_u at 20 K	Q_u at 77 K	Q_u at 300 K
Alumina	704,000	269,000	40,000
BaMgTaO (BMT)	87,000	37,000	21,000
ZrSnTiO	56,000	17,000	5300
TiO$_2$	77,000	30,000	4400

Any conventional dielectric resonator filter can be potentially designed to operate at cryogenic temperatures. The key design considerations are the proper selection of dielectric materials and resonator mechanical structure.

16.7 HYBRID DIELECTRIC/SUPERCONDUCTOR FILTERS

For the conventional dielectric resonator shown in Figure 16.30a, the tangential electric field of the HEE$_{11}$ mode vanishes at the plane $z = 0$. As a consequence,

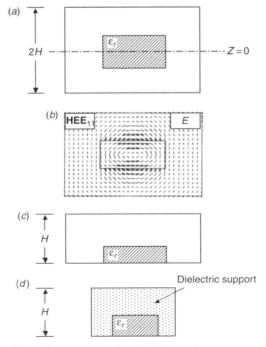

Figure 16.30 Transition from conventional DR cavity to an image DR cavity: (a) conventional DR resonator; (b) field distribution; (c) Image DR resonator; (d) Image DR resonator with a dielectric support [28].

introducing a conducting wall at this plane will slightly perturb the field distribution of this mode. Thus the size of the dielectric resonator filter operating in the HEE mode may be reduced by supporting the resonator directly on a conducting surface (image plate) as shown in Figure 16.30c. With the use of dielectric support that fills the whole cavity as shown in Figure 16.30d, further size reduction can be achieved.

The use of a normal-conducting image-plate, however, degrades the Q of the resonator, which in turn makes the approach unfeasible for high-Q applications. On the other hand, if the image plate is replaced by a surface made of high-temperature superconductor (HTS) materials, the original unloaded resonator Q will be only slightly affected, allowing the use of this resonator structure to realize compact size filters with a superior loss performance for cryogenic applications. The reduction in size is attributed to the use of the image plate, while the improvement in loss performance is attributed to the use of the HTS image plate as well as to the increase in the unloaded Q of the dielectric resonator as the temperature is lowered from 300 to 77 K.

Several hybrid four-pole/eight-pole DR/HTS filters, which utilize the concept illustrated in Figure 16.30, have been built and tested [28,29,30]. Figure 16.31 shows a detailed configuration of an eight-pole hybrid DR/HTS filter [30]. The filter consists of four dielectric resonators operating in image-type dual modes. The resonators are supported inside the filter housing using blocks made of

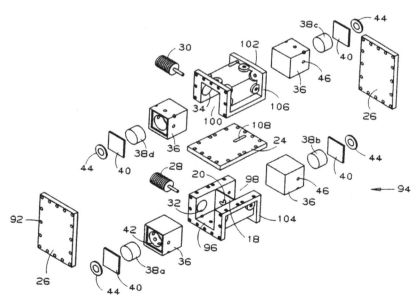

Figure 16.31   A hybrid DR/HTS filter. Components: 38s—dielectric resonators; 40s—HTS wafers diced from a standard 2-in. wafer; 6s—dielectric supports; 44s—spring washers; 26s–24, 102, 104—elements of filter housing; 28s—input/output probes [30].

low-loss low-dielectric-constant ceramic materials. The HTS material is in the form of small wafers (0.5 × 0.5 in. at C band), which are diced from a standard 2-in. wafer. The HTS films are kept in contact with the dielectric resonators using springs and metallic plates bolted to the filter housing. Both the housing and the ceramic blocks have holes to allow the filter to be tuned using standard metallic screws. The insertion loss of this type of filter is determined mainly by the loss tangent of the dielectric resonators and the ceramic blocks as well as the surface resistance of the cavity walls. Both the conductivity of the cavity walls and the loss tangent of the dielectric material improve at lower temperatures, which helps to lower the insertion loss of the filter. A detailed description of this type of filter is given in [30].

The major drawback of this class of filter is its mechanical design complexity. However, hybrid DR/HTS filters have several advantages over conventional HTS thin-film planar filters:

1. No etching is required for the HTS material. Etching may reduce the power handling capability of the HTS material and degrade its surface resistance. The design also eliminates the need to deposit gold contacts, which are required for thin-film filters. Also, there is no need to generate circuit and contact masks.

2. In patterned microstrip thin-film filters, a good portion of the HTS materials is wasted in the etching process, while in hybrid DR/HTS filters the wafer is diced into small pieces, used as "shorting plates." The wafer is used efficiently.

3. The filters can be easily tuned using conventional tuning screws.

4. The filter has a good spurious performance. The image plate helps not only reduce the filter size but also eliminate spurious modes, making it possible to design filters of this type with a spurious free window of more than 2 GHz width centered at a frequency of 4 GHz.

Figure 16.32 A layout of an eight-pole dual-mode hybrid DR/HTS filter at 4 GHz. The filter volume is one-eighth the size of a conventional dual-mode DR filter [28,30].

With metal plates replacing the HTS wafers, the filter can be used in conventional room-temperature applications. The advantages in this case are significant size reduction in comparison with traditional dual-mode filters. The unloaded filter Q is certainly much lower than that of traditional DR filters resonators, but such filters are useful in applications where the emphasis is on size reduction. Figure 16.32 illustrates an eight-pole filter realized using hybrid DR/HTS resonators [28,30]. The filter volume is one-eighth the volume of a conventional dual-mode DR filter.

SUMMARY

Dielectric resonator filters are widely employed in wireless and satellite applications. Continuing advances in the quality of dielectric materials is a good indication of the growing application of this technology. This chapter has been devoted to the design of dielectric resonator filters in a variety of configurations. Mode charts along with plots illustrating the field distribution of the first four modes in dielectric resonators are included. Commercial software packages such as HFSS and CST Microwave Studio can be readily utilized to calculate the resonant frequency, field distribution, and resonator Q of dielectric resonators having any arbitrary shape.

Major design considerations of dielectric resonator filters illustrating two different approaches for improving their spurious performance have been described. Several unconventional designs for dielectric resonator filters are included. These designs provide a tradeoff between size reduction and achievable Q-value. This tradeoff bridges the gap in terms of size and Q-value between conventional dielectric resonator filters and conventional combine coaxial filters. Design configuration of a dielectric resonator filter made of high K-ceramic, substrates is also included. The structure is attractive since the fabrication, assembly, and integration of the dielectric resonators is much simpler than that of conventional DR filters. The structure combines the low cost feature of the combine coaxial filters and the high Q feature of dielectric resonator filters. It promises to be useful in mass production of filters for wireless base station applications.

The chapter also addresses the improvement in the Q value of dielectric resonator filters when cooled to cryogenic temperatures. A hybrid dielectric/superconductor filter that offers a significant size reduction compared to conventional design is presented. The structure uses the image concept and can be employed for room temperature applications by replacing the image superconductor plates with normal metallic plates.

REFERENCES

1. R. D. Richtmyer, Dielectric resonators, *J. Appl. Phys.* **10**, 391–398 (June 1939).
2. S. B. Cohn and E. N. Targow, *Investigation of Microwave Dielectric Resonator Filters*, Rantec Div., Emerson Electric Co., internal report, 1965.

3. S. B. Cohn, Microwave bandpass filters containing high Q dielectric resonators, *IEEE Trans. Microwave Theory Tech.* **MTT-16**, 218–227 (1967).

4. W. H. Harrison, Microwave bandpass filters containing high-Q dielectric resonators, *IEEE Trans. Microwave Theory Tech.* **MTT-16**, 218–227 (1968).

5. Trans-Tech, http://www.trans-techinc.com.

6. K. A. Zaki and C. Chunming, New results in dielectric-loaded resonators, *IEEE Trans. Microwave Theory Tech.* **34**(7), 815–824 (July 1986).

7. L. Xiao-Peng and K. A. Zaki, Modeling of cylindrical dielectric resonators in rectangular waveguides and cavities, *IEEE Trans. Microwave Theory Tech.* **41**(12), 2174–2181 (Dec. 1993).

8. Ansoft HFSS, http://www.ansoft.com.

9. D. Kajfez and P. Guillon, *Dielectric Resonators*, Artech House, Norwood, MA, 1986.

10. CST, Microwave Studio, www.cst.com.

11. K. W. Zaki, Univ. Maryland, College Park, MD, USA.

12. D. Pozar, *Microwave Engineering*, 2nd ed., Wiley, New York, 1998.

13. J. Salefi, *Design Engineering Fourth-Year Design Project Report*, Univ. Waterloo, 2004.

14. J. Fiedziuzko et al., Dielectric materials, devices and circuits, *IEEE Trans. Microwave Theory Tech.* **MTT-50**, 706–720 (March 2002).

15. C. Kudsia, R. Cameron, and W. C. Tang, Innovations in microwave filters and multiplexing networks for communications satellite systems, *IEEE Trans. Microwave Theory Tech.* **40**(6), 1133–1149 (June 1992).

16. T. Noshikawa, Comparative filter technologies for communications systems, *Proc. 1998 MTT-S Workshop Comparative Filter Technologies for Communications Systems*, Baltimore, June 1998.

17. L. Ji-Fuh and B. William, High-Q TE01 mode DR filters for PCS wireless base stations, *IEEE Trans. Microwave Theory Tech.* **MTT-46**, 2493–2500 (1998).

18. R. R. Mansour et al., Quasi dual-mode resonators, *IEEE Trans. Microwave Theory Tech.* **MTT-48**(12), 2476–2482 (Dec. 2000).

19. C. Wang, C. K. A. Zaki, A. E. Atia, and T. Dolan, Dielectric combline resonators and filters. *IEEE Trans. Microwave Theory Tech.* **3**(7–12), 1315–1318 (June 1998).

20. R. R. Mansour, Filter technologies for wireless base station filters, *IEEE Microwave Mag.* **5**(1), 68–74 (March 2004).

21. V. Walker and I. C. Hunter, Design of triple mode TE_{01} resonator transmission filters, *IEEE Microwave Wireless Compon. Lett.* **12**, 215–217 (June 2002).

22. J. Hattori et al., 2 GHz band quadruple mode dielectric resonator filter for cellular base stations, *IEEE-MTT IMS-2003 Digest.*, Philadelphia, June 2003.

23. R. R. Mansour, Dual-mode dielectric resonator filters with improved spurious performance, *IEEE MTT Int. Microwave Symp. Digest*, 1993, pp. 439–442.

24. T. Nishikawa et al., Dielectric high-power bandpass filter using quarter-cut TE_{01} image resonator for cellular base stations, *IEEE Trans. Microwave Theory Tech.* **MTT-35**, 1150–1154 (Dec. 1987).

25. R. Zhang and R. R. Mansour, Dielectric resonator filters fabricated from high-K ceramic substrates, *IEEE-IMS*, June 2006.

26. S. Penn and N. Alford, *High Dielectric Constant, Low Loss Dielectric Resonator Materials*, EPSRC final report GR/K70649, EEIE, South Bank Univ., London.

27. Murata, http://www.murata.com.

28. R. R. Mansour, Design considerations of superconductive input multiplexers for satellite applications, *IEEE Trans. Microwave Theory Tech.* 1213–1228 (July 1996).

29. R. R. Mansour, Microwave superconductivity, *IEEE Trans. Microwave Theory Tech.* (Special issue—50th Anniversary of *IEEE MTT Trans.* **MTT-48**(12), 2476–2481 (March 2002).

30. R. R. Mansour and V. Dokas, *Miniaturized Dielectric Resonator Filters and Method of Operation Thereof at Cryogenic Temperatures*, U.S. Patent 0549,8771 (March 1996).

CHAPTER 17

ALLPASS PHASE AND GROUP DELAY EQUALIZER NETWORKS

The ideal characteristics for a filter network call for a brick-wall amplitude response and a linear phase across the band of interest. Such a response is not feasible, even with idealized components. However, it can be approximated with tradeoffs between the amplitude and the phase responses as a function of the order and thus complexity of the filter network.

From a communication system perspective, filters are required to channelize a given frequency band into a number of RF channels. This must be accomplished with a minimum of guard band between the channels in order to achieve the highest efficiency in the usage of the available bandwidth. As a result, most filter designs are based on the minimum phase networks since they provide the highest rate of attenuation. The filter response is optimized on the basis of amplitude response, providing a specified level of attenuation outside its passband. The phase response (and group delay) of such a filter is uniquely related to the filter's amplitude response. This implies that once the amplitude response has been optimized, the resulting phase and the group delay response must be accepted as is. For many applications, this tradeoff is acceptable.

However, it is possible to improve the phase and group delay response of a filter without sacrificing the amplitude performance. This can be accomplished by either incorporating internal equalization (linear phase filters) or using allpass external phase/group delay equalizer networks. Either method adds complexity and increases the passband loss for the overall filter network. In this chapter, we deal with the external allpass phase and group delay equalizers. It also includes a section in which we discuss the practical tradeoffs between the linear phase and externally equalized filter networks.

Microwave Filters for Communication Systems: Fundamentals, Design, and Applications,
by Richard J. Cameron, Chandra M. Kudsia, and Raafat R. Mansour
Copyright © 2007 John Wiley & Sons, Inc.

17.1 CHARACTERISTICS OF ALLPASS NETWORKS

A *network* is defined as an allpass network if it transmits all the incident power and reflects none. Such a network implies a transmission coefficient of unity and the reflection coefficient of zero. The pole–zero configuration of the network is determined by analyzing the scattering matrix of a two-port network, given by

$$[S] = \begin{bmatrix} S_{11} & S_{12} \\ S_{12} & S_{22} \end{bmatrix} \tag{17.1}$$

The network is defined as an allpass network if

$$S_{11} = S_{22} = 0 \quad \text{and} \quad |S_{12}| = 1 \text{ evaluated at } s = j\omega \tag{17.2}$$

The transfer function S_{12} of a lossless two-port network is represented by

$$S_{12}(s) = \frac{N(s)}{E(s)} \tag{17.3}$$

where $E(s)$ is a strict Hurwitz polynomial and $N(s)$ is a polynomial with real coefficients whose order does not differ from that of $E(s)$ by more than 1. For an equalizer network, polynomials $N(s)$ and $E(s)$ must be equal in magnitude by virtue of equation (17.2). Furthermore, for an allpass network to be able to manipulate the phase response, the two polynomials cannot have an identical phase response, because it leads to the trivial solution representing just a matched transmission network. The only other possibility within the constraint of equal magnitude is for the two polynomials to have an equal and opposite phase. Thus, a two-port lossless allpass network is represented by

$$S_{12}(s) = \frac{E(-s)}{E(s)} \tag{17.4}$$

The Hurwitz polynomial $E(s)$ is formed by three factors:

$$\begin{aligned}
&(s + a), & &a \text{ real and positive} \\
&(s^2 + b^2), & &b \text{ real} \\
&(s^2 + 2cs + c^2 + d^2), & &c \text{ real and positive}; \ d \text{ real}
\end{aligned} \tag{17.5}$$

It is readily seen that for the allpass functions, the allowable locations for the zeros of Hurwitz polynomial are given by

$$(s + a) \quad \text{and} \quad (s^2 + 2cs + c^2 + d^2) \tag{17.6}$$

The zeros pertaining to $(s^2 + b^2)$ cancel out, and are thus excluded from the allpass functions. This leads to the conclusion that allpass networks consist of a cascade of

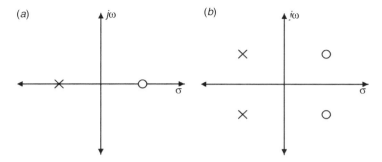

Figure 17.1 Pole–zero location of lumped-element allpass networks: (*a*) first-order network; (*b*) second-order network.

networks that can be represented by a symmetric pair of pole–zero locations on the real axis or a complex quad, as depicted in Figure 17.1. Therefore, the general form of an allpass function consisting of k first-order sections and m second-order sections is given by

$$E(s) = \prod_{a=1}^{k} (s + \sigma_a) \prod_{b=1}^{m} (s^2 + 2s\sigma_b + |s_b|^2), \quad \text{where } s_b = \sigma_b + j\omega_b \quad (17.7)$$

and σ_a, σ_b, ω_b are positive and real numbers.

The magnitude and phase response for such a network is given by

Magnitude: $\alpha = |S_{12}(s)| = 1$

$$\text{Phase}: \quad \beta = -\sum_{a=1}^{k} 2 \tan^{-1} \frac{\omega}{\sigma_a} - \sum_{b=1}^{m} 2 \tan^{-1} \frac{2\sigma_b \omega}{|s_b|^2 - \omega^2} \quad (17.8)$$

The group delay response is given by

$$\tau = -\frac{d\beta}{d\omega} = 2 \sum_{a=1}^{k} \frac{\sigma_a}{\sigma_a^2 + \omega^2} + 2 \sum_{b=1}^{m} \frac{2\sigma_b(|s_b|^2 + \omega^2)}{\omega^4 + 2\omega^2(\sigma_b^2 - \omega_b^2) + |s_b|^4} \quad (17.9)$$

17.2 LUMPED-ELEMENT ALLPASS NETWORKS

The lumped-element allpass networks are thoroughly covered by many authors [1,2]. Here, we present a summary of such networks as a background to the design of microwave allpass networks.

Typically, lumped-element allpass networks are realized by symmetric lattice structures. The general form of a lattice structure is shown in Figure 17.2*a*, and its equivalent bridge network is shown in Figure 17.2*b*.

The two alternative representations are identical. However, the bridge representation lends itself easily for the analysis of a single lattice section. On the other hand,

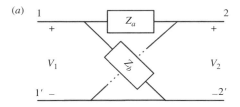

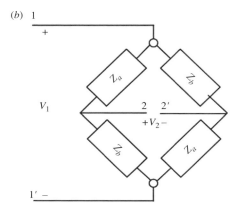

Figure 17.2 Lumped-element symmetric lattice networks: (*a*) standard representation of symmetric lattice; (*b*) bridge form of symmetric lattice.

if a number of lattice networks are connected in tandem, then the general series-type representation is a more practical description. The open-circuit impedance of the lattice network, as shown in Figure 17.2*b*, is given by

$$z_{11} = z_{22} = \frac{1}{2}(Z_a + Z_b) \tag{17.10}$$

This simply represents the impedance of two identical branches in parallel. The open-circuit transfer impedance z_{12} is determined as follows.

Let V_1 and V_2 be the voltages across terminals 1 and 2, and let I_1 and I_2 be the corresponding currents. The voltage of terminal 2 with respect to 1 is

$$V_2 = \frac{Z_b V_1}{Z_a + Z_b} - \frac{Z_a V_1}{Z_a + Z_b} \tag{17.11}$$

The voltage V_1 is related to current I_1 with terminals 2–2' open according to equation (17.10):

$$V_1 = I_1 \left[\frac{1}{2}(Z_a + Z_b) \right] \tag{17.12}$$

From equations (17.11) and (17.12), we obtain

$$\frac{V_2}{I_1} = z_{12} = \frac{1}{2}(Z_b - Z_a) \tag{17.13}$$

As a result, the impedances Z_a and Z_b are determined to be

$$Z_a = z_{11} - z_{12}$$
$$Z_b = z_{11} + z_{12} \tag{17.14}$$

In a similar manner

$$y_{11} = y_{22} = \frac{1}{2}(Y_a + Y_b)$$
$$y_{12} = \frac{1}{2}(Y_b - Y_a) \tag{17.15}$$

where

$$Y_a = \frac{1}{Z_a}, \qquad Y_b = \frac{1}{Z_b}$$

Many different methods are used to synthesize open-circuit lattice networks according to the relationships in equations (17.14) and (17.15). However, our interest lies in networks that can be used to equalize the phase and group delay of filters that are terminated in a matched load. Consequently, we next examine lattice networks that are terminated in a resistance.

17.2.1 Resistively Terminated Symmetric Lattice Networks

Let us start with a general network terminated in a resistive load as shown in Figure 17.3. The equations for the network, in terms of its open-circuit functions, are given by

$$V_1 = z_{11}I_1 + z_{12}I_2$$
$$V_2 = z_{12}I_1 + z_{22}I_2 \tag{17.16}$$

For an allpass network, the driving-point impedance must equal the terminating impedance of the filter network; therefore

$$Z(s) = \frac{V_1}{I_1} = R \tag{17.17}$$

Figure 17.3 General network terminated in the resistor, R.

By substituting $V_2 = -RI_2$ into equation (17.16), it is possible to determine $Z(s)$ in terms of the open-circuit parameters. Assuming a symmetric lattice (i.e., $z_{11} = z_{22}$) and a network normalized with $R = 1$, equation (17.17) is evaluated as

$$Z(s) = \frac{z_{11}^2 - z_{12}^2 + z_{11}}{z_{11} + 1} = 1 \quad \text{or} \quad z_{11}^2 = 1 + z_{12}^2 \tag{17.18}$$

Combining equation (17.18) with equations (17.10) and (17.13), we obtain

$$Z_a Z_b = 1 \tag{17.19}$$

If the lattice is terminated in R ohms, then we have the more general form $Z_a Z_b = R^2$. Thus, Z_a and Z_b need to be reciprocal impedances for the input impedance to be a constant value. For the network terminated in R, the transfer impedance $Z_{12}(=V_2/I_1)$ is related to Z_a and Z_b as follows [2]:

$$Z_{12} = R\left(\frac{R - Z_a}{R + Z_a}\right) \tag{17.20}$$

For the network normalized with $R = 1$, we obtain

$$Z_{12} = \left(\frac{1 - Z_a}{1 + Z_a}\right), \qquad Z_a = \frac{1 - Z_{12}}{1 + Z_{12}}, \qquad Z_b = \frac{1}{Z_a} \tag{17.21}$$

This rather important relationship makes it possible to find Z_a and Z_b from a given Z_{12} specification. As always, Z_{12} must satisfy the conditions of physical realizability, which means that Z_{12} has no poles in the right half-plane or on the imaginary axis, and that $|Z_{12}(j\omega)| \le 1$.

Let us now consider lattice networks composed entirely of lossless inductors and capacitors except for the termination. For such a network, evaluated at real frequencies, we have

$$|Z_{12}(j\omega)| = \left|\frac{1 - Z_a(j\omega)}{1 + Z_a(j\omega)}\right| = \left|\frac{1 - jX_a}{1 + jX_a}\right| = 1 \tag{17.22}$$

Thus, a symmetric lattice network, composed of L and C, and terminated in a matched resistor, represents an allpass network and satisfies the condition $|Z_{12}(j\omega)| = 1$ for all ω. The condition $|Z_{12}(j\omega)| < 1$ implies a loss in the lattice structure.

There is one more important property of the constant-resistance lattice network. Given a single lattice terminated in R ohms, its input impedance is also R ohms; thus, this terminated lattice can serve as a proper termination of R ohms for another lattice. This argument can be repeated any number of times for the cascade connection of lattices in Figure 17.4. For lattice networks with the same constant-R specification,

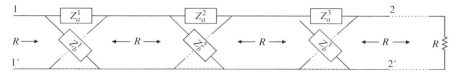

Figure 17.4 Cascade connection of constant resistance lumped-element lattice networks. (From Ref. 2.)

the input impedance of the cascade connection of any number of lattices is still R ohms. This reflects the physical reality of the nature of allpass functions.

17.2.2 Network Realizations

We know that any allpass function can be represented as a cascade of the first- and second-order allpass functions. Furthermore, an allpass function can be realized as a cascade of individual lattice networks. Thus, all we need to do is to relate the transfer function of a first- and a second-order allpass function to the impedances of a single-lattice network, shown in Figure 17.5.

The transfer function $H(s)$ is defined by the output to input voltage ratio V_2/V_1. For the allpass network, $V_1 = I_1 R$, and therefore, the transfer function is given by

$$H(s) = \frac{Z_{12}}{R} = \frac{R - Z_a}{R + Z_a} \tag{17.23}$$

First-Order Lattice Section For the first-order section, $Z_a = sL$, and substituting this value in equation (17.23) yields

$$H(s) = \frac{R/L - s}{R/L + s}$$

Using equation (17.7), we obtain

$$L = \frac{R}{\sigma_a}$$

$$Z_b = \frac{R^2}{Z_a} = \frac{R^2}{sL} = \frac{1}{sC} \tag{17.24}$$

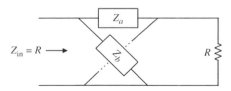

Figure 17.5 Constant-resistance lumped-element lattice network.

The value of the capacitor is

$$C = \frac{L}{R^2} = \frac{1}{\sigma_a R} \tag{17.25}$$

Thus, the reactive elements of the lattice are determined in terms of the pole–zero location of the first-order allpass function. Note that L and C resonate at σ_a.

Second-Order Lattice Section   For the second-order section, Z_a is the impedance of the antiresonant circuit:

$$Z_a = \frac{Ls}{LCs^2 + 1} \tag{17.26}$$

Substitution into equation (17.23) yields

$$H(s) = \frac{s^2 - (1/RC)s + 1/LC}{s^2 + (1/RC)s + 1/LC} \tag{17.27}$$

Comparing this with equation (17.7), we have

$$\frac{1}{RC} = 2\sigma_b$$

$$\frac{1}{LC} = |s_b|^2 = \sigma_b^2 + \omega_b^2 \tag{17.28}$$

The values L and C are therefore given by

$$L = \frac{2\sigma_b R}{|s_b|}$$

$$C = \frac{1}{s\sigma_b R |s_b|} \tag{17.29}$$

Since $Z_a Z_b = R^2$ and $Z_a = sL_1/(s^2 L_1 C_1 + 1)$, the series resonant circuit Z_b is given by

$$Z_b = \frac{R^2(s^2 L_1 C_1 + 1)}{sL_1}$$

$$= sR^2 C_1 + \frac{1}{s(L_1/R^2)}$$

$$= sL_2 + \frac{1}{sC_2} \tag{17.30}$$

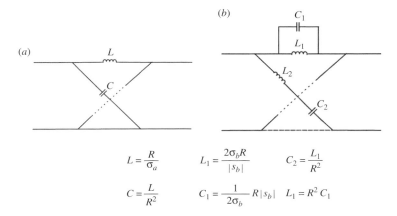

$$L = \frac{R}{\sigma_a} \qquad L_1 = \frac{2\sigma_b R}{|s_b|} \qquad C_2 = \frac{L_1}{R^2}$$

$$C = \frac{L}{R^2} \qquad C_1 = \frac{1}{2\sigma_b} R|s_b| \quad L_1 = R^2 C_1$$

Figure 17.6 Lumped-element lattice forms of (a) first-order and (b) second-order allpass network.

This is the impendence of a series circuit, composed of L_2 and C_2, where

$$L_2 = R^2 C_1$$
$$C_2 = \frac{L_1}{R^2} \tag{17.31}$$

The design equations for the first- and second-order lattice networks are summarized in Figure 17.6.

17.3 MICROWAVE ALLPASS NETWORKS

An ideal allpass network is characterized by a constant amplitude response and a variable phase or group delay response over an infinite bandwidth. This implies that the amplitude and phase responses are completely independent of each other. Furthermore, owing to the network's unique amplitude characteristics and infinite bandwidth, there is no need to develop lowpass, lumped-element prototype models as we did for the filter networks. Instead, the design can proceed directly at the required frequency and bandwidth over which equalization is required. Consequently, microwave allpass networks are dealt with directly in terms of the distributed frequency variable for their analysis and the synthesis.

Scanlon and Rhodes [3] show that any TEM allpass commensurate network must have a transfer function that has the form

$$S_{12} = \frac{E(-t)}{E(t)} \tag{17.32}$$

The distributed frequency variable t is given by

$$t = \tanh \gamma\ell = \Sigma + j\Omega \tag{17.33}$$

where γ represents the complex propagation constant and ℓ is the commensurate length of the network. This result is very similar to that of the transfer function of lumped-element allpass networks, where the real frequency variable is replaced by the distributed frequency variable t. For the lossless case, $\Sigma = 0$ and

$$t = j\Omega = j\tan\theta, \tag{17.34}$$

where $\theta = (\omega\ell/v) = (2\pi/\lambda)\ell$, θ is electrical line length, ℓ is shortest commensurate-length line, v is velocity of propagation, ω is real frequency, and λ is wavelength of the signal. If n is the effective number of unit elements in the network, the expression for the transfer function is [3]

$$S_{12}(t) = \left(\frac{1-t}{1+t}\right)^{n/2} \frac{E(-t)}{E(t)} \tag{17.35}$$

The appearance of factor $[(1-t)/(1+t)]^{n/2}$ in the transfer function is equivalent to adding a delay of a transmission line $n/4$ radians long.

The phase β of this transfer function is

$$\beta = -j\,\ell n\,S_{12}(t) \tag{17.36}$$

The group delay τ is given by

$$\tau = -\frac{d\beta}{d\omega}\bigg|_{t=j\Omega} = -\frac{1}{2\pi f_0}\frac{1}{dt}\frac{d\beta}{dt}\frac{dt}{dF}\bigg|_{t=j\Omega} \tag{17.37}$$

Here, $F = f/f_0$ is the normalized frequency variable and f_0 is the center frequency. In the physical realization of allpass networks, the commensurate-line length is usually chosen to be a quarter-length wavelength with respect to the center frequency. As a consequence, the variable t is expressed as

$$t = j\tan\frac{2\pi}{\lambda}\frac{\lambda_{\text{cf}}}{4}$$

$$= j\tan\frac{\pi}{2}\frac{\lambda_{\text{cf}}}{\lambda} \tag{17.38}$$

where λ_{cf} is the wavelength corresponding to the center frequency. For TEM structures, we obtain

$$\lambda_{\text{cf}} = \lambda_0 = \frac{c}{f_0} \tag{17.39}$$

where λ_0 is the free-space wavelength at the band center frequency of f_0. Therefore

$$t = j \tan \frac{\pi}{2} \frac{\lambda_0}{\lambda}$$

$$= j \tan \frac{\pi}{2} \frac{f}{f_0} \tag{17.40}$$

For waveguide networks, we must deal with the guide wavelength λ_g derived from

$$t = j \tan \frac{2\pi}{\lambda_g} \ell, \qquad \lambda_g = \frac{\lambda}{\sqrt{1 - \left(\dfrac{\lambda}{\lambda_{co}}\right)^2}} \tag{17.41}$$

where λ is the free-space wavelength and λ_{co} is the cutoff wavelength of the waveguide. By assuming ℓ to be a quarter of a wavelength at the center frequency, we obtain

$$t = j \tan \frac{2\pi}{\lambda} \sqrt{1 - \left(\frac{\lambda}{\lambda_{co}}\right)^2} \frac{\lambda_{g0}}{4} \tag{17.42}$$

where λ_{g0} is the guide wavelength at the center frequency f_0, given by

$$\lambda_{g0} = \frac{\lambda_0}{\sqrt{1 - \left(\dfrac{\lambda_0}{\lambda_{co}}\right)^2}} \tag{17.43}$$

Therefore

$$t = j \tan \frac{\pi}{2} \frac{\lambda_0}{\lambda} \frac{\sqrt{1 - \left(\dfrac{\lambda}{\lambda_{co}}\right)^2}}{\sqrt{1 - \left(\dfrac{\lambda_0}{\lambda_{co}}\right)^2}} \tag{17.44}$$

By substituting

$$F = \frac{f}{f_0} = \frac{\lambda_0}{\lambda}$$

$$F_0 = \frac{f_{c0}}{f_0} = \frac{\lambda_0}{\lambda_{c0}}$$

we obtain

$$t = j \tan \frac{\pi}{2} \sqrt{\frac{F^2 - F_0^2}{1 - F_0^2}}$$

$$\frac{dt}{dF} = j \frac{\pi}{2} (1 - t^2) \frac{F}{\left[(F^2 - F_0^2)(1 - F_0^2)\right]^{1/2}} \tag{17.45}$$

For the TEM-mode transmission, we obtain

$$f_{c0} = 0$$

$$t = j \tan \frac{\pi}{2} F = j \tan \frac{\pi}{2} \frac{f}{f_0} \tag{17.46}$$

$$\frac{dt}{dF} = j \frac{\pi}{2} (1 - t^2)$$

Next, let us consider the general form of the Hurwitz polynomial $H(t)$ for the allpass networks given by

$$E(t) = \prod_{i=1}^{k} (t + \sigma_i) \prod_{j=1}^{m} (t^2 + 2\sigma_j t + |t_j|^2) \tag{17.47}$$

where σ_i, $\sigma_j > 0$ and $t_j = \sigma_j + j\omega_j$. Note that this equation is similar to (17.7) except that s is replaced by the distributed frequency variable t. The integer k represents the number of real zeros and m represents the number of complex zeros; $k + m$ gives the order of the allpass network, and subscripts i and j refer to the ith and jth zeros, respectively. In microwave terminology, the first-order allpass network with a pole–zero location on the real axis is referred to as a *C section*, and the second-order network with complex conjugate pairs of pole–zero locations is referred to as a *D section*. Figure 17.7 describes the pole–zero distribution of a C/D-section allpass network.

On the basis of the relationships described in equations (17.36)–(17.47), the phase and normalized group delay for a C-section (first-order) allpass equalizer are given by

$$\beta = -2 \tan^{-1} \frac{\Omega}{\sigma_c}$$

$$\tau_N = \tau f_0 = \frac{1}{2} \frac{\sigma_c(1 + \Omega^2)}{\sigma_c^2 + \Omega^2} S(F, F_0) \tag{17.48}$$

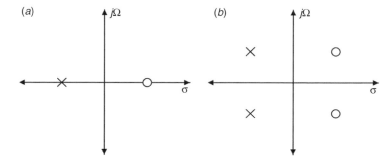

Figure 17.7 Pole–zero location of distributed allpass networks: (*a*) first-order allpass C section; (*b*) second-order allpass D section.

where

$$S(F,F_0) = \frac{F}{[F^2 - F_0^2]^{1/2}[1 - F_0^2]^{1/2}}$$

The phase and normalized time delay for a D-section (second-order) all-pass equalizer are

$$\beta = -2\tan^{-1}\frac{2\sigma_d\Omega}{|t_d|^2 - \Omega^2}$$

$$\tau_N = \sigma_d(1 + \Omega^2)\frac{\Omega^2 + |t_d|^2}{\Omega 4 + 2\Omega^2(\sigma_d^2 - \omega_d^2) + |t_d|^4}S(F,F_0) \tag{17.49}$$

$$t_d = \sigma_d + j\omega_d$$

Thus, in its most general form, the phase and normalized group delay of microwave allpass network, consisting of a cascade of k first-order and m second-order networks, is calculated by

$$\beta = -n\tan^{-1}\Omega - 2\left[\sum_{i=1}^{k}\tan^{-1}\frac{\Omega}{\sigma_i} + \sum_{j=1}^{m}\tan^{-1}\frac{2\sigma_j\Omega}{|t_j|^2 - \Omega^2}\right]$$

$$\tau_N = \frac{n}{4} + S(F,F_0)\left[\frac{1}{2}\sum_{i=1}^{k}\frac{\sigma_i(1 + \Omega^2)}{\sigma_i^2 + \Omega^2} + \sum_{j=1}^{m}\sigma_i(1 + \Omega^2)\frac{\Omega^2 + |t_j|^2}{\Omega^4 + 2(\sigma_j^2 - t_j^2) + |t_j|^4}\right]$$

$$\tag{17.50}$$

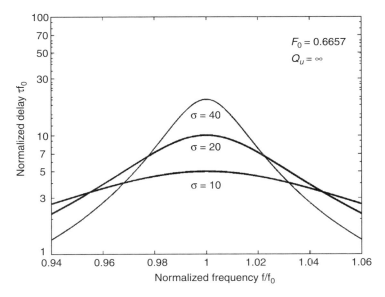

Figure 17.8 Normalized time delay of C-section TEM equalizer.

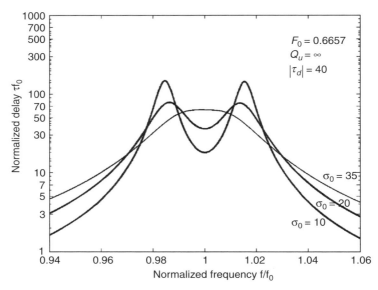

Figure 17.9   Normalized time delay of D-section waveguide equalizer.

From these relationships, it is possible to compute the phase and group delays responses of any allpass network. The normalized delays of C and D sections for a range of values of pole–zero locations are shown in Figures 17.8 and 17.9, respectively.

The shape of the group delays of allpass networks is nearly opposite that of the group delay of filter networks. Therefore, it is possible to equalize the delay of filters by cascading them with allpass networks. This implies that the location of the zeros of the allpass networks must be optimized to achieve the desired level of the overall phase and the group delay responses.

17.4   PHYSICAL REALIZATION OF ALLPASS NETWORKS

A filter or an allpass equalizer is a two-port network. In lumped circuits, an equalizer is realized as a lattice structure, described in Section 17.2. At microwave frequencies, the equivalent of a lattice structure is complicated, and often impractical. On the other hand, it is possible to realize allpass networks by making use of nonreciprocal ferrite circulators and isolators widely available at microwave frequencies. Such devices permit cascading of microwave circuits very effectively. As a result, microwave allpass networks can be realized directly as a two-port network, referred to as a *transmission-type equalizer*. The alternative is to realize the allpass network via the three-port circulator, with one of its ports terminated in a short-circuited reactance, referred to as a *reflection-type equalizer*.

17.4.1 Transmission-Type Equalizers

Many authors have dealt with two-port transmission-type equalizer designs in TEM [5–7] and waveguide structures [4,8]. A key requirement of such a network is that it maintain a near-perfect match over the band of interest, while its phase and group delay response is optimized to neutralize the filter delay. This tends to make the structure sensitive, and very costly. As a result, such equalizers have found limited practical use. On the other hand, reflection-type equalizers are much simpler to implement, and are described in the following section.

17.4.2 Reflection-Type Allpass Networks

The simplest configuration for an allpass microwave network is described in Figure 17.10.

This network consists of an ideal nonreciprocal three-port ferrite circulator, where port 1 is the input, port 2 is terminated with a lossless reactance network, and port 3 is the output. All the transmission lines and the circulator have the same characteristic impedance. A brief description of circulators is provided in Section 1.8.2. Due to the property of nonreciprocity and a virtual short circuit at port 2, all the energy incident at port 1 exits at port 2, providing an intrinsic allpass configuration. By designing the reactance network at port 2 as a cascade of commensurate transmission lines, we can realize an allpass network of any order. Ferrite circulators are widely available in the microwave frequency bands, in planar, coaxial, and waveguide structures. This changes the task of designing allpass networks to that of synthesizing one-port reactance networks. Thus, the design and implementation of allpass networks is much simpler and more cost-effective. For some applications, the use of ferrite circulators may not be suitable owing to the requirements of a very low loss (<0.1 dB), wide operating temperature range ($>30–40°C$ variation), high power ($>$tens or hundreds of watts), or any combination thereof. An alternative configuration that obviates the need of a nonreciprocal ferrite devices is displayed in Figure 17.11.

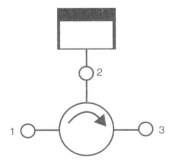

Figure 17.10 Circulator-coupled reflection-type allpass equalizer network.

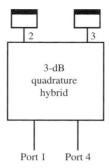

Figure 17.11 Hybrid-coupled reflection-type equalizer network.

The hybrid-coupled reflection-type equalizer network consists of a 3- dB quadrature hybrid, where port 1 is the input, port 4 is the output, and the coupled ports 2 and 3 terminated in identical reactance networks. In such a scheme (assuming an ideal hybrid), all the power incident at port 1 exits at port 4, or vice versa, yielding an inherent allpass and reciprocal network. The design of hybrids is well established in a variety of microwave structures. In planar technology, use of hybrids offers the advantage of a compact structure, which is easy to implement. Furthermore, this configuration does not require ferrite elements and thus is not limited by the properties of ferrite materials. However, the configuration requires the use of two identical reactance networks, incurring extra hardware and a higher sensitivity in the implementation of such designs.

17.5 SYNTHESIS OF REFLECTION-TYPE ALLPASS NETWORKS

Synthesis of the reflection-type microwave equalizer has been studied by several authors [5–11]. For low-order ($n \leq 3$) allpass networks and narrow bandwidths (<5%), Cristal [8] provides an excellent description and formulas for both waveguide and TEM structures. Such a range covers the majority of applications with adequate accuracy. This technique can be readily extended to higher-order networks by using modern synthesis techniques. The material presented in this section is extracted primarily from Ref. 8.

The synthesis of reflection type equalizers is simplified by assuming an ideal circulator. An isolation of 30–40 dB or greater over the band of interest is quite adequate to justify such an assumption. Such circulators are readily available at microwave frequencies. The same holds true for the quadrature hybrid-coupled realization of an allpass network. It then boils down to the synthesis of a short-circuited reactance (one-port) network. More specifically, we need to synthesize the input reflection coefficient so that its delay characteristic has the desired response. The reflection coefficient of the reactance network is given by

$$\Gamma(j\Omega) = \frac{Z_{in} - Z_0}{Z_{in} + Z_0} = \frac{z_{in} - 1}{z_{in} + 1} \qquad (17.51)$$

where Z_0 is the characteristic impedance of the circulator, Z_{in} is the input impedance of the reactance network, and z_{in} is the normalized input impedance Z_{in}/Z_0. The next step is to relate the poles and zeros of the allpass function to the input impedance of the reactance network. It is well known that any reactance function is expressed as the ratio of the odd to even or even to odd parts of an associated Hurwitz polynomial. We may therefore express the normalized impedance as

$$z(j\Omega) = \frac{\text{even } E(j\Omega)}{\text{odd } E(j\Omega)} = \frac{E(j\Omega) - E(-j\Omega)}{E(j\Omega) + E(-j\Omega)} \tag{17.52}$$

Substituting (17.52) into (17.51), we obtain

$$\Gamma(j\Omega) = -\frac{E(-j\Omega)}{E(j\Omega)} \tag{17.53}$$

The negative sign suggests a phase change of π radians and presents no difficulty in realizing the equalizer network. The same results are derived by the dual formulation in terms of the admittance of the reactance networks. This result is identical to that of the formulation of allpass commensurate transmission-line allpass networks as described by equation (17.32). It is the mathematical proof that any arbitrary transmission-line allpass transfer function can be realized by the circulator-coupled reactance network. Once function $E(t)$ is optimized to achieve the desired delay characteristic, the $E(t)$ parameters are related to the reactance of the networks by equation (17.52).

The impedance Z_{in} of a two-port network in the distributed frequency variable t in terms of *ABCD* parameters is given by [see equation (12.18)]

$$Z_{in} = \frac{A(t)Z_L + B(t)}{C(t)Z_L + D(t)} \tag{17.54}$$

For a short-circuited reactance network, $Z_L = 0$, and

$$Z_{in} = \frac{B(t)}{D(t)} \tag{17.55}$$

Making use of equation (17.52), we have

$$Z_{in}(j\Omega) = \frac{B(t)}{D(t)} = Z_0 \frac{E(j\Omega) - E(-j\Omega)}{E(j\Omega) + E(-j\Omega)} \tag{17.56}$$

Since the optimized value of $E(t)$ is known, the input impedance, in terms of the [*ABCD*] parameters, is given by equation (17.56). Following the procedure described in Chapter 11, one can synthesize a reactance network of any order.

The input impedance of C-, D-, and CD-section reactance networks in terms of its *ABCD* parameters is evaluated as follows:

C-Section Network:

$$E(t) = t + \sigma_c$$

$$Z_{\text{in}}(j\Omega) = \frac{B(t)}{D(t)} = Z_0 \frac{E(j\Omega) - E(-j\Omega)}{E(j\Omega) + E(-j\Omega)} \qquad (17.57)$$

$$= Z_0 \frac{j\Omega}{\sigma_c}$$

D-Section Network:

$$E(t) = t^2 + 2\sigma_D t + |t_D|^2 \qquad \text{where } t_D = \sigma_D + j\omega_D \text{ and}$$

$$Z_{\text{in}}(j\Omega) = \frac{B(t)}{D(t)} = Z_0 \frac{j2\sigma_D \Omega}{|t_D|^2 - \Omega^2} \qquad (17.58)$$

CD-Section Network:

$$E(t) = (t + \sigma_c)(t^2 + 2\sigma_D t + |t_D|^2)$$

$$Z_{\text{in}}(j\Omega) = \frac{B(t)}{D(t)} = Z_0 j\Omega \frac{(2\sigma_C \sigma_D + |t_D|^2) - \Omega^2}{\sigma_C |t_D|^2 - (\sigma_C + 2\sigma_D)\Omega^2} \qquad (17.59)$$

An alternative synthesis procedure to optimize the parameters of the reactance network directly is described by Hsu et al. [11]. The procedure is based on expressing the input group delay in terms of the parameters of the reactance network and then optimizing the delay by varying these parameters. Consequently, the step of synthesizing the reactance network from the optimized Hurwitz polynomial is bypassed. The number of variables required for the optimization is the same for both methods. The first method is more general, and keeps the optimization of the group delay and the physical realization separate. The first method also has the merit of being capable of generating a variety of reactance networks from the optimized Hurwitz polynomial to suit specific applications as well as exploiting the property that any allpass function is simply a cascade of first- and second-order allpass functions. However, for the specific realization of a reactance network, such as multiple coupled cavity resonators, it might be more expedient to design the equalizer directly [11].

17.6   PRACTICAL NARROWBAND REFLECTION-TYPE ALLPASS NETWORKS

As described in Chapter 3, the group delay response of filters varies inversely with absolute passband bandwidth. This implies that the group delay variation is large for filters with narrower absolute bandwidths and small for filters with larger bandwidths, irrespective of their center frequency. Typically, commercial communication systems employ narrowband channels to optimize the traffic flow and communication capacity. As a consequence, the bulk of the applications for microwave allpass networks require that the phase and group delays be equalized over narrow bandwidths, usually in the range of 0.05–5% of center frequency. The typical bandwidths of the RF channels in a communication system vary from a few megahertz up to 100 MHz. Such narrowband filters exhibit large group delays, especially at the bandedges. It is, therefore, not possible to equalize the delay over the entire bandwidth of the filter without incurring a large group delay ripple. Consequently, typical specifications call for equalization of the group delay to be within a ripple of 1–3 ns in the middle 80% of the bandwidth, allowing higher group delay toward the band edges.

In a practical system, the operating thermal environment has a significant impact on the equalized delay. This is due to the fact that any misalignment between the filter and equalizer can result in a large deviation in the equalized delay. The higher the order of the equalizer, the greater is its sensitivity, and thus the greater the misalignment of the group delays of the filter and equalizer over the operating temperature range. As a result, most practical systems require no more than a third-order allpass network, with the bulk of the applications requiring a first- or second-order equalizer. For such applications, closed-form formulas are derived for narrowband one-port reactance networks [8]. A summary of these derivations for a narrowband C- and a D-section reactance networks is summarized as follows.

17.6.1 C-Section Allpass Equalizer in Waveguide Structure

The Hurwitz polynomial and the required input reactance for a single C-section equalizer are derived as follows:

$$E(t) = t + \sigma_c$$

$$z_{in} = j\left(\frac{\Omega}{\sigma_c}\right) = \frac{t}{\sigma_c} \tag{17.60}$$

The input reactance of the network, as shown in Figure 17.12, is given by

$$Z_1 = jZ_C \tan 2\theta,$$

$$Z_2 = \frac{K_1^2}{jZ_0 \tan 2\theta} \tag{17.61}$$

$$z_{in} = \frac{Z_2}{Z_0} = \left(\frac{K_1}{Z_0}\right)^2 \frac{1}{j \tan 2\theta}$$

The waveguide cavities are π radians at the band center, and this is reflected in the choice of 2θ, as opposed to $\pi/2$ radians for cascaded transmission-line networks.

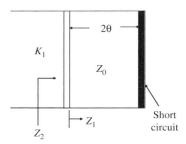

Figure 17.12 One-cavity reactance network.

For the frequency variable t, $\tan 2\theta$ corresponds to $2t/1 + t^2$, and hence

$$z_{in} = \frac{K_1^2}{Z_0^2}\left(\frac{1+t^2}{2t}\right) = \left(\frac{K_1}{Z_0}\right)^2\left[\frac{1}{2t}+\frac{t}{2}\right] \tag{17.62}$$

In the vicinity of $t \to \infty$, near the resonant frequency, we obtain

$$z_{in} = \frac{1}{2}\left(\frac{K_1}{Z_0}\right)^2 t \tag{17.63}$$

Equating this impedance to that obtained in terms of σ_c for a C-section network in equation (17.60), we obtain

$$\frac{K_1}{Z_0} = \sqrt{2\sigma_c} \tag{17.64}$$

The normalized reactance X_i/Z_0 and cavity length θ_i are related to the normalized values of the impedance inverters by [12]

$$\frac{X_i}{Z_0} = \frac{K_i/Z_0}{1-(K_i/Z_0)^2}, \qquad i = 1, 2, \ldots$$

$$\theta_i = \pi - \frac{1}{2}\left[\tan^{-1}\left(2\frac{X_i}{Z_0}\right)+\tan^{-1}\left(2\frac{X_{i-1}}{Z_0}\right)\right] \tag{17.65}$$

The reactance and cavity length is therefore evaluated as follows:

$$\frac{X_1}{Z_0} = \frac{\sqrt{2\sigma_c}}{\sigma_c - 2}$$

$$\theta_1 = \pi - \frac{1}{2}\tan^{-1}\left(2\frac{X_1}{Z_0}\right) \tag{17.66}$$

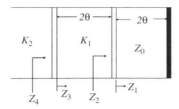

Figure 17.13   Two-cavity reactance network.

17.6.2   D-Section Allpass Equalizer in Waveguide Structure

Figure 17.13 shows a two-section reactance network in a waveguide structure. For the second-order allpass network, we have

$$E(t) = t^2 + 2\sigma_d t + |t_d|^2 \tag{17.67}$$

where $t_d = \sigma_d + j\omega_d$.

Following the same procedure as that of the C-section equalizer, the following approximate relationships for the D-section equalizer network are obtained [8].

$$
\begin{aligned}
\frac{X_1}{Z_0} &= \frac{2|t_d|}{|t_d|^2 - 4}, & \frac{K_1}{Z_0} &= \frac{2}{|t_0|} \\[2mm]
\frac{X_2}{Z_0} &= \frac{2\sqrt{\sigma_d} \cdot |t_d|}{|t_d|^2 - 4\sigma_d}, & \frac{K_2}{Z_0} &= 2\frac{\sqrt{\sigma_d}}{|t_d|} \\[2mm]
\theta_1 &= \pi - \frac{1}{2}\tan^{-1} 2\frac{X_1}{Z_0} \\[2mm]
\theta_2 &= \pi - \frac{1}{2}\left[\tan^{-1}\left(2\frac{X_2}{Z_0}\right) + \tan^{-1}\left(2\frac{X_1}{Z_0}\right)\right]
\end{aligned}
\tag{17.68}
$$

17.6.3   Narrowband TEM Reactance Networks

For narrowband TEM network applications, the series capacitive coupled resonator network in Figure 17.14 is useful, particularly with printed circuits and microstrip, since grounding the strips is not required. Also, it is particularly attractive to replace the circulator by a quadrature hybrid, thereby making the entire circuit relatively compact and easy to manufacture. This circuit is virtually the dual of the waveguide shunt inductive coupled cavity circuit described earlier. Therefore, equations (17.66) and (17.68) apply, provided that X_i/Z_0 is replaced by B_i/Y_0, and K_i/Z_0 is replaced by J_i/Y_0. B_i/Y_0 is the normalized series capacitive coupling suspectance, and J_i/Y_0 is the ideal admittance inverter, corresponding to the impedance inverter discussed above.

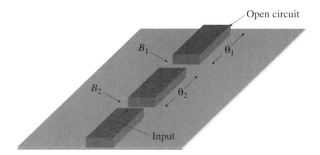

Figure 17.14 Capacitive coupled two-resonator reactance network for a second-order TEM transmission-line equalizer.

Numerous other types of TEM reactance networks are possible by using various distributed coupled line configurations. The design procedure for a cascade of commensurate transmission lines is well established. Cristal [8] describes an iterative relationship for determining the admittances of the cascaded reactance network, as well as a variety of TEM structures for arbitrary pole–zero locations of allpass networks.

17.7 OPTIMIZATION CRITERIA FOR ALLPASS NETWORKS

Allpass networks are required to equalize the group delay of filters in a communication system. An ideal group delay response calls for a flat relative delay across the entire passband of the RF channel. However, such a response is not feasible. The best that can be achieved by an allpass network is an equiripple group delay over the entire passband. However, in such an approximation, the ripple can be quite large with a slow time period, introducing a large amount of distortion in the channel. An alternative is to achieve a lower value of the group delay ripple over the middle portion of the channel and let the value rise near the band edges. This allows a tradeoff between the flat portion of the delay, usually symmetric about the band center, and the delay variation at the band edges. This channel characteristic is suitable as it tends to follow the energy spectrum of most modulation schemes. Thus, the optimization constraints typically call for the total system relative group delay to lie within a prescribed region. This allows the tradeoff for flatness of group delay (i.e., control of group delay ripple) in the central region, at the expense of the increased delay toward passband edges. A typical situation in which a communication channel requires equalization is depicted in Figure 17.15.

Let the curve $C_1(f_i, \tau_i, i = 1, 2, \ldots, N)$ be the set of points that describe the absolute group delay versus frequency characteristics of the filter, where f_i is the ith frequency point and τ_i is the corresponding absolute group delay. Let τ_c and

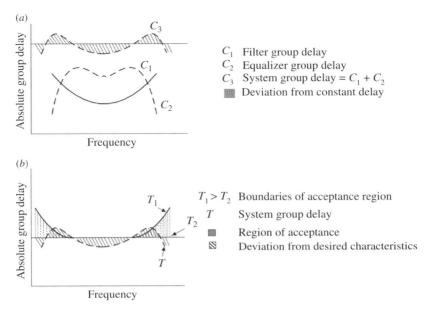

Figure 17.15 Optimization criteria for the equalizer: (*a*) when the system group delay is constant across the desired band and the constant itself either has a preassigned value or is allowed to float; (*b*) when the system group delay lies between the curves T_1 and T_2 designated as the region of acceptance.

τ_D be the group delays, generated at a particular frequency f_i by a cascade of a C- and D-section equalizer. Then, curve $C_2(f_i) = (\tau_C)_i + (\tau_D)_i$, where

$$(\tau_C)_i + (\tau_D)_i = \sum_{k=1}^{k} \tau_k + \sum_{m=1}^{m} \tau_m \qquad (17.69)$$

represents the group delay of the equalizer consisting of k C sections and m D sections.

The overall group delay C is then given by

$$C = C_1 + C_2 \qquad (17.70)$$

Defining C_0 as the overall delay at band center, the relative delay C_R of the equalized network is

$$C_R = C - C_0, \quad C_0 = (C_1 + C_2) \quad \text{at} \quad f = f_0 \qquad (17.71)$$

The minimization of C_R yields an equiripple overall delay over the desired bandwidth. Going a step further, we let the curves T_1 and T_2, as shown in Figure 17.15*b*, describe the prescribed region for the overall relative delay. To meet such a criterion requires the minimization of function F, derived from

$$F = \sum_{i=1}^{N}(T_{1i} - C_{Ri})^2 + \sum_{i=1}^{N}(T_{2i} - C_{Ri})^2$$

$$T_{1i} < C_{Ri} \qquad\qquad T_{2i} > C_{Ri} \qquad\qquad (17.72)$$

The minimization of F leads to determination of the optimum locations of the zeros of the C and D sections. The total number of variables involved is $(k + 2m)$. Function F is minimized by a variety of optimization procedures well documented in the literature.

Example 17.1   In this example, we determine the tradeoffs and design parameters to equalize the relative group delay of a channelizing filter in a $6/4$-GHz communication satellite.

Channelizing Filter Parameters.   Typical filter parameters to channelize 36-MHz transponders are as follows:

Order of the filter	9
Filter type	Chebyshev
Center frequency	4000 MHz
Passband bandwidth	38 MHz
Return loss	22 dB

Computed amplitude and group delay response of this filter is described in Figure 17.16.

Equalizer Tradeoffs.   In order to generate group delay tradeoffs, we will consider a C-, D-, and CD-section allpass networks to equalize the delay of the channelizing

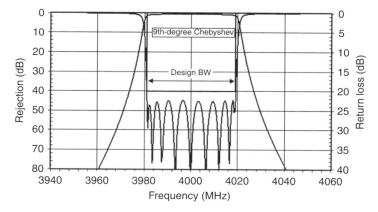

Figure 17.16   Amplitude response of a nine-pole Chebyshev filter.

filter. Furthermore, we designate the criterion of the equalization to be within 2 ns or less. It is, of course, not possible to equalize the delay to within 2 ns over the entire bandwidth of the filter. The intent is to maximize the equalized bandwidth to be within 2 ns for the first-, second-, and third-order equalizer networks.

Equalizer Parameters. We assume the WR229 waveguide for the realization of the reactance networks. Let us assume the initial parameters for the equalizers to be $\sigma_c = 100$ for the C section and $\sigma_d = 80$, $\omega_d = 50$ for the D section. By using optimization procedure described in Section 17.6, the optimized values of the equalizer networks are computed as $\sigma_c = 75.0$ for the C section, $\sigma_d = 57.3$, $\omega_d = 37.4$ for the D section, and $\sigma_c = 48.4$, $\sigma_d = 46.3$, $\omega_d = 57.9$ for the CD section.

The optimized group delay for the three equalizers, along with the filter delay, are displayed in Figure 17.17. For most applications, a D-section equalizer provides the most efficient practical equalizer network. The gain in the equalized bandwidth diminishes rapidly beyond that, while the complexity tends to increase.

Physical Realization. From the optimized locations of the pole–zero configuration of the allpass network, the input impedance of the reactance networks is determined by employing the relationships described in equations (17.57–17.59). Based on these relationships, the reactance networks can be evaluated using the [*ABCD*] synthesis technique as explained in Chapter 12. Alternatively, one can use the coupling matrix synthesis technique to determine the reactance values [11]. For the present example, it is most expedient to invoke the direct

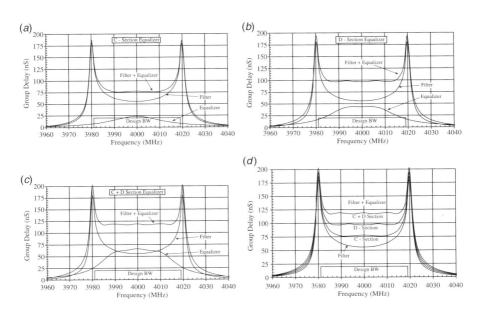

Figure 17.17 Optimized group delay of the equalized filter (*a*) using C-section equalizer; (*b*) using D-section equalizer; (*c*) using CD-section equalizer; and (*d*) comparison of the equalized group delays.

relationships described in Section 17.6 for C- and D-section allpass equalizers. The normalized values of the reactance networks are computed as follows.

For C section, $X_1/Z_0 = 0.1678$.

For D section, $X_1/Z_0 = 0.02925$, $X_2/Z_0 = 0.2326$.

The corresponding values of the resonator lengths are:

For C section

$$\theta_1 = \pi - \frac{1}{2}\tan^{-1} 2\frac{X_1}{Z_0}$$

$$= 180^\circ - 9.3^\circ = 170.7^\circ$$

For the D section, $\theta_1 \approx 178.3^\circ$ and $\theta_2 \approx 165.9^\circ$.

From the normalized values of reactance, circuit theory approximation can be used to compute the dimensions for a variety of physical structures, including circular or rectangular inductive posts or thin capacitive irises as described in Ref. 9. The physical dimensions can be refined using techniques described in Chapters 14 and 15.

17.8 EFFECT OF DISSIPATION

The dissipation of allpass networks is determined by following the same procedure as for the filter networks described in Section 3.9.1.

17.8.1 Dissipation Loss of a Lumped-Element First-Order Allpass Equalizer

Assuming a uniform dissipation factor δ, the frequency variable s is modified to $s + \delta$ in order to evaluate the effect of the dissipation of a network. By substituting this value, the amplitude response of a first-order equalizer network is given by

$$\alpha = -20\log e\left[\ell n|\sigma - (s + \delta)| - \ell n|\sigma + (s + \delta)|\right]$$

$$= -20\log e\,\ell n\,\frac{(1 - \delta')^2 + \omega'^2}{(1 + \delta)^2 + \omega'^2} \quad \text{dB} \tag{17.73}$$

where $\delta' = \delta/\sigma$ and $\omega' = \omega/\sigma$. At band center ($\omega = 0$) and for $\delta \ll 1$

$$\alpha \cong -20\log e\,(-2\delta') \cong 20\,\log e\,\delta\tau_0 \tag{17.74}$$

where τ_0 is the group delay at zero frequency given by $2/\sigma$. This result is similar to that obtained for the filter networks.

17.8.2 Dissipation Loss of a Second-Order Lumped Equalizer

Following the same procedure as for a C-section allpass network, it is easily shown that for $\delta \ll 1$, the dissipation at zero frequency is related to the group delay τ_0 by

$$\alpha \cong 20 \log e \; \delta \; \tau_0 \qquad (17.75)$$

Thus, the dissipation of allpass networks is proportional to the absolute group delay, a result similar to that for filter networks.

17.8.3 Effect of Dissipation in Distributed Allpass Networks

For distributed networks with dissipation loss, we obtain

$$t = \Sigma + j\Omega = \tanh\left(\delta' + j\frac{\pi}{2}\frac{\omega}{\omega_0}\right) \qquad (17.76)$$

where

$$\delta' = \frac{\pi}{2}\frac{1}{\omega_0}\delta$$

where δ' represents the normalized dissipation factor for lossy TEM structures. For resonant circuits, $\delta = \omega_0/2Q_0$ [see equation (3.123)] and

$$\delta' = \frac{\pi}{4Q_0} \qquad (17.77)$$

This represents the displacement of zeros of the Hurwitz polynomial, along the real axis for the TEM structures. For the waveguide structures, we have

$$t = \delta' + j\frac{\pi}{2}\frac{\lambda_{g0}}{\lambda_g}$$

and the dissipation factor δ' is therefore given by

$$\begin{aligned}
\delta' &= \frac{\pi}{2} \cdot \frac{1}{Q_0} \cdot \frac{\lambda_{g0}}{\lambda_g} \\
&= \frac{\pi}{4} \cdot \frac{1}{Q_0} \cdot \frac{1}{[1 - F_0^2]^{1/2}}
\end{aligned} \qquad (17.78)$$

where $F_0 = \lambda_0/\lambda_{co}$, λ_0 is the free-space wavelength at the center frequency and λ_{co} is the cutoff wavelength of the waveguide. It is evident that for the lumped allpass equalizers and filters, the dissipation loss is proportional to the absolute group delay, and it has a negligible impact on the delay itself. Intuitively, there is no reason why such an approximation should not hold for distributed networks as well. This is verified by the simulations. Consequently, a good approximation for the midband loss for a microwave equalizer network is given by

$$\alpha_E = 20 \log e \, \frac{\pi}{Q_E} \cdot \tau_E \tag{17.79}$$

where τ_E is the midband group delay for the equalizer and Q_E is the unloaded Q. From equations (17.48) and (17.49), the normalized midband delay of a C section is calculated by $\sigma/2(1 - F_0^2)$ and for a D section by $\sigma/(1 - F_0^2)$, respectively. The total midband delay of an equalizer is obtained by summing the contributions from the individual C and D sections. This approximation holds for the entire passband, that is, the loss variation tracks the group delay variation, where the constant of proportionality is $20 \log e \, \pi/Q_E$ or $27.3/Q_E$.

17.9 EQUALIZATION TRADEOFFS

The primary application of allpass networks is to enhance the passband performance of a filter network without altering its amplitude response. However, the use of allpass networks increases the complexity, loss, mass, volume, and cost. The use of allpass networks is, therefore, dictated by overall system level tradeoffs.

Equalization of the channel is accomplished by using external equalizers or self-equalized (linear phase) filters. Linear phase filters represent an attractive alternative to employing external equalizer networks, but the penalty involves an increase in the order of the filter and its design complexity. Such designs are sensitive to physical tolerances, and difficult to tune to achieve the desired response. This is especially true for narrowband filters, where the bulk of the applications are. On the plus side, linear phase filters represent the minimum mass and volume solution to achieve a given level of performance. Their sensitivity to the operating environment is similar to that of minimum phase filters, since self-equalization implies additional resonators. However, external equalizers offer simplicity in the design and implementation to achieve the desired overall (equalized) channel response. This is especially true for reflective-type equalizers. Such external equalizer networks are completely independent of the filter network. The equalizers can be employed during the final phases of integration to equalize the filter delays as well as any residual delays in the system.

Also, the networks do not need to be integrated with filters and can be separated by transmission lines or waveguide media. The design of equalizer networks requires reactance networks that are simple to design and tune to meet a desired response. For a given overall order of filter–equalizer network, external equalization typically offers better channel characteristics than do the equivalent-order linear phase filters.

In addition, external equalization offers the degree of freedom to change the unloaded Q of the reactance network, thereby providing a measure of control to equalize the passband amplitude with little or no impact on the group delay equalization. However, external equalization incurs extra mass and volume of the nonreciprocal (ferrite circulator) or reciprocal (3-dB quadrature hybrid) devices used to couple the reactance networks. The external equalizer incurs a higher loss owing to the losses associated with the coupling devices. For low-power applications, this extra loss is quite small and has a minimum impact on the system design. Another drawback of reflection equalizers is their sensitivity to the operating temperature range, and that can be problematic beyond $\pm 20^{\circ}$C. Generally, this is due to the properties of the ferrite materials used in circulators.

From the foregoing discussion, it is clear that the detailed system level tradeoffs are necessary to determine the best filter network design to achieve the desired channel characteristics in a communication system.

SUMMARY

This chapter describes the transfer functions of allpass networks, composed of lossless lumped elements, as well as distributed elements. An allpass network is characterized as a cascade of the first-order (C-section) and second-order (D-section) allpass networks. The poles and zeros of allpass networks are located symmetrically on the real axis for C section and distributed as a complex quad for the D section. These locations can be optimized to achieve group delays to equalize the delay of a filter over a given bandwidth. This allows the computation of the overall channel performance and tradeoffs, independent of the physical realization of the equalizer network. For microwave applications, it is most convenient to realize the allpass network by using a three-port nonreciprocal circulator whose one port is terminated in a short-circuited reactance network. At microwave frequencies, circulators with 30–40 dB isolation are readily available. For practical applications, such an isolation between the input and output ports of a circulator is adequate to realize a near-ideal allpass network. This reduces the synthesis procedure for such reflection-type equalizers to the realization of an appropriate reactance network. This is accomplished by forming the reactance function from the optimized values of the pole–zero locations of the allpass network. For a given reactance function, the synthesis is carried out by using the $[ABCD]$ matrix or the coupling matrix synthesis technique presented in Chapters 7 and 8.

The effect of dissipation can be incorporated, knowing the quality factor (unloaded Q) of the reactance network. The amplitude response of a lossy allpass network follows its group delay response. Thus, it is possible to equalize the amplitude response of the filter network by an appropriate choice of the equalizer unloaded Q. The chapter concludes with a brief discussion of the relative merits of external equalizer networks, compared with those of the equivalent linear phase filters.

REFERENCES

1. H. J. Blinchkoff and A. I. Zverev, *Filtering in the Time and Frequency Domains*, Wiley, New York, 1976.
2. M. E. Van Valkenburg, *Modern Network Synthesis*, Wiley, New York, 1960.

3. J. O. Scanlon and J. D. Rhodes, Microwave all-pass networks Part 1 and Part 2, *IEEE Trans. Microwave Theory Tech.* **MTT-16**, 62–79 (Feb. 1968).

4. K. Woo, An adjustable microwave delay equalizer, *IEEE Trans. Microwave Theory Tech.* **MTT-13**, 224–232 (March 1965).

5. D. Merlo, Development of group-delay equalizers for 4GC, *Proc. IEE (Lond.)* **112**, 289–295 (Feb. 1965).

6. E. G. Cristal, Analysis and exact synthesis of cascaded commensurate transmission-line C-section all-pass networks, *IEEE Trans. Microwave Theory Tech.* **MTT-14**, 285–291 (June 1966). [see also addendum to this article, ibid., **MTT-14**, 498–499 (Oct. 1966)].

7. T. A. Abele and H. C. Wang, An adjustable narrow band microwave delay equalizer, *IEEE Trans. Microwave Theory Tech.* **MTT-15**, 566–574 (Oct. 1967).

8. E. G. Cristal, Theory and design of transmission line all-pass equalizers, *IEEE Trans. Microwave Theory Tech.* **MTT-17**(1) (Jan. 1969).

9. C. M. Kudsia, Synthesis of optimum reflection type microwave equalizers, *RCA Rev.* **31**(3) (Sept. 1970).

10. M. H. Chen, The design of a multiple cavity equalizer, *IEEE Trans. Microwave Theory Tech.* **MTT-30**, 1380–1383 (Sept. 1982).

11. H. T. Hsu et al., Synthesis of coupled-resonators group-delay equalizers, *IEEE Trans. Microwave Theory Tech.* **50**(8), 1960–1967 (Aug. 2002).

12. S. B. Cohn, Direct-coupled resonator filters, *Proc. IRE* **45** (Feb. 1957).

CHAPTER 18

MULTIPLEXER THEORY AND DESIGN

Previous chapters have dealt with the design of microwave filters as an individual component of a communication network. In a multiuser enviror.ment, communication systems employ a range of microwave filters as individual components, as well as part of a multiport network required to separate or combine a number of RF channels. This is commonly referred to as a *multiplexing network* or simply a *multiplexer*. This chapter is devoted to the design and tradeoffs of multiplexers for a variety of practical applications.

18.1 BACKGROUND

Multiplexers (MUXs) are used in communication system applications, where there is a need to separate a wideband signal into a number of narrowband signals (RF channels). Channelization of the allocated frequency band allows flexibility for the flow of communication traffic in a multiuser environment. Amplification of individual channels also eases the requirements on the high-power amplifiers (HPAs), enabling them to operate at relatively high efficiency with an acceptable degree of nonlinearity. Multiplexers are also employed to provide the opposite function, that is, to combine several narrowband channels into a single wideband composite signal for transmission via a common antenna. Multiplexers are, therefore, referred to as *channelizers* or *combiners*. Due to the reciprocity of filter networks, a MUX can also be configured to separate the transmit and receive frequency bands in a common

Microwave Filters for Communication Systems: Fundamentals, Design, and Applications,
by Richard J. Cameron, Chandra M. Kudsia, and Raafat R. Mansour
Copyright © 2007 John Wiley & Sons, Inc.

device, referred to as a *duplexer* or *diplexer*. Multiplexers have many applications such as in satellite payloads, wireless systems, and electronic warfare (EW) systems. A more detailed rationale for the channelization of an allocated frequency band and an overview of multiplexing schemes and their impact on channel characteristics are presented in Chapter 1.

Although the principles of combining or separating frequency-diverse signals (channels) for interfacing with a single port of an antenna system have been known for many years, it was the advent of satellite communication systems in the 1970s that motivated major advances in this field. Figure 18.1 illustrates a simplified block diagram of a conventional satellite payload system. It consists of a receive and transmit antenna, a low-noise receiver, input and output multiplexers, and high-power amplifiers (HPAs). The payload behaves as an orbiting repeater that receives, amplifies, and transmits signals in the allocated frequency bands. The practical constraints on HPAs require the channelization of the received signal into a number of RF channels by a multiplexer (IMUX). After the relatively narrowband channels are amplified separately by the HPAs, the channels are recombined by an output multiplexer (OMUX) for transmission back to the ground via a common antenna. In satellite payloads, the input and output multiplexers govern the characteristics of RF channels and hence have a substantial impact on the performance of the overall payload. The typical number of filters required for input and output multiplexing networks ranges from 48 to well over 100.

Other multiplexer structures that have found widespread applications are the transmit/receive diplexers which are used in the base stations of mobile telephony systems. These devices have the most demanding electrical and build-standard requirements of all, accommodating both high- and low-power signals in close proximity within one housing. In some cases, these diplexers are located at the top of transmitter masts, where diplexers are subjected to the worst of climatic extremes. Figure 18.2 shows a simplified block diagram of the front-end of a cellular base station. The purpose of the receive filter is to reject the out-of-band interference prior to the low-noise amplification and downconversion. The transmit filter is used

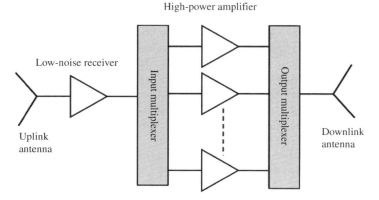

Figure 18.1 A simplified block diagram of a satellite payload.

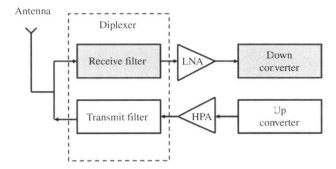

Figure 18.2 A simplified block diagram of the front end in base stations.

primarily to limit the out-of-band signals generated by the transmit portion of the base station. The transmit filter must also have a very high level of rejection in the receive band in order to eliminate the possibility of intermodulation products being fed into the receiver through the common antenna.

Multiplexers are also used in wireless applications, where the base station may need to transmit various frequency channels in different directions by using directive antennas. In this case, a multiplexer is needed to separate the overall band into separate channels, radiated in various directions. Another possible application of multiplexers in wireless base stations is in cases where the base station provides service to a number of independent operators that are licensed to operate only in specific channels within the frequency band covered by the base station.

In EW systems, multiplexers are used to construct switched filter banks, which are essential building blocks in wideband receivers that operate in a hostile signal-dense environment. In this case, switches are integrated into the multiplexer to allow the selection of a particular channel, effectively allowing the realization of a tunable filter with a variable bandwidth and a variable center frequency.

18.2 MULTIPLEXER CONFIGURATIONS

Over the past three decades (i.e., since the mid-1970s), there have been many advances in the design and implementation of multiplexing networks [1–20]. The most commonly used configurations are hybrid-coupled multiplexers, circulator-coupled multiplexers, directional filter multiplexers, and manifold-coupled multiplexers. A summary of the advantages and disadvantages of these multiplexer configurations is given in Table 18.1 [5].

18.2.1 Hybrid Coupled Approach

Figure 18.3 shows a layout of a hybrid-coupled multiplexer. Each channel consists of two identical filters and two identical 90° hybrids. The main advantage of the

TABLE 18.1 Comparison among Various Multiplexer Configurations

Hybrid Coupled MUX	Circulator Coupled MUX	Directional Filter MUX	Manifold Multiplexer
Advantages	Advantages	Advantages	Advantages
Amenable to modular concept	Requires one filter per channel	Requires one filter per channel	Requires one filter per channel
Simple to tune, no interaction between channel filters	Employs standard design of filters	Simple to tune, no interaction between channel filters	Most compact design
Total power in transmission modes as well as reflection mode is divided by the hybrid so that only 50% of the power is incident on each filter; power handling is thus increased and susceptibility to voltage breakdown is reduced	Simple to tune; no interaction between channel filters	Amenable to modular concept	Capable of realizing optimum performance for absolute insertion loss, amplitude and group delay response
	Amenable to a modular concept	Disadvantages	Disadvantages
Disadvantages	Disadvantages	Restricted to realize all-pole functions such as Butterworth and Chebyshev	Complex design
Two identical filters and two hybrids are required for each channel.	Signals must pass in succession through circulators, incurring extra loss per trip	Difficult to realize bandwidths greater than 1%	Tuning of multiplexer can be time-consuming and expensive
Line lengths between hybrids and filters require precise balancing to preserve circuit directivity	Low-loss, high-power ferrite circulators are expensive		Not amenable to a flexible frequency plan; i.e., change of a channel frequency would require a new multiplexer design
Physical size and weight of the multiplexer is greater than in other approaches	Higher level of passive intermodulation (PIM) products than in other configurations		

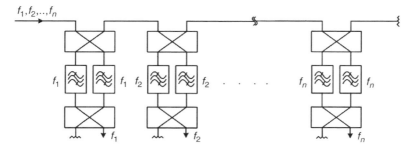

Figure 18.3 Layout of a hybrid-coupled multiplexer.

hybrid-coupled approach is its directional property, which minimizes the interaction among the channel filters. As a consequence, the hybrid-coupled multiplexer is amenable to a modular concept. It also allows the integration of additional channels at a later date without disrupting the existing multiplexer design, which is a requirement in some systems. Another key advantage of this approach is that only half of the input power goes through each filter. Thus, the filter design can be relaxed when using this type of multiplexer in high-power applications. On the other hand, its large size is a disadvantage, since two filters and two hybrids per channel are required. Another design consideration of such multiplexers is the phase deviation between the two filter paths that the two signals undergo before they add constructively at the channel output. As a result, the structure must be fabricated with tight tolerances to minimize the phase deviation particularly in planar circuit applications, where it is difficult to use tuning elements to balance the two paths.

18.2.2 Circulator-Coupled Approach

Each channel, in this approach, consists of a channel-dropping circulator and one filter, as shown in Figure 18.4. The unidirectional property of the circulator provides the same advantages as does the hybrid-coupled approach in terms of amenability to modular integration and ease of design and assembly. The insertion loss of the first channel is the sum of the insertion loss of the channel filter and the insertion loss of the circulator. The subsequent channels exhibit a relatively higher loss due to the insertion loss incurred during each trip through the channel dropping circulators. Figure 18.5 is a photo of a four-channel multiplexer [17]. It consists of a stack of four circulators connected to a stack of four filters by cables.

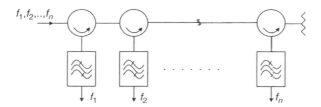

Figure 18.4 A circulator-coupled multiplexer.

Figure 18.5   A four-channel circulator-coupled multiplexer [17].

18.2.3   Directional Filter Approach

Figure 18.6 illustrates a layout of a multiplexer, realized by connecting directional filters in series. A direction filter is a four-port device where one port is terminated in a load. The other three ports of the directional filter essentially act as a circulator connected to a bandpass filter. The power incident at one port emerges at the second port with a bandpass frequency response, while the reflected power from the filter emerges at the third port. However, directional filters do not require the use of ferrite circulators. Figure 18.7 illustrates two possible approaches for realizing directional filters in waveguide and microstrip, respectively [1]. The waveguide directional filter is realized by coupling rectangular waveguides, operating in TE_{10} mode, to a circular waveguide filter operating in TE_{11} modes. In the microstrip version, each $360°$ wavelength ring resonator is coupled to another wavelength ring resonator and to two transmission lines. This multiplexing approach has the same advantages as the hybrid-coupled and circulator-coupled approaches. It is, however, limited to narrowband applications.

18.2.4   Manifold-Coupled Approach

The manifold-coupled approach, shown in Figure 18.8, is viewed as the optimum choice as far as miniaturization and absolute insertion loss are concerned; this

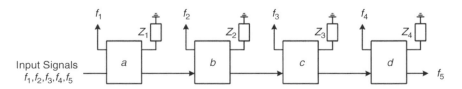

Figure 18.6   A directional filter multiplexer.

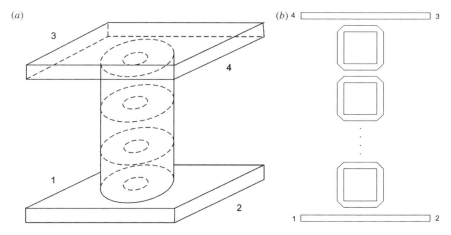

Figure 18.7 (*a*) A waveguide directional filter; (*b*) microstrip directional filter.

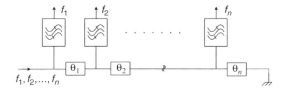

Figure 18.8 A manifold-coupled multiplexer.

type of MUX requires the presence of all the channel filters at the same time so that the effect of channel interactions can be compensated for in the design process. This implies that a manifold-coupled MUX is not amenable to a flexible frequency plan. Any change in the allocation of the channels requires a new multiplexer design. Moreover, as the number of channels increases, this approach becomes more difficult to implement. The manifold coupled multiplexer shown in Figure 18.8 acts as a channelizer. The same configuration can be used as a combiner. Figure 18.9 is a photo of a four-channel multiplexer, employing a waveguide manifold and four dual-mode circular waveguide filters. The same figure also shows a miniature multiplexer with a coaxial manifold and superconductive filters [20]. The manifold-coupled concept can be implemented as well in planar circuits as shown in Figure 18.10. Here, three microstrip filters are integrated with a microstrip manifold [17]. In this particular case, one of the channels is connected directly to the manifold.

There are three distinct categories of multiplexing networks required by communication systems, namely, RF channelizers, RF combiners, and transmit–receive diplexers. The design of each type of MUX is dictated by its application and system constraints.

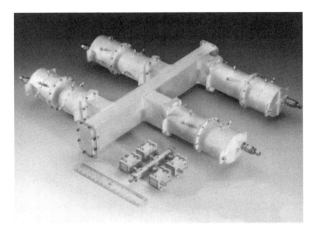

Figure 18.9 A C-band Four-channel waveguide multiplexer compared with a Four-channel multiplexer implemented using superconductor technology [20].

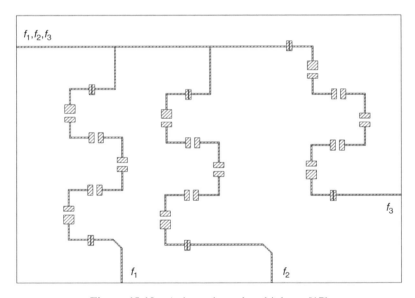

Figure 18.10 A three-channel multiplexer [17].

18.3 RF CHANNELIZERS (DEMULTIPLEXERS)

For broadband communication systems, the available frequency spectrum is split into a number of smaller frequency bands or RF channels. This is done to optimize the system performance, as explained in Chapter 1. The function is carried out by a multiplexer, referred to as an *input mux* (IMUX) or *demultiplexer* or *RF channelizer*.

Such a MUX is required at each receiving station in the system. Design constraints on such a MUX are:

1. It must be able to split a frequency band into a number of discrete contiguous RF channels
2. Each RF channel must exhibit a near ideal response, that is, a minimum of amplitude and group delay variation across its passband and a large attenuation outside of it. This is required to minimize the interference from other channels and to provide HPA with a clean signal for amplification. In other words, a near-ideal filter at this stage prevents the amplification of the interference that may be present outside the passband of the RF channel.

In a communication system, RF channelization is carried out after amplification of the broadband signal by an LNA. As a consequence, absolute insertion loss is not a constraint. This feature is exploited in the design of such multiplexers. It should be noted that excessive loss can become a liability if it incurs the requirement of an extra amplifier. The most common designs for IMUX are the hybrid branching network and the circulator-coupled MUX.

18.3.1 Hybrid Branching Network

Figure 18.11 shows the typical arrangement for a hybrid branching network. The figure depicts the case for the demultiplexing of a group of five contiguous channels. Each 3 dB hybrid in the branch splits the incoming wideband signal into two equal paths, until the number of paths equals the number of channels. At the end of each path channel filters are located, each with an isolator at its input.

The entire spectrum of the channel group is presented to the input port of each filter, and the channel corresponding to the passband of the filter passes through; the rest are reflected and absorbed in the isolator loads. There is no restriction in the placement of filters corresponding to any channel at the end of any of the paths.

The advantages of the hybrid branching network are summarized here;

- Design is simple, requiring standard hybrids and channel filters.
- The branching network can be very lightweight, compact, and reliable, using integrated microstrip technology.
- The second port of the input hybrid can be used for the input from a redundant low-noise amplifier (LNA), without the need for an RF switch.
- There is no channel interaction, since each filter receives the broadband signal directly at its input port. Each channel filter operates independently and is tuned separately without regard to the other filters in the overall MUX.
- The input return loss of the overall IMUX is good over a wide bandwidth.

A disadvantage is that insertion loss is high, typically 3.5 dB per hybrid. The worst-case insertion loss of the IMUX shown in Figure 18.11 is estimated to be

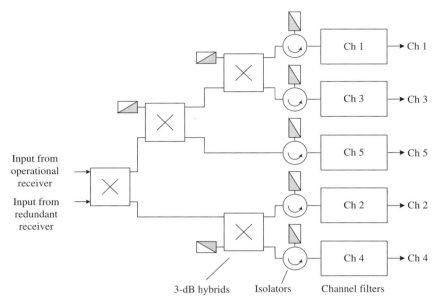

Figure 18.11 Hybrid branching input multiplexer (IMUX).

approximately 12 dB at Ku band, if the interconnection and isolator losses are included, but not the channel filter losses.

18.3.2 Circulator-Coupled MUX

The circulator-coupled IMUX is sometimes referred to as a "channel dropping" multiplexer, which gives a better indication of how it operates [15]. An example of a circulator-coupled IMUX for demultiplexing a group of five contiguous channels is shown in Figure 18.12. The hybrid at the input divides the power of the five-channel group into two branches. In each branch, the first circulator directs the group to the input of the first channel filter in the branch. The channel corresponding to the filter's passband drops through the filter to its output, and the other channels, in out-of-band position relative to this filter, are reflected at its input and are directed by the circulator toward the second circulator/channel filter in the chain. Here, the second channel drops through, the rest are reflected, and so on, until all the remaining channels are absorbed in the terminating load at the end of the chain.

This design approach provides an excellent way to retain nearly all the advantages of the hybrid branching scheme with a much lower insertion loss. The use of a single hybrid provides two alternative paths to extract the RF channels. Availability of the two paths allows the use of alternate (noncontiguous) channel filters in each path. Once the channels are extracted from each path, energy of all other channels, including those assigned to the other path, are absorbed by the load that terminates the port of the last circulator. Owing to the large frequency gap among

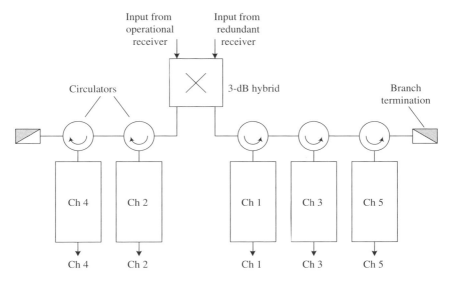

Figure 18.12 Circulator-coupled or channel dropping input demultiplexer.

the nonadjacent channels, the degradation due to reflections of channels is minimal. It can be readily seen that without the use of the input hybrid, we are forced to extract contiguous channels in the same path, simply because there is no alternative path anymore. Under such a constraint, there is significant degradation to the channel performance. The degradation, caused by reflections, is sometimes referred to as "en passant" distortion. It is dealt with in the next section.

Figure 18.13 Circulator-coupled input demultiplexer subsystem. (Photo courtesy COM DEV Ltd.)

The advantages of circulator-coupled MUX are similar to those of the hybrid branching MUX; namely, the circulator chains can be rendered very lightweight, compact, and reliable, and the channel filters can be designed and tuned individually. Their integration into the MUX subsystem has a negligible effect on their individual performance characteristics.

The losses are considerably less than those of the hybrid branching IMUX. Figure 18.13 presents a typical Ku-band waveguide circulator-coupled IMUX subsystem for a communication satellite. In this case, the filters are dual-mode bandpass with circulator-coupled external group delay equalizers.

18.3.3 En Passant Distortion

In the circulator-coupled multiplexing scheme, the transfer characteristics of the individual channel filters are not affected by their integration into the MUX subsystem. The channels themselves are affected by the reflections from the inputs of the noncorresponding filters in their path, until they reach their own channel filter. The deviations introduced by such reflections in the channel characteristics are sometimes referred to as *en passant distortion*. The word *en passant* is a very apt description of such a distortion and is being gradually adopted. It is computed as follows.

The channel corresponding to the first filter in the chain (e.g., channel 1 in the right-hand branch of Figure 18.12) drops directly through to that filter's output without incurring any reflections from other filters in the chain. However, the second and subsequent channels are reflected off the input to the channel 1 filter, and acquire the return loss characteristic of filter 1 before dropping through their own filter. This is expressed, approximately, by

$$
\begin{aligned}
S_{1c}(s) &= S_{21}^{(1)}(s) \text{ (dB)} &&\text{(for } n = 1\text{)} \\
S_{nc}(s) &= S_{21}^{(n)}(s) + \sum_{i=1}^{n-1} S_{11}^{(i)}(s) \text{ (dB)} &&\text{(for } n = 2, 3, \ldots\text{)}
\end{aligned}
\tag{18.1}
$$

where

$s = j\omega$

$n =$ physical filter position in chain; $n = 1$ is nearest to the common input, $n = 2$ is second along the chain, and so on

$S_{nc}(s)$ (dB) = overall transfer function of path from common input to the output of nth filter in chain

$S_{21}^{(n)}(s)$ (dB) = transfer function of nth filter in chain

$S_{11}^{(i)}(s)$ (dB) = return loss function of ith filter in chain.

This formulation assumes that the circulators in the MUX are perfect, that is, that they possess infinite directivity. This implies a transmission of signals in one direction and an infinite isolation in the reverse direction. The directivity of the circulators in the range of 30–40 dB is quite adequate for most practical multiplexers to justify this assumption. The reflection response depends on the channel filter formulation characteristics, as well as the frequency separation between the transmitted and reflected channels.

En Passant Distortion for a Noncontiguous Multiplexing Scheme

The IMUX configuration, described in Figure 18.12, requires noncontiguous channels in the two paths of the MUX. Taking the right-hand branch (odd-numbered channels) as an example, the channel 1 (Ch1) transfer function is that of the channel filter itself. Channel 3, en route from the common input, encounters a reflection from the channel 1 filter before it is extracted by the channel 3 filter. Thus, the transfer characteristic for the channel 3 signal through the IMUX is the return loss characteristic of the channel 1 filter plus the transfer characteristic of the channel 3 filter, expressed by

$$S_{3c}(s) = S_{21}^{(2)}(s) + S_{11}^{(1)}(s) \tag{18.2}$$

Channel 1 is unaffected by the presence of the other channel filters in the chain, but the downstream channels; Ch3 and Ch5 pick up the return loss characteristics of channel 1 and channel 1 + channel 3, respectively, in addition to their own transfer characteristics. This is illustrated in Figure 18.14. For the noncontiguous case, the addition of the return loss characteristics of the other downstream channels in the chain has little effect on the in-band amplitude and group delay characteristics.

En Passant Distortion with Contiguous Channels

The reflection response for a contiguous channel is calculated in the same way as for that of the noncontiguous channels. We compute the reflection response over the frequency band, corresponding to the contiguous channel. This is described in Figure 18.15, where channels 1, 3, and 5 are contiguous. Channel 1 is not affected by reflection, whereas channel 3 is reflected off channel 1. The lower edge of channel 3 is in close proximity to channel 1, and hence it suffers a large deviation in amplitude and group delay near the lower band edge as shown in Figure 18.15. The upper edge of channel 3 is far away from channel 1, and hence is relatively unaffected. Similarly, channel 5 suffers an asymmetric reflection from channel 3, and so on.

It should be noted that if the sequence of channels in the example is reversed, it is the upper edge of the channels that are adversely affected. The large deviations introduced by such reflections affect only one band edge, whereas the other edge remains relatively unaffected. As a consequence, the channel response is rendered asymmetric. Another way of looking at this is in terms of the rejection provided by the channel filters to the reflected channels. If the rejection at a given frequency is greater than 40–50 dB, there is virtually no impact on the reflected response. That is representative of noncontiguous channels. If the rejection is in the range of 5–10 dB, it can cause large deviations in the amplitude and group delay, as in the case of contiguous channels. As will be described in more detail in the following section on hybrid-coupled directional filter module multiplexers, these distortions can be countered to a certain extent with asymmetric filter designs.

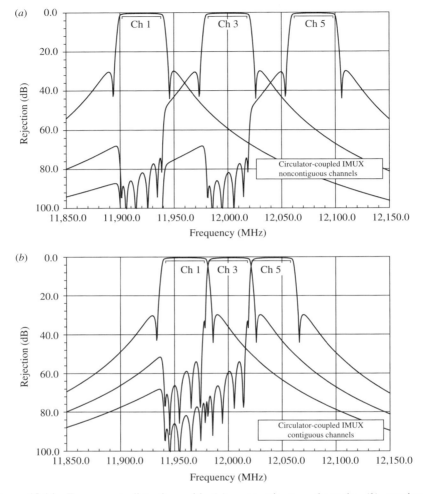

Figure 18.14 En passant distortion with (*a*) noncontiguous channels; (*b*) contiguous channels.

18.4 RF COMBINERS

RF channel combiners, often referred to as *output MUX* (OMUX) or simply *MUX*, provide a function that is opposite that of RF channelizers. Such a MUX is required at each transmit station in the system. The design constraints on such a MUX are as follows:

1. The MUX must be able to combine the powers of a number of RF channels, which are contiguous or noncontiguous.

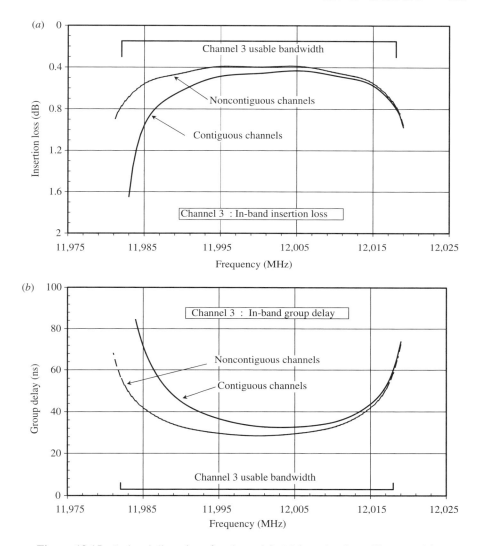

Figure 18.15 In-band distortions for channel 3: (*a*) insertion loss; (*b*) group delay.

2. Insertion loss is a critical parameter, since it has a direct impact on the power radiated and thus the traffic capacity.

3. The equipment must be capable of handling high power in terrestrial or space environments.

4. The MUX must provide a high isolation in the receive band and suppress the harmonics generated in the high-power amplifiers.

The primary aim of an OMUX is to combine the power of a number of RF channels into a single port that can then be fed to an antenna. An antenna with a single input port is simple to implement and simple to optimize its performance. MUX design depends on the number of channels handled by it, and whether the channels are contiguous or noncontiguous. Above all, it is the power handling capability of a MUX and its operating environment that dictate the MUX design and implementation. Although all of the MUX configurations summarized in Table 18.1 are applicable for combining RF channels, it is the manifold coupled or hybrid coupled multiplexer that is used in most systems.

18.4.1   Circulator-Coupled MUX

Owing to the low-loss requirement, this configuration is viable for only a two-channel system with low to moderate power levels (tens to low hundreds of watts). It represents the simplest and cheapest configuration but it does depend on the availability of a low-loss circulator capable of handling high power and low loss. The limitations of this configuration can be overcome by the hybrid-coupled network.

18.4.2   Hybrid-Coupled Filter Combiner Module (HCFM) Multiplexer

Instead of circulators, the HCFM derives its directional properties from an assembly of two identical bandpass filters and two 90° (quadrature) hybrids, shown schematically in Figure 18.16. In view of the scattering matrix of the 90° hybrid [21], we can readily show the following properties for such a module:

- Any out-of-band channels entering at port 3 are directed by the main-path hybrid toward the two-channel filters, are reflected off by the inputs to the two channel filters, and are then recombined in the hybrid to emerge from port 4.
- Residual power that is not reflected passes diagonally through the module to port 2 and is absorbed by the termination at this port.
- In-band channel power entering at port 1 traverses the module diagonally and emerges at port 4, combining with any out-of-band channels that enter at port 3.

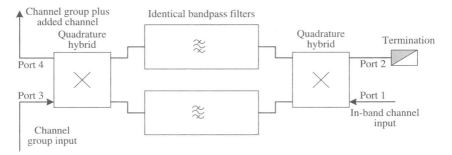

Figure 18.16   Hybrid-coupled filter module (HCFM).

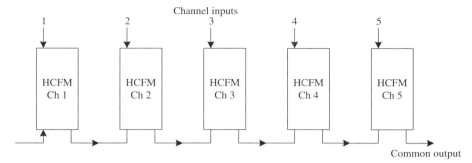

Figure 18.17   Output multiplexer with hybrid-coupled filter modules.

Port 2 of the HCFM can be used as a second channel input port instead of being terminated. The in-band power entering at port 2 emerges at port 3, and propagates away from the module in the direction opposite that of the power input to port 1, toward another antenna in a frequency reuse system for example.

The HCFMs are interconnected by arbitrary lengths of transmission line to form a multiplexer as shown in Figure 18.17. Within each module, the power entering at port 1 is shared by the twin filters, halving the thermal dissipation that each filter has to handle and reducing the peak voltages in the cavities by $1/\sqrt{2}$.

In waveguide networks, hybrids are typically short-slot E-plane couplers that are simple, compact, and internally open structures. They are ideally suited for the handling of the high peak voltages of the combined channels in the common regions of the modules. In addition, broadwall branchline couplers can be used, when the orientation of the input and output ports of the filters are at right angles to each other for example, 6th degree dual mode filters. Figure 18.18 gives a possible internal structure of HCFMs with extended box filters (dual-mode realization) and with phase-coupled extracted pole filters (manufactured in a common housing).

En Passant Distortion   A multiplexer consisting of HCFMs is similar in concept to the circulator-coupled MUX module. Here, the hybrids plus the twin filters perform the same function as the circulator and the filter. This being so, equation (18.1) may be used to characterize the transfer characteristics of each channel path, within the constraint that the hybrids used are perfect, that is, that their directivity is infinite. This corresponds to a similar assumption we made for the circulator-coupled MUX, where we assumed that the directivity of the circulator is infinite. A directivity of 30–40 dB for the hybrids is sufficient to justify such an assumption for practical systems. The transfer characteristic of each path is dominated by that of the channel filters of the module itself. However, the return loss characteristics of any other channel that may be physically located downstream from the module are also superimposed on the transfer characteristic. If it is required to combine the noncontiguous channels, this scheme provides an excellent way to accomplish it with a high power handling capability. However, if there are any contiguous channels to be combined, then there are in-band insertion loss and group

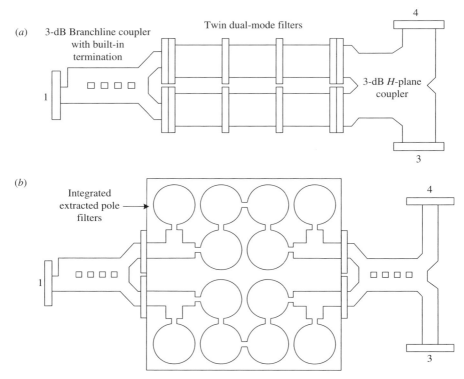

Figure 18.18 Possible HCFM configurations: (*a*) using dual-mode filters; (*b*) using extracted pole filters.

delay distortions near the band edges of the channels. The computed response for such distortions for contiguous channels is illustrated in Figure 18.15.

The large deviations in the insertion loss and group delay near the band edges renders this approach unacceptable for a practical system. For these reasons, it has been very rare to find contiguous or even semicontiguous channels in the same branch of an HCFM multiplexer. In satellite earth stations, even-numbered channels are combined in one branch and the odd-numbered channels in a second branch. Then the two branches are combined with a formidable hybrid whose termination must be water- or forced-air-cooled, capable of dissipating many kilowatts of RF energy. Such an approach incurs an extra loss of 3 dB due to the hybrid, and that represents a big waste of energy since we are dealing with kilowatts of wasted power [22].

There are two ways to overcome the limitations of this design approach for contiguous channels. One way is to use manifold multiplexers, described in the following section. Another way is to modify the design of the channel filters to provide a degree of compensation for the degradation of channels near the band edges and, at

TABLE 18.2   HCFM Filter Planning

Downstream Channels	HCF Module Filter Solution
None contiguous, or none at all	Regular symmetric with transmission zeros
One contiguous on one side of the passband	Asymmetric design with one or two Tx zeros, on side opposite that of downstream neighbor
Two contiguous on both sides	All-pole Chebyshev. downstream neighbors provide selectivity

the same time, restore symmetry to the channel response. This approach is described as follows.

Knowledge of filter types and design bandwidths of the downstream modules makes it possible to tailor the design of the filters in the modules, building in asymmetry where it is needed to compensate for a downstream neighbor. Table 18.2 provides a summary of these requirements.

The idea here is to try to keep the overall transfer characteristic as symmetric as possible by not providing built-in Tx zeros in the channel filters, when the return loss characteristics of the downstream neighbors are effectively providing "virtual" Tx zeros for it.

Figure 18.19 shows the overall transfer function of an HCFM (Ch1) with a contiguous downstream neighbor (Ch2) and two others (Ch3 and Ch4) on the upper side of its passband. The filters of the HCFMs for Ch1–Ch3 are designed with one TZ each on the lower side, while the HCFM for Ch4, which has no downstream neighbors, has two symmetric TZs. With contiguous downstream neighbors, it can be seen that a certain amount of symmetry is achieved for the close-to-band rejection characteristics of Ch1–Ch3. A high selectivity has been provided on the lower side by these filters' own built-in TZ, and on the upper side by the return loss characteristic of the downstream neighbor. The in-band insertion loss and group delay characteristics for Ch1 are shown in Figure 18.19b and 18.19c, including a comparison with the performance of an equivalent symmetric filter (with two built-in symmetric TZs) and no downstream neighbor.

Figure 18.19d shows the effect on the rejection of Ch1 after we remove the Ch2 HCFM. In the absence of the downstream contiguous neighbor, the rejection characteristic of Ch1 tends to "collapse" into the gap left by the missing module. This demonstrates Ch1's reliance on the presence of the neighboring channel for its own proper performance.

18.4.3   Directional Filter Combiner

The configuration of a multiplexer with directional filters is described in Section 18.2. Its advantages and disadvantages are summarized in Table 18.1. Because of the sensitivity of this structure and bandwidth limitations, this type of MUX is

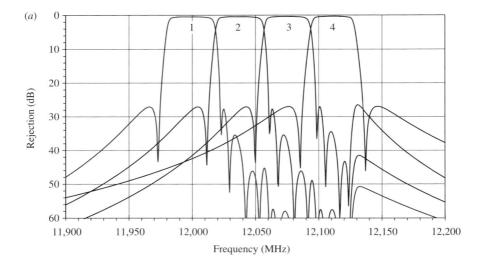

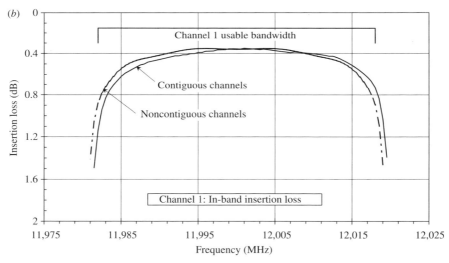

Figure 18.19   HCFM directional multiplexer with four contiguous channels: (*a*) overall transfer characteristics; (*b*) in-band insertion loss characteristic; (*c*) in-band group delay characteristic; (*d*) effect on channel 1 rejection characteristic of removing channel 2 module.

rarely used for high-power combining networks. An interesting aspect of this design is that the directional property of the filters can be switched by changing the polarization of the input to the filters. This feature has been exploited in satellite systems for narrowband multiplexing networks requiring switching between two antenna beams [6].

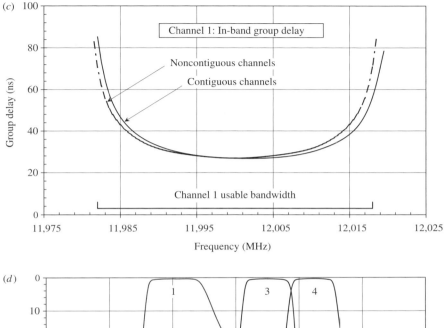

Figure 18.19 *Continued.*

18.4.4 Manifold Multiplexer

A manifold MUX is characterized by channel filters connected to a common output port, by either a common junction or a common manifold. These two types are very similar in concept and design. The principal difference is the presence of interchannel manifold phase lengths that separate the channel launch points in the manifold MUX. This provides the manifold MUX with extra degrees of freedom to optimize the design. Moreover, it is easier to separate the inputs to the common output transmission line, thereby rendering the structure less sensitive. In the star connection, the power of all the channels congregates at the common junction, which can cause very high voltages and a hotspot, whereas the power in the manifold of a manifold MUX tends to be more evenly distributed along the manifold. Often, the star connection is used where there are a relatively small number of channels, up to four, and is particularly favored for coaxial filter diplexing at lower frequencies, where the junction can be built very compactly into the volume of the filter output cavities themselves.

Star Junction Multiplexer Some star combining schemes are shown in Figure 18.20: a four-channel multiplexer with a regular star junction, a scheme where the channels are combined in stages, and the compact internal combination scheme of a Rx/Tx coaxial filter diplexer.

Manifold-Coupled Multiplexer Three common manifold multiplexer configurations are shown in Figure 18.21, where the channel filters are connected to one side of the manifold (comb), to both sides (herringbone), and end-fed (applicable to either of the first two).

The design techniques for manifold multiplexers underwent rapid development in the 1970s and 1980s, when it was apparent that they were ideal for communication satellite payloads [3–6]. From the electrical perspective, design techniques have advanced to the point where it is possible to combine an arbitrary number of channels, regardless of their bandwidths and channel separations. In addition, there are no restrictions on the design and implementation of the channel filters in the manifold.

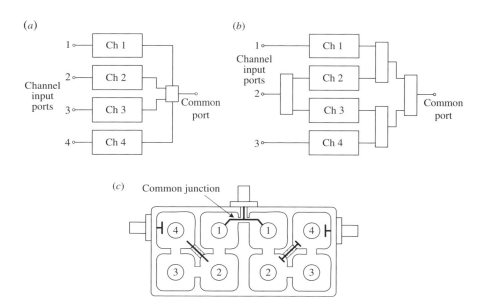

Figure 18.20 Types of star combining junctions: (*a*) basic star combining junction; (*b*) staged star junction; (*c*) internal star junction in coaxial diplexer housing.

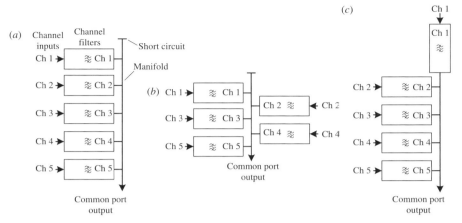

Figure 18.21 Common configurations for manifold multiplexers: (*a*) comb; (*b*) herringbone; (*c*) one filter feeding directly into the manifold.

The manifold itself is a transmission line, either coaxial line, a rectangular waveguide, or some other low-loss structure. It is possible to achieve a channel performance in the multiplexer configuration which is close to that can be obtained from a channel filter by itself. No other MUX configuration can match this performance.

From the mechanical perspective, the MUX structure can be very lightweight and compact, but at the same time, rugged enough to withstand the vibrations and other rigors of a launch into space. By using special materials, the structure may be made to be electrically stable in the presence of large environmental temperature fluctuations, and can conduct the dissipated RF thermal energy efficiently to a cooling baseplate.

Figure 18.22 shows a model of a typical C-band output multiplexer intended for space application. The short-circuited manifold supports five dual-mode channel filters, arranged in a herringbone fashion (Figure 18.21*b*).

Design Methodology Since there are no directional or isolating elements in the manifold multiplexer (circulators, hybrids), all the channel filters are electrically connected to each other through the near-lossless manifold. Moreover, the design of the manifold MUX needs to be considered as a whole, not as individual channels. In the early days, a number of ingenious techniques were invented [1–4,7,8] to design the individual filters such that they would properly interact with other filters on the same manifold and function virtually as if operating into a matched termination. However, with the dramatic increase in computer power in recent years, multiplexer design procedures have utilized optimization methods to achieve the final design, in preference for the more limited analytic design techniques. Optimization techniques for multiplexer design have been developed to a high degree of sophistication, and worthy of note here is the space mapping technique developed

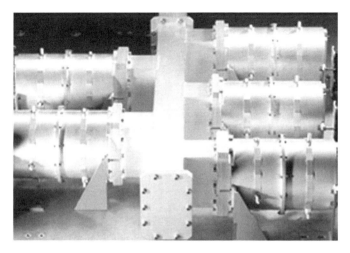

Figure 18.22 Five-channel C-band manifold output multiplexer with conventional dual-mode filters. (Photo courtesy COM DEV Ltd.)

by Bandler [23], which involves optimizing a rough model (circuit or hybrid circuit model with embedded EM-modeled components) and connecting that coarse model to a fine model (e.g., full-wave electromagnetic modeling). Because of the large number of optimization variables, the fine model EM simulator takes a formidable amount of computer CPU time and has to be used sparingly during the optimization process. On the other hand, the rough model can be optimized by a circuit simulator very rapidly, especially if the S parameters of fixed and noninteracting elements (e.g., the manifold waveguide junctions) are modeled in advance by EM techniques and are available to the main program either as lookup tables or with the call of a rapid specialist routine. It therefore provides a means to simulate the manifold multiplexer with a good degree of accuracy and efficiency.

It is not the intention in this chapter to discuss the details of the EM simulation and optimization, which is thoroughly covered by Bandler [23]. In the following sections some of the basic techniques needed for building the circuit model ready for optimization are outlined, followed by a description of the optimization strategy itself.

Analysis of Common Port Return Loss and Channel Transfer Characteristics

At the heart of any circuit optimizer is an efficient analysis routine. The optimization calls the analysis routine many times at different frequencies as it progresses with the task of optimizing the various parameters [10]. In general, the two parameters from which the overall cost function is built are the multiplexer common port return loss and the individual channel transfer characteristics.

Sampling the common port return loss at frequency points corresponding to the in-band return loss zeros (points of perfect transmission) and return loss maxima (points of maximum in-band insertion loss ripple) of the original filtering function of each filter, with the aim of restoring them to their original RF frequency positions

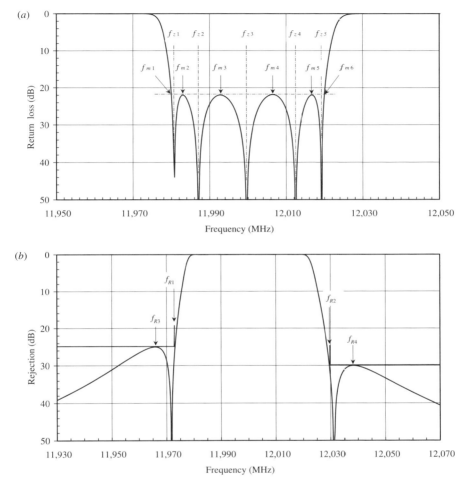

Figure 18.23 Typical sampling points for the evaluation of the optimization cost function: (*a*) return loss zeros and maxima; (*b*) close-to-band rejection points.

while the filter is on the manifold and is interacting with the other filters on the manifold, has been found to be effective in achieving a good overall common port return loss. In addition, the two band-edge RL points are usually added, giving a total of $2N_k + 1$ frequency sampling points for each filter, where N_k is the degree of the kth filter on the manifold. Sometimes when close-to-band rejection is critical, two or four rejection corner points are added to the cost function. This necessitates that the transfer and reflection characteristics for each channel be known. Figure 18.23 shows the location of these sample points for a typical (5–2) quasi-elliptic filter.

During the optimization process, the subroutines to calculate common port return loss and channel transfer characteristics at each frequency sampling point is called many times. Although it is possible to construct an admittance matrix for the entire multiplexer circuit and analyze it at each frequency point to obtain the desired transfer and reflection data, this matrix is quite large for the MUX, and the large number of channel filters requires a significant amount of CPU time to invert the MUX even though it is relatively sparse. Moreover, much data is acquired that is not needed for the optimization.

Later on in this chapter, a piecewise optimization strategy, which is quite efficient in terms of computer CPU time, is described [12,13]. Part of the strategy involves optimizing the channel filters one after the other in a repeated cycle. As the optimization parameters of each filter in turn are optimized, it is necessary to calculate only the transfer characteristic of this filter. Others are relatively unaffected by the changes (assumed small) being made to the optimized filter.

To speed up the overall optimization, it is more efficient to analyze each filter's input-to-common port transfer characteristic individually, and the common port return loss. The manifold of the multiplexer can be represented as an open-wire circuit, with a short circuit at one end and three-port E-plane or H-plane waveguide junctions spaced along the manifold's length. The channel filters are located at the third port of each junction, and are separated from the junction by short lengths of transmission lines (stubs), as shown in Figure 18.24.

If the manifold is waveguide, the junctions are E-plane or H-plane. Since their intrinsic parameters do not change during the optimization, the junctions are best characterized with three-port S parameters precalculated by a mode matching routine and stored over a range of frequencies that covers the bandwidths of all

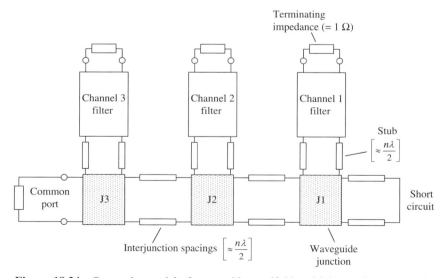

Figure 18.24   Open-wire model of waveguide manifold multiplexer (three-channel).

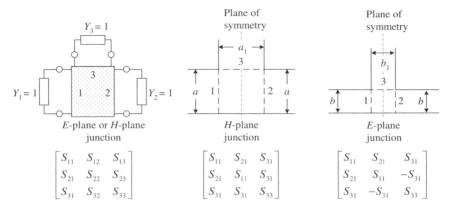

Figure 18.25 *E*-plane and *H*-plane waveguide junctions and *S*-parameter matrix representation.

the channel filters, or specifically, at the frequencies of the sampling points. Typically, the junctions are symmetric about ports 1 and 2, leaving the possibility for port 3 to be of a different size as described in Figure 18.25.

Because of reciprocity and the symmetry of the junctions about the vertical axis in Figure 18.25, $S_{22} = S_{11}$ and $S_{32} = S_{31}$ (*H*-plane), and $S_{32} = -S_{31}$ (*E*-plane). This means that only four parameters—S_{11}, S_{21}, S_{31}, and S_{33}—are needed to characterize the electrical performance of the junctions. The *S*-parameter matrices are calculated by assuming a matched termination at each port. However, if the termination at one port is arbitrary, the two-port *S* parameters between the other two ports can be defined:

1. *Admittance Y_{L2} ($\neq 1$) at port 2:*

$$\begin{bmatrix} S'_{11} & S'_{13} \\ S'_{31} & S'_{33} \end{bmatrix} = \begin{bmatrix} S_{11} & S_{13} \\ S_{31} & S_{33} \end{bmatrix} + \frac{\Gamma_2}{1 - \Gamma_2 S_{11}} \begin{bmatrix} S_{21}^2 & kS_{21}S_{31} \\ kS_{21}S_{31} & S_{31}^2 \end{bmatrix} \quad (18.3a)$$

where $\Gamma_2 = (1 - Y_{L2})/(1 + Y_{L2})$ and Y_{L2} is the admittance at port 2 of the junction and $k = 1$ for *H*-plane junctions, and $k = -1$ for *E*-plane junctions. This modified *S* matrix also applies if an admittance Y_{L1} is terminating port 1. It is necessary only to change the subscripts from 2 to 1 and vice versa.

2. *Admittance Y_{L3} ($\neq 1$) at port 3:*

$$\begin{bmatrix} S'_{11} & S'_{13} \\ S'_{31} & S'_{33} \end{bmatrix} = \begin{bmatrix} S_{11} & S_{13} \\ S_{31} & S_{33} \end{bmatrix} + \frac{\Gamma_3 S_{31}^2}{1 - \Gamma_3 S_{33}} \begin{bmatrix} 1 & k \\ k & 1 \end{bmatrix} \quad (18.3b)$$

where $\Gamma_3 = (1 - Y_{L3})/(1 + Y_{L3})$, Y_{L3} is the admittance at port 3 of the junction and k has the same meaning as before.

Common Port Return Loss   The common port return loss (CPRL) computation at a given frequency point (e.g., at a sample point) proceeds as follows:

1. The channel filter input admittances Y_{F1}, Y_{F2}, and Y_{F3} are determined at the frequency sample point and stored.
2. With the known admittances at ports 3 of each junction, the transfer and reflection S parameters $[S'_{21}, S'_{11};$ see Equation (18.3b)] for each junction are calculated and converted to $[ABCD]$ matrices. From the short-circuit end, the manifold phases and the junctions are cascaded at each stage for finding and storing the along-manifold admittances $Y_{M1}, Y_{M2}, \ldots$ as shown in Figure 18.26

$$Y_{Mi} = \frac{1 + S_{11i}}{1 - S_{11i}} \qquad (18.4)$$

where $i = 1, 2, \ldots, n + 1$ and n is the number of channels on the manifold.

The final admittance (Y_{M4} in Fig. 18.26) may be used to calculate the CPRL as

$$\Gamma_{CP} = \frac{1 + Y_{Mn+1}}{1 - Y_{Mn+1}}$$

$$RL_{CP} = -20 \log_{10} \Gamma_{CP} \text{ (dB)} \qquad (18.5)$$

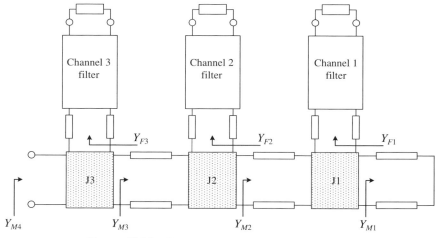

Figure 18.26   Common port return loss computation.

If only the manifold spacings θ_{M1}, θ_{M2}, ... are being optimized, then the filter input admittances Y_{F1}, Y_{F2}, ... need to be calculated only once at each sample frequency and stored for optimization of the CPRL.

Channel Transfer Characteristics   If optimization is now focused on an individual channel filter and the transfer function between its input and the common port is needed (e.g., for a rejection sample point), the following procedure may be used. Let us consider channel 2 as an example, as shown in Figure 18.27:

1. Calculate the new Y_{F2} for channel 2 with its newly updated parameters.
2. The previously stored value of Y_{M2} is used in conjunction with equation 18.3a to calculate S_{31} and S_{11} for junction 2 with Y_{M2} at its port 2.
3. The [ABCD] matrices for the channel 2 filter and the S_{31} path of J2 and then the rest of the manifold toward the common port are cascaded to calculate the transfer characteristic for channel 2. The process can be repeated for the other channels.

For both common port and the transfer characteristics, the computations are conducted by assuming lossless (purely reactive) components for the filter plus the stub networks. Then, noncomplex arithmetic is carried out for the matrix multiplications, inversions, and other calculations, speeding up the process considerably.

Channel Filter Design   Being purely reactive (i.e., with no resistive elements between the channel inputs and the common port output) the channel filters of the manifold multiplexer react with each other through the manifold itself. If the

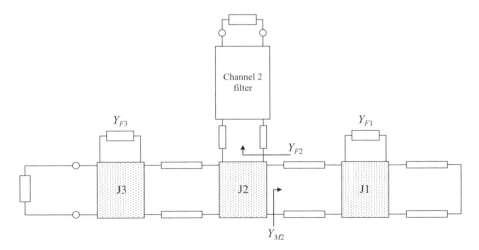

Figure 18.27   Channel 2 transfer function calculation.

channels are widely spaced, the channel interreactions are quite low, because over one filter's passband, the other filters are well into their reject regions and present short circuits at the ports closest to the manifold. The channel filters are designed as doubly terminated networks, separately from the other filters and the manifold. The filters are integrated on the manifold; all that is needed are adjustments to the spacings and stub lengths along the manifold, and minor adjustments to the first three or four elements of the filter, to recover a good common port return loss.

However, as the guard bands between the channel filters decrease toward contiguity, the filters begin to interreact strongly along the manifold. Now significant adjustments to the filter parameters are needed in addition to the manifold and stub phase lengths to reach an acceptable CPRL. Although it is possible to optimize doubly terminated filters to operate in a contiguous channel environment, a starting point much closer to the final optimal result is obtained if singly terminated (ST) filter prototypes are used in these conditions. The design methods for ST filter prototypes are outlined in Chapter 7 [1,24]. The ST filter network is useful for the design of contiguous channel manifold multiplexers, because the contiguous ST channel filters, along the manifold, tend to interact in a mutually beneficial manner, providing a conjugate match for each other at their launch points. This "natural multiplexing" effect can be explained by studying the special characteristics of the input admittance of the ST circuit, looking in at the port opposite to the terminated port.

The real part of the input admittance has the same characteristic as the filter's own power transfer characteristic. It is close to unity over the filter's passband, dropping to near zero in the out-of-band regions. As the frequency increases from minus infinity, the imaginary part of the admittance increases from zero toward a positive peak near the lower passband edge, then traverses the passband with a negative slope toward a negative peak near the upper passband edge. It then slowly returns to zero with a positive slope as the frequency goes to infinity. Figure 18.28 illustrates the variations of the real and imaginary parts of the admittance for singly and doubly terminated prototype filters.

If the contiguous band singly terminated filters are connected to the manifold with the "zero impedance" termination closest to the manifold junction, then the negative in-band slope of the imaginary part of the admittance tends to cancel with the positive slopes of the two contiguous neighbors that extend into this filter's passband. The results are illustrated in Figure 18.29, which shows that the imaginary parts of the source (the ST filter) and the load (the other filters on the manifold) have been mostly canceled, whereas the unity-valued real part of the source mostly sees the unity load at the common port of the manifold. This is due to the impedances at the input ports of the other channel filters being almost isolated by their own rejection characteristics.

Thus, a conjugate match between the source and the load is partially achieved. Although it is not a perfect conjugate match, it is better than that if doubly terminated filters were used, which makes it a good point at which to start the optimization process. The channels at the edges of the contiguous group on the manifold, with a neighbor on one side only, have some small in-band and rejection asymmetries after the optimization process.

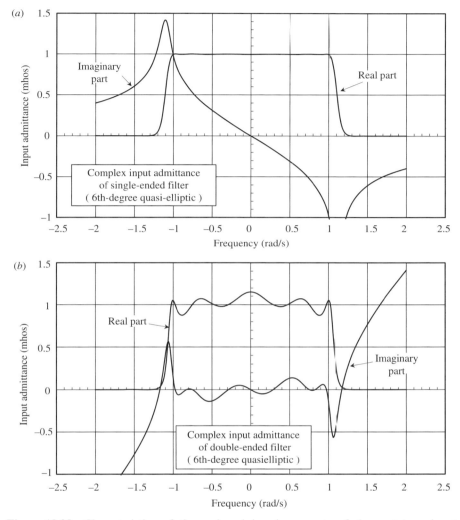

Figure 18.28 Characteristics of the real and imaginary parts of the prototype input admittance: (*a*) singly terminated filter; (*b*) doubly terminated filter.

Optimization Strategy The networks that model even a moderate-sized wave-guide manifold multiplexer tend to be quite complex. An open-wire equivalent circuit of a six-channel manifold multiplexer with 6th-degree quasielliptic dual-mode filters has in the order of 90 frequency sampling points and 100 electrical elements of varying sensitivities and different constraints. They all need to be correctly valued before the overall multiplexer can operate to specification.

If all these parameters are optimized simultaneously, not only the amount of CPU time is enormous but also there is little likelihood that the global

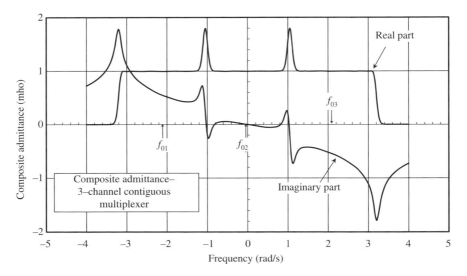

Figure 18.29 Single-terminated filter with two contiguous neighbors—composite admittances as seen from the common port of the manifold.

optimum can be attained; there will be a myriad of shallow local optimum solutions. With manifold multiplexers routinely incorporating 20 channels, and perhaps up to 30 in the future, global optimization with so many variables is clearly unsuitable.

For these reasons, most of the major satellite OMUX designers and manufacturers have developed more efficient proprietary methods for manifold multiplexer optimization. Among these is the "piecewise" approach, optimizing parts of the multiplexer separately in repeated cycles, while converging upon an optimal solution. The "parts" or "parameter groups" referred to here might include the first five elements of each channel filter (narrowband domain), or all the manifold stub lengths (wideband domain). It is usual practice to commence the optimization process with the wideband sections first (parameters relating to the manifold and stubs), followed by a shift in emphasis to the narrowband sections (filter parameters) as the common port return loss begins to take shape. A typical design optimization project can proceed as follows.

Design
1. Design the channel filter transfer/reflection functions to meet the individual in band and rejection specifications.
2. Synthesize the corresponding coupling matrices, doubly terminated if the channel filter design bandwidths (DBW) are separated by guard bands greater than about 25% of the DBWs; singly terminated if otherwise. If singly terminated prototypes are employed, a more practical design results if the initial prototype is generated with a low return loss to bring

the value of the termination at the end that is opposite to the manifold as close to unity as possible.

3. Set the initial manifold spacings between the *E-* or *H*-plane junctions at $m\lambda_g/2$, where m is as low as possible for a convenient mechanical layout. Set the initial manifold short circuit/first junction spacing at $\lambda_g/4$ (*H*-plane) or $\lambda_g/2$ (*E*-plane). λ_g is the wavelength in the manifold waveguide at the center frequency of the filter nearest to the length of the waveguide, in the direction of the common port.

4. Set the initial manifold junction with filter stub lengths at $n\lambda_g/2$. Again, n should be as small as possible.

Optimization

1. Wideband components. Optimize the spacings between the junctions and between the short circuit and the first junction, and the stub lengths. This often has the most dramatic effect in terms of an improvement in the common port return loss.

2. Optimize the stubs and first three or four parameters of the channel 1 filter [M_{S1} (filter manifold coupling), M_{11} (first resonance tuning), and M_{12} (resonance 1 to resonance 2 coupling)].

3. Repeat the cycle for all the channels, possibly omitting the stub lengths since these do not change much, until the improvements in the cost function are beginning to become negligible.

Refinement

1. Repeat the optimization of manifold and stub lengths.

2. Reoptimize with a fine step on all of each channel filter's parameters, most lightly on the elements farthest away from the manifold, and not at all on the final coupling M_{LN} (i.e., the input coupling to the multiplexer filter).

Demonstration of the Piecewise Optimization Process and Stages in the Optimization of a Four-Channel Contiguous OMUX The piecewise optimization process is demonstrated through the optimization of a four-contigous-channel Ku-band waveguide manifold multiplexer. The channel electrical specifications are satisfied with $(5-2)$ quasielliptic filters with 30 dB rejection lobe levels and design bandwidths of 38 MHz, and 40 MHz center frequency spacings that fall within the definition of contiguity.

Designing the filters as singly terminated prototypes and attaching them to the manifold with the initial manifold spacings and stub lengths results in a rather disappointing performance as shown in Figure 18.30. The CPRL is poor, and one of the channels is unrecognizable.

Figure 18.31 shows the dramatic improvement that results from optimization of the manifold spacings. Now the channel rejection characteristics are close to design, and the average CPRL is in the order of 10 dB.

Further improvement in the CPRL is obtained by optimizing next, the first four parameters (first coupling iris M_{S1}, first cavity tuning state M_{11}, second coupling iris M_{12}, second cavity tuning state M_{22}) of each filter. The computed response following this procedure is shown in Figure 18.32. The inner rejection lobe levels are close to the design of 30 dB, but the outer lobe levels, because there are no neighbors on the outer side of the group, have risen to about 24 dB.

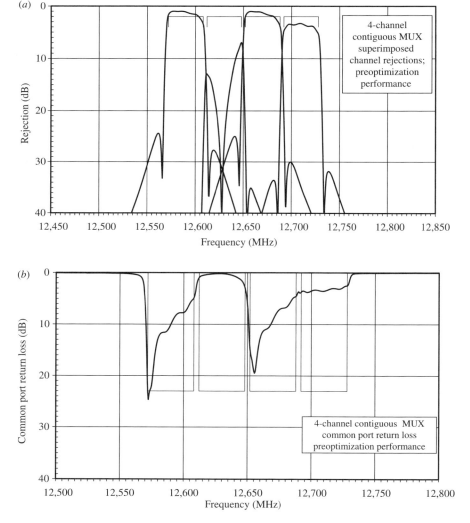

Figure 18.30   Four-channel manifold multiplexer preoptimization performance: (*a*) superimposed channel transfer characteristics; (*b*) common port return loss.

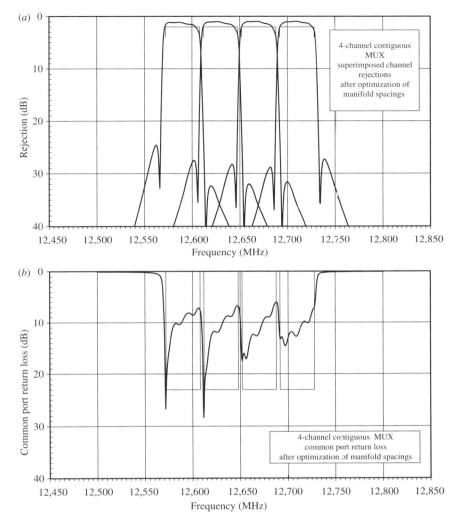

Figure 18.31   Four-channel manifold multiplexer after optimization of manifold lengths: (*a*) superimposed channel transfer characteristics; (*b*) common port return loss.

Then, the refinement optimization cycles are carried out, and the final result is shown in Figure 18.33. Now the CPRL is above 23 dB over all the channel bandwidths. The effect of not having a contiguous neighbor on one side is evident in the rejection responses of channels 1 and 4, and the lobe levels. The rejection slopes on the outer sides are not as steep as the sides with a contiguous neighbor. This causes a small amount of asymmetric in-band distortion to the group delay and insertion loss.

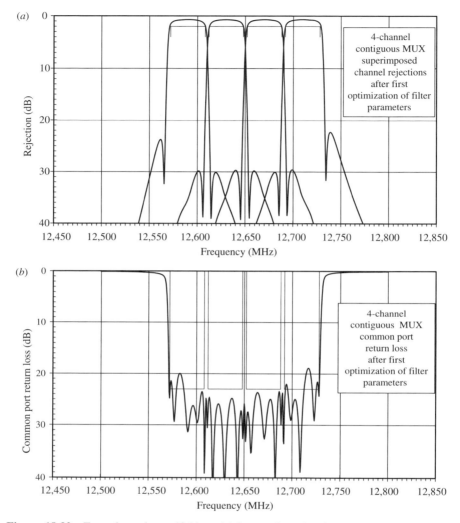

Figure 18.32 Four-channel manifold multiplexer after the first optimization of filter parameters: (*a*) superimposed channel transfer characteristics; (*b*) common port return loss.

The simulations have all been conducted by assuming a Q_u of 12,000, but the optimizations were carried out with a lossless network to speed up the process. With a moderate-speed PC (700 MHz), the whole optimization process took approximately 2 min.

This optimization strategy appears to work well even for a larger number of channels. Figure 18.34 shows the common port return loss and rejection characteristics of a manifold multiplexer with 20 contiguous channels at the C band (4th-degree filters), which was designed using the piecewise optimization process.

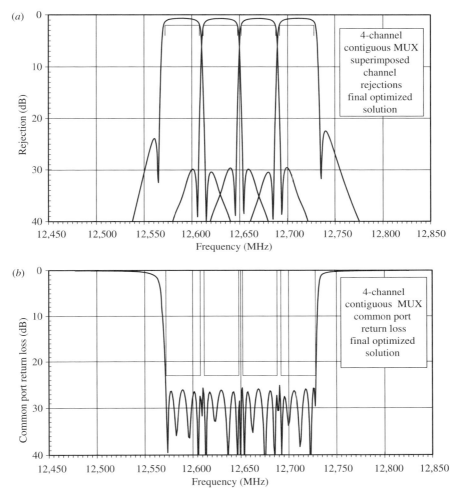

Figure 18.33   Four-channel manifold multiplexer and final performance: (*a*) superimposed channel transfer characteristics; (*b*) common port return loss.

18.5   TRANSMIT–RECEIVE DIPLEXERS

The terms *diplexer* and *duplexer* have been used interchangeably by RF engineers for many years. The prefix *di* is defined as twice or double, while the prefix *du* is defined as two or dual. The term *plex* is generated from the Latin word *plexus*, which means an interwoven combination of parts or elements. Thus, there is no basic difference in the meaning of the two terms. Some RF engineers prefer to use the term *duplexer* for the case where the antenna is shared between the receive and transmit signals through the use of either a switch or a circulator. Others use the term *diplexer* for the same device. In view of IEEE publications

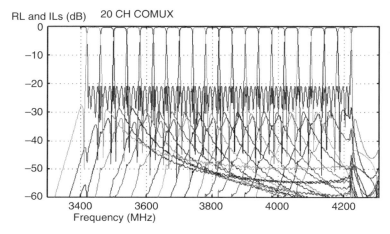

Figure 18.34   20-channel manifold multiplexer superimposed channel transfer characteristics and common port return loss (RL—return loss; IL—insertion loss). (Courtesy Dr. Yu, COM DEV Ltd.)

on the subject over the past 20 years, one can clearly see that the two terms are often used interchangeably. In this chapter, we will use the term *diplexer*. A clear distinction exists between a diplexer, which combines the receive and transmit functions, and a simple two-channel MUX, which may be required for separating or combining two channels.

A simple form of a diplexer is shown in Figure 18.35. It consists of two 90° hybrids and two identical filters. They are designed to operate within the receive band with enough isolation to reject the transmit band. The diplexer allows the use of one antenna for the receive (Rx) and transmit (Tx) signals. In view of the scattering matrix of the 90° hybrid [21], we can readily show that the signal received from the antenna port is sent to the receive port, while the signal from the transmit port is reflected from the filters and directed to the antenna port. This diplexer configuration is bulky, requiring the use of two hybrids. However, it offers design simplicity.

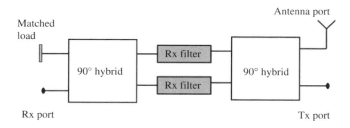

Figure 18.35   A diplexer implemented using the 90° hybrid approach.

Another approach to realizing the diplexer is to utilize one filter for the receive channel and another one for the transmit channel. The two filters are then combined by a T junction. Figure 18.36a illustrates two coaxial filters, combined with a wire T junction, while Figure 18.36b shows two iris waveguide bandpass filters combined together by a waveguide T junction. The interaction between the two filters, in this case, is very strong, requiring that the phase at the junction arms be adjusted to ensure that the receive and transmit signals are integrated properly. In some applications, such as in radar systems, the receive and transmit signals do not exist simultaneously. A fast-acting switch is used to replace the T junction, allowing the antenna to be connected to either the receive or the transmit circuits.

In diplexer applications, the insertion loss of the receive filter is an important factor, since the loss contributes directly to the overall noise figure of the receiver that follows the diplexer. The insertion loss of the transmit filter is equally important, since the loss impacts the power transmitted and the efficiency of the transmit system. A tradeoff exists between the loss, rejection, and the size of the filter. High rejection with minimal effect on insertion loss and size can be achieved by using filters with asymmetric response. This allows the allocation of several attenuation poles in the appropriate band to increase the filter rejection. Typically, the insertion loss of the 90° hybrid design is slightly higher than that of the T-junction design. However, the 90° hybrid approach offers a higher power handling capability. The reason is that the power handling capability of the T-junction approach is determined by the power handling capability of the transmit filter alone, whereas in the hybrid approach, the power is handled by the reflection of the two identical filters.

Tx/Rx diplexers are found in those systems where the antenna is used for transmitting and receiving signals simultaneously (as opposed to *half-duplexing* or a *simplex* system, where the receiving and transmitting are separated in time). In radar systems the Tx and Rx signals are seldom present simultaneously as they are in communication systems, and a fast-acting electronic switch (switching ferrite circulator) is often used to isolate the sensitive low-noise receiver from the high-power transmitter. In satellite communication transponders, a wideband Tx/

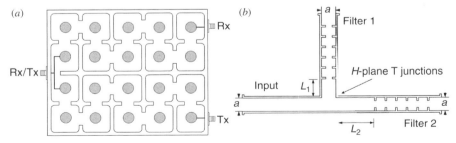

Figure 18.36 Transmit–receive diplexer configurations: (a) implemented using two coaxial filters and a wire T junction; (b) implemented using two waveguide iris filters and T waveguide junction.

Rx diplexer is sometimes found to cover the bandwidths of the Tx channel group in one filter, and the Rx channel group in the other filter for up/downlinking by one antenna system. However, the most demanding specifications are placed on the Tx/Rx diplexers in the base stations of cellular telephony systems, where the Tx and Rx channel groups are very closely spaced.

When diplexing for a Tx/Rx system, it is essential that a very high isolation be maintained between the Tx and Rx paths, perhaps up to 120 dB, to prevent the Tx power, its harmonics, or intermodulation (IM) products from reaching the sensitive LNA units. This means not only that there must be a very high intrinsic rejection by the Tx filter over the Rx bandwidth (and vice versa) but also that a very high build standard must be imposed in order to prevent passive IM products (PIMs) from being generated in the high-power regions of the diplexer. Since the rejection requirements are usually less severe on the outer sides of the filters, filters with asymmetric characteristics are often used when the Tx and Rx bands are very close, minimizing the insertion losses and overall mass, volume, and complexity. To attain a 100 dB isolation between the closely spaced receive and transmit bands of a cellular telephony system, 12th-degree coaxial filters with three TZs have been used for the filters of the base station diplexers, Figure 18.37 is a sketch of such a diplexer, and Figure 18.38 gives its measured response. Resonators 1–6 form a 6th-degree extended box section that realizes two of the TZs, and resonators 9–12 form a single box section realizing one more TZ. There are still some

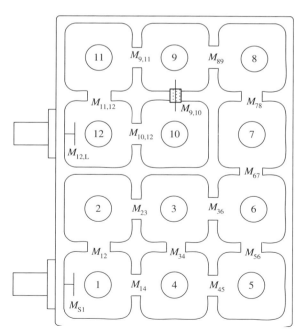

Figure 18.37 Outline sketch of a (12–3) filter realized with one 6th-degree extended box section (two TZs) and one 4th-degree box section (one TZ).

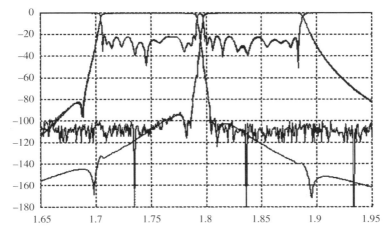

Figure 18.38 Simulated and measured rejection characteristics of a transmit/receive diplexer (duplexer) of a cellular telephony base station. (Courtesy of COM DEV Ltd.)

negative couplings, but these are "straight" rather than diagonal (e.g., $M_{9,10}$). Elimination of the diagonal cross-couplings benefits the volume production process and the eventual reliability of the device.

At L- and S-band frequencies (most common for mobile communications), the reentrant coaxial resonator is usually employed for construction of the filter resonators with the star-connected common junction located within the first cavities of the two filters. This leads to a very compact and efficient combination structure with line lengths quite short compared with the wavelengths. Figure 18.20 gives an outline sketch of a coaxial diplexer, consisting of two asymmetric (4–2) filters, and showing the combining junction.

18.5.1 Internal Voltage Levels in Tx/Rx Diplexer Filters

It is important to know the peak voltage levels within the resonators of a high-power microwave filter, if a proper design, avoiding damaging electrical discharges, is to be carried out. Such discharge phenomena include arcing in the presence of a gas (e.g., in high-power terrestrial TV broadcast systems), corona in partial-pressure conditions (e.g., for equipment in missiles or operating in a rocket launcher), or multipactor effect in vacuum (in spacecraft). These high-power effects are discussed in Chapter 20.

The peak voltages in the resonators of a microwave filter can be estimated at the prototype design stage, ensuring a proper mechanical design later with the dimensions of the gaps in the resonator cavities sufficient to withstand the peak voltages during operation. One procedure for determining the resonator voltages based on the $N \times N$ or $N + 2$ coupling matrix for the filter may be summarized as follows:

- Build up the $N \times N$ open-circuit impedance matrix $[z]$ for the filter network as detailed in Section 8.1.2, that is, $[z'] = R + [jM + sI]$, where $[M]$ is the filter coupling matrix [including frequency offsets (self couplings) on the diagonal, if any], sI is the diagonal matrix containing the frequency variable s (plus the factor $\delta_k = f_0/(\text{BW} \cdot Q_{uk})$, $k = 1, 2, \ldots, N$, accounting for noninfinite resonator Q_u factors), and $[R]$ is the termination matrix. Adding an inverter at either end of the network to normalize the terminations to unity converts the $N \times N$ impedance matrix $[z]$ to the $N+2$ admittance matrix $[y]$ (dual network theorem).

- Invert the $N+2$ admittance matrix $[y]$ to give $[y']^{-1} = [z'']$. Designating the input and output nodes of the circuit as 0 and $N+1$ respectively, use equation (18.6) to evaluate the voltage v_k in the kth resonator and the voltage v_0 at the input to the network ($k = 0$ in equation (18.6)):

$$v_k = \left(z_{k,0} - \frac{z_{k,N+1} \, z_{N+1,0}}{1 + z_{N+1,N+1}} \right) i_0 \qquad (18.6)$$

where the elements $z_{i,j}$ are taken from the impedance matrix $[z]$. The ratio $v_k(1 + S_{11}(s))/v_0$ then gives the voltage in the kth resonator referred to the voltage incident at the input of the network [25].

- Multiply the node voltages by $\sqrt{R_s}$. If $R_S = R_L = 1\,\Omega$ and the generator voltage $e_g = 2$ V, then the voltages v_k will be referred to 1 volt, which corresponds to the maximum available power from the generator P_i of 1 W (see Section 20.5.5).

The RF voltage values for a bandpass filter are found by multiplying the voltages v_k at the internal resonator nodes by a factor F_v, the value of which depends on the technology that the filter is constructed from. For example, for a waveguide bandpass filter

$$F_v = \frac{\lambda_0}{\lambda_{g0}} \sqrt{\frac{2f_0}{n\pi \text{BW}}} \qquad (18.7)$$

where n is the number of half-wavelengths of the waveguide resonators, f_0 is the center frequency, BW is the design bandwidth, and λ_{g0} and λ_0 are the wavelengths at f_0 in the waveguide and in free space, respectively [25]. For a given incident power P_i watts, the peak voltage gradient across the narrow b dimension along the centerline of the cavities of a rectangular waveguide TE_{01} mode filter are found by multiplying equation (18.6) by a further factor

$$\hat{E} = \sqrt{\frac{480\pi \lambda_g P_i}{ab\lambda}} \quad \text{V/cm} \qquad (18.8)$$

where a and b are the broad and narrow dimensions, respectively, of the waveguide in centimeters, and λ_g and λ are the wavelengths in the waveguide and free space, respectively.

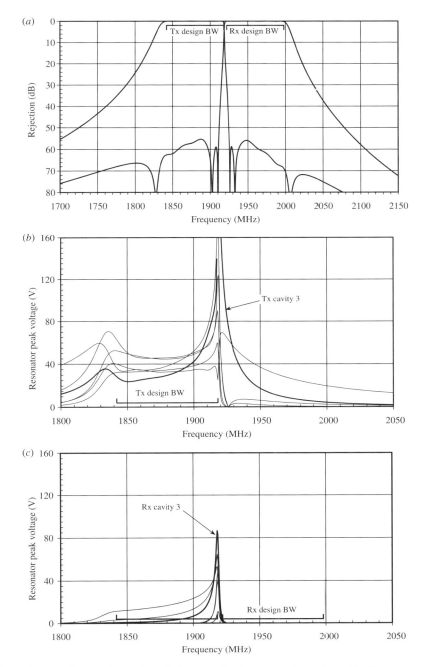

Figure 18.39   Transmit–receive diplexer with two complementary (6–2) asymmetric filters: (*a*) rejection characteristics; (*b*) voltage characteristics in transmit filter cavities; (*c*) voltage characteristics in receive filter cavities.

For cavities with a more complicated geometry, such as coaxial cavities, it is more realistic to use a stored energy approach, analyzing the voltage patterns and gradients in a single resonator with a unit energy by using finite-element EM methods. Then the absolute voltage values in each resonator of the filter are estimated by scaling by the voltage found as above for each resonator of the prototype network, and by the square root of the incident power [26].

Figure 18.39 shows the rejection characteristics and the estimated voltage levels within the cavities of the two (6−2) filters of an Rx/Tx diplexer, where the high-power filter is lower in frequency. The transmission zeros of these filters were realized with two trisections each. The voltage levels are referred to 1 W of input power, and the plot, shown in bold, is the cavity with the highest peak voltage, the third cavity from the input port in the case of the Tx filter. The plots in Figure 18.39 show that quite high voltages are also experienced by the cavities in the receive filter. This is caused by the power at the upper edge of the Tx filter bandwidth. The highest is also in the third cavity from the common port, meaning that the resonators of the receive filter need to also be designed to withstand high voltages. For a coaxial cavity resonator, the highest voltage gradients tend to concentrate around the zone where the tuning screw is close to the top of the resonator rod [27,28].

These voltage profiles and their levels are dependent on the filter's coupling topology. A different topology (e.g., folded instead of trisections) that gives the same rejection characteristics may give different voltage levels. Some work has been done to investigate topologies that give a more evenly distributed but lower average voltage level. However, the configurations that are best for lower voltages tend to be quite complex. [29]. On the other hand, the cavity arrangements that have the minimum number of couplings, such as the cul-de-sac filter, have the worst voltage levels. It seems that simple coupling topologies are inconsistent with low internal peak voltage levels.

SUMMARY

The previous chapters deal with the design of a microwave filter as an individual component of a communication network. In a multiuser environment, communication systems employ a range of microwave filters as individual components, as well as parts of a multiport network required to separate or combine a number of RF channels. This is commonly referred to as a *multiplexing network*, or simply, a *multiplexer* (MUX). This chapter is devoted to the design and tradeoffs of multiplexers for a variety of practical applications.

Multiplexers are used in communication system applications where there is a need to separate a wideband signal into a number of narrowband signals (RF channels). The channelization of the allocated frequency band allows flexibility for the flow of communication traffic in a multiuser environment. This also eases the requirements on the high-power amplifiers (HPAs) so that they can operate at relatively high efficiency with an acceptable degree of nonlinearity. Multiplexers are also employed to provide the opposite function, that of combining several narrowband channels into a single wideband composite signal for transmission via a common antenna. Owing to the reciprocity of filter networks, a MUX can also be configured to separate transmit and receive frequency bands in a common device, referred to as a *duplexer* or *diplexer*. Since multiplexers deal with the individual RF channels, they essentially control their characteristics. Wideband devices, such as amplifiers, switching networks, and antennas, have little impact over the narrow bandwidths of RF channels. Consequently, performance of multiplexers is critical in the efficient utilization of the frequency spectrum and hence the channel capacity. Multiplexers find widespread use in satellite and cellular communication systems.

The starting point of this chapter is a discussion of the tradeoffs among the various types of multiplexing networks. They include circulator-coupled, hybrid-coupled, and manifold-coupled multiplexers employing single-mode or dual-mode filters. It also includes multiplexers based on using directional filters. This is followed by trade-off and design considerations for each type of multiplexer. The design methodology and optimization strategy are dealt with in depth for the manifold-coupled multiplexer, by far the most complex microwave network. Numerous examples and photographs are included to illustrate the designs. The chapter concludes with a discussion of the high-power requirements and their impact on multiplexing networks.

REFERENCES

1. G. Matthaei, L. Young, and M. T. Jones, *Microwave Filters, Impedance Matching Networks and Coupling Structures*, Artech House, Norwood, MA, 1985.

2. E. G. Cristal and G. L. Matthaei, A technique for the design of multiplexers having contiguous channels, *IEEE Trans. Microwave Theory Tech.* **MTT-12**, 88–93 (Jan. 1964).

3. A. E. Atia, Computer aided design of waveguide multiplexers, *IEEE Trans. Microwave Theory Tech.* **MTT-22**, 322–336 (March 1974).

4. M. H. Chen, F. Assal, and C. Mahle, A contiguous band multiplexer, *COMSAT Tech. Rev.* **6**, 285–307 (Fall 1976).

5. C. M. Kudsia et al., A new type of low loss 14 GHz high power combining network, *Proc. 9th European Microwave Conf.*, London, Sept. 1979.

6. C. M. Kudsia, K. R. Ainsworth, and M. V. O'Donovan, Microwave filters and multiplexing networks for communication satellites in the 1980s, *Proc. AIAA 8th Communications Satellite Systems Conf.*, April 1980.

7. J. D. Rhodes and R. Levy, Design of general manifold multiplexers, *IEEE Trans. Microwave Theory Tech.* **MTT-27**, 111–123 (1979).

8. J. D. Rhodes and R. Levy, A generalized multiplexer theory, *IEEE Trans. Microwave Theory Tech.* **MTT-27**, 99–110 (1979).

9. D. Doust et al., Satellite multiplexing using dielectric resonator filters, *Microwave J.* **32**(12), 93–166 (Dec. 1989).

10. J. Bandler, S. Daijavad, and Q.-J. Zhang, Exact simulation and sensitivity analysis of multiplexing networks, *IEEE Trans. Microwave Theory Tech.* **MTT-34**, 111–102 (Jan. 1986).

11. D. S. Levinson and R. L. Bennett, Multiplexing with high performance directional filters, *Microwave J.* 92–112 (June 1989).

12. D. Rosowsky, Design of manifold multiplexers, *Proc. ESA Workshop Microwave Filters*, June 1990, pp. 145–156.

13. U. Rosenberg, D. Wolk, and H. Zeh, High performance output multiplexers for Ku-band satellites, *Proc. 13th AIAA Int. Communication Satellite Conf.*, Los Angeles, March 1990, pp. 747–752.

14. C. Kudsia, R. Cameron, and W. C. Tang, Innovation in microwave filters and multiplexing networks for communication satellite systems, *IEEE Trans. Microwave Theory Tech.* **MTT-40**, 1133–1149 (June 1992).

15. J. Uher, J. Bornemann, and U. Rosenberg, *Waveguide Components for Antenna Feed Systems—Theory and CAD*, Artech House, Norwood, MA, 1993.

16. S. Ye and R. R. Mansour, Design of manifold-coupled multiplexers using superconductive lumped element filters, *IEEE MTT-IMS*, 1994, pp. 191–194.

17. R. R. Mansour et al., Design considerations of superconductive input multiplexers for satellite applications, *IEEE Trans. Microwave Theory Tech.* **MTT-44**, 1213–1228 (July 1996).

18. G. Matthaei, S. Rohlfing, and R. Forse, Design of HTS lumped element manifold-type microwave multiplexers, *IEEE Trans. Microwave Theory Tech.* **MTT-44**, 1313–1320 (July 1996).

19. R. R. Mansour et al., A 60 channel superconductive input multiplexer integrated with pulse-tube cryocoolers, *IEEE Trans. Microwave Theory Tech.* **MTT-48(7)**, 1171–1180 (July 2000).

20. R. R. Mansour, S. Ye, V. Dokas, B. Jolley, W. C. Tang, and C. Kudsia, System integration issues of high power HTS output multiplexers, *IEEE Trans. Microwave Theory Tech.* **MTT-48**, 1199–1208 (July 2000).

21. D. Pozar, *Microwave Engineering*, Wiley, New York, 1998.

22. C. M. Kudsia, High power contiguous combiners for satellite earth terminals, *Proc. Canadian Satellite User Conf.*, Ottawa, Ontario, Canada, May 25–28, 1987.

23. J. Bandler, R. Biernacki, S. Chen, P. Grobelny, and R. Hemmers, Space mapping technique for electromagnetic optimization, *IEEE Trans. Microwave Theory Tech.* **MTT-42**, 2536–2544 (Dec. 1994).

24. M. H. Chen, Singly-terminated pseudo-elliptic function filter, *COMSAT Tech. Rev.* **7**, 527–541 (Fall 1977).

25. A. Sivadas, M. Yu, and R. J. Cameron, A simplified analysis for high power microwave bandpass filter structures, *IEEE MTT-S Int. Microwave Symp. Digest*, Boston, 2000, pp. 1771–1774.

26. C. Ernst and V. Postoyalko, Comparison of the stored energy distributions in a QC-type and a TC-type prototype with the same power transfer function, *IEEE MTT-S Int. Microwave Symp. Digest* Anaheim, CA, 1999, pp. 1339–1342.

27. A. R. Harish, J. S. Petit, and R. J. Cameron, Generation of high equivalent peak powers in coaxial filter cavities, *Proc. 31st European Microwave Conf.*, London, Sept. 2001, pp. 289–292.

28. C. Ernst and V. Postoyalko, Prediction of peak internal fields in direct-coupled filters, *IEEE Trans. Microwave Theory Tech.* **MTT-51**, 64–73 (Jan. 2003).

29. B. S. Senior, *Optimized Network Topologies for High Power Filter Applications*, Ph.D. thesis, Univ. Leeds, 2004.

CHAPTER 19

COMPUTER-AIDED DIAGNOSIS AND TUNING OF MICROWAVE FILTERS

As a result of manufacturing and material tolerances, filter tuning is an essential postproduction process. Traditionally, filters have been tuned manually by skilled technologists. The tuning process is not only time-consuming but also expensive, particularly for high-order narrowband filters with stringent requirements. It is a fact that almost all filters used for wireless base stations and satellite applications must be subjected to a postproduction tuning process. For example, in typical satellite applications, channel filters that operate at 4 GHz have very stringent in-band and out-of-band requirements with a design margin of 300 kHz, that is, <0.01%. Such a design margin is needed to accommodate the drift in filter frequency due to temperature, leaving almost a zero margin for design imperfections due to manufacturing tolerances. The specifications for wireless base station filters are as stringent as those for satellite applications.

The complexity of the tuning process depends on the technology that is employed and the filter configuration. For example, tuning is a must for narrowband dielectric resonator filters and narrowband planar filters, since the dielectric constant of the dielectric resonator or the substrate of planar structures can vary from batch to batch. For dielectric resonators with $\varepsilon_r = 38$, a deviation of ±0.5 in the dielectric constant at 4 GHz translates into a frequency shift close to 25 MHz. In some applications, such a shift in the center frequency exceeds the filter bandwidth itself. The filter topologies and filter functions can also contribute to the complexity of the tuning process. Dual-mode elliptic filters and self-equalized filters are usually much more difficult to tune than single-mode filters with a Chebyshev response.

Microwave Filters for Communication Systems: Fundamentals, Design, and Applications,
by Richard J. Cameron, Chandra M. Kudsia, and Raafat R. Mansour
Copyright © 2007 John Wiley & Sons, Inc.

Typically, manual tuning is performed as a real-time iterative optimization process. To allow for tuning, filters are constructed with tuning screws or other forms of tuning elements to allow the technologist to change the resonance frequency of the filter resonators and the interresonator couplings. The technologist monitors the filter performance on the vector network analyzer (VNA) and tweaks the tuning screws iteratively, until the filter meets the specification requirements. For many technologists, the manual tuning process has been more of an art than a science. Therefore, the manual tuning of complex filter and multiplexer structures is usually performed by well-experienced technologists.

Tuning is a major factor in the overall filter production cost. Also, tuning has a significant impact on project schedules. The availability of computer-aided tuning techniques that can guide the technologists during the tuning process can be a major factor in reducing the tuning time. The use of robots has the potential to eliminate the need to use human operators, which can further reduce production costs and scheduling.

Although the concept of computer-aided tuning has been known for years [1–14], the low-cost and short-term delivery requirements of wireless base station filters in the mid-1990s has contributed to the advances and innovations in computer-aided filter tuning technology. Since then, several papers have been published on the computer-aided tuning of microwave filters, employing different techniques. These techniques can be divided into five categories:

1. Sequential tuning of coupled resonator filters
2. Computer-aided tuning based on circuit model parameter extraction
3. Computer-aided tuning using poles and zeros of the input reflection coefficient
4. Time-domain tuning
5. Fuzzy logic tuning

The following sections present details of these five approaches.

19.1 SEQUENTIAL TUNING OF COUPLED RESONATOR FILTERS

Ness [15] has shown that the group delay of the input reflection coefficient of sequentially tuned resonators contains all the necessary information to tune filters. The paper demonstrates that the group delay value at the center frequency of the

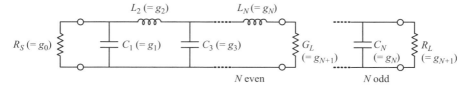

Figure 19.1 Lowpass prototype circuit.

filter can be written quite simply in terms of the prototype lowpass g_k values. Consider the lowpass prototype filter shown in Figure 19.1. The Q_e and M_{ij} of the coupled matrix model are given in terms of the g_k lowpass prototype values, filter center frequency f_0, and filter bandwidth BW as

$$M_{j,j+1} = \frac{1}{\sqrt{g_j g_{j+1}}}, \quad j = 1, 2, \ldots, N-1 \tag{19.1}$$

$$R_1 = \frac{1}{g_0 \, g_1}, \quad R_2 = \frac{1}{g_N g_{N+1}} \tag{19.2}$$

$$Q_e = \frac{f_0}{R_1 \cdot \text{BW}} = \frac{g_0 \, g_1 f_0}{\text{BW}} \tag{19.3}$$

$$k_{j,j+1} = \frac{\text{BW}}{f_0} M_{j,j+1} = \frac{\text{BW}}{f_0} \frac{1}{\sqrt{g_j g_{j+1}}} \tag{19.4}$$

The calculation from the lowpass prototype, with the proper lowpass-to-bandpass transformation, enables the group delay of the reflected signal to be expressed simply and directly in terms of the g_k values and the normalized bandwidth of the bandpass filter. Using equations (19.1)–(19.4), the external Q_e and the coupling k_{ij} can be determined from the filter specification. These coupling parameters can also be related to group delay Γ_d of the reflected signal S_{11} as successive resonators are tuned to resonance. The group delay of the $S_{11}(\omega)$ is defined as

$$\Gamma_d(\omega) = -\frac{\partial \varphi}{\partial \omega} \tag{19.5}$$

For a standard lowpass-to-bandpass transformation, we obtain

$$\omega' \longrightarrow \frac{\omega_0}{(\omega_2 - \omega_1)} \left(\frac{\omega}{\omega_0} - \frac{\omega_0}{\omega} \right) \tag{19.6}$$

where ω' is the angular frequency of the lowpass prototype, ω_0 is the center frequency of the bandpass filter, and ω_1 and ω_2 are the lower edge and the upper edge frequencies respectively, of the bandpass filter. Whereas $\omega_0 = (\omega_1 \, \omega_2)^{1/2}$, $\Gamma_d(\omega)$ is then written as

$$\Gamma_d(\omega) = -\frac{\partial \varphi}{\partial \omega'} \frac{\partial \omega'}{\partial \omega} \tag{19.7}$$

$$\Gamma_d(\omega) = -\frac{\omega^2 + \omega_0^2}{\omega^2 (\omega_2 - \omega_1)} \frac{\partial \varphi}{\partial \omega'} \tag{19.8}$$

Consider the case where all the resonators are short-circuited ("shorted") (detuned) with the exception of the first resonator. This corresponds to the case where the elements $g_2, g_3, \ldots$ are disconnected from the lowpass prototype circuit.

Input impedance Z_{in} and reflection coefficient S_{11} are then given by

$$Z_{in} = -\frac{j}{\omega' g_1}, \quad Z_0 = g_0, \tag{19.9}$$

$$S_{11} = \frac{Z_{in} - Z_0}{Z_{in} + Z_0} \tag{19.10}$$

$$S_{11} = \frac{\omega' g_1 g_0 + j}{-\omega' g_1 g_0 + j} \tag{19.11}$$

$$\phi = 2 \tan^{-1} \frac{1}{\omega' g_1 g_0} \tag{19.12}$$

We define $\Gamma_{d1}(\omega)$ as group delay, when all the resonators, except resonator 1, are shorted (detuned). Thus, $\Gamma_{d1}(\omega)$ can be written as

$$\Gamma_{d1}(\omega) = -\frac{2(\omega^2 + \omega_0^2)}{\omega^2(\omega_2 - \omega_1)} \frac{g_0 g_1}{1 + (g_0 g_1 \omega')^2} \tag{19.13}$$

At the center frequency ω_0, $\Gamma_{d1}(\omega_0)$ is given by

$$\Gamma_{d1}(\omega_0) = \frac{4 g_0 g_1}{(\omega_2 - \omega_1)} \tag{19.14}$$

In view of equation (19.3), we can relate the group delay $\Gamma_{d1}(\omega_0)$ to Q_e as

$$\Gamma_{d1}(\omega_0) = \frac{4Q_e}{\omega_0} \tag{19.15}$$

We then consider the case where all resonators of the bandpass filters are shorted (detuned) with the exception of the first and second resonators. In the prototype lowpass circuit, this corresponds to the case where the second element g_2 is shorted to the ground. We can then find the input impedance and use equations (19.8)–(19.10) to derive $\Gamma_{d2}(\omega_0)$, which is given as

$$\Gamma_{d2}(\omega_0) = \frac{4 g_2}{g_0(\omega_2 - \omega_1)} \tag{19.16}$$

In view of equations (19.1)–(19.4), $\Gamma_{d2}(\omega_0)$ is related to K_{12} by

$$\Gamma_{d2}(\omega_0) = \frac{16}{\omega_0^2 k_{12}^2 \Gamma_{d1}(\omega_0)} = \frac{4}{\omega_0 Q_e k_{12}^2} \tag{19.17}$$

We can follow a similar procedure to derive the group delay as the successive resonators are shorted (detuned). Table 19.1 provides the group delay and phase of the input reflection coefficient up to $N = 6$ [15]. It is interesting to note that at ω_0, the group delay of S_{11}, is determined only by the shunt elements for an odd number of resonators and by the series elements for an even number of resonators. The reverse applies if the dual circuit for the lowpass filter is selected.

TABLE 19.1   Group Delay at Center Frequency ω_0 as Successive Resonators are Tuned

$n = 1$	$\Gamma_{d1}(\omega_0) = \dfrac{4g_0g_1}{(\omega_2 - \omega_1)}$	$\Gamma_{d1}(\omega_0) = \dfrac{4Q_e}{\omega_0}$	$\phi \to \pm 180^\circ$
$n = 2$	$\Gamma_{d2}(\omega_0) = \dfrac{4g_2}{g_0(\omega_2 - \omega_1)}$	$\Gamma_{d2}(\omega_0) = \dfrac{4}{\omega_0 Q_e k_{12}^2}$	$\phi \to 0^\circ$
$n = 3$	$\Gamma_{d3}(\omega_0) = \dfrac{4g_0(g_1 + g_3)}{(\omega_2 - \omega_1)}$	$\Gamma_{d3}(\omega_0) = \Gamma_{d1} + \dfrac{4Q_e k_{12}^2}{\omega_0 k_{23}^2}$	$\phi \to \pm 180^\circ$
$n = 4$	$\Gamma_{d4}(\omega_0) = \dfrac{4(g_2 + g_4)}{g_0(\omega_2 - \omega_1)}$	$\Gamma_{d4}(\omega_0) = \Gamma_{d2} - \dfrac{4k_{23}^2}{\omega_0 Q_e k_{12}^2 k_{34}^2}$	$\phi \to 0^\circ$
$n = 5$	$\Gamma_{d5}(\omega_0) = \dfrac{4g_0(g_1 + g_3 + g_5)}{(\omega_2 - \omega_1)}$	$\Gamma_{d5}(\omega_0) = \Gamma_{d3} + \dfrac{4Q_e k_{23}^2 k_{34}^2}{\omega_0 k_{23}^2 k_{45}^2}$	$\phi \to \pm 180^\circ$
$n = 6$	$\Gamma_{d6}(\omega_0) = \dfrac{4(g_2 + g_4 + g_6)}{g_0(\omega_2 - \omega_1)}$	$\Gamma_{d6}(\omega_0) = \Gamma_{d4} + \dfrac{4k_{45}^2 k_{23}^2}{\omega_0 Q_e k_{12}^2 k_{34}^2 k_{56}^2}$	$\phi \to 0^\circ$

We can easily show that for narrow band filters the group delay $\Gamma_{d1}(\omega)$, $\Gamma_{d2}(\omega), \ldots$, and $\Gamma_{dn}(\omega)$ can be approximated as even functions of ω around ω_0. It can be also demonstrated that the phase of S_{11} passes through 180° and 0° crossing at ω_0, as the successive resonators are tuned. Figures 19.2 and 19.3 show sketches for the phase and group delay around the center frequency, respectively, as the resonators are successively tuned for a filter, designed at a center frequency of 12 GHz. The location of the $180^\circ/0^\circ$ phase crossing deviates slightly from the center frequency. Note that the group delay is not perfectly symmetric. The slight unsymmetry is also attributed to the fact that the results shown in Figures 19.2 and 19.3 are obtained from coupling matrix model M, where the values of diagonal elements M_{ii} determine how well the resonators are tuned ($M_{ii} = 0$ represents the case of perfectly tuned resonators). The results given in these two figures represent the case where resonator detuning is obtained by using a large value for the diagonal elements M_{ii}, rather than perfectly shorting out (short-circuiting) the resonators. This, in fact, represents a real-case scenario since shorting the resonator is not readily achieved in most practical applications.

Tuning Steps

Step 1. For the given filter specifications, calculate the group delay values Γ_{d1}, $\Gamma_{d2}, \ldots, \Gamma_{dN}$ in terms of the lowpass g_k values or the coupling values k_{ij}.

Step 2. Short (detune) all resonators except resonator 1. Adjust the input coupling and resonance frequency of resonator 1 to set the group delay to the specified value Γ_{d1} at the filter center frequency.

Step 3. Short (detune) all the resonators except resonators 1 and 2. Adjust the resonance frequency of resonator 2 and the coupling between resonators 1 and 2 to get a symmetric group delay response about the filter center frequency

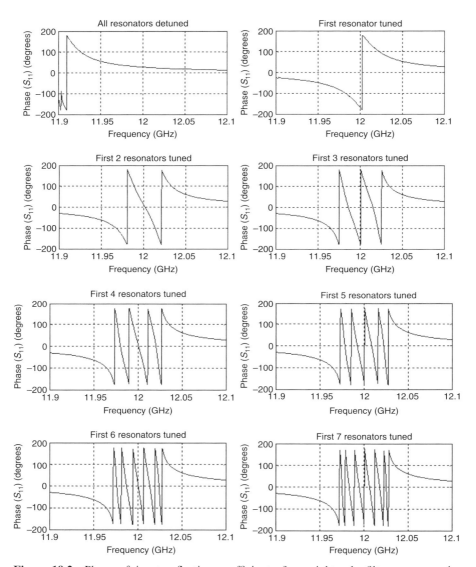

Figure 19.2 Phase of input reflection coefficient of an eight-pole filter as successive resonators are tuned.

and with the specified value Γ_{d2}. To maintain symmetry, it may be necessary to readjust resonator 1, if the coupling between resonators 1 and 2 is so strong to the extent that it detunes resonator 1.

Step 4. Progress through the filter by tuning each subsequent resonator as in step 3 to maintain group delay symmetry and to set the group delay at the filter center frequency to the specified values $\Gamma_{d3}, \Gamma_{d4}, \ldots, \Gamma_{dN-1}$. In each case, a slight

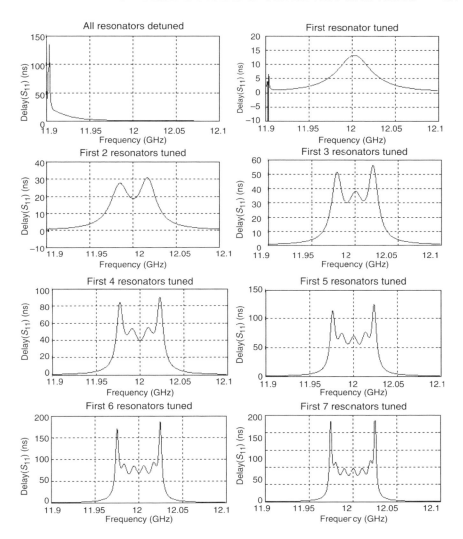

Figure 19.3 Group delay of input reflection coefficient of an eight-pole filter as successive resonators are tuned.

readjustment of the resonance frequency of the preceding resonator may be required to maintain the group delay symmetry.

Step 5. When the last resonator is reached and the filter output is properly terminated, observe the amplitude response of S_{11} and tune the last resonator and the output coupling to obtain the specified return loss.

Precise results are obtained only if the resonators are properly shorted rather than simply detuned. Nevertheless, the tuning procedure presents a simple approach to tune the filter so that its response is close to the required response. A minimum fine-tuning may be required at the end of step 5 in order to meet the exact filter specifications.

In this tuning procedure [15], each step involves the simultaneous adjustment of two parameters, the resonance frequency and interresonator coupling, requiring the group delay symmetry. Also, since the phase passes through the $0°/180°$ crossing at the center frequency, as the resonators are detuned, these two parameters can be successively tuned rather than simultaneously tuned. After we tune the resonator center frequency to achieve either $0°$ or $180°$ crossing at the center frequency, we adjust the interresonator coupling to set the group delay as close as possible to the specified value as feasible. At this point, both the resonator center frequency and the interresonator coupling can be fine-tuned to set the group delay exactly to the specified value. However, in practical applications, filters are usually attached to input/output transmission lines, which affect the location where the $180°/0°$ phase crossing takes place, and the use of phase information requires knowledge of the input reference plane. Several approaches are available to determine the input reference planes. A simple approach is to apply step 2 to the first resonator to get group delay symmetry and to set the group delay to the value Γ_{d1}. The frequency at which this occurs directly gives the frequency shift needed to accommodate the shift in the reference plane. Thus, the use of $180°/0°$ phase crossing can be used by itself to rough tune the filter to the proper frequency band.

19.2 COMPUTER-AIDED TUNING BASED ON CIRCUIT MODEL PARAMETER EXTRACTION

This approach incorporates a circuit model for the filter with an element optimization routine. The idea in this technique is to optimize the parameters of the circuit model such that the response from the circuit model best fits the measured response of the filter. After that, a comparison between the extracted parameters and the parameters of the ideal filter is used to identify the filter elements that need tuning. The technique was first introduced by Thal [7]. Several modifications of this technique have been reported [16–22]. Computer-aided tuning software tools based on parameter extraction were developed by COM DEV [17] and other filter suppliers in early 1990s. These software tools were used to guide technologists through the tuning process of microwave filters.

Although several circuit models are available that can be employed in this technique, the advantages of using the coupling matrix circuit model is that the coupling elements are directly related to the position of the physically tunable elements. Once one of these coupling elements is identified and found to deviate from the desired ideal value, the coupling element can be easily adjusted back to the desired value by turning the corresponding tuning element.

Consider the generalized filter network shown in Figure 19.4. The circuit model consists of n coupled lossless resonators, where M_{iji} denotes the frequency-independent coupling between resonators i and j. The generalized matrix is given

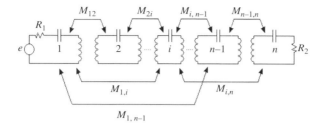

Figure 19.4   A generalized model for coupled resonator filters.

in equation (19.18). For given filter specifications, the coupling matrix is synthesized, as explained in Chapters 8–10.

$$
M = \begin{bmatrix} m_{11} & m_{12} & \cdots & m_{1n} \\ m_{21} & m_{22} & \cdots & m_{2n} \\ \vdots & & & \vdots \\ m_{n1} & m_{n2} & \cdots & m_{nn} \end{bmatrix}
$$
(19.18)

The filter scattering parameters in terms of the coupling elements matrix and input and output coupling R_1 and R_2 are given as:

$$
S_{21} = -2j\sqrt{R_1 R_2}[A^{-1}]_{n1}
$$
(19.19)

$$
S_{11} = 1 + 2jR_1[A^{-1}]_{11}
$$
(19.20)

$$
A = \lambda I - jR + M
$$
(19.21)

$$
\lambda = \frac{f_0}{\text{BW}}\left(\frac{f}{f_0} - \frac{f_0}{f}\right)
$$
(19.22)

where I is the unity matrix and R is a diagonal matrix with all the elements zero, except $R_{11} = R_1$ and $R_{nn} = R_2$.

The S parameters of the measured filter are described by the same circuit model. Here, the coupling elements are related to a particular tuning element. For example, the tuning of the filter resonance frequencies correspond to varying the diagonal elements M_{ii}, while tuning of the coupling between the resonators corresponds to varying coupling elements M_{ij}. Thus, we can extract the coupling matrix, associated with the measured results, by optimizing the elements of the coupling matrix such that the S parameters of the circuit model match the measured S parameters. A comparison of the coupling matrix extracted from the measured results with the coupling matrix of the ideal filter reveals the elements that need tuning. The tuning procedure can then be summarized as follows:

Step 1. Measure the performance of the filter being tuned.

Step 2. Use the coupling matrix circuit model and extract the coupling elements M_{ij}, R_1, and R_2 by optimization for best fit with measured data. To construct the objective function, the measured performance of the filter is sampled at several frequency points within the band, and the objective function to be minimized can be defined as

$$\varphi = \sum_{\text{freq}} \sum_{i=1}^{2} \sum_{j=1}^{2} \left(abs\left(S_{ij}^{\text{model}} \right) - abs\left(S_{ij}^{\text{measured}} \right) \right), \tag{19.23}$$

where S_{ij}^{model}, $i = 1, 2$ and $j = 1, 2$ are the S parameters, calculated from the coupling matrix circuit model at specific frequency points, whereas S_{ij}^{measured} $i = 1, 2, j = 1, 2$ are the measured parameters at the same frequency points. The variables to be optimized are the coupling elements M_{ij} and input/output coupling R_1 and R_2.

Step 3. Compare the coupling matrix elements, extracted from the experimental results, with the ideal coupling matrix elements. Adjust the tuning elements accordingly.

Step 4. Repeat steps 1–3 iteratively until the filter is tuned. At this point, the extracted coupling elements from the measured results should reasonably match the ideal coupling elements.

One problem associated with this technique is that very good initial values for the variables to be optimized are usually needed to reduce the possibility of running into a local minimum or failing to converge to the proper solution. The problem is particularly pronounced in the case of highly detuned filters. To overcome this problem, two approaches were proposed, one is based on sequential parameter extraction and the other is based on the fuzzy logic technique. The latter technique is explained in Section 19.6.

The sequential parameter extraction procedure [22] starts with detuning all the resonators (i.e., shorting) by setting the resonance frequencies of the corresponding resonators to values outside the frequency range. Then one resonator after the other is tuned (unshorted). Thus for a filter of degree n, we obtain a sequence of n subfilters, with the subfilter i characterized by $n - i$ shorted resonators and i unshorted resonators. Then the parameter extraction is carried out at each step to extract the coupling elements of the subfilter. At step i, after unshorting resonator i, the filter return loss S_{11} is measured, and the parameters of the corresponding subfilter are extracted by minimizing the following error function:

$$\varepsilon(i) = \sum_{j} \left| S_{11}^{\text{meas}}(\omega_j) - S_{11}^{\text{mod}}(\omega_j)_{(i)} \right|^2 \tag{19.24}$$

Equation (19.24) represents the difference between the measured and the modeled responses for each subfilter. Once the subfilter parameters are extracted, the actual filter elements are tuned to minimize the deviation between the ideal parameters and the actual subfilter parameters before proceeding to the next step.

Let the coupling matrix of the ideal complete filter be denoted as $[M]$, the reflection coefficient of the subfilter I, $S_{11(i)}^{\text{mod}}$ can be calculated from the physical circuit shown in Figure 19.4 by shorting $n - i$ last resonators. Alternatively, $S_{11(i)}^{\text{mod}}$ can be calculated from equations (19.19)–(19.22) by using a different coupling matrix for the subfilter at each step. Let $[M]^{(1)}, [M]^{(2)}, \ldots, [M]^{(n-1)}$ be the coupling matrices associated with subfilters $1, 2, \ldots, n-1$. The M-matrices of the subfilters can be obtained by simply setting the corresponding diagonal elements to a very large value, which is equivalent to resonator detuning. These matrices are obtained as follows:

Subfilter 1. All the resonators with the exception of resonator 1 are shorted. The parameters to be optimized are R_1 and M_{11}. Then $[M]^{(1)} = [M]$ with M_{22}, $M_{33}, \ldots, M_{nn}$ are set to a very large value.

Subfilter 2. All the resonators with the exception of resonators 1 and 2 are shorted. The parameters to be optimized are M_{12} and M_{22}, whereas R_1 and M_{11} are fixed at the values extracted from step 1, but can also be included in the optimization for fine-tuning to achieve the best fit with the measured results of subfilter 2. then $[M]^{(2)} = [M]$ with $M_{33}, M_{44}, \ldots, M_{nn}$ are set to a very large value.

Subfilter $n-1$. Resonator n is shorted. The parameters to be optimized are $M_{n-2, \, n-1}$ and $M_{n-1, \, n-1}$. Then $[M]^{(n-1)} = [M]$ with M_{nn} is set to a very large value.

The Complete Filter. The parameters to be optimized are M_{nn} and R_2. The other coupling elements of the coupling matrix can be set to the values obtained from the previous steps or included for the optimization as fine-tuning for the best fit with the measured results of the whole filter.

This approach provides the flexibility of controlling the number of variables to be optimized. The parameters that are extracted and tuned up to step i may be assumed fixed in the model during the next parameter extraction step, $i + 1$. Alternatively, since bringing the resonator $i + 1$ in-band by removing the detuning may have an impact on adjacent couplings, some of the parameters extracted in step i may be assumed to be variables in the parameter extraction step $i + 1$. However, the initial values for these variables will be the values extracted in step i rather than ideal values, which in turn will speed up the optimization process.

This kind of sequential parameter extraction, although it requires several optimization steps, helps eliminate the convergence problems associated with optimizing all filter parameters at once. The number of parameter extraction processes (optimization) may be lowered by reducing the number of subfilters. As an example, if we take a 10-pole filter, we may use three subfilters: subfilter 1 with the last 7 resonators detuned; subfilter 2, with the last 3 resonators detuned; and subfilter 3, as the whole filter, with no resonator detuned. The use of sequential parameter extraction technique to tune complex filter structures has been demonstrated [22], including the tuning of a 12-pole self-equalized dual-mode waveguide filter.

At each step in the tuning process for the subfilters, a certain number of filter parameters that deviate from the ideal parameters need to be adjusted. Thus, for a particular stage, where several of the tuning elements need tuning, one of these elements must be selected for the very next tuning action. In some cases, the convergence of the tuning process depends strongly on the decision as to which tuning elements should be adjusted first. A common selection criterion is to base the action on the difference between the actual coupling matrix and ideal coupling matrix, where the elements that have a large deviation need to be tuned first. However, this approach tends to give preference to the elements that have large values such as sequential coupling over the elements with small values such as cross-coupling and resonance frequencies (the diagonal elements of the coupling matrix). Furthermore, even when using normalized values, it is difficult to compare the mistuning level of the coupling with the mistuning level of the resonators. An approach based on sensitivity analysis of the equivalent circuit has been proposed in [22] to identify the elements that need to be tuned first.

Up to this point, the parameter extraction technique has been presented as a diagnostic tool, where it is used to extract the coupling matrix so it can be compared with the ideal matrix in order to identify the elements (coupling and resonance frequencies) that need adjustment. The tool can guide the technologists to select the proper elements during the tuning process, and with repeated iterations, they can align the filter to the desired response. The question is how the deviation in coupling value (M_{ij}, R_1, R_2) or resonance frequency (M_{ii}) is translated to the penetration depth of the tuning element (screw). The information is needed to automate the parameter extraction process to construct a robot that can replace the technologist.

A sensitivity analysis is needed to determine the relationship between the deviation in the coupling value and the tuning screw, or, the position of the tuning elements. The steps needed for this sensitivity analysis [19] are listed next:

Step 1. Rough-tune the filter within the tuning range.

Step 2. Measure the S parameters for this basis position.

Step 3. Measure S parameters with one screw turned at a time.

Step 4. Extract coupling matrix elements for the basis position and for each measurement when one screw is turned at a time.

Assume that the number of screws is m and that the screw turns are the parameters that must be controlled, when the entire process is automated. For m tuning screws, let $n_1, n_2, \ldots, n_m$ be the number of turns associated with screws $1, 2, \ldots, m$. A comparison between the extracted coupling matrix of the basis position and the extracted coupling matrices provides ΔR_1, ΔR_2, Δm_{11}, and Δm_{ij}, which are associated with the m steps $\Delta n_1, \Delta n_2, \ldots, \Delta n_m$. We can then build the following sensitivity matrix:

$$\text{Sensitivity matrix} = \begin{bmatrix} \dfrac{\partial R_1}{\partial n_1} & \dfrac{\partial R_1}{\partial n_2} & \cdots & \dfrac{\partial R_1}{\partial n_m} \\[2mm] \dfrac{\partial m_{11}}{\partial n_1} & \dfrac{\partial m_{11}}{\partial n_2} & \cdots & \dfrac{\partial m_{11}}{\partial n_m} \\[2mm] \dfrac{\partial m_{12}}{\partial n_1} & \dfrac{\partial m_{12}}{\partial n_2} & \cdots & \dfrac{\partial m_{12}}{\partial n_n} \\[1mm] \vdots & \vdots & \cdots & \vdots \\[1mm] \dfrac{\partial m_{nn}}{\partial n_1} & \dfrac{\partial m_{nn}}{\partial n_2} & \cdots & \dfrac{\partial m_{nn}}{\partial n_m} \\[2mm] \dfrac{\partial R_2}{\partial n_1} & \dfrac{\partial R_2}{\partial n_2} & \cdots & \dfrac{\partial R_2}{\partial n_m} \end{bmatrix}. \qquad (19.25)$$

The relation between the deviation in the coupling value versus the screw position can then be written as follows:

$$\Delta m_{ij} = \sum_{k=1}^{k=m} \frac{\partial m_{ij}}{\partial n_k} \Delta n_k$$

The sensitivity matrix can be used to construct a robot to automate the tuning process (see Section 19.7). In most filter structures, each tuning element affects not only the corresponding element in the coupling matrix but also nearby elements. The extent of the influence of each tuning screw on the other coupling elements is also found from the sensitivity matrix. Thus, this sensitivity matrix is beneficial to both the automation process and the diagnosis process.

The technique presented in this section has been successfully used in the diagnosis and automated computer-aided tuning of microwave filters. Several modifications of this technique have been reported offering the use of other filter circuit models or an alternative approach to extract the coupling elements. For example, the following objective function has been used [18] to replace the objective function given in equation (19.23)

$$\varphi = \sum_{i=1}^{N} \left| S_{11}^{model}(P_i) \right|^2 + \sum_{i=1}^{M} \left| S_{21}^{model}(Q_i) \right|^2 \tag{19.26}$$

where P_i and Q_i are the locations of the reflection and transmission zeros, respectively, which are generated from the measured response. The idea here is to construct polynomials generated from sample points from the measured filter response using the model-based parameter estimation technique [23]. The location of the reflection and transmission zeros are then extracted from these polynomials. Details on this approach are given in [18].

19.3 COMPUTER-AIDED TUNING BASED ON POLES AND ZEROS OF THE INPUT REFLECTION COEFFICIENT

A general model for the determination of individual resonance frequencies and coupling coefficients of a system with cascaded coupled resonators based on knowledge of the zeros and poles of the filter is described in this section. The key idea in this approach [24] is that measurement or computation of the phase of the reflection coefficient provides all the information necessary to evaluate the couplings and resonance frequencies. To understand the concept, we consider the two-pole filter shown in Figure 19.5 with the output shorted. The loop equations are given by

$$e_1 = i_1 \left(j\omega L_1 + \frac{1}{j\omega C_1} \right) - jM_{12}i_2 \tag{19.27}$$

$$0 = i_2 \left(j\omega L_2 + \frac{1}{j\omega C_2} \right) - jM_{12}i_1 \tag{19.28}$$

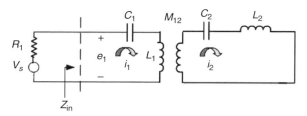

Figure 19.5 Equivalent circuit of two cascaded resonators terminated in a shortcircuit.

If we define m_{12} as

$$m_{12}^2 = \omega_{01}\omega_{02}\frac{M_{12}^2}{Z_{01}Z_{02}} \tag{19.29}$$

where $\omega_{0i} = 1/(\sqrt{L_i C_i})$ and $Z_{0i} = \sqrt{L_i/C_i}$, then Z_{in} is written as

$$Z_{in} = \frac{e_1}{i_1} = j\frac{Z_{01}}{\omega\omega_{01}}\frac{(\omega^2 - \omega_{01}^2)(\omega^2 - \omega_{02}^2) - \omega^2 m_{12}^2}{(\omega^2 - \omega_{02}^2)} \tag{19.30}$$

and Z_{in} can be written in a polynomial form as

$$Z_{in} = j\frac{Z_{01}}{\omega\omega_{01}}\frac{(\omega^2 - \omega_{z1}^2)(\omega^2 - \omega_{z2}^2)}{(\omega^2 - \omega_{p1}^2)} \tag{19.31}$$

where

$$\omega_{01}^2 = \frac{\omega_{z1}^2\omega_{z2}^2}{\omega_{p1}^2}, \quad \omega_{02}^2 = \omega_{p1}^2 \tag{19.32}$$

$$m_{12}^2 = \omega_{z1}^2 + \omega_{z2}^2 - \omega_{p1}^2 - \omega_{01}^2 \tag{19.33}$$

This analysis demonstrates that the zeros and poles of the input reflection coefficient with output port terminated in a short circuit are related to individual resonance frequencies and coupling coefficients.

Consider the generalized coupled resonator model shown in Figure 19.6 with resonator n terminated in a short circuit at reference plane TT. The input impedance at loop i can be written as

$$Z_{in}^{(i)} = j\frac{Z_{0i}}{\omega\omega_{0i}}\frac{P_i(\omega^2)}{Q_i(\omega^2)}, \quad i = 1, 2, \ldots, n \tag{19.34}$$

where $P_i(\omega^2)$ and $Q_i(\omega^2)$ are polynomials of orders $(n - i + 1)$ and $(n - 1)$ respectively and Z_{0i} and ω_{0i} are the characteristic impedance and resonance frequency of

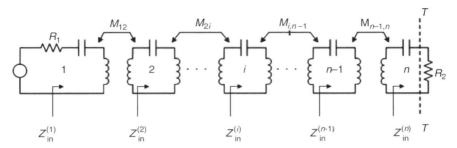

Figure 19.6   The equivalent circuit of N-cascaded resonators terminated in a short circuit.

resonator i. These two polynomials are expressed as

$$P_i(\omega^2) = \prod_{t=1}^{n-i+1} (\omega^2 - \omega_{zt}^{(i)^2}) \qquad i = 1, 2, \ldots, n \tag{19.35}$$

$$Q_i(\omega^2) = \prod_{q=1}^{n-i} (\omega^2 - \omega_{pq}^{(i)^2}), \qquad i = 1, 2, \ldots, n \tag{19.36}$$

where $\omega_{zt}^{(i)2}$ ($t = 1, 2, \ldots, n - i + 1$) and $\omega_{qt}^{(i)2}$ ($q = 1, 2, \ldots, n - i$) are the zeros of $P_i(\omega^2)$ and $Q_i(\omega^2)$ corresponding respectively to the zeros and poles of the input impedance of the one-port network at loop i.

It can be proved [24] that the resonance frequencies and couplings are related to the zeros and poles by the following equations

$$\omega_{0i}^2 = \frac{\prod_{t=1}^{n-i+1} \omega_{zt}^{(i)2}}{\prod_{q=1}^{n-i} \omega_{pq}^{(i)^2}}, \qquad i = 1, 2, \ldots, n \tag{19.37}$$

$$m_{i,i+1}^2 = \sum_{t=1}^{n-i+1} \omega_{zt}^{(i)^2} - \sum_{q=1}^{n-i} \omega_{pq}^{(i)^2} - \omega_{0i}^2, \qquad i = 1, 2, \ldots, n - 1 \tag{19.38}$$

$$r_{1,n} = \left| \frac{\prod_{i=1}^{n} (\omega_R^2 - \omega_{zi}^{(1)2})}{\omega_R \prod_{i=1}^{n} (\omega_R^2 - \omega_{pq}^{(1)2})} \right| \tag{19.39}$$

where ω_R is the frequency corresponding to a $\pm 90°$ phase of the input reflection coefficient. Equations (19.37) and (19.38) provide a simple and direct correlation between the filter parameters (the resonance frequencies of the individual resonators and the interresonator couplings) and the zeros and poles of the short-circuited network. These zeros and poles can be obtained by either direct or indirect measurement. In the case of direct measurement, where the reference plan is clearly defined, the frequencies, corresponding to a phase value of $\pm 180°$ and $0°$, are the corresponding zeros and poles of the input reflection coefficient (input impedance), while in the case of indirect measurement, the locations of zeros and poles can be determined through the calculation of phase derivatives with respect to frequency. The frequencies at which the phase derivative is maximum (minimum) are the corresponding zeros (poles) of the input impedance.

This technique is based on having a short-circuited termination directly at the end of the last resonator. However, in practice, the actual short-circuit reference plane for the last resonator is seldom accessible. In most cases, the last resonator is loaded by a transmission line and/or connector, which represents a loading on the last resonator. Hsu et al. [24] proposed an efficient systematic method to remove the loading effect on the last resonator. The loading affects only the resonance frequency of the last resonator; therefore, the effect of the loading may not be an important step for the initial tuning stage. The reader may refer to the original paper [24] for details on removal of loading effects on the last resonator.

The tuning procedure in this approach relies on the fact that there is a direct relation between the poles and zeros of the input impedance as one set and the filter resonance frequency and interresonator coupling as another set; in other words, we can extract (extrapolate) one set from the knowledge of the other set. Let us define the desired theoretical set of interresonator couplings and resonator center frequencies as $M_{12}, M_{23}, \ldots, \dot{M}_{n-1,n}, F_{01}, F_{02}, \ldots, F_{0n}$. The tuning steps are summarized here:

1. Calibrate to determine the proper location for the reference plane for accurate phase measurement. In practice, the procedure starts by inserting all the filter tuning screws deep into the cavities so that their resonance frequencies are far from the desired resonance frequencies of the individual cavities (the detuned condition). This condition is observed on the polar display of a network analyzer when the input reflection coefficient appears as a single spot, while the frequency is being swept several times the bandwidth around the desired center frequency. The phase reference plane is then adjusted to achieve the "best" spot on the sweep, which will be used to correspond to the 0.

2. The measurement and adjustment of R_1 (input coupling) is performed by bringing the first resonator into resonance, while all the remaining resonators in the filter are still in the detuned condition. The input coupling R_1 can be determined using the procedure reported in [3] or [15]. The same approach is applied for the measurement of the output coupling R_2.

3. Terminate the output port with a short circuit and look at the phase of the input reflection coefficient. Bring all the resonators into resonance by adjusting the tuning screws until all the zeros and poles are shown on the display of VNA. The ones at the left correspond to the frequencies of zeros ($\pm 180°$), and the ones at the right correspond to the poles ($0°$).

4. On the VNA, record the frequencies of all the zeros and poles (n zeros and $n - 1$ poles). Next, extract the interresonator couplings and resonance frequencies (n resonance frequencies and $n - 1$ couplings) [24]. These

extracted couplings and resonance frequencies differ from the desired ones at this point. Let us identify this set as $\{M'_{12}, M'_{23}, \ldots, M'_{n-1,n}, F'_{01}, F'_{02}, \ldots, F'_{0n})$.

5. In the extracted set of interresonator coupling and resonator center frequencies in step 4, replace M'_{12} by the desired coupling value M_{12}. The new set $\{M_{12}, M'_{23}, \ldots, M'_{n-1,n}, F'_{01}, F'_{02}, \ldots, F'_{0n})$ is then used to synthesize the zeros and poles. The obtained set of zeros and poles are used as criterion for the adjustment of the coupling between resonators 1 and 2. The idea here is that by adjusting the tuning screw that corresponds to M_{12}, such that the zeros and poles are placed to the recorded position, the coupling between resonators 1 and 2 will be set to the desired value.

6. Repeat step 5 for all interresonator couplings. For example, for the coupling between resonators 2 and 3, the set $\{M_{12}, M_{23}, \ldots, M'_{n-1,n}, F'_{01}, F'_{02}, \ldots, F'_{0n}\}$ is used to synthesize the zeros and poles. Step 5 is therefore repeated $n-1$ times, until all the interresonator couplings are set to the desired value.

7. Apply step 5 to the resonance frequencies of the individual resonators; that is, for the first resonator, use the set $\{M_{12}, M_{23}, \ldots, M_{n-1,n}, F_{01}, F'_{02}, \ldots, F'_{0n}\}$ to synthesize the new set of zeros and poles. The tuning screw for the first resonator is adjusted until the measured poles and zeros match the ones that are generated by synthesis using the above set. For the resonance frequencies step 5 is therefore repeated n times.

Overall, the number of tuning steps is $(2n-1)$ after the input and output couplings have been set correctly. Examples on the application of this approach are given in [24].

19.4 TIME-DOMAIN TUNING

This technique was first reported by Agilent [25,26]. The idea in this tuning technique is that certain features of the time-domain response of the input reflection coefficient S_{11} correspond exactly to the resonator center frequencies and the interresonator coupling elements. Figure 19.7 shows the frequency response and the S_{11} time-domain response of a perfectly tuned five-pole Chebyshev filter. The time-domain response is the inverse Fourier transform of the frequency response of S_{11}. There are five distinct dips in the time-domain response for $t > 0$. We demonstrate that each of these five dips corresponds to a resonator in the filter. The first dip, with the shorter time delay, corresponds to the first resonator, the second dip corresponds to the second resonator, and so on. Also, we demonstrate that the peaks between the dips relate to the interresonator couplings. The time domain response is basically an indication of the reflected energy; thus, the dips and peaks for $t < 0$ are not meaningful and have nothing to do with the filter elements. A detailed Agilent application note on the subject is given in [27].

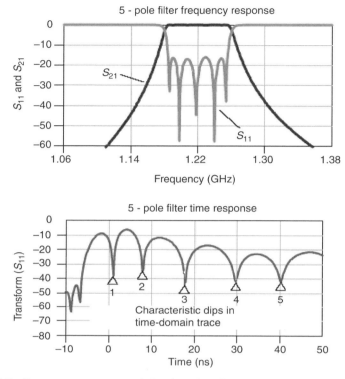

Figure 19.7 Frequency response and S_{11} time-domain response of a five-pole bandpass filter [27].

19.4.1 Time-Domain Tuning of Resonator Frequencies

In order to demonstrate the relationship between the dips in the time domain response and the individual resonators, we detune each resonator and monitor both the simulated frequency response and the time-domain response. Figure 19.8

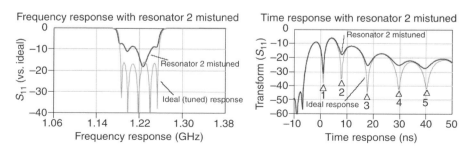

Figure 19.8 Frequency response and S_{11} time-domain response when resonator 2 is detuned [27] (the lighter trace is the ideal response; the darker trace is after detuning) [27].

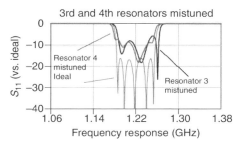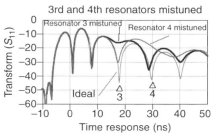

Figure 19.9 The Frequency response and S_{11} time-domain response when resonator 3 is detuned (lighter trace indicates ideal response; darker trace, after detuning) [27].

shows the frequency response and S_{11} time-domain response for the case of detuning of resonator 2 only, whereas Figure 19.9 illustrates the case of detuning resonator 3 only. The perfectly tuned ideal response of the filter is also included in these two figures. It is noted that, by detuning resonator 2, the first dip does not change very much; however, the second dip is no longer minimized, and neither are the following dips. In the case where resonator 3 is detuned, the first and second dips do not change, and the third, fourth, and fifth dips are no longer minimized. The same pattern is observed by detuning the other resonators in the filter. This demonstrates the relationship between the dips and the individual resonators. Detuning one resonator does not affect the dips associated with the preceding resonators. This example also illustrates the fact that the dips in the S_{11} time-domain response are minimized only when the resonators are perfectly tuned. Detuning in either direction causes the dips to deviate from the minimum value.

Alignment of the resonator frequencies is achieved by tuning the resonators successively, starting from resonator 1, until each dip is minimized (placed as low as possible). The adjustments of the individual resonators are mainly independent; however, for highly detuned filters, tuning a succeeding resonator may cause the dip of the previous resonator to deviate slightly from its minimum. Thus, after completing the tuning process of all the resonators, we may need to go through another round, starting from resonator 1, to fine-tune the dips to their minimum position. In high-order filters, the dips associated with the last few resonators may be not appear as distinct as those associated with the first few resonators. In this case, we use the time-domain response of S_{22} and follow the same tuning process starting from the last resonator and moving toward the middle. The first dip in the S_{22} time domain is associated in this case with the resonance frequency of the last resonator.

19.4.2 Time-Domain Tuning of Interresonator Coupling

To demonstrate the effects of interresonator couplings on the time-domain response of S_{11}, we examine what happens to the time-domain response of the five-pole filter when the first and second interresonator coupling elements M_{12} and M_{23} are varied. Figure 19.10 shows the frequency response and S_{11} time-domain response, where the

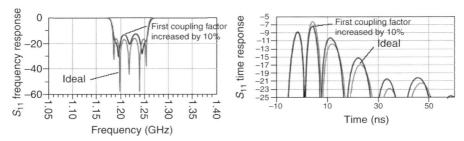

Figure 19.10   The frequency response and S_{11} time-domain response when M_{12} is increased by 10% (the lighter trace indicates the ideal response before increasing M_{12}) [27].

first interresonator coupling M_{12} is increased by 10%. The ideal response is also shown in this figure in the lighter trace. We see that the filter bandwidth is slightly wider and the return loss has changed when the input coupling M_{12} is increased. This is expected since increasing the coupling means more energy passing through the filter, resulting in a wider bandwidth. The time-domain response exhibits different peaks between the dips. The ones of interest to us are the peaks that are located at $t > 0$, namely, the peaks that are located between the dips are associated with the individual resonators.

It is noteworthy that increasing the coupling M_{12} causes the magnitude of the first peak (the one that is located after the dip associated with the first resonator) to decrease. The reduction in the level of this peak, when the interresonator coupling M_{12} is increased, makes sense, and can be explained as: Increasing the coupling means that more energy is coupled to the next resonator; thus less energy is reflected, and the peak corresponding to reflected energy from that coupling element should decrease. It is also noted that the level of the peaks that follow are higher in comparison with the ideal case. When energy is being coupled through the first coupling aperture, there is more energy to reflect off the remaining coupling apertures [27].

Figure 19.11 illustrates the case where the second interresonator coupling element M_{23} is reduced by 10%. We can see that although reducing M_{23} has no effect on the first peak, it causes the level of the second peak to increase

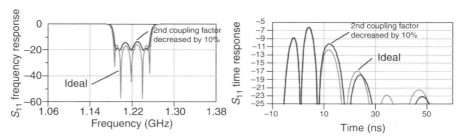

Figure 19.11   The frequency response and S_{11} time-domain response when M_{23} is reduced by 10% (the lighter trace indicates the ideal response before decreasing M_{23}) [27].

(the peak that is located after the dip associated with the second resonator). This is expected since more energy is being reflected as a result of the decreased coupling M_{23}. The peaks that follow have a lower level, since the amount of energy coupled to the following apertures is reduced.

This example demonstrates how well the time-domain response separates the effects of changing each interresonator coupling, allowing the couplings to be individually adjusted. It is important, however, to recognize that changing the first interresonator coupling M_{12} does affect the responses of all the following peaks. Consequently, we begin with the tuning of the first interresonator coupling and move toward the middle of the filter for the successive tuning of the other interresonator couplings. An improperly tuned coupling near the input can mask the real response of the inner coupling elements. As in the case of tuning the resonance frequency, we can use the time-domain response of S_{11} to tune the coupling elements that are close to the input and use the time-domain response of S_{22} to tune those elements that are located near the output.

19.4.3 Time-Domain Response of a Golden Filter

It is obvious that tuning the interresonator couplings requires the availability of a target time-domain response to use as a reference to adjust the peak level. This target time-domain response can be obtained by taking the Fourier transform of the S_{11} frequency response of an ideal filter. The simulated time-domain response is then downloaded to the vector network analyzer (VNA) and used as a template. Also, the target response can be obtained by measuring the time-domain response of an experimental filter (a "golden" filter) that is properly tuned. Each subsequent filter can be tuned to obtain the same response. However, it should be mentioned that the peak levels in the time-domain response depend on the ratio of the filter bandwidth to the frequency sweep used to compute the time-domain transform. The larger the frequency sweep/filter bandwidth ratio, the more total energy is reflected, which in turn will result in increased peak levels. Thus, during the tuning process, we need to maintain the same frequency sweep used to generate the target response. It is also essential to set the center frequency of the analyzer's frequency sweep during tuning equal to the desired center frequency of the filter. Typically a frequency span that is $2-5$ times the bandwidth of the filter is needed. A span that is too narrow does not provide sufficient resolution to differentiate between the individual sections of the filter, whereas too wide a span causes too much energy to be reflected, reducing the tuning sensitivity.

For the network analyzer time-domain setup [27], the bandpass mode must be used. The start and stop times are set so that the individual resonators can be seen. For most filters, the start time should be set slightly before zero time, and the stop time should be set somewhat longer than twice the group delay of the filter. The correct settings can be approximated by setting the start time at $t = -(2/\pi BW)$ and the stop time at $t = (2N+1)/(\pi BW)$, where BW is the filter's bandwidth and N is the number of filter sections. This should give a little extra time-domain response before the start of the filter and after the end of the filter time response. If both S_{11} and S_{22} responses are used in time-domain tuning of the filter, one can set the stop time to a smaller value, since the S_{22} response will be used to tune the resonators that are farther out in time (and closer to the output port). The format to use for viewing the time-domain response is log magnitude (dB). It is usually helpful to set the top of the screen at 0 dB.

It should be mentioned that tuning of the resonator frequencies can be accomplished without the need for the golden filter target time-domain response, since only one needs to minimize the dips. The golden filter time-domain response is needed only when tuning interresonator couplings. The time-domain technique does not deal with adjusting the input/output coupling elements R_1 and R_2. We need to use other tuning techniques to tune these elements. The technique is also not applicable for handling elliptic filters that have coupling between nonadjacent resonators. Detailed examples on time-domain tuning are given in [27].

19.5   FILTER TUNING BASED ON FUZZY LOGIC TECHNIQUES

The idea of using fuzzy logic in filter tuning was introduced for the first time to the microwave community by Miraftab and Mansour [28–30]. The idea stems from the fact that during the manual tuning process, technologists use the concept of sets to describe the various coupling elements. Nevertheless, they converge to the proper solution and manage to align the filter perfectly. Examples of these sets for the filter coupling elements are: very small, small, large, very large, and so on. Some of these technologists have the expertise to look at the measured filter response and conclude that a specific coupling element, for example, is very small, and accordingly they take the action of adjusting the tuning screw to increase the coupling of this particular element. This process is repeated several times during the tuning process. While they do not have any means to find the exact value of the coupling elements, associated with a particular measured filter response, they manage to tune the filter and converge to the desired response. This type of human thought process can be duplicated by the fuzzy logic technique, which also uses the concept of sets.

The computer-aided tuning techniques described in Sections 19.2–19.4 are based on implementing a mathematical model to guide the interpretation of the measured data. The fuzzy logic technique is capable of combining mathematical models, measured data, and human experience into one single model. This is possible since fuzzy logic systems interpret and process numerical information and linguistic (expert) information the same way.

The fuzzy logic (FL) technique itself was conceived by Zadeh [31] in 1965. It is now well adapted in many engineering fields. It has become a standard technology in industrial automation, pattern recognition, and medical diagnosis systems. Nevertheless, microwave researchers have been hesitant to use this technique because of the perceived fuzziness aspect of the concept. In fact, there is nothing "fuzzy" about this technique. It is simply a "function approximator"—actually a very good function approximator. It has been shown in [32] that a FLS can approximate any real continuous nonlinear function to an arbitrary degree of accuracy.

19.5.1 Description of Fuzzy Logic Systems

In classical Boolean logic, sets are defined in a crisp manner; an element either belongs to a set or does not belong to it. In fuzzy logic, a membership value between binary 0 and 1 is assigned to each element of the set. A value of 0 means that the element does not belong to the set at all, whereas a value of 1 means the element totally belongs to that set. Fuzzy logic interprets the numerical data as linguistic rules. Then the extracted rules will be used as a kind of system specification to calculate the output values of the system.

The fuzzy logic system can be described as a function approximator as shown in Figure 19.12. Overall, the FL systems map crisp inputs into crisp outputs. The system has four components: fuzzifier, rules, inference engine, and defuzzifier. Once the rules are established, the FL system can be viewed to perform the function $y = f(x)$. Rules can be expert rules, or rules extracted from numerical data. In either case, the rules are expressed as a collection of IF-THEN statements.

The fuzzifier maps crisp input numbers into fuzzy sets. The purpose of the inference engine is to generate the output fuzzy sets using the rules and the input fuzzy sets. There are many different types of fuzzy logic inferential procedures. However, a small number of them are usually being used in engineering applications. This is similar to the human method of making decisions, where different types of inferential procedures are used to understand things. The defuzzifier maps fuzzy output sets into crisp output numbers. This step is necessary since we need to obtain a crisp number in most of the engineering applications.

As an example of fuzzy sets, consider Figure 19.13, which shows different membership functions for the variable "age." We can interpret "age" as a variable X. This variable can be decomposed into the following sets of terms: $X(\text{age}) = \{\text{very young, young, middle age, old, very old}\}$, where each term in $X(\text{age})$ is characterized by a fuzzy set in the universe of discourse $U = \{0, 80\}$. For example, as can be seen from Figure 19.13, we can interpret a person with the

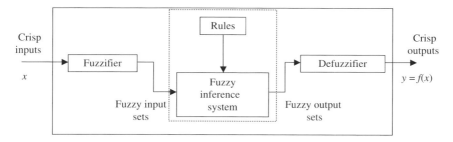

Figure 19.12 A block diagram of the fuzzy logic system.

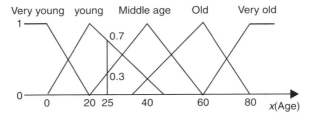

Figure 19.13   Functions for X(age) = {very young, young, middle age, old, very old}.

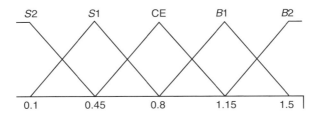

Figure 19.14   Potential membership functions for the elements of the coupling matrix.

age 25 to belong to both fuzzy sets of "young" and "middle age," but with different membership functions of 0.7 and 0.3, respectively. This example demonstrates that in fuzzy logic an element can reside in more than one set with different degrees of similarity. We can treat coupling values in a similar fashion; for example, we can select membership functions for the coupling element M_{ij}, as shown in Figure 19.14. The sets can be identified as $S2$, $S1$, CE, $B1$, and $B2$, covering the values that the coupling values M_{ij} might take.

The most commonly used shapes for the membership functions are triangular, trapezoidal piecewise linear, and Gaussian. The membership functions are designed by using an optimization procedure [30], but they are usually based on the user's experience for a particular problem. The number of membership functions is up to the designer. By choosing more membership functions, we can achieve a better resolution at the price of more computational complexity.

19.5.2   Steps in Building the FL System

To illustrate the main steps involved in the fuzzy logic technique, we consider the following example. Assume that we have the set of data given in Table 19.2. We show in this example how fuzzy logic can be used as a function approximator $y = f(x)$. Thus, the problem is: Given these sets of data pair points, use the fuzzy logic technique to calculate the value of the output y for $x_1 = 0.5$, $x_2 = 0.3$. The steps for building the fuzzy logic are as follows:

Step 1. Assigning the membership functions for all variables. Details on selecting and optimizing the shape of the membership functions are found in the literature

TABLE 19.2   Example for a Set of Data

x_1	x_2	y
0.3	0.7	0.5
0.6	0.5	0.1
0.1	0.2	0.9

[30–33]. To illustrate the concept, let us select the simplest form of membership function, which is triangular, as shown in Figures 19.15–19.17. Five membership functions ($S2$, $S1$, CE, $B2$, $B1$) are selected for variable x_1, three membership functions are selected for variable x_2, and three membership functions ($S1$, CE, $B1$) are selected for the variable y.

Step 2. Creating the IF-THEN fuzzy rules. Many approaches are available for generating fuzzy rules from numerical data [32–34]. In this section we use the method proposed in [33], since it allows the combination of both numerical and linguistic information into a common simple framework. The rules are created by using the data pairs given in Table 19.2 as well as the membership functions. The construction of the IF-THEN rule associated with the first data pair (for $x_1 = 0.3$ and $x_2 = 0.7$, $y = 0.5$) is also illustrated in Figures 19.15– 19.17. As shown in Figure 19.15, the point $x_1 = 0.7$ belongs to the set $S1$ with a membership value 0.8 and to the set CE with a membership value of 0.2. Since its membership value to set $S1$ is higher than its membership value to set CE, it is assumed that it belongs to set $S1$ (i.e., x_1 is $S1$). Similarly, one can see from Figures 19.16 and 19.17 that the point $x_2 = 0.5$ belongs to set CE and the point $y = 0.5$ belongs to the set CE. Thus the rule associated with the first data pair ($x_1 = 0.3$, $x_2 = 0.7$ and $y = 0.5$) is

> *Rule 1. IF (x_1 is $S1$) and (x_2 is CE) THEN (y is CE).*
>
> Similarly, the two other data pairs in Table 19.2 can be used to generate rules 2 and 3. Using the same concept, we can readily show that these two rules are as follows.

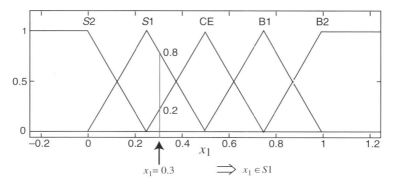

Figure 19.15   The membership function for variable x_1.

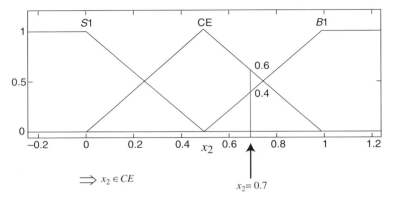

$\Longrightarrow x_2 \in CE$ $\qquad\qquad\qquad x_2 = 0.7$

Figure 19.16   The membership function for variable x_2.

Rule 2. IF $(x_1$ is $CE)$ and $(x_2$ is $CE)$ THEN $(y$ is $S1)$.

Rule 3. IF $(x_1$ is $S2)$ and $(x_2$ is $S1)$ THEN $(y$ is $B1)$.

Notice that in this step, we effectively converted the three crisp data pairs given in Table 19.2 into three IF-THEN rules.

Step 3. Fuzzy inference and defuzzification process. In this step we find the output y for the given inputs $x_1 = 0.5$ and $x_2 = 0.3$ using the centriod defuzzification theorem [29].

$$y_i = \frac{\sum_{j=1}^{K} m_j y_i^{\,j}}{\sum_{j=1}^{K} m_j}$$

where $m_j = m_j(x_1)\, m_j(x_2), \ldots,$ where y_i^j denotes the center value of the fuzzy set corresponding to rule j and output y_i. The term $m_j(x_k)$ is the membership

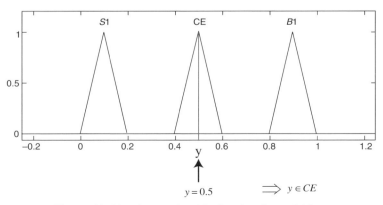

$y = 0.5$ $\qquad\qquad \Longrightarrow y \in CE$

Figure 19.17   The membership function for variable y.

value of x_k to the fuzzy set corresponding to a rule j and input x_k. K is the number of rules. Note that as long as the output membership functions are symmetric, y_i^j is simply the center of each membership function. If we chose a nonsymmetric membership, we would need to calculate the center of gravity of each membership function as y_i^j. By choosing symmetric membership functions, less computation is required. In this example, $m_1(x_1)$ and $m_1(x_2)$ are calculated by looking at rule 1 and finding the membership values of $x_1 = 0.4$ and $x_2 = 0.3$. In view of rule 1 and Figures 19.15 and 19.16, we can readily show that the point $x_1 = 0.4$ belongs to the set $S1$ by a membership value of 0.4 [i.e., $m_{S1}(0.4) = 0.4$] while point $x_2 = 0.3$ belongs to the set CE with a membership value of 0.6 [i.e., $m_{CE}(0.3) = 0.6$]. Thus we have the following set of rules:

Rule 1. IF $(x_1$ is $S1)$ and $(x_2$ is $CE)$ THEN (y is CE):

$$m_1 = m_{S1}(0.4)\, m_{CE}(0.3) = 0.4 \times 0.6 = 0.24, \quad \bar{y}_1 = 0.5$$

Similarly, one can show that for rules 2 and 3 we have

Rule 2. IF $(x_1$ is $CE)$ and $(x_2$ is $CE)$ THEN (y is $S1$):

$$m_2 = m_{CE}(0.4\, rm_{CE}(0.3) = 0.6 \times 0.6 = 0.36, \quad \bar{y}_2 = 0.1$$

Rule 3. IF $(x_1$ is $S2)$ and $(x_2$ is $S1)$ THEN (y is $B1$):

$$m_3 = m_{S2}(0.4)\, m_{S1}(0.3) = 0 \times 0.4 = 0, \quad \bar{y}_3 = 0.9$$

The value of y corresponding to $x_1 = 0.4$ and $x_2 = 0.3$ is as follows:

$$y_i = \frac{\sum_{j=1}^{3} m_j y_i^j}{\sum_{j=1}^{3} m_j} = \frac{m_1 \bar{y}_1 + m_2 \bar{y}_2 + m_3 \bar{y}_3}{m_1 + m_2 + m_3}$$

$$= \frac{0.24 \times 0.5 + 0.36 \times 0.1 + 0 \times 0.9}{0.24 + 0.36 + 0} = 0.26$$

Note that rules 1 and 2 are fired while rule 3 is not fired. In this example, the fuzzy logic system acts as a function approximator. The solution for $x_1 = 0.4$ and $x_2 = 0.3$ is $y = 0.26$.

19.5.3 Comparison Between Boolean Logic and Fuzzy Logic

In this section, we apply both fuzzy logic and Boolean logic to approximate the simple function $y = x^2 + 1$ as an example to illustrate that the logic system is in fact a good function approximator [38]. A function approximator based on Boolean logic divides all the input values into groups, and each group is associated with an output value. Since each input value must belong to a single output value

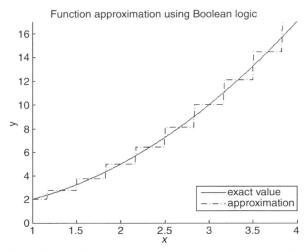

Figure 19.18   Function approximation using Boolean logic.

group, the output curve will often appear in "staircase" shape. The accuracy of the approximator depends on how the input values are grouped together and how many groups are used. However, the "staircase" nature of the approximator cannot be avoided by any means (since it's Boolean), as shown in Figure 19.18.

In this example, the Boolean logic function approximator divides the input values into 10 groups. The ranges of x values that each group covers are the same for design simplicity. The output value of each group is derived by taking the middle value of each group and calculates the corresponding output value using the function itself ($y = x^2 + 1$). In the example, 10 output values are used. The input/output values used are (1, 2), (1.33, 2.78), (1.67, 3.78), (2, 5), (2.33, 6.44), (2.67, 8.11), (3, 10), (3.33, 12.11), (3.67, 14.44), and (4, 17).

With the availability of a set of data pairs, we can follow an approach similar to that outlined in Section 19.5.2 to build a fuzzy logic system to approximate this function. On the other hand, fuzzy logic tools are available in MATLAB [36], allowing users to easily build fuzzy logic systems. Users can simply define the number of inputs and outputs of the system, and then define the number of membership functions for each input and output. MATLAB fuzzy logic tools offer a number of membership function types, such as the triangular function and the Gaussian function, for users to choose from. The evaluation of fuzzy logic system is extremely easy once all parameters (inputs, outputs, and rules) of the fuzzy inference system (FIS) are set. MATLAB's fuzzy logic tools take the input values and return the output values. All fuzzification and defuzzification processes are internal to MATLAB fuzzy logic tools.

By adjusting the shape of the input membership function, the number of rules, and the shape of the output membership function, we can develop a highly accurate function approximator using just a few data points. Figure 19.19

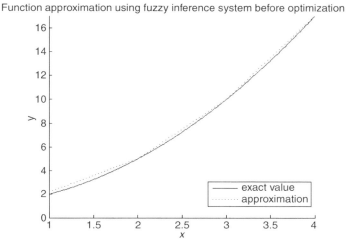

Figure 19.19 Function approximation using fuzzy logic.

shows the output of a fuzzy logic function approximator developed by the four data points.

The fuzzy logic system used to construct the approximator is a Sugeno-type fuzzy system [35] that has one input variable (x). The input x contains four triangular membership functions centered at 1, 2, 3, and 4. The four data pairs used are (1, 2), (2, 5), (3, 10), and (4, 17). Four IF-THEN rules are generated from these data points. Triangular membership functions were chosen to simplify calculation.

The fuzzy logic approximator can be improved by using Gaussin membership functions. MATLAB fuzzy logic tools are used to optimize these Gaussin membership functions. In this case more data pair points are needed. Some of these data pairs are used to optimize the Gaussian membership functions and the output values such that the output of the system will closely match the given data pairs. The process is called "training the fuzzy logic systems" and is described in [30]. In the optimized approximator, 10 data points (the same data points used to construct the Boolean logic approximator) are used to train and build the fuzzy system.

Figure 19.20 shows the output of the optimized function approximator, whereas Figure 19.21 shows the percentage error of the Boolean logic solution and the non-optimized and optimized fuzzy logic solution. As shown in Figure 19.21, the error exists in using the Boolean logic function approximator, which fluctuates between a high value (9–17%) and 0. The fuzzy logic function approximator has a maximum error of 14% and can be reduced to less than 2% with an optimized fuzzy logic system. Although increasing the number of output groups in the Boolean function approximator can improve its accuracy, the fuzzy logic function approximator is shown to be more accurate given the same number of data pairs.

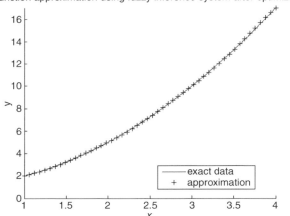

Figure 19.20   Function approximation using fuzzy logic with optimization.

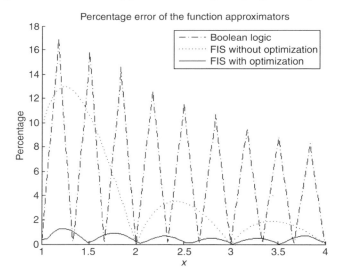

Figure 19.21   Percentage error of three different function approximators.

19.5.4   Applying Fuzzy Logic to Filter Tuning

Three fuzzy logic systems for tuning of microwave filters have been developed by Miraftab and Mansour [28–30,39,40] for filter tuning:

1. An objective fuzzy logic system where the IF-THEN rules are constructed from data pairs generated from a mathematical model such as the filter circuit model [29,30].

2. An objective FL system where the IF-THEN rules are created by tracking a human expert actions during manual tuning [39].
3. A subjective FL system where the IF-THEN rules are generated directly from a human expert is proposed in [40]. This paper also demonstrates the feasibility of using the concept to construct a robot to carry out the tuning process [40].

Details of these three techniques are given in [41]. In this chapter we briefly describe the first approach where the IF-THEN rules are generated from a mathematical model [29,30].

The mathematical model used in [29] and [30] is the coupling matrix circuit model. The fuzzy logic system in this case identifies the coupling matrix that is associated with a particular experimental filter response. The FL system acts as a function approximator $y = f(x)$, where x is the experimental performance and y is the coupling matrix. In Section 19.3 the same problem is solved by applying an optimization routine to the coupling matrix circuit model. The optimization involves many iterations and is considered an indirect approach method. The fuzzy logic technique is a direct approach. Efforts are taken initially in building the fuzzy logic system by using data pairs generated from the coupling matrix model. Once the system is constructed, it can be used directly to extract the coupling matrix associated with a particular experimental response.

We consider the M-matrix coupling coefficients as outputs and the S parameters of the filter at different frequencies as inputs. Suppose that we have p frequency sampling points, that is, p inputs and q unknown coupling coefficients as outputs. We can extract the input information from either S_{21} or S_{11}. The inputs will then be in the form $S(f_1) \cdots S(f_p)$, which can be written in the form $x_1, x_2, \ldots, x_p$ for simplicity. The outputs, which are the coupling coefficients, could also be written in the form $y_1, y_2, \ldots, y_q$ for simplicity. The coupling matrix model [equations (19.19)–(19.22)] is used here to generate the initial data pairs. We select values for each coupling coefficient around the ideal design value and generate a number of input/output data pairs:

$$(x_1^{(1)}, x_2^{(1)}, \ldots, x_p^{(1)};\ y_1^{(1)}, y_2^{(1)}, \ldots, y_q^{(1)})$$

$$(x_1^{(2)}, x_2^{(2)}, \ldots, x_p^{(2)};\ y_1^{(2)}, y_2^{(2)}, \ldots, y_q^{(2)})$$

$$\cdots$$

$$(x_1^{(n)}, x_2^{(n)}, \ldots, x_p^{(n)};\ y_1^{(n)}, y_2^{(n)}, \ldots, y_q^{(n)})$$

Once the data pairs are generated, we can employ the simple three steps described in Section 19.5.2 to build the fuzzy logic system or use commercial fuzzy logic software tools such as those of MATLAB. The IF-THEN rules are in the following form:

$$IF(x_1 \text{ is } fs_{x1}) \text{ and } (x_2 \text{ is } fs_{x2}) \ldots$$
$$and \ (x_p \text{ is } fs_{xp}), \ THEN$$
$$(y_1 \text{ is } fs_{y1}) \ldots \quad and \ (y_q \text{ is } fs_{yq})$$

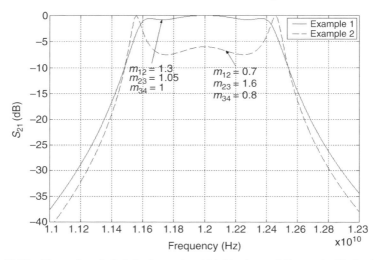

Figure 19.22   Examples of slightly detuned and highly detuned Four-pole Chebyshev filter characteristics [29].

where f_s is a fuzzy set among the fuzzy sets of each input/output variable. To illustrate the concept, we consider the tuning of a four-pole bandpass Chebyshev filter. The coupling matrix (*M*-matrix) is a symmetric 4×4 matrix ($m_{ij} = m_{ji}$) with all elements zero except m_{12}, m_{23}, m_{34}. These are the variables that need tuning; R_1 and R_2 are assumed fixed in this example. The ideal coupling values for this filter are $m_{12} = 1.2$, $m_{23} = 0.95$, $m_{34} = 1.2$. Figure 19.22 depicts S_{21} versus the frequency of two detuned filters, one with a slight deviation and the other with a high deviation from the ideal filter performance.

These two examples represent the experimental data of two detuned filters, each with two different deviations. In order to use the tuning procedure, we need to extract the *M*-matrix elements associated with the experimental results. Then, with the knowledge of the ideal coupling matrix, we can identify the elements that caused the detuning. In assigning the membership functions or fuzzy sets, we choose five fuzzy sets for each input variable. The membership functions for the output variables are selected as depicted in Figure 19.23 [28].

In this example the inputs are selected to be the magnitude of S_{21} at nine frequencies with seven frequencies inside the passband and the other two outside the passband. The fuzzy logic system in this example has nine inputs and three outputs. Seventy IF THEN rules are generated from data pairs obtained using the coupling matrix model [equations (19.19)–(19.22)]. The extracted response of the two filters (slightly detuned and highly detuned) are shown in Figures 19.24 and 19.25. The coupling elements are also shown in the same figures. The extracted responses are fairly close to the assumed experimental filter response.

The fuzzy logic technique is able to integrate theoretical models, measured data, and human experience into one model. In particular, its main advantages lies in

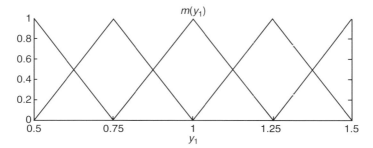

Figure 19.23 Output membership functions for the Four-pole filter.

the ability to translate human experience into a model. Details on using human experience are given in [39–41].

19.6 AUTOMATED SETUPS FOR FILTER TUNING

Figure 19.26 shows a schematic diagram for an automated filter tuning station (robot). It consists of a vector network analyzer (VNA), a computer, a motor with arms, and the filter under test with tuning screws. The filter measurements data are taken by the VNA and are transferred to the computer. The data are then processed by the tuning algorithm running on the computer, which in turn interfaces with the motor to provide the necessary tuning. Any of the five computer-aided tuning techniques described in this chapter can be automated to construct a robot for tuning microwave filters.

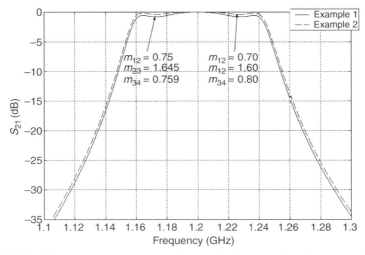

Figure 19.24 A Comparison between experimental and extracted performance using fuzzy logic for the slightly detuned filter.

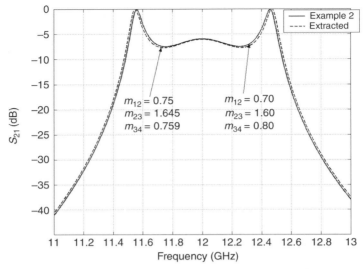

Figure 19.25   A comparison between experimental and extracted performance using fuzzy logic for the highly detuned filter.

COM DEV, one of the world's leading suppliers of filters and multiplexers has reported the first robotic system for filter tuning [42,43]. Figure 19.27 is a photograph of COM DEV's RoboCat system [43]. The system has a replaceable coaxial screw/nut driver head that allows simultaneous independent servo mechanically driven control of a tuning screw and its locknut. This capability is crucial in allowing the robot to perform the final tuning screw lockdown on sensitive products where the lockdown procedure can affect the device performance. The system is able to predict the effect of tightening the nut, and make microscale adjustments to the tuning screw position to counteract the associated detuning. The screw/nut driver can be easily swapped to handle different screw/nut combinations, and has been used on screws of various sizes. The system is fully compatible with self-locking

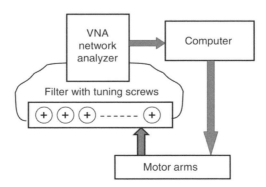

Figure 19.26   The automated tuning station.

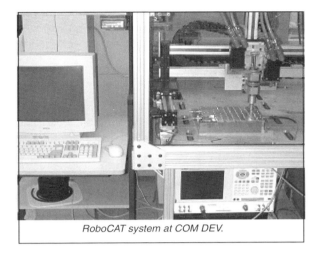

RoboCAT system at COM DEV.

Coaxial
screw/nut
driver
handles
standard
tuning screw
and lock nut.

Figure 19.27   COM DEV's RoboCat [43].

Figure 19.28   An automated filter tuning setup [40].

screws and other nutless screw designs. It is capable of tuning coaxial diplexers for wireless base stations in a few minutes. More details on the system are given in [43].

Figure 19.28 shows a robotic system for filter tuning employing several motors [40]. The system is capable of tuning filters where the tuning screws are scattered over the filter body. The screws are connected to servo/step motors by custom-designed flexible leads to connect the screws to motor shafts. Universal mounting brackets are designed to arbitrarily hold and position the motors on a main plate. This enables adjustability in three dimensions to target any position on a particular filter or multiplexer. More details on this system are given in [40].

SUMMARY

From a theoretical standpoint, the physical dimensions of a microwave filter can be perfected using EM based techniques with near arbitrary accuracy. In practice, the use of EM based tools can be very time consuming and prohibitively so for higher order filters and multiplexing networks. Moreover, a perfect EM based design cannot provide measured results that exactly match theoretical results. This is due to manufacturing tolerances and variations in the characteristics of the materials used to construct microwave filters. These problems are furthers exacerbated by the very stringent performance requirements for applications in the wireless and satellite communication systems. For these reasons, the vast majority of practical microwave filters require tuning to achieve the desired response. This chapter is devoted to the computer-aided techniques for tuning microwave filters.

By starting with a good EM-based design (Chapter 14 and Chapter 15), tuning can be invoked to achieve a measured response that meets all design requirements. Tuning is usually carried out by experienced technologists, and is considered a key bottleneck in filter production. For many decades, tuning has been considered more of an art than a science. Lately, several computer aided tuning techniques have emerged for the purpose of augmenting the science part of the tuning process.

This chapter presents five different techniques for computer-aided tuning of microwave filters. These include filter tuning procedures based on sequentially tuned resonators, circuit model extraction parameters, pole-zero configuration of the filter function, time domain response of filters, and fuzzy logic systems. The sequential and the time domain tuning techniques are useful in guiding the technologists to tune the resonators and coupling elements, one by one. Such techniques have been successfully implemented in tuning Chebyshev filters. The circuit model parameter extraction tuning technique has several variations. This technique has been successfully employed to tune large order elliptic function filters as well as diplexers and multiplexers. The fuzzy logic technique is investigated to deal with both the science and the art parts of the tuning process. The technique has been successfully employed to rough tune Chebyshev filters. An outline of other concepts on computer-aided tuning is given in [44].

This chapter also includes an outline for two automated filter tuning step-ups. The first set-up has a built-in mechanism making it possible to lock the screws once the filter is tuned. The second set-up is built with flexible arms which make it possible to tune microwave filters and multiplexers with tuning screws scattered over the 3-dimensional filter body. It also allows the tuning of several screws simultaneously, which can help considerably in speeding up the tuning process.

REFERENCES

1. M. Dishal, Alignment and adjustment procedure of synchronously tuned multiple resonant circuit filters, *Proc. IRE* **39**, 1448–1455 (Nov. 1951).

2. H. A. Wheeler, Tuning of waveguide filters by perturbing of individual sections, in *Proc. Symp. Modern Advances in Microwave Technology*, 1954, pp. 343–353.

3. A. E. Atia and A. E. Williams, Nonminimum phase optimum amplitude band pass waveguide filters, *IEEE Trans. Microwave Theory Tech.* **MTT-22**, 425–431 (April 1974).

4. A. E. Atia and A. E. Williams, Measurement of intercavity couplings, *IEEE Trans. Microwave Theory Tech.* **MTT-23**, 519–522 (June 1975).

5. A. E. Williams, R. G. Egri, and R. R. Johnson, Automatic measurement of filter coupling parameters, *IEEE MTT-S Int. Microwave Symp. Digest*, 1983, pp. 418–420.

6. L. Accatino, Computer-aided tuning of microwave filters, in *IEEE MTT-S Int. Microwave Symp. Digest*, 1986, pp. 249–252.

7. H. L. Thal, Computer aided filter alignment and diagnosis, *IEEE Trans. Microwave Theory Tech.* **MTT-26**, 958–963, (Dec. 1978).

8. T. Ishizaki, H. Ikeda, T. Uwano, M. Hatanaka, and H. Miyake, A computer aided accurate adjustment of cellular radio RF filters, *IEEE MTT-S Int. Microwave Symp. Digest*, 1990, pp. 139–142.

9. A. R. Mirzai, C. F. N. Cowan, and T. M. Crawford, Intelligent alignment of waveguide filters using a machine learning approach, *IEEE Trans. Microwave Theory Tech.* **37**, 166–173 (Jan. 1989).

10. K. Antreich, E. Gleissner, and G. Muller, Computer aided tuning of electrical circuits, *Nachrichtentech. Z.*, **28**(6), 200–206 (1975).

11. R. L. Adams and V. K. Manaktala, An optimization algorithm suitable for computer-assisted network tuning, *Proc. IEEE Int. Symp. Circuits Systems*, Newton, MA, 1975, pp. 210–212.

12. J. W. Bandler and A. E. Salama, Functional approach to microwave postproduction tuning, *IEEE Trans. Microwave Theory Tech.* **MTT-33**, 302–310 (1985).

13. P. V. Lopresti, Optimum design of linear tuning algorithms, *IEEE Trans. Circuits Syst.* **CAS-24**, 144–151 (1977).

14. C. J. Alajajian, T. N. Trick, and E. I. El-Masry, On the design of an efficient tuning algorithm, *Proc. IEEE Int. Symp. Circuits Systems*, Houston, TX. 1980, pp. 807–811.

15. J. B. Ness, A unified approach to the design, measurement, and tuning of coupled-resonator filters, *IEEE Trans. Microwave Theory Tech.* **MTT-46**, 343–351, (1998).

16. L. Accatino, Computer-aided tuning of microwave filters, *IEEE MTT-S Int. Microwave Symp. Digest*, 1986, pp. 249–252.

17. M. Yu, *Computer Aided Tuning*, COM DEV internal report, 1994.

18. M. Kahrizi, S. Safavi-Naeini, and S. K. Chaudhuri, Computer diagnosis and tuning of microwave filters using model-based parameter estimation and multi-level optimization, *IEEE MTTS Int. Microwave Symp. Digest*, 2000, pp. 1641–1644.

19. P. Harscher and R. Vahldieck, Automated computer-controlled tuning of waveguide filters using adaptive network models, *IEEE Trans. Microwave Theory Tech.* **49**(11), 2125–2130 (Nov. 2001).

20. P. Harscher, S. Amari, and R. Vahldieck, Automated filter tuning using generalized low pass prototype networks and gradient-based parameter extraction, *IEEE Trans. Microwave Theory Tech.* **49**, 2532–2538 (Dec. 2001).

21. M. Yu, Simulation/design techniques for microwave filters—an engineering perspective, *Proc. Workshop WSA: State-of-the-Art Filter Design Using EM and Circuit Simulation Techniques, Int. Symposium IEEE Microwave Theory and Techniques*, Phoenix, May 2001.

22. G. Pepe, F. J. Gortz, and H. Chaloupka, Computer-aided tuning and diagnosis of microwave filters using sequential parameter extraction, *IEEE MTT-S Int. Microwave Symp. Digest*, 2004.

23. E. K. Miller and G. J. Burke, Using model-based parameter estimation to increase the physical interpretability and numerical efficiency of computational electromagnetics, *Comput. Phys. Commun.* **69**, 43–75 (1991).

24. H. Hsu, H. Yao, K. A. Zaki, and A. E. Atia, Computer-aided diagnosis and tuning of cascaded coupled resonators filters, *IEEE Trans. Microwave Theory Tech.* **50**, 1137–1145 (April 2002).

25. J. Dunsmore, Simplify filter tuning using time domain transformers, *Microwaves RF*, (March 1999).

26. J. Dunsmore, Tuning band pass filters in the time domain, *IEEE MTT-S Int. Microwave Symp. Digest*, 1999, pp. 1351–1354.

27. Agilent application note, Agilent AN 1287–8.

28. V. Miraftab and R. R. Mansour, Computer-aided tuning of microwave filters using fuzzy logic, *IEEE MTT-S Int. Microwave Symp. Digest*, 2002, vol. 2, pp. 1117–1120.

29. V. Miraftab and R. R. Mansour, Computer-aided tuning of microwave filters using fuzzy logic *IEEE Trans. Microwave Theory Tech.* **50**, (12), 2781–2788 (Dec. 2002).

30. V. Miraftab and R. R. Mansour, A robust fuzzy technique for computer-aided diagnosis of microwave filters," *IEEE Trans. Microwave Theory Tech.* **52**(1), 450–456 (2004).

31. L. A. Zadeh, Fuzzy sets, *Inform. Control* **8**, 338–353 (1965).

32. Y. M. Kim and J. M. Mendel, Fuzzy basis functions: Comparisons with other basis functions, *IEEE Trans. Fuzzy Syst.* **43**, 1663–1676 (1995).

33. L. X. Wang and J. Mendel, Fuzzy basis functions, universal approximation, and orthogonal least-squares learning, *IEEE Trans. Neural Networks* **3**(5), (1992).

34. G. C. Mouzouris and J. M. Mendel, Non-singleton fuzzy logic systems, *Proc. 1994 IEEE Conf. Fuzzy Systems*, Orlando, FL, June 1994.

35. M. Sugeno and K. Tanaka, Successive identification of a fuzzy model and its applications to prediction of a complex system, *Fuzzy Sets Syst.* **42**(3), 315–334 (1991).

36. J. M. Mendel, Fuzzy logic systems for engineering, *Proc. IEEE* (Special Issue on Engineering Applications of Fuzzy Logic) **83**(3), (March 1995).

37. P. M. Larsen, Industrial applications of fuzzy logic control, *Int. J. Man Machine Studies.* **12**(1), 3–10 (1980).

38. K. T. Lau, *Function Approximation Using Fuzzy Logic*, technical report, Univ. Waterloo, 2005.

39. V. Miraftab and R. R. Mansour, tuning of microwave filters by extracting human experience using fuzzy logic, *IEEE MTT-S Int. Microwave Symp. Digest*, June 12–17, 2005, pp. 1605–1608.

40. V. Miraftab and R. R. Mansour, Automated microwave filter tuning by extracting human experience in terms of linguistic rules using fuzzy controllers, *IEEE MTT-S Int. Microwave Symp. Digest*, San Francisco, CA, June 2006, pp. 1439–1442.

41. V. Miraftab, *Computer Aided Tuning and Design of Microwave Circuits Using Fuzzy Logic*, Ph.D. thesis, Jan. 2006.

42. M. Yu and W. C. Tang, A fully automated filter tuning robots for wireless basestation diplexers, *Workshop Computer Aided Filter Tuning*, *IEEE Int. Microwave Symp.* Philadelphia, June 8–13, 2003.

43. M. Yu, Robotic computer-aided tuning, *Microwave J.* 136–138 (March 2006).

44. M. Mongiardo and D. Swanson, *Workshop on Computer Aided Filter Tuning*, *IEEE Int. Microwave Symp.*, Philadelphia, June 8–13, 2003.

CHAPTER 20

HIGH-POWER CONSIDERATIONS IN MICROWAVE FILTER NETWORKS

This last chapter is devoted to an overview of high-power considerations for microwave filters and multiplexing networks in terrestrial and space applications. Emphasis is placed on the practical aspects in the design of high-power filter networks.

20.1 BACKGROUND

Microwave breakdown in gases under various physical conditions is a well-known and well-studied problem. A classical summary of relevant results is given by MacDonald [1]. Another phenomenon, which is dormant under normal pressures, shows up when the gas pressure is reduced to values approaching near vacuum. This is called *multipaction*. A summary of this is presented by Kudsia et al. [2]. Just as significant is the phenomenon of passive intermodulation (PIM) in high-power equipment. It depends on the materials used in the equipment, mechanical design features, and the standards of work quality. Novel microwave structures and topologies, and continuing demands for higher power, give rise to new situations and parameter regimes where previous results are no longer applicable. The problem is further compounded by the use of new materials in the implementation of microwave components. Above all, when it comes to the practical aspects of high-power breakdown, passive intermodulation and the operating environment, it seems that each generation of microwave engineers tends to repeat the mistakes of the past before coming to grips with such issues. In this chapter, we briefly

Microwave Filters for Communication Systems: Fundamentals, Design, and Applications,
by Richard J. Cameron, Chandra M. Kudsia, and Raafat R. Mansour
Copyright © 2007 John Wiley & Sons, Inc.

review the phenomenon of breakdown and then describe in more detail the practical considerations in the design of high-power filter networks.

There are two application areas that require high power for wireless systems at microwave frequencies:

- Radar systems
- Communications systems

Radars make use of pulsed power with low duty cycles and require low to moderate bandwidths. Another feature of radars is the requirement of frequency agility, and the ability to cope with jamming. Typically, peak pulsed power requirements are in the kilowatt range, and CW (continuous-wave) power is in the 100s of watts range. Various pulse compression techniques are employed to achieve a good range and Doppler resolutions. High-power equipment is optimized for pulse operation and the bulk of high-power radar systems are for military applications. Since the main focus of this book is on nonmilitary systems, we focus on the high-power equipment required for communication systems.

20.2 HIGH-POWER REQUIREMENTS IN WIRELESS SYSTEMS

As described in Section 1.3, the capacity of a communication channel is governed by the fundamental Shannon–Hartley theorem, which relates the capacity to the bandwidth and the signal-to-noise ratio (SNR) within the channel bandwidth. For a given allocation of bandwidth, the communication capacity is determined by the achievable SNR. The minimization of noise is limited by the ever-present thermal noise; therefore, the ultimate capacity of a system depends on the available power for radio transmission. Does this mean that we can indefinitely increase the capacity by simply increasing the level of RF power? There are two problems associated with that: (1) the generation of RF power is expensive and (2) the physical aperture of an antenna imposes a limit on the amount of energy that can be focused toward a distant receiver. This implies that only a limited amount of radiated energy reaches the intended destination; the rest is lost in space. What this means is that there is an inherent inefficiency in the radio transmission of energy. In addition, the spillage of energy causes interference in other competing systems. The level of such interference limits the power that can be used by a system. As discussed in Section 1.2.2, frequency allocation and the allowable levels of interference are strictly controlled by government agencies to ensure a viable multiple communication system environment. Thus, generation and transmission of RF power for communication systems is both costly and a source of interference. In addition, there are always limitations of technology in the efficient generation of RF power. For handheld mobile phones and other devices, there are health-related concerns for the potential harmful effects of microwave radiation on human

TABLE 20.1 High-Power Requirements for Terrestrial Systems[a]

System Type	Typical Power Requirements (W)
PCS	0.1–1
Cellular—handheld	0.6
Mobile unit	4
Cell sites	60
Line of sight (LOS)	1–10
CATV	100

[a]Frequency allocations are in the 400, 800, 900, 1700, 1800, and 1900 MHz frequency bands. Table 1.3 includes more details.

TABLE 20.2 High-Power Requirements for Typical Satellite Systems

Satellite System (GHz)	Transponder Bandwidths (MHz)	Typical Power Requirements (per Transponder)	
		Uplink (Earth Station) (W)	Downlink (Satellite) (W)
1.5	10s of kHz Uplink 0.5–5 MHz Downlink	1–10	150–220
6/4	36, 54, 72	500–3000	10–100
14/11	27, 54, 72	500–3000	20–200
30/20	36, 72, 112	100–600	5–80

beings. The typical power levels of high-power transmitters in use for terrestrial and for space systems are described in Tables 20.1 and 20.2, respectively.

20.3 HIGH-POWER AMPLIFIERS (HPAs)

In a microwave system, the output amplifier can be a solid-state power amplifier (SSPA), a traveling-wave-tube amplifier (TWTA), or a klystron. They differ in the maximum RF power delivered, as plotted in Figure 20.1 [3,4].

SSPAs are widely used for cellular and line-of-sight (LOS) wireless systems, whereas TWTAs dominate satellite systems. Klystrons are used when very high power is required. Earth stations use klystrons for high power uplinks, and ground-based radar systems use them in a pulsed mode operation.

The efficiencies of TWTAs and klystron amplifiers are in the range of 30–50%. For space applications, TWTAs have been developed for efficiencies as high as 60%. Typically, SSPAs are limited to efficiencies of 20–35%. Even with efficiencies in the 50% range, HPAs dissipate a great deal of heat. This represents a major challenge for the design of high-power equipment for airborne and satellite applications.

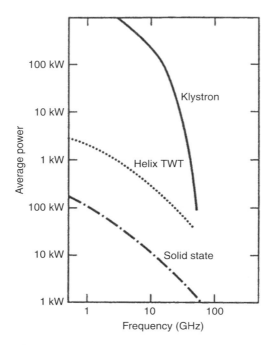

Figure 20.1   Approximate limits of microwave power sources: solid state, TWTA, and klystron. (From Ref. 4.)

20.4   HIGH-POWER BREAKDOWN PHENOMENA

High-power microwave equipment consists of filter networks, circulator devices, couplers, power dividers and combiners, feed elements, and transmission lines. All of these components have one thing in common—they make use of air or other gaseous materials inside the structures to prevent electrical breakdown. For some applications, solid dielectric materials are used to reduce the size of the equipment. For space applications, enclosed structures are either vented to achieve a near-vacuum condition or hermetically sealed with some inert gas. For many applications, gaseous breakdown plays a critical role in the design of high-power microwave equipment.

The phenomena associated with electrical discharges in gases have attracted a great deal of attention, which has led to many discoveries. Initial studies dealt with dc voltages, or relatively low ac voltages with applications in power generation and distribution. The advent of radar and communication systems has required the use of high power at much higher frequencies. Use of relatively high frequencies (MHz–GHz range) at high-power levels lead to some new and unique effects. This section reviews the high-power breakdown phenomenon at microwave frequencies, and the impact on the design of high-power passive microwave components.

20.4.1 Gaseous Breakdown

Any gas consists of atoms and molecules that are neutral particles. Also, there are a small number of electrons and charged particles present in any gaseous volume because of ionization by cosmic rays or some other phenomena, such as the photo-electric effect. When an electric field is applied to a gas, electrons and charged particles are accelerated in the direction of the field, colliding with neutral particles in their path and the sides of the container. Electron collision constitutes the dominant effect. Ions are much heavier, and slow to accelerate, resulting in fewer collisions and imparting far less energy than electrons in such collisions.

Microwave breakdown in gases is explored in great detail by MacDonald [1]. The ionization of air and other gases at microwave frequencies is described in terms of collision frequency (mean free path), diffusion, attachment, and the nature of the gas.

20.4.2 Mean Free Path

The mean free path concept is based on the classical kinetic theory of gases. It assumes that electrons, atoms, molecules, and ions are rigid elastic spheres of very small dimensions moving freely and in a random manner throughout a gas. These particles collide with each other like billiard balls. The distances between these collisions are the free paths, and the average of the lengths of these paths is the mean free path. Since we are dealing with millions and billions of such particles in an enclosed volume of gas, it is possible to compute the collision probability and frequency with a high degree of accuracy by using statistical mechanics. It is assumed that all the particles are moving at the same speed, and that all the directions are equally probable. The mean free path inherently depends on the concentration of the atoms and molecules in the gas, determined by its pressure.

There are two types of collisions: elastic and inelastic. In an elastic collision, the electrons or ions bounce off an atom with an exchange of momentum but no change in the state of the atom. Free electrons move in the direction of the field, constituting a small current. The gas envelope retains its mix of largely neutral particles. In an inelastic collision, the energy of the electron is sufficiently high to cause a change in the internal state of the atom at the expense of its kinetic energy. An excited atom normally returns to its ground state very soon after collision, and the energy that it has acquired is radiated. If the electric field is high enough, some of the electrons, on collision, knock off an electron from the atom, producing a secondary electron and a positive ion. When the electric field is strong enough for a sustained period of time, such collisions will occur often and the production rate of electrons will become greater than the loss rate due to diffusion or recombination, and breakdown will occur. The excess production rate needs to be only infinitesimally greater than the loss rate for the concentration to increase very rapidly, leading to breakdown. Figure 20.2 illustrates the variation of the electron concentration as a function of the applied electric field.

It is evident that when the electric field is gradually increased, the gas initially appears to obey Ohm's law until it starts to produce secondary electrons. When that happens, the electron concentration can vary widely as shown by the flat

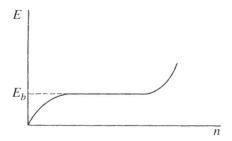

Figure 20.2   Variation of electron concentration versus electric field. (From Ref. 1.)

portion of the curve, and the gas becomes conductive. Thereafter, even a minute change in the electric field can produce orders-of-magnitude change in the electron concentration and flow of current, leading to glow and breakdown.

20.4.3 Diffusion

In a gaseous medium, any gradient in particle concentration or velocity causes the flow of particles in directions that reduce such gradients. This flow of particles is called *diffusion*. In ionized gas, electrons are scattered by the atoms and molecules as they move in the direction of the applied field. The surfaces of the structure containing the gas eventually absorb the scattered electrons. This loss of electrons is attributed to the process of diffusion. The rate of diffusion depends on the electron concentration, field gradients, the rate at which the electrons are produced, and the geometry, size, and surface conditions of the container. In addition, the rate depends on the interaction between the electrons and ions, although the dominant factor is the free diffusion of electrons.

20.4.4 Attachment

An electron can become attached to neutral particles of gas. Once that happens, the attached electrons cease to play any significant role in the ionization process. The reason is that the neutral particle is at least 2000 times heavier than an electron, and therefore, its velocity is much lower than that of free electrons. It is tantamount to a loss of an electron. It should be noted that this loss of an electron is different from the diffusion process, which removes an electron from the region in which the field exists. The attachment rate varies for the various gases and is relatively high for oxygen. The nature of atoms and molecules of the gas governs the attachment process.

20.4.5 Breakdown in Air

There has been a longstanding view that the breakdown strength of air is 29 kV/cm. It is possible that this number originated from power engineering and is more appropriate to dc and low-frequency RF engineering practices. Up until the 1960s, it was possible to find this value stated in well-known microwave texts [5–9].

It has taken a long time for this rule of thumb to be eradicated from the toolsets of microwave engineers in favor of a more accurate description that takes into account the interaction between the microwave field and an air molecule. Since the early 1960s, with the publication of works by MacDonald [1], there has been a revision of the 29-kV/cm rule derived from better theoretical understanding and more experimental data. In a given volume of gas, the electron concentration is given by [1]

$$n = n_0 \, e^{((v_i - v_a) - D/\Lambda^2)t} \tag{20.1}$$

where n = accumulated electron concentration

$\quad n_0$ = initial concentration

$\quad v_i$ = ionization rate

$\quad v_a$ = attachment rate

$\quad D$ = diffusion coefficient

$\quad \Lambda$ = characteristic diffusion length

The condition where the net rate $(v_i - v_a)$ is just balanced by the rate of diffusion, D/Λ^2, represents a threshold below which breakdown cannot occur. If, on the other hand, the net ionization rate, that is, the exponent of equation (20.1), becomes greater than zero, the electrons accumulate rapidly within the volume and breakdown follows.

An analysis of the breakdown data for air [1] indicates that in a high-pressure regime, which means above one-tenth of the atmospheric pressure (i.e., 76 Torr, where 1 Torr = 1 mm Hg) or greater, the electric field for breakdown is virtually independent of frequency and the characteristic diffusion length, and is represented by the value

$$\frac{E}{p} = 30 \, \text{V}/(\text{cm-Torr})$$

Figure 20.3 demonstrates that at 9.4 GHz, the influence of the characteristic diffusion length Λ diminishes as the pressure is increased, whereas the breakdown strength of air converges at 30 V/(cm-Torr). The same conclusions are drawn from the data taken at 992 MHz. From this, it is inferred that breakdown is an attachment-controlled process. At the normal atmospheric pressure of 760 Torr, the breakdown strength of air E_b becomes

$$E_b = 30 \times 760 = 22.8 \, \text{V/cm}$$

It is interesting to see that this threshold level is significantly lower than the commonly used value of 29 kV/cm. Figures 20.4 and 20.5 describe the experimental data on the breakdown voltages at 992 MHz and 9.4 GHz for air, oxygen, and nitrogen.

20.4.6 Critical Pressure

At low pressures, there are fewer collisions between electrons and gas molecules. The electrons oscillate freely, and mostly, out of phase with the applied field. As the pressure is increased, the number of collisions also increases. This tends to reduce the phase lag between the electrons and the applied field owing to the increased collisions, encountered by the out-of-phase electrons. The resulting

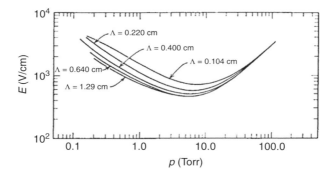

Figure 20.3 Breakdown strength of air at 9.4 GHz [1].

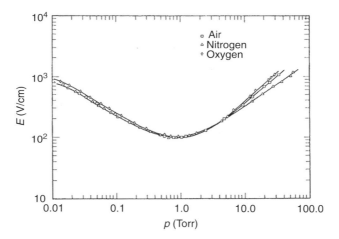

Figure 20.4 CW breakdown in air, oxygen, and nitrogen at 992 MHz [1].

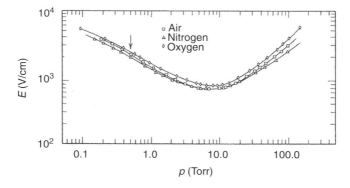

Figure 20.5 CW breakdown in air, nitrogen, and oxygen at 9.4 GHz [1].

motion provides a more efficient transfer of energy. The transition from a few collisions per oscillation of the field to many collisions per oscillation is pivotal, and occurs when the electron collision frequency equals the frequency of the applied RF. The minimum value of the breakdown field strength occurs at a pressure where the radio frequency is equal to the gas collision frequency. For air at microwave frequencies, this pressure varies linearly from a pressure of nearly 1mm Hg at 1 GHz to about 10 mm Hg at 10 GHz, as illustrated in Figures 20.4 and 20.5, respectively. At lower pressures, the breakdown field strength increases rapidly. High-power microwave equipment is often designed to withstand critical pressure even though a communication system is never intended to operate under such conditions.

20.4.7 Power Rating of Waveguides and Coaxial Transmission Lines

The high-power rating of microwave equipment depends on two factors, namely, voltage breakdown and the permissible rise in temperature. Under CW operation, either one of these two conditions dominate, depending on the peak and average power, and the operating environment of the equipment. For air-filled waveguides, the voltage gradient at which breakdown occurs is approximately 22.8 kV/cm, as described in the previous section, and is based on the following assumptions [1,9]:

1. The distance between the walls of the waveguide is much greater than the mean free path of the electrons.
2. The air inside the waveguide is still and dry.
3. The air pressure is one atmosphere.
4. Only the dominant mode is present under matched conditions [i.e., the voltage standing-wave ratio (VSWR) is unity].

Under these conditions, the theoretical power handling capability of rectangular and circular waveguides, operating in their respective dominant modes, is given by [9]

$$P_{max}(TE_{10}) = 0.34 \ a \ b \ \frac{\lambda}{\lambda_g} MW \quad \text{and} \quad P_{max}(TE_{11}) = 0.26 \ D^2 \frac{\lambda}{\lambda_g} \ MW \quad (20.2)$$

where a and b are dimensions of the rectangular waveguide and D is the diameter of the circular waveguide in centimeteres; λ and λ_g represent the free space and guide wavelengths. The power handling capability of a matched coaxial line is derived from [9]

$$P = \frac{E_m^2}{480} \left(\frac{b}{2.54} \right)^2 \frac{\ln b/d}{(b/d)^2} \ W$$

$$= 0.17 \ d^2 \ln \left(\frac{b}{d} \right) MW \quad (20.3)$$

E_m represents the breakdown voltage of 22.8 kV/cm; d and b are the inner and outer diameters of the coaxial line in centimeters.

20.4.8   Derating Factors

The power carrying capacity of microwave equipment depends on a number of factors, none of which can be precisely assessed [10]. The basic phenomenon of gaseous breakdown is not spontaneous but has to undergo the process of creation of sufficient electrons, which are accelerated with enough energy to produce ionizing collisions. An important stage of this process is the creation of effective electrons. The number of electrons created in a breakdown gap by cosmic rays and natural radioactivity is of the order of 10 per minute in a microwave component. Some of these electrons are captured by molecules or by the walls of the components before they can be accelerated to produce new electrons.

Whether an electron is effective depends on its initial velocity, position, and phase at the time of the electron's creation. All of these factors are random. Even under CW power, it is not possible to place a precise value on either the breakdown power or the time necessary to achieve breakdown with a given power. Besides these uncertainties, we need to take into account variables such as VSWR, moisture and other impurities inside the equipment, air pressure (altitude), the presence of any spurious modes, and any discontinuities in the structure of the components. All of these factors tend to degrade the power handling capability. The cumulative effect can reduce the capacity to 5–10% of the theoretical maximum value. A practical guide of derating factors is shown in Figure 20.6.

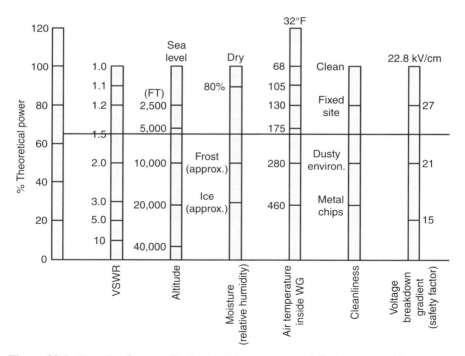

Figure 20.6   Derating factors affecting the high-power capability in waveguide structures (VSWR—voltage standing-wave ratio). (From Ref. 10.)

Breakdown margins can be improved by employing pressurized systems with the concomitant penalty of increased cost and decreased reliability. The use of electronegative gases such as Freon, with high attachment coefficients, can provide a much higher breakdown voltage. However, such gases are toxic and are not used in commercial systems.

20.4.9 Impact of Thermal Dissipation on Power Rating

Under CW power, the transmission lines and microwave components dissipate significant amounts of heat. As a consequence, the permissible rise of temperature is an important criterion in specifying the power handling capability. The thermal design must ensure an acceptable increase in the temperature of high-power equipment. This implies consideration of techniques to cool the equipment, including radiation cooling by using enlarged cooling surfaces, or by employing air or water cooled systems in a terrestrial environment. The theoretical values of power ratings for waveguides for terrestrial application with temperature rise as a criterion are given by King [8] and illustrated in Figure 20.7. VSWR and imperfections in the materials create a concentration of energy dissipation or hotspots, degrading the power handling capability. More in-depth coverage of the derating factors for

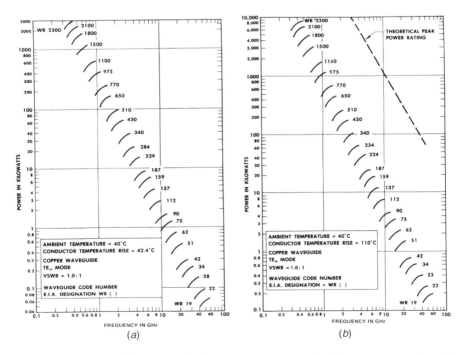

Figure 20.7 Theoretical curves of the average power rating for a copper rectangular waveguide operating in the TE_{10} mode with unity VSWR at an ambient temperature of 40°C: (*a*) for a temperature rise of 62°C; (*b*) temperature rise of 110°C. (From Ref. 8.)

waveguides and other microwave components are found in the literature [8,9]. One conclusion that can be drawn from these data is that the CW power handling capability for ground based equipment is governed by the specified temperature rise of the equipment rather than the limitation imposed by voltage breakdown.

20.5 HIGH-POWER BANDPASS FILTERS

High-power equipment is designed to have low loss to conserve power and, at the same time, minimize thermal dissipation. The requirement of low loss necessitates use of high-Q microwave structures for such equipment. For bandpass filters, the problem is further exacerbated owing to the requirement of relatively narrow bandwidths in most communication systems. Narrow bandwidths imply higher losses and electric fields inside the filter cavities. As a consequence, the power handling capability of high-Q, narrowband filters represents the most severe problem in the design of high-power equipment. Microwave filters are therefore limited in their power handling capacity by the high fields generated inside a filter as well as the heating of the cavities due to thermal dissipation. Cohn [11] was the first to discuss general design considerations for high-power filters. He provided a simple analytic expression for the peak electric fields inside the various cavities of a "narrowband" direct-coupled waveguide filter in terms of its lowpass prototype parameters. These formulas described the variation of electric field as a function of frequency in each cavity. Cohn [11] described how the variation of peak internal fields between the cavities can be minimized by enlarging certain cavities at the expense of an increased number of spurious resonances.

Young [12] introduced an alternative approach by analyzing the power ratios in a filter cavity in terms of the group delay of the filter. This analysis is based on the assumption that a direct-coupled cavity filter resembles a "periodic structure." With that assumption, it can be readily shown that the stored energy in a filter cavity is proportional to its group delay. This analysis extended the computation of electric fields to direct-coupled "wideband" filters. This approach has been resurrected and extended in depth by Ernst [13]. Rigorous expressions for the time-averaged stored energy (TASE) in passive lossless two-port networks have been derived and shown to be proportional to the group delay for Butterworth, Chebyshev, and elliptic function filters. This allows a rigorous justification for using the group delay as an indicator of the power handling capability of a filter. This approach highlights the sensitivity of high-power to a range of filter parameters, and specifically, to the selectivity of filters. Peak internal fields are very sensitive to the bandwidth and selectivity requirements of filters. The smaller the absolute bandwidth, the higher the group delay and electric fields inside it. Similarly, the higher the selectivity by virtue of having transmission zeros close to the passband, or a higher order of filter, the sharper the rise in group delays and peak internal fields. Stored energy in a resonator can be computed using circuit models derived from the lumped lowpass prototype filter. The stored energy in practical bandpass filters can be related to the original prototype filter by frequency transformation. For direct coupled ladder filters, TASE of individual cavities can be readily computed in terms of the lowpass prototype parameters [13]. The next step is the computation of electrical fields in terms of the stored energy. For standard rectangular or circular waveguides, this can be done analytically. This approach simplifies the analysis of power handling capability of filters realized in standard topologies, such as ladder networks. For more complex structures involving discontinuities in the filter cavities, a more rigorous EM analysis may be required.

A general procedure to estimate the voltage in filters based on the lumped element prototype without any restrictions has been introduced by [14,15]. It highlights the fact that it is possible to realize a variety of practical topologies (in theory, an infinite number if all cross coupling are available) having the same filter response. The total stored energy remains the same in each topology; it simply gets redistributed among the various resonators. It leads to the possibility of a tradeoff between the topology of the filter and the peak electrical fields inside the resonators [13,16]. In practice, the topology tradeoff must involve design simplicity and the sensitivity of the structure over the operating environment and that puts a limiting constraint on the design options. This analysis allows computation of peak voltages in a variety of filter structures, including diplexers and multiplexers.

20.5.1 Bandpass Filters Limited by Thermal Dissipation

The power of ground based transmitters is typically in the range of 100s of watts to kilowatts of RF power. Such high-power generates lots of heat due to losses in the equipment and, especially, in the narrow band channel filters. Thermal dissipation is minimized by using high Q structures. The Q of a cavity depends on its physical dimensions as well as the propagation mode. Use of larger cavities and low loss propagation modes is typical in the design of high-power narrow band bandpass filters. However, lower loss is achieved at the expense of a larger number of spurious modes. Suppression of spurious modes is therefore an integral part of the design. Another design feature is to locate the hot spots in the filter and provide an efficient path for the conduction of heat to the thermal radiators, within the constraints of maintaining good thermal stability of the overall filter. This can be achieved by constructing the filter cavities out of thermally stable silver plated invar. The hot spots are centered at the coupling irises. By making irises of thick copper provides an efficient means to remove the heat away from the filter with a minimal impact on the overall thermal stability of the filter. The copper irises can be readily integrated with cooling fins. Such design features are typical in the design of high power filters and mulitiplexers for satellite earth stations [17,18]. For Cellular base stations operating in the 800 MHz range, dielectric loaded cavities are preferred in order to reduce the size without too much of a sacrifice in the unloaded Q of the filter. This poses a challenge for the removal of heat under high-power. Use of TM110 mode dielectric resonators and a monoblock construction has been used to develop thermally stable high-power dielectric filters [19]. The monoblock resonator is made up of a high dielectric constant low loss ceramic. It is encased in a surrounding ceramic which is silver plated on the outside providing a smooth contact with the metal housing. Such a construction provides a continuous thermal diffusion path from the resonators to the housing. At times it is necessary to use fans to conduct the heat away or even employ water cooled systems. For space applications, the only means available to get rid of the heat is via conduction or radiation. However, the power requirement is typically much lower than that required for the ground based equipment.

20.5.2 Bandpass Filters Limited by Voltage Breakdown

For space based applications, filters and multiplexers are subjected to the phenomenon of multipaction, described in the subsequent Section 20.6. It severely limits the breakdown voltages to much lower values if the equipment is to operate under conditions of hard vacuum. In addition, heat due to thermal dissipation must be carried away via conduction and radiation. As a consequence, evaluation of peak electric fields inside the filters, especially the narrow bandwidth channel filters is a critical requirement. This analysis is equally applicable for high-power ground equipment such as diplexers used in Cellular base stations (see Section 18.5.1).

An exact analysis of the voltage and power distribution inside a filter involves the 3D analysis of the filter structure using methods such as the finite element method (FEM) or the finite difference time domain (FDTD) method. However, such analysis is time consuming and impractical if the filter in question is very narrow band owing to the extreme sensitivity of the filter performance on dimensions. A good alternative is to compute voltages inside the filter based on the lumped element prototype network (coupling matrix representation) of a band pass filter. This is an efficient procedure and often sufficient to evaluate the peak voltages. If necessary, the cavities with the highest energy can be analyzed more rigorously by using commercially available FEM software. It allows an accurate assessment of the hot spots owing to the specific geometry and coupling structures.

20.5.3 Filter Prototype Network

Bandpass filters of narrow to moderate bandwidth can be represented by a prototype network consisting of N resonators coupled through admittance invertors [20,21]. This technique to analyze and synthesize filter networks is described extensively in Chapter 8. Here, we provide an adaptation of the technique for the evaluation of voltage distribution inside filters and multiplexers.

Figure 20.8 shows the generic lowpass filter representation wherein there are N unit capacitances cross coupled through frequency invariant admittance invertors of admittances M_{ij}. A bandpass filter with center f_0 and bandwidth BW can be obtained by using the lowpass to bandpass transformation namely $s \leftarrow \frac{f_0}{BW}\left(\frac{s}{\omega_0} + \frac{\omega_0}{s}\right)$, where $s = j\omega$ is the complex frequency variable and $\omega_0 = 2\pi f_0$. Thus, the bandpass filter network would consist of N lumped LC resonators with $N + 2$ nodes, wherein

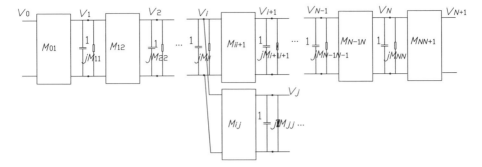

Figure 20.8 General admittance invertor cross coupled lowpass filter prototype.

the nodes designated 0 and $N + 1$ are the input and output respectively. The frequency invariant shunt susceptance M_{ij} across each resonator represents the tuning state of the resonator. Losses can be introduced into the network by shunting each resonator with a conductance $(f_0/BW)/Q_i$, where Q_i is the unloaded quality factor of the resonator. This network problem can be solved by writing down the nodal system of equations in the form

$$\mathbf{Y}\mathbf{v} = \mathbf{i} \tag{20.4}$$

where

$$Y_{pp} = s + jM_{pp}, \quad p = 1, \ldots, N$$

$$Y_{pp} = 0, \quad p = 0, \quad N + 1$$

$$Y_{pq} = jM_{pq}, \quad p \neq q$$

$$\mathbf{v} = [V_0, V_1, \ldots, V_N, V_{N+1}]^T$$

$$\mathbf{i} = [I_0 \quad 0 \cdots 0 \quad I_{N+1}]^T$$

where I_0 and I_{N+1} are the filter port excitation currents. The two-port S parameters of the filter can be computed from the two-port impedance matrix $\mathbf{Z}_f$ of the filter. This can be done by first evaluating the impedance matrix of $\mathbf{Z} = \mathbf{Y}^{-1}$ of the system and then selecting the corner elements of the $\mathbf{Z}$ matrix to form the filter two-port impedance matrix $\mathbf{Z}_f$,

$$\mathbf{Z}_f = \begin{bmatrix} Z_{0,0} & Z_{0,N+1} \\ Z_{N+1,0} & Z_{N+1,N+1} \end{bmatrix} \tag{20.5}$$

20.5.4 Lumped To Distributed Scaling

In practice, the resonators are realized as a waveguide or coaxial transmission line cavity and therefore an equivalence between a lumped and a distributed resonator has to be established for voltage computation. Here, we assume that the resonator is realized as an $n\lambda_{g0}/2$ long shorted cavity, where n is the number of half cycle variations along the axis of the cavity and λ_{g0} is the guide wavelength at f_0. Although both the prototype LC resonator and the transmission line cavity are resonant at f_0, their susceptance slopes are respectively $(\pi\,\mathrm{BW})^{-1}$ and $\frac{n}{2}(\lambda_{g0}/\lambda_0)^2/f_0$. Therefore, if the susceptance slope of the lumped resonator can be made equal to that of the cavity resonator, the lumped element prototype analysis becomes equivalent to that of a distributed transmission line prototype. As a consequence, the unit capacitances in the lowpass prototype have to be scaled by a factor $T = n\frac{\pi}{2}\left(\frac{BW}{f_0}\right)\left(\frac{\lambda_{g0}}{\lambda_0}\right)^2$. This scaling of the unit capacitance values in the prototype is carried out in such a way that the two-terminal filter response of the system in equation (20.4) is unchanged. This can be done by multiplying all the rows and columns in the **Y** matrix except the first and last by $\sqrt{T}$. This scales the internal (resonator) voltages V_1 to V_N by $1/\sqrt{T}$. Therefore, the resonator voltages in a cavity filter can be obtained by first solving the lumped prototype problem from equation (20.4) and then by scaling the lumped resonator voltages V_1 to V_N by $1/\sqrt{T}$.

20.5.5   Resonator Voltages from Prototype Network

The general case of a filter excitation can be treated by considering a terminating impedance Z_L at the output port when the input is driven by a unit current. On treating the network as an $N + 2$ port with an impedance matrix of $\mathbf{Z} = \mathbf{Y}^{-1}$, the voltage at the kth resonator is given by

$$V_k = I_0\left(Z_{k,0} - \frac{Z_{k,N+1}\,Z_{N+1,0}}{(Z_L + Z_{N+1,N+1})}\right) \qquad (20.6)$$

Similarly V_0 may be determined as:

$$V_0 = I_0\left(Z_{0,0} - \frac{Z_{0,N+1}\,Z_{N+1,0}}{(Z_L + Z_{N+1,N+1})}\right) \qquad (20.7)$$

thus allowing the ratio V_k/V_0 to be evaluated in terms of the elements of the impedance matrix [equation (20.5)]. The generator voltage V_{gen} and the incident voltage V_0 are related as follows [see equation (8.26a)]:

$$V_0 = V_{\mathrm{gen}}(1 + S_{11})/2 \qquad (20.8)$$

If the generator impedance $Z_S = Z_L = 1\ \Omega$, then at a frequency of perfect transmission through the lossless purely reactive network the source and load will be matched ($S_{11} = 0$), and the maximum power available from the generator is transferred to the load. From equation (20.8), in the special case where $S_{11} = 0$,

$V_0 \rightarrow V_{0\mathrm{max}} = V_{\mathrm{gen}}/2$. If V_{gen} is set to 2 volts, then the incident voltage $V_{0\mathrm{max}} = 1$ V, and the power transferred to the load is the maximum possible 1 W. The internal voltage V_k at any frequency, normalized to the voltage corresponding to an incident power of 1 W, then becomes:

$$\frac{V_k}{V_{0\mathrm{max}}} = \frac{V_k}{V_0}(1 + S_{11}) \tag{20.9}$$

which becomes $\frac{V_k}{V_0}(1 + S_{11})\frac{1}{\sqrt{T}}$ which **scaled** to the transmission line environment.

20.5.6 Example and Verification Via FEM Simulation

In order to verify the proposed method, the voltage distribution inside a 5 pole H plane iris filter (see Fig. 20.9) with Chebychev response is computed using *Ansoft HFSS* FEM package. The responses of the filter using the prototype and by using FEM are compared in Figure 20.10. The voltages of the five resonators are computed as a function of the frequency using the prototype filter and plotted in Figure 20.11. The voltage corresponding to the E_z component of the electric field (E_z integrated between the broad walls) is computed along the center line of the filter at the band center (12026 MHz) and at the 3 dB band edge (11737 MHz) in Figure 20.12. The peak values in the voltage distribution are compared against the resonator voltages obtained from the prototype (equation 20.7). It can be seen that there is close correlation between the prototype values and the peak voltage values inside a filter.

It is interesting to note that (i) the middle resonator has the highest voltage at band center, (ii) the second resonator has the highest voltage at band edge, and (iii) in the out off band region, the first resonator gets an appreciable amount of voltage due to the reflected power.

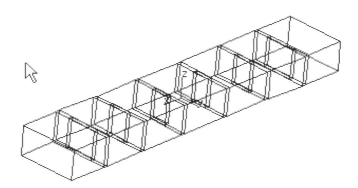

Figure 20.9 5-Pole H plane iris filter (WR75).

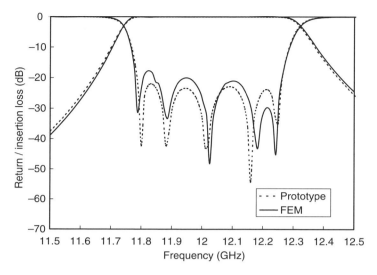

Figure 20.10 5-Pole H plane iris filter response using FEM and prototype, $f_0 = 12026$ MHz, BW = 470 MHz, Couplings: $M_{01} = M_{56} = 1.078$, $M_{12} = M_{45} = 0.928$, $M_{23} = M_{34} = 0.662$.

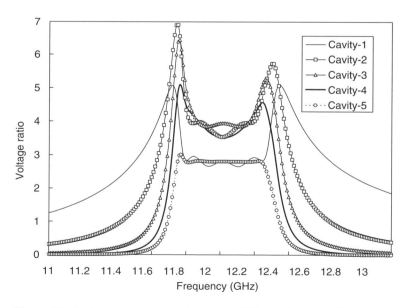

Figure 20.11 Resonator voltages (equation 20.7) as a function of frequency.

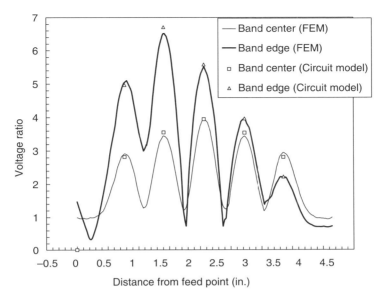

Figure 20.12 Voltage distribution along the center line of the filter using FEM and comparison of the peak values to the values obtained using the prototype.

20.5.7 Example of High Voltages in a Multiplexer

Multiplexers for satellite applications typically comprise of channels realized as 4 pole circular waveguide dual mode filters with two transmission zeros. A typical example is a filter with a cavity diameter of 1.07 inch at 12100 MHz with 36 MHz bandwidth operating in the TE113 mode. The voltage ratios in the 4 resonators at the 3 salient frequency points are given in Table 20.3. It can be seen that the peak voltage appears on the second mode at the band edge. Since power is proportional to the square of the voltage, the peak power in the second resonator can often reach up to 120 times the input power at the band edge.

TABLE 20.3 **Voltage Ratios for the 4 Modes Inside a 4 Pole Typical Dual Mode Filter**[a]

Frequency (MHz)	Resonator Voltage Ratio			
	Mode 1	Mode 2	Mode 3	Mode 4
12100, midband	6.16	7.18	6.48	6.39
12082, bandedge	7.29	11.45	10.57	6.10
12020, out of band	3.29	0.67	0.09	0.150

[a]The couplings are $M_{12} = M_{34} = 0.8745$, $M_{23} = 0.8024$, $M_{14} = -0.2363$, $M_{01} = M_{45} = 1.0429$, $f_0 = 12100$ MHz, BW $= 36$ MHz.

20.6 MULTIPACTION BREAKDOWN

Multipaction is an RF breakdown phenomenon. It takes place under vacuum conditions when the mean free path of the electrons is larger than the gap between the walls guiding the flow of the RF power. The applied RF voltage accelerates the free electrons across the gap between the walls. Let us suppose that the transmit time is such that when the electrons hit the walls, the field is reversed and the electrons are accelerated toward the opposite wall, creating an electron resonance. If the intensity of the applied field is such that the electrons bombarding the walls cause continuous release of secondary electrons, an RF breakdown results. It depends on the electron multiplication via the secondary emission of the electrons from the walls and is often referred to as the *secondary electron resonance phenomenon.* The generation of multipaction depends on the following constraints:

1. Vacuum condition
2. Applied RF voltage
3. Frequency of operation in conjunction with geometry of RF components ($f \times d$ product)
4. Surface conditions

Brief description of each of these constraints and its impact on the generation and sustenance of multiplication follow.

20.6.1 Dependence on Vacuum Environment

The condition for multipaction to occur is that the mean free path of electrons must be long enough to permit the electrons to be accelerated between the emitting surfaces with a low probability of collision with the ambient atoms or molecules. It has been shown [22] that for pressure of 10^{-3} Torr or less, the mean free path of electrons is in the tens of centimeters range when they are surrounded by typical gas molecules such as nitrogen, oxygen, and helium. For RF equipment, typical gap sizes are in the millimeter range. Consequently, it is safe to assume that multipaction can occur under pressures of 10^{-3} Torr or less.

20.6.2 Dependence on Applied RF Voltage

The electrons are accelerated by the electric field created by the applied RF voltage. Under optimum conditions, the electron motion must be in phase with the field. However, the breakdown does not require the optimum conditions in order to occur. There is a fairly broad range of fields and frequencies over which the phenomenon can be observed. A condition for multipaction is that the incident electrons on a wall must produce secondary electrons such that the ratio (δ) of the secondary to incident electrons is greater than one. If not, the generation of electrons subsides quickly and multipaction is not sustained.

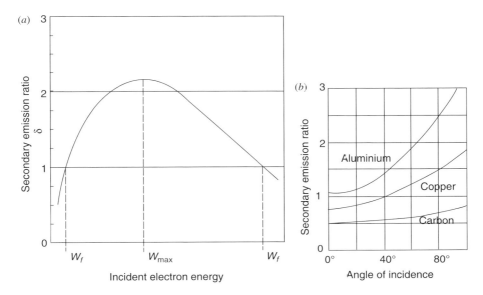

Figure 20.13 Secondary emission parameters: (*a*) δ versus *W*; (*b*) δ versus θ. (From NASA Report CR-488.)

For a given material, the value of δ is a function of the energy and angle of incidence of the primary electrons [22], as illustrated in Figure 20.13. At low incident energies, the primary electrons are unable to liberate many secondary electrons and multipaction is not initiated. At a very high incident energy, the primary electrons penetrate so deeply into the surface that the secondary electrons produced are trapped in the substance and do not reach the surface. The energy of the primary electrons must lie between these minimum and maximum values to sustain multipaction breakdown as depicted in Figure 20.13*a*. This energy depends on the phase angle of the electrons with respect to the applied RF field. As a consequence, there is a phase window within which the incident electrons can produce secondary electrons, causing multipaction, as shown in Figure 20.13*b*. These upper and lower bounds on the energy and phase permit multipaction to occur over a broad range of frequencies and applied voltages.

20.6.3 Dependence on *f* × *d* Product

Another condition of multipaction is that the gap size should be a multiple of the half-cycle of the applied RF voltage in order to satisfy the condition of electron resonance. The simplest geometry to determine breakdown as a function of an *f* × *d* product is that of a parallel-plate geometry. Many researchers [22–25] have made extensive measurements in such a way as to obtain both the maximum and minimum breakdown field strengths at various excitation frequencies. A typical depiction of this relationship is shown in Figure 20.14.

There is a multipaction curve for each odd multiple of the gap width as the curve corresponds to the half-cycle of the RF voltage. The primary mode corresponds to the multiple unity. Other modes are referred to as higher-order multipaction modes. Each multipaction curve has a minimum value and a maximum value of energy or peak voltage to initiate and sustain multipaction, as explained in the previous section. Also, each multipaction curve has a width that corresponds to the angles of arrival within which the secondary emission coefficient is greater than unity to sustain the multipaction. Such curves are available for a number of pure materials.

The geometry of high-power RF equipment rarely corresponds to simple parallel-plate conductors. Generally, discontinuities and complex geometries are required to meet the electrical performance requirements. In the vicinity of discontinuities and sharp edges, the electric fields can be much higher than those based on simple parallel geometry. As a result, multipaction can occur at RF voltages seemingly lower than those based on parallel-plate geometry in Figure 20.14.

20.6.4 Dependence on Surface Conditions of Materials

The secondary emission coefficient δ is very sensitive to the relative purity of materials. At low energy levels, there are large variations in δ caused by surface impurities. Most contaminants increase the value of δ and enhance the possibility of a multipaction occurrence. Contamination-free implementation and upkeep of high-power equipment are critical for space applications.

20.6.5 Detection and Prevention of Multipaction

The types of detection methods fall into two categories: local and global. Local detection methods are suitable in the vicinity close to the point of the actual discharge, for example, in an individual component. For system testing, global detection methods indicate that a discharge is present somewhere in the high-power assembly. The detection methods most often used are the noise floor, forward/reverse power, and balanced phase methods [25,26]. Earlier in this section, we learned that a number of constraints are essential to initiate and sustain multipaction discharge. The violation of any of those constraints should, therefore, provide the means to suppress multipaction. Possible ways to prevent the occurrence of this phenomenon are enumerated as follows:

- Control of the frequency–gap product
- Dielectric filling
- Pressurization
- Magnetic or dc bias
- Reduction of the surface potential by treatment of materials

The characteristics and relative advantages of these methods are summarized in Table 20.4.

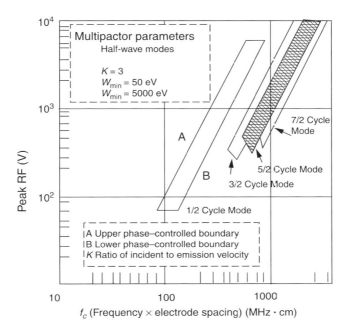

Figure 20.14 Possible regions of multipaction between parallel plates.

20.6.6 Design Margins in Multipaction

Design margins are a sensitive issue for customers, prime contractors, and suppliers of high-power components and subsystems. Because of the variability of multipaction breakdown, its dependence on work quality standards, sensitivity to contaminants, complexity of component geometries, and change of personnel every few years, all contribute to the tendency by the customer to overspecify multipaction margins and test requirements. The prime contractor often has very little choice but to accept such specifications for competitive reasons. The supplier of high-power equipment is invariably faced with the difficult task of having to meet such specifications within tight schedules and cost constraints. The optimum choice for all concerned is to adopt certain guidelines for specifying design margins on the basis of theoretical calculations and experimental verifications of a repeatable nature. This aspect was discussed quite extensively at an MTT Workshop [27], and its outcome is summarized here.

There are two categories of specifications:

1. For individual components such as filters, couplers, power dividers, circulators, and transmission lines, which handle power of a single RF channel
2. For equipment designed to handle the combined power of a number of RF channels such as multiplexers, transmissions lines, and beamforming networks

TABLE 20.4 Multipaction Prevention Methods

Method	Description	Advantages	Disadvantages
Large gap size	Electron transit time is made nonresonant such that it requires much higher voltage to initiate and sustain multipaction	Simplest mechanical structure	Limitation on RF design and achievable performance
Gap size below cutoff	Electron transit time is made nonresonant; electrons are scattered and absorbed, thus preventing onset of multipaction	No power handling limitations exist due to multipaction breakdown	RF design unsuitable for most practical applications; achievable Q values are much lower
Dielectric filling	Mean free path of electrons is reduced by filling empty spaces of discharge region with foam or solid dielectric	Simple in theory	Lowers effective Q of the devices; difficult to apply above 12 GHz
Pressurization	Mean free path of electrons is rendered close to zero by pressurization using inert gas	Removal of multipactor mechanism	Possible failure mode if leakage exists; increase in system weight; system must use hermetically sealed connectors; a source of potential PIM
Direct-current or magnetic bias	A dc bias voltage or applied magnetic field can alter the electron resonance condition to allow higher power handling	Power handling margin improved by ~ 1 dB or more; arching/discharge prevented across the gap	Surface erosion is still possible, creating wideband noise and other performance degradation; performance improvement not significant
Reduced surface potential	Secondary emission ratio of surfaces is spoiled by use of coating of low emission ratio	Simple to apply	Limited applicability due to higher loss

Individual high-power components are designed by using the methods summarized in Table 20.4. For most applications, a 3 dB margin is specified for the power handling capability. It is relatively easy to incorporate a 3 dB margin at the expense of increased volume and mass and other design features, all adding to extra complexity and cost. Moreover, testing at these elevated power levels also adds to the cost, especially at higher frequencies. However, the most complex and limiting operating condition is for equipment that is required to handle the combined powers of a number of RF channels.

High Power with Multiple RF Carriers   The worst-case scenario occurs when the equipment must handle the combined power of a large number of RF carriers, where all the carriers are in phase. In such a case, the total peak power is $n^2 P_{in}$, where n is the number of carriers and P_{in} is the power of an individual carrier. Assuming a 3 dB margin, the specification for such equipment can be as high as $2 \times n^2 P_{in}$. If we assume six carriers, a power handling specification of 72 times the value of an individual carrier is implied! On the contrary, if the carriers are completely random in phase, the combined power is usually $n \times P_{in}$. The challenge lies in determining the duration of the peak power and whether that duration is long enough to initiate multipactor breakdown. For ground-based equipment, we rarely require more than a few carriers at a satellite earth station or a cellular site. Moreover, the mass and volume of the equipment is not so critical. A specification of $2 \times n^2 P_{in}$ can be met by placing equipment in hermetically sealed enclosures containing some inert gas. For space applications with a relatively larger number of carriers, such a specification can be very costly to design, and often, beyond the range of high-power equipment to test it. The problem becomes more complicated because mass, volume, and reliability are critical issues as well. Equipment has to operate for 10–15 years, trouble-free, in the hard vacuum of space. Hermetically sealed units are deemed too risky for such requirements because of the potential leakage of the inert gas over time. Besides, such an option is too costly. Consequently, one has to look critically at the realistic scenarios and design margins for multipaction breakdown in a muticarrier environment.

This problem can be analyzed by making the following assumptions [26]: (1) there are n equally spaced carriers of equal magnitudes, and (2) over a short duration of time ($\sim$5 ns), the relative phase of the incident carriers at the spacecraft remains unchanged. This condition implies good short-term stability for the local oscillators generating the individual carriers.

Under these assumptions, the computed response for the case of 12 equally spaced carriers, 50 MHz apart, is shown in Figure 20.15. The reference for this combined power level is chosen to be at a time when all the carriers are in phase. Table 20.5 lists the computed power levels for the case of 6, 12, and 100 carriers over discrete dwell-time intervals.

Next, we compare these results with the time it takes to initiate and sustain multipaction. As discussed in Section 20.4.2, electrons require a minimum energy level to initiate multipaction. This energy level is built up as the electrons travel back and forth between the walls, when the resonant condition is satisfied. Thus, a finite-transit-time envelope is required for the electrons to gather enough energy to initiate and sustain multipaction. On the basis of experimental evidence, it is reasonable to assume that for materials with a secondary coefficient of 2 or less, it takes more than 15 cycles for multipaction to initiate. For the first-order multipaction, this translates into overall transit times of 7.5 ns at 1 GHz, 1.9 ns at 4 GHz, and 0.6 ns at 12 GHz. These transit-time envelopes can now be compared with

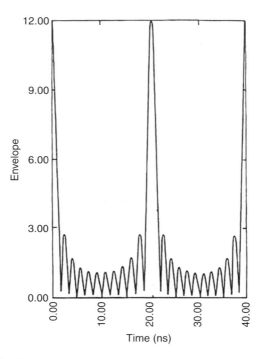

Figure 20.15 Computed power envelope of 12 carriers with 50 MHz spacing.

TABLE 20.5 Computed Power Levels of Carriers with 50 MHz Spacing

Number of Carriers	Dwell-Time Envelope (ns)	Percent of Maximum Power (%)
6	0	100
	0.5	93
	2	0.26
	20	100
12	0	100
	0.5	74
	2	0.025
	20	100
100	0	100
	0.5	0.00
	20	100

Note: The reference for the dwell time of 0 is when all the carriers are in phase. The envelope repeats itself every 20 ns since the carriers are chosen to be 50 MHz apart.

the "dwell" times of multicarrier powers as described in Table 20.5. This analysis shows the following:

1. For n equally spaced carriers, the dwell time of the phased peak power goes down rapidly as the number of carriers increases. Depending on the actual number of carriers and the frequency of operation, the dwell time can be less than the time it takes to initiate multipaction. Typical specifications based on an additional 3–6 dB margin over $n^2 P_{in}$ (P_{in} is the power per carrier) are, therefore, not realistic.

2. For multicarrier operations when the number of carriers is six or greater, the test levels for multipaction should be specified at power levels no greater than $n^2 P_{in}$. For a smaller number of carriers, the dwell time and frequency of operation should be factored in when the margins for design and testing are specified.

20.6.7 Multipactor Breakdown Levels

Multipaction breakdown has been observed at very low power (of the order of a few watts) levels, and more frequently, at higher power levels. High-Q circuits such as filters and multiplexers are especially susceptible to this breakdown. This is due to

TABLE 20.6   Examples of Flight-Qualified High-Power Equipment

Satellites	Description	Design Approach for Multipactor Prevention	Power Handing (Peak)
INTELSATs VI and VII	C-band harmonic filter	Dielectric-filled Large gap separation	>1.5 kW
INTELSAT VII	Ku-band transmit–reject filter	Large gap separation	4 kW
STC DBS	Output assembly (12 GHz)		
	Multiplexer	TE$_{114}$ filters	>2 kW
	Harmonic filter	Stepped impedance	>2 kW
ERS-1	Output Assembly (5 GHz)		
	Circulator	Dielectric-filled	26 kW
	Harmonic filter	Large gaps	26 kW
	Mechanical switch	Large gaps	45 kW
Mobile Satcom	800-MHz diplexer	Dielectric-coated probe Coax-coupled design	1 kW
SCS-1	Ku-band diplexer	Large gap separation	>4 kW
	Ku- and Ka-band isolators	Using ridges and posts	>4 kW
Aussat "B"	L-band diplexer	Coax-coupled design	400 W
Satcom "K"	12-GHz VPDs	Half-wave polarizers	>500 W
Olympus	18-GHz diplexer	Large gaps	180 W
	18-GHz isolator	Planar design	
NIMIQ 4	8-CH Ku-band mux	TE113, Large gaps	19.2 kW

Source: These data have been provided by COM DEV, Cambridge, Ontario, Canada.

Figure 20.16 Five channel high power (100 W per channel) S-band multiplexer assembly. (Courtesy COM DEV.)

the compact structure of these devices, field enhancement at discontinuities, sharp mismatches outside the band of interest, and the choice of materials. Table 20.6 summarizes of the high-power measured data for a range of microwave equipment. The measured power levels represent state-of-the-art high-power output circuits required for communications and remote sensing satellites.

Figure 20.16 is the photograph of a high-power 5-channel S-band multiplexer, with a power rating of 100 watts per channel. An 8-channel high-power multiplexer at Ku-band, rated at 300 watts per channel is shown in Figure 20.17.

Figure 20.17 An eight channel high-power (300 W per channel), low PIM (−140 dBm) Ku-band multiplexer. (Courtesy COM DEV.)

20.7 PASSIVE INTERMODULATION (PIM) CONSIDERATION FOR HIGH-POWER EQUIPMENT

It is well known that active devices produce intermodulation (IM) products because of their inherent nonlinearity, as described in Section 1.5.6. However, what has not been well understood is that even passive devices such as filters and antennas generate lower-level IM products, referred to as *passive intermodulation* (PIM) products [28]. In space, PIM was first observed in the late 1960s in the Lincoln Laboratories satellites LES-5 and LES-6.

Since then, PIM was discovered in other satellites, high-power earth station equipment, and terrestrial microwave systems. Continual advances in spacecraft technology, requiring a higher number of transponders and power levels, have provided a strong motivation to focus on the problem of PIM and, in general, on the nonlinear behavior of passive components. In satellites, all the high-power RF channels are in close proximity with each other, as well as with the receive equipment, as described in Section 1.8.4. Moreover, the difference in power levels between the transmit portion and the receive section is of the order of 120 dB, implying that the PIM levels need to be 140–150 dB lower than the power of the transmit channels. This represents the most severe environment for PIM specifications. As a consequence, PIM is a factor in determining the frequency plans of satellite systems (to avoid 3rd order PIM falling in the receive band) and achievable specifications for high-power equipment. Similar problems exist for high-power ground-based transmitters. However, for satellite earth stations, the problem is minimized by physically separating the transmit and the receive portions of the earth station. For cellular stations, PIM is not a significant issue owing to the relatively smaller difference in power levels between the transmit and receive portions of the station.

The sources of PIM are divided into two categories: contact nonlinearity and material nonlinearity. PIM results primarily from the formation of very thin oxide layers on metal surfaces, mechanical imperfections in the joints, or both. For metal surfaces separated by 10–40 Å thick oxide layers, nonlinear electron tunneling through the barrier occurs, giving rise to low-level PIM. Nonlinearities also occur as a result of microcracks or voids in metal structures (bulk effects), or dirt or metal particles on the surfaces. The buildup of microvoltages at these locations produces microcurrents that are manifested as low-level PIM. All interfaces and joints are susceptible to generate low level PIMs. The observed values of PIM represent the summation of all of these effects. Magnetic and dielectric materials have a degree of inherent nonlinearity and therefore are capable of producing significant levels of PIM. It should be noted that PIM reduction strategies can only mitigate, and not cure, the effects. Although experience has taught us that some methods are better than others, PIM reduction must be carried out experimentally. Smooth and clean surfaces, free of dirt and oxide layers, are a good baseline.

20.7.1 PIM Measurement

The PIM specifications for high-power communication systems, such as direct broadcast satellites or military satellite systems, can be as low as −140 dBm. The measuring equipment must therefore be at least 5–10 dB better than the specified level of PIM. This is a tall order for standard measurement instrumentation. Invariably, it requires custom measurement set ups, suited for specific measurement accuracy and system design. Figure 20.18 describes a typical set up for PIM measurement.

It consists of two high-power input signals at frequencies F1 and F2, the device under test (DUT) combiner and the low power measuring equipment. Signals are generated using frequency synthesizers, and amplified to the desired power levels and followed by narrow band filters to retain the purity of the signals. Power level can be adjusted via attenuators in the path prior to amplification of the signals. Directional couplers in the path of the signals allow measurement of the incident (FWD) and reflected (RVS) powers at the DUT interface.

DUT consists of a 3 dB hybrid and the device or component to be tested for PIM. The two input ports of the hybrid are connected to the signals F1 and F2. Each of the two output ports receives half the power of each carrier. One of the output ports is connected to the device to be tested and the other port to a load. Half of the power of the carriers is dissipated in the load whereas the PIM power, generated by the

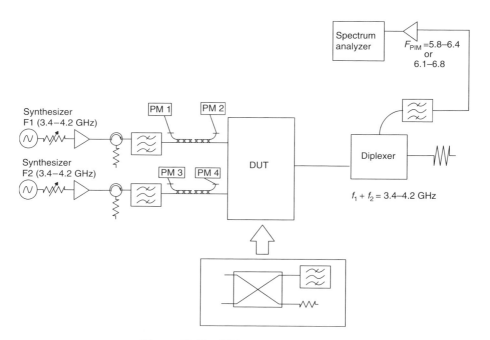

Figure 20.18 PIM measurement set-up.

TABLE 20.7   **State-of-the Art Low PIM Measurement Systems**

Frequency (MHz)	Total Power (W)	3rd-Order PIM (dBm)	dBc
300	60	−120 to −130	−167 to −177
8000	2000	−150	−213
12000	40	−140	−186

TABLE 20.8   **State of the Art: Intermodulation in Passive Components, Passive IM—3rd-Order Level for Two +45-dBm Carriers**

Component	Conventional Units (dBm)	State-of-the-Art Units (dBm)
Diplexer	−70	−120
Multiplexer	−60	−120
Hybrid	−60	−140
Circulators	−70	−110
Coaxial connectors	−60 to −100	−140
OrthoMode transducers	−100	−140

component under test, is connected to a Diplexer. The Diplexer separates the carrier signals from the PIM products. The remaining carrier power is dissipated in the load connected to the Diplexer whereas the PIM product is filtered, amplified by an LNA and then fed to the spectrum analyzer. The measurement sequence involves calibration of the noise floor and any PIM generated by the measuring equipment. A high-power signal at the input of the diplexer (consisting of F1 and F2) is used to evaluate and minimize any PIM contribution from the load or the diplexer. A low level signal (PIM frequency) at the input of the diplexer is used to calibrate the noise floor of the system.

From DUT onwards, each component and interface in the measurement set up is capable of generating low level PIM which must be maintained at power levels of 5–10 dB lower than the specification on the device. For a specification of −140 dBm, this implies PIM levels of −145 to −150 dBm. Identification of the sources of PIM and controlling their levels over time and the operating temperature range (i.e., repeatability) at such low power levels puts a severe constraint on the measuring equipment, especially at the various interfaces. The problem is at its worst when coaxial interfaces are involved. This is one area where experience plays a critical role in establishing precision set ups for the measurement of low level PIMs. Table 20.7 describes the state of the art for PIM measurement.

20.7.2 PIM Control Guidelines

The nature of PIM is such that it is not amenable to theoretical design. It depends on the quality of materials almost at the molecular level and the way they are joined together, surface conditions, work standards, and the subsequent handling of the equipment. The most cost-effective approach to protection against or minimization of PIM is to enforce a set of guidelines to be followed at all stages of the program. Such guidelines are summarized as follows [25,26].

- Avoid the use of ferromagnetic material in high-power equipment.
- Use metals that tend to exhibit low levels of PIM, such as, gold, silver, copper, and brass. Aluminum and steel tend to generate higher levels of PIM.
- Reduce the regions of high current density by incorporating designs with larger dimensions of RF components.
- Pay special attention to the choice of materials, connectors, and interfaces. All contacts and joints must be composed of the same material and be rigidly mounted. Cleanliness and avoidance of surface contamination are critical to minimizing PIM.
- Minimize or eliminate the use of tuning screws.
- Take special care of electroplated surfaces and avoid nickel plating.
- Avoid sharp edges that can concentrate current flow.

Table 20.8 lists the state of the art in achievable performance for a range of components that require a low PIM performance.

SUMMARY

High-power microwave equipment for terrestrial applications uses air or other gaseous materials inside the equipment. It is cost-effective but requires a good understanding of the phenomenon of gaseous breakdown at microwave frequencies.

In space, enclosed structures are either vented to ensure a hard vacuum or hermetically sealed with an inert gas. The employment of hermetically sealed equipment is rare because of the very stringent requirement on the leakage of gas in the hard vacuum of space. Equipment operating in a vacuum environment is subjected to the phenomenon of multipaction breakdown, which becomes viable under pressures of 10^{-3} Torr or less. In a geostationary orbit, the vacuum is of the order of 10^{-17} Torr. As a consequence, multipaction represents a major constraint for the design of high-power equipment for space applications.

This chapter provides an overview of the phenomena of microwave gaseous breakdown. It addresses the issue of voltage breakdown as a function of pressure, including the critical pressure. It highlights the importance of derating factors that can severely degrade the performance of high-power equipment. The phenomenon of multipaction is described in some depth. The topic of design margins, especially when the equipment must handle a number of high-power carriers, is discussed in detail. Methods to prevent multipaction are highlighted.

Of all the high-power components, filters and multiplexers are especially susceptible to breakdown owing to the high electric fields generated inside narrowband and high-Q filters. The topic of electric fields in resonators and filters is dealt with in terms of group delay inside each cavity of a coupled resonator filter. A more general approach, based on a lumped element prototype using general matrix topology is also described. This allows lowpass prototype filter parameters to be directly related to the peak electric fields generated inside the filters. This approach provides an excellent guide to the power handling capability of practical filters. It allows the computation of peak internal fields in a variety of filter structures, including diplexers and multiplexers. Resonators with the highest energy can then be analyzed more rigorously by using commercially available finite-element modeling (FEM) software.

Another phenomenon that becomes significant in designing high-power equipment is that of passive intermodulation (PIM), a significant issue in high-power satellite systems. PIM is difficult to analyze and depends on the choice of materials and work quality standards. Guidelines to measure and minimize PIM in the design of high-power equipment are included.

REFERENCES

1. A. D. MacDonald, *Microwave Breakdown in Gases*, Wiley, New York, 1966.
2. C. Kudsia, R. Cameron, and W. C. Tang, Innovations in microwave filters and multiplexing networks for communications satellite systems, *IEEE Trans. Microwave Theory Tech.* (June 1992).
3. R. Strauss, J. Bretting, and R. Metivier, Traveling wave tubes for communication satellites, *Proc. IEEE* **65**, 387–400 (March 1977).
4. G. D. Gordon and W. L. Morgan, *Principles of Communications Satellites*, Wiley, 1993.
5. G. L. Ragan, *Microwave Transmission Circuits*, MIT Radiation Lab. Series, Vol. 9, McGraw-Hill, New York, 1946.
6. T. Moreno, *Microwave Transmission Design Data*, McGraw-Hill, New York, 1948.
7. *Reference Data for Radio Engineers*, 4th ed., ITT, 1961.
8. H. E. King, Rectangular waveguide theoretical CW average power rating, *IRE Trans. Microwave Theory Tech.* (July 1961).
9. G. L. Matthaei, L. Young, and E. M. T. Jones, *Microwave Filters, Impedance Matching Networks and Coupling Structures*, McGraw-Hill, New York, 1964.
10. J. Ciavolella, Take the hassle out of high power design, *Microwaves* (June 1972).
11. S. B. Cohn, Design considerations for high-power microwave filters, *IRE Trans. Microwave Theory Tech.* (Jan. 1959).
12. L. Young, Peak internal fields in direct-coupled filters, *IRE Trans. Microwave Theory Tech.* (Nov. 1960).
13. C. Ernst, *Energy Storage in Microwave Cavity Filter Networks*, Ph.D. thesis, Univ. Leeds, School of Electronic and Electrical Engineering, 2000.
14. A. Sivadas, M. Yu, and R. Cameron, A simplified analysis for high power microwave bandpass filter structures, *IEEE MTT-S Microwave Symp. Digest* (June 2000).
15. C. Wang and K. A. Zaki, Analysis of power handling capability of bandpass filters, *IEEE MTT-S Microwave Symp. Digest* (June 2001).
16. Ben S. Senior et al., Optimum network topologies for high power microwave filters, *IEEE MTT-S Digest* 53–58 (2001).

17. Ali E. Atia, A 14-GHz high-power filter, *IEEE MTT-S Digest* (1979).

18. C. M. Kudsia et al., A New Type of Low Loss 14 GHz High Power Combining Networks for Satellite Earth Terminals, 9th European Microwave Conference, 1979, Brighton, England.

19. T. Nishikawa et al., 800 MHz high-power bandpass filter using TM110 Mode dielectric resonators for cellular base stations, *IEEE MTT-S Digest* (1988).

20. A. E. Atia and Williams, Narrow band waveguide filters, *IEEE Trans. Microwave Theory Tech.* **MTT-20**, 258–265 (Apr. 1972).

21. R. J. Cameron, General prototype network-synthesis methods for microwave filters, *ESA Journal* **6**, 193–206 (1982).

22. *The Study of Multipactor Breakdown in Space Electronic System*, NASA CR-488, Goddard Space Flight Center, 1966.

23. A. J. Hatch and H. J. B. Williams, The secondary electron resonance mechanism of low pressure-high frequency breakdown, *J. Appl. Phys.* **25**, 417–423 (1954).

24. E. F. Vance and J. E. Nanevicz, *Multipactor discharge experiments*, ARCRL-68-0063, SRI Project 5359, Stanford Research Institute, Palo Alto, CA, Dec. 1967.

25. *Multipaction Design and Test*, European Cooperation for Space Standardisation ESA-ESTEC Document, ECSS-E-20-01A, May 2003.

26. W. C. Tang and C. M. Kudsia, Multipactor Breakdown and Passive Intermodulation in Microwave Equipment for Satellite Application, *IEEE Military Communication Conference Proceedings*, Monterey, CA, Oct. 1990.

27. C. M. Kudsia (COMDEV) and J. Fiedzuisko (LORAL), High power passive equipment for satellite applications, *1989 IEEE MTT-S Workshop Proceeding.*, Long Beach, CA, June 13–15, 1989.

28. R. C. Chapman et al., Hidden threat: Multicarrier passive component IM generation, Paper 76-296, *AIAA/CASI 6th Communications Satellite Systems Conf.*, April 5–8 1976, Montreal.

APPENDIX A

APPENDIX A.   Physical Constants

Speed of light in free-space, $c = 2.998 \times 10^8$ m/s
Permittivity of free-space, $\varepsilon_0 = 8.854 \times 10^{-12}$ F/m
Permeability of free-space, $\mu_0 = 4\pi \times 10^{-7}$ H/m
Impedance of free-space, $\eta_0 = 376.7$ Ω
Charge of electron, $e = 1.602 \times 10^{-19}$ C
Mass of electron, $m = 9.107 \times 10^{-31}$ kg
Boltzmann's constant, $k = 1.380 \times 10^{-23}$ J/K

Microwave Filters for Communication Systems: Fundamentals, Design, and Applications,
by Richard J. Cameron, Chandra M. Kudsia, and Raafat R. Mansour
Copyright © 2007 John Wiley & Sons, Inc.

APPENDIX B

APPENDIX B. Conductivities of Metals

Material	Conductivity S/m (20 °C)	Material	Conductivity S/m (20 °C)
Aluminium	3.816×10^7	Mercury	1.04×10^6
Brass	2.564×10^7	Lead	4.56×10^6
Bronze	1.00×10^7	Nickel	1.449×10^7
Chromium	3.846×10^7	Platinum	9.52×10^6
Copper	5.813×10^7	Silver	6.173×10^7
Germanium	2.2×10^6	Steel (stainless)	1.1×10^6
Gold	4.098×10^7	Solder	7.0×10^6
Graphite	7.0×10^4	Tungsten	1.825×10^7
Iron	1.03×10^7	Zinc	1.67×10^7

Microwave Filters for Communication Systems: Fundamentals, Design, and Applications,
by Richard J. Cameron, Chandra M. Kudsia, and Raafat R. Mansour
Copyright © 2007 John Wiley & Sons, Inc.

APPENDIX C

APPENDIX C.   Dielectric Constants and Loss Tangents of some Materials

Material	Frequency	ε_r	$\tan \delta$
Alumina	10 GHz	9.7–10	0.0002
Fused quartz	10 GHz	3.78	0.0001
Gallium arsenide	10 GHz	13	0.0016
Glass (pyrex)	3 GHz	4.82	0.0054
Glazed ceramic	10 GHz	7.2	0.008
Plexiglass	3 GHz	2.60	0.0057
Polyethylene	10 GHz	2.25	0.0004
Polystyrene	10 GHz	2.54	0.00033
Porcelain	100 GHz	5.04	0.0078
Rexolite	3 GHz	2.54	0.00048
RT/duriod 5880	10 GHz	2.2	0.0009
RT/duriod 6002	10 GHz	2.94	0.0012
RT/duriod 6006	10 GHz	6.15	0.0019
RT/duriod 6010	10 GHz	10.8	0.0023
Silicon	10 GHz	11.9	0.004
Styrofoam	3 GHz	1.03	0.0001
Teflon	10 GHz	2.08	0.0004
Vaseline	10 GHz	2.16	0.001
Water (distilled)	3 GHz	76.7	0.157

Microwave Filters for Communication Systems: Fundamentals, Design, and Applications,
by Richard J. Cameron, Chandra M. Kudsia, and Raafat R. Mansour
Copyright © 2007 John Wiley & Sons, Inc.

APPENDIX D

APPENDIX D. **Rectangular Waveguide Designation**

Designation WR-XX	Recommended Frequency Range (GHz)	TE_{10} Cutoff Frequency (GHz)	Inside Dimensions Inches (cm)
WR-650	1.12–1.70	0.908	6.500 × 3.250 (16.51 × 8.255)
WR-430	1.70–2.60	1.372	4.300 × 2.150 (10.922 × 5.461)
WR-284	2.60–3.95	2.078	2.840 × 1.340 (7.214 × 3.404)
WR-187	3.95–5.85	3.152	1.872 × 0.872 (4.755 × 2.215)
WR-137	5.85–8.20	4.301	1.372 × 0.622 (3.485 × 1.580)
WR-112	7.05–10.0	5.259	1.122 × 0.497 (2.850 × 1.262)
WR-90	8.20–12.4	6.557	0.900 × 0.400 (2.286 × 1.016)
WR-62	12.4–18.0	9.486	0.622 × 0.311 (1.580 × 0.790)
WR-42	18.0–26.5	14.047	0.420 × 0.170 (1.07 × 0.43)
WR-28	26.5–40.0	21.081	0.280 × 0.140 (0.711 × 0.356)
WR-22	33.0–50.5	26.342	0.224 × 0.112 (0.57 × 0.28)
WR-19	40.0–60.0	31.357	0.188 × 0.094 (0.48 × 0.24)
WR-15	50.0–75.0	39.863	0.148 × 0.074 (0.38 × 0.19)
WR-12	60.0–90.0	48.350	0.122 × 0.061 (0.31 × 0.015)
WR-10	75.0–110.0	59.010	0.100 × 0.050 (0.254 × 0.127)
WR-8	90.0–140.0	73.840	0.080 × 0.040 (0.203 × 0.102)
WR-6	110.0–170.0	90.854	0.065 × 0.0325 (0.170 × 0.083)
WR-5	140.0–220.0	115.750	0.051 × 0.0255 (0.130 × 0.0648)

Microwave Filters for Communication Systems: Fundamentals, Design, and Applications,
by Richard J. Cameron, Chandra M. Kudsia, and Raafat R. Mansour
Copyright © 2007 John Wiley & Sons, Inc.

INDEX

Microwave Filters for Communication Systems: Fundamentals, Design, and Applications,
by Richard J. Cameron, Chandra M. Kudsia, and Raafat R. Mansour
Copyright © 2007 John Wiley & Sons, Inc.